PATTERN
CLASSIFICATION

PATTERN CLASSIFICATION

Second Edition

Richard O. Duda

Peter E. Hart

David G. Stork

A Wiley-Interscience Publication

JOHN WILEY & SONS, INC.

New York ■ Chichester ■ Weinheim ■ Brisbane ■ Singapore ■ Toronto

For ordering and customer service, call 1-800-CALL-WILEY.

Library of Congress Cataloging-in-Publication Data:
Duda, Richard O.
    Pattern classification / Richard O. Duda, Peter
E. Hart [and] David G. Stork. — 2nd ed.
      p. cm.
    "A Wiley-Interscience Publication."
    Includes bibliographical references and index.
    Partial Contents: Part 1. Pattern classification.
    ISBN 0-471-05669-3 (alk. paper)
    1. Pattern recognition systems.   2. Statistical decision.
I. Hart, Peter E.   II. Stork, David G.   III. Title.
Q327.D83   2000
006.4—dc21   99-29981
                                         CIP

22

To
C. A. Rosen
and
C. W. Stork

CONTENTS

3 MAXIMUM-LIKELIHOOD AND BAYESIAN PARAMETER ESTIMATION   84

4   NONPARAMETRIC TECHNIQUES   161

5	LINEAR DISCRIMINANT FUNCTIONS	215

6   MULTILAYER NEURAL NETWORKS   282

7 STOCHASTIC METHODS 350

8 NONMETRIC METHODS 394

9   ALGORITHM-INDEPENDENT MACHINE LEARNING   453

10   UNSUPERVISED LEARNING AND CLUSTERING   517

A | MATHEMATICAL FOUNDATIONS
601

INDEX   637

PREFACE

Our purpose in writing this second edition—more than a quarter century after the original—remains the same: to give a systematic account of the major topics in pattern recognition, based whenever possible on fundamental principles. We believe that this provides the required foundation for solving problems in more specialized application areas such as speech recognition, optical character recognition, or signal classification. Readers of the first edition often asked why we combined in one book a Part I on pattern classification with a Part II on scene analysis. At the time, we could reply that classification theory was the most important domain-independent theory of pattern recognition, and that scene analysis was the only important application domain. Moreover, in 1973 it was still possible to provide an exposition of the major topics in pattern classification and scene analysis without being superficial. In the intervening years, the explosion of activity in both the theory and practice of pattern recognition has made this view untenable. Knowing that we had to make a choice, we decided to focus our attention on classification theory, leaving the treatment of applications to the books that specialize on particular application domains. Since 1973, there has been an immense wealth of effort, and in many cases progress, on the topics we addressed in the first edition. The pace of progress in algorithms for learning and pattern recognition has been exceeded only by the improvements in computer hardware. Some of the outstanding problems acknowledged in the first edition have been solved, whereas others remain as frustrating as ever. Taken with the manifest usefulness of pattern recognition, this makes the field extremely vigorous and exciting.

While we wrote then that pattern recognition might appear to be a rather specialized topic, it is now abundantly clear that pattern recognition is an immensely broad subject, with applications in fields as diverse as handwriting and gesture recognition, lipreading, geological analysis, document searching, and the recognition of bubble chamber tracks of subatomic particles; it is central to a host of human-machine interface problems, such as pen-based computing. The size of the current volume is a testament to the body of established theory. Whereas we expect that most of our readers will be interested in developing pattern recognition systems, perhaps a few will be active in understanding existing pattern recognition systems, most notably human and animal nervous systems. To address the biological roots of pattern recognition would of course be beyond the scope of this book. Nevertheless, because neurobiologists and psychologists interested in pattern recognition in the natural world continue to rely on more advanced mathematics and theory, they too may profit from the material presented here.

Despite the existence of a number of excellent books that focus on a small set of specific *techniques*, we feel that there is still a strong need for a book such as ours, which takes a somewhat different approach. Rather than focus on a specific technique such as neural networks, we address a specific *class of problems*—pattern recognition problems—and consider the wealth of different techniques that can be applied to it. Students and practitioners typically have a particular problem and need to know which technique is best suited for their needs and goals. In contrast, books that focus on neural networks may not explain decision trees, or nearest-neighbor methods, or many other classifiers to the depth required by the pattern recognition practitioner who must decide among the various alternatives. To avoid this problem, we often discuss the relative strengths and weaknesses of various classification techniques.

These developments demanded a unified presentation in an updated edition of Part I of the original book. We have tried not only to expand but also to improve the text in a number of ways:

New Material. The text has been brought up to date with chapters on pattern recognition topics that have, over the past decade or so, proven to be of value: neural networks, stochastic methods, and some topics in the theory of learning, to name a few. While the book continues to stress methods that are statistical at root, for completeness we have included material on syntactic methods as well. "Classical" material has been included, such as Hidden Markov models, model selection, combining classifiers, and so forth.

Examples. Throughout the text we have included worked examples, usually containing data and methods simple enough that no tedious calculations are required, yet complex enough to illustrate important points. These are meant to impart intuition, clarify the ideas in the text, and to help students solve the homework problems.

Algorithms. Some pattern recognition or learning techniques are best explained with the help of algorithms, and thus we have included several throughout the book. These are meant for clarification, of course; they provide only the skeleton of structure needed for a full computer program. We assume that every reader is familiar with such pseudocode, or can understand it from context here.

Starred Sections. The starred sections (*) are a bit more specialized, and they are typically expansions upon other material. Starred sections are generally not needed to understand subsequent unstarred sections, and thus they can be skipped on first reading.

Computer Exercises. These are not specific to any language or system, and thus can be done in the language or style the student finds most comfortable.

Problems. New homework problems have been added, organized by the earliest section where the material is covered. In addition, in response to popular demand, a Solutions Manual has been prepared to help instructors who adopt this book for courses.

Chapter Summaries. Chapter summaries are included to highlight the most important concepts covered in the rest of the text.

Graphics. We have gone to great lengths to produce a large number of high-quality figures and graphics to illustrate our points. Some of these required extensive calculations, selection, and reselection of parameters to best illustrate the concepts at hand. Study the figures carefully! The book's illustrations are available in Adobe Acrobat format that can be used by faculty adopting this book for courses to create presentations for lectures. The files can be accessed through a standard web browser or an ftp client program at the Wiley STM ftp area at:

ftp://ftp.wiley.com/public/sci_tech_med/pattern/

The files can also be accessed from a link on the Wiley Electrical Engineering software supplements page at:

http://www.wiley.com/products/subject/engineering/electrical/
software_supplem_elec_eng.html

Mathematical Appendixes. It comes as no surprise that students do not have the same mathematical background, and for this reason we have included mathematical appendixes on the foundations needed for the book. We have striven to use clear notation throughout—rich enough to cover the key properties, yet simple enough for easy readability. The list of symbols in the Appendix should help those readers who dip into an isolated section that uses notation developed much earlier.

This book surely contains enough material to fill a two-semester upper-division or graduate course; alternatively, with careful selection of topics, faculty can fashion a one-semester course. A one-semester course could be based on Chapters 1–6, 9 and 10 (most of the material from the first edition, augmented by neural networks and machine learning), with or without the material from the starred sections.

Because of the explosion in research developments, our historical remarks at the end of most chapters are necessarily cursory and somewhat idiosyncratic. Our goal has been to stress important references that help the reader rather than to document the complete historical record and acknowledge, praise, and cite the established researcher. The Bibliography sections contain some valuable references that are not explicitly cited in the body of the text. Readers should also scan through the titles in the Bibliography sections for references of interest.

This book could never have been written without the support and assistance of several institutions. First and foremost is of course Ricoh Innovations (DGS and PEH). Its support of such a long-range and broadly educational project as this book— amidst the rough and tumble world of industry and its never-ending need for products and innovation—is proof positive of a wonderful environment and a rare and enlightened leadership. The enthusiastic support of Morio Onoe, who was Director of Research, Ricoh Company Ltd. when we began our writing efforts, is gratefully acknowledged. Likewise, San Jose State University (ROD), Stanford University (Departments of Electrical Engineering, Statistics and Psychology), The University of California, Berkeley Extension, The International Institute of Advanced Scientific Studies, the Niels Bohr Institute, and the Santa Fe Institute (DGS) all provided a temporary home during the writing of this book. Our sincere gratitude goes to all.

Deep thanks go to Stanford graduate students Regis Van Steenkiste, Chuck Lam and Chris Overton who helped immensely on figure preparation and to Sudeshna Adak, who helped in solving homework problems. Colleagues at Ricoh aided in numerous ways; Kathrin Berkner, Michael Gormish, Maya Gupta, Jonathan Hull and Greg Wolff deserve special thanks, as does research librarian Rowan Fairgrove, who efficiently found obscure references, including the first names of a few authors. The book has been used in manuscript form in several courses at Stanford University and San Jose State University, and the feedback from students has been invaluable. Numerous faculty and scientific colleagues have sent us many suggestions and caught many errors. The following such commentators warrant special mention: Leo Breiman, David Cooper, Lawrence Fogel, Gary Ford, Isabelle Guyon, Robert Jacobs, Dennis Kibler, Scott Kirkpatrick, Daphne Koller, Benny Lautrup, Nick Littlestone, Amir Najmi, Art Owen, Rosalind Picard, J. Ross Quinlan, Cullen Schaffer, and David Wolpert. Specialist reviewers—Alex Pentland (1), Giovanni Parmigiani (2), Peter Cheeseman (3), Godfried Toussaint (4), Padhraic Smyth (5), Yann Le Cun (6), Emile Aarts (7), Horst Bunke (8), Tom Dietterich (9), Anil Jain (10), and Rao Vemuri (Appendix)—focused on single chapters (as indicated by the numbers in parentheses); their perceptive comments were often enlightening and improved the text in numerous ways. (Nevertheless, we are responsible for any errors that remain.) George Telecki, our editor, gave the needed encouragement and support, and he refrained from complaining as one manuscript deadline after another passed. He, and indeed all the folk at Wiley, were extremely helpful and professional. Finally, deep thanks go to Nancy, Alex, and Olivia Stork for understanding and patience.

DAVID G. STORK
RICHARD O. DUDA
PETER E. HART

Menlo Park, California
August, 2000

PATTERN CLASSIFICATION

INTRODUCTION

The ease with which we recognize a face, understand spoken words, read handwritten characters, identify our car keys in our pocket by feel, and decide whether an apple is ripe by its smell belies the astoundingly complex processes that underlie these acts of pattern recognition. Pattern recognition—the act of taking in raw data and making an action based on the "category" of the pattern—has been crucial for our survival, and over the past tens of millions of years we have evolved highly sophisticated neural and cognitive systems for such tasks.

1.1 MACHINE PERCEPTION

It is natural that we should seek to design and build machines that can recognize patterns. From automated speech recognition, fingerprint identification, optical character recognition, DNA sequence identification, and much more, it is clear that reliable, accurate pattern recognition by machine would be immensely useful. Moreover, in solving the myriad problems required to build such systems, we gain deeper understanding and appreciation for pattern recognition systems in the natural world—most particularly in humans. For some problems, such as speech and visual recognition, our design efforts may in fact be influenced by knowledge of how these are solved in nature, both in the algorithms we employ and in the design of special-purpose hardware.

1.2 AN EXAMPLE

To illustrate the complexity of some of the types of problems involved, let us consider the following imaginary and somewhat fanciful example. Suppose that a fish-packing plant wants to automate the process of sorting incoming fish on a conveyor belt according to species. As a pilot project it is decided to try to separate sea bass from salmon using optical sensing. We set up a camera, take some sample images, and begin to note some physical differences between the two types of fish—length, lightness, width, number and shape of fins, position of the mouth, and so on—and these suggest *features* to explore for use in our classifier. We also notice noise or

FEATURE

1

variations in the images—variations in lighting, position of the fish on the conveyor, even "static" due to the electronics of the camera itself.

MODEL

Given that there truly are differences between the population of sea bass and that of salmon, we view them as having different *models*—different descriptions, which are typically mathematical in form. The overarching goal and approach in pattern classification is to hypothesize the class of these models, process the sensed data to eliminate noise (not due to the models), and for any sensed pattern choose the model that corresponds best. Any techniques that further this aim should be in the conceptual toolbox of the designer of pattern recognition systems.

Our prototype system to perform this very specific task might well have the form shown in Fig. 1.1. First the camera captures an image of the fish. Next, the camera's signals are *preprocessed* to simplify subsequent operations without losing relevant information. In particular, we might use a *segmentation* operation in which the images of different fish are somehow isolated from one another and from the background. The information from a single fish is then sent to a *feature extractor*, whose purpose is to reduce the data by measuring certain "features" or "properties."

PREPROCESSING
SEGMENTATION

FEATURE
EXTRACTION

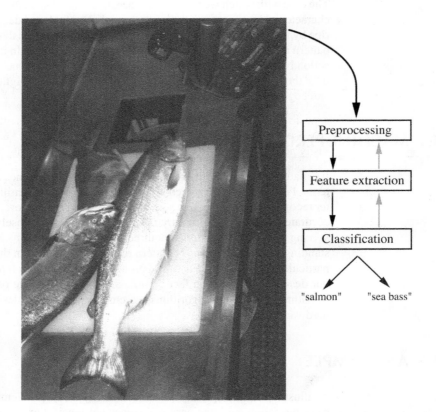

FIGURE 1.1. The objects to be classified are first sensed by a transducer (camera), whose signals are preprocessed. Next the features are extracted and finally the classification is emitted, here either "salmon" or "sea bass." Although the information flow is often chosen to be from the source to the classifier, some systems employ information flow in which earlier levels of processing can be altered based on the tentative or preliminary response in later levels (gray arrows). Yet others combine two or more stages into a unified step, such as simultaneous segmentation and feature extraction.

These features (or, more precisely, the values of these features) are then passed to a *classifier* that evaluates the evidence presented and makes a final decision as to the species.

The preprocessor might automatically adjust for average light level, or threshold the image to remove the background of the conveyor belt, and so forth. For the moment let us pass over how the images of the fish might be segmented and consider how the feature extractor and classifier might be designed. Suppose somebody at the fish plant tells us that a sea bass is generally longer than a salmon. These, then, give us our tentative *models* for the fish: Sea bass have some typical length, and this is greater than that for salmon. Then length becomes an obvious feature, and we might attempt to classify the fish merely by seeing whether or not the length l of a fish exceeds some critical value l^*. To choose l^* we could obtain some *design* or *training* samples of the different types of fish, make length measurements, and inspect the results.

TRAINING
SAMPLES

Suppose that we do this and obtain the histograms shown in Fig. 1.2. These disappointing histograms bear out the statement that sea bass are somewhat longer than salmon, on average, but it is clear that this single criterion is quite poor; no matter how we choose l^*, we cannot reliably separate sea bass from salmon by length alone.

Discouraged, but undeterred by these unpromising results, we try another feature, namely the average lightness of the fish scales. Now we are very careful to eliminate variations in illumination, because they can only obscure the models and corrupt our new classifier. The resulting histograms and critical value x^*, shown in Fig. 1.3, are much more satisfactory: The classes are much better separated.

So far we have tacitly assumed that the consequences of our actions are equally costly: Deciding the fish was a sea bass when in fact it was a salmon was just as undesirable as the converse. Such a symmetry in the *cost* is often, but not invariably, the case. For instance, as a fish-packing company we may know that our customers easily accept occasional pieces of tasty salmon in their cans labeled "sea bass," but they object vigorously if a piece of sea bass appears in their cans labeled "salmon." If we want to stay in business, we should adjust our decisions to avoid antagonizing our customers, even if it means that more salmon makes its way into the cans of

COST

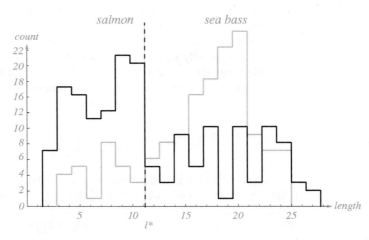

FIGURE 1.2. Histograms for the length feature for the two categories. No single threshold value of the length will serve to unambiguously discriminate between the two categories; using length alone, we will have some errors. The value marked *l** will lead to the smallest number of errors, on average.

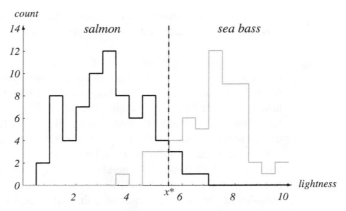

FIGURE 1.3. Histograms for the lightness feature for the two categories. No single threshold value x^* (decision boundary) will serve to unambiguously discriminate between the two categories; using lightness alone, we will have some errors. The value x^* marked will lead to the smallest number of errors, on average.

sea bass. In this case, then, we should move our decision boundary to smaller values of lightness, thereby reducing the number of sea bass that are classified as salmon (Fig. 1.3). The more our customers object to getting sea bass with their salmon (i.e., the more costly this type of error) the lower we should set the decision threshold x^* in Fig. 1.3.

Such considerations suggest that there is an overall single cost associated with our decision, and our true task is to make a decision rule (i.e., set a decision boundary) so as to minimize such a cost. This is the central task of *decision theory* of which pattern classification is perhaps the most important subfield.

Even if we know the costs associated with our decisions and choose the optimal critical value x^*, we may be dissatisfied with the resulting performance. Our first impulse might be to seek yet a different feature on which to separate the fish. Let us assume, however, that no other single visual feature yields better performance than that based on lightness. To improve recognition, then, we must resort to the use of *more* than one feature at a time.

In our search for other features, we might try to capitalize on the observation that sea bass are typically wider than salmon. Now we have two features for classifying fish—the lightness x_1 and the width x_2. If we ignore how these features might be measured in practice, we realize that the feature extractor has thus reduced the image of each fish to a point or *feature vector* **x** in a two-dimensional *feature space*, where

$$\mathbf{x} = \begin{pmatrix} x_1 \\ x_2 \end{pmatrix}.$$

Our problem now is to partition the feature space into two regions, where for all points in one region we will call the fish a sea bass, and for all points in the other we call it a salmon. Suppose that we measure the feature vectors for our samples and obtain the scattering of points shown in Fig. 1.4. This plot suggests the following rule for separating the fish: Classify the fish as sea bass if its feature vector falls above the *decision boundary* shown, and as salmon otherwise.

This rule appears to do a good job of separating our samples and suggests that perhaps incorporating yet more features would be desirable. Besides the lightness

DECISION
THEORY

DECISION
BOUNDARY

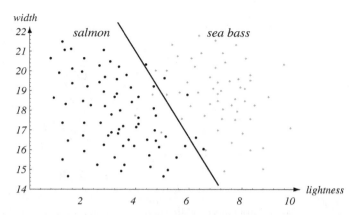

FIGURE 1.4. The two features of lightness and width for sea bass and salmon. The dark line could serve as a decision boundary of our classifier. Overall classification error on the data shown is lower than if we use only one feature as in Fig. 1.3, but there will still be some errors.

and width of the fish, we might include some shape parameter, such as the vertex angle of the dorsal fin, or the placement of the eyes (as expressed as a proportion of the mouth-to-tail distance), and so on. How do we know beforehand which of these features will work best? Some features might be redundant. For instance, if the eye color of all fish correlated perfectly with width, then classification performance need not be improved if we also include eye color as a feature. Even if the difficulty or computational cost in attaining more features is of no concern, might we ever have *too many* features—is there some "curse" for working in very high dimensions?

Suppose that other features are too expensive to measure, or provide little improvement (or possibly even degrade the performance) in the approach described above, and that we are forced to make our decision based on the two features in Fig. 1.4. If our models were extremely complicated, our classifier would have a decision boundary more complex than the simple straight line. In that case all the training patterns would be separated perfectly, as shown in Fig. 1.5. With such a "solution," though, our satisfaction would be premature because the central aim of designing a classifier is to suggest actions when presented with *novel* patterns, that is, fish not yet seen. This is the issue of *generalization*. It is unlikely that the complex decision boundary in Fig. 1.5 would provide good generalization—it seems to be "tuned" to the particular training samples, rather than some underlying characteristics or true model of all the sea bass and salmon that will have to be separated.

GENERALIZATION

Naturally, one approach would be to get more training samples for obtaining a better estimate of the true underlying characteristics, for instance the probability distributions of the categories. In some pattern recognition problems, however, the amount of such data we can obtain easily is often quite limited. Even with a vast amount of training data in a continuous feature space though, if we followed the approach in Fig. 1.5 our classifier would give a horrendously complicated decision boundary—one that would be unlikely to do well on novel patterns.

Rather, then, we might seek to "simplify" the recognizer, motivated by a belief that the underlying models will not require a decision boundary that is as complex as that in Fig. 1.5. Indeed, we might be satisfied with the slightly poorer performance on the training samples if it means that our classifier will have better performance

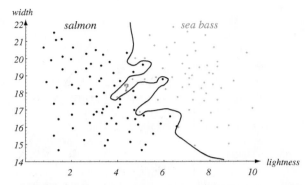

FIGURE 1.5. Overly complex models for the fish will lead to decision boundaries that are complicated. While such a decision may lead to perfect classification of our training samples, it would lead to poor performance on future patterns. The novel test point marked **?** is evidently most likely a salmon, whereas the complex decision boundary shown leads it to be classified as a sea bass.

on novel patterns.* But if designing a very complex recognizer is unlikely to give good generalization, precisely how should we quantify and favor simpler classifiers? How would our system automatically determine that the simple curve in Fig. 1.6 is preferable to the manifestly simpler straight line in Fig. 1.4 or the complicated boundary in Fig. 1.5? Assuming that we somehow manage to optimize this tradeoff, can we then *predict* how well our system will generalize to new patterns? These are some of the central problems in *statistical pattern recognition*.

For the same incoming patterns, we might need to use a drastically different task or cost function, and this will lead to different actions altogether. We might, for instance, wish instead to separate the fish based on their sex—all females (of either species) from all males—if we wish to sell roe. Alternatively, we might wish to cull

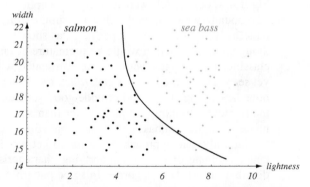

FIGURE 1.6. The decision boundary shown might represent the optimal tradeoff between performance on the training set and simplicity of classifier, thereby giving the highest accuracy on new patterns.

---

*The philosophical underpinnings of this approach derive from William of Occam (1284–1347?), who advocated favoring *simpler* explanations over those that are needlessly complicated: *Entia non sunt multiplicanda praeter necessitatem* ("Entities are not to be multiplied without necessity"). Decisions based on overly complex models often lead to lower accuracy of the classifier.

the damaged fish (to prepare separately for cat food), and so on. Different decision tasks may require features and yield boundaries quite different from those useful for our original categorization problem.

This makes it quite clear that our decisions are fundamentally task- or cost-specific, and that creating a single *general purpose* artificial pattern recognition device—that is, one capable of acting accurately based on a wide variety of tasks—is a profoundly difficult challenge. This, too, should give us added appreciation of the ability of humans to switch rapidly and fluidly between pattern recognition tasks.

Because classification is, at base, the task of recovering the model that generated the patterns, different classification techniques are useful depending on the type of candidate models themselves. In statistical pattern recognition we focus on the statistical properties of the patterns (generally expressed in probability densities), and this will command most of our attention in this book. Here the model for a pattern may be a single specific set of features, though the actual pattern sensed has been corrupted by some form of random noise. Occasionally it is claimed that *neural* pattern recognition (or neural network pattern classification) should be considered its own discipline, but despite its somewhat different intellectual pedigree, we will consider it a close descendant of statistical pattern recognition, for reasons that will become clear. If instead the model consists of some set of crisp logical rules, then we employ the methods of *syntactic* pattern recognition, where rules or grammars describe our decision. For example, we might wish to classify an English sentence as grammatical or not. Here crisp rules, rather than statistical descriptions of word frequencies or correlations, are appropriate.

It was necessary in our fish example to choose our features carefully, and hence achieve a *representation* (as in Fig. 1.6) that enabled reasonably successful pattern classification. A central aspect in virtually every pattern recognition problem is that of achieving such a "good" representation, one in which the structural relationships among the components are simply and naturally revealed, and one in which the true (unknown) model of the patterns can be expressed. In some cases, patterns should be represented as vectors of real-valued numbers, in others ordered lists of attributes, in yet others descriptions of parts and their relations, and so forth. We seek a representation in which the patterns that lead to the same action are somehow "close" to one another, yet "far" from those that demand a different action. The extent to which we create or learn a proper representation and how we quantify near and far apart will determine the success of our pattern classifier. A number of additional characteristics are desirable for the representation. We might wish to favor a small number of features, which might lead to (a) simpler decision regions, and (b) a classifier easier to train. We might also wish to have features that are robust—that is, relatively insensitive to noise or other errors. In practical applications we may need the classifier to act *quickly*, or use few electronic components, memory or processing steps.

A central technique, when we have insufficient training data, is to incorporate knowledge of the problem domain. Indeed, the less the training data, the more important is such knowledge—for instance, how the patterns themselves were produced. ANALYSIS BY SYNTHESIS One method that takes this notion to its logical extreme is that of *analysis by synthesis*, where in the ideal case one has a model of how each pattern is generated. Consider speech recognition. Amidst the manifest acoustic variability among the possible "dee"s that might be uttered by different people, one thing they have in common is that they were all produced by lowering the jaw slightly, opening the mouth, placing the tongue tip against the roof of the mouth after a certain delay, and so on. We might assume that "all" the acoustic variation is due to the happenstance of whether the talker is male or female, old or young, with different overall pitches, and so forth.

At some deep level, such a "physiological" model (or so-called "motor" model) for production of the "dee" utterances is appropriate, and different (say) from that for "doo" and indeed all other utterances. *If* this underlying model of production can be determined from the sound (and that is a very big *if*), then we can classify the utterance by how it was produced. That is to say, the production representation may be the "best" representation for classification. Our pattern recognition systems should then analyze (and hence classify) the input pattern based on how one would have to synthesize that pattern. The trick is, of course, to recover the generating parameters from the sensed pattern.

Consider the difficulty in making a recognizer of all types of chairs—standard office chair, contemporary living room chair, beanbag chair, and so forth—based on an image. Given the astounding variety in the number of legs, material, shape, and so on, we might despair of ever finding a representation that reveals the unity within the class of chair. Perhaps the only such unifying aspect of chairs is *functional*: A chair is a stable artifact that supports a human sitter, including back support. Thus we might try to deduce such functional properties from the image; and the property "can support a human sitter" is very indirectly related to the orientation of the larger surfaces, and it would need to be answered in the affirmative even for a beanbag chair. Of course, this requires some reasoning about the properties and naturally touches upon computer vision rather than pattern recognition proper.

Without going to such extremes, many real-world pattern recognition systems seek to incorporate at least *some* knowledge about the method of production of the patterns or their functional use in order to ensure a good representation, though of course the goal of the representation is classification, not reproduction. For instance, in optical character recognition (OCR) one might confidently assume that handwritten characters are written as a sequence of strokes and might first try to recover a stroke representation from the sensed image and then deduce the character from the identified strokes.

1.2.1 Related Fields

Pattern classification differs from classical statistical *hypothesis testing*, wherein the sensed data are used to decide whether or not to reject a *null hypothesis* in favor of some alternative hypothesis. Roughly speaking, if the probability of obtaining the data given some null hypothesis falls below a "significance" threshold, we reject the null hypothesis in favor of the alternative. Hypothesis testing is often used to determine whether a drug is effective, where the null hypothesis is that it has no effect. Hypothesis testing might be used to determine whether the fish on the conveyor belt belong to a single class (all salmon, for instance)—the null hypothesis—or instead from two classes (the alternative).

IMAGE
PROCESSING

Pattern classification differs, too, from *image processing*. In image processing, the input is an image and the output is an image. Image processing steps often include rotation, contrast enhancement, and other transformations which preserve all the original information. Feature extraction, such as finding the peaks and valleys of the intensity, loses information (but hopefully preserves everything relevant to the task at hand).

ASSOCIATIVE
MEMORY

As just described, *feature extraction* takes in a pattern and produces feature values. The number of features is virtually always chosen to be fewer than the total necessary to describe the complete target of interest, and this leads to a loss in information. In acts of *associative memory*, the system takes in a pattern and emits another pattern which is representative of a general group of patterns. It thus reduces the information somewhat, but rarely to the extent that pattern classification does. In short, because of the crucial role of a *decision* in pattern recognition information, it is fundamentally an information reduction process. You cannot reconstruct a pattern given only its category membership. The classification step represents an even more radical loss of information, reducing the original several thousand bits representing all the color of each of several thousand pixels down to just a few bits representing the chosen category (a single bit in our fish example.)

REGRESSION

Three closely interrelated fields, which are often employed in pattern recognition research, are regression, interpolation, and density estimation. In *regression*, we seek to find some functional description of data, often with the goal of predicting values for new input. Linear regression—in which the function is linear in the input variables—is by far the most popular and well studied form of regression. We might, for instance, feel that the length of a salmon varies linearly with its age or with weight, and take measurements of the age and length of many typical salmon and then use linear regression to find the coefficients.

INTERPOLATION

In *interpolation* we know or can easily deduce the function for certain ranges of input; the problem is then to infer the function for intermediate ranges of input. Thus we might know how the length of a salmon varies as age in the first two weeks of life, and above two years of age. We might then use any of a variety of interpolatation methods to infer how the length depends upon age between two weeks and two years

DENSITY ESTIMATION

of age. *Density estimation* is the problem of estimating the density (or probability) that a member of a certain category will be found to have particular features.

These fields are often employed—explicitly or implicitly—as first steps in pattern recognition. For instance, we shall see several methods for estimating the densities of different categories; an unknown pattern is then classified according to which category is the most probable. While these fields are highly developed and useful, we shall only indirectly address them as they relate to pattern classification.

1.3 PATTERN RECOGNITION SYSTEMS

In describing our hypothetical fish classification system, we distinguished between the three different operations of preprocessing, feature extraction and classification (see Fig. 1.1). Figure 1.7 shows a slightly more elaborate diagram of the components of a typical pattern recognition system. To understand the problem of designing such a system, we must understand the problems that each of these components must solve. Let us consider the operations of each component in turn, and reflect on the kinds of problems that can arise.

1.3.1 Sensing

The input to a pattern recognition system is often some kind of a transducer, such as a camera or a microphone array. The difficulty of the problem may well depend on the characteristics and limitations of the transducer—its bandwidth, resolution, sensitivity, distortion, signal-to-noise ratio, latency, etc. As important as it is in practice, the design of sensors for pattern recognition is beyond the scope of this book.

1.3.2 Segmentation and Grouping

In our fish example, we tacitly assumed that each fish was isolated, separate from others on the conveyor belt, and could easily be distinguished from the conveyor belt.

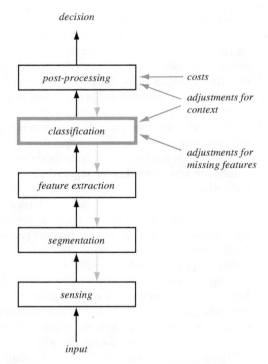

FIGURE 1.7. Many pattern recognition systems can be partitioned into components such as the ones shown here. A sensor converts images or sounds or other physical inputs into signal data. The segmentor isolates sensed objects from the background or from other objects. A feature extractor measures object properties that are useful for classification. The classifier uses these features to assign the sensed object to a category. Finally, a post processor can take account of other considerations, such as the effects of context and the costs of errors, to decide on the appropriate action. Although this description stresses a one-way or "bottom-up" flow of data, some systems employ feedback from higher levels back down to lower levels (gray arrows).

In practice, the fish would often be abutting or overlapping, and our system would have to determine where one fish ends and the next begins—the individual patterns have to be *segmented*. If we have already recognized the fish then it would be easier to segment their images. But how can we segment the images before they have been categorized, or categorize them before they have been segmented? It seems we need a way to know when we have switched from one model to another, or to know when we just have background or "no category." How can this be done?

Segmentation is one of the deepest problems in pattern recognition. In auto-mated speech recognition, we might seek to recognize the individual sounds (e.g., phonemes, such as "ss," "k," ...) and then put them together to determine the word. But consider two nonsense words, "sklee" and "skloo." Speak them aloud and notice that for "skloo" you push your lips forward (so-called "rounding" in anticipation of the upcoming "oo") *before* you utter the "ss." Such rounding influences the sound of the "ss," lowering the frequency spectrum compared to the "ss" sound in "sklee"—a phenomenon known as anticipatory coarticulation. Thus, the "oo" phoneme reveals its presence in the "ss" *earlier* than the "k" and "l" which nominally occur *before* the "oo" itself! How do we segment the "oo" phoneme from the others when they are so manifestly intermingled? Or should we even try? Perhaps we are focusing on groupings of the wrong size, and that the most useful unit for recognition is somewhat larger.

Closely related to the problem of segmentation is the problem of recognizing or grouping together the various parts of a composite object. The letter **i** or the symbol = have two connected components, but we see them as one symbol. We effortlessly read a simple word such as **BEATS**. But consider this: Why didn't we read instead *other* words that are perfectly good subsets of the full pattern, such as **BE**, **BEAT**, **EAT**, **AT**, and **EATS**? Why don't they enter our minds, unless explicitly brought to our attention? Or when we saw the **B** why didn't we read a **P** or an **I**, which are "there" within the **B**? Conversely, how is it that we can read the two unsegmented words in **POLOPONY**—without placing the *entire* input into a single word category?

MEREOLOGY

This is the problem of *subsets and supersets*—formally part of *mereology*, the study of part/whole relationships. It appears as though the best classifiers try to incorporate as much of the input into the categorization as "makes sense," but not too much. How can this be done automatically?

1.3.3 Feature Extraction

The conceptual boundary between feature extraction and classification proper is somewhat arbitrary: An ideal feature extractor would yield a representation that makes the job of the classifier trivial; conversely, an omnipotent classifier would not need the help of a sophisticated feature extractor. The distinction is forced upon us for practical rather than theoretical reasons.

The traditional goal of the feature extractor is to characterize an object to be recognized by measurements whose values are very similar for objects in the same category, and very different for objects in different categories. This leads to the idea of

INVARIANT
FEATURES

seeking *distinguishing features* that are *invariant* to irrelevant transformations of the input. In our fish example, the absolute location of a fish on the conveyor belt is irrelevant to the category, and thus our representation should also be insensitive to the absolute location of the fish. Ideally, in this case we want the features to be invari-

TRANSLATION
ROTATION

ant to translation, whether horizontal or vertical. Because rotation is also irrelevant for classification, we would also like the features to be invariant to rotation. Finally, the size of the fish may not be important—a young, small salmon is still a salmon.

SCALE

Thus, we may also want the features to be invariant to scale. In general, features that describe properties such as shape, color and many kinds of texture are invariant to translation, rotation and scale.

The problem of finding rotation invariant features from an overhead image of a fish on a conveyor belt is simplified by the fact that the fish is likely to be lying flat, and the axis of rotation is always parallel to the camera's line of sight. A more general invariance would be for rotations about an arbitrary line in three dimensions. The image of even such a "simple" object as a coffee cup undergoes radical variation as the cup is rotated to an arbitrary angle: The handle may become *occluded*—that is,

OCCLUSION

hidden by another part. The bottom of the inside volume come into view, the circular lip appear oval or a straight line or even obscured, and so forth. Furthermore, if the distance between the cup and the camera can change, the image is subject to

PROJECTIVE
DISTORTION

projective distortion. How might we ensure that the features are invariant to such complex transformations? Or should we define different subcategories for the image of a cup and achieve the rotation invariance at a higher level of processing?

In speech recognition, we want features that are invariant to translations in time and to changes in the overall amplitude. We may also want features that are insensitive to the duration of the word, i.e., invariant to the *rate* at which the pattern evolves. Rate variation is a serious problem in speech recognition. Not only do different people talk at different rates, but even a single talker may vary in rate, causing the speech signal to change in complex ways. Likewise, cursive handwriting varies in complex ways as the writer speeds up—the placement of dots on the **i**'s, and cross bars on the **t**'s and **f**'s, are the first casualties of rate increase, while the appearance of **l**'s and **e**'s are relatively inviolate. How can we make a recognizer that changes its representations for some categories *differently* from that for others under such rate variation?

A large number of highly complex transformations arise in pattern recognition, and many are domain specific. We might wish to make our handwritten optical character recognizer insensitive to the overall thickness of the pen line, for instance. Far
more severe are transformations such as *nonrigid deformations* that arise in three-dimensional object recognition, such as the radical variation in the image of your hand as you grasp an object or snap your fingers. Similarly, variations in illumination or the complex effects of cast shadows may need to be taken into account.

As with segmentation, the task of feature extraction is much more problem- and domain-dependent than is classification proper, and thus requires knowledge of the domain. A good feature extractor for sorting fish would probably be of little use for identifying fingerprints, or classifying photomicrographs of blood cells. However, some of the principles of pattern classification can be used in the design of the feature extractor. Although the pattern classification techniques presented in this book cannot substitute for domain knowledge, they can be helpful in making the feature
values less sensitive to noise. In some cases, they can also be used to select the most valuable features from a larger set of candidate features.

1.3.4 Classification

The task of the classifier component proper of a full system is to use the feature vector provided by the feature extractor to assign the object to a category. Most of this book is concerned with the design of the classifier. Because perfect classification performance is often impossible, a more general task is to determine the probability for each of the possible categories. The abstraction provided by the feature-vector representation of the input data enables the development of a largely domain-independent theory of classification.

The degree of difficulty of the classification problem depends on the variability in the feature values for objects in the same category relative to the difference between feature values for objects in different categories. The variability of feature values for
objects in the same category may be due to complexity, and may be due to *noise*. We define noise in very general terms: any property of the sensed pattern which is not due to the true underlying model but instead to randomness in the world or the sensors. All nontrivial decision and pattern recognition problems involve noise in some form. What is the best way to design a classifier to cope with this variability? What is the best performance that is possible?

One problem that arises in practice is that it may not always be possible to determine the values of all of the features for a particular input. In our hypothetical system for fish classification, for example, it may not be possible to determine the width of the fish because of occlusion by another fish. How should the categorizer compensate? Since our two-feature recognizer never had a single-variable criterion value x^* determined in anticipation of the possible absence of a feature (cf. Fig. 1.3), how shall it make the best decision using only the feature present? The naïve method, of merely assuming that the value of the missing feature is zero or the average of the values for the patterns already seen, is provably nonoptimal. Likewise, how should we train a classifier or use one when some features are missing?

1.3.5 Post Processing

A classifier rarely exists in a vacuum. Instead, it is generally to be used to recommend actions (put this fish in this bucket, put that fish in that bucket), each action having an associated cost. The post-processor uses the output of the classifier to decide on the recommended action.

ERROR RATE

Conceptually, the simplest measure of classifier performance is the classification error rate—the percentage of new patterns that are assigned to the wrong category. Thus, it is common to seek minimum-error-rate classification. However, it may be much better to recommend actions that will minimize the total expected cost, which

RISK

is called the *risk*. How do we incorporate knowledge about costs and how will they affect our classification decision? Can we estimate the total risk and thus tell whether our classifier is acceptable even before we field it? Can we estimate the lowest possible risk of *any* classifier, to see how close ours meets this ideal, or whether the problem is simply too hard overall?

CONTEXT

The post-processor might also be able to exploit *context*—input-dependent information other than from the target pattern itself—to improve system performance. Suppose in an optical character recognition system we encounter a sequence that looks like T/~\E C/~\T. Even though the system may be unable to classify each /~\ as an isolated character, in the context of English it is clear that the first instance should be an H and the second an A. Context can be highly complex and abstract. The utterance "jeetyet?" may seem nonsensical, unless you hear it spoken by a friend in the context of the cafeteria at lunchtime—"did you eat yet?" How can such a visual and temporal context influence your recognition of speech?

In our fish example we saw how using multiple features could lead to improved recognition. We might imagine that we could also do better if we used multiple clas-

MULTIPLE
CLASSIFIERS

sifiers, each classifier operating on different aspects of the input. For example, we might combine the results of acoustic recognition and lip reading to improve the performance of a speech recognizer.

If all of the classifiers agree on a particular pattern, there is no difficulty. But suppose they disagree. How should a "super" classifier *pool the evidence* from the component recognizers to achieve the best decision? Imagine calling in ten experts for determining whether or not a particular fish is diseased. While nine agree that the fish is healthy, one expert does not. Who is right? It may be that the lone dissenter is the only one familiar with the particular very rare symptoms in the fish, and is in fact correct. How would the "super" categorizer know when to base a decision on a minority opinion, even from an expert in one small domain who is not well-qualified to judge throughout a broad range of problems?

We have asked more questions in this section than we have answered. Our purpose was to emphasize the complexity of pattern recognition problems and to dispel naïve hope that any single approach has the power to solve all pattern recognition problems. The methods presented in this book are primarily useful for the classification step. We shall see that they also have relevance to those segmentation, feature extraction and post-processing problems that are not highly domain-dependent. However, performance on difficult pattern recognition problems generally requires exploiting domain-specific knowledge.

1.4 THE DESIGN CYCLE

The design of a pattern recognition system usually entails the repetition of a number of different activities: data collection, feature choice, model choice, training, and evaluation. In this section we present an overview of this design cycle (Fig. 1.8) and consider some of the problems that frequently arise.

1.4.1 Data Collection

Data collection can account for surprisingly large part of the cost of developing a pattern recognition system. It may be possible to perform a preliminary feasibility study with a small set of "typical" examples, but much more data will usually be needed to assure good performance in the fielded system. How do we know when we have collected an adequately large and representative set of examples for training and testing the system?

1.4.2 Feature Choice

The choice of the distinguishing features is a critical design step and depends on the characteristics of the problem domain. Having access to example data, such as

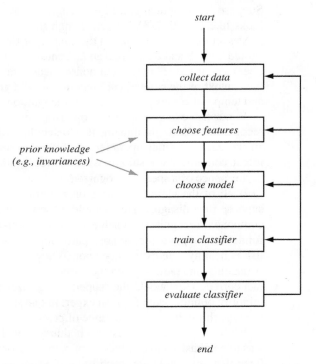

FIGURE 1.8. The design of a pattern recognition system involves a design cycle similar to the one shown here. Data must be collected, both to train and to test the system. The characteristics of the data affect both the choice of appropriate discriminating features and the choice of models for the different categories. The training process uses some or all of the data to determine the system parameters. The results of evaluation may call for repetition of various steps in this process in order to obtain satisfactory results.

PRIOR
KNOWLEDGE pictures of fish on the conveyor belt, will certainly be valuable for choosing a feature set. However, *prior knowledge* also plays a major role.

In our hypothetical fish-classification example, prior knowledge about the lightness of the different fish categories helped in the design of a classifier by suggesting a promising feature. Incorporating prior knowledge can be far more subtle and difficult. In some applications the knowledge ultimately derives from information about the production of the patterns, as we saw in analysis-by-synthesis. In others the knowledge may be about the *form* of the underlying categories, or specific attributes of the patterns, such as the fact that a face has two eyes, one nose, and so on.

In selecting or designing features, we obviously would like to find features that are simple to extract, invariant to irrelevant transformations, insensitive to noise, and useful for discriminating patterns in different categories. How do we combine prior knowledge and empirical data to find relevant and effective features?

1.4.3 Model Choice

We might have been unsatisfied with the performance of our fish classifier in Figs. 1.4 and 1.5, and thus jumped to an entirely different class of model, for instance one based on some function of the number and position of the fins, the color of the eyes, the weight, shape of the mouth, and so on. How do we know when a hypothesized model differs significantly from the true model underlying our patterns, and thus a new model is needed? In short, how are we to know to reject a class of models and try another one? Are we as designers reduced to random and tedious trial and error in model selection, never really knowing whether we can expect improved performance? Or might there be principled methods for knowing when to jettison one class of models and invoke another?

1.4.4 Training

In general, the process of using data to determine the classifier is referred to as *training* the classifier. Much of this book is concerned with the many different procedures for training classifiers and choosing models.

We have already seen many problems that arise in the design of pattern recognition systems. No universal methods have been found for solving all of these problems. However, the repeated experience of the last quarter century has been that the most effective methods for developing classifiers involve learning from example patterns. Throughout this book, we shall see again and again how methods of learning relate to these central problems, and how they are essential in the engineering of pattern recognition systems.

1.4.5 Evaluation

When we went from the use of one feature to two in our fish classification problem, it was essentially the result of an evaluation that the error rate we could obtain with one feature was inadequate, and that it was possible to do better. When we went from the simple straight-line classifier in Fig. 1.4 to the more complicated model illustrated in Fig. 1.5, it was again the result of an evaluation that it was possible to do still better. Evaluation is important both to measure the performance of the system and to identify the need for improvements in its components.

OVERFITTING

While an overly complex system may allow perfect classification of the training samples, it is unlikely perform well on new patterns. This situation is known as *overfitting*. One of the most important areas of research in statistical pattern classification is determining how to adjust the complexity of the model—not so simple that it cannot explain the differences between the categories, yet not so complex as to give poor classification on novel patterns. Are there principled methods for finding the best (intermediate) complexity for a classifier?

1.4.6 Computational Complexity

Some pattern recognition problems can be "solved" using algorithms that are highly impractical. For instance, we might try to hand label all possible 20×20 binary pixel images with a category label for optical character recognition, and use table lookup to classify incoming patterns. Although we might in theory achieve error-free recognition, the labeling time and storage requirements would be quite prohibitive since it would require a labeling each of $2^{20 \times 20} \approx 10^{120}$ patterns. Thus the computational resources necessary and the computational complexity of different algorithms are of considerable practical importance.

In more general terms, we may ask how an algorithm scales as a function of the number of feature dimensions, or the number of patterns or the number of categories. What is the tradeoff between computational ease and performance? In some problems we know we can design an excellent recognizer, but not within the engineering constraints. How can we optimize *within* such constraints? We are typically less concerned with the complexity of learning, which is done in the laboratory, than with the complexity of making a decision, which is done with the fielded application. While computational complexity generally correlates with the complexity of the hypothesized model of the patterns, these two notions are conceptually different.

1.5 LEARNING AND ADAPTATION

In the broadest sense, any method that incorporates information from training samples in the design of a classifier employs learning. Because nearly all practical or interesting pattern recognition problems are so hard that we cannot guess the best classification decision ahead of time, we shall spend the great majority of our time here considering learning. Creating classifiers then involves positing some general form of model, or form of the classifier, and using training patterns to learn or estimate the unknown parameters of the model. Learning refers to some form of algorithm for reducing the error on a set of training data. A range of *gradient descent* algorithms that alter a classifier's parameters in order to reduce an error measure now permeate the field of statistical pattern recognition, and these will demand a great deal of our attention. Learning comes in several general forms.

1.5.1 Supervised Learning

In supervised learning, a teacher provides a category label or cost for each pattern in a training set, and seeks to reduce the sum of the costs for these patterns. How can we be sure that a particular learning algorithm is powerful enough to learn the solution to a given problem and that it will be stable to parameter variations? How can we determine if it will converge in finite time or if it will scale reasonably with the number of training patterns, the number of input features or the number of categories? How can we ensure that the learning algorithm appropriately favors "simple" solutions (as in Fig. 1.6) rather than complicated ones (as in Fig. 1.5)?

1.5.2 Unsupervised Learning

In *unsupervised learning* or *clustering* there is no explicit teacher, and the system forms clusters or "natural groupings" of the input patterns. "Natural" is always defined explicitly or implicitly in the clustering system itself; and given a particular set of patterns or cost function, different clustering algorithms lead to different clusters. Often the user will set the hypothesized number of different clusters ahead of time, but how should this be done? How do we avoid inappropriate representations?

1.5.3 Reinforcement Learning

CRITIC

The most typical way to train a classifier is to present an input, compute its tentative category label, and use the known target category label to improve the classifier. For instance, in optical character recognition, the input might be an image of a character, the actual output of the classifier the category label "R," and the desired output a "B." In *reinforcement learning* or *learning with a critic*, no desired category signal is given; instead, the only teaching feedback is that the tentative category is right or wrong. This is analogous to a critic who merely states that something is right or wrong, but does not say specifically *how* it is wrong. In pattern classification, it is most common that such reinforcement is binary—either the tentative decision is correct or it is not. How can the system learn from such nonspecific feedback?

1.6 CONCLUSION

At this point the reader may be overwhelmed by the number, complexity, and magnitude of the subproblems of pattern recognition. Furthermore, these subproblems are rarely addressed in isolation and they are invariably interrelated. Thus for instance in seeking to reduce the complexity of our classifier, we might affect its ability to deal with invariance. We point out, however, that the good news is at least threefold: (1) There is an "existence proof" that many of these problems can indeed be solved— as demonstrated by humans and other biological systems, (2) mathematical theories solving some of these problems have in fact been discovered, and finally (3) there remain many fascinating unsolved problems providing opportunities for progress.

SUMMARY BY CHAPTERS

This book first addresses those cases where a great deal of information about the models is known (such as the probability densities, category labels,...) and moves, chapter by chapter, toward problems where the form of the distributions are unknown and even the category membership of training patterns is unknown. We begin in Chapter 2 (Bayesian Decision Theory) by considering the ideal case in which the probability structure underlying the categories is known perfectly. While this sort of situation rarely occurs in practice, it permits us to determine the optimal (Bayes) classifier against which we can compare all other classifiers. Moreover, in some problems it enables us to predict the error we will get when we generalize to novel patterns. In Chapter 3 (Maximum-Likelihood and Bayesian Parameter Estimation) we address the case when the full probability structure underlying the categories is not known, but the general *forms* of their distributions *are* known. Thus the uncertainty about a probability distribution is represented by the values of some unknown parameters, and we seek to determine these parameters to attain the best categorization. In Chapter 4 (Nonparametric Techniques) we move yet further from the Bayesian ideal, and assume that we have *no* prior parameterized knowledge about the underlying probability structure; in essence our classification will be based on information provided by training samples alone. Classic techniques such as the nearest-neighbor algorithm and potential functions play an important role here.

Then in Chapter 5 (Linear Discriminant Functions) we return somewhat toward the general approach of parameter estimation. We shall assume that the so-called "discriminant functions" are of a very particular form—namely linear—in order to derive a class of incremental training rules. Next, in Chapter 6 (Multilayer Neural Networks) we see how some of the ideas from such linear discriminants can be extended to a class of very powerful algorithms for training multilayer neural networks; these neural techniques have a range of useful properties that have made them a mainstay in contemporary pattern recognition research. In Chapter 7 (Stochastic Methods) we discuss simulated annealing, the Boltzmann learning algorithm and other stochastic methods which can avoid some of the estimation problems that plague other neural methods. Chapter 8 (Nonmetric Methods) moves beyond models that are statistical in nature to ones that can be best described by logical rules. Here we discuss tree-based algorithms such as CART (which can also be applied to statistical data) and syntactic-based methods based on grammars.

Chapter 9 (Algorithm-Independent Machine Learning) is both the most important chapter and the most difficult one in the book. Some of the results described there—those related to bias and variance, degrees of freedom, the desire for "simple" classifiers, and computational complexity—are subtle but nevertheless crucial both theoretically and practically. In some sense, the other chapters can only be fully understood (or used) in light of the results presented here.

We conclude in Chapter 10 (Unsupervised Learning and Clustering) by addressing the case when input training patterns are not labeled, and where our recognizer must determine the cluster structure. We also treat a related problem, that of learning with a critic, in which the teacher provides only a single bit of information during the presentation of a training pattern—"yes," that the classification provided by the recognizer is correct, or "no," it isn't.

BIBLIOGRAPHICAL AND HISTORICAL REMARKS

Classification is among the first crucial steps in making sense of the blooming buzzing confusion of sensory data that intelligent systems confront. In the Western world, the foundations of pattern recognition can be traced to Plato [2], which were later extended by Aristotle [1], who distinguished between an "essential property" (which would be shared by all members in a class or "natural kind" as he put it) from an "accidental property" (which could differ among members in the class). Pattern recognition can be cast as the problem of finding such essential properties of a category. In the Eastern world, the first Zen patriarch, Bodhidharma, would point at things and demand students to answer "What is that?" as a way of confronting the deepest issues in mind, the identity of objects, and the nature of classification and decision [3]. It has been a central theme in the discipline of philosophical epistemology, the study of the nature of knowledge. A more modern treatment of some philosophical problems of pattern recognition, relating to the technical matter in the current book, can be found in references [22, 4] and [18]. A delightful and particularly insightful book on the foundations of artificial intelligence, including pattern recognition, is reference [10].

There are a number of overviews and reference books that can be recommended, including references [5] and [6]. The modern literature on decision theory and pattern recognition is now overwhelming, and comprises dozens of journals, thousands of books and conference proceedings and innumerable articles; it continues to grow rapidly. While some disciplines such as statistics [8], machine learning [17], and neural networks [9], expand the foundations of pattern recognition, others, such as computer vision [7, 19] and speech recognition [16], rely on it heavily. Perceptual Psychology, Cognitive Science [13], Psychobiology [21], and Neuroscience [11] analyze how pattern recognition is performed by humans and other animals. The extreme view that everything in human cognition—including rule-following and logic—can be reduced to pattern recognition is presented in reference [14]. Pattern recognition techniques have been applied in virtually every scientific and technical discipline.

BIBLIOGRAPHY

[1] Aristotle, Robin Waterfield, and David Bostock. *Physics*. Oxford University Press, Oxford, UK, 1996.

[2] Allan Bloom. *The Republic of Plato*. Basic Books, New York, second edition, 1991.

[3] Bodhidharma. *The Zen Teachings of Bodhidharma*. North Point Press, San Francisco, CA, 1989.

[4] Mikhail M. Bongard. *Pattern Recognition*. Spartan Books, Washington, D.C., 1970.

[5] Chi-hau Chen, Louis François Pau, and Patrick S. P. Wang, editors. *Handbook of Pattern Recognition & Computer Vision*. World Scientific, Singapore, second edition, 1993.

[6] Luc Devroye, László Györfi, and Gábor Lugosi. *A Probabilistic Theory of Pattern Recognition*. Springer, New York, 1996.

[7] Marty Fischler and Oscar Firschein. *Readings in Computer Vision: Issues, Problems, Principles and Paradigms*. Morgan Kaufmann, San Mateo, CA, 1987.

[8] Keinosuke Fukunaga. *Introduction to Statistical Pattern Recognition*. Academic Press, New York, second edition, 1990.

[9] John Hertz, Anders Krogh, and Richard G. Palmer. *Introduction to the Theory of Neural Computation*. Addison-Wesley, Redwood City, CA, 1991.

[10] Douglas Hofstadter. *Gödel, Escher, Bach: An Eternal Golden Braid*. Basic Books, New York, 1979.

[11] Eric R. Kandel and James H. Schwartz. *Principles of Neural Science*. Elsevier, New York, second edition, 1985.

[12] Immanuel Kant. *Critique of Pure Reason*. Prometheus Books, New York, 1990.

[13] George F. Luger. *Cognitive Science: The Science of Intelligent Systems*. Academic Press, New York, 1994.

[14] Howard Margolis. *Patterns, Thinking, and Cognition: A Theory of Judgement*. University of Chicago Press, Chicago, IL, 1987.

[15] Karl Raimund Popper. *Popper Selections*. Princeton University Press, Princeton, NJ, 1985.

[16] Lawrence Rabiner and Biing-Hwang Juang. *Fundamentals of Speech Recognition*. Prentice-Hall, Englewood Cliffs, NJ, 1993.

[17] Jude W. Shavlik and Thomas G. Dietterich, editors. *Readings in Machine Learning*. Morgan Kaufmann, San Mateo, CA, 1990.

[18] Brian Cantwell Smith. *On the Origin of Objects*. MIT Press, Cambridge, MA, 1996.

[19] Louise Stark and Kevin Bowyer. *Generic Object Recognition Using Form & Function*. World Scientific, River Edge, NJ, 1996.

[20] Donald R. Tveter. *The Pattern Recognition Basis of Artificial Intelligence*. IEEE Press, New York, 1998.

[21] William R. Uttal. *The Psychobiology of Sensory Coding*. HarperCollins, New York, 1973.

[22] Satoshi Watanabe. *Knowing and Guessing: A Quantitative Study of Inference and Information*. Wiley, New York, 1969.

BAYESIAN DECISION THEORY

2.1 INTRODUCTION

Bayesian decision theory is a fundamental statistical approach to the problem of pattern classification. This approach is based on quantifying the tradeoffs between various classification decisions using probability and the costs that accompany such decisions. It makes the assumption that the decision problem is posed in probabilistic terms, and that all of the relevant probability values are known. In this chapter we develop the fundamentals of this theory and we show how it can be viewed as being simply a formalization of common-sense procedures; in subsequent chapters we will consider the problems that arise when the probabilistic structure is not completely known.

While we will give a quite general, abstract development of Bayesian decision theory in Section 2.2, we begin our discussion with a specific example. Let us re-consider the hypothetical problem posed in Chapter 1 of designing a classifier to separate two kinds of fish: sea bass and salmon. Suppose that an observer watching fish arrive along the conveyor belt finds it hard to predict what type will emerge next and that the sequence of types of fish appears to be random. In decision-theoretic terminology we would say that as each fish emerges nature is in one or the other of the two possible states: Either the fish is a sea bass or the fish is a salmon. We let ω

STATE OF NATURE

denote the *state of nature*, with $\omega = \omega_1$ for sea bass and $\omega = \omega_2$ for salmon. Because the state of nature is so unpredictable, we consider ω to be a variable that must be described probabilistically.

If the catch produced as much sea bass as salmon, we would say that the next fish is equally likely to be sea bass or salmon. More generally, we assume that there is

PRIOR

some *a priori probability* (or simply *prior*) $P(\omega_1)$ that the next fish is sea bass, and some prior probability $P(\omega_2)$ that it is salmon. If we assume there are no other types of fish relevant here, then $P(\omega_1)$ and $P(\omega_2)$ sum to one. These prior probabilities reflect our prior knowledge of how likely we are to get a sea bass or salmon before the fish actually appears. It might, for instance, depend upon the time of year or the choice of fishing area.

Suppose for a moment that we were forced to make a decision about the type of fish that will appear next without being allowed to see it. For the moment, we shall assume that any incorrect classification entails the same cost or consequence, and that the only information we are allowed to use is the value of the prior probabilities. If a decision must be made with so little information, it seems logical to use the following *decision rule*: Decide ω_1 if $P(\omega_1) > P(\omega_2)$; otherwise decide ω_2.

DECISION RULE

This rule makes sense if we are to judge just one fish, but if we are to judge many fish, using this rule repeatedly may seem a bit strange. After all, we would always make the same decision even though we know that *both* types of fish will appear. How well it works depends upon the values of the prior probabilities. If $P(\omega_1)$ is very much greater than $P(\omega_2)$, our decision in favor of ω_1 will be right most of the time. If $P(\omega_1) = P(\omega_2)$, we have only a fifty-fifty chance of being right. In general, the probability of error is the smaller of $P(\omega_1)$ and $P(\omega_2)$, and we shall see later that under these conditions no other decision rule can yield a larger probability of being right.

In most circumstances we are not asked to make decisions with so little information. In our example, we might for instance use a lightness measurement x to improve our classifier. Different fish will yield different lightness readings, and we express this variability in probabilistic terms; we consider x to be a continuous random variable whose distribution depends on the state of nature and is expressed as $p(x|\omega)$.* This is the *class-conditional probability density* function, the probability density function for x given that the state of nature is ω. (It is also sometimes called state-conditional probability density.) Then the difference between $p(x|\omega_1)$ and $p(x|\omega_2)$ describes the difference in lightness between populations of sea bass and salmon (Fig. 2.1).[†]

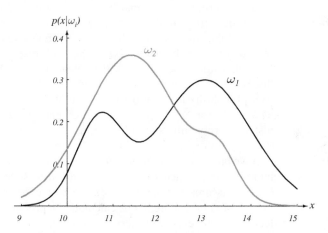

FIGURE 2.1. Hypothetical class-conditional probability density functions show the probability density of measuring a particular feature value x given the pattern is in category ω_i. If x represents the lightness of a fish, the two curves might describe the difference in lightness of populations of two types of fish. Density functions are normalized, and thus the area under each curve is 1.0.

---

*We generally use an uppercase $P(\cdot)$ to denote a probability mass function and use a lowercase $p(\cdot)$ to denote a probability density function.

[†]Strictly speaking, the probability density function $p(x|\omega)$ should be written as $p_x(x|\omega)$ to indicate that we are speaking about a particular density function for the random variable X. This more elaborate subscripted notation makes it clear that $p_x(\cdot)$ and $p_y(\cdot)$ denote two different functions, a fact that is obscured when writing $p(x)$ and $p(y)$. Because this potential confusion rarely arises in practice, we have elected to adopt the simpler notation. Readers who are unsure of our notation or who would like to review probability theory should see Section A.4 of the Appendix.

Suppose that we know both the prior probabilities $P(\omega_j)$ and the conditional densities $p(x|\omega_j)$ for $j = 1, 2$. Suppose further that we measure the lightness of a fish and discover that its value is x. How does this measurement influence our attitude concerning the true state of nature—that is, the category of the fish? We note first that the (joint) probability density of finding a pattern that is in category ω_j *and* has feature value x can be written in two ways: $p(\omega_j, x) = P(\omega_j|x)p(x) = p(x|\omega_j)P(\omega_j)$. Rearranging these leads us to the answer to our question, which is called *Bayes formula*:

$$P(\omega_j|x) = \frac{p(x|\omega_j)P(\omega_j)}{p(x)}, \tag{1}$$

where in this case of two categories

$$p(x) = \sum_{j=1}^{2} p(x|\omega_j)P(\omega_j). \tag{2}$$

Bayes formula can be expressed informally in English by saying that

$$posterior = \frac{likelihood \times prior}{evidence}. \tag{3}$$

POSTERIOR

LIKELIHOOD

EVIDENCE

Bayes formula shows that by observing the value of x we can convert the prior probability $P(\omega_j)$ to the *a posteriori* probability (or *posterior*) $P(\omega_j|x)$—the probability of the state of nature being ω_j given that feature value x has been measured. We call $p(x|\omega_j)$ the *likelihood* of ω_j with respect to x, a term chosen to indicate that, other things being equal, the category ω_j for which $p(x|\omega_j)$ is large is more "likely" to be the true category. Notice that it is the product of the likelihood and the prior probability that is most important in determining the posterior probability; the *evidence* factor, $p(x)$, can be viewed as merely a scale factor that guarantees that the posterior probabilities sum to one, as all good probabilities must. The variation of $P(\omega_j|x)$ with x is illustrated in Fig. 2.2 for the case $P(\omega_1) = 2/3$ and $P(\omega_2) = 1/3$.

If we have an observation x for which $P(\omega_1|x)$ is greater than $P(\omega_2|x)$, we would naturally be inclined to decide that the true state of nature is ω_1. Conversely, if $P(\omega_2|x)$ is greater than $P(\omega_1|x)$, we would be inclined to choose ω_2. To justify this decision procedure, let us calculate the probability of error whenever we make a decision. Whenever we observe a particular x, the probability of error is

$$P(error|x) = \begin{cases} P(\omega_1|x) & \text{if we decide } \omega_2 \\ P(\omega_2|x) & \text{if we decide } \omega_1. \end{cases} \tag{4}$$

Clearly, for a given x we can minimize the probability of error by deciding ω_1 if $P(\omega_1|x) > P(\omega_2|x)$ and ω_2 otherwise. Of course, we may never observe exactly the same value of x twice. Will this rule minimize the average probability of error? Yes, because the average probability of error is given by

$$P(error) = \int_{-\infty}^{\infty} P(error, x)\, dx = \int_{-\infty}^{\infty} P(error|x)p(x)\, dx \tag{5}$$

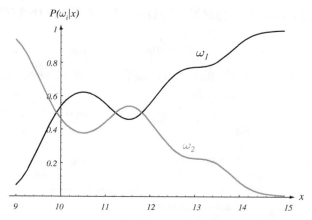

FIGURE 2.2. Posterior probabilities for the particular priors $P(\omega_1) = 2/3$ and $P(\omega_2) = 1/3$ for the class-conditional probability densities shown in Fig. 2.1. Thus in this case, given that a pattern is measured to have feature value $x = 14$, the probability it is in category ω_2 is roughly 0.08, and that it is in ω_1 is 0.92. At every x, the posteriors sum to 1.0.

BAYES
DECISION RULE

and if for every x we ensure that $P(error|x)$ is as small as possible, then the integral must be as small as possible. Thus we have justified the following *Bayes decision rule* for minimizing the probability of error:

$$\text{Decide } \omega_1 \text{ if } P(\omega_1|x) > P(\omega_2|x); \text{ otherwise decide } \omega_2. \tag{6}$$

Under this rule Eq. 4 becomes

$$P(error|x) = \min[P(\omega_1|x), P(\omega_2|x)]. \tag{7}$$

EVIDENCE

This form of the decision rule emphasizes the role of the posterior probabilities. By using Eq. 1 we can instead express the rule in terms of the conditional and prior probabilities. First note that the *evidence*, $p(x)$, in Eq. 1 is unimportant as far as making a decision is concerned. It is basically just a scale factor that states how frequently we will actually measure a pattern with feature value x; as mentioned above, its presence in Eq. 1 assures us that $P(\omega_1|x) + P(\omega_2|x) = 1$. By eliminating this scale factor, we obtain the following completely equivalent decision rule:

$$\text{Decide } \omega_1 \text{ if } p(x|\omega_1)P(\omega_1) > p(x|\omega_2)P(\omega_2); \quad \text{otherwise decide } \omega_2. \tag{8}$$

Some additional insight can be obtained by considering a few special cases. If for some x we have $p(x|\omega_1) = p(x|\omega_2)$, then that particular observation gives us no information about the state of nature; in this case, the decision hinges entirely on the prior probabilities. On the other hand, if $P(\omega_1) = P(\omega_2)$, then the states of nature are equally probable; in this case the decision is based entirely on the likelihoods $p(x|\omega_j)$. In general, both of these factors are important in making a decision, and the Bayes decision rule combines them to achieve the minimum probability of error.

2.2 BAYESIAN DECISION THEORY—CONTINUOUS FEATURES

We shall now formalize the ideas just considered, and generalize them in four ways:

- By allowing the use of more than one feature
- By allowing more than two states of nature
- By allowing actions other than merely deciding the state of nature
- By introducing a loss function more general than the probability of error

FEATURE SPACE

These generalizations and their attendant notational complexities should not obscure the central points illustrated in our simple example. Allowing the use of more than one feature merely requires replacing the scalar x by the *feature vector* $\mathbf{x}$, where $\mathbf{x}$ is in a d-dimensional Euclidean space $\mathbf{R}^d$, called the *feature space*. Allowing more than two states of nature provides us with a useful generalization for a small notational expense. Allowing actions other than classification primarily allows the possibility of rejection—that is, of refusing to make a decision in close cases; this is a

LOSS FUNCTION

useful option if being indecisive is not too costly. Formally, the *loss function* states exactly how costly each action is, and is used to convert a probability determination into a decision. Cost functions let us treat situations in which some kinds of classification mistakes are more costly than others, although we often discuss the simplest case, where all errors are equally costly. With this as a preamble, let us begin the more formal treatment.

Let $\{\omega_1, \ldots, \omega_c\}$ be the finite set of c states of nature ("categories") and let $\{\alpha_1, \ldots, \alpha_a\}$ be the finite set of a possible actions. The loss function $\lambda(\alpha_i|\omega_j)$ describes the loss incurred for taking action α_i when the state of nature is ω_j. Let the feature vector $\mathbf{x}$ be a d-component vector-valued random variable and let $p(\mathbf{x}|\omega_j)$ be the state-conditional probability density function for $\mathbf{x}$, with the probability density function for $\mathbf{x}$ conditioned on ω_j being the true state of nature. As before, $P(\omega_j)$ describes the prior probability that nature is in state ω_j. Then the posterior probability $P(\omega_j|\mathbf{x})$ can be computed from $p(\mathbf{x}|\omega_j)$ by Bayes formula:

$$P(\omega_j|\mathbf{x}) = \frac{p(\mathbf{x}|\omega_j)P(\omega_j)}{p(\mathbf{x})}, \tag{9}$$

where the evidence is now

$$p(\mathbf{x}) = \sum_{j=1}^{c} p(\mathbf{x}|\omega_j)P(\omega_j). \tag{10}$$

Suppose that we observe a particular $\mathbf{x}$ and that we contemplate taking action α_i. If the true state of nature is ω_j, by definition we will incur the loss $\lambda(\alpha_i|\omega_j)$. Because $P(\omega_j|\mathbf{x})$ is the probability that the true state of nature is ω_j, the expected loss associated with taking action α_i is merely

$$R(\alpha_i|\mathbf{x}) = \sum_{j=1}^{c} \lambda(\alpha_i|\omega_j)P(\omega_j|\mathbf{x}). \tag{11}$$

RISK

In decision-theoretic terminology, an expected loss is called a *risk*, and $R(\alpha_i|\mathbf{x})$ is called the *conditional risk*. Whenever we encounter a particular observation $\mathbf{x}$, we can minimize our expected loss by selecting the action that minimizes the conditional risk. We shall now show that this *Bayes decision procedure* actually provides the optimal performance.

Stated formally, our problem is to find a decision rule against $P(\omega_j)$ that minimizes the overall risk. A general *decision rule* is a function $\alpha(\mathbf{x})$ that tells us which action to take for every possible observation. To be more specific, for every $\mathbf{x}$ the *decision function* $\alpha(\mathbf{x})$ assumes one of the a values $\alpha_1, \ldots, \alpha_a$. The overall risk R is the expected loss associated with a given decision rule. Because $R(\alpha_i|\mathbf{x})$ is the conditional risk associated with action α_i and because the decision rule specifies the action, the overall risk is given by

$$R = \int R(\alpha(\mathbf{x})|\mathbf{x})p(\mathbf{x}) \, d\mathbf{x}, \tag{12}$$

where $d\mathbf{x}$ is our notation for a d-space volume element and where the integral extends over the entire feature space. Clearly, if $\alpha(\mathbf{x})$ is chosen so that $R(\alpha_i(\mathbf{x}))$ is as small as possible for every $\mathbf{x}$, then the overall risk will be minimized. This justifies the following statement of the *Bayes decision rule*: To minimize the overall risk, compute the conditional risk

$$R(\alpha_i|\mathbf{x}) = \sum_{j=1}^{c} \lambda(\alpha_i|\omega_j)P(\omega_j|\mathbf{x}) \tag{13}$$

for $i = 1, \ldots, a$ and then select the action α_i for which $R(\alpha_i|\mathbf{x})$ is minimum.* The resulting minimum overall risk is called the *Bayes risk*, denoted R^*, and is the best performance that can be achieved.

2.2.1 Two-Category Classification

Let us consider these results when applied to the special case of two-category classification problems. Here action α_1 corresponds to deciding that the true state of nature is ω_1, and action α_2 corresponds to deciding that it is ω_2. For notational simplicity, let $\lambda_{ij} = \lambda(\alpha_i|\omega_j)$ be the loss incurred for deciding ω_i when the true state of nature is ω_j. If we write out the conditional risk given by Eq. 13, we obtain

$$R(\alpha_1|\mathbf{x}) = \lambda_{11}P(\omega_1|\mathbf{x}) + \lambda_{12}P(\omega_2|\mathbf{x}) \tag{14}$$

$$R(\alpha_2|\mathbf{x}) = \lambda_{21}P(\omega_1|\mathbf{x}) + \lambda_{22}P(\omega_2|\mathbf{x}). \tag{15}$$

There are a variety of ways of expressing the minimum-risk decision rule, each having its own minor advantages. The fundamental rule is to decide ω_1 if $R(\alpha_1|\mathbf{x}) < R(\alpha_2|\mathbf{x})$. In terms of the posterior probabilities, we decide ω_1 if

$$(\lambda_{21} - \lambda_{11})P(\omega_1|\mathbf{x}) > (\lambda_{12} - \lambda_{22})P(\omega_2|\mathbf{x}). \tag{16}$$

Ordinarily, the loss incurred for making an error is greater than the loss incurred for being correct, and both of the factors $\lambda_{21}-\lambda_{11}$ and $\lambda_{12}-\lambda_{22}$ are positive. Thus in practice, our decision is generally determined by the more likely state of nature, although

---

*Note that if more than one action minimizes $R(\alpha|\mathbf{x})$, it does not matter which of these actions is taken, and any convenient tie-breaking rule can be used.

we must scale the posterior probabilities by the loss differences. By employing Bayes formula, we can replace the posterior probabilities by the prior probabilities and the conditional densities. This results in the equivalent rule, to decide ω_1 if

$$(\lambda_{21} - \lambda_{11})p(\mathbf{x}|\omega_1)P(\omega_1) > (\lambda_{12} - \lambda_{22})p(\mathbf{x}|\omega_2)P(\omega_2), \tag{17}$$

and otherwise decide ω_2.

Another alternative, which follows at once under the reasonable assumption that $\lambda_{21} > \lambda_{11}$, is to decide ω_1 if

$$\frac{p(\mathbf{x}|\omega_1)}{p(\mathbf{x}|\omega_2)} > \frac{\lambda_{12} - \lambda_{22}}{\lambda_{21} - \lambda_{11}} \frac{P(\omega_2)}{P(\omega_1)}. \tag{18}$$

LIKELIHOOD
RATIO

This form of the decision rule focuses on the $\mathbf{x}$-dependence of the probability densities. We can consider $p(\mathbf{x}|\omega_j)$ a function of ω_j (i.e., the likelihood function) and then form the *likelihood ratio* $p(\mathbf{x}|\omega_1)/p(\mathbf{x}|\omega_2)$. Thus the Bayes decision rule can be interpreted as calling for deciding ω_1 if the likelihood ratio exceeds a threshold value that is independent of the observation $\mathbf{x}$.

2.3 MINIMUM-ERROR-RATE CLASSIFICATION

In classification problems, each state of nature is usually associated with a different one of the c classes, and the action α_i is usually interpreted as the decision that the true state of nature is ω_i. If action α_i is taken and the true state of nature is ω_j, then the decision is correct if $i = j$ and in error if $i \neq j$. If errors are to be avoided, it is natural to seek a decision rule that minimizes the probability of error, that is, the *error rate*.

The loss function of interest for this case is hence the so-called *symmetrical* or *zero-one* loss function,

ZERO-ONE LOSS

$$\lambda(\alpha_i|\omega_j) = \begin{cases} 0 & i = j \\ 1 & i \neq j \end{cases} \qquad i, j = 1, \ldots, c. \tag{19}$$

This loss function assigns no loss to a correct decision, and assigns a unit loss to any error; thus, all errors are equally costly.* The risk corresponding to this loss function is precisely the average probability of error because the conditional risk is

$$R(\alpha_i|\mathbf{x}) = \sum_{j=1}^{c} \lambda(\alpha_i|\omega_j) P(\omega_j|\mathbf{x})$$

$$= \sum_{j \neq i} P(\omega_j|\mathbf{x})$$

$$= 1 - P(\omega_i|\mathbf{x}) \tag{20}$$

---

*We note that other loss functions, such as quadratic and linear difference, find greater use in regression tasks where there is a natural ordering on the predictions and we can meaningfully penalize predictions that are "more wrong" than others.

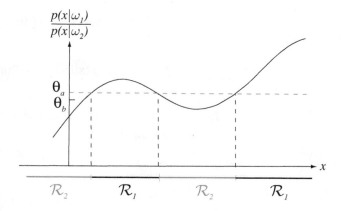

FIGURE 2.3. The likelihood ratio $p(x|\omega_1)/p(x|\omega_2)$ for the distributions shown in Fig. 2.1. If we employ a zero-one or classification loss, our decision boundaries are determined by the threshold θ_a. If our loss function penalizes miscategorizing ω_2 as ω_1 patterns more than the converse, we get the smaller threshold θ_b, and hence $\mathcal{R}_1$ becomes smaller.

and $P(\omega_i|\mathbf{x})$ is the conditional probability that action α_i is correct. The Bayes decision rule to minimize risk calls for selecting the action that minimizes the conditional risk. Thus, to minimize the average probability of error, we should select the i that *maximizes* the posterior probability $P(\omega_i|\mathbf{x})$. In other words, for *minimum error rate*:

$$\text{Decide } \omega_i \text{ if } P(\omega_i|\mathbf{x}) > P(\omega_j|\mathbf{x}) \qquad \text{for all } j \neq i. \tag{21}$$

This is the same rule as in Eq. 6. The region in the input space where we decide ω_i is denoted $\mathcal{R}_i$; such a region need not be simply connected.

We saw in Fig. 2.2 some class-conditional probability densities and the posterior probabilities; Fig. 2.3 shows the likelihood ratio $p(x|\omega_1)/p(x|\omega_2)$ for the same case. In general, this ratio can range between zero and infinity. The threshold value θ_a marked is from the same prior probabilities but with a zero-one loss function. Notice that this leads to the same decision boundaries as in Fig. 2.2, as it must. If we penalize mistakes in classifying ω_1 patterns as ω_2 more than the converse (i.e., $\lambda_{21} > \lambda_{12}$), then Eq. 18 leads to the threshold θ_b marked. Note that the range of x values for which we classify a pattern as ω_1 gets larger, as it should.

⋆2.3.1 Minimax Criterion

Sometimes we must design our classifier to perform well over a *range* of prior probabilities. For instance, in our fish categorization problem we can imagine that whereas the physical properties of lightness and width of each type of fish remain constant, the prior probabilities might vary widely and in an unpredictable way, or alternatively we want to use the classifier in a different plant where we do not know the prior probabilities. A reasonable approach is then to design our classifier so that the *worst* overall risk for any value of the priors is as small as possible—that is, minimize the maximum possible overall risk.

In order to understand this, we let $\mathcal{R}_1$ denote that (as yet unknown) region in feature space where the classifier decides ω_1 and likewise for $\mathcal{R}_2$ and ω_2, and then we write our overall risk Eq. 12 in terms of conditional risks:

$$R = \int_{\mathcal{R}_1} [\lambda_{11} P(\omega_1)\, p(\mathbf{x}|\omega_1) + \lambda_{12} P(\omega_2)\, p(\mathbf{x}|\omega_2)]\, d\mathbf{x}$$

$$+ \int_{\mathcal{R}_2} [\lambda_{21} P(\omega_1)\, p(\mathbf{x}|\omega_1) + \lambda_{22} P(\omega_2)\, p(\mathbf{x}|\omega_2)]\, d\mathbf{x}. \tag{22}$$

We use the fact that $P(\omega_2) = 1 - P(\omega_1)$ and that $\int_{\mathcal{R}_1} p(\mathbf{x}|\omega_1)\, d\mathbf{x} = 1 - \int_{\mathcal{R}_2} p(\mathbf{x}|\omega_1)\, d\mathbf{x}$ to rewrite the risk as:

$$R(P(\omega_1)) = \overbrace{\lambda_{22} + (\lambda_{12} - \lambda_{22}) \int_{\mathcal{R}_1} p(\mathbf{x}|\omega_2)\, d\mathbf{x}}^{= R_{mm}, \text{ minimax risk}} \tag{23}$$

$$+ P(\omega_1) \underbrace{\left[(\lambda_{11} - \lambda_{22}) + (\lambda_{21} - \lambda_{11}) \int_{\mathcal{R}_2} p(\mathbf{x}|\omega_1)\, d\mathbf{x} - (\lambda_{12} - \lambda_{22}) \int_{\mathcal{R}_1} p(\mathbf{x}|\omega_2)\, d\mathbf{x} \right]}_{= 0 \text{ for minimax solution}}.$$

This equation shows that once the decision boundary is set (i.e., $\mathcal{R}_1$ and $\mathcal{R}_2$ determined), the overall risk is linear in $P(\omega_1)$. If we can find a boundary such that the constant of proportionality is 0, then the risk is independent of priors. This is the *minimax solution*, and the *minimax risk*, R_{mm}, can be read from Eq. 23:

MINIMAX RISK

$$R_{mm} = \lambda_{22} + (\lambda_{12} - \lambda_{22}) \int_{\mathcal{R}_1} p(\mathbf{x}|\omega_2)\, d\mathbf{x}$$

$$= \lambda_{11} + (\lambda_{21} - \lambda_{11}) \int_{\mathcal{R}_2} p(\mathbf{x}|\omega_1)\, d\mathbf{x}. \tag{24}$$

Figure 2.4 illustrates the approach. Briefly stated, we search for the prior for which the Bayes risk is *maximum*, and the corresponding decision boundary then gives the minimax solution. The value of the minimax risk, R_{mm}, is hence equal to the worst Bayes risk. In practice, finding the decision boundary for minimax risk may be difficult, particularly when distributions are complicated. Nevertheless, in some cases the boundary can be determined analytically (Problem 4).

The minimax criterion finds greater use in game theory than it does in traditional pattern recognition. In game theory you have a hostile opponent who can be expected to take an action maximally detrimental to you. Thus it makes great sense for you to take an action (e.g., make a classification) where your costs—due to your opponent's subsequent actions—are minimized.

⋆2.3.2 Neyman-Pearson Criterion

In some problems, we may wish to minimize the overall risk subject to a constraint; for instance, we might wish to minimize the total risk subject to the constraint $\int R(\alpha_i|\mathbf{x})\, d\mathbf{x} < constant$ for some particular i. Such a constraint might arise when there is a fixed resource that accompanies one particular action α_i, or when we must not misclassify a pattern from a particular state of nature ω_i at more than some limited frequency. For instance, in our fish example, there might be some government

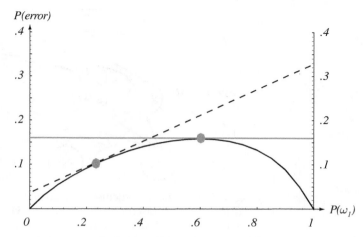

FIGURE 2.4. The curve at the bottom shows the minimum (Bayes) error as a function of prior probability $P(\omega_1)$ in a two-category classification problem of fixed distributions. For each value of the priors (e.g., $P(\omega_1) = 0.25$) there is a corresponding optimal decision boundary and associated Bayes error rate. For any (fixed) such boundary, if the priors are then changed, the probability of error will change as a linear function of $P(\omega_1)$ (shown by the dashed line). The maximum such error will occur at an extreme value of the prior, here at $P(\omega_1) = 1$. To minimize the maximum of such error, we should design our decision boundary for the maximum Bayes error (here $P(\omega_1) = 0.6$), and thus the error will not change as a function of prior, as shown by the solid red horizontal line.

regulation that we must not misclassify more than 1% of salmon as sea bass. We might then seek a decision that minimizes the chance of classifying a sea bass as a salmon subject to this condition.

We generally satisfy such a *Neyman-Pearson criterion* by adjusting decision boundaries numerically. However, for Gaussian and some other distributions, Neyman-Pearson solutions can be found analytically (Problems 6 and 7). We shall have cause to mention Neyman-Pearson criteria again in Section 2.8.3 on operating characteristics.

2.4 CLASSIFIERS, DISCRIMINANT FUNCTIONS, AND DECISION SURFACES

2.4.1 The Multicategory Case

There are many different ways to represent pattern classifiers. One of the most useful is in terms of a set of *discriminant functions* $g_i(\mathbf{x})$, $i = 1, \ldots, c$. The classifier is said to assign a feature vector $\mathbf{x}$ to class ω_i if

$$g_i(\mathbf{x}) > g_j(\mathbf{x}) \qquad \text{for all } j \neq i. \tag{25}$$

Thus, the classifier is viewed as a network or machine that computes c discriminant functions and selects the category corresponding to the largest discriminant. A network representation of a classifier is illustrated in Fig. 2.5.

A Bayes classifier is easily and naturally represented in this way. For the general case with risks, we can let $g_i(\mathbf{x}) = -R(\alpha_i|\mathbf{x})$, because the maximum discriminant

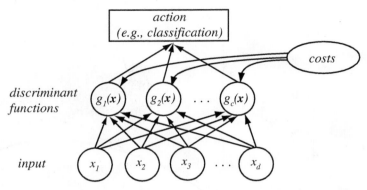

FIGURE 2.5. The functional structure of a general statistical pattern classifier which includes d inputs and c discriminant functions $g_i(\mathbf{x})$. A subsequent step determines which of the discriminant values is the maximum, and categorizes the input pattern accordingly. The arrows show the direction of the flow of information, though frequently the arrows are omitted when the direction of flow is self-evident.

function will then correspond to the minimum conditional risk. For the minimum-error-rate case, we can simplify things further by taking $g_i(\mathbf{x}) = P(\omega_i|\mathbf{x})$, so that the maximum discriminant function corresponds to the maximum posterior probability.

Clearly, the choice of discriminant functions is not unique. We can always multiply all the discriminant functions by the same positive constant or shift them by the same additive constant without influencing the decision. More generally, if we replace every $g_i(\mathbf{x})$ by $f(g_i(\mathbf{x}))$, where $f(\cdot)$ is a monotonically increasing function, the resulting classification is unchanged. This observation can lead to significant analytical and computational simplifications. In particular, for minimum-error-rate classification, any of the following choices gives identical classification results, but some can be much simpler to understand or to compute than others:

$$g_i(\mathbf{x}) = P(\omega_i|\mathbf{x}) = \frac{p(\mathbf{x}|\omega_i)P(\omega_i)}{\sum\limits_{j=1}^{c} p(\mathbf{x}|\omega_j)P(\omega_j)} \qquad (26)$$

$$g_i(\mathbf{x}) = p(\mathbf{x}|\omega_i)P(\omega_i) \qquad (27)$$

$$g_i(\mathbf{x}) = \ln p(\mathbf{x}|\omega_i) + \ln P(\omega_i), \qquad (28)$$

where ln denotes natural logarithm.

Even though the discriminant functions can be written in a variety of forms, the decision rules are equivalent. The effect of any decision rule is to divide the feature space into c *decision regions*, $\mathcal{R}_1, \ldots, \mathcal{R}_c$. If $g_i(\mathbf{x}) > g_j(\mathbf{x})$ for all $j \neq i$, then $\mathbf{x}$ is in $\mathcal{R}_i$, and the decision rule calls for us to assign $\mathbf{x}$ to ω_i. The regions are separated by *decision boundaries*, surfaces in feature space where ties occur among the largest discriminant functions (Fig. 2.6).

DECISION
REGION

2.4.2 The Two-Category Case

While the two-category case is just a special instance of the multicategory case, it has traditionally received separate treatment. Indeed, a classifier that places a pattern in one of only two categories has a special name—a *dichotomizer*.* Instead of using

DICHOTOMIZER

_____

*A classifier for more than two categories is called a polychotomizer.

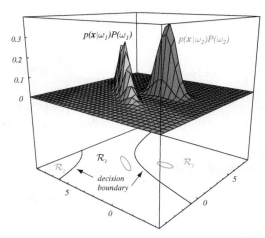

FIGURE 2.6. In this two-dimensional two-category classifier, the probability densities are Gaussian, the decision boundary consists of two hyperbolas, and thus the decision region $\mathcal{R}_2$ is not simply connected. The ellipses mark where the density is $1/e$ times that at the peak of the distribution.

two discriminant functions g_1 and g_2 and assigning $\mathbf{x}$ to ω_1 if $g_1 > g_2$, it is more common to define a single discriminant function

$$g(\mathbf{x}) \equiv g_1(\mathbf{x}) - g_2(\mathbf{x}), \tag{29}$$

and to use the following decision rule: Decide ω_1 if $g(\mathbf{x}) > 0$; otherwise decide ω_2. Thus, a dichotomizer can be viewed as a machine that computes a single discriminant function $g(\mathbf{x})$, and classifies $\mathbf{x}$ according to the algebraic sign of the result. Of the various forms in which the minimum-error-rate discriminant function can be written, the following two (derived from Eqs. 26 and 28) are particularly convenient:

$$g(\mathbf{x}) = P(\omega_1|\mathbf{x}) - P(\omega_2|\mathbf{x}) \tag{30}$$

$$g(\mathbf{x}) = \ln \frac{p(\mathbf{x}|\omega_1)}{p(\mathbf{x}|\omega_2)} + \ln \frac{P(\omega_1)}{P(\omega_2)}. \tag{31}$$

2.5 THE NORMAL DENSITY

The structure of a Bayes classifier is determined by the conditional densities $p(\mathbf{x}|\omega_i)$ as well as by the prior probabilities $P(\omega_i)$. Of the various density functions that have been investigated, none has received more attention than the multivariate normal or Gaussian density. To a large extent this attention is due to its analytical tractability. However, the multivariate normal density is also an appropriate model for an important situation, namely, the case where the feature vectors $\mathbf{x}$ for a given class ω_i are continuous-valued, randomly corrupted versions of a single typical or prototype vector $\boldsymbol{\mu}_i$. In this section we provide a brief exposition of the multivariate normal density, focusing on the properties of greatest interest for classification problems.

EXPECTATION
First, recall the definition of the *expected value* of a scalar function $f(x)$, defined for some density $p(x)$:

$$\mathcal{E}[f(x)] \equiv \int_{-\infty}^{\infty} f(x)p(x)\,dx. \tag{32}$$

If the values of the feature x are restricted to points in a discrete set $\mathcal{D}$, we must sum over all samples as

$$\mathcal{E}[f(x)] = \sum_{x \in \mathcal{D}} f(x)P(x), \tag{33}$$

where $P(x)$ is the probability mass at x. We shall occassionally need to calculate expected values by these and analogous equations defined in higher dimensions (see Appendix Sections A.4.2, A.4.5 and A.4.9).*

2.5.1 Univariate Density

We begin with the continuous univariate normal or Gaussian density,

$$p(x) = \frac{1}{\sqrt{2\pi}\sigma} \exp\left[-\frac{1}{2}\left(\frac{x - \mu}{\sigma}\right)^2 \right], \tag{34}$$

for which the *expected value* of x (an average, here taken over the feature space) is

$$\mu \equiv \mathcal{E}[x] = \int_{-\infty}^{\infty} xp(x)\,dx, \tag{35}$$

VARIANCE

and where the expected squared deviation or *variance* is

$$\sigma^2 \equiv \mathcal{E}[(x - \mu)^2] = \int_{-\infty}^{\infty} (x - \mu)^2 p(x)\,dx. \tag{36}$$

MEAN

The univariate normal density is completely specified by two parameters: its mean μ and variance σ^2. For simplicity, we often abbreviate Eq. 34 by writing $p(x) \sim N(\mu, \sigma^2)$ to say that x is distributed normally with mean μ and variance σ^2. Samples from normal distributions tend to cluster about the mean, with a spread related to the standard deviation σ (Fig. 2.7).

ENTROPY

There is a deep relationship between the normal distribution and *entropy*. We discuss entropy in greater detail in Appendix Section A.7, but for now we merely state that the entropy of a distribution is given by

$$H(p(x)) = -\int p(x)\,\ln p(x)\,dx, \tag{37}$$

NAT
BIT

and measured in *nats*; if a $\log_2$ is used instead, the unit is the *bit*. The entropy measures the fundamental uncertainty in the values of points selected randomly from a

---

*We will often use somewhat loose engineering terminology and refer to a single point as a "sample." Statisticians, however, always refer to a sample as a *collection* of points, and they discuss "a sample of size n." When taken in context, there are rarely ambiguities in such usage.

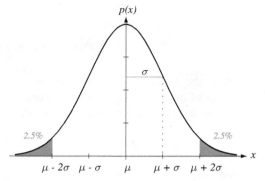

FIGURE 2.7. A univariate normal distribution has roughly 95% of its area in the range $|x - \mu| \leq 2\sigma$, as shown. The peak of the distribution has value $p(\mu) = 1/\sqrt{2\pi}\sigma$.

distribution. It can be shown that the normal distribution has the maximum entropy of all distributions having a given mean and variance (Problem 20). Moreover, as stated by the *Central Limit Theorem*, the aggregate effect of the sum of a large number of small, independent random disturbances will lead to a Gaussian distribution (Computer exercise 5). Because many patterns—from fish to handwritten characters to some speech sounds—can be viewed as some ideal or prototype pattern corrupted by a large number of random processes, the Gaussian is often a good model for the actual probability distribution.

CENTRAL LIMIT
THEOREM

2.5.2 Multivariate Density

The general multivariate normal density in d dimensions is written as

$$p(\mathbf{x}) = \frac{1}{(2\pi)^{d/2}|\mathbf{\Sigma}|^{1/2}}\exp\left[-\frac{1}{2}(\mathbf{x} - \boldsymbol{\mu})^t\mathbf{\Sigma}^{-1}(\mathbf{x} - \boldsymbol{\mu})\right], \tag{38}$$

where $\mathbf{x}$ is a d-component column vector, $\boldsymbol{\mu}$ is the d-component *mean vector*, $\mathbf{\Sigma}$ is the d-by-d *covariance matrix*, and $|\mathbf{\Sigma}|$ and $\mathbf{\Sigma}^{-1}$ are its determinant and inverse, respectively. Further, we let $(\mathbf{x} - \boldsymbol{\mu})^t$ denote the transpose of $\mathbf{x} - \boldsymbol{\mu}$.* Our notation for the *inner product* is

COVARIANCE
MATRIX

INNER PRODUCT

$$\mathbf{a}^t\mathbf{b} = \sum_{i=1}^{d} a_i b_i, \tag{39}$$

which is often called a *dot product*. For simplicity, we often abbreviate Eq. 38 as $p(\mathbf{x}) \sim N(\boldsymbol{\mu}, \mathbf{\Sigma})$.

Formally, we have

$$\boldsymbol{\mu} \equiv \mathcal{E}[\mathbf{x}] = \int \mathbf{x} p(\mathbf{x}) \, d\mathbf{x} \tag{40}$$

---

*The mathematical expressions for the multivariate normal density are greatly simplified by employing the concepts and notation of linear algebra. Readers who are unsure of our notation or who would like to review linear algebra should see Appendix Section A.2.

and

$$\boldsymbol{\Sigma} \equiv \mathcal{E}[(\mathbf{x} - \boldsymbol{\mu})(\mathbf{x} - \boldsymbol{\mu})^t] = \int (\mathbf{x} - \boldsymbol{\mu})(\mathbf{x} - \boldsymbol{\mu})^t \, p(\mathbf{x}) \, d\mathbf{x}, \tag{41}$$

where the expected value of a vector or a matrix is found by taking the expected values of its components. In other words, if x_i is the ith component of $\mathbf{x}$, μ_i the ith component of $\boldsymbol{\mu}$, and σ_{ij} the ijth component of $\boldsymbol{\Sigma}$, then

$$\mu_i = \mathcal{E}[x_i] \tag{42}$$

and

$$\sigma_{ij} = \mathcal{E}[(x_i - \mu_i)(x_j - \mu_j)]. \tag{43}$$

The covariance matrix $\boldsymbol{\Sigma}$ is always symmetric and positive semidefinite. We shall restrict our attention to the case in which $\boldsymbol{\Sigma}$ is positive definite, so that the determinant of $\boldsymbol{\Sigma}$ is strictly positive.* The diagonal elements σ_{ii} are the variances of the respective x_i (i.e., σ_i^2), and the off-diagonal elements σ_{ij} are the *covariances* of x_i and x_j. We would expect a positive covariance for the length and weight features of a population of fish, for instance. If x_i and x_j are *statistically independent*, then $\sigma_{ij} = 0$. If all the off-diagonal elements are zero, $p(\mathbf{x})$ reduces to the product of the univariate normal densities for the components of $\mathbf{x}$.

COVARIANCE

STATISTICAL
INDEPENDENCE

Linear combinations of jointly normally distributed random variables, independent or not, are normally distributed. In particular, if $p(\mathbf{x}) \sim N(\boldsymbol{\mu}, \boldsymbol{\Sigma})$, $\mathbf{A}$ is a d-by-k matrix and $\mathbf{y} = \mathbf{A}^t\mathbf{x}$ is a k-component vector, then $p(\mathbf{y}) \sim N(\mathbf{A}^t\boldsymbol{\mu}, \mathbf{A}^t\boldsymbol{\Sigma}\mathbf{A})$, as illustrated in Fig. 2.8. In the special case where $k = 1$ and $\mathbf{A}$ is a unit-length vector $\mathbf{a}$, $y = \mathbf{a}^t\mathbf{x}$ is a scalar that represents the projection of $\mathbf{x}$ onto a line in the direction of $\mathbf{a}$; in that case $\mathbf{a}^t\boldsymbol{\Sigma}\mathbf{a}$ is the variance of the projection of $\mathbf{x}$ onto $\mathbf{a}$. In general then, knowledge of the covariance matrix allows us to calculate the dispersion of the data in any direction, or in any subspace.

It is sometimes convenient to perform a coordinate transformation that converts an arbitrary multivariate normal distribution into a spherical one—that is, one having a covariance matrix proportional to the identity matrix $\mathbf{I}$. If we define $\boldsymbol{\Phi}$ to be the matrix whose columns are the orthonormal eigenvectors of $\boldsymbol{\Sigma}$, and $\boldsymbol{\Lambda}$ the diagonal matrix of the corresponding eigenvalues, then the transformation

$$\mathbf{A}_w = \boldsymbol{\Phi}\boldsymbol{\Lambda}^{-1/2} \tag{44}$$

applied to the coordinates ensures that the transformed distribution has covariance matrix equal to the identity matrix. In signal processing, $\mathbf{A}_w$ yields a so-called *whitening* transform, because it makes the spectrum of eigenvalues of the transformed distribution uniform.

WHITENING
TRANSFORM

The multivariate normal density is completely specified by $d + d(d + 1)/2$ parameters, namely the elements of the mean vector $\boldsymbol{\mu}$ and the independent elements of the covariance matrix $\boldsymbol{\Sigma}$. Samples drawn from a normal population tend to fall in a single cloud or cluster (Fig. 2.9); the center of the cluster is determined by the mean vector, and the shape of the cluster is determined by the covariance matrix. It

---

*If sample vectors are drawn from a linear subspace, $|\boldsymbol{\Sigma}| = 0$ and $p(\mathbf{x})$ is degenerate. This occurs, for example, when one component of $\mathbf{x}$ has zero variance, or when two components are identical or multiples of one another.

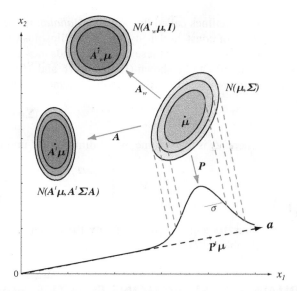

FIGURE 2.8. The action of a linear transformation on the feature space will convert an arbitrary normal distribution into another normal distribution. One transformation, **A**, takes the source distribution into distribution $N(\mathbf{A}^t\boldsymbol{\mu}, \mathbf{A}^t\boldsymbol{\Sigma}\mathbf{A})$. Another linear transformation—a projection **P** onto a line defined by vector **a**—leads to $N(\mu, \sigma^2)$ measured along that line. While the transforms yield distributions in a different space, we show them superimposed on the original $x_1 x_2$-space. A whitening transform, $\mathbf{A}_w$, leads to a circularly symmetric Gaussian, here shown displaced.

follows from Eq. 38 that the loci of points of constant density are hyperellipsoids for which the quadratic form $(\mathbf{x} - \boldsymbol{\mu})^t \boldsymbol{\Sigma}^{-1} (\mathbf{x} - \boldsymbol{\mu})$ is constant. The principal axes of these hyperellipsoids are given by the eigenvectors of $\boldsymbol{\Sigma}$ (described by $\boldsymbol{\Phi}$); the eigenvalues (described by $\boldsymbol{\Lambda}$) determine the lengths of these axes. The quantity

$$r^2 = (\mathbf{x} - \boldsymbol{\mu})^t \boldsymbol{\Sigma}^{-1} (\mathbf{x} - \boldsymbol{\mu}) \tag{45}$$

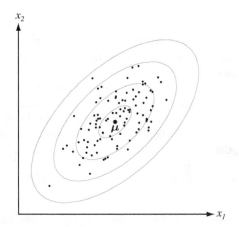

FIGURE 2.9. Samples drawn from a two-dimensional Gaussian lie in a cloud centered on the mean $\boldsymbol{\mu}$. The ellipses show lines of equal probability density of the Gaussian.

is sometimes called the squared *Mahalanobis distance* from $\mathbf{x}$ to $\boldsymbol{\mu}$. Thus, the contours of constant density are hyperellipsoids of constant Mahalanobis distance to $\boldsymbol{\mu}$ and the volume of these hyperellipsoids measures the scatter of the samples about the mean. It can be shown (Problems 15 and 16) that the volume of the hyperellipsoid corresponding to a Mahalanobis distance r is given by

$$V = V_d |\boldsymbol{\Sigma}|^{1/2} r^d, \tag{46}$$

where V_d is the volume of a d-dimensional unit hypersphere:

$$V_d = \begin{cases} \pi^{d/2}/(d/2)! & d \text{ even} \\ 2^d \pi^{(d-1)/2} (\frac{d-1}{2})!/d! & d \text{ odd.} \end{cases} \tag{47}$$

Thus, for a given dimensionality, the scatter of the samples varies directly with $|\boldsymbol{\Sigma}|^{1/2}$ (Problem 17).

2.6 DISCRIMINANT FUNCTIONS FOR THE NORMAL DENSITY

In Section 2.4.1 we saw that the minimum-error-rate classification can be achieved by use of the discriminant functions

$$g_i(\mathbf{x}) = \ln p(\mathbf{x}|\omega_i) + \ln P(\omega_i). \tag{48}$$

This expression can be readily evaluated if the densities $p(\mathbf{x}|\omega_i)$ are multivariate normal—that is, if $p(\mathbf{x}|\omega_i) \sim N(\boldsymbol{\mu}_i, \boldsymbol{\Sigma}_i)$. In this case, then, from Eq. 38 we have

$$g_i(\mathbf{x}) = -\frac{1}{2}(\mathbf{x} - \boldsymbol{\mu}_i)^t \boldsymbol{\Sigma}_i^{-1}(\mathbf{x} - \boldsymbol{\mu}_i) - \frac{d}{2}\ln 2\pi - \frac{1}{2}\ln|\boldsymbol{\Sigma}_i| + \ln P(\omega_i). \tag{49}$$

Let us examine this discriminant function and resulting classification for a number of special cases.

2.6.1 Case 1: $\boldsymbol{\Sigma}_i = \sigma^2 \mathbf{I}$

The simplest case occurs when the features are statistically independent and when each feature has the same variance, σ^2. In this case the covariance matrix is diagonal, being merely σ^2 times the identity matrix $\mathbf{I}$. Geometrically, this corresponds to the situation in which the samples fall in equal-size hyperspherical clusters, the cluster for the ith class being centered about the mean vector $\boldsymbol{\mu}_i$. The computation of the determinant and the inverse of $\boldsymbol{\Sigma}_i$ is particularly easy: $|\boldsymbol{\Sigma}_i| = \sigma^{2d}$ and $\boldsymbol{\Sigma}_i^{-1} = (1/\sigma^2)\mathbf{I}$. Because both $|\boldsymbol{\Sigma}_i|$ and the $(d/2)\ln 2\pi$ term in Eq. 49 are independent of i, they are unimportant additive constants that can be ignored. Thus we obtain the simple discriminant functions

$$g_i(\mathbf{x}) = -\frac{\|\mathbf{x} - \boldsymbol{\mu}_i\|^2}{2\sigma^2} + \ln P(\omega_i), \tag{50}$$

where $\|\cdot\|$ denotes the *Euclidean norm*, that is,

$$\|\mathbf{x} - \boldsymbol{\mu}_i\|^2 = (\mathbf{x} - \boldsymbol{\mu}_i)^t (\mathbf{x} - \boldsymbol{\mu}_i). \tag{51}$$

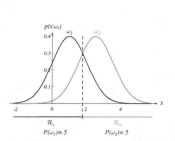

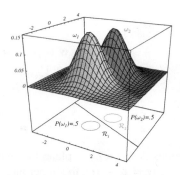

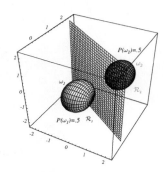

FIGURE 2.10. If the covariance matrices for two distributions are equal and proportional to the identity matrix, then the distributions are spherical in d dimensions, and the boundary is a generalized hyperplane of $d - 1$ dimensions, perpendicular to the line separating the means. In these one-, two-, and three-dimensional examples, we indicate $p(\mathbf{x}|\omega_i)$ and the boundaries for the case $P(\omega_1) = P(\omega_2)$. In the three-dimensional case, the grid plane separates $\mathcal{R}_1$ from $\mathcal{R}_2$.

If the prior probabilities are not equal, then Eq. 50 shows that the squared distance $\|\mathbf{x} - \boldsymbol{\mu}\|^2$ must be normalized by the variance σ^2 and offset by adding $\ln P(\omega_i)$; thus, if $\mathbf{x}$ is equally near two different mean vectors, the optimal decision will favor the a priori more likely category.

Regardless of whether the prior probabilities are equal or not, it is not actually necessary to compute distances. Expansion of the quadratic form $(\mathbf{x} - \boldsymbol{\mu}_i)^t (\mathbf{x} - \boldsymbol{\mu}_i)$ yields

$$g_i(\mathbf{x}) = -\frac{1}{2\sigma^2}[\mathbf{x}^t\mathbf{x} - 2\boldsymbol{\mu}_i^t\mathbf{x} + \boldsymbol{\mu}_i^t\boldsymbol{\mu}_i] + \ln P(\omega_i), \tag{52}$$

which appears to be a quadratic function of $\mathbf{x}$. However, the quadratic term $\mathbf{x}^t\mathbf{x}$ is the same for all i, making it an ignorable additive constant. Thus, we obtain the equivalent *linear discriminant functions*

LINEAR
DISCRIMINANT

$$g_i(\mathbf{x}) = \mathbf{w}_i^t\mathbf{x} + w_{i0}, \tag{53}$$

where

$$\mathbf{w}_i = \frac{1}{\sigma^2}\boldsymbol{\mu}_i \tag{54}$$

and

$$w_{i0} = \frac{-1}{2\sigma^2}\boldsymbol{\mu}_i^t\boldsymbol{\mu}_i + \ln P(\omega_i). \tag{55}$$

THRESHOLD
BIAS

We call w_{i0} the *threshold* or *bias* for the ith category.

LINEAR
MACHINE

A classifier that uses linear discriminant functions is called a *linear machine*. This kind of classifier has many interesting theoretical properties, some of which will be discussed in Chapter 5. At this point we merely note that the decision surfaces for a linear machine are pieces of hyperplanes defined by the linear equations $g_i(\mathbf{x}) = g_j(\mathbf{x})$ for the two categories with the highest posterior probabilities. For our particular case, this equation can be written as

$$\mathbf{w}^t(\mathbf{x} - \mathbf{x}_0) = 0, \tag{56}$$

where

$$\mathbf{w} = \boldsymbol{\mu}_i - \boldsymbol{\mu}_j \tag{57}$$

and

$$\mathbf{x}_0 = \frac{1}{2}(\boldsymbol{\mu}_i + \boldsymbol{\mu}_j) - \frac{\sigma^2}{\|\boldsymbol{\mu}_i - \boldsymbol{\mu}_j\|^2} \ln \frac{P(\omega_i)}{P(\omega_j)}(\boldsymbol{\mu}_i - \boldsymbol{\mu}_j). \tag{58}$$

These equations define a hyperplane through the point $\mathbf{x}_0$ and orthogonal to the vector $\mathbf{w}$. Because $\mathbf{w} = \boldsymbol{\mu}_i - \boldsymbol{\mu}_j$, the hyperplane separating $\mathcal{R}_i$ and $\mathcal{R}_j$ is orthogonal to the line linking the means. If $P(\omega_i) = P(\omega_j)$, the second term on the right of Eq. 58 vanishes, and thus the point $\mathbf{x}_0$ is halfway between the means, and the hy-

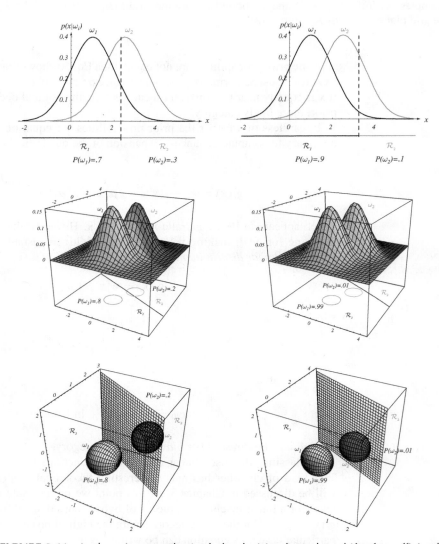

FIGURE 2.11. As the priors are changed, the decision boundary shifts; for sufficiently disparate priors the boundary will not lie between the means of these one-, two- and three-dimensional spherical Gaussian distributions.

perplane is the perpendicular bisector of the line between the means (Fig. 2.11). If $P(\omega_i) \neq P(\omega_j)$, the point $\mathbf{x}_0$ shifts away from the more likely mean. Note, however, that if the variance σ^2 is small relative to the squared distance $\|\boldsymbol{\mu}_i - \boldsymbol{\mu}_j\|^2$, then the position of the decision boundary is relatively insensitive to the exact values of the prior probabilities.

If the prior probabilities $P(\omega_i)$ are the same for all c classes, then the $\ln P(\omega_i)$ term becomes another unimportant additive constant that can be ignored. When this happens, the optimum decision rule can be stated very simply: To classify a feature vector $\mathbf{x}$, measure the Euclidean distance $\|\mathbf{x} - \boldsymbol{\mu}_i\|$ from each $\mathbf{x}$ to each of the c
MINIMUM-
DISTANCE
CLASSIFIER
mean vectors, and assign $\mathbf{x}$ to the category of the nearest mean. Such a classifier is called a *minimum-distance classifier*. If each mean vector is thought of as being an ideal prototype or template for patterns in its class, then this is essentially a *template-*
TEMPLATE-
MATCHING
matching procedure (Fig. 2.10), a technique we will consider again in Chapter 4 on the nearest-neighbor algorithm.

2.6.2 Case 2: $\Sigma_i = \Sigma$

Another simple case arises when the covariance matrices for all of the classes are identical but otherwise arbitrary. Geometrically, this corresponds to the situation in which the samples fall in hyperellipsoidal clusters of equal size and shape, the cluster for the ith class being centered about the mean vector $\boldsymbol{\mu}_i$. Because both $|\Sigma_i|$ and the $(d/2) \ln 2\pi$ term in Eq. 49 are independent of i, they can be ignored as superfluous additive constants. This simplification leads to the discriminant functions

$$g_i(\mathbf{x}) = -\frac{1}{2}(\mathbf{x} - \boldsymbol{\mu}_i)^t \Sigma^{-1}(\mathbf{x} - \boldsymbol{\mu}_i) + \ln P(\omega_i). \qquad (59)$$

If the prior probabilities $P(\omega_i)$ are the same for all c classes, then the $\ln P(\omega_i)$ term can be ignored. In this case, the optimal decision rule can once again be stated very simply: To classify a feature vector $\mathbf{x}$, measure the squared Mahalanobis distance $(\mathbf{x} - \boldsymbol{\mu}_i)^t \Sigma^{-1}(\mathbf{x} - \boldsymbol{\mu}_i)$ from $\mathbf{x}$ to each of the c mean vectors, and assign $\mathbf{x}$ to the category of the nearest mean. As before, unequal prior probabilities bias the decision in favor of the a priori more likely category.

Expansion of the quadratic form $(\mathbf{x} - \boldsymbol{\mu}_i)^t \Sigma^{-1}(\mathbf{x} - \boldsymbol{\mu}_i)$ results in a sum involving a quadratic term $\mathbf{x}^t \Sigma^{-1} \mathbf{x}$ which here is independent of i. After this term is dropped from Eq. 59, the resulting discriminant functions are again linear:

$$g_i(\mathbf{x}) = \mathbf{w}_i^t \mathbf{x} + w_{i0}, \qquad (60)$$

where

$$\mathbf{w}_i = \Sigma^{-1} \boldsymbol{\mu}_i \qquad (61)$$

and

$$w_{i0} = -\frac{1}{2} \boldsymbol{\mu}_i^t \Sigma^{-1} \boldsymbol{\mu}_i + \ln P(\omega_i). \qquad (62)$$

Because the discriminants are linear, the resulting decision boundaries are again hyperplanes (Fig. 2.10). If $\mathcal{R}_i$ and $\mathcal{R}_j$ are contiguous, the boundary between them has the equation

$$\mathbf{w}^t(\mathbf{x} - \mathbf{x}_0) = 0, \tag{63}$$

where

$$\mathbf{w} = \Sigma^{-1}(\mu_i - \mu_j) \tag{64}$$

and

$$\mathbf{x}_0 = \frac{1}{2}(\mu_i + \mu_j) - \frac{\ln[P(\omega_i)/P(\omega_j)]}{(\mu_i - \mu_j)^t \Sigma^{-1}(\mu_i - \mu_j)}(\mu_i - \mu_j). \tag{65}$$

Because $\mathbf{w} = \Sigma^{-1}(\mu_i - \mu_j)$ is generally not in the direction of $\mu_i - \mu_j$, the hyperplane separating $\mathcal{R}_i$ and $\mathcal{R}_j$ is generally not orthogonal to the line between the means. However, it does intersect that line at the point $\mathbf{x}_0$; if the prior probabilities are equal then $\mathbf{x}_0$ is halfway between the means. If the prior probabilities are not equal, the optimal boundary hyperplane is shifted away from the more likely mean (Fig. 2.12). As before, with sufficient bias the decision plane need not lie between the two mean vectors.

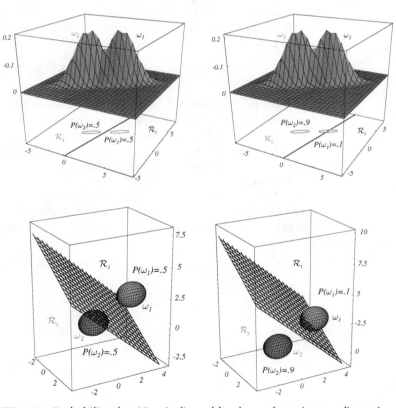

FIGURE 2.12. Probability densities (indicated by the surfaces in two dimensions and ellipsoidal surfaces in three dimensions) and decision regions for equal but asymmetric Gaussian distributions. The decision hyperplanes need not be perpendicular to the line connecting the means.

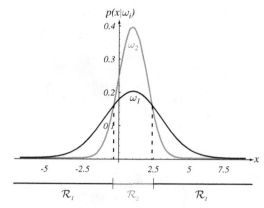

FIGURE 2.13. Non-simply connected decision regions can arise in one dimensions for Gaussians having unequal variance, as shown in this case with $P(\omega_1) = P(\omega_2)$.

2.6.3 Case 3: Σ_i = arbitrary

In the general multivariate normal case, the covariance matrices are different for each category. The only term that can be dropped from Eq. 49 is the $(d/2) \ln 2\pi$ term, and the resulting discriminant functions are inherently quadratic:

$$g_i(\mathbf{x}) = \mathbf{x}^t \mathbf{W}_i \mathbf{x} + \mathbf{w}_i^t \mathbf{x} + w_{i0}, \tag{66}$$

where

$$\mathbf{W}_i = -\frac{1}{2}\Sigma_i^{-1}, \tag{67}$$

$$\mathbf{w}_i = \Sigma_i^{-1}\boldsymbol{\mu}_i \tag{68}$$

and

$$w_{i0} = -\frac{1}{2}\boldsymbol{\mu}_i^t\Sigma_i^{-1}\boldsymbol{\mu}_i - \frac{1}{2}\ln|\Sigma_i| + \ln P(\omega_i). \tag{69}$$

HYPERQUADRIC   In the two-category case, the decision surfaces are *hyperquadrics*, and they can assume any of the general forms: hyperplanes, pairs of hyperplanes, hyperspheres, hyperellipsoids, hyperparaboloids, and hyperhyperboloids of various types (Problem 30). Even in one dimension, for arbitrary variance the decision regions need not be simply connected (Fig. 2.13). The two- and three-dimensional examples in Figs. 2.14 and 2.15 indicate how these different forms can arise.

The extension of these results to more than two categories is straightforward though here we need to keep clear which two of the total c categories are responsible for any boundary segment. Figure 2.16 shows the decision surfaces for a four-category case made up of Gaussian distributions. Of course, if the distributions are more complicated, the decision regions can be even more complex, though the same underlying theory holds there too.

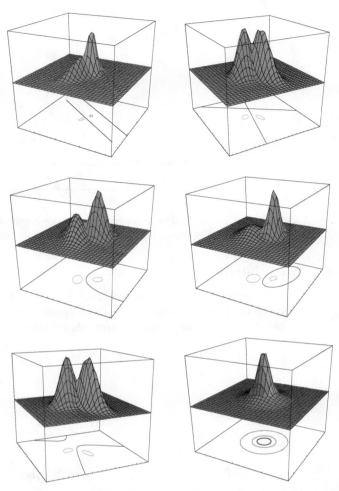

FIGURE 2.14. Arbitrary Gaussian distributions lead to Bayes decision boundaries that are general hyperquadrics. Conversely, given any hyperquadric, one can find two Gaussian distributions whose Bayes decision boundary is that hyperquadric. These variances are indicated by the contours of constant probability density.

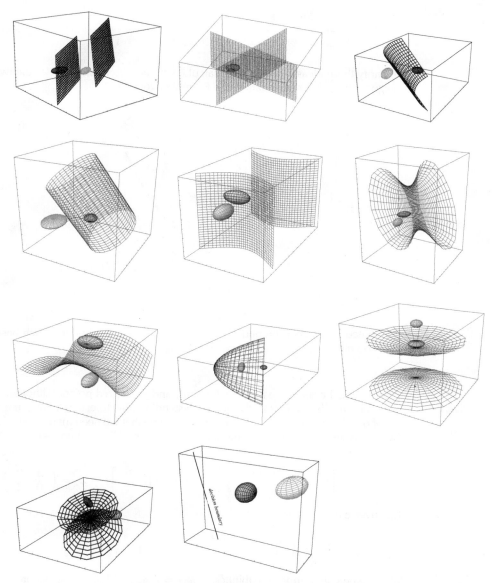

FIGURE 2.15. Arbitrary three-dimensional Gaussian distributions yield Bayes decision boundaries that are two-dimensional hyperquadrics. There are even degenerate cases in which the decision boundary is a line.

EXAMPLE 1   Decision Regions for Two-Dimensional Gaussian Data

---

To clarify these ideas, we explicitly calculate the decision boundary for the two-category two-dimensional data in the Example figure.

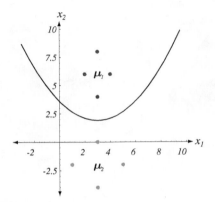

The computed Bayes decision boundary for two Gaussian distributions, each based on four data points.

Let ω_1 be the set of the four black points, and ω_2 the red points. Although we will spend much of the next chapter understanding how to estimate the parameters of our distributions, for now we simply assume that we need merely calculate the means and covariances by the discrete versions of Eqs. 40 and 41; they are found to be:

$$\boldsymbol{\mu}_1 = \begin{bmatrix} 3 \\ 6 \end{bmatrix}; \quad \boldsymbol{\Sigma}_1 = \begin{pmatrix} 1/2 & 0 \\ 0 & 2 \end{pmatrix} \text{ and } \boldsymbol{\mu}_2 = \begin{bmatrix} 3 \\ -2 \end{bmatrix}; \quad \boldsymbol{\Sigma}_2 = \begin{pmatrix} 2 & 0 \\ 0 & 2 \end{pmatrix}.$$

The inverse matrices are then,

$$\boldsymbol{\Sigma}_1^{-1} = \begin{pmatrix} 2 & 0 \\ 0 & 1/2 \end{pmatrix} \text{ and } \boldsymbol{\Sigma}_2^{-1} = \begin{pmatrix} 1/2 & 0 \\ 0 & 1/2 \end{pmatrix}.$$

We assume equal prior probabilities, $P(\omega_1) = P(\omega_2) = 0.5$, and substitute these into the general form for a discriminant, Eqs. 66–69, setting $g_1(\mathbf{x}) = g_2(\mathbf{x})$ to obtain the decision boundary:

$$x_2 = 3.514 - 1.125x_1 + 0.1875x_1^2.$$

This equation describes a parabola with vertex at $\binom{3}{1.83}$. Note that despite the fact that the variance in the data along the x_2 direction for both distributions is the same, the decision boundary does not pass through the point $\binom{3}{2}$, midway between the means, as we might have naively guessed. This is because for the ω_1 distribution, the probability distribution is "squeezed" in the x_1-direction more so than for the ω_2 distribution. The ω_1 distribution is increased along the x_2 direction (relative to that for the ω_2 distribution). Thus the decision boundary lies slightly lower than the point midway between the two means, as can be seen in the decision boundary.

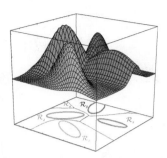

FIGURE 2.16. The decision regions for four normal distributions. Even with such a low number of categories, the shapes of the boundary regions can be rather complex.

★2.7 ERROR PROBABILITIES AND INTEGRALS

We can obtain additional insight into the operation of a general classifier—Bayes or otherwise—if we consider the sources of its error. Consider first the two-category case, and suppose the dichotomizer has divided the space into two regions $\mathcal{R}_1$ and $\mathcal{R}_2$ in a possibly nonoptimal way. There are two ways in which a classification error can occur; either an observation $\mathbf{x}$ falls in $\mathcal{R}_2$ and the true state of nature is ω_1, or $\mathbf{x}$ falls in $\mathcal{R}_1$ and the true state of nature is ω_2. Because these events are mutually exclusive and exhaustive, the probability of error is

$$
\begin{aligned}
P(error) &= P(\mathbf{x} \in \mathcal{R}_2, \omega_1) + P(\mathbf{x} \in \mathcal{R}_1, \omega_2) \\
&= P(\mathbf{x} \in \mathcal{R}_2|\omega_1)P(\omega_1) + P(\mathbf{x} \in \mathcal{R}_1|\omega_2)P(\omega_2) \\
&= \int_{\mathcal{R}_2} p(\mathbf{x}|\omega_1)P(\omega_1)\,d\mathbf{x} + \int_{\mathcal{R}_1} p(\mathbf{x}|\omega_2)P(\omega_2)\,d\mathbf{x}.
\end{aligned} \tag{70}
$$

This result is illustrated in the one-dimensional case in Fig. 2.17. The two integrals in Eq. 70 represent the pink and the gray areas in the tails of the functions $p(\mathbf{x}|\omega_i)P(\omega_i)$. Because the decision point x^* (and hence the regions $\mathcal{R}_1$ and $\mathcal{R}_2$) were chosen arbitrarily for that figure, the probability of error is not as small as it might be. In particular, the triangular area marked "reducible error" can be eliminated if the decision boundary is moved to x_B. This is the Bayes optimal decision boundary and gives lowest probability of error. In general, if $p(\mathbf{x}|\omega_1)P(\omega_1) > p(\mathbf{x}|\omega_2)P(\omega_2)$, it is advantageous to classify $\mathbf{x}$ as in $\mathcal{R}_1$ so that the smaller quantity will contribute to the error integral; this is exactly what the Bayes decision rule achieves.

In the multicategory case, there are more ways to be wrong than to be right, and it is simpler to compute the probability of being correct. Clearly,

$$
\begin{aligned}
P(correct) &= \sum_{i=1}^{c} P(\mathbf{x} \in \mathcal{R}_i, \omega_i) \\
&= \sum_{i=1}^{c} P(\mathbf{x} \in \mathcal{R}_i|\omega_i)P(\omega_i) \\
&= \sum_{i=1}^{c} \int_{\mathcal{R}_i} p(\mathbf{x}|\omega_i)P(\omega_i)\,d\mathbf{x}.
\end{aligned} \tag{71}
$$

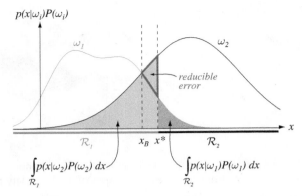

FIGURE 2.17. Components of the probability of error for equal priors and (nonoptimal) decision point x^*. The pink area corresponds to the probability of errors for deciding ω_1 when the state of nature is in fact ω_2; the gray area represents the converse, as given in Eq. 70. If the decision boundary is instead at the point of equal posterior probabilities, x_B, then this reducible error is eliminated and the total shaded area is the minimum possible; this is the Bayes decision and gives the Bayes error rate.

The general result of Eq. 71 depends neither on how the feature space is partitioned into decision regions nor on the form of the underlying distributions. The Bayes classifier maximizes this probability by choosing the regions so that the integrand is maximal for all $\mathbf{x}$; no other partitioning can yield a smaller probability of error.

*2.8 ERROR BOUNDS FOR NORMAL DENSITIES

The Bayes decision rule guarantees the lowest average error rate, and we have seen how to calculate the decision boundaries for normal densities. However, these results do not tell us what the probability of error actually *is*. The full calculation of the error for the Gaussian case would be quite difficult, especially in high dimensions, because of the discontinuous nature of the decision regions in the integral in Eq. 71. However, in the two-category case the general error integral of Eq. 5 can be approximated analytically to give us an upper bound on the error.

2.8.1 Chernoff Bound

To derive a bound for the error, we need the following inequality:

$$\min[a, b] \le a^\beta b^{1-\beta} \quad \text{for } a, b \ge 0 \text{ and } 0 \le \beta \le 1. \tag{72}$$

To understand this inequality we can, without loss of generality, assume $a \ge b$. Thus we need only show that $b \le a^\beta b^{1-\beta} = (a/b)^\beta b$. But this inequality is manifestly valid, because $(a/b)^\beta \ge 1$. Using Eqs. 7 and 1, we apply this inequality to the vector form of Eq. 5 and get the bound:

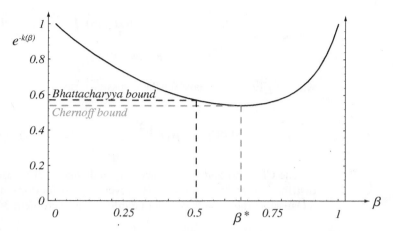

FIGURE 2.18. The Chernoff error bound is never looser than the Bhattacharyya bound. For this example, the Chernoff bound happens to be at $\beta^* = 0.66$, and is slightly tighter than the Bhattacharyya bound ($\beta = 0.5$).

$$P(error) \leq P^{\beta}(\omega_1) P^{1-\beta}(\omega_2) \int p^{\beta}(\mathbf{x}|\omega_1) p^{1-\beta}(\mathbf{x}|\omega_2) \, d\mathbf{x} \qquad \text{for } 0 \leq \beta \leq 1. \quad (73)$$

Note especially that this integral is over *all* feature space—we do not need to impose integration limits corresponding to decision boundaries.

If the conditional probabilities are normal, the integral in Eq. 73 can be evaluated analytically (Problem 36), yielding

$$\int p^{\beta}(\mathbf{x}|\omega_1) p^{1-\beta}(\mathbf{x}|\omega_2) \, d\mathbf{x} = e^{-k(\beta)} \qquad (74)$$

where

$$k(\beta) = \frac{\beta(1-\beta)}{2} (\boldsymbol{\mu}_1 - \boldsymbol{\mu}_2)^t [(1-\beta)\boldsymbol{\Sigma}_1 + \beta\boldsymbol{\Sigma}_2]^{-1} (\boldsymbol{\mu}_1 - \boldsymbol{\mu}_2)$$

$$+ \frac{1}{2} \ln \frac{(1-\beta)\boldsymbol{\Sigma}_1 + \beta\boldsymbol{\Sigma}_2}{|\boldsymbol{\Sigma}_1|^{1-\beta}|\boldsymbol{\Sigma}_2|^{\beta}}. \qquad (75)$$

The graph in Fig. 2.18 shows a typical example of how $e^{-k(\beta)}$ varies with β. The *Chernoff bound* on $P(error)$ is found by analytically or numerically finding the value of β that minimizes $P^{\beta}(\omega_1) P^{1-\beta}(\omega_2) e^{-k(\beta)}$ and then substituting this β into Eq. 73. The key benefit here is that this optimization is in the one-dimensional β space, despite the fact that the distributions themselves might be in a space of arbitrarily high dimension.

2.8.2 Bhattacharyya Bound

The general dependence of the Chernoff bound upon β shown in Fig. 2.18 is typical of a wide range of problems; The bound is loose for extreme values (i.e., $\beta \to 1$ and $\beta \to 0$), and it is tighter for intermediate ones. While the precise value of the optimal β depends upon the parameters of the distributions and the prior probabilities, a computationally simpler but slightly less tight bound can be derived by simply setting $\beta = 1/2$. This gives the so-called *Bhattacharyya bound* on the error, where Eq. 73 then has the form

$$P(error) \leq \sqrt{P(\omega_1)P(\omega_2)} \int \sqrt{p(\mathbf{x}|\omega_1)p(\mathbf{x}|\omega_2)}\, d\mathbf{x}$$

$$= \sqrt{P(\omega_1)P(\omega_2)}\, e^{-k(1/2)}, \tag{76}$$

where by Eq. 75 we have for the Gaussian case:

$$k(1/2) = 1/8(\boldsymbol{\mu}_2 - \boldsymbol{\mu}_1)^t \left[\frac{\boldsymbol{\Sigma}_1 + \boldsymbol{\Sigma}_2}{2}\right]^{-1} (\boldsymbol{\mu}_2 - \boldsymbol{\mu}_1) + \frac{1}{2} \ln \frac{\left|\frac{\boldsymbol{\Sigma}_1 + \boldsymbol{\Sigma}_2}{2}\right|}{\sqrt{|\boldsymbol{\Sigma}_1||\boldsymbol{\Sigma}_2|}}. \tag{77}$$

The Chernoff and Bhatacharyya bounds may still be used even if the underlying distributions are not Gaussian. However, for distributions that deviate markedly from a Gaussian, the bounds will not be informative (Problem 34).

EXAMPLE 2   **Error Bounds for Gaussian Distributions**

It is a straightforward matter to calculate the Bhattacharyya bound for the two-dimensional data sets of Example 1. Substituting the means and covariances of Example 1 into Eq. 77, we find $k(1/2) = 4.11157$, and thus by Eqs. 76 and 77 the Bhattacharyya bound on the error is $P(error) \leq 0.008191$.

A slightly tighter bound on the error can be approximated by searching numerically for the Chernoff bound of Eq. 75, which for this problem gives 0.008190. Numerical integration of Eq. 5 gives an error rate of 0.0021; thus the bounds here are not particularly tight. Such numerical integration is often impractical for Gaussians in higher than two or three dimensions.

2.8.3 Signal Detection Theory and Operating Characteristics

Another measure of distance between two Gaussian distributions has found great use in experimental psychology, radar detection and other fields. Suppose we are interested in detecting a single weak pulse, such as a dim flash of light or a weak radar reflection. Our model is, then, that at some point in the detector there is an internal signal (such as a voltage) x, whose value has mean μ_2 when the external signal (pulse) is present, and mean μ_1 when it is not present. Because of random noise—within and outside the detector itself—the actual value is a random variable. We assume the distributions are normal with different means but the same variance—that is, $p(x|\omega_i) \sim N(\mu_i, \sigma^2)$—as shown in Fig. 2.19.

The detector (classifier) employs a threshold value x^* for determining whether the external pulse is present, but suppose we, as experimenters, do not have access to this value (nor to the means and standard deviations of the distributions). We seek to find some measure of the ease of discriminating whether the pulse is present or not, in a form independent of the choice of x^*. Such a measure is the *discriminability*, which describes the inherent and unchangeable properties due to noise and the strength of the external signal, but not on the decision strategy (i.e., the actual choice of x^*). This discriminability is defined as

DISCRIMIN-
ABILITY

$$d' = \frac{|\mu_2 - \mu_1|}{\sigma}. \tag{78}$$

A high d' is of course desirable.

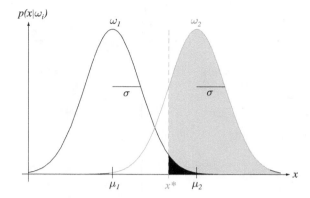

FIGURE 2.19. During any instant when no external pulse is present, the probability density for an internal signal is normal, that is, $p(x|\omega_1) \sim N(\mu_1, \sigma^2)$; when the external signal is present, the density is $p(x|\omega_2) \sim N(\mu_2, \sigma^2)$. Any decision threshold x^* will determine the probability of a hit (the pink area under the ω_2 curve, above x^*) and of a false alarm (the black area under the ω_1 curve, above x^*).

While we do not know μ_1, μ_2, σ or x^*, we assume here that we know the state of nature and the decision of the system. Such information allows us to find d'. To this end, we consider the following four probabilities:

- $P(x > x^*|x \in \omega_2)$: a *hit*—the probability that the internal signal is above x^* given that the external signal is present
- $P(x > x^*|x \in \omega_1)$: a *false alarm*—the probability that the internal signal is above x^* despite there being no external signal present
- $P(x < x^*|x \in \omega_2)$: a *miss*—the probability that the internal signal is below x^* given that the external signal is present
- $P(x < x^*|x \in \omega_1)$: a *correct rejection*—the probability that the internal signal is below x^* given that the external signal is not present.

If we have a large number of trials (and we can assume x^* is fixed, albeit at an unknown value), we can determine these probabilities experimentally, in particular the hit and false alarm rates. We plot a point representing these rates on a two-dimensional graph. If the densities are fixed but the threshold x^* is changed, then our hit and false alarm rates will also change. Thus we see that for a given discriminability d', our point will move along a smooth curve—a *receiver operating characteristic* or ROC curve (Fig. 2.20).

RECEIVER OPERATING CHARACTER-ISTIC

The great benefit of this signal detection framework is that we can distinguish operationally between *discriminability* and *decision bias*: While the former is an inherent property of the detector system, the latter is due to the receiver's implied but changeable loss matrix. Through any pair of hit and false alarm rates passes one and only one ROC curve; thus, so long as neither rate is exactly 0 or 1, we can determine the discriminability from these rates (Problem 39). Moreover, if the Gaussian assumption holds, a determination of the discriminability (from an arbitrary x^*) allows us to calculate the Bayes error rate—the most important property of any classifier. If the actual error rate differs from the Bayes rate inferred in this way, we should alter the threshold x^* accordingly.

It is a simple matter to generalize the above discussion and apply it to two categories having arbitrary multidimensional distributions, Gaussian or not. Suppose we

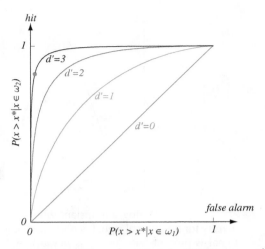

FIGURE 2.20. In a receiver operating characteristic (ROC) curve, the abscissa is the probability of false alarm, $P(x > x^* | x \in \omega_1)$, and the ordinate is the probability of hit, $P(x > x^* | x \in \omega_2)$. From the measured hit and false alarm rates (here corresponding to x^* in Fig. 2.19 and shown as the red dot), we can deduce that $d' = 3$.

have two distributions $p(\mathbf{x}|\omega_1)$ and $p(\mathbf{x}|\omega_2)$ which overlap, and thus have nonzero Bayes classification error. Just as we saw above, any pattern actually from ω_2 could be properly classified as ω_2 (a "hit") or misclassified as ω_1 (a "false alarm"). Unlike the one-dimensional case above, however, there may be *many* decision boundaries that correspond to a particular hit rate, each with a different false alarm rate. Clearly here we cannot determine a fundamental measure of discriminability without knowing more about the underlying decision rule than just the hit and false alarm rates.

In a rarely attainable ideal, we can imagine that our measured hit and false alarm rates are *optimal*—for example, that of all the decision rules giving the measured hit rate, the rule that is actually used is the one having the minimum false alarm rate. If we constructed a multidimensional classifier—regardless of the distributions used—we might try to characterize the problem in this way, though it would probably require great computational resources to search for such optimal hit and false alarm rates.

In practice, instead we forgo optimality, and simply vary a single control parameter for the decision rule and plot the resulting hit and false alarm rates—a curve called merely an *operating characteristic*. It is traditional to use a control parameter that can yield, at extreme values, either a vanishing false alarm or a vanishing hit rate, just as can be achieved with a very large or a very small x^* in an ROC curve. We should note that since the distributions can be arbitrary, the operating characteristic need not be symmetric (Fig. 2.21); in rare cases it need not even be concave down at all points.

Classifier operating curves are of value for problems where the loss matrix λ_{ij} might be changed. If the operating characteristic has been determined as a function of the control parameter ahead of time, it is a simple matter, when faced with a new loss function, to deduce the control parameter setting that will minimize the expected risk (Problem 39).

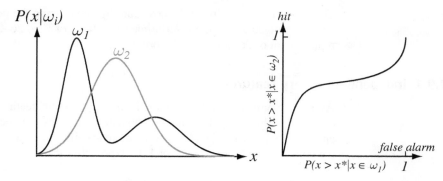

FIGURE 2.21. In a general operating characteristic curve, the abscissa is the probability of false alarm, $P(x \in \mathcal{R}_2 | x \in \omega_1)$, and the ordinate is the probability of hit, $P(x \in \mathcal{R}_2 | x \in \omega_2)$. As illustrated here, operating characteristic curves are generally not symmetric, as shown at the right.

2.9 BAYES DECISION THEORY—DISCRETE FEATURES

Until now we have assumed that the feature vector $\mathbf{x}$ could be any point in a d-dimensional Euclidean space, $\mathbf{R}^d$. However, in many practical applications the components of $\mathbf{x}$ are binary-, ternary-, or higher-integer-valued, so that $\mathbf{x}$ can assume only one of m discrete values $\mathbf{v}_1, \ldots, \mathbf{v}_m$. In such cases, the probability density function $p(\mathbf{x}|\omega_j)$ becomes singular; integrals of the form

$$\int p(\mathbf{x}|\omega_j)\, d\mathbf{x} \tag{79}$$

must then be replaced by corresponding sums, such as

$$\sum_{\mathbf{x}} P(\mathbf{x}|\omega_j), \tag{80}$$

where we understand that the summation is over all values of $\mathbf{x}$ in the discrete distribution.* Bayes formula then involves probabilities, rather than probability densities:

$$P(\omega_j|\mathbf{x}) = \frac{P(\mathbf{x}|\omega_j)P(\omega_j)}{P(\mathbf{x})}, \tag{81}$$

where

$$P(\mathbf{x}) = \sum_{j=1}^{c} P(\mathbf{x}|\omega_j)P(\omega_j). \tag{82}$$

The definition of the conditional risk $R(\alpha|\mathbf{x})$ is unchanged, and the fundamental Bayes decision rule remains the same: To minimize the overall risk, select the action α_i for which $R(\alpha_i|\mathbf{x})$ is minimum, or stated formally,

$$\alpha^* = \arg \min_{i} R(\alpha_i|\mathbf{x}). \tag{83}$$

---

*Technically speaking, Eq. 80 should be written as $\sum_k P(\mathbf{v}_k|\omega_j)$ where $P(\mathbf{v}_k|\omega_j)$ is the conditional probability that $\mathbf{x} = \mathbf{v}_k$, given that the state of nature is ω_j.

The basic rule to minimize the error rate by maximizing the posterior probability is also unchanged as are the discriminant functions of Eqs. 26–28, given the obvious replacement of densities $p(\cdot)$ by probabilities $P(\cdot)$.

2.9.1 Independent Binary Features

As an example of a classification involving discrete features, consider the two-category problem in which the components of the feature vector are binary-valued and conditionally independent. To be more specific we let $\mathbf{x} = (x_1, \ldots, x_d)^t$, where the components x_i are either 0 or 1, with probabilities

$$p_i = \Pr[x_i = 1|\omega_1] \tag{84}$$

and

$$q_i = \Pr[x_i = 1|\omega_2]. \tag{85}$$

This is a model of a classification problem in which each feature gives us a yes/no answer about the pattern. If $p_i > q_i$, we expect the ith feature to give a "yes" answer more frequently when the state of nature is ω_1 than when when it is ω_2. (As an example, consider two factories each making the same automobile, each of whose d components could be functional or defective. If it were known how the factories differed in their reliabilities for making each component, then this model could be used to judge which factory manufactured a given automobile based on the knowledge of which features are functional and which defective.) By assuming conditional independence we can write $P(\mathbf{x}|\omega_i)$ as the product of the probabilities for the components of $\mathbf{x}$. Given this assumption, a particularly convenient way of writing the class-conditional probabilities is as follows:

$$P(\mathbf{x}|\omega_1) = \prod_{i=1}^{d} p_i^{x_i}(1-p_i)^{1-x_i} \tag{86}$$

and

$$P(\mathbf{x}|\omega_2) = \prod_{i=1}^{d} q_i^{x_i}(1-q_i)^{1-x_i}. \tag{87}$$

Then the likelihood ratio is given by

$$\frac{P(\mathbf{x}|\omega_1)}{P(\mathbf{x}|\omega_2)} = \prod_{i=1}^{d} \left(\frac{p_i}{q_i}\right)^{x_i} \left(\frac{1-p_i}{1-q_i}\right)^{1-x_i} \tag{88}$$

and consequently Eq. 31 yields the discriminant function

$$g(\mathbf{x}) = \sum_{i=1}^{d} \left[x_i \ln \frac{p_i}{q_i} + (1-x_i) \ln \frac{1-p_i}{1-q_i}\right] + \ln \frac{P(\omega_1)}{P(\omega_2)}. \tag{89}$$

We note especially that this discriminant function is linear in the x_i and thus we can write

$$g(\mathbf{x}) = \sum_{i=1}^{d} w_i x_i + w_0, \tag{90}$$

where

$$w_i = \ln \frac{p_i(1 - q_i)}{q_i(1 - p_i)} \qquad i = 1, \ldots, d \tag{91}$$

and

$$w_0 = \sum_{i=1}^{d} \ln \frac{1 - p_i}{1 - q_i} + \ln \frac{P(\omega_1)}{P(\omega_2)}. \tag{92}$$

Let us examine these results to see what insight they can give. Recall first that we decide ω_1 if $g(\mathbf{x}) > 0$ and ω_2 if $g(\mathbf{x}) \leq 0$. We have seen that $g(\mathbf{x})$ is a weighted combination of the components of $\mathbf{x}$. The magnitude of the weight w_i indicates the relevance of a "yes" answer for x_i in determining the classification. If $p_i = q_i$, x_i gives us no information about the state of nature, and $w_i = 0$, just as we might expect. If $p_i > q_i$, then $1 - p_i < 1 - q_i$ and w_i is positive. Thus in this case a "yes" answer for x_i contributes w_i votes for ω_1. Furthermore, for any fixed $q_i < 1$, w_i gets larger as p_i gets larger. On the other hand, if $p_i < q_i$, w_i is negative and a "yes" answer contributes $|w_i|$ votes for ω_2.

The condition of feature independence leads to a very simple (linear) classifier; of course if the features were not independent, a more complicated classifier would be needed. We shall come across this again for systems with continuous features but note here that the more independent we can make the features, the simpler the classifier can be.

The prior probabilities $P(\omega_i)$ appear in the discriminant only through the threshold weight w_0. Increasing $P(\omega_1)$ increases w_0 and biases the decision in favor of ω_1, whereas decreasing $P(\omega_1)$ has the opposite effect. Geometrically, the possible values for $\mathbf{x}$ appear as the vertices of a d-dimensional hypercube; the decision surface defined by $g(\mathbf{x}) = 0$ is a hyperplane that separates ω_1 vertices from ω_2 vertices.

EXAMPLE 3   **Bayesian Decisions for Three-Dimensional Binary Data**

Consider a two-class problem having three independent binary features with known feature probabilities. Let us construct the Bayesian decision boundary if $P(\omega_1) = P(\omega_2) = 0.5$ and the individual components obey $p_i = 0.8$ and $q_i = 0.5$ for $i = 1, 2, 3$. By Eqs. 91 and 92 we have that the weights are

$$w_i = \ln \frac{.8(1 - .5)}{.5(1 - .8)} = 1.3863$$

and the bias value is

$$w_0 = \sum_{i=1}^{3} \ln \frac{1 - .8}{1 - .5} + \ln \frac{.5}{.5} = -2.75.$$

The surface $g(\mathbf{x}) = 0$ from Eq. 90 is shown on the left of the figure. Indeed, as we might have expected, the boundary places points with two or more "yes" answers into category ω_1, because that category has a higher probability of having any feature take value 1.

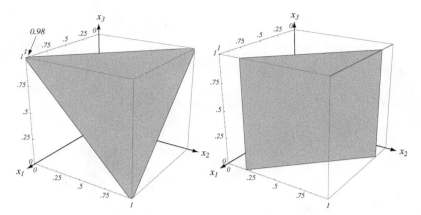

The decision boundary for the example involving three-dimensional binary features. On the left we show the case $p_i = .8$ and $q_i = .5$. On the right we use the same values except $p_3 = q_3$, which leads to $w_3 = 0$ and a decision surface parallel to the x_3 axis.

Suppose instead that while the prior probabilities remained the same, our individual components obeyed $p_1 = p_2 = 0.8$, $p_3 = 0.5$ and $q_1 = q_2 = q_3 = 0.5$. In this case feature x_3 gives us no predictive information about the categories, and hence the decision boundary is parallel to the x_3 axis. Note that in this discrete case there is a large range in positions of the decision boundary that leaves the categorization unchanged, as is particularly clear in the figure on the right.

*2.10 MISSING AND NOISY FEATURES

If we know the full probability structure of a problem, we can construct the (optimal) Bayes decision rule. Suppose we develop a Bayes classifier using uncorrupted data, but our input (test) data are then corrupted in particular known ways. How can we classify such corrupted inputs to obtain a minimum error now?

There are two analytically solvable cases of particular interest: when some of the features are *missing*, and when they are corrupted by a *noise source* with known properties. In each case our basic approach is to recover as much information about the underlying distribution as possible and use the Bayes decision rule.

2.10.1 Missing Features

Suppose we have a Bayesian (or other) recognizer for a problem using two features, but that for a particular pattern to be classified, one of the features is missing. For example, we can easily imagine that the lightness can be measured from a portion of a fish, but the width cannot because of occlusion by another fish.

We can illustrate with four categories a somewhat more general case (Fig. 2.22). Suppose that for a particular test pattern the feature x_1 is missing, and the measured value of x_2 is $\hat{x}_2$. Clearly if we assume that the missing value is the *mean* of all the x_1 values (i.e., $\bar{x}_1$), we will classify the pattern as ω_3. However, if the priors are equal, ω_2 would be a better decision, because the figure implies that $p(\hat{x}_2|\omega_2)$ is the largest of the four likelihoods.

To clarify our derivation we let $\mathbf{x} = [\mathbf{x}_g, \mathbf{x}_b]$, where $\mathbf{x}_g$ represents the known or "good" features and $\mathbf{x}_b$ represents the "bad" ones—that is, either unknown or missing. We seek the Bayes rule given the good features, and for that the posterior

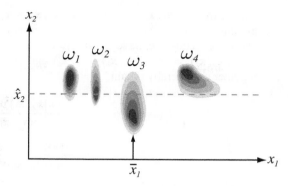

FIGURE 2.22. Four categories have equal priors and the class-conditional distributions shown. If a test point is presented in which one feature is missing (here, x_1) and the other is measured to have value $\hat{x}_2$ (red dashed line), we want our classifier to classify the pattern as category ω_2, because $p(\hat{x}_2|\omega_2)$ is the largest of the four likelihoods.

probabilities are needed. In terms of the good features the posteriors are

$$P(\omega_i|\mathbf{x}_g) = \frac{p(\omega_i, \mathbf{x}_g)}{p(\mathbf{x}_g)} = \frac{\int p(\omega_i, \mathbf{x}_g, \mathbf{x}_b)\, d\mathbf{x}_b}{p(\mathbf{x}_g)}$$

$$= \frac{\int P(\omega_i|\mathbf{x}_g, \mathbf{x}_b)\,p(\mathbf{x}_g, \mathbf{x}_b)\, d\mathbf{x}_b}{p(\mathbf{x}_g)}$$

$$= \frac{\int g_i(\mathbf{x})\,p(\mathbf{x})\, d\mathbf{x}_b}{\int p(\mathbf{x})\, d\mathbf{x}_b}, \tag{93}$$

where $g_i(\mathbf{x}) = g_i(\mathbf{x}_g, \mathbf{x}_b) = P(\omega_i|\mathbf{x}_g, \mathbf{x}_b)$ is one form of our discriminant function.

MARGINAL We refer to $\int p(\omega_i, \mathbf{x}_g, \mathbf{x}_b)\, d\mathbf{x}_b$, as a *marginal distribution*; we say the full joint distribution is marginalized over the variable $\mathbf{x}_b$. In short, Eq. 93 shows that we must integrate (marginalize) the posterior probability over the bad features. Finally we use the Bayes decision rule on the resulting posterior probabilities, that is, choose ω_i if $P(\omega_i|\mathbf{x}_g) > P(\omega_j|\mathbf{x}_g)$ for all i and j. In Chapter 3 we shall consider the Expectation-Maximization (EM) algorithm, which addresses a related problem involving missing features.

2.10.2 Noisy Features

It is a simple matter to generalize the results of Eq. 93 to the case where a particular feature has been corrupted by statistically independent noise. For instance, in our fish classification example, we might have a reliable measurement of the length, while variability of the light source might degrade the measurement of the lightness. We assume we have uncorrupted (good) features $\mathbf{x}_g$, as before, and a *noise model*, expressed as $p(\mathbf{x}_b|\mathbf{x}_t)$. Here we let $\mathbf{x}_t$ denote the true value of the observed $\mathbf{x}_b$ features, that is, without the noise present; in short, the $\mathbf{x}_b$ are observed instead of the true $\mathbf{x}_t$. We assume that if $\mathbf{x}_t$ were known, $\mathbf{x}_b$ would be independent of ω_i and $\mathbf{x}_g$. From such an assumption we get:

$$P(\omega_i|\mathbf{x}_g, \mathbf{x}_b) = \frac{\int p(\omega_i, \mathbf{x}_g, \mathbf{x}_b, \mathbf{x}_t)\, d\mathbf{x}_t}{p(\mathbf{x}_g, \mathbf{x}_b)}. \tag{94}$$

Now $p(\omega_i, \mathbf{x}_g, \mathbf{x}_b, \mathbf{x}_t) = P(\omega_i|\mathbf{x}_g, \mathbf{x}_b, \mathbf{x}_t)p(\mathbf{x}_g, \mathbf{x}_b, \mathbf{x}_t)$, but by our independence assumption, if we know $\mathbf{x}_t$, then $\mathbf{x}_b$ does not provide any additional information about ω_i. Thus we have $P(\omega_i|\mathbf{x}_g, \mathbf{x}_b, \mathbf{x}_t) = P(\omega_i|\mathbf{x}_g, \mathbf{x}_t)$. Similarly, we have $p(\mathbf{x}_g, \mathbf{x}_b, \mathbf{x}_t) = p(\mathbf{x}_b|\mathbf{x}_g, \mathbf{x}_t)p(\mathbf{x}_g, \mathbf{x}_t)$ and $p(\mathbf{x}_b|\mathbf{x}_g, \mathbf{x}_t) = p(\mathbf{x}_b|\mathbf{x}_t)$. We put these together and thereby obtain

$$
P(\omega_i|\mathbf{x}_g, \mathbf{x}_b) = \frac{\int P(\omega_i|\mathbf{x}_g, \mathbf{x}_t)p(\mathbf{x}_g, \mathbf{x}_t)p(\mathbf{x}_b|\mathbf{x}_t)\,d\mathbf{x}_t}{\int p(\mathbf{x}_g, \mathbf{x}_t)p(\mathbf{x}_b|\mathbf{x}_t)\,d\mathbf{x}_t}
$$

$$
= \frac{\int g_i(\mathbf{x})p(\mathbf{x})p(\mathbf{x}_b|\mathbf{x}_t)\,d\mathbf{x}_t}{\int p(\mathbf{x})p(\mathbf{x}_b|\mathbf{x}_t)\,d\mathbf{x}_t}, \tag{95}
$$

which we use as discriminant functions for classification in the manner dictated by Bayes.

Equation 95 differs from Eq. 93 solely by the fact that the integral is weighted by the noise model. In the extreme case where $p(\mathbf{x}_b|\mathbf{x}_t)$ is uniform over the entire space (and hence provides no predictive information for categorization), the equation reduces to the case of missing features—a satisfying result.

★2.11 BAYESIAN BELIEF NETWORKS

The methods we have described up to now are fairly general—all that we assumed, at base, was that we could parameterize the probability distributions by a vector $\boldsymbol{\theta}$. If we had prior information about $\boldsymbol{\theta}$, this too could be used. Sometimes our knowledge about a distribution is not directly expressed by a parameter vector, but instead about the statistical dependencies (or independencies) or the causal relationships among the component variables. (Recall that for some multidimensional distribution $p(\mathbf{x})$, if for two features we have $p(x_i, x_j) = p(x_i)p(x_j)$, we say those variables are statistically independent, as illustrated in Fig. 2.23.)

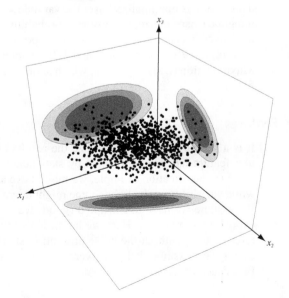

FIGURE 2.23. A three-dimensional distribution which obeys $p(x_1, x_3) = p(x_1)p(x_3)$; thus here x_1 and x_3 are statistically independent but the other feature pairs are not.

There are many such cases where we know—or can safely assume—which variables are or are not causally related, even if it may be more difficult to specify the precise probabilistic relationships among those variables. Suppose, for instance, we are describing the state of an automobile: temperature of the engine, pressure of the brake fluid, pressure of the air in the tires, voltages in the wires, and so on. Our basic knowledge of cars includes the fact that the oil pressure in the engine and the air pressure in a tire are *not* causally related while the engine temperature and oil temperature *are* causally related. Furthermore, we may know *several* variables that might influence another: The coolant temperature is affected by the engine temperature, the speed of the radiator fan (which blows air over the coolant-filled radiator), and so on. We shall now exploit this structural information when reasoning about the system and its variables.

We represent these causal dependencies graphically by means of *Bayesian belief nets*, also called *causal networks*, or simply *belief nets*. While these nets can represent continuous multidimensional distributions over their variables, they have enjoyed greatest application and success for discrete variables. For this reason, and because the calculations are simpler, we shall concentrate on the discrete case.

NODE

Each *node* (or unit) represents one of the system components, and here it takes on discrete values. We label nodes **A**, **B**, ... and their variables by the corresponding lowercase letter. Thus, while there are a discrete number of possible values of node **A**—for instance two, a_1 and a_2—there may be continuous-valued *probabilities* on these discrete states. For example, if node **A** represents the automobile ignition switch—$a_1 = on$, $a_2 = off$—we might have $P(a_1) = 0.739$, $P(a_2) = 0.261$, or indeed any other probabilities. Each link in the net is directional and joins two nodes; the link represents the causal influence of one node upon another. Thus in the net in Fig. 2.24 **A** directly influences **D**. While **B** also influences **D**, such influence is indirect, through **C**. In considering a single node in a net, it is useful to distinguish the set of nodes immediately *before* that node—called its *parents*—and the set of those immediately *after* it—called its *children*. Thus in Fig. 2.24 the parents of **D** are **A** and **C** while the child of **D** is **E**. Suppose we have a belief net, complete with causal dependencies indicated by the topology of the links. Through a direct application of Bayes rule, we can determine the probability of any configuration of variables in

PARENT
CHILD

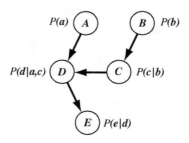

FIGURE 2.24. A belief network consists of nodes (labeled with uppercase bold letters) and their associated discrete states (in lowercase). Thus node **A** has states $\{a_1, a_2, \ldots\}$, which collectively are denoted simply **a**; node **B** has states $\{b_1, b_2, \ldots\}$, denoted **b**, and so forth. The links between nodes represent direct causal influence. For example the link from **A** to **D** represents the direct influence of **A** upon **D**. In this network, the variables at **B** may influence those at **D**, but only indirectly through their effect on **C**. Simple probabilities are denoted $P(\mathbf{a})$ and $P(\mathbf{b})$, and conditional probabilities $P(\mathbf{c}|\mathbf{b})$, $P(\mathbf{d}|\mathbf{a}, \mathbf{c})$ and $P(\mathbf{e}|\mathbf{d})$.

the joint distribution. To proceed, though, we also need the *conditional probability tables*, which give the probability of any variable at a node for each conditioning event—that is, for the values of the variables in the parent nodes. Each row in a conditional probability table sums to 1, as its entries describe all possible cases for the variable. If a node has no parents, then the table just contains the prior probabilities of the variables. (There are sophisticated algorithms for learning the entries in such a table based on a data set of variable values. We shall not address such learning here as our main concern is how to represent and reason about this probabilistic information.) Since the network and conditional probability tables contain all the information of the problem domain, we can use them to calculate any entry in the joint probability distribution, as illustrated in Example 4.

EXAMPLE 4   Belief Network for Fish

Consider again the problem of classifying fish, but now we want to incorporate more information than the measurements of the lightness and width. Imagine that a human expert has constructed the simple belief network in the figure, where node **A** represents the time of year and can have four values: $a_1 = winter$, $a_2 = spring$, $a_3 = summer$ and $a_4 = autumn$. Node **B** represents the locale where the fish was caught: $b_1 = north\ Atlantic$ and $b_2 = south\ Atlantic$. Node **X**, which represents the fish, has just two possible values: $x_1 = salmon$ and $x_2 = sea\ bass$. **A** and **B** are the parents of the **X**. Similarly, our expert tells us that the children nodes of **X** represent lightness, **C**, with $c_1 = dark$, $c_2 = medium$ and $c_3 = light$, as well as thickness, **D**, with $d_1 = thick$ and $d_2 = thin$. Thus the season and the locale determine directly what kind of fish is likely to be caught; the season and locale also determine the fish's lightness and thickness, but only indirectly through their effect on **X**.

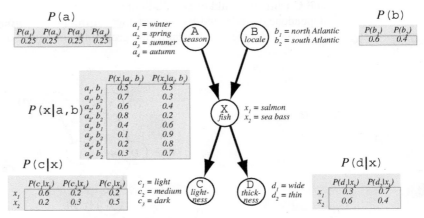

A simple belief net for the fish example. The season and the locale (and many other stochastic variables) determine directly the probability of the two different types of fish caught. The type of fish directly affects the lightness and thickness measured. The conditional probability tables quantifying these relationships are shown in pink.

Imagine that fishing boats go out throughout the year; then the probability distribution on the variables at **A** is uniform. Imagine, too, that boats generally spend more time in the north than the south Atlantic areas, specifically the probabilities that any fish came from those areas are 0.6 and 0.4, respectively. The other conditional probabilities are similarly given in the tables.

Now we can determine the value of any entry in the joint probability, for instance the probability that the fish was caught in the summer in the north Atlantic and is a sea bass that is dark and thin:

$$P(a_3, b_1, x_2, c_3, d_2) = P(a_3)P(b_1)P(x_2|a_3, b_1)P(c_3|x_2)P(d_2|x_2)$$

$$= 0.25 \times 0.6 \times 0.6 \times 0.5 \times 0.4$$

$$= 0.018.$$

Note how the topology of the net is captured by the probabilities in the expression. Specifically, since **X** is the only node to have two parents, only the $P(x_2|\cdot, \cdot)$ term has two conditioning variables; the other conditional probabilities have just one each. The product of these probabilities corresponds to the assumption of statistical independence.

We now illustrate more fully how to exploit the causal structure in a Bayes belief net when determining the probability of its variables. Suppose we wish to determine the probability distribution over the variables $d_1, d_2, \ldots$ at **D** in the left network of Fig. 2.25 using the conditional probability tables and the network topology. We evaluate this by summing the full joint distribution, $P(\mathbf{a}, \mathbf{b}, \mathbf{c}, \mathbf{d})$, over all the variables other than **d**:

$$P(\mathbf{d}) = \sum_{\mathbf{a},\mathbf{b},\mathbf{c}} P(\mathbf{a}, \mathbf{b}, \mathbf{c}, \mathbf{d}) \tag{96}$$

$$= \sum_{\mathbf{a},\mathbf{b},\mathbf{c}} P(\mathbf{a})P(\mathbf{b}|\mathbf{a})P(\mathbf{c}|\mathbf{b})P(\mathbf{d}|\mathbf{c})$$

$$= \sum_{\mathbf{c}} P(\mathbf{d}|\mathbf{c}) \underbrace{\sum_{\mathbf{b}} P(\mathbf{c}|\mathbf{b}) \underbrace{\sum_{\mathbf{a}} P(\mathbf{b}|\mathbf{a})P(\mathbf{a})}_{P(\mathbf{b})}}_{P(\mathbf{c})}.$$

$$\underbrace{\phantom{= \sum_{\mathbf{c}} P(\mathbf{d}|\mathbf{c}) \sum_{\mathbf{b}} P(\mathbf{c}|\mathbf{b}) \sum_{\mathbf{a}} P(\mathbf{b}|\mathbf{a})P(\mathbf{a})}}_{P(\mathbf{d})}$$

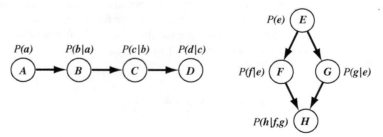

FIGURE 2.25. Two simple belief networks. The one on the left is a simple linear chain, the one on the right a simple loop. The conditional probability tables are indicated, for instance, as $P(\mathbf{h}|\mathbf{f}, \mathbf{g})$.

In Eq. 96 the summation variables can be split simply, and the intermediate terms have simple interpretations, as indicated. If we wanted the probability of a *particular* value of $\mathbf{D}$, for instance d_2, we would compute

$$P(d_2) = \sum_{\mathbf{a},\mathbf{b},\mathbf{c}} P(\mathbf{a}, \mathbf{b}, \mathbf{c}, d_2), \tag{97}$$

and proceed as above. In either case, the conditional probabilities are simple because of the simple linear topology of the network.

Now consider computing the probabilities of the variables at $\mathbf{H}$ in the network with the loop on the right of Fig. 2.25. Here we find

$$P(\mathbf{h}) = \sum_{\mathbf{e},\mathbf{f},\mathbf{g}} P(\mathbf{e}, \mathbf{f}, \mathbf{g}, \mathbf{h}) \tag{98}$$

$$= \sum_{\mathbf{e},\mathbf{f},\mathbf{g}} P(\mathbf{e}) P(\mathbf{f}|\mathbf{e}) P(\mathbf{g}|\mathbf{e}) P(\mathbf{h}|\mathbf{f}, \mathbf{g})$$

$$= \sum_{\mathbf{e}} P(\mathbf{e}) P(\mathbf{f}|\mathbf{e}) P(\mathbf{g}|\mathbf{e}) \sum_{\mathbf{f},\mathbf{g}} P(\mathbf{h}|\mathbf{f}, \mathbf{g}).$$

Note particularly that the expansion of the full sum differs somewhat from that in Eq. 96 because of the $P(\mathbf{h}|\mathbf{f}, \mathbf{g})$ term, which itself arises from the loop topology of the network.

Bayes belief nets are most useful in the case where are given the values of some of the variables—the *evidence*—and we seek to determine some particular configuration of other variables. Thus in our fish example we might seek to determine the probability that a fish came from the north Atlantic, given that it is springtime, and that the fish is a light salmon. (Notice that even here we may not be given the values of some variables such as the width of the fish.) In that case, the probability we seek is $P(b_1|a_2, x_1, c_1)$. In practice, we determine the values of several query variables (denoted collectively $\mathbf{x}$) given the evidence of all other variables (denoted $\mathbf{e}$) by

$$P(\mathbf{x}|\mathbf{e}) = \frac{P(\mathbf{x}, \mathbf{e})}{P(\mathbf{e})} = \alpha P(\mathbf{x}, \mathbf{e}), \tag{99}$$

where α is a constant of proportionality.

As an example, suppose we are given the Bayes belief net and conditional probability tables in Example 4. Suppose we know that a fish is light (c_1) and caught in the south Atlantic (b_2), but we do not know what time of year the fish was caught nor it thickness. How shall we classify the fish for minimum expected classification error? Of course we must compute the probability it is a salmon, and also the probability it is sea bass. We focus first on the relative probability the fish is a salmon given this evidence:

$$P(x_1|c_1, b_2) = \frac{P(x_1, c_1, b_2)}{P(c_1, b_2)} \tag{100}$$

$$= \alpha \sum_{\mathbf{a},\mathbf{d}} P(x_1, \mathbf{a}, b_2, c_1, \mathbf{d})$$

$$= \alpha \sum_{a,d} P(\mathbf{a})P(b_2)P(x_1|\mathbf{a}, b_2)P(c_1|x_1)P(\mathbf{d}|x_1)$$

$$= \alpha P(b_2)P(c_1|x_1)$$

$$\times \left[\sum_a P(\mathbf{a})P(x_1|\mathbf{a}, b_2)\right]\left[\sum_d P(\mathbf{d}|x_1)\right]$$

$$= \alpha P(b_2)P(c_1|x_1)$$

$$\times [P(a_1)P(x_1|a_1, b_2) + P(a_2)P(x_1|a_2, b_2)$$

$$+ P(a_3)P(x_1|a_3, b_2) + P(a_4)P(x_1|a_4, b_2)]$$

$$\times \underbrace{[P(d_1|x_1) + P(d_2|x_1)]}_{=1}$$

$$= \alpha(0.4)(0.6)[(0.25)(0.7) + (0.25)(0.8) + (0.25)(0.1)$$

$$+ (0.25)(0.3)]1.0$$

$$= \alpha \, 0.114.$$

Note that in this case,

$$\sum_d P(\mathbf{d}|x_1) = 1, \tag{101}$$

that is, if we do not measure information corresponding to node **D**, the conditional probability table at **D** does not affect our results. A computation similar to that in Eq. 100 shows $P(x_2|c_1, b_2) = \alpha \, 0.042$. We normalize these probabilities (and hence eliminate α) and find $P(x_1|c_1, b_2) = 0.73$ and $P(x_2|c_1, b_2) = 0.27$. Thus given this evidence, we should classify this fish as a salmon.

When the dependency relationships among the features used by a classifier are unknown, we generally proceed by taking the simplest assumption, namely, that the features are conditionally independent given the category, that is,

$$P(\mathbf{a}, \mathbf{b}|\mathbf{x}) = P(\mathbf{a}|\mathbf{x})P(\mathbf{b}|\mathbf{x}). \tag{102}$$

NAIVE BAYES'
RULE

In practice, this so-called *naive Bayes' rule* or *idiot Bayes' rule* often works quite well in practice, despite its manifest simplicity. Other approaches are to assume some functional form of conditional probability tables.

Belief nets have found increasing use in complicated problems such as medical diagnosis. Here the uppermost nodes (ones without their own parents) represent a fundamental biological agent such as the presence of a virus or bacteria. Intermediate nodes then describe diseases, such as flu or emphysema, and the lowermost nodes describe the symptoms, such as high temperature or coughing. A physician enters measured values into the net and finds the most likely disease or cause.

*2.12 COMPOUND BAYESIAN DECISION THEORY AND CONTEXT

Let us reconsider our introductory example of designing a classifier to sort two types of fish. Our original assumption was that the sequence of types of fish was so unpredictable that the state of nature looked like a random variable. Without abandoning this attitude, let us consider the possibility that the consecutive states of nature might not be statistically independent. We should be able to exploit such statistical dependence to gain improved performance. This is one example of the use of *context* to aid decision making.

The way in which we exploit such context information is somewhat different when we can wait for n fish to emerge and then make all n decisions jointly than when we must decide as each fish emerges. The first problem is a *compound decision problem*, and the second is a *sequential compound decision problem*. The former case is conceptually simpler and is the one we shall examine here.

To state the general problem, let $\boldsymbol{\omega} = (\omega(1), \ldots, \omega(n))^t$ be a vector denoting the n states of nature, with $\omega(i)$ taking on one of the c values $\omega_1, \ldots, \omega_c$. Let $P(\boldsymbol{\omega})$ be the prior probability for the n states of nature. Let $\mathbf{X} = (\mathbf{x}_1, \ldots, \mathbf{x}_n)$ be a matrix giving the n observed feature vectors, with $\mathbf{x}_i$ being the feature vector obtained when the state of nature was $\omega(i)$. Finally, let $p(\mathbf{X}|\boldsymbol{\omega})$ be the conditional probability density function for $\mathbf{X}$ given the true set of states of nature $\boldsymbol{\omega}$. Using this notation we see that the posterior probability of $\boldsymbol{\omega}$ is given by

$$P(\boldsymbol{\omega}|\mathbf{X}) = \frac{p(\mathbf{X}|\boldsymbol{\omega})P(\boldsymbol{\omega})}{p(\mathbf{X})} = \frac{p(\mathbf{X}|\boldsymbol{\omega})P(\boldsymbol{\omega})}{\sum_{\boldsymbol{\omega}} p(\mathbf{X}|\boldsymbol{\omega})P(\boldsymbol{\omega})}. \tag{103}$$

In general, one can define a loss matrix for the compound decision problem and seek a decision rule that minimizes the compound risk. The development of this theory parallels our discussion for the simple decision problem, and concludes that the optimal procedure is to minimize the compound conditional risk. In particular, if there is no loss for being correct and if all errors are equally costly, then the procedure reduces to computing $P(\boldsymbol{\omega}|\mathbf{X})$ for all $\boldsymbol{\omega}$ and selecting the $\boldsymbol{\omega}$ for which this posterior probability is maximum.

While this provides the theoretical solution, in practice the computation of $P(\boldsymbol{\omega}|\mathbf{X})$ can easily prove to be an enormous task. If each component $\omega(i)$ can have one of c values, there are c^n possible values of $\boldsymbol{\omega}$ to consider. Some simplification can be obtained if the distribution of the feature vector $\mathbf{x}_i$ depends only on the corresponding state of nature $\omega(i)$, not on the values of the other feature vectors or the other states of nature. In this case the joint density $p(\mathbf{X}|\boldsymbol{\omega})$ is merely the product of the component densities $p(\mathbf{x}_i|\omega(i))$:

$$p(\mathbf{X}|\boldsymbol{\omega}) = \prod_{i=1}^{n} p(\mathbf{x}_i|\omega(i)). \tag{104}$$

While this simplifies the problem of computing $p(\mathbf{X}|\boldsymbol{\omega})$, there is still the problem of computing the prior probabilities $P(\boldsymbol{\omega})$. This joint probability is central to the compound Bayes decision problem, because it reflects the interdependence of the states of nature. Thus it is unacceptable to simplify the problem of calculating $P(\boldsymbol{\omega})$ by assuming that the states of nature are independent. In addition, practical applications usually require some method of avoiding the computation of $P(\boldsymbol{\omega}|\mathbf{X})$ for all c^n possible values of $\boldsymbol{\omega}$. We shall find some solutions to this problem in Chapter 3.

SUMMARY

The basic ideas underlying Bayes decision theory are very simple. To minimize the overall risk, one should always choose the action that minimizes the conditional risk $R(\alpha|\mathbf{x})$. In particular, to minimize the probability of error in a classification problem, one should always choose the state of nature that maximizes the posterior probability $P(\omega_j|\mathbf{x})$. Bayes formula allows us to calculate such probabilities from the prior probabilities $P(\omega_j)$ and the conditional densities $p(\mathbf{x}|\omega_j)$. If there are different penalties for misclassifying patterns from ω_i as if from ω_j, the posteriors must be first weighted according to such penalties before taking action.

If the underlying distributions are multivariate Gaussian, the decision boundaries will be hyperquadrics, whose form and position depends upon the prior probabilities, means and covariances of the distributions in question. The true expected error can be bounded above by the Chernoff and computationally simpler Bhattacharyya bounds. If an input (test) pattern has missing or corrupted features, we should form the marginal distributions by integrating over such features and then using Bayes decision procedure on the resulting distributions. Receiver operating characteristic curves describe the inherent and unchangeable properties of a classifier and can be used, for example, to determine the Bayes rate.

Bayesian belief nets allow the designer to specify, by means of connection topology, the functional dependencies and independencies among model variables. When any subset of variables is clamped to some known values, each node comes to a probability of its value through a Bayesian inference calculation. Parameters representing conditional dependencies can be set by an expert.

For many pattern classification applications, the chief problem in applying these results is that the conditional densities $p(\mathbf{x}|\omega_j)$ are not known. In some cases we may know the form these densities assume, but we may not know characterizing parameter values. The classic case occurs when the densities are known to be, or can assumed to be, multivariate normal, but the values of the mean vectors and the covariance matrices are not known. More commonly, even less is known about the conditional densities, and procedures that are less sensitive to specific assumptions about the densities must be used. Most of the remainder of this book will be devoted to various procedures that have been developed to attack such problems.

BIBLIOGRAPHICAL AND HISTORICAL REMARKS

The power, coherence, and elegance of Bayesian theory in pattern recognition make it among the most beautiful formalisms in science. Its foundations go back to the Reverend Bayes himself, of course [3], but he stated his theorem (Eq. 1) for the case of uniform priors. It was Laplace [29] who first stated it for the more general (but discrete) case. There are several modern and clear descriptions of the ideas—in pattern recognition and general decision theory—that can be recommended [6, 7, 15, 17, 30, 31]. Because Bayesian theory rests on an axiomatic foundation, it is guaranteed to have quantitative coherence; some other classification methods do not. Wald presents a non-Bayesian perspective on these topics that can be highly recommended [41], and the philosophical foundations of Bayesian and non-Bayesian methods are explored in reference [18]. Neyman and Pearson provided some of the most important pioneering work in hypothesis testing, and they used the probability of error as the criterion [32]; Wald extended this work by introducing the notions of loss and risk [40]. Certain conceptual problems have always attended the use of loss functions and prior probabilities. In fact, the Bayesian approach is avoided by many statisticians, partly because there are problems for which a decision is made only once, and partly because there may be no reasonable way to determine the prior probabilities. Neither of these difficulties seems to present a serious drawback in typical pattern recognition applications: For nearly all important pattern recognition problems we will have training data and we will use our recognizer more than once. For these reasons, the Bayesian approach will continue to be of great use in pattern recognition. The single most important drawback of the Bayesian approach is the difficulty of determining and computing the conditional density functions. The multivariate Gaussian model may provide an adequate approximation to the true density, but there are many problems for which the densitites are far from Gaussian. Even when the Gaussian model is satisfactory, we shall see in the next chapter that estimating the unknown parameters from data may not be a trivial task. Subsequent chapters will investigate what can be done when the Gaussian model is not adequate.

Chow was among the earliest to use Bayesian decision theory for pattern recognition [12], and he later established fundamental relations between error and reject rate [13]. Error rates for Gaussians have been explored in reference [20], and the Chernoff and Bhattacharyya bounds were first presented in references [11] and [8], respectively; the bounds are explored in a number of statistics texts, such as reference [19]. Computational approximations for bounding integrals for Bayesian probability of error (the source for one of the homework problems) appears in reference [2]. Neyman and Pearson also worked on classification given constraints [32], and the analysis of minimax estimators for multivariate normals is presented in references [4, 5] and [16]. Signal detection theory and receiver operating characteristics are fully explored in reference [22]; a brief overview, targeting experimental psychologists, is presented in reference [39]. Our discussion of the missing feature problem follows closely the work of Ahmad and Tresp [1], while the definitive book on missing features, including a great deal beyond our discussion here, is reference [35].

The origins of Bayesian belief nets can be traced back to reference [43], and a thorough literature review can be found in reference [10]; excellent modern books [27, 33] and tutorials [9] can be recommended. An important dissertation on the theory of belief nets, with an application to medical diagnosis, is reference [25], and a summary of work on diagnosis of machine faults is given in reference [24]. While we have focussed on directed acyclic graphs, belief nets are of broader use, and they even allow loops or arbitrary topologies—a topic that would lead us far afield here but which is treated in reference [27].

Entropy was the central concept in the foundation of information theory [36], and the relation of Gaussians to entropy is explored in reference [38]. Readers requiring a review of information theory [14], linear algebra [28], calculus and continuous mathematics [37, 44], probability [34], calculus of variations and Lagrange multipliers [21], should consult these texts and those listed in our Appendix.

PROBLEMS

Section 2.1

1. In the two-category case, under the Bayes decision rule the conditional error is given by Eq. 7. Even if the posterior densities are continuous, this form of the conditional error virtually always leads to a discontinuous integrand when calculating the full error by Eq. 5.

   (a) Show that for arbitrary densities, we can replace Eq. 7 by $P(error|x) = 2P(\omega_1|x)P(\omega_2|x)$ in the integral and get an upper bound on the full error.

   (b) Show that if we use $P(error|x) = \alpha P(\omega_1|x)P(\omega_2|x)$ for $\alpha < 2$, then we are not guaranteed that the integral gives an upper bound on the error.

   (c) Analogously, show that we can use instead $P(error|x) = P(\omega_1|x)P(\omega_2|x)$ and get a lower bound on the full error.

   (d) Show that if we use $P(error|x) = \beta P(\omega_1|x)P(\omega_2|x)$ for $\beta > 1$, then we are not guaranteed that the integral gives an lower bound on the error.

Section 2.2

2. Suppose two equally probable one-dimensional densities are of the form $p(x|\omega_i) \propto e^{-|x-a_i|/b_i}$ for $i = 1, 2$ and $0 < b_i$.

   (a) Write an analytic expression for each density, that is, normalize each function for arbitrary a_i and positive b_i.

   (b) Calculate the likelihood ratio as a function of your four variables.

   (c) Sketch a graph of the likelihood ratio $p(x|\omega_1)/p(x|\omega_2)$ for the case $a_1 = 0$, $b_1 = 1$, $a_2 = 1$ and $b_2 = 2$.

Section 2.3

3. Consider minimax criterion for the zero-one loss function, that is, $\lambda_{11} = \lambda_{22} = 0$ and $\lambda_{12} = \lambda_{21} = 1$.

   (a) Prove that in this case the decision regions will satisfy

   $$\int_{\mathcal{R}_2} p(\mathbf{x}|\omega_1)\,d\mathbf{x} = \int_{\mathcal{R}_1} p(\mathbf{x}|\omega_2)\,d\mathbf{x}.$$

   (b) Is this solution always unique? If not, construct a simple counterexample.

4. Consider the minimax criterion for a two-category classification problem.

   (a) Fill in the steps of the derivation of Eq. 23.

   (b) Explain why the overall Bayes risk must be concave down as a function of the prior $P(\omega_1)$, as shown in Fig. 2.4.

   (c) Assume we have one-dimensional Gaussian distributions $p(x|\omega_i) \sim N(\mu_i, \sigma_i^2)$, $i = 1, 2$, but completely unknown prior probabilities. Use the minimax criterion to find the optimal decision point x^* in terms of μ_i and σ_i under a zero-one risk.

(d) For the decision point x^* you found in (c), what is the overall minimax risk? Express this risk in terms of an error function $\text{erf}(\cdot)$.

(e) Assume $p(x|\omega_1) \sim N(0, 1)$ and $p(x|\omega_2) \sim N(1/2, 1/4)$, under a zero-one loss. Find x^* and the overall minimax loss.

(f) Assume $p(x|\omega_1) \sim N(5, 1)$ and $p(x|\omega_2) \sim N(6, 1)$. Without performing any explicit calculations, determine x^* for the minimax criterion. Explain your reasoning.

5. Generalize the minimax decision rule in order to classify patterns from three categories having triangle densities as follows:

$$p(x|\omega_i) = T(\mu_i, \delta_i) \equiv \begin{cases} (\delta_i - |x - \mu_i|)/\delta_i^2 & \text{for } |x - \mu_i| < \delta_i \\ 0 & \text{otherwise,} \end{cases}$$

where $\delta_i > 0$ is the half-width of a distribution ($i = 1, 2, 3$). Assume for convenience that $\mu_1 < \mu_2 < \mu_3$, and make some minor simplifying assumptions about the δ_i's as needed, to answer the following:

(a) In terms of the priors $P(\omega_i)$, means and half-widths, find the optimal decision points x_1^* and x_2^* under a zero-one (categorization) loss.

(b) Generalize the minimax decision rule to *two* decision points, x_1^* and x_2^*, for such triangular distributions.

(c) Let $\{\mu_i, \delta_i\} = \{0, 1\}, \{.5, .5\}$, and $\{1, 1\}$. Find the minimax decision rule (i.e., x_1^* and x_2^*) for this case.

(d) What is the minimax risk for part (c)?

6. Consider the Neyman-Pearson criterion for two univariate normal distributions: $p(x|\omega_i) \sim N(\mu_i, \sigma_i^2)$ and $P(\omega_i) = 1/2$ for $i = 1, 2$. Assume a zero-one error loss, and for convenience let $\mu_2 > \mu_1$.

(a) Suppose the maximum acceptable error rate for classifying a pattern that is actually in ω_1 as if it were in ω_2 is E_1. Determine the single-point decision boundary in terms of the variables given.

(b) For this boundary, what is the error rate for classifying ω_2 as ω_1?

(c) What is the overall error rate under zero-one loss?

(d) Apply your results to the specific case $p(x|\omega_1) \sim N(-1, 1)$ and $p(x|\omega_2) \sim N(1, 1)$ and $E_1 = 0.05$.

(e) Compare your result to the Bayes error rate (i.e., without the Neyman-Pearson conditions).

7. Consider Neyman-Pearson criteria for two Cauchy distributions in one dimension:

$$p(x|\omega_i) = \frac{1}{\pi b} \cdot \frac{1}{1 + \left(\frac{x - a_i}{b}\right)^2}, \qquad i = 1, 2.$$

Assume a zero-one error loss, and for simplicity $a_2 > a_1$, the same "width" b, and equal priors.

(a) Suppose the maximum acceptable error rate for classifying a pattern that is actually in ω_1 as if it were in ω_2 is E_1. Determine the single-point decision boundary in terms of the variables given.

(b) For this boundary, what is the error rate for classifying ω_2 as ω_1?

(c) What is the overall error rate under zero-one loss?

(d) Apply your results to the specific case $b = 1$ and $a_1 = -1$, $a_2 = 1$ and $E_1 = 0.1$.

(e) Compare your result to the Bayes error rate (i.e., without the Neyman-Pearson conditions).

8. Let the conditional densities for a two-category one-dimensional problem be given by the Cauchy distribution described in Problem 7.

(a) By explicit integration, check that the distributions are indeed normalized.

(b) Assuming $P(\omega_1) = P(\omega_2)$, show that $P(\omega_1|x) = P(\omega_2|x)$ if $x = (a_1 + a_2)/2$, that is, the minimum error decision boundary is a point midway between the peaks of the two distributions, regardless of b.

(c) Plot $P(\omega_1|x)$ for the case $a_1 = 3$, $a_2 = 5$ and $b = 1$.

(d) How do $P(\omega_1|x)$ and $P(\omega_2|x)$ behave as $x \to -\infty$? $x \to +\infty$? Explain.

9. Use the conditional densities given in Problem 7, and assume equal prior probabilities for the categories.

(a) Show that the minimum probability of error is given by

$$P(error) = \frac{1}{2} - \frac{1}{\pi}\tan^{-1}\left|\frac{a_2 - a_1}{2b}\right|.$$

(b) Plot this as a function of $|a_2 - a_1|/b$.

(c) What is the maximum value of $P(error)$ and under which conditions can this occur? Explain.

10. Consider the following decision rule for a two-category one-dimensional problem: Decide ω_1 if $x > \theta$; otherwise decide ω_2.

(a) Show that the probability of error for this rule is given by

$$P(error) = P(\omega_1) \int_{-\infty}^{\theta} p(x|\omega_1)\,dx + P(\omega_2) \int_{\theta}^{\infty} p(x|\omega_2)\,dx.$$

(b) By differentiating, show that a necessary condition to minimize $P(error)$ is that θ satisfies

$$p(\theta|\omega_1)P(\omega_1) = p(\theta|\omega_2)P(\omega_2).$$

(c) Does this equation define θ uniquely?

(d) Give an example where a value of θ satisfying the equation actually *maximizes* the probability of error.

11. Suppose that we replace the deterministic decision function $\alpha(\mathbf{x})$ with a *randomized rule*, namely, one giving the probability $P(\alpha_i|\mathbf{x})$ of taking action α_i upon observing $\mathbf{x}$.

(a) Show that the resulting risk is given by

$$R = \int \left[\sum_{i=1}^{a} R(\alpha_i|\mathbf{x}) P(\alpha_i|\mathbf{x}) \right] p(\mathbf{x}) \, d\mathbf{x}.$$

(b) In addition, show that R is minimized by choosing $P(\alpha_i|\mathbf{x}) = 1$ for the action α_i associated with the minimum conditional risk $R(\alpha_i|\mathbf{x})$, thereby showing that no benefit can be gained from randomizing the best decision rule.

(c) Can we benefit from randomizing a suboptimal rule? Explain.

12. Let $\omega_{max}(\mathbf{x})$ be the state of nature for which $P(\omega_{max}|\mathbf{x}) \geq P(\omega_i|\mathbf{x})$ for all i, $i = 1, \ldots, c$.

(a) Show that $P(\omega_{max}|\mathbf{x}) \geq 1/c$.

(b) Show that for the minimum-error-rate decision rule the average probability of error is given by

$$P(error) = 1 - \int P(\omega_{max}|\mathbf{x}) p(\mathbf{x}) \, d\mathbf{x}.$$

(c) Use these two results to show that $P(error) \leq (c-1)/c$.

(d) Describe a situation for which $P(error) = (c-1)/c$.

Section 2.4

13. In many pattern classification problems one has the option either to assign the pattern to one of c classes, or to *reject* it as being unrecognizable. If the cost for rejects is not too high, rejection may be a desirable action. Let

$$\lambda(\alpha_i|\omega_j) = \begin{cases} 0 & i = j \quad i, j = 1, \ldots, c \\ \lambda_r & i = c+1 \\ \lambda_s & \text{otherwise}, \end{cases}$$

where λ_r is the loss incurred for choosing the $(c+1)$th action, rejection, and λ_s is the loss incurred for making any substitution error. Show that the minimum risk is obtained if we decide ω_i if $P(\omega_i|\mathbf{x}) \geq P(\omega_j|\mathbf{x})$ for all j and if $P(\omega_i|\mathbf{x}) \geq 1 - \lambda_r/\lambda_s$, and reject otherwise. What happens if $\lambda_r = 0$? What happens if $\lambda_r > \lambda_s$?

14. Consider the classification problem with rejection option.

(a) Use the results of Problem 13 to show that the following discriminant functions are optimal for such problems:

$$g_i(\mathbf{x}) = \begin{cases} p(\mathbf{x}|\omega_i) P(\omega_i) & i = 1, \ldots, c \\ \frac{\lambda_s - \lambda_r}{\lambda_s} \sum_{j=1}^{c} p(\mathbf{x}|\omega_j) P(\omega_j) & i = c+1. \end{cases}$$

(b) Plot these discriminant functions and the decision regions for the two-category one-dimensional case having

- $p(x|\omega_1) \sim N(1, 1)$,
- $p(x|\omega_2) \sim N(-1, 1)$,
- $P(\omega_1) = P(\omega_2) = 1/2$, and
- $\lambda_r/\lambda_s = 1/4$.

(c) Describe qualitatively what happens as λ_r/λ_s is increased from 0 to 1.

(d) Repeat for the case having

- $p(x|\omega_1) \sim N(1, 1)$,
- $p(x|\omega_2) \sim N(0, 1/4)$,
- $P(\omega_1) = 1/3$, $P(\omega_2) = 2/3$, and
- $\lambda_r/\lambda_s = 1/2$.

Section 2.5

15. Confirm Eq. 47 for the volume of a d-dimensional hypersphere as follows:

   (a) Verify that the equation is correct for a line segment ($d = 1$).

   (b) Verify that the equation is correct for a disk ($d = 2$).

   (c) Integrate the volume of a line over appropriate limits to obtain the volume of a disk.

   (d) Consider a general d-dimensional hypersphere. Integrate its volume to obtain a formula (involving the ratio of gamma functions, $\Gamma(\cdot)$) for the volume of a $(d + 1)$-dimensional hypersphere.

   (e) Apply your formula to find the volume of a hypersphere in an odd-dimensional space by integrating the volume of a hypersphere in the lower even-dimensional space, and thereby confirm Eq. 47 for odd dimensions.

   (f) Repeat the above but for finding the volume of a hypersphere in even dimensions.

16. Derive the formula for the volume of a d-dimensional hypersphere in Eq. 47 as follows:

   (a) State by inspection the formula for V_1.

   (b) Follow the general procedure outlined in Problem 15 and integrate twice to find V_{d+2} as a function of V_d.

   (c) Assume that the functional form of V_d is the same for all odd dimensions (and likewise for all even dimensions). Use your integration results to determine the formula for V_d for d odd.

   (d) Use your intermediate integration results to determine V_d for d even.

   (e) Explain why we should expect the functional form of V_d to be different in even and in odd dimensions.

17. Derive the formula (Eq. 46) for the volume V of a hyperellipsoid of constant Mahalanobis distance r (Eq. 45) for a Gaussian distribution having covariance Σ.

18. Consider two normal distributions in one dimension: $N(\mu_1, \sigma_1^2)$ and $N(\mu_2, \sigma_2^2)$. Imagine that we choose two random samples x_1 and x_2, one from each of the normal distributions and calculate their sum $x_3 = x_1 + x_2$. Suppose we do this repeatedly.

(a) Consider the resulting distribution of the values of x_3. Show that x_3 possesses the requisite statistical properties and thus its distribution is normal.

(b) What is the mean, μ_3, of your new distribution?

(c) What is the variance, σ_3^2?

(d) Repeat the above with two distributions in a multi-dimensional space, i.e., $N(\boldsymbol{\mu}_1, \boldsymbol{\Sigma}_1)$ and $N(\boldsymbol{\mu}_2, \boldsymbol{\Sigma}_2)$.

19. Starting from the definition of entropy (Eq. 37), derive the general equation for the maximum-entropy distribution given constraints expressed in the general form

$$\int b_k(x) p(x)\, dx = a_k, \quad k = 1, 2, \dots, q$$

as follows:

(a) Use Lagrange undetermined multipliers $\lambda_1, \lambda_2, \dots, \lambda_q$ and derive the synthetic function:

$$H_s = -\int p(x) \left[\ln p(x) - \sum_{k=0}^{q} \lambda_k b_k(x) \right] dx - \sum_{k=0}^{q} \lambda_k a_k.$$

State why we know $a_0 = 1$ and $b_0(x) = 1$ for all x.

(b) Take the derivative of H_s with respect to $p(x)$. Set the integrand to zero, and thereby prove that the minimum-entropy distribution obeys

$$p(x) = \exp\left[\sum_{k=0}^{q} \lambda_k b_k(x) - 1 \right],$$

where the $q + 1$ parameters are determined by the constraint equation.

20. Use the final result from Problem 19 for the following.

(a) Suppose we know solely that a distribution is nonzero only in the range $x_l \leq x \leq x_u$. Prove that the maximum entropy distribution is uniform in that range, that is,

$$p(x) \sim U(x_l, x_u) = \begin{cases} 1/|x_u - x_l| & x_l \leq x \leq x_u \\ 0 & \text{otherwise.} \end{cases}$$

(b) Suppose we know solely that a distribution is nonzero only for $x \geq 0$ and that its mean is μ. Prove that the maximum entropy distribution is

$$p(x) = \begin{cases} \frac{1}{\mu} e^{-x/\mu} & \text{for } x \geq 0 \\ 0 & \text{otherwise.} \end{cases}$$

(c) Now suppose we know solely that the distribution is normalized, has mean μ, and standard deviation σ, and thus from Problem 19 our maximum entropy distribution must be of the form

$$p(x) = \exp[\lambda_0 - 1 + \lambda_1 x + \lambda_2 x^2].$$

Write out the three constraints and solve for λ_0, λ_1, and λ_2 and thereby prove that the maximum entropy solution is a Gaussian, that is,

$$p(x) = \frac{1}{\sqrt{2\pi}\sigma} \exp\left[\frac{-(x - \mu)^2}{2\sigma^2}\right].$$

21. Three distributions—a Gaussian, a uniform distribution, and a triangle distribution (cf. Problem 5)—each have mean zero and standard deviation σ. Use Eq. 37 to calculate and compare their entropies.

22. Calculate the entropy of a multidimensional Gaussian $p(\mathbf{x}) \sim N(\boldsymbol{\mu}, \boldsymbol{\Sigma})$.

23. Consider the three-dimensional normal distribution $p(\mathbf{x}|\omega) \sim N(\boldsymbol{\mu}, \boldsymbol{\Sigma})$, where

$$\boldsymbol{\mu} = \begin{pmatrix} 1 \\ 2 \\ 2 \end{pmatrix} \text{ and } \boldsymbol{\Sigma} = \begin{pmatrix} 1 & 0 & 0 \\ 0 & 5 & 2 \\ 0 & 2 & 5 \end{pmatrix}.$$

(a) Find the probability density at the point $\mathbf{x}_0 = (.5, 0, 1)^t$.

(b) Construct the whitening transformation $\mathbf{A}_w$ (Eq. 44). Compute the matrices representing eigenvectors and eignvalues, $\boldsymbol{\Phi}$ and $\boldsymbol{\Lambda}$. Next, convert the distribution to one centered on the origin with covariance matrix equal to the identity matrix, $p(\mathbf{x}|\omega) \sim N(\mathbf{0}, \mathbf{I})$.

(c) Apply the same overall transformation to $\mathbf{x}_0$ to yield a transformed point $\mathbf{x}_w$.

(d) By explicit calculation, confirm that the Mahalanobis distance from $\mathbf{x}_0$ to the mean $\boldsymbol{\mu}$ in the original distribution is the same as for $\mathbf{x}_w$ to $\mathbf{0}$ in the transformed distribution.

(e) Does the probability density remain unchanged under a general linear transformation? In other words, is $p(\mathbf{x}_0|N(\boldsymbol{\mu}, \boldsymbol{\Sigma})) = p(\mathbf{T}^t\mathbf{x}_0|N(\mathbf{T}^t\boldsymbol{\mu}, \mathbf{T}^t\boldsymbol{\Sigma}\mathbf{T}))$ for some linear transform $\mathbf{T}$? Explain.

(f) Prove that a general whitening transform $\mathbf{A}_w = \boldsymbol{\Phi}\boldsymbol{\Lambda}^{-1/2}$ when applied to a Gaussian distribution ensures that the final distribution has covariance proportional to the identity matrix $\mathbf{I}$. Check whether normalization is preserved by the transformation.

24. Consider the multivariate normal density with mean $\boldsymbol{\mu}$, $\sigma_{ij} = 0$ and $\sigma_{ii} = \sigma_i^2$, that is, the covariance matrix is diagonal: $\boldsymbol{\Sigma} = diag(\sigma_1^2, \sigma_2^2, \ldots, \sigma_d^2)$.

(a) Show that the evidence is

$$p(\mathbf{x}) = \frac{1}{\prod\limits_{i=1}^{d} \sqrt{2\pi}\sigma_i} \exp\left[-\frac{1}{2}\sum_{i=1}^{d}\left(\frac{x_i - \mu_i}{\sigma_i}\right)^2\right].$$

(b) Plot and describe the contours of constant density.

(c) Write an expression for the Mahalanobis distance from $\mathbf{x}$ to $\boldsymbol{\mu}$.

Section 2.6

25. Fill in the steps in the derivation from Eq. 59 to Eqs. 60–65.

26. Let $p(\mathbf{x}|\omega_i) \sim N(\boldsymbol{\mu}_i, \boldsymbol{\Sigma})$ for $i = 1, 2$ in a two-category d-dimensional problem with the same covariances but arbitrary means and prior probabilities. Consider

the squared Mahalanobis distance

$$r_i^2 = (\mathbf{x} - \boldsymbol{\mu}_i)^t \boldsymbol{\Sigma}^{-1}(\mathbf{x} - \boldsymbol{\mu}_i).$$

(a) Show that the gradient of r_i^2 is given by

$$\nabla r_i^2 = 2\boldsymbol{\Sigma}^{-1}(\mathbf{x} - \boldsymbol{\mu}_i).$$

(b) Show that at any position on a given line through $\boldsymbol{\mu}_i$ the gradient ∇r_i^2 points in the same direction. Must this direction be parallel to that line?

(c) Show that ∇r_1^2 and ∇r_2^2 point in opposite directions along the line from $\boldsymbol{\mu}_1$ to $\boldsymbol{\mu}_2$.

(d) Show that the optimal separating hyperplane is tangent to the constant probability density hyperellipsoids at the point that the separating hyperplane cuts the line from $\boldsymbol{\mu}_1$ to $\boldsymbol{\mu}_2$.

(e) True or False: For a two-category problem involving normal densities with arbitrary means and covariances, and $P(\omega_1) = P(\omega_2) = 1/2$, the Bayes decision boundary consists of the set of points of equal Mahalanobis distance from the respective sample means. Explain.

27. Suppose we have two normal distributions with the same covariances but different means: $N(\boldsymbol{\mu}_1, \boldsymbol{\Sigma})$ and $N(\boldsymbol{\mu}_2, \boldsymbol{\Sigma})$. In terms of their prior probabilities $P(\omega_1)$ and $P(\omega_2)$, state the condition that the Bayes decision boundary *not* pass between the two means.

28. Two random variables $\mathbf{x}$ and $\mathbf{y}$ are called statistically independent if $p(\mathbf{x}, \mathbf{y}|\omega) = p(\mathbf{x}|\omega)p(\mathbf{y}|\omega)$.

(a) Prove that if $x_i - \mu_i$ and $x_j - \mu_j$ are statistically independent (for $i \neq j$), then σ_{ij} as defined in Eq. 43 is 0.

(b) Prove that the converse is true for the Gaussian case.

(c) Show by counterexample that this converse is *not* true in the general case.

29. Figure 2.15 shows that it is possible for a decision boundary for two three-dimensional Gaussians to be a line segment. Explain how this can arise by analyzing a simpler one-dimensional case as follows.

(a) Consider two one-dimensional Gaussians whose means differ and whose variances differ. Explain why for this case we can always find prior probabilities such that the decision boundary is a single point.

(b) Use your result to explain how the three-dimensional two-Gaussian case can yield a line segment decision boundary.

30. Consider the Bayes decision boundary for two-category classification in d dimensions.

(a) Prove that for any arbitrary hyperquadric in d dimensions, there exist normal distributions $p(\mathbf{x}|\omega_i) \sim N(\boldsymbol{\mu}_i, \boldsymbol{\Sigma}_i)$ and priors $P(\omega_i)$, $i = 1, 2$, that possess this hyperquadric as their Bayes decision boundary.

(b) Is your answer to part (a) true if the priors are held fixed and nonzero, e.g., $P(\omega_1) = P(\omega_2) = 1/2$?

Section 2.7

31. Let $p(x|\omega_i) \sim N(\mu_i, \sigma^2)$ for a two-category one-dimensional problem with $P(\omega_1) = P(\omega_2) = 1/2$.

(a) Show that the minimum probability of error is given by

$$P_e = \frac{1}{\sqrt{2\pi}} \int_a^\infty e^{-u^2/2} \, du,$$

where $a = |\mu_2 - \mu_1|/(2\sigma)$.

(b) Use the inequality

$$P_e = \frac{1}{\sqrt{2\pi}} \int_a^\infty e^{-t^2/2} \, dt \leq \frac{1}{\sqrt{2\pi}a} e^{-a^2/2}$$

to show that P_e goes to zero as $|\mu_2 - \mu_1|/\sigma$ goes to infinity.

32. Let $p(\mathbf{x}|\omega_i) \sim N(\boldsymbol{\mu}_i, \sigma^2 \mathbf{I})$ for a two-category d-dimensional problem with $P(\omega_1) = P(\omega_2) = 1/2$.

(a) Show that the minimum probability of error is given by

$$P_e = \frac{1}{\sqrt{2\pi}} \int_a^\infty e^{-u^2/2} \, du,$$

where $a = \|\boldsymbol{\mu}_2 - \boldsymbol{\mu}_1\|/(2\sigma)$.

(b) Let $\boldsymbol{\mu}_1 = \mathbf{0}$ and $\boldsymbol{\mu}_2 = (\mu_1, \ldots, \mu_d)^t \neq \mathbf{0}$. Use the inequality from Problem 31 to show that P_e approaches zero as the dimension d approaches infinity.

(c) Express the meaning of this result in words.

33. Suppose we know exactly two arbitrary distributions $p(\mathbf{x}|\omega_i)$ and priors $P(\omega_i)$ in a d-dimensional feature space.

(a) Prove that the true error cannot decrease if we first project the distributions to a lower-dimensional space and then classify them.

(b) Despite this fact, suggest why in an actual pattern recognition application we might not want to include an arbitrarily high number of feature dimensions.

Section 2.8

34. Show that if the densities in a two-category classification problem differ significantly from Gaussian, the Chernoff and Bhattacharyya bounds are not likely to be informative by considering the following one-dimensional examples. Consider a number of problems in which the mean and variance are the same (and thus the Chernoff bound and the Bhattacharyya bound remain the same), but nevertheless have a wide range in Bayes error. For definiteness, assume $P(\omega_1) = P(\omega_2) = 0.5$ and the distributions have means at $\mu_1 = -\mu$ and $\mu_2 = +\mu$, and $\sigma_1^2 = \sigma_2^2 = \mu^2$.

(a) Use the equations in the text to calculate the Chernoff and the Bhattacharyya bounds on the error.

(b) Suppose the distributions are both Gaussian. Calculate explicitly the Bayes error. Express it in terms of an error function erf($\cdot$) and as a numerical value.

(c) Now consider another case, in which half the density for ω_1 is concentrated at a point $x = -2\mu$ and half at $x = 0$; likewise (symmetrically) the density for ω_2 has half its mass at $x = +2\mu$ and half at $x = 0$. Show that the means and variance remain as desired, but that now the Bayes error is 0.25.

(d) Now consider yet another case, in which half the density for ω_1 is concentrated near $x = -2$ and half at $x = -\epsilon$, where ϵ is an infinitessimally small positive distance; likewise (symmetrically) the density for ω_2 has half its mass near $x = +2\mu$ and half at $+\epsilon$. Show that by making ϵ sufficiently small, the means and variances can be made arbitrarily close to μ and μ^2, respectively. Show, too, that now the Bayes error is zero.

(e) Compare your errors in (b), (c), and (d) to your Chernoff and Bhattacharyya bounds of (a) and explain in words why those bounds are unlikely to be of much use if the distributions differ markedly from Gaussians.

35. Show for nonpathological cases that if we include more feature dimensions in a Bayesian classifier for multidimensional Gaussian distributions then the Bhattacharyya bound decreases. Do this as follows: Let $P_d(P(\omega_1), \boldsymbol{\mu}_1, \boldsymbol{\Sigma}_1, P(\omega_2), \boldsymbol{\mu}_2, \boldsymbol{\Sigma}_2)$, or simply P_d, be the Bhattacharyya bound if we consider the distributions restricted to d dimensions.

(a) Using general properties of a covariance matrix, prove that $k(1/2)$ of Eq. 77 must increase as we increase from d to $d + 1$ dimensions, and hence the error bound must decrease.

(b) Explain why this general result does or does not depend upon *which* dimension is added.

(c) What is a "pathological" case in which the error bound does *not* decrease, that is, for which $P_{d+1} = P_d$?

(d) Is it ever possible that the *true* error—that is, not just the *bound*—could *increase* as we go to higher dimension?

(e) Prove that as $d \to \infty$, $P_d \to 0$ for nonpathological distributions. Describe pathological distributions for which this infinite limit does not hold.

(f) Given that the Bhattacharyya bound decreases for the inclusion of a particular dimension, does this guarantee that the *true* error will decrease? Explain.

36. Derive Eqs. 74 and 75 from Eq. 73 by the following steps:

(a) Substitute the normal distributions into the integral and gather the terms dependent upon $\mathbf{x}$ and those that are not dependent upon $\mathbf{x}$.

(b) Factor the term independent of $\mathbf{x}$ from the integral.

(c) Integrate explicitly the term dependent upon $\mathbf{x}$.

37. Consider a two-category classification problem in two dimensions with

$$p(\mathbf{x}|\omega_1) \sim N(\mathbf{0}, \mathbf{I}), \ p(\mathbf{x}|\omega_2) \sim N\left(\begin{pmatrix}1\\1\end{pmatrix}, \mathbf{I}\right), \text{ and } P(\omega_1) = P(\omega_2) = 1/2.$$

(a) Calculate the Bayes decision boundary.

(b) Calculate the Bhattacharyya error bound.

(c) Repeat the above for the same prior probabilities, but

$$p(\mathbf{x}|\omega_1) \sim N\left(\mathbf{0}, \begin{pmatrix}2 & .5\\.5 & 2\end{pmatrix}\right) \text{ and } p(\mathbf{x}|\omega_2) \sim N\left(\begin{pmatrix}1\\1\end{pmatrix}, \begin{pmatrix}5 & 4\\4 & 5\end{pmatrix}\right).$$

38. Derive the Bhattacharyya error bound without the need for first examining the Chernoff bound. Do this as follows:

(a) If a and b are nonnegative numbers, show directly that $\min[a,b] \leq \sqrt{ab}$.

(b) Use this to show that the error rate for a two-category Bayes classifier must satisfy

$$P(error) \leq \sqrt{P(\omega_1)P(\omega_2)}\ \rho \leq \rho/2,$$

where ρ is the so-called *Bhattacharyya coefficient*

$$\rho = \int \sqrt{p(\mathbf{x}|\omega_1)\ p(\mathbf{x}|\omega_2)}\ d\mathbf{x}.$$

39. Use signal detection theory, as well as the notation and basic Gaussian assumptions described in the text, to address the following.

(a) Prove that $P(x > x^*|x \in \omega_2)$ and $P(x > x^*|x \in \omega_1)$, taken together, uniquely determine the discriminability d'.

(b) Use error functions erf$(\cdot)$ to express d' in terms of the hit and false alarm rates. Estimate d' if $P(x > x^*|x \in \omega_1) = 0.8$ and $P(x > x^*|x \in \omega_2) = 0.3$. Repeat for $P(x > x^*|x \in \omega_1) = 0.7$ and $P(x > x^*|x \in \omega_1) = 0.4$.

(c) Given that the Gaussian assumption is valid, calculate the Bayes error for both the cases in (b).

(d) Using a trivial one-line computation determine which case has the higher d':

Case A: $P(x > x^*|x \in \omega_1) = 0.8$, $P(x > x^*|x \in \omega_2) = 0.3$ or

Case B: $P(x > x^*|x \in \omega_1) = 0.3$, $P(x > x^*|x \in \omega_2) = 0.7$.

Explain your logic.

40. Suppose in our signal detection framework we had two Gaussians, but with different variances (cf. Fig. 2.20)—that is, $p(x|\omega_1) \sim N(\mu_1, \sigma_1^2)$ and $p(x|\omega_2) \sim N(\mu_2, \sigma_2^2)$ for $\mu_2 > \mu_1$ and $\sigma_2^2 \neq \sigma_1^2$. In that case the resulting ROC curve would no longer be symmetric.

(a) Suppose in this asymmetric case we modified the definition of the discriminability to be $d'_a = |\mu_2 - \mu_1|/\sqrt{\sigma_1\sigma_2}$. Show by nontrivial counterexample or analysis that one cannot determine d'_a uniquely based on a single pair of hit and false alarm rates.

(b) Assume we measure the hit and false alarm rates for two different, but unknown, values of the threshold x^*. Derive a formula for d'_a based on such measurements.

(c) State and explain all pathological values for which your formula does not give a meaningful value for d'_a.

(d) Plot several ROC curves for the case $p(x|\omega_1) \sim N(0, 1)$ and $p(x|\omega_2) \sim N(1, 2)$.

41. Consider two one-dimensional triangle distributions having different means, but the same width:

$$p(x|\omega_i) = T(\mu_i, \delta) = \begin{cases} (\delta - |x - \mu_i|)/\delta^2 & \text{for } |x - \mu_i| < \delta \\ 0 & \text{otherwise,} \end{cases}$$

with $\mu_2 > \mu_1$. We define a new discriminability here as $d'_T = (\mu_2 - \mu_1)/\delta$.

(a) Write an analytic function, parameterized by d'_T, for the operating characteristic curves.

(b) Plot these novel operating characteristic curves for $d'_T = \{.1, .2, \ldots, 1.0\}$. Interpret your answer for the case $d'_T = 1.0$ and 2.0.

(c) Suppose we measure $P(x > x^*|x \in \omega_2) = 0.4$ and $P(x > x^*|x \in \omega_1) = 0.7$. What is d'_T? What is the Bayes error rate?

(d) Infer the decision rule employed in part (c). That is, express x^* in terms of the variables given in the problem.

(e) Suppose we measure $P(x > x^*|x \in \omega_2) = 0.3$ and $(x > x^*|x \in \omega_1) = 0.9$. What is d'_T? What is the Bayes error rate?

(f) Infer the decision rule employed in part (e). That is, express x^* in terms of the variables given in the problem.

42. Equation 72 can be used to obtain an upper bound on the error. One can also derive tighter analytic bounds in the two-category case—both upper and lower bounds—analogous to Eq. 73 for general distributions. If we let $p \equiv p(x|\omega_1)$, then we seek tighter bounds on $\min[p, 1 - p]$ (which has discontinuous derivative).

(a) Prove that

$$b_L(p) = \frac{1}{\beta} \ln \left[\frac{1 + e^{-\beta}}{e^{-\beta p} + e^{-\beta(1-p)}} \right]$$

for any $\beta > 0$ is a lower bound on $\min[p, 1 - p]$.

(b) Prove that one can choose β in (a) to give an arbitrarily tight lower bound.

(c) Repeat (a) and (b) for the upper bound given by

$$b_U(p) = b_L(p) + [1 - 2b_L(0.5)]b_G(p)$$

where $b_G(p)$ is any upper bound that obeys

$$b_G(p) \geq \min[p, 1 - p]$$
$$b_G(p) = b_G(1 - p)$$
$$b_G(0) = b_G(1) = 0$$
$$b_G(0.5) = 0.5.$$

(d) Confirm that $b_G(p) = 1/2 \sin[\pi p]$ obeys the conditions in (c).

(e) Let $b_G(p) = 1/2 \sin[\pi p]$, and plot your upper and lower bounds as a function of p, for $0 \leq p \leq 1$ and $\beta = 1, 10$, and 50.

Section 2.9

43. Let the components of the vector $\mathbf{x} = (x_1, \ldots, x_d)^t$ be binary-valued (0 or 1), and let $P(\omega_j)$ be the prior probability for the state of nature ω_j and $j = 1, \ldots, c$. Now define

$$p_{ij} = \Pr[x_i = 1|\omega_j] \qquad \begin{matrix} i = 1, \ldots, d \\ j = 1, \ldots, c, \end{matrix}$$

with the components of x_i being statistically independent for all $\mathbf{x}$ in ω_j.

(a) Interpret in words the meaning of p_{ij}.

(b) Show that the minimum probability of error is achieved by the following decision rule: Decide ω_k if $g_k(\mathbf{x}) \geq g_j(\mathbf{x})$ for all j and k, where

$$g_j(\mathbf{x}) = \sum_{i=1}^{d} x_i \ln \frac{p_{ij}}{1 - p_{ij}} + \sum_{i=1}^{d} \ln(1 - p_{ij}) + \ln P(\omega_j).$$

44. Let the components of the vector $\mathbf{x} = (x_1, \ldots, x_d)^t$ be ternary valued (1, 0 or -1), with

$$p_{ij} = \Pr[x_i = 1 \,|\omega_j]$$
$$q_{ij} = \Pr[x_i = 0 \,|\omega_j]$$
$$r_{ij} = \Pr[x_i = -1|\omega_j],$$

and with the components of x_i being statistically independent for all $\mathbf{x}$ in ω_j.

(a) Show that a minimum probability of error decision rule can be derived that involves discriminant functions $g_j(\mathbf{x})$ that are quadratic function of the components x_i.

(b) Suggest a generalization to more categories of your answers to this and Problem 43.

45. Let $\mathbf{x}$ be distributed as in Problem 43 with $c = 2$, d odd, and

$$\begin{aligned} p_{i1} &= p > 1/2 & i &= 1, \ldots, d \\ p_{i2} &= 1 - p & i &= 1, \ldots, d, \end{aligned}$$

and $P(\omega_1) = P(\omega_2) = 1/2$.

(a) Show that the minimum-error-rate decision rule becomes

$$\text{Decide } \omega_1 \text{ if } \sum_{i=1}^{d} x_i > d/2 \text{ and } \omega_2 \text{ otherwise.}$$

(b) Show that the minimum probability of error is given by

$$P_e(d, p) = \sum_{k=0}^{(d-1)/2} \binom{d}{k} p^k (1 - p)^{d-k}.$$

where $\binom{d}{k} = d!/(k!(d - k)!)$ is the binomial coefficient.

(c) What is the limiting value of $P_e(d, p)$ as $p \to 1/2$? Explain.

(d) Show that $P_e(d, p)$ approaches zero as $d \to \infty$. Explain.

46. Under the natural assumption concerning losses, i.e., that $\lambda_{21} > \lambda_{11}$ and $\lambda_{12} > \lambda_{22}$, show that the general minimum risk discriminant function for the independent binary case described in Section 2.9.1 is given by $g(\mathbf{x}) = \mathbf{w}^t \mathbf{x} + w_0$, where $\mathbf{w}$ is unchanged, and

$$w_0 = \sum_{i=1}^{d} \ln \frac{1 - p_i}{1 - q_i} + \ln \frac{P(\omega_1)}{P(\omega_2)} + \ln \frac{\lambda_{21} - \lambda_{11}}{\lambda_{12} - \lambda_{22}}.$$

47. The Poisson distribution for a discrete variable $x = 0, 1, 2, \ldots$ and real parameter λ is

$$P(x|\lambda) = e^{-\lambda} \frac{\lambda^x}{x!}.$$

(a) Prove that the mean of such a distribution is $\mathcal{E}[x] = \lambda$.

(b) Prove that the variance of such a distribution is $\mathcal{E}[x - \bar{x}] = \lambda$.

(c) The *mode* of a distribution is the value of x that has the maximum probability. Prove that the mode of a Poisson distribution is the greatest integer that does not exceed λ. That is, prove that the mode is $\lfloor \lambda \rfloor$, read "floor of lambda." (If λ is an integer, then both λ and $\lambda - 1$ are modes.)

(d) Consider two equally probable categories having Poisson distributions but with differing parameters; assume for definiteness $\lambda_1 > \lambda_2$. What is the Bayes classification decision?

(e) What is the Bayes error rate?

Section 2.10

48. Suppose we have three categories in two dimensions with the following underlying distributions:

- $p(\mathbf{x}|\omega_1) \sim N(\mathbf{0}, \mathbf{I})$

- $p(\mathbf{x}|\omega_2) \sim N\left(\begin{pmatrix} 1 \\ 1 \end{pmatrix}, \mathbf{I}\right)$

- $p(\mathbf{x}|\omega_3) \sim \frac{1}{2}N\left(\begin{pmatrix} .5 \\ .5 \end{pmatrix}, \mathbf{I}\right) + \frac{1}{2}N\left(\begin{pmatrix} -.5 \\ .5 \end{pmatrix}, \mathbf{I}\right)$

with $P(\omega_i) = 1/3, i = 1, 2, 3$.

(a) By explicit calculation of posterior probabilities, classify the point $\mathbf{x} = \begin{pmatrix} .3 \\ .3 \end{pmatrix}$ for minimum probability of error.

(b) Suppose that for a particular test point the first feature is missing. That is, classify $\mathbf{x} = \begin{pmatrix} * \\ .3 \end{pmatrix}$.

(c) Suppose that for a particular test point the second feature is missing. That is, classify $\mathbf{x} = \begin{pmatrix} .3 \\ * \end{pmatrix}$.

(d) Repeat all of the above for $\mathbf{x} = \begin{pmatrix} .2 \\ .6 \end{pmatrix}$.

49. Show that Eq. 95 reduces to Bayes rule when the true feature is $\boldsymbol{\mu}_i$ and $p(\mathbf{x}_b|\mathbf{x}_t) \sim N(\mathbf{x}_t, \boldsymbol{\Sigma})$. Interpret this answer in words.

Section 2.11

50. Use the conditional probability matrices in Example 4 to answer the following separate problems.

(a) Suppose it is December 20—the end of autumn and the beginning of winter—and thus let $P(a_1) = P(a_4) = 0.5$. Furthermore, it is known that the fish was caught in the north Atlantic, that is, $P(b_1) = 1$. Suppose the lightness has not been measured but it is known that the fish is thin, that is, $P(d_2) = 1$. Classify the fish as salmon or sea bass. What is the expected error rate?

(b) Suppose all we know is that a fish is thin and medium lightness. What season is it now, most likely? What is your probability of being correct?

(c) Suppose we know a fish is thin and medium lightness and that it was caught in the north Atlantic. What season is it, most likely? What is the probability of being correct?

51. Consider a Bayesian belief net with several nodes having unspecified values. Suppose that one such node is selected at random, with the probabilities of its nodes computed by the formulas described in the text. Next, another such node is chosen at random (possibly even a node already visited), and the probabilities are similarly updated. Prove that this procedure will converge to the desired probabilities throughout the full network.

Section 2.12

52. Suppose we have three categories with $P(\omega_1) = 1/2$, $P(\omega_2) = P(\omega_3) = 1/4$ and the following distributions

- $p(x|\omega_1) \sim N(0, 1)$
- $p(x|\omega_2) \sim N(.5, 1)$
- $p(x|\omega_3) \sim N(1, 1)$,

and that we sample the following four points: $x = 0.6, 0.1, 0.9, 1.1$.

(a) Calculate explicitly the probability that the sequence actually came from $\omega_1, \omega_3, \omega_3, \omega_2$. Be careful to consider normalization.

(b) Repeat for the sequence $\omega_1, \omega_2, \omega_2, \omega_3$.

(c) Find the sequence having the maximum probability.

COMPUTER EXERCISES

Several of the computer exercises will rely on the following data.

sample	ω_1			ω_2			ω_3		
	x_1	x_2	x_3	x_1	x_2	x_3	x_1	x_2	x_3
1	−5.01	−8.12	−3.68	−0.91	−0.18	−0.05	5.35	2.26	8.13
2	−5.43	−3.48	−3.54	1.30	−2.06	−3.53	5.12	3.22	−2.66
3	1.08	−5.52	1.66	−7.75	−4.54	−0.95	−1.34	−5.31	−9.87
4	0.86	−3.78	−4.11	−5.47	0.50	3.92	4.48	3.42	5.19
5	−2.67	0.63	7.39	6.14	5.72	−4.85	7.11	2.39	9.21
6	4.94	3.29	2.08	3.60	1.26	4.36	7.17	4.33	−0.98
7	−2.51	2.09	−2.59	5.37	−4.63	−3.65	5.75	3.97	6.65
8	−2.25	−2.13	−6.94	7.18	1.46	−6.66	0.77	0.27	2.41
9	5.56	2.86	−2.26	−7.39	1.17	6.30	0.90	−0.43	−8.71
10	1.03	−3.33	4.33	−7.50	−6.32	−0.31	3.52	−0.36	6.43

Section 2.5

1. You may need the following procedures for several exercises below.

(a) Write a procedure to generate random samples according to a normal distribution $N(\mu, \Sigma)$ in d dimensions.

(b) Write a procedure to calculate the discriminant function (of the form given in Eq. 49) for a given normal distribution and prior probability $P(\omega_i)$.

(c) Write a procedure to calculate the Euclidean distance between two arbitrary points.

(d) Write a procedure to calculate the Mahalanobis distance between the mean μ and an arbitrary point $\mathbf{x}$, given the covariance matrix Σ.

2. Refer to Computer exercise 1 (b) and consider the problem of classifying 10 samples from the table above. Assume that the underlying distributions are normal.

(a) Assume that the prior probabilities for the first two categories are equal ($P(\omega_1) = P(\omega_2) = 1/2$ and $P(\omega_3) = 0$) and design a dichotomizer for those two categories using only the x_1 feature value.

(b) Determine the empirical training error on your samples, that is, the percentage of points misclassified.

(c) Use the Bhattacharyya bound to bound the error you will get on novel patterns drawn from the distributions.

(d) Repeat all of the above, but now use *two* feature values, x_1 and x_2.

(e) Repeat, but use all *three* feature values.

(f) Discuss your results. In particular, is it ever possible for a finite set of data that the empirical error might be *larger* for more data dimensions?

3. Repeat Computer exercise 2 but for categories ω_1 and ω_3.

4. Consider the three categories in Computer exercise 2, and assume $P(\omega_i) = 1/3$.

(a) What is the Mahalanobis distance between each of the following test points and each of the category means in Computer exercise 2: $(1, 2, 1)^t$, $(5, 3, 2)^t$, $(0, 0, 0)^t$, $(1, 0, 0)^t$.

(b) Classify those points.

(c) Assume instead that $P(\omega_1) = 0.8$, and $P(\omega_2) = P(\omega_3) = 0.1$ and classify the test points again.

5. Illustrate the fact that the average of a large number of independent random variables will approximate a Gaussian by the following:

(a) Write a program to generate n random integers from a uniform distribution $U(x_l, x_u)$. (Some computer systems include this as a single, compiled function call.)

(b) Now write a routine to choose x_l and x_u randomly, in the range $-100 \leq x_l < x_u \leq +100$, and n (the number of samples) randomly in the range $0 < n \leq 1000$.

(c) Generate and plot a histogram of the accumulation of 10^4 points sampled as just described.

(d) Calculate the mean and standard deviation of your histogram, and plot it

(e) Repeat the above for 10^5 and for 10^6. Discuss your results.

Section 2.8

6. Explore how the empirical error does or does not approach the Bhattacharyya bound as follows:

(a) Write a procedure to generate sample points in d dimensions with a normal distribution having mean $\boldsymbol{\mu}$ and covariance matrix $\boldsymbol{\Sigma}$.

(b) Consider the normal distributions

$$p(\mathbf{x}|\omega_1) \sim N\left(\begin{pmatrix}1\\0\end{pmatrix}, \mathbf{I}\right) \text{ and } p(\mathbf{x}|\omega_2) \sim N\left(\begin{pmatrix}-1\\0\end{pmatrix}, \mathbf{I}\right)$$

with $P(\omega_1) = P(\omega_2) = 1/2$. By inspection, state the Bayes decision boundary.

(c) Generate $n = 100$ points (50 for ω_1 and 50 for ω_2) and calculate the empirical error.

(d) Repeat for increasing values of n, $100 \leq n \leq 1000$, in steps of 100 and plot your empirical error.

(e) Discuss your results. In particular, is it ever possible that the empirical error is greater than the Bhattacharyya or Chernoff bound?

7. Consider two one-dimensional normal distributions $p(x|\omega_1) \sim N(-.5, 1)$ and $p(x|\omega_2) \sim N(+.5, 1)$ and $P(\omega_1) = P(\omega_2) = 0.5$.

(a) Calculate the Bhattacharyya bound for the error of a Bayesian classifier.

(b) Express the true error rate in terms of an error function, $\text{erf}(\cdot)$.

(c) Evaluate this true error to four significant figures by numerical integration (or other routine).

(d) Generate 10 points each for the two categories and determine the empirical error using your Bayesian classifier. (You should recalculate the decision boundary for each of your data sets.)

(e) Plot the empirical error as a function of the number of points from either distribution by repeating the previous part for 50, 100, 200, 500 and 1000 sample points from each distribution. Compare your asymptotic empirical error to the true error and the Bhattacharyya error bound.

8. Repeat Computer exercise 7 with the following conditions:

(a) $p(x|\omega_1) \sim N(-.5, 2)$ and $p(x|\omega_2) \sim N(.5, 2)$, $P(\omega_1) = 2/3$ and $P(\omega_2) = 1/3$.

(b) $p(x|\omega_1) \sim N(-.5, 2)$ and $p(x|\omega_2) \sim N(.5, 2)$ and $P(\omega_1) = P(\omega_2) = 1/2$.

(c) $p(x|\omega_1) \sim N(-.5, 3)$ and $p(x|\omega_2) \sim N(.5, 1)$ and $P(\omega_1) = P(\omega_2) = 1/2$.

Section 2.11

9. Write a program to evaluate the Bayesian belief net for fish in Example 3, including the information in $P(x_i|a_j)$, $P(x_i|b_j)$, $P(c_i|x_j)$, and $P(d_i|x_j)$. Test your program on the calculation given in the Example. Apply your program to the following cases, and state any assumptions you need to make.

(a) A dark, thin fish is caught in the north Atlantic in summer. What is the probability it is a salmon?

(b) A thin, medium fish is caught in the north Atlantic. What is the probability it is winter? spring? summer? autumn?

(c) A light, wide fish is caught in the autumn. What is the probability it came from the north Atlantic?

BIBLIOGRAPHY

[1] Subutai Ahmad and Volker Tresp. Some solutions to the missing feature problem in vision. In Stephen J. Hanson, Jack D. Cowan, and C. Lee Giles, editors, *Advances in Neural Information Processing Systems*, volume 5, pages 393–400, Morgan Kaufmann San Mateo, CA, 1993.

[2] Hadar Avi-Itzhak and Thanh Diep. Arbitrarily tight upper and lower bounds on the Bayesian probability of error. *IEEE Transactions on Pattern Analysis and Machine Intelligence*, PAMI-18(1):89–91, 1996.

[3] Thomas Bayes. An essay towards solving a problem in the doctrine of chances. *Philosophical Transactions of the Royal Society (London)*, 53:370–418, 1763.

[4] James O. Berger. Minimax estimation of a multivariate normal mean under arbitrary quadratic loss. *Journal of Multivariate Analysis*, 6(2):256–264, 1976.

[5] James O. Berger. Selecting a minimax estimator of a multivariate normal mean. *Annals of Statistics*, 10(1):81–92, 1982.

[6] James O. Berger. *Statistical Decision Theory and Bayesian Analysis*. Springer-Verlag, New York, second edition, 1985.

[7] José M. Bernardo and Adrian F. M. Smith. *Bayesian Theory*. Wiley, New York, 1996.

[8] Anil Bhattacharyya. On a measure of divergence between two statistical populations defined by their probability distributions. *Bulletin of the Calcutta Mathematical Society*, 35:99–110, 1943.

[9] Wray L. Buntine. Operations for learning with graphical models. *Journal of Artificial Intelligence Research*, 2:159–225, 1994.

[10] Wray L. Buntine. A guide to the literature on learning probabilistic networks from data. *IEEE Transactions on Knowledge and Data Engineering*, 8(2):195–210, 1996.

[11] Herman Chernoff. A measure of asymptotic efficiency for tests of a hypothesis based on the sum of observations. *Annals of Mathematical Statistics*, 23:493–507, 1952.

[12] Chao K. Chow. An optimum character recognition system using decision functions. *IRE Transactions*, pages 247–254, 1957.

[13] Chao K. Chow. On optimum recognition error and reject tradeoff. *IEEE Transactions on Information Theory*, IT-16:41–46, 1970.

[14] Thomas M. Cover and Joy A. Thomas. *Elements of Information Theory*. Wiley-Interscience, New York, 1991.

[15] Morris H. DeGroot. *Optimal Statistical Decisions*. McGraw-Hill, New York, 1970.

[16] Bradley Efron and Carl Morris. Families of minimax estimators of the mean of a multivariate normal distribution. *Annals of Statistics*, 4:11–21, 1976.

[17] Thomas S. Ferguson. *Mathematical Statistics: A Decision Theoretic Approach*. Academic Press, New York, 1967.

[18] Simon French. *Decision Theory: An Introduction to the Mathematics of Rationality*. Halsted Press, New York, 1986.

[19] Keinosuke Fukunaga. *Introduction to Statistical Pattern Recognition*. Academic Press, New York, second edition, 1990.

[20] Keinosuke Fukunaga and Thomas F. Krile. Calculation of Bayes recognition error for two multivariate Gaussian distributions. *IEEE Transactions on Computers*, C-18:220–229, 1969.

[21] Izrail M. Gelfand and Sergei Vasilevich Fomin. *Calculus of Variations*. Prentice-Hall, Englewood Cliffs, NJ, translated from the Russian by Richard A. Silverman, 1963.

[22] David M. Green and John A. Swets. *Signal Detection Theory and Psychophysics*. Wiley, New York, 1974.

[23] David J. Hand. *Construction and Assessment of Classification Rules*. Wiley, New York, 1997.

[24] Peter E. Hart and Jamey Graham. Query-free information retrieval. *IEEE Expert: Intelligent Systems and Their Application*, 12(5):32–37, 1997.

[25] David Heckerman. *Probabilistic Similarity Networks*. ACM Doctoral Dissertation Award Series. MIT Press, Cambridge, MA, 1991.

[26] Anil K. Jain. On an estimate of the Bhattacharyya distance. *IEEE Transactions on Systems, Man and Cybernetics*, SMC-16(11):763–766, 1976.

[27] Michael I. Jordan, editor. *Learning in Graphical Models*. MIT Press, Cambridge, MA, 1999.

[28] Bernard Kolman. *Elementary Linear Algebra*. Macmillan, New York, fifth edition, 1991.

[29] Pierre Simon Laplace. *Théorie Analytique des Probabiltiés*. Courcier, Paris, France, 1812.

[30] Peter M Lee. *Bayesian Statistics: An Introduction*. Edward Arnold, London, 1989.

[31] Dennis V. Lindley. *Making Decisions*. Wiley, New York, 1991.

[32] Jerzy Neyman and Egon S. Pearson. On the problem of the most efficient tests of statistical hypotheses. *Philosophical Transactions of the Royal Society, London*, 231:289–337, 1928.

[33] Judea Pearl. *Probabilistic Reasoning in Intelligent Systems: Networks of Plausible Inference*. Morgan Kaufmann, San Mateo, CA, 1988.

[34] Sheldon M. Ross. *Introduction to Probability and Statistics for Engineers*. Wiley, New York, 1987.

[35] Donald B. Rubin and Roderick J. A. Little. *Statistical Analysis with Missing Data*. Wiley, New York, 1987.

[36] Claude E. Shannon. A mathematical theory of communication. *Bell Systems Technical Journal*, 27:379–423, 623–656, 1948.

[37] George B. Thomas, Jr. and Ross L. Finney. *Calculus and Analytic Geometry*. Addison-Wesley, New York, ninth edition, 1996.

[38] Julius T. Tou and Rafael C. Gonzalez. *Pattern Recognition Principles*. Addison-Wesley, New York, 1974.

[39] William R. Uttal. *The Psychobiology of Sensory Coding*. HarperCollins, New York, 1973.

[40] Abraham Wald. Contributions to the theory of statistical estimation and testing of hypotheses. *Annals of Mathematical Statistics*, 10:299–326, 1939.

[41] Abraham Wald. *Statistical Decision Functions*. Wiley, New York, 1950.

[42] Charles T. Wolverton and Terry J. Wagner. Asymptotically optimal discriminant functions for pattern classifiers. *IEEE Transactions on Information Theory*, IT-15(2):258–265, 1969.

[43] Sewal Wright. Correlation and causation. *Journal of Agricultural Research*, 20(7):557–585, 1921.

[44] C. Ray Wylie and Louis C. Barrett. *Advanced Engineering Mathematics*. McGraw-Hill, New York, sixth edition, 1995.

MAXIMUM-LIKELIHOOD AND BAYESIAN PARAMETER ESTIMATION

3.1 INTRODUCTION

In Chapter 2 we saw how we could design an optimal classifier if we knew the prior probabilities $P(\omega_i)$ and the class-conditional densities $p(\mathbf{x}|\omega_i)$. Unfortunately, in pattern recognition applications we rarely, if ever, have this kind of complete knowledge about the probabilistic structure of the problem. In a typical case we merely have some vague, general knowledge about the situation, together with a number of *design samples* or *training data*—particular representatives of the patterns we want to classify. The problem, then, is to find some way to use this information to design or train the classifier.

TRAINING DATA

One approach to this problem is to use the samples to estimate the unknown probabilities and probability densities, and then use the resulting estimates as if they were the true values. In typical supervised pattern classification problems, the estimation of the prior probabilities presents no serious difficulties (Problem 3). However, estimation of the class-conditional densities is quite another matter. The number of available samples always seems too small, and serious problems arise when the dimensionality of the feature vector $\mathbf{x}$ is large. If we know the number of parameters in advance and our general knowledge about the problem permits us to parameterize the conditional densities, then the severity of these problems can be reduced significantly. Suppose, for example, that we can reasonably assume that $p(\mathbf{x}|\omega_i)$ is a normal density with mean $\boldsymbol{\mu}_i$ and covariance matrix $\boldsymbol{\Sigma}_i$, although we do not know the exact values of these quantities. This knowledge simplifies the problem from one of estimating an unknown *function* $p(\mathbf{x}|\omega_i)$ to one of estimating the *parameters* $\boldsymbol{\mu}_i$ and $\boldsymbol{\Sigma}_i$.

MAXIMUM-
LIKELIHOOD

BAYESIAN
ESTIMATION

The problem of parameter estimation is a classical one in statistics, and it can be approached in several ways. We shall consider two common and reasonable procedures, namely, *maximum-likelihood* estimation and *Bayesian* estimation. Although the results obtained with these two procedures are frequently nearly identical, the approaches are conceptually quite different. Maximum-likelihood and several other methods view the parameters as quantities whose values are fixed but unknown. The best estimate of their value is defined to be the one that maximizes the probability of obtaining the samples actually observed. In contrast, Bayesian methods view the parameters as random variables having some known prior distribution. Observation of the samples converts this to a posterior density, thereby revising our opinion about the true values of the parameters. In the Bayesian case, we shall see that a typical effect of observing additional samples is to sharpen the *a posteriori* density function, causing it to peak near the true values of the parameters. This phenomenon is known as *Bayesian learning*. In either case, we use the posterior densities for our classification rule, as we have seen before.

BAYESIAN
LEARNING

It is important to distinguish between supervised learning and unsupervised learning. In both cases, samples **x** are assumed to be obtained by selecting a state of nature ω_i with probability $P(\omega_i)$, and then independently selecting **x** according to the probability law $p(\mathbf{x}|\omega_i)$. The distinction is that with supervised learning we know the state of nature (class label) for each sample, whereas with unsupervised learning we do not. As one would expect, the problem of unsupervised learning is the more difficult one. In this chapter we shall consider only the supervised case, deferring consideration of unsupervised learning to Chapter 10.

Finally, there are nonparametric procedures for transforming the feature space in the hope that it may be possible to employ parametric methods in the transformed space. These discriminant analysis methods include the Fisher linear discriminant, which provides an important link between the parametric techniques of Chapter 3 and the adaptive techniques of Chapters 5 and 6 and some methods in feature selection described in Chapter 10.

3.2 MAXIMUM-LIKELIHOOD ESTIMATION

Maximum-likelihood estimation methods have a number of attractive attributes. First, they nearly always have good convergence properties as the number of training samples increases. Furthermore, maximum-likelihood estimation often can be simpler than alternative methods, such as Bayesian techniques or other methods presented in subsequent chapters.

3.2.1 The General Principle

Suppose that we separate a collection of samples according to class, so that we have c data sets, $\mathcal{D}_1, \ldots, \mathcal{D}_c$, with the samples in $\mathcal{D}_j$ having been drawn independently according to the probability law $p(\mathbf{x}|\omega_j)$. We say such samples are *i.i.d.*—independent and identically distributed random variables. We assume that $p(\mathbf{x}|\omega_j)$ has a known parametric form, and is therefore determined uniquely by the value of a parameter vector $\boldsymbol{\theta}_j$. For example, we might have $p(\mathbf{x}|\omega_j) \sim N(\boldsymbol{\mu}_j, \boldsymbol{\Sigma}_j)$, where $\boldsymbol{\theta}_j$ consists of the components of $\boldsymbol{\mu}_j$ and $\boldsymbol{\Sigma}_j$. To show the dependence of $p(\mathbf{x}|\omega_j)$ on $\boldsymbol{\theta}_j$ explicitly, we write $p(\mathbf{x}|\omega_j)$ as $p(\mathbf{x}|\omega_j, \boldsymbol{\theta}_j)$. Our problem is to use the information provided by the training samples to obtain good estimates for the unknown parameter vectors $\boldsymbol{\theta}_1, \ldots, \boldsymbol{\theta}_c$ associated with each category.

I.I.D.

To simplify treatment of this problem, we shall assume that samples in $\mathcal{D}_i$ give no information about $\boldsymbol{\theta}_j$ if $i \neq j$; that is, we shall assume that the parameters for the different classes are functionally independent. This permits us to work with each class separately and to simplify our notation by deleting indications of class distinctions. With this assumption we thus have c separate problems of the following form: Use a set $\mathcal{D}$ of training samples drawn independently from the probability density $p(\mathbf{x}|\boldsymbol{\theta})$ to estimate the unknown parameter vector $\boldsymbol{\theta}$.

Suppose that $\mathcal{D}$ contains n samples, $\mathbf{x}_1, \ldots, \mathbf{x}_n$. Then, because the samples were drawn independently, we have

$$p(\mathcal{D}|\boldsymbol{\theta}) = \prod_{k=1}^{n} p(\mathbf{x}_k|\boldsymbol{\theta}). \tag{1}$$

Recall from Chapter 2 that, viewed as a function of $\boldsymbol{\theta}$, $p(\mathcal{D}|\boldsymbol{\theta})$ is called the *likelihood* of $\boldsymbol{\theta}$ with respect to the set of samples. The *maximum-likelihood estimate* of $\boldsymbol{\theta}$ is, by definition, the value $\hat{\boldsymbol{\theta}}$ that maximizes $p(\mathcal{D}|\boldsymbol{\theta})$. Intuitively, this estimate corresponds to the value of $\boldsymbol{\theta}$ that in some sense best agrees with or supports the actually observed training samples (Fig. 3.1).

For analytical purposes, it is usually easier to work with the logarithm of the likelihood than with the likelihood itself. Because the logarithm is monotonically increasing, the $\hat{\boldsymbol{\theta}}$ that maximizes the log-likelihood also maximizes the likelihood. If $p(\mathcal{D}|\boldsymbol{\theta})$ is a well-behaved, differentiable function of $\boldsymbol{\theta}$, $\hat{\boldsymbol{\theta}}$ can be found by the standard methods of differential calculus. If the number of parameters to be estimated is

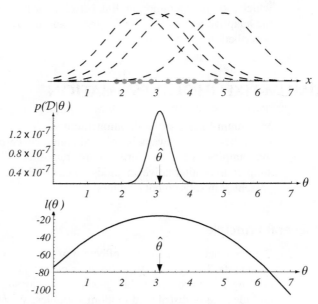

FIGURE 3.1. The top graph shows several training points in one dimension, known or assumed to be drawn from a Gaussian of a particular variance, but unknown mean. Four of the infinite number of candidate source distributions are shown in dashed lines. The middle figure shows the likelihood $p(\mathcal{D}|\theta)$ as a function of the mean. If we had a very large number of training points, this likelihood would be very narrow. The value that maximizes the likelihood is marked $\hat{\theta}$; it also maximizes the logarithm of the likelihood—that is, the log-likelihood $l(\theta)$, shown at the bottom. Note that even though they look similar, the likelihood $p(\mathcal{D}|\theta)$ is shown as a function of θ whereas the conditional density $p(x|\theta)$ is shown as a function of x. Furthermore, as a function of θ, the likelihood $p(\mathcal{D}|\theta)$ is not a probability density function and its area has no significance.

p, then we let $\boldsymbol{\theta}$ denote the p-component vector $\boldsymbol{\theta} = (\theta_1, \ldots, \theta_p)^t$, and we let $\boldsymbol{\nabla}_{\boldsymbol{\theta}}$ be the gradient operator

$$\boldsymbol{\nabla}_{\boldsymbol{\theta}} \equiv \begin{bmatrix} \dfrac{\partial}{\partial \theta_1} \\ \vdots \\ \dfrac{\partial}{\partial \theta_p} \end{bmatrix}. \tag{2}$$

LOG-LIKELIHOOD

We define $l(\boldsymbol{\theta})$ as the *log-likelihood* function*

$$l(\boldsymbol{\theta}) \equiv \ln p(\mathcal{D}|\boldsymbol{\theta}). \tag{3}$$

We can then write our solution formally as the argument $\boldsymbol{\theta}$ that maximizes the log-likelihood, that is,

$$\hat{\boldsymbol{\theta}} = \arg \max_{\boldsymbol{\theta}} l(\boldsymbol{\theta}), \tag{4}$$

where the dependence on the data set $\mathcal{D}$ is implicit. Thus we have from Eq. 1

$$l(\boldsymbol{\theta}) = \sum_{k=1}^{n} \ln p(\mathbf{x}_k|\boldsymbol{\theta}) \tag{5}$$

and

$$\boldsymbol{\nabla}_{\boldsymbol{\theta}} l = \sum_{k=1}^{n} \boldsymbol{\nabla}_{\boldsymbol{\theta}} \ln p(\mathbf{x}_k|\boldsymbol{\theta}). \tag{6}$$

Thus, a set of necessary conditions for the maximum-likelihood estimate for $\boldsymbol{\theta}$ can be obtained from the set of p equations

$$\boldsymbol{\nabla}_{\boldsymbol{\theta}} l = \mathbf{0}. \tag{7}$$

A solution $\hat{\boldsymbol{\theta}}$ to Eq. 7 could represent a true global maximum, a *local* maximum or minimum, or (rarely) an inflection point of $l(\boldsymbol{\theta})$. One must be careful, too, to check if the extremum occurs at a boundary of the parameter space, which might not be apparent from the solution to Eq. 7. If all solutions are found, we are guaranteed that one represents the true maximum, though we might have to check each solution individually (or calculate second derivatives) to identify which is the global optimum. Of course, we must bear in mind that $\hat{\boldsymbol{\theta}}$ is an estimate; it is only in the limit of an infinitely large number of training points that we can expect that our estimate will equal to the true value of the generating function.

MAXIMUM A POSTERIORI

We note in passing that a related class of estimators—*maximum a posteriori* or MAP estimators—find the value of $\boldsymbol{\theta}$ that maximizes $l(\boldsymbol{\theta}) + \ln p(\boldsymbol{\theta})$, where $p(\boldsymbol{\theta})$ describes the prior probability of different parameter values. Thus a maximum-likelihood estimator is a MAP estimator for the uniform or "flat" prior. As such,

MODE

a MAP estimator finds the peak, or *mode* of a posterior density. The drawback of MAP estimators is that if we choose some arbitrary nonlinear transformation of the

---

*Of course, the base of the logarithm can be chosen for convenience, and in most analytic problems, base e is most appropriate. For that reason we will generally use ln rather than log or $\log_2$.

parameter space (e.g., an overall rotation), the density will change, and our MAP solution need no longer be appropriate (Section 3.5.2).

3.2.2 The Gaussian Case: Unknown μ

To see how maximum-likelihood methods apply to a specific case, suppose that the samples are drawn from a multivariate normal population with mean μ and covariance matrix Σ. For simplicity, consider first the case where only the mean is unknown. Under this condition, we consider a sample point $\mathbf{x}_k$ and find

$$\ln p(\mathbf{x}_k|\mu) = -\frac{1}{2}\ln\left[(2\pi)^d|\Sigma|\right] - \frac{1}{2}(\mathbf{x}_k - \mu)^t\Sigma^{-1}(\mathbf{x}_k - \mu) \tag{8}$$

and

$$\nabla_\mu \ln p(\mathbf{x}_k|\mu) = \Sigma^{-1}(\mathbf{x}_k - \mu). \tag{9}$$

Identifying θ with μ, we see from Eqs. 6, 7 and 9 that the maximum-likelihood estimate for μ must satisfy

$$\sum_{k=1}^{n} \Sigma^{-1}(\mathbf{x}_k - \hat{\mu}) = 0. \tag{10}$$

Multiplying by Σ and rearranging, we obtain

$$\hat{\mu} = \frac{1}{n}\sum_{k=1}^{n} \mathbf{x}_k. \tag{11}$$

SAMPLE MEAN

This is a very satisfying result. It says that the maximum-likelihood estimate for the unknown population mean is just the arithmetic average of the training samples— the *sample mean*, sometimes written $\hat{\mu}_n$ to clarify its dependence on the number of samples. Geometrically, if we think of the n samples as a cloud of points, the sample mean is the centroid of the cloud. The sample mean has a number of desirable statistical properties as well, and one would be inclined to use this rather obvious estimate even without knowing that it is the maximum-likelihood solution.

3.2.3 The Gaussian Case: Unknown μ and Σ

In the more general (and more typical) multivariate normal case, neither the mean μ nor the covariance matrix Σ is known. Thus, these unknown parameters constitute the components of the parameter vector θ. Consider first the univariate case with $\theta_1 = \mu$ and $\theta_2 = \sigma^2$. Here the log-likelihood of a single point is

$$\ln p(x_k|\theta) = -\frac{1}{2}\ln 2\pi\theta_2 - \frac{1}{2\theta_2}(x_k - \theta_1)^2 \tag{12}$$

and its derivative is

$$\nabla_\theta l = \nabla_\theta \ln p(x_k|\theta) = \begin{bmatrix} \dfrac{1}{\theta_2}(x_k - \theta_1) \\ -\dfrac{1}{2\theta_2} + \dfrac{(x_k - \theta_1)^2}{2\theta_2^2} \end{bmatrix}. \tag{13}$$

Applying Eq. 7 to the full log-likelihood leads to the conditions

$$\sum_{k=1}^{n} \frac{1}{\hat{\theta}_2}(x_k - \hat{\theta}_1) = 0 \tag{14}$$

and

$$-\sum_{k=1}^{n} \frac{1}{\hat{\theta}_2} + \sum_{k=1}^{n} \frac{(x_k - \hat{\theta}_1)^2}{\hat{\theta}_2^2} = 0, \tag{15}$$

where $\hat{\theta}_1$ and $\hat{\theta}_2$ are the maximum-likelihood estimates for θ_1 and θ_2, respectively. By substituting $\hat{\mu} = \hat{\theta}_1$ and $\hat{\sigma}^2 = \hat{\theta}_2$ and doing a little rearranging, we obtain the following maximum-likelihood estimates for μ and σ^2:

$$\hat{\mu} = \frac{1}{n} \sum_{k=1}^{n} x_k \tag{16}$$

and

$$\hat{\sigma}^2 = \frac{1}{n} \sum_{k=1}^{n} (x_k - \hat{\mu})^2. \tag{17}$$

While the analysis of the multivariate case is basically very similar, considerably more manipulations are involved (Problem 6). Just as we would predict, however, the result is that the maximum-likelihood estimates for μ and Σ are given by

$$\hat{\mu} = \frac{1}{n} \sum_{k=1}^{n} \mathbf{x}_k \tag{18}$$

and

$$\hat{\Sigma} = \frac{1}{n} \sum_{k=1}^{n} (\mathbf{x}_k - \hat{\mu})(\mathbf{x}_k - \hat{\mu})^t. \tag{19}$$

Thus, once again we find that the maximum-likelihood estimate for the mean vector is the sample mean. The maximum-likelihood estimate for the covariance matrix is the arithmetic average of the n matrices $(\mathbf{x}_k - \hat{\mu})(\mathbf{x}_k - \hat{\mu})^t$. Because the true covariance matrix is the expected value of the matrix $(\mathbf{x} - \hat{\mu})(\mathbf{x} - \hat{\mu})^t$, this is also a very satisfying result.

3.2.4 Bias

BIAS

The maximum-likelihood estimate for the variance σ^2 is *biased*; that is, the expected value over all data sets of size n of the sample variance is not equal to the true variance:*

$$\mathcal{E}\left[\frac{1}{n} \sum_{i=1}^{n} (x_i - \bar{x})^2 \right] = \frac{n-1}{n}\sigma^2 \neq \sigma^2. \tag{20}$$

---

*The word "bias" is often used to refer generally to an offset. The bias in a statistical estimate is unrelated to the bias weights in a discriminant function or a multilayer neural network.

We shall return to a more general consideration of bias in Chapter 9, but for the moment we can verify Eq. 20 for an underlying distribution with nonzero variance, σ^2, in the extreme case of $n = 1$, in which the expectation value is given by $\mathcal{E}[\cdot] = 0 \neq \sigma^2$. The maximum-likelihood estimate of the covariance matrix is similarly biased.

An elementary *unbiased* estimator for $\boldsymbol{\Sigma}$ is given by

$$\mathbf{C} = \frac{1}{n-1} \sum_{k=1}^{n} (\mathbf{x}_k - \hat{\boldsymbol{\mu}})(\mathbf{x}_k - \hat{\boldsymbol{\mu}})^t, \tag{21}$$

SAMPLE
COVARIANCE

ABSOLUTELY
UNBIASED

ASYMPTOT-
ICALLY
UNBIASED

where $\mathbf{C}$ is the so-called *sample covariance matrix*, as explored in Problem 30. If an estimator is unbiased for *all* distributions, as for example the variance estimator in Eq. 21, then it is called *absolutely unbiased*. If the estimator tends to become unbiased as the number of samples becomes very large, as for instance Eq. 20, then the estimator is *asymptotically unbiased*. In many pattern recognition problems with large training data sets, asymptotically unbiased estimators are acceptable.

Clearly, $\widehat{\boldsymbol{\Sigma}} = [(n-1)/n]\mathbf{C}$, and $\widehat{\boldsymbol{\Sigma}}$ is asymptotically unbiased; these two estimates are essentially identical when n is large. However, the existence of two similar but nevertheless distinct estimates for the covariance matrix may be disconcerting, and it is natural to ask which one is "correct." Of course, for $n > 1$ the answer is that these estimates are neither right nor wrong—they are just different. What the existence of two actually shows is that no single estimate possesses all of the properties we might desire. For our purposes, the most desirable property is rather complex—we want the estimate that leads to the best classification performance. While it is usually both reasonable and sound to design a classifier by substituting the maximum-likelihood estimates for the unknown parameters, we might well wonder if other estimates might not lead to better performance. Below we address this question from a Bayesian viewpoint.

If we have a reliable model for the underlying distributions and their dependence upon the parameter vector $\boldsymbol{\theta}$, the maximum-likelihood classifier can give excellent results. But what if our model is wrong? Do we nevertheless get the best classifier in our assumed set of models? For instance, what if we assume that a distribution comes from $N(\mu, 1)$ but instead it actually comes from $N(\mu, 10)$? Will the value we find for $\theta = \mu$ by maximum-likelihood yield the best of all classifiers of the form derived from $N(\mu, 1)$? Unfortunately, the answer is "no," and an illustrative counterexample is given in Problem 7 where the so-called *model error* is large indeed. This points out the need for reliable information concerning the models: If the assumed model is very poor, we cannot be assured that the classifier we derive is the best, even among our model set. We shall return to the problem of choosing among candidate models in Chapter 9.

3.3 BAYESIAN ESTIMATION

We now consider the Bayesian estimation or Bayesian learning approach to pattern classification problems. Although the answers we get by this method will generally be nearly identical to those obtained by maximum-likelihood, there is a conceptual difference: Whereas in maximum-likelihood methods we view the true parameter vector we seek, $\boldsymbol{\theta}$, to be fixed, in Bayesian learning we consider $\boldsymbol{\theta}$ to be a random variable, and training data allow us to convert a distribution on this variable into a posterior probability density.

3.3.1 The Class-Conditional Densities

The computation of posterior probabilities $P(\omega_i|\mathbf{x})$ lies at the heart of Bayesian classification. Bayes formula allows us to compute these probabilities from the prior probabilities $P(\omega_i)$ and the class-conditional densities $p(\mathbf{x}|\omega_i)$, but how can we proceed when these quantities are unknown? The general answer to this question is that the best we can do is to compute $P(\omega_i|\mathbf{x})$ using all of the information at our disposal. Part of this information might be prior knowledge, such as knowledge of the functional forms for unknown densities and ranges for the values of unknown parameters. Part of this information might reside in a set of training samples. If we again let $\mathcal{D}$ denote the set of samples, then we can emphasize the role of the samples by saying that our goal is to compute the posterior probabilities $P(\omega_i|\mathbf{x}, \mathcal{D})$. From these probabilities we can obtain the Bayes classifier.

Given the sample $\mathcal{D}$, Bayes formula then becomes

$$P(\omega_i|\mathbf{x}, \mathcal{D}) = \frac{p(\mathbf{x}|\omega_i, \mathcal{D}) P(\omega_i|\mathcal{D})}{\sum\limits_{j=1}^{c} p(\mathbf{x}|\omega_j, \mathcal{D}) P(\omega_j|\mathcal{D})}. \tag{22}$$

As this equation suggests, we can use the information provided by the training samples to help determine both the class-conditional densities and the prior probabilities.

Although we could maintain this generality, we shall henceforth assume that the true values of the prior probabilities are known or obtainable from a trivial calculation; thus we substitute $P(\omega_i) = P(\omega_i|\mathcal{D})$. Furthermore, because we are treating the supervised case, we can separate the training samples by class into c subsets $\mathcal{D}_1, \ldots, \mathcal{D}_c$, with the samples in $\mathcal{D}_i$ belonging to ω_i. As we mentioned when addressing maximum-likelihood methods, in most cases of interest (and in all of the cases we shall consider) the samples in $\mathcal{D}_i$ have no influence on $p(\mathbf{x}|\omega_j, \mathcal{D})$ if $i \neq j$. This has two simplifying consequences. First, it allows us to work with each class separately, using only the samples in $\mathcal{D}_i$ to determine $p(\mathbf{x}|\omega_i, \mathcal{D})$. Used in conjunction with our assumption that the prior probabilities are known, this allows us to write Eq. 22 as

$$P(\omega_i|\mathbf{x}, \mathcal{D}) = \frac{p(\mathbf{x}|\omega_i, \mathcal{D}_i) P(\omega_i)}{\sum\limits_{j=1}^{c} p(\mathbf{x}|\omega_j, \mathcal{D}_j) P(\omega_j)}. \tag{23}$$

Second, because each class can be treated independently, we can dispense with needless class distinctions and simplify our notation. In essence, we have c separate problems of the following form: Use a set $\mathcal{D}$ of samples drawn independently according to the fixed but unknown probability distribution $p(\mathbf{x})$ to determine $p(\mathbf{x}|\mathcal{D})$. This is the central problem of Bayesian learning.

3.3.2 The Parameter Distribution

Although the desired probability density $p(\mathbf{x})$ is unknown, we assume that it has a known parametric form. The only thing assumed unknown is the value of a parameter vector $\boldsymbol{\theta}$. We shall express the fact that $p(\mathbf{x})$ is unknown but has known parametric form by saying that the function $p(\mathbf{x}|\boldsymbol{\theta})$ is completely known. Any information we might have about $\boldsymbol{\theta}$ prior to observing the samples is assumed to be contained in a *known* prior density $p(\boldsymbol{\theta})$. Observation of the samples converts this to a posterior density $p(\boldsymbol{\theta}|\mathcal{D})$, which, we hope, is sharply peaked about the true value of $\boldsymbol{\theta}$.

Note that we have converted our problem of learning a probability density function to one of estimating a parameter vector. To this end, our basic goal is to compute $p(\mathbf{x}|\mathcal{D})$, which is as close as we can come to obtaining the unknown $p(\mathbf{x})$. We do this by integrating the joint density $p(\mathbf{x}, \boldsymbol{\theta}|\mathcal{D})$ over $\boldsymbol{\theta}$. That is,

$$p(\mathbf{x}|\mathcal{D}) = \int p(\mathbf{x}, \boldsymbol{\theta}|\mathcal{D}) \, d\boldsymbol{\theta}, \tag{24}$$

where the integration extends over the entire parameter space. Now we can always write $p(\mathbf{x}, \boldsymbol{\theta}|\mathcal{D})$ as the product $p(\mathbf{x}|\boldsymbol{\theta}, \mathcal{D})p(\boldsymbol{\theta}|\mathcal{D})$. Because the selection of $\mathbf{x}$ and that of the training samples in $\mathcal{D}$ is done independently, the first factor is merely $p(\mathbf{x}|\boldsymbol{\theta})$. That is, the distribution of $\mathbf{x}$ is known completely once we know the value of the parameter vector. Thus, Eq. 24 can be rewritten as

$$p(\mathbf{x}|\mathcal{D}) = \int p(\mathbf{x}|\boldsymbol{\theta})p(\boldsymbol{\theta}|\mathcal{D}) \, d\boldsymbol{\theta}. \tag{25}$$

This key equation links the desired class-conditional density $p(\mathbf{x}|\mathcal{D})$ to the posterior density $p(\boldsymbol{\theta}|\mathcal{D})$ for the unknown parameter vector. If $p(\boldsymbol{\theta}|\mathcal{D})$ peaks very sharply about some value $\hat{\boldsymbol{\theta}}$, we obtain $p(\mathbf{x}|\mathcal{D}) \simeq p(\mathbf{x}|\hat{\boldsymbol{\theta}})$, i.e., the result we would obtain by substituting the estimate $\hat{\boldsymbol{\theta}}$ for the true parameter vector. This result rests on the assumption that $p(\mathbf{x}|\boldsymbol{\theta})$ is smooth, and that the tails of the integral are not important. These conditions are typically but not invariably the case. In general, if we are less certain about the exact value of $\boldsymbol{\theta}$, this equation directs us to average $p(\mathbf{x}|\boldsymbol{\theta})$ over the possible values of $\boldsymbol{\theta}$. Thus, when the unknown densities have a known parametric form, the samples exert their influence on $p(\mathbf{x}|\mathcal{D})$ through the posterior density $p(\boldsymbol{\theta}|\mathcal{D})$. We should also point out that in practice, the integration in Eq. 25 can be performed numerically, for instance by Monte-Carlo simulation.

3.4 BAYESIAN PARAMETER ESTIMATION: GAUSSIAN CASE

In this section we use Bayesian estimation techniques to calculate the *a posteriori* density $p(\boldsymbol{\theta}|\mathcal{D})$ and the desired probability density $p(\mathbf{x}|\mathcal{D})$ for the case where $p(\mathbf{x}|\boldsymbol{\mu}) \sim N(\boldsymbol{\mu}, \boldsymbol{\Sigma})$.

3.4.1 The Univariate Case: $p(\mu|\mathcal{D})$

Consider the case where μ is the only unknown parameter. For simplicity we treat first the univariate case, that is,

$$p(x|\mu) \sim N(\mu, \sigma^2), \tag{26}$$

where the only unknown quantity is the mean μ. We assume that whatever prior knowledge we might have about μ can be expressed by a *known* prior density $p(\mu)$. Later we shall make the further assumption that

$$p(\mu) \sim N(\mu_0, \sigma_0^2), \tag{27}$$

where both μ_0 and σ_0^2 are known. Roughly speaking, μ_0 represents our best prior guess for μ, and σ_0^2 measures our uncertainty about this guess. The assumption that the prior distribution for μ is normal will simplify the subsequent mathematics. However, the crucial assumption is not so much that the prior distribution for μ is normal, but that it is known.

Having selected the prior density for μ, we can view the situation as follows. Imagine that a value is drawn for μ from a population governed by the probability law $p(\mu)$. Once this value is drawn, it becomes the true value of μ and completely determines the density for x. Suppose now that n samples $x_1, \ldots, x_n$ are independently drawn from the resulting population. Letting $\mathcal{D} = \{x_1, \ldots, x_n\}$, we use Bayes formula to obtain

$$p(\mu|\mathcal{D}) = \frac{p(\mathcal{D}|\mu)p(\mu)}{\int p(\mathcal{D}|\mu)p(\mu)\, d\mu}$$

$$= \alpha \prod_{k=1}^{n} p(x_k|\mu)p(\mu), \tag{28}$$

where α is a normalization factor that depends on $\mathcal{D}$ but is independent of μ. This equation shows how the observation of a set of training samples affects our ideas about the true value of μ; it relates the prior density $p(\mu)$ to an *a posteriori* density $p(\mu|\mathcal{D})$. Because $p(x_k|\mu) \sim N(\mu, \sigma^2)$ and $p(\mu) \sim N(\mu_0, \sigma_0^2)$, we have

$$p(\mu|\mathcal{D}) = \alpha \prod_{k=1}^{n} \overbrace{\frac{1}{\sqrt{2\pi}\sigma} \exp\left[-\frac{1}{2}\left(\frac{x_k - \mu}{\sigma}\right)^2\right]}^{p(x_k|\mu)} \overbrace{\frac{1}{\sqrt{2\pi}\sigma_0} \exp\left[-\frac{1}{2}\left(\frac{\mu - \mu_0}{\sigma_0}\right)^2\right]}^{p(\mu)}$$

$$= \alpha' \exp\left[-\frac{1}{2}\left(\sum_{k=1}^{n}\left(\frac{\mu - x_k}{\sigma}\right)^2 + \left(\frac{\mu - \mu_0}{\sigma_0}\right)^2\right)\right]$$

$$= \alpha'' \exp\left[-\frac{1}{2}\left[\left(\frac{n}{\sigma^2} + \frac{1}{\sigma_0^2}\right)\mu^2 - 2\left(\frac{1}{\sigma^2}\sum_{k=1}^{n}x_k + \frac{\mu_0}{\sigma_0^2}\right)\mu\right]\right], \tag{29}$$

where factors that do not depend on μ have been absorbed into the constants α, α', and α''. Thus, $p(\mu|\mathcal{D})$ is an exponential function of a quadratic function of μ, i.e., is again a normal density. Because this is true for any number of training samples, $p(\mu|\mathcal{D})$ remains normal as the number n of samples is increased, and $p(\mu|\mathcal{D})$ is said to be a *reproducing density* and $p(\mu)$ is said to be a *conjugate prior*. If we write $p(\mu|\mathcal{D}) \sim N(\mu_n, \sigma_n^2)$, then μ_n and σ_n^2 can be found by equating coefficients in Eq. 29 with corresponding coefficients in the generic Gaussian of the form

REPRODUCING
DENSITY

$$p(\mu|\mathcal{D}) = \frac{1}{\sqrt{2\pi}\sigma_n} \exp\left[-\frac{1}{2}\left(\frac{\mu - \mu_n}{\sigma_n}\right)^2\right]. \tag{30}$$

Identifying coefficients in this way yields

$$\frac{1}{\sigma_n^2} = \frac{n}{\sigma^2} + \frac{1}{\sigma_0^2} \tag{31}$$

and

$$\frac{\mu_n}{\sigma_n^2} = \frac{n}{\sigma^2}\hat{\mu}_n + \frac{\mu_0}{\sigma_0^2}, \tag{32}$$

where $\hat{\mu}_n$ is the sample mean

$$\hat{\mu}_n = \frac{1}{n} \sum_{k=1}^{n} x_k. \tag{33}$$

We solve explicitly for μ_n and σ_n^2 and obtain

$$\mu_n = \left(\frac{n\sigma_0^2}{n\sigma_0^2 + \sigma^2} \right) \hat{\mu}_n + \frac{\sigma^2}{n\sigma_0^2 + \sigma^2} \mu_0 \tag{34}$$

and

$$\sigma_n^2 = \frac{\sigma_0^2 \sigma^2}{n\sigma_0^2 + \sigma^2}. \tag{35}$$

These equations show how the prior information is combined with the empirical information in the samples to obtain the *a posteriori* density $p(\mu|\mathcal{D})$. Roughly speaking, μ_n represents our best guess for μ after observing n samples, and σ_n^2 measures our uncertainty about this guess. Because σ_n^2 decreases monotonically with n—approaching σ^2/n as n approaches infinity—each additional observation decreases our uncertainty about the true value of μ. As n increases, $p(\mu|\mathcal{D})$ becomes more and more sharply peaked, approaching a Dirac delta function as n approaches infinity. This behavior is commonly known as *Bayesian learning* (Fig. 3.2).

BAYESIAN
LEARNING

In general, μ_n is a linear combination of $\hat{\mu}_n$ and μ_0, with coefficients that are nonnegative and sum to one. Thus μ_n always lies somewhere between $\hat{\mu}_n$ and μ_0. If $\sigma_0 \neq 0$, μ_n approaches the sample mean as n approaches infinity. If $\sigma_0 = 0$, we have a degenerate case in which our prior certainty that $\mu = \mu_0$ is so strong that no number of observations can change our opinion. At the other extreme, if $\sigma_0 \gg \sigma$, we are so uncertain about our prior guess that we take $\mu_n = \hat{\mu}_n$, using only the samples

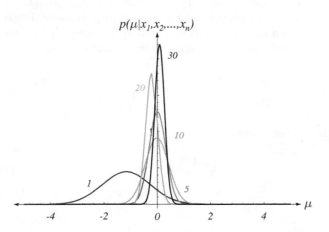
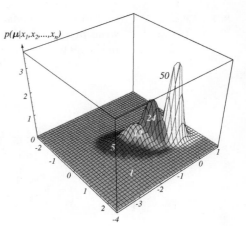

FIGURE 3.2. Bayesian learning of the mean of normal distributions in one and two dimensions. The posterior distribution estimates are labeled by the number of training samples used in the estimation.

to estimate μ. In general, the relative balance between prior knowledge and empirical data is set by the ratio of σ^2 to σ_0^2, which is sometimes called the *dogmatism*. If the dogmatism is not infinite, after enough samples are taken the exact values assumed for μ_0 and σ_0^2 will be unimportant, and μ_n will converge to the sample mean.

3.4.2 The Univariate Case: $p(x|\mathcal{D})$

Having obtained the *a posteriori* density for the mean, $p(\mu|\mathcal{D})$, all that remains is to obtain the "class-conditional" density for $p(x|\mathcal{D})$.* From Eqs. 25, 26 and 30 we have

$$
p(x|\mathcal{D}) = \int p(x|\mu)p(\mu|\mathcal{D})\,d\mu
$$

$$
= \int \frac{1}{\sqrt{2\pi}\sigma} \exp\left[-\frac{1}{2}\left(\frac{x-\mu}{\sigma}\right)^2\right] \frac{1}{\sqrt{2\pi}\sigma_n} \exp\left[-\frac{1}{2}\left(\frac{\mu-\mu_n}{\sigma_n}\right)^2\right]\,d\mu
$$

$$
= \frac{1}{2\pi\sigma\sigma_n} \exp\left[-\frac{1}{2}\frac{(x-\mu_n)^2}{\sigma^2+\sigma_n^2}\right] f(\sigma,\sigma_n), \tag{36}
$$

where

$$
f(\sigma,\sigma_n) = \int \exp\left[-\frac{1}{2}\frac{\sigma^2+\sigma_n^2}{\sigma^2\sigma_n^2}\left(\mu - \frac{\sigma_n^2 x + \sigma^2\mu_n}{\sigma^2+\sigma_n^2}\right)^2\right]\,d\mu.
$$

That is, as a function of x, $p(x|\mathcal{D})$ is proportional to $\exp[-(1/2)(x-\mu_n)^2/(\sigma^2+\sigma_n^2)]$, and hence $p(x|\mathcal{D})$ is normally distributed with mean μ_n and variance $\sigma^2 + \sigma_n^2$:

$$
p(x|\mathcal{D}) \sim N(\mu_n, \sigma^2 + \sigma_n^2). \tag{37}
$$

In other words, to obtain the class-conditional density $p(x|\mathcal{D})$, whose parametric form is known to be $p(x|\mu) \sim N(\mu, \sigma^2)$, we merely replace μ by μ_n and σ^2 by $\sigma^2 + \sigma_n^2$. In effect, the conditional mean μ_n is treated as if it were the true mean, and the known variance is increased to account for the additional uncertainty in x resulting from our lack of exact knowledge of the mean μ. This, then, is our final result: the density $p(x|\mathcal{D})$ is the desired class-conditional density $p(x|\omega_j, \mathcal{D}_j)$, and together with the prior probabilities $P(\omega_j)$ it gives us the probabilistic information needed to design the classifier. This is in contrast to maximum-likelihood methods that only make point estimates for $\hat{\mu}$ and $\hat{\sigma}^2$, rather that estimate a *distribution* for $p(x|\mathcal{D})$.

3.4.3 The Multivariate Case

The treatment of the multivariate case in which $\mathbf{\Sigma}$ is known but $\boldsymbol{\mu}$ is not is a direct generalization of the univariate case. For this reason we shall only sketch the derivation. As before, we assume that

$$
p(\mathbf{x}|\boldsymbol{\mu}) \sim N(\boldsymbol{\mu}, \mathbf{\Sigma}) \quad \text{and} \quad p(\boldsymbol{\mu}) \sim N(\boldsymbol{\mu}_0, \mathbf{\Sigma}_0), \tag{38}
$$

where $\mathbf{\Sigma}$, $\mathbf{\Sigma}_0$, and $\boldsymbol{\mu}_0$ are assumed to be known. After observing a set $\mathcal{D}$ of n independent samples $\mathbf{x}_1, \ldots, \mathbf{x}_n$, we use Bayes formula to obtain

---

*Recall that for simplicity we dropped class distinctions, but that all samples here come from the same class, say ω_i, and hence $p(\mathbf{x}|\mathcal{D})$ is really $p(\mathbf{x}|\omega_i, \mathcal{D}_i)$.

$$p(\boldsymbol{\mu}|\mathcal{D}) = \alpha \prod_{k=1}^{n} p(\mathbf{x}_k|\boldsymbol{\mu}) p(\boldsymbol{\mu}) \tag{39}$$

$$= \alpha' \exp\left[-\frac{1}{2}\left(\boldsymbol{\mu}^t \left(n\boldsymbol{\Sigma}^{-1} + \boldsymbol{\Sigma}_0^{-1}\right)\boldsymbol{\mu} - 2\boldsymbol{\mu}^t \left(\boldsymbol{\Sigma}^{-1}\sum_{k=1}^{n}\mathbf{x}_k + \boldsymbol{\Sigma}_0^{-1}\boldsymbol{\mu}_0\right)\right)\right],$$

which has the form

$$p(\boldsymbol{\mu}|\mathcal{D}) = \alpha'' \exp\left[-\frac{1}{2}(\boldsymbol{\mu} - \boldsymbol{\mu}_n)^t \boldsymbol{\Sigma}_n^{-1}(\boldsymbol{\mu} - \boldsymbol{\mu}_n)\right]. \tag{40}$$

Thus, $p(\boldsymbol{\mu}|\mathcal{D}) \sim N(\boldsymbol{\mu}_n, \boldsymbol{\Sigma}_n)$, and once again we have a reproducing density. Equating coefficients, we obtain the analogs of Eqs. 34 and 35,

$$\boldsymbol{\Sigma}_n^{-1} = n\boldsymbol{\Sigma}^{-1} + \boldsymbol{\Sigma}_0^{-1} \tag{41}$$

and

$$\boldsymbol{\Sigma}_n^{-1}\boldsymbol{\mu}_n = n\boldsymbol{\Sigma}^{-1}\hat{\boldsymbol{\mu}}_n + \boldsymbol{\Sigma}_0^{-1}\boldsymbol{\mu}_0, \tag{42}$$

where $\hat{\boldsymbol{\mu}}_n$ is the sample mean

$$\hat{\boldsymbol{\mu}}_n = \frac{1}{n}\sum_{k=1}^{n}\mathbf{x}_k. \tag{43}$$

The solution of these equations for $\boldsymbol{\mu}_n$ and $\boldsymbol{\Sigma}_n$ is simplified by knowledge of the matrix identity

$$(\mathbf{A}^{-1} + \mathbf{B}^{-1})^{-1} = \mathbf{A}(\mathbf{A} + \mathbf{B})^{-1}\mathbf{B} = \mathbf{B}(\mathbf{A} + \mathbf{B})^{-1}\mathbf{A}, \tag{44}$$

which is valid for any pair of nonsingular, d-by-d matrices $\mathbf{A}$ and $\mathbf{B}$. After a little manipulation (Problem 16), we obtain the final results:

$$\boldsymbol{\mu}_n = \boldsymbol{\Sigma}_0 \left(\boldsymbol{\Sigma}_0 + \frac{1}{n}\boldsymbol{\Sigma}\right)^{-1}\hat{\boldsymbol{\mu}}_n + \frac{1}{n}\boldsymbol{\Sigma}\left(\boldsymbol{\Sigma}_0 + \frac{1}{n}\boldsymbol{\Sigma}\right)^{-1}\boldsymbol{\mu}_0 \tag{45}$$

(which, as in the univariate case, is a linear combination of $\hat{\boldsymbol{\mu}}_n$ and $\boldsymbol{\mu}_0$) and

$$\boldsymbol{\Sigma}_n = \boldsymbol{\Sigma}_0 \left(\boldsymbol{\Sigma}_0 + \frac{1}{n}\boldsymbol{\Sigma}\right)^{-1}\frac{1}{n}\boldsymbol{\Sigma}. \tag{46}$$

The proof that $p(\mathbf{x}|\mathcal{D}) \sim N(\boldsymbol{\mu}_n, \boldsymbol{\Sigma} + \boldsymbol{\Sigma}_n)$ can be obtained as before by performing the integration

$$p(\mathbf{x}|\mathcal{D}) = \int p(\mathbf{x}|\boldsymbol{\mu}) p(\boldsymbol{\mu}|\mathcal{D})\, d\boldsymbol{\mu}. \tag{47}$$

However, this result can be obtained with less effort by observing that $\mathbf{x}$ can be viewed as the sum of two mutually independent random variables, a random vector $\boldsymbol{\mu}$ with $p(\boldsymbol{\mu}|\mathcal{D}) \sim N(\boldsymbol{\mu}_n, \boldsymbol{\Sigma}_n)$ and an independent random vector $\mathbf{y}$ with $p(\mathbf{y}) \sim$

$N(\mathbf{0}, \mathbf{\Sigma})$. Because the sum of two independent, normally distributed vectors is again a normally distributed vector whose mean is the sum of the means and whose co-variance matrix is the sum of the covariance matrices (Chapter 2, Problem 18), we have

$$p(\mathbf{x}|\mathcal{D}) \sim N(\boldsymbol{\mu}_n, \mathbf{\Sigma} + \mathbf{\Sigma}_n), \tag{48}$$

and the generalization is complete.

3.5 BAYESIAN PARAMETER ESTIMATION: GENERAL THEORY

We have just seen how the Bayesian approach can be used to obtain the desired density $p(\mathbf{x}|\mathcal{D})$ in a special case—the multivariate Gaussian. This approach can be generalized to apply to any situation in which the unknown density can be parameterized. The basic assumptions are summarized as follows:

- The form of the density $p(\mathbf{x}|\boldsymbol{\theta})$ is assumed to be known, but the value of the parameter vector $\boldsymbol{\theta}$ is not known exactly.
- Our initial knowledge about $\boldsymbol{\theta}$ is assumed to be contained in a known prior density $p(\boldsymbol{\theta})$.
- The rest of our knowledge about $\boldsymbol{\theta}$ is contained in a set $\mathcal{D}$ of n samples $\mathbf{x}_1, \ldots, \mathbf{x}_n$ drawn independently according to the unknown probability density $p(\mathbf{x})$.

The basic problem is to compute the posterior density $p(\boldsymbol{\theta}|\mathcal{D})$, because from this we can use Eq. 25 to compute $p(\mathbf{x}|\mathcal{D})$:

$$p(\mathbf{x}|\mathcal{D}) = \int p(\mathbf{x}|\boldsymbol{\theta}) p(\boldsymbol{\theta}|\mathcal{D}) \, d\boldsymbol{\theta}. \tag{49}$$

By Bayes formula we have

$$p(\boldsymbol{\theta}|\mathcal{D}) = \frac{p(\mathcal{D}|\boldsymbol{\theta}) p(\boldsymbol{\theta})}{\int p(\mathcal{D}|\boldsymbol{\theta}) p(\boldsymbol{\theta}) \, d\boldsymbol{\theta}}, \tag{50}$$

and by the independence assumption

$$p(\mathcal{D}|\boldsymbol{\theta}) = \prod_{k=1}^{n} p(\mathbf{x}_k|\boldsymbol{\theta}). \tag{51}$$

This constitutes the formal solution to the problem, and Eqs. 50 and 51 illuminate its relation to the maximum-likelihood solution. Suppose that $p(\mathcal{D}|\boldsymbol{\theta})$ reaches a sharp peak at $\boldsymbol{\theta} = \hat{\boldsymbol{\theta}}$. If the prior density $p(\boldsymbol{\theta})$ is not zero at $\boldsymbol{\theta} = \hat{\boldsymbol{\theta}}$ and does not change much in the surrounding neighborhood, then $p(\boldsymbol{\theta}|\mathcal{D})$ also peaks at that point. Thus, Eq. 49 shows that $p(\mathbf{x}|\mathcal{D})$ will be approximately $p(\mathbf{x}|\hat{\boldsymbol{\theta}})$, the result one would obtain by using the maximum-likelihood estimate as if it were the true value. If the peak of $p(\mathcal{D}|\boldsymbol{\theta})$ is very sharp, then the influence of prior information on the uncertainty in the true value of $\boldsymbol{\theta}$ can be ignored. In this and even the more general case, however, the Bayesian solution tells us how to use *all* the available information to compute the desired density $p(\mathbf{x}|\mathcal{D})$.

While we have obtained the formal Bayesian solution to the problem, a number of interesting questions remain. One concerns the difficulty of carrying out these computations. Another concerns the convergence of $p(\mathbf{x}|\mathcal{D})$ to $p(\mathbf{x})$. We shall discuss the matter of convergence briefly, and we will later turn to the computational question.

To indicate explicitly the number of samples in a set for a single category, we shall write $\mathcal{D}^n = \{\mathbf{x}_1, \dots, \mathbf{x}_n\}$. Then from Eq. 51, if $n > 1$ we obtain

$$p(\mathcal{D}^n|\boldsymbol{\theta}) = p(\mathbf{x}_n|\boldsymbol{\theta})p(\mathcal{D}^{n-1}|\boldsymbol{\theta}). \tag{52}$$

Substituting this in Eq. 50 and using Bayes formula, we see that the posterior density satisfies the recursion relation

$$p(\boldsymbol{\theta}|\mathcal{D}^n) = \frac{p(\mathbf{x}_n|\boldsymbol{\theta})p(\boldsymbol{\theta}|\mathcal{D}^{n-1})}{\int p(\mathbf{x}_n|\boldsymbol{\theta})p(\boldsymbol{\theta}|\mathcal{D}^{n-1})\,d\boldsymbol{\theta}}. \tag{53}$$

RECURSIVE
BAYES
INCREMENTAL
LEARNING

With the understanding that $p(\boldsymbol{\theta}|\mathcal{D}^0) = p(\boldsymbol{\theta})$, repeated use of this equation produces the sequence of densities $p(\boldsymbol{\theta})$, $p(\boldsymbol{\theta}|\mathbf{x}_1)$, $p(\boldsymbol{\theta}|\mathbf{x}_1, \mathbf{x}_2)$, and so forth. (It should be obvious from Eq. 53 that $p(\boldsymbol{\theta}|\mathcal{D}^n)$ depends only on the points in $\mathcal{D}^n$, not the sequence in which they were selected.) This is called the *recursive Bayes* approach to parameter estimation. This is our first example of an *incremental* or *on-line* learning method, where learning goes on as the data are collected. When this sequence of densities converges to a Dirac delta function centered about the true parameter value, we have *Bayesian learning* (Example 1). We shall come across many other, nonincremental learning schemes, where all the training data must be present before learning can take place.

In principle, Eq. 53 requires that we preserve all the training points in $\mathcal{D}^{n-1}$ to calculate $p(\boldsymbol{\theta}|\mathcal{D}^n)$. However, for some distributions, just a few estimates of parameters associated with $p(\boldsymbol{\theta}|\mathcal{D}^{n-1})$ contain all the information needed. Such parameters are the *sufficient statistics* of those distributions, as we shall see in Section 3.6. Some authors reserve the term recursive learning to apply to only those cases where the sufficient statistics are retained—not the training data—when incorporating the information from a new training point. We could call this more restrictive usage *true recursive Bayes learning*.

EXAMPLE 1   **Recursive Bayes Learning**

Suppose we believe our one-dimensional samples come from a uniform distribution

$$p(x|\theta) \sim U(0, \theta) = \begin{cases} 1/\theta & 0 \le x \le \theta \\ 0 & \text{otherwise,} \end{cases}$$

but initially we know only that our parameter is bounded. In particular we assume $0 < \theta \le 10$ (a noninformative or "flat prior" we shall discuss in Section 3.5.2). We will use recursive Bayes methods to estimate θ and the underlying densities from the data $\mathcal{D} = \{4, 7, 2, 8\}$, which were selected randomly from the underlying distribution. Before any data arrive, then, we have $p(\theta|\mathcal{D}^0) = p(\theta) = U(0, 10)$. When our first data point $x_1 = 4$ arrives, we use Eq. 53 to get an improved estimate:

$$p(\theta|\mathcal{D}^1) \propto p(x|\theta)p(\theta|\mathcal{D}^0) = \begin{cases} 1/\theta & \text{for } 4 \le \theta \le 10 \\ 0 & \text{otherwise,} \end{cases}$$

where throughout we will ignore the normalization. When the next data point $x_2 = 7$ arrives, we have

$$p(\theta|\mathcal{D}^2) \propto p(x|\theta)p(\theta|\mathcal{D}^1) = \begin{cases} 1/\theta^2 & \text{for } 7 \leq \theta \leq 10 \\ 0 & \text{otherwise,} \end{cases}$$

and similarly for the remaining sample points. It should be clear that because each successive step introduces a factor of $1/\theta$ into $p(x|\theta)$, and the distribution is nonzero only for x values above the largest data point sampled, the general form of our solution is $p(\theta|\mathcal{D}^n) \propto 1/\theta^n$ for $\max_x[\mathcal{D}^n] \leq \theta \leq 10$, as shown in the figure.

Given our full data set, the maximum-likelihood solution here is clearly $\hat{\theta} = 8$, and this implies a uniform $p(x|\mathcal{D}) \sim U(0, 8)$.

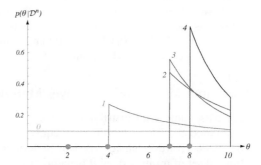

The posterior $p(\theta|\mathcal{D}^n)$ for the model and n points in the data set in this Example. For $n = 0$, the posterior starts our as a flat, uniform density from 0 to 10, denoted $p(\theta) \sim U(0, 10)$. As more points are incorporated, it becomes increasingly peaked at the value of the highest data point.

According to our Bayesian methodology, which requires the integration in Eq. 49, the density is uniform up to $x = 8$, but has a tail at higher values—an indication that the influence of our prior $p(\theta)$ has not yet been swamped by the information in the training data.

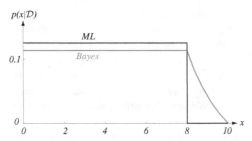

Given the full set of four points, the distribution based on the maximum-likelihood solution is $p(x|\hat{\theta}) \sim U(0, 8)$, whereas the distribution derived from Bayesian methods has a small tail above $x = 8$, reflecting the prior information that values of x near 10 are possible.

Whereas the maximum-likelihood approach estimates a *point* in $\boldsymbol{\theta}$ space, the Bayesian approach instead estimates a *distribution*. Technically speaking, then, we cannot directly compare these estimates. It is only when the second stage of inference is done—that is, when we compute the distributions $p(x|\mathcal{D})$, as shown in the above figure—that the comparison is fair.

For most of the typically encountered probability densities $p(\mathbf{x}|\boldsymbol{\theta})$, the sequence of posterior densities does indeed converge to a delta function. Roughly speaking, this implies that with a large number of samples there is only one value for $\boldsymbol{\theta}$ that causes $p(\mathbf{x}|\boldsymbol{\theta})$ to fit the data, that is, that $\boldsymbol{\theta}$ can be determined uniquely from $p(\mathbf{x}|\boldsymbol{\theta})$. When this is the case, $p(\mathbf{x}|\boldsymbol{\theta})$ is said to be *identifiable*. A rigorous proof of convergence under these conditions requires a precise statement of the properties required of $p(\mathbf{x}|\boldsymbol{\theta})$ and $p(\boldsymbol{\theta})$ and considerable care, but presents no serious difficulties (Problem 21).

There are occasions, however, when more than one value of $\boldsymbol{\theta}$ may yield the same value for $p(\mathbf{x}|\boldsymbol{\theta})$. In such cases, $\boldsymbol{\theta}$ cannot be determined uniquely from $p(\mathbf{x}|\boldsymbol{\theta})$, and $p(\mathbf{x}|\mathcal{D}^n)$ will peak near all of the values of $\boldsymbol{\theta}$ that explain the data. Fortunately, this ambiguity is erased by the integration in Eq. 25, because $p(\mathbf{x}|\boldsymbol{\theta})$ is the same for all of these values of $\boldsymbol{\theta}$. Thus, $p(\mathbf{x}|\mathcal{D}^n)$ will typically converge to $p(\mathbf{x})$ whether or not $p(\mathbf{x}|\boldsymbol{\theta})$ is identifiable. While this might make the problem of identifiabilty appear to be moot, we shall see in Chapter 10 that identifiability presents a genuine problem in the case of unsupervised learning.

3.5.1 When Do Maximum-Likelihood and Bayes Methods Differ?

For reasonable prior distributions that do not preclude the true solution, maximum-likelihood and Bayes solutions are equivalent in the asymptotic limit of infinite training data. However since practical pattern recognition problems invariably have a limited set of training data, it is natural to ask when maximum-likelihood and Bayes solutions may be expected to differ, and then which we should prefer.

There are several criteria that will influence our choice. One is computational complexity (Section 3.7.2), and here maximum-likelihood methods are often to be preferred because they require merely differential calculus techniques or gradient search for $\hat{\boldsymbol{\theta}}$, rather than a possibly complex multidimensional integration needed in Bayesian estimation. This leads to another consideration: interpretability. In many cases the maximum-likelihood solution will be easier to interpret and understand because it returns the single best model from the set the designer provided (and presumably understands). In contrast, Bayesian methods give a weighted average of models (parameters), often leading to solutions more complicated and harder to understand than those provided by the designer. The Bayesian approach reflects the remaining uncertainty in the possible models.

Another consideration is our confidence in the prior information, such as in the *form* of the underlying distribution $p(\mathbf{x}|\boldsymbol{\theta})$. A maximum-likelihood solution $p(\mathbf{x}|\hat{\boldsymbol{\theta}})$ must of course be of the assumed parametric form; not so for the Bayesian solution. We saw this difference in Example 1, where the Bayes solution was not of the parametric form originally assumed, that is, a uniform $p(x|\mathcal{D})$. In general, through their use of the full $p(\boldsymbol{\theta}|\mathcal{D})$ distribution Bayesian methods use more of the information brought to the problem than do maximum-likelihood methods. (For instance, in Example 1 the addition of the third training point did not change the maximum-likelihood solution but did refine the Bayesian estimate.) If such information is reliable, Bayes methods can be expected to give better results. Furthermore, general Bayesian methods with a "flat" or uniform prior (i.e., where no prior information is explicitly imposed) are more similar to maximum-likelihood methods. If there is much data, leading to a strongly peaked $p(\boldsymbol{\theta}|\mathcal{D})$, and the prior $p(\boldsymbol{\theta})$ is uniform or flat, then the MAP estimate is essentially the same as the maximum-likelihood estimate.

When $p(\boldsymbol{\theta}|\mathcal{D})$ is broad, or asymmetric around $\hat{\boldsymbol{\theta}}$, the methods are quite likely to yield $p(\mathbf{x}|\mathcal{D})$ distributions that differ from one another. Such a strong asymmetry (when not due to rare statistical irregularities in the selection of the training data) generally conveys some information about the distribution, just as did the asymmetric role of the threshold θ in Example 1. Bayes methods would exploit such information, whereas maximum-likelihood methods would not (at least not directly). Furthermore, Bayesian methods make more explicit the crucial problem of bias and variance tradeoffs—roughly speaking, the balance between the accuracy of the estimation and its variance, which depend upon the amount of training data. This important matter was irrelevant in Chapter 2, where there was no notion of a finite training set, but it will be crucial in our considerations of the theory of machine learning in Chapter 9.

When designing a classifier by either of these methods, we determine the posterior densities for each category and also classify a test point by the maximum posterior. (If there are costs, summarized in a cost matrix, these can be incorporated as well.) There are three sources of classification error in our final system:

Bayes or Indistinguishability Error: The error due to overlapping densities $p(\mathbf{x}|\omega_i)$ for different values of i. This error is an inherent property of the problem and can never be eliminated.

Model Error: The error due to having an incorrect model. This error can only be eliminated if the designer specifies a model that includes the true model which generated the data. Designers generally choose the model based on knowledge of the problem domain rather than on the subsequent estimation method, and thus the model error in maximum-likelihood and Bayes methods rarely differ.

Estimation Error: The error arising from the fact that the parameters are estimated from a finite sample. This error can best be reduced by increasing the training data.

The relative contributions of these sources depend upon problem, of course. In the limit of infinite training data, the estimation error vanishes, and the total classification error will be the same for both maximum-likelihood and Bayes methods.

In summary, there are strong theoretical and methodological arguments supporting Bayesian estimation, though in practice maximum-likelihood estimation is simpler and, when used for designing classifiers, can lead to classifiers nearly as accurate.

3.5.2 Noninformative Priors and Invariance

Generally speaking, the information about the prior $p(\boldsymbol{\theta})$ derives from the designer's knowledge of the problem domain and as such is beyond our study of the design of classifiers. Nevertheless, in some cases we have guidance in how to create priors that do not impose structure when we believe none exists, and this leads us to the notion of noninformative priors.

Recall our discussion of the role of prior category probabilities in Chapter 2, where, in the absence of other information, we assumed each of c categories equally likely. Analogously, in a Bayesian framework we can have a "noninformative" prior over a parameter for a single category's distribution. Suppose for instance that we are using Bayesian methods to infer from data some position and scale parameters, which we denote μ and σ, which might be the mean and standard deviation of a Gaussian, or the position and width of a triangle distribution, and so on. What prior might we put on these parameters? Consider first the location parameter. Surely the prior distribution should not depend upon our arbitrary choice of origin, that is, we demand *translation invariance*. The only prior to have this property is the uniform prior, that is, uniform over the entire one-dimensional space. Of course, such a prior is *improper*, that is, $\int p(\mu)d\mu = \infty$.

What about the prior over the scale parameter, $p(\sigma)$? Surely the unit of spatial measurement—meters, feet, inches—is irrelevant to the functional form of the prior, that is, we demand *scale invariance*. Consider a new variable, $\tilde{\sigma} = \ln\sigma$. If σ is rescaled by some positive constant a, that is, $\sigma \to a\sigma$, then our new variable experiences a shift, i.e.,

$$\tilde{\sigma} \to \ln a + \ln\sigma = \underbrace{\ln a}_{shift} + \tilde{\sigma}. \tag{54}$$

Thus, just as with the location parameter μ, we demand that $\tilde{\sigma}$ should have a uniform distribution through all possible values. A uniform prior on $\tilde{\sigma}$ implies that the noninformative prior for σ itself should be (Problem 20)

$$p(\sigma) = 1/\sigma. \tag{55}$$

Of course, this prior is also improper.

In general, then, if there is known or assumed invariance—such as translation, or for discrete distributions invariance to the sequential order of data selection—there will be constraints on the form of the prior. If we can find a prior that satisfies such constraints, the resulting prior is "noninformative" with respect to that invariance.

It is tempting to assert that the use of noninformative priors is somehow "objective" and lets the data speak for themselves, but such a view is a bit naive. For example, we may seek a noninformative prior when estimating the standard deviation σ of a Gaussian. But this requirement might not lead to the noninformative prior for estimating the variance, σ^2. Which should we use? In fact, the greatest benefit of this approach is that it forces the designer to acknowledge and be clear about the assumed invariance—the choice of which generally lies outside our methodology. It may be more difficult to accommodate such arbitrary transformations in a maximum *a posteriori* (MAP) estimator (Section 3.2.1), and hence considerations of invariance are of greatest use in Bayesian estimation, or when the posterior is very strongly peaked and the mode not influenced by transformations of the density (Problem 19).

3.5.3 Gibbs Algorithm

Given the assumptions described above, the Bayes optimal classifier will give the optimum classification performance. Unfortunately, the integration described by Eq. 25 can be very difficult, and a simple alternative is to pick a parameter vector $\boldsymbol{\theta}$ according to $p(\boldsymbol{\theta}|\mathcal{D})$ and use the single value as if it were the true value. Given weak assumptions, this so-called *Gibbs algorithm* gives a misclassification error that is at most twice the expected error of the Bayes optimal classifier (Problem 22).

*3.6 SUFFICIENT STATISTICS

From a practical viewpoint, the formal solution provided by Eqs. 49–51 is not computationally attractive. In pattern recognition applications it is not unusual to have dozens or hundreds of parameters and thousands of training samples, which makes the direct computation and tabulation of $p(\mathcal{D}|\boldsymbol{\theta})$ or $p(\boldsymbol{\theta}|\mathcal{D})$ quite out of the question.

We shall see in Chapter 6 how neural network methods avoid many of the difficulties of setting such a large number of parameters in a classifier, but for now we note that the only hope for an analytic, computationally feasible maximum-likelihood solution lies in being able to find a parametric form for $p(\mathbf{x}|\boldsymbol{\theta})$ that on the one hand matches the characteristics of the problem and on the other hand allows a reasonably tractable solution.

Consider the simplification that occurred in the problem of learning the parameters of a multivariate Gaussian density. The basic data processing required was merely the computation of the sample mean and sample covariance. These easily computed and easily updated statistics contained all the information in the samples relevant to estimating the unknown population mean and covariance. One might suspect that this simplicity is just one more happy property of the normal distribution, and that such good fortune is not likely to occur in other cases. While this is largely true, there are distributions for which computationally feasible solutions can be obtained, and the key to their simplicity lies in the notion of a sufficient statistic.

To begin with, any function of the samples is a statistic. Roughly speaking, a *sufficient statistic* is a (possibly vector-valued) function $\mathbf{s}$ of the samples $\mathcal{D}$ that contains all of the information relevant to estimating some parameter $\boldsymbol{\theta}$.* Intuitively, one might expect the definition of a sufficient statistic to involve the requirement that $p(\boldsymbol{\theta}|\mathbf{s}, \mathcal{D}) = p(\boldsymbol{\theta}|\mathbf{s})$. However, this would require treating $\boldsymbol{\theta}$ as a random variable, limiting the definition to a Bayesian domain. To avoid such a limitation, the conventional definition is as follows: A statistic $\mathbf{s}$ is said to be *sufficient* for $\boldsymbol{\theta}$ if $p(\mathcal{D}|\mathbf{s}, \boldsymbol{\theta})$ is independent of $\boldsymbol{\theta}$. If we think of $\boldsymbol{\theta}$ as a random variable, we can write

$$p(\boldsymbol{\theta}|\mathbf{s}, \mathcal{D}) = \frac{p(\mathcal{D}|\mathbf{s}, \boldsymbol{\theta})\, p(\boldsymbol{\theta}|\mathbf{s})}{p(\mathcal{D}|\mathbf{s})}, \tag{56}$$

whereupon it becomes evident that $p(\boldsymbol{\theta}|\mathbf{s}, \mathcal{D}) = p(\boldsymbol{\theta}|\mathbf{s})$ if $\mathbf{s}$ is sufficient for $\boldsymbol{\theta}$. Conversely, if $\mathbf{s}$ is a statistic for which $p(\boldsymbol{\theta}|\mathbf{s}, \mathcal{D}) = p(\boldsymbol{\theta}|\mathbf{s})$ and if $p(\boldsymbol{\theta}|\mathbf{s}) \neq 0$, it is easy to show that $p(\mathcal{D}|\mathbf{s}, \boldsymbol{\theta})$ is independent of $\boldsymbol{\theta}$ (Problem 28). Thus, the intuitive and the conventional definitions are basically equivalent. As one might expect, for a Gaussian distribution the sample mean and covariance, taken together, represent a sufficient statistic for the true mean and covariance; if these are known, all other statistics such as the mode, range, higher-order moments, number of data points, and so on, are superfluous when estimating the true mean and covariance.

A fundamental theorem concerning sufficient statistics is the *Factorization Theorem*, which states that $\mathbf{s}$ is sufficient for $\boldsymbol{\theta}$ if and only if $p(\mathcal{D}|\boldsymbol{\theta})$ can be factored into the product of two functions—one depending only on $\mathbf{s}$ and $\boldsymbol{\theta}$, and the other depending only on the training samples. The virtue of the Factorization Theorem is that it allows us to shift our attention from the rather complicated density $p(\mathcal{D}|\mathbf{s}, \boldsymbol{\theta})$, used to define a sufficient statistic, to the simpler function

$$p(\mathcal{D}|\boldsymbol{\theta}) = \prod_{k=1}^{n} p(\mathbf{x}_k|\boldsymbol{\theta}). \tag{57}$$

In addition, the Factorization Theorem makes it clear that the characteristics of a sufficient statistic are completely determined by the density $p(\mathbf{x}|\boldsymbol{\theta})$, and have nothing to do with a felicitous choice of a prior density $p(\boldsymbol{\theta})$. A proof of the Factorization Theorem in the continuous case is somewhat tricky because degenerate situations are

---

*When we must distinguish between the function and its value, we shall write $\mathbf{s} = \boldsymbol{\varphi}(\mathcal{D})$.

involved. Because the proof has some intrinsic interest, however, we include one for the simpler discrete case.

■ **Theorem 3.1.** (**Factorization**) A statistic $\mathbf{s}$ is sufficient for $\boldsymbol{\theta}$ if and only if the probability $P(\mathcal{D}|\boldsymbol{\theta})$ can be written as the product

$$P(\mathcal{D}|\boldsymbol{\theta}) = g(\mathbf{s}, \boldsymbol{\theta})h(\mathcal{D}), \tag{58}$$

for some functions $g(\cdot, \cdot)$ and $h(\cdot)$.

Proof. (**a**) We begin by showing the "only if" part of the theorem. Suppose first that $\mathbf{s}$ is sufficient for $\boldsymbol{\theta}$, so that $P(\mathcal{D}|\mathbf{s}, \boldsymbol{\theta})$ is independent of $\boldsymbol{\theta}$. Because we want to show that $P(\mathcal{D}|\boldsymbol{\theta})$ can be factored, our attention is directed toward computing $P(\mathcal{D}|\boldsymbol{\theta})$ in terms of $P(\mathcal{D}|\mathbf{s}, \boldsymbol{\theta})$. We do this by summing the joint probability $P(\mathcal{D}, \mathbf{s}|\boldsymbol{\theta})$ over all values of $\mathbf{s}$:

$$P(\mathcal{D}|\boldsymbol{\theta}) = \sum_{\mathbf{s}} P(\mathcal{D}, \mathbf{s}|\boldsymbol{\theta})$$

$$= \sum_{\mathbf{s}} P(\mathcal{D}|\mathbf{s}, \boldsymbol{\theta})P(\mathbf{s}|\boldsymbol{\theta}). \tag{59}$$

But because $\mathbf{s} = \boldsymbol{\varphi}(\mathcal{D})$ for some $\boldsymbol{\varphi}(\cdot)$, there is only one possible value for $\mathbf{s}$ for the given data, and thus

$$P(\mathcal{D}|\boldsymbol{\theta}) = P(\mathcal{D}|\mathbf{s}, \boldsymbol{\theta})P(\mathbf{s}|\boldsymbol{\theta}). \tag{60}$$

Moreover, because by hypothesis $P(\mathcal{D}|\mathbf{s}, \boldsymbol{\theta})$ is independent of $\boldsymbol{\theta}$, the first factor depends only on $\mathcal{D}$. Identifying $P(\mathbf{s}|\boldsymbol{\theta})$ with $g(\mathbf{s}, \boldsymbol{\theta})$, we see that $P(\mathcal{D}|\boldsymbol{\theta})$ factors, as desired.

(**b**) We now consider the "if" part of the theorem. To show that the ability to factor $P(\mathcal{D}|\boldsymbol{\theta})$ as the product $g(\mathbf{s}, \boldsymbol{\theta})h(\mathcal{D})$ implies that $\mathbf{s}$ is sufficient for $\boldsymbol{\theta}$, we must show that such a factoring implies that the conditional probability $P(\mathcal{D}|\mathbf{s}, \boldsymbol{\theta})$ is independent of $\boldsymbol{\theta}$. Because $\mathbf{s} = \boldsymbol{\varphi}(\mathcal{D})$, specifying a value for $\mathbf{s}$ constrains the possible sets of samples to some set $\bar{\mathcal{D}}$. Formally, $\bar{\mathcal{D}} = \{\mathcal{D}|\boldsymbol{\varphi}(\mathcal{D}) = \mathbf{s}\}$. If $\bar{\mathcal{D}}$ is empty, no assignment of values to the samples can yield that value of $\mathbf{s}$, and $P(\mathbf{s}|\boldsymbol{\theta}) = 0$. Excluding such cases (i.e., considering only values of $\mathbf{s}$ that can arise), we have

$$P(\mathcal{D}|\mathbf{s}, \boldsymbol{\theta}) = \frac{P(\mathcal{D}, \mathbf{s}|\boldsymbol{\theta})}{P(\mathbf{s}|\boldsymbol{\theta})}. \tag{61}$$

The denominator can be computed by summing the numerator over all values of $\mathcal{D}$. Because the numerator will be zero if $\mathcal{D} \notin \bar{\mathcal{D}}$, we can restrict the summation to $\mathcal{D} \in \bar{\mathcal{D}}$. That is,

$$P(\mathcal{D}|\mathbf{s}, \boldsymbol{\theta}) = \frac{P(\mathcal{D}, \mathbf{s}|\boldsymbol{\theta})}{\sum\limits_{\mathcal{D} \in \bar{\mathcal{D}}} P(\mathcal{D}, \mathbf{s}|\boldsymbol{\theta})} = \frac{P(\mathcal{D}|\boldsymbol{\theta})}{\sum\limits_{\mathcal{D} \in \bar{\mathcal{D}}} P(\mathcal{D}|\boldsymbol{\theta})} = \frac{g(\mathbf{s}, \boldsymbol{\theta})h(\mathcal{D})}{\sum\limits_{\mathcal{D} \in \bar{\mathcal{D}}} g(\mathbf{s}, \boldsymbol{\theta})h(\mathcal{D})} = \frac{h(\mathcal{D})}{\sum\limits_{\mathcal{D} \in \bar{\mathcal{D}}} h(\mathcal{D})}, \tag{62}$$

which is independent of $\boldsymbol{\theta}$. Thus, by definition, $\mathbf{s}$ is sufficient for $\boldsymbol{\theta}$.

It should be pointed out that there are trivial ways of constructing sufficient statistics. For example we can define $\mathbf{s}$ to be a vector whose components are the n samples themselves: $\mathbf{x}_1, \ldots, \mathbf{x}_n$. In that case $g(\mathbf{s}, \boldsymbol{\theta}) = p(\mathcal{D}|\boldsymbol{\theta})$ and $h(\mathcal{D}) = 1$. One can even produce a scalar sufficient statistic by the trick of interleaving the digits in the decimal expansion of the components of the n samples. Sufficient statistics such as these are of little interest, because they do not provide us with simpler results. The ability to factor $p(\mathcal{D}|\boldsymbol{\theta})$ into a product $g(\mathbf{s}, \boldsymbol{\theta})h(\mathcal{D})$ is interesting only when the function g and the sufficient statistic $\mathbf{s}$ are simple. It should be noted that sufficiency is an integral notion. That is, if $\mathbf{s}$ is a sufficient statistic for $\boldsymbol{\theta}$, this does not necessarily imply that their corresponding components are sufficient, i.e., that s_1 is sufficient for θ_1, or s_2 for θ_2, and so on (Problem 27).

An obvious fact should also be mentioned: The factoring of $p(\mathcal{D}|\boldsymbol{\theta})$ into $g(\mathbf{s}, \boldsymbol{\theta})h(\mathcal{D})$ is not unique. If $f(\mathbf{s})$ is any function of $\mathbf{s}$, then $g'(\mathbf{s}, \boldsymbol{\theta}) = f(\mathbf{s})g(\mathbf{s}, \boldsymbol{\theta})$ and $h'(\mathcal{D}) = h(\mathcal{D})/f(\mathbf{s})$ are equivalent factors. This kind of ambiguity can be eliminated by defining the *kernel density*

KERNEL DENSITY

$$\bar{g}(\mathbf{s}, \boldsymbol{\theta}) = \frac{g(\mathbf{s}, \boldsymbol{\theta})}{\int g(\mathbf{s}, \boldsymbol{\theta}) \, d\boldsymbol{\theta}}, \tag{63}$$

which is invariant to such scaling.

What is the importance of sufficient statistics and kernel densities for parameter estimation? The general answer is that the most practical applications of classical parameter estimation to pattern classification involve density functions that possess simple sufficient statistics and simple kernel densities. Moreover, it can be shown that for any classification rule, we can find another based solely on sufficient statistics that has equal or better performance. Thus—in principle at least—we need only consider decisions based on sufficient statistics. It is, in essence, the ultimate in data reduction: We can reduce an extremely large data set down to a few numbers—the sufficient statistics—confident that all relevant information has been preserved. This means, too, that we can always create the Bayes classifier from sufficient statistics, as for example our Bayes classifiers for Gaussian distributions were functions solely of the sufficient statistics, estimates of $\boldsymbol{\mu}$ and $\boldsymbol{\Sigma}$.

In the case of maximum-likelihood estimation, when searching for a value of $\boldsymbol{\theta}$ that maximizes $p(\mathcal{D}|\boldsymbol{\theta}) = g(\mathbf{s}, \boldsymbol{\theta})h(\mathcal{D})$, we can restrict our attention to $g(\mathbf{s}, \boldsymbol{\theta})$. In this case, the normalization provided by Eq. 63 is of no particular value unless $\bar{g}(\mathbf{s}, \boldsymbol{\theta})$ is simpler than $g(\mathbf{s}, \boldsymbol{\theta})$. The significance of the kernel density is revealed, however, in the Bayesian case. If we substitute $p(\mathcal{D}|\boldsymbol{\theta}) = g(\mathbf{s}, \boldsymbol{\theta})h(\mathcal{D})$ in Eq. 50, we obtain

$$p(\boldsymbol{\theta}|\mathcal{D}) = \frac{g(\mathbf{s}, \boldsymbol{\theta})p(\boldsymbol{\theta})}{\int g(\mathbf{s}, \boldsymbol{\theta})p(\boldsymbol{\theta}) \, d\boldsymbol{\theta}}. \tag{64}$$

If our prior knowledge of $\boldsymbol{\theta}$ is very vague, $p(\boldsymbol{\theta})$ will tend to be uniform, or changing very slowly as a function of $\boldsymbol{\theta}$. For such an essentially uniform $p(\boldsymbol{\theta})$, Eq. 64 shows that $p(\boldsymbol{\theta}|\mathcal{D})$ is approximately the same as the kernel density. Roughly speaking, the kernel density is the posterior distribution of the parameter vector when the prior distribution is uniform. Even when the prior distribution is far from uniform, the kernel density typically gives the asymptotic distribution of the parameter vector. In particular, when $p(\mathbf{x}|\boldsymbol{\theta})$ is identifiable and when the number of samples is large, $g(\mathbf{s}, \boldsymbol{\theta})$ usually peaks sharply at some value $\boldsymbol{\theta} = \hat{\boldsymbol{\theta}}$. If the prior density $p(\boldsymbol{\theta})$ is continuous at $\boldsymbol{\theta} = \hat{\boldsymbol{\theta}}$ and if $p(\hat{\boldsymbol{\theta}})$ is not zero, $p(\boldsymbol{\theta}|\mathcal{D})$ will approach the kernel density $\bar{g}(\mathbf{s}, \boldsymbol{\theta})$.

3.6.1 Sufficient Statistics and the Exponential Family

To see how the Factorization Theorem can be used to obtain sufficient statistics, consider once again the familiar d-dimensional normal case with fixed covariance but unknown mean, that is, $p(\mathbf{x}|\boldsymbol{\theta}) \sim N(\boldsymbol{\theta}, \boldsymbol{\Sigma})$. Here we have

$$
\begin{aligned}
p(\mathcal{D}|\boldsymbol{\theta}) &= \prod_{k=1}^{n} \frac{1}{(2\pi)^{d/2}|\boldsymbol{\Sigma}|^{1/2}} \exp\left[-\frac{1}{2}(\mathbf{x}_k - \boldsymbol{\theta})^t \boldsymbol{\Sigma}^{-1}(\mathbf{x}_k - \boldsymbol{\theta})\right] \\
&= \frac{1}{(2\pi)^{nd/2}|\boldsymbol{\Sigma}|^{n/2}} \exp\left[-\frac{1}{2}\sum_{k=1}^{n}\left(\boldsymbol{\theta}^t\boldsymbol{\Sigma}^{-1}\boldsymbol{\theta} - 2\boldsymbol{\theta}^t\boldsymbol{\Sigma}^{-1}\mathbf{x}_k + \mathbf{x}_k^t\boldsymbol{\Sigma}^{-1}\mathbf{x}_k\right)\right] \\
&= \exp\left[-\frac{n}{2}\boldsymbol{\theta}^t\boldsymbol{\Sigma}^{-1}\boldsymbol{\theta} + \boldsymbol{\theta}^t\boldsymbol{\Sigma}^{-1}\left(\sum_{k=1}^{n}\mathbf{x}_k\right)\right] \\
&\quad \times \frac{1}{(2\pi)^{nd/2}|\boldsymbol{\Sigma}|^{n/2}} \exp\left[-\frac{1}{2}\sum_{k=1}^{n}\mathbf{x}_k^t\boldsymbol{\Sigma}^{-1}\mathbf{x}_k\right].
\end{aligned}
\tag{65}
$$

This factoring isolates the $\boldsymbol{\theta}$ dependence of $p(\mathcal{D}|\boldsymbol{\theta})$ in the first term, and hence from the Factorization Theorem we conclude that $\sum_{k=1}^{n}\mathbf{x}_k$ is sufficient for $\boldsymbol{\theta}$. Of course, any one-to-one function of this statistic is also sufficient for $\boldsymbol{\theta}$; in particular, the sample mean

$$
\hat{\boldsymbol{\mu}}_n = \frac{1}{n}\sum_{k=1}^{n}\mathbf{x}_k
\tag{66}
$$

is also sufficient for $\boldsymbol{\theta}$. Using this statistic, we can write

$$
g(\hat{\boldsymbol{\mu}}_n, \boldsymbol{\theta}) = \exp\left[-\frac{n}{2}\left(\boldsymbol{\theta}^t\boldsymbol{\Sigma}^{-1}\boldsymbol{\theta} - 2\boldsymbol{\theta}^t\boldsymbol{\Sigma}^{-1}\hat{\boldsymbol{\mu}}_n\right)\right].
\tag{67}
$$

From using Eq. 63 or by completing the square, we can obtain the kernel density:

$$
\bar{g}(\hat{\boldsymbol{\mu}}_n, \boldsymbol{\theta}) = \frac{1}{(2\pi)^{d/2}|\frac{1}{n}\boldsymbol{\Sigma}|^{1/2}} \exp\left[-\frac{1}{2}(\boldsymbol{\theta} - \hat{\boldsymbol{\mu}}_n)^t \left(\frac{1}{n}\boldsymbol{\Sigma}\right)^{-1}(\boldsymbol{\theta} - \hat{\boldsymbol{\mu}}_n)\right].
\tag{68}
$$

These results make it immediately clear that $\hat{\boldsymbol{\mu}}_n$ is the maximum-likelihood estimate for $\boldsymbol{\theta}$. The Bayesian posterior density can be obtained from $\bar{g}(\hat{\boldsymbol{\mu}}_n, \boldsymbol{\theta})$ by performing the integration indicated in Eq. 64. If the prior density is essentially uniform, then $p(\boldsymbol{\theta}|\mathcal{D}) = \bar{g}(\hat{\boldsymbol{\mu}}_n, \boldsymbol{\theta})$.

This same general approach can be used to find sufficient statistics for other density functions. In particular, it applies to any member of the *exponential family*, a group of probability and probability density functions that possess simple sufficient statistics. Members of the exponential family include the Gaussian, exponential, Rayleigh, Poisson, and many other familiar distributions. They can all be written in the form

$$
p(\mathbf{x}|\boldsymbol{\theta}) = \alpha(\mathbf{x}) \exp\left[\mathbf{a}(\boldsymbol{\theta}) + \mathbf{b}(\boldsymbol{\theta})^t\mathbf{c}(\mathbf{x})\right].
\tag{69}
$$

If we multiply n terms of the form in Eq. 69 we find

$$p(\mathcal{D}|\boldsymbol{\theta}) = \exp\left[n\mathbf{a}(\boldsymbol{\theta}) + \mathbf{b}(\boldsymbol{\theta})^t \sum_{k=1}^{n} \mathbf{c}(\mathbf{x}_k)\right] \prod_{k=1}^{n} \alpha(\mathbf{x}_k) = g(\mathbf{s}, \boldsymbol{\theta})h(\mathcal{D}), \qquad (70)$$

where we can take

$$\mathbf{s} = \frac{1}{n}\sum_{k=1}^{n} \mathbf{c}(\mathbf{x}_k),$$

$$g(\mathbf{s}, \boldsymbol{\theta}) = \exp\left[n\{\mathbf{a}(\boldsymbol{\theta}) + \mathbf{b}(\boldsymbol{\theta})^t\mathbf{s}\}\right],$$

and

$$h(\mathcal{D}) = \prod_{k=1}^{n} \alpha(\mathbf{x}_k).$$

The distributions, sufficient statistics, and unnormalized kernels for a number of commonly encountered members of the exponential family are given in Table 3.1. It is a fairly routine matter to derive maximum-likelihood estimates and Bayesian *a posteriori* distributions from these solutions. With two exceptions, the solutions given are for univariate cases, though they can be used in multivariate situations if statistical independence can be assumed. Note that a few well-known probability distributions, such as the Cauchy, do *not* have sufficient statistics, so that the sample mean can be a very poor estimator of the true mean (Problem 29).

3.7 PROBLEMS OF DIMENSIONALITY

In practical multicategory applications, it is not at all unusual to encounter problems involving fifty or a hundred features, particularly if the features are binary valued. We might typically believe that each feature is useful for at least some of the discriminations; while we may doubt that each feature provides independent information, intentionally superfluous features have not been included. There are two issues that must be confronted. The most important is how classification accuracy depends upon the dimensionality (and amount of training data); the second is the computational complexity of designing the classifier.

3.7.1 Accuracy, Dimension, and Training Sample Size

If the features are statistically independent, there are some theoretical results that suggest the possibility of excellent performance. For example, consider the two-class multivariate normal case with the same covariance, i.e., where $p(\mathbf{x}|\omega_j) \sim N(\boldsymbol{\mu}_j, \boldsymbol{\Sigma})$, $j = 1, 2$. If the prior probabilities are equal, then it is not hard to show (Problem 31 in Chapter 2) that the Bayes error rate is given by

$$P(e) = \frac{1}{\sqrt{2\pi}} \int_{r/2}^{\infty} e^{-u^2/2} \, du, \qquad (71)$$

where r^2 is the squared Mahalanobis distance (Chapter 2, Section 2.5):

$$r^2 = (\boldsymbol{\mu}_1 - \boldsymbol{\mu}_2)^t \boldsymbol{\Sigma}^{-1}(\boldsymbol{\mu}_1 - \boldsymbol{\mu}_2). \qquad (72)$$

Table 3.1. Common Exponential Distributions and Their Sufficient Statistics

Name	Distribution	Domain		s	$|g(\mathbf{s}, \boldsymbol{\theta})|^{1/n}$
Normal	$p(x|\boldsymbol{\theta}) = \sqrt{\dfrac{\theta_2}{2\pi}}\, e^{-(1/2)\theta_2(x-\theta_1)^2}$	$\theta_2 > 0$		$\begin{bmatrix} \dfrac{1}{n}\sum_{k=1}^{n} x_k \\[1em] \dfrac{1}{n}\sum_{k=1}^{n} x_k^2 \end{bmatrix}$	$\sqrt{\theta_2}\, e^{\frac{1}{2}\theta_2(s_2 - 2\theta_1 s_1 + \theta_1^2)}$
Multivariate Normal	$p(\mathbf{x}|\boldsymbol{\theta}) = \dfrac{|\boldsymbol{\Theta}_1|^{1/2}}{(2\pi)^{d/2}}\, e^{-(1/2)(\mathbf{x}-\boldsymbol{\theta}_1)^t \boldsymbol{\Theta}_2 (\mathbf{x}-\boldsymbol{\theta}_1)}$	$\boldsymbol{\Theta}_2$ positive definite		$\begin{bmatrix} \dfrac{1}{n}\sum_{k=1}^{n} \mathbf{x}_k \\[1em] \dfrac{1}{n}\sum_{k=1}^{n} \mathbf{x}_k \mathbf{x}_k^t \end{bmatrix}$	$|\boldsymbol{\Theta}_2|^{1/2}\, e^{-\frac{1}{2}[\operatorname{tr}\boldsymbol{\Theta}_2 \mathbf{s}_2 - 2\boldsymbol{\theta}_1^t \boldsymbol{\Theta}_2 \mathbf{s}_1 + \boldsymbol{\theta}_1^t \boldsymbol{\Theta}_2 \boldsymbol{\theta}_1]}$
Exponential	$p(x|\theta) = \begin{cases} \theta e^{-\theta x} & x \geq 0 \\ 0 & \text{otherwise} \end{cases}$	$\theta > 0$		$\dfrac{1}{n}\sum_{k=1}^{n} x_k$	$\theta e^{-\theta s}$
Rayleigh	$p(x|\theta) = \begin{cases} 2\theta x e^{-\theta x^2} & x \geq 0 \\ 0 & \text{otherwise} \end{cases}$	$\theta > 0$		$\dfrac{1}{n}\sum_{k=1}^{n} x_k^2$	$\theta e^{-\theta s}$
Maxwell	$p(x|\theta) = \begin{cases} \dfrac{4}{\sqrt{\pi}} \theta^{3/2} x^2 e^{-\theta x^2} & x \geq 0 \\ 0 & \text{otherwise} \end{cases}$	$\theta > 0$		$\dfrac{1}{n}\sum_{k=1}^{n} x_k^2$	$\theta^{3/2} e^{-\theta s}$
Gamma	$p(x|\boldsymbol{\theta}) = \begin{cases} \dfrac{\theta_2^{\theta_1+1}}{\Gamma(\theta_1+1)} x^{\theta_1} e^{-\theta_2 x} & x \geq 0 \\ 0 & \text{otherwise} \end{cases}$	$\theta_1 > -1$ $\theta_2 > 0$		$\begin{bmatrix} \left(\prod_{k=1}^{n} x_k\right)^{1/n} \\[1em] \dfrac{1}{n}\sum_{k=1}^{n} x_k \end{bmatrix}$	$\dfrac{\theta_2^{\theta_1+1}}{\Gamma(\theta_1+1)} s_1^{\theta_1} e^{-\theta_2 s_2}$

	$p(x\mid\boldsymbol{\theta})$				
Beta	$p(x\mid\boldsymbol{\theta}) = \begin{cases} \dfrac{\Gamma(\theta_1+\theta_2+2)}{\Gamma(\theta_1+1)\Gamma(\theta_2+1)}x^{\theta_1}(1-x)^{\theta_2} & 0 \le x \le 1 \\ 0 & \text{otherwise} \end{cases}$	$\begin{aligned}\theta_1 &> -1 \\ \theta_2 &> -1\end{aligned}$		$\left[\begin{array}{c}\left(\displaystyle\prod_{k=1}^{n}x_k\right)^{1/n} \\ \left(\displaystyle\prod_{k=1}^{n}(1-x_k)\right)^{1/n}\end{array}\right]$	$\dfrac{\Gamma(\theta_1+\theta_2+2)}{\Gamma(\theta_1+1)\Gamma(\theta_2+1)}s_1^{\theta_1}s_2^{\theta_2}$
Poisson	$P(x\mid\theta) = \dfrac{\theta^x}{x!}e^{-\theta} \qquad x = 0,1,2,\dots$	$\theta > 0$		$\dfrac{1}{n}\displaystyle\sum_{k=1}^{n}x_k$	$\theta^s e^{-\theta}$
Bernoulli	$P(x\mid\theta) = \theta^x(1-\theta)^{1-x} \qquad x = 0,1$	$0 < \theta < 1$		$\dfrac{1}{n}\displaystyle\sum_{k=1}^{n}x_k$	$\theta^s(1-\theta)^{1-s}$
Binomial	$P(x\mid\theta) = \dfrac{m!}{x!(m-x)!}\theta^x(1-\theta)^{m-x} \qquad \begin{array}{l} x_i = 0,1,\dots,m \\ \displaystyle\sum_{i=1}^{d}x_i = m\end{array}$	$0 < \theta < 1$		$\dfrac{1}{n}\displaystyle\sum_{k=1}^{n}x_k$	$\theta^s(1-\theta)^{m-s}$
Multinomial	$P(\mathbf{x}\mid\boldsymbol{\theta}) = \dfrac{m!\displaystyle\prod_{i=1}^{d}\theta_i^{x_i}}{\displaystyle\prod_{i=1}^{d}x_i!}$	$\begin{aligned}0 &< \theta_i < 1 \\ \displaystyle\sum_{i=1}^{d}\theta_i &= 1\end{aligned}$		$\dfrac{1}{n}\displaystyle\sum_{k=1}^{n}\mathbf{x}_k$	$\displaystyle\prod_{i=1}^{d}\theta_i^{s_i}$

Thus, the probability of error decreases as r increases, approaching zero as r approaches infinity. In the conditionally independent case, $\mathbf{\Sigma} = diag(\sigma_1^2, \ldots, \sigma_d^2)$ and

$$r^2 = \sum_{i=1}^{d} \left(\frac{\mu_{i1} - \mu_{i2}}{\sigma_i} \right)^2. \tag{73}$$

This shows how each feature contributes to reducing the probability of error. Naturally, the most useful features are the ones for which the difference between the means is large relative to the standard deviations. However, no feature is useless if its means for the two classes differ. An obvious way to reduce the error rate further is to introduce new, independent features. Each new feature need not add much, but if r can be increased without limit, the probability of error can be made arbitrarily small.

In general, if the performance obtained with a given set of features is inadequate, it is natural to consider adding new features, particularly ones that will help separate the class pairs most frequently confused. Although increasing the number of features increases the cost and complexity of both the feature extractor and the classifier, it is often reasonable to believe that the performance will improve. After all, if the probabilistic structure of the problem were completely known, the Bayes risk could not possibly be increased by adding new features. At worst, the Bayes classifier would ignore the new features, but if the new features provide any additional information, the performance must improve (Fig. 3.3).

Unfortunately, it has frequently been observed in practice that, beyond a certain point, the inclusion of additional features leads to worse rather than better performance. This apparent paradox presents a genuine and serious problem for classifier design. The basic source of the difficulty can always be traced to the fact that we have the wrong model (e.g., the Gaussian assumption or conditional assumption is

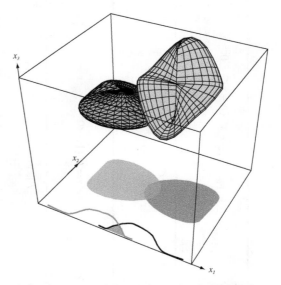

FIGURE 3.3. Two three-dimensional distributions have nonoverlapping densities, and thus in three dimensions the Bayes error vanishes. When projected to a subspace—here, the two-dimensional $x_1 - x_2$ subspace or a one-dimensional x_1 subspace—there can be greater overlap of the projected distributions, and hence greater Bayes error.

wrong) or the number of design or training samples is finite and thus the distributions are not estimated accurately. However, analysis of the problem is both challenging and subtle. Simple cases do not exhibit the experimentally observed phenomena, and more realistic cases are difficult to analyze (Chapter 9).

3.7.2 Computational Complexity

ORDER
BIG OH

We have mentioned that one consideration affecting our design methodology is that of the computational difficulty, and here the technical notion of computational complexity can be useful. First, we will will need to understand the notion of the *order* of a function $f(x)$: we say that the $f(x)$ is "of the order of $h(x)$"—written $f(x) = O(h(x))$ and generally read "big oh of $h(x)$"—if there exist constants c and x_0 such that $|f(x)| \leq c|h(x)|$ for all $x > x_0$. This means simply that for sufficiently large x, an upper bound on the function grows no worse than $h(x)$. For instance, suppose $f(x) = a_0 + a_1 x + a_2 x^2$; in that case we have $f(x) = O(x^2)$ because for sufficiently large x the constant, linear, and quadratic terms can be "overcome" by proper choice of c and x_0. The generalization to functions of two or more variables is straightforward. It should be clear that by the definition above, the big oh order of a function is not unique. For instance, we can describe our particular $f(x)$ as being $O(x^2)$, $O(x^3)$, $O(x^4)$, or $O(x^2 \ln x)$.

Because of the nonuniqueness of the big oh notation, we occasionally need to be more precise in describing the order of a function. We say that $f(x) = \Theta(h(x))$ "big theta of $h(x)$" if there are constants x_0, c_1, and c_2 such that for $x > x_0$, $f(x)$ always lies between $c_1 h(x)$ and $c_2 h(x)$. Thus our simple quadratic function above would obey $f(x) = \Theta(x^2)$, but would *not* obey $f(x) = \Theta(x^3)$. (A fuller explanation is provided in Section A.8 of the Appendix.)

In describing the computational complexity of an algorithm we are generally interested in the number of basic mathematical operations, such as additions, multiplications, and divisions it requires, or in the time and memory needed on a computer. To illustrate this concept we consider the complexity of a maximum-likelihood estimation of the parameters in a classifier for Gaussian priors in d dimensions, with n training samples for each of c categories. For each category it is necessary to calculate the discriminant function of Eq. 74. The computational complexity of finding the sample mean $\hat{\boldsymbol{\mu}}$ is $O(nd)$, because for each of the d dimensions we must add n component values. The required division by n in the mean calculation is a single computation, independent of the number of points, and hence does not affect this complexity. For each of the $d(d + 1)/2$ independent components of the sample covariance matrix $\widehat{\boldsymbol{\Sigma}}$ there are n multiplications and additions (Eq. 19), giving a complexity of $O(d^2 n)$. Once $\widehat{\boldsymbol{\Sigma}}$ has been computed, its determinant is an $O(d^2)$ calculation, as we can easily verify by counting the number of operations in matrix "sweep" methods. The inverse can be calculated in $O(d^3)$ calculations, for instance by Gaussian elimination.* The complexity of estimating $P(\omega)$ is of course $O(n)$. Equation 74 illustrates these individual components for the problem of setting the parameters of normal distributions via maximum-likelihood:

$$g(\mathbf{x}) = -\frac{1}{2}(\mathbf{x} - \overset{O(dn)\ O(nd^3)}{\hat{\boldsymbol{\mu}}})^t \overbrace{\widehat{\boldsymbol{\Sigma}}^{-1}} (\mathbf{x} - \hat{\boldsymbol{\mu}}) - \overbrace{\frac{d}{2} \ln 2\pi}^{O(1)} - \overbrace{\frac{1}{2} \ln |\widehat{\boldsymbol{\Sigma}}|}^{O(d^3)} + \overbrace{\ln P(\omega)}^{O(n)}. \qquad (74)$$

---

*We mention for the aficionado that there are more complex matrix inversion algorithms that are $O(d^{2.376\cdots})$, and there may be algorithms with even lower complexity yet to be discovered.

Naturally we assume that $n > d$ (otherwise our covariance matrix will not have a well defined inverse), and thus for large problems the overall complexity of calculating an individual discriminant function is dominated by the $O(d^2 n)$ term in Eq. 74. This is done for each of the categories, and hence our overall computational complexity for learning in this Bayes classifier is $O(cd^2 n)$. Since c is typically a constant much smaller than d^2 or n, we can call our complexity $O(d^2 n)$. We saw in Section 3.7 that it was generally desirable to have more training data from a larger-dimensional space; our complexity analysis shows the steep cost in so doing.

We next reconsider the matter of estimating a covariance matrix in a bit more detail. This requires the estimation of $d(d + 1)/2$ parameters—the d diagonal elements and $d(d - 1)/2$ independent off-diagonal elements. We observe first that the appealing maximum-likelihood estimate

$$\widehat{\boldsymbol{\Sigma}} = \frac{1}{n} \sum_{k=1}^{n} (\mathbf{x}_k - \mathbf{m}_n)(\mathbf{x}_k - \mathbf{m}_n)^t \tag{75}$$

which is an $O(nd^2)$ calculation; it is the sum of $n - 1$ independent d-by-d matrices of rank one, and thus is guaranteed to be singular if $n \leq d$. Because we must invert $\widehat{\boldsymbol{\Sigma}}$ to obtain the discriminant functions, we have an algebraic requirement for at least $d + 1$ samples. To smooth our statistical fluctuations and obtain a really good estimate, it would not be surprising if several times that number of samples were needed.

The computational complexity for *classification* is less, of course. Given a test point $\mathbf{x}$ we must compute $(\mathbf{x} - \hat{\boldsymbol{\mu}})$, an $O(d)$ calculation. Moreover, for each of the categories we must multiply the inverse covariance matrix by the separation vector, an $O(d^2)$ calculation. The $\max_i g_i(\mathbf{x})$ decision is a separate $O(c)$ operation. For small c then, classification is an $O(d^2)$ operation. Here, as throughout virtually all pattern classification, classification is much simpler (and faster) than learning. The complexity of the corresponding case for Bayesian learning, summarized in Eq. 48, yields the same computational complexity as in maximum-likelihood. More generally, however, Bayesian learning has higher complexity as a consequence of integrating over model parameters $\boldsymbol{\theta}$.

Such a rough analysis did not tell us the constants of proportionality. For a finite size problem it is possible (though not particularly likely) that a particular $O(n^3)$ algorithm is simpler than a particular $O(n^2)$ algorithm, and it is occasionally necessary for us to determine these constants to find which of several implementations is the simplest. Nevertheless, big oh and big theta analyses, as just described, are generally the best way to describe the computational complexity of an algorithm.

SPACE COMPLEXITY

TIME COMPLEXITY

Sometimes we stress space and time complexities, which are particularly relevant when contemplating parallel implementations. For instance, the sample mean of a category could be calculated with d separate processors, each adding n sample values. Thus we can describe this implementation as $O(d)$ in *space* (i.e., the amount of memory or possibly the number of processors) and $O(n)$ in *time* (i.e., number of sequential steps). Of course for any particular algorithm there may be a number of time-space tradeoffs, for instance using a single processor many times, or using many processors in parallel for a shorter time. Such tradeoffs must be considered carefully and can be important in neural network implementations, as we shall see in Chapter 6.

A common qualitative distinction is made between *polynomially* complex and *exponentially* complex algorithms—$O(a^k)$ for some constant a and aspect or variable k of the problem. Exponential algorithms are generally so complex that for reasonable size cases we avoid them altogether and resign ourselves to approximate solutions that can be found by polynomially complex algorithms.

3.7.3 Overfitting

It frequently happens that the number of available samples is inadequate, and the question of how to proceed arises. One possibility is to reduce the dimensionality, either by redesigning the feature extractor, by selecting an appropriate subset of the existing features, or by combining the existing features in some way (Chapter 10). Another possibility is to assume that all c classes share the same covariance matrix, and to pool the available data. Yet another alternative is to look for a better estimate for $\boldsymbol{\Sigma}$. If any reasonable prior estimate $\boldsymbol{\Sigma}_0$ is available, a Bayesian or pseudo-Bayesian estimate of the form $\lambda\boldsymbol{\Sigma}_0+(1-\lambda)\widehat{\boldsymbol{\Sigma}}$ might be employed. If $\boldsymbol{\Sigma}_0$ is diagonal, this diminishes the troublesome effects of "accidental" correlations. Alternatively, one can remove chance correlations heuristically by thresholding the sample covariance matrix. For example, one might assume that all covariances for which the magnitude of the correlation coefficient is not near unity are actually zero. An extreme of this approach is to assume statistical independence, thereby making all the off-diagonal elements be zero, regardless of empirical evidence to the contrary—an $O(nd)$ calculation. Even though such assumptions are almost surely incorrect, the resulting heuristic estimates sometimes provide better performance than the maximum-likelihood estimate of the full parameter space.

Here we have another apparent paradox. The classifier that results from assuming independence is almost certainly suboptimal. It is understandable that it will perform better if it happens that the features actually are independent, but how can it provide better performance when this assumption is untrue? The answer again involves the problem of insufficient data, and some insight into its nature can be gained from considering an analogous problem in curve fitting. Figure 3.4 shows a set of ten data points and two candidate curves for fitting them. The data points were obtained by adding zero-mean, independent noise to a parabola. Thus, of all the possible polynomials, presumably a parabola would provide the best fit, assuming that we are interested in fitting data obtained in the future as well as the points at hand. Even a straight line could fit the training data fairly well. The parabola provides a better fit, but one might wonder whether the data are adequate to fix the curve. The best parabola for a larger data set might be quite different, and over the interval shown the straight line could easily be superior. The tenth-degree polynomial fits the given data perfectly. However, we do not expect that a tenth-degree polynomial is required here. In general, reliable interpolation or extrapolation cannot be obtained unless the solution is overdetermined, that is, there are more points than function parameters to be set.

In fitting the points in Fig. 3.4, then, we might consider beginning with a high-order polynomial (e.g., 10th order), and then successively smoothing or simplifying our model by eliminating the highest-order terms. While this would in virtually all cases lead to greater error on the "training data," we might expect the generalization to improve.

Analogously, there are a number of heuristic methods that can be applied in the Gaussian classifier case. For instance, suppose we wish to design a classifier for distributions $N(\boldsymbol{\mu}_1, \boldsymbol{\Sigma}_1)$ and $N(\boldsymbol{\mu}_2, \boldsymbol{\Sigma}_2)$ and we have reason to believe that we have insufficient data for accurately estimating the parameters. We might make the simplification that they have the same covariance—that is, $N(\boldsymbol{\mu}_1, \boldsymbol{\Sigma})$ and $N(\boldsymbol{\mu}_2, \boldsymbol{\Sigma})$—and

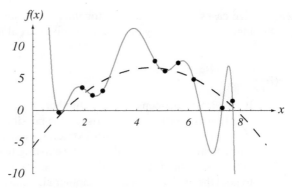

FIGURE 3.4. The "training data" (black dots) were selected from a quadratic function plus Gaussian noise, i.e., $f(x) = ax^2 + bx + c + \epsilon$ where $p(\epsilon) \sim N(0, \sigma^2)$. The 10th-degree polynomial shown fits the data perfectly, but we desire instead the second-order function $f(x)$, because it would lead to better predictions for new samples.

estimate Σ accordingly. Such estimation requires proper normalization of the data (Problem 37).

SHRINKAGE

An intermediate approach is to assume a weighted combination of the equal and individual covariances, a technique known as *shrinkage* (also called regularized discriminant analysis), because the individual covariances "shrink" toward a common one. If i is an index on the c categories in question, we have

$$\Sigma_i(\alpha) = \frac{(1 - \alpha)n_i\Sigma_i + \alpha n\Sigma}{(1 - \alpha)n_i + \alpha n}, \tag{76}$$

for $0 < \alpha < 1$. Additionally, we could "shrink" the estimate of the (assumed) common covariance matrix toward the identity matrix, as

$$\Sigma(\beta) = (1 - \beta)\Sigma + \beta\mathbf{I}, \tag{77}$$

for $0 < \beta < 1$ (Computer exercise 8). (Such methods for simplifying classifiers have counterparts in regression, generally known as *ridge regression*.)

Our short, intuitive discussion here will have to suffice until Chapter 9, where we will explore the crucial issue of controlling the complexity or expressive power of a classifier for optimum performance.

*3.8 COMPONENT ANALYSIS AND DISCRIMINANTS

One approach to coping with the problem of excessive dimensionality is to reduce the dimensionality by combining features. Linear combinations are particularly attractive because they are simple to compute and analytically tractable. In effect, linear methods project the high-dimensional data onto a lower dimensional space. There are two classical approaches to finding effective linear transformations. One approach—known as Principal Component Analysis or PCA—seeks a projection that best represents the data in a least-squares sense. Another approach—known as Multiple Discriminant Analysis or MDA—seeks a projection that best separates the data in a least-squares sense. We consider each of these approaches in turn.

3.8.1 Principal Component Analysis (PCA)

We begin by considering the problem of representing all of the vectors in a set of n d-dimensional samples $\mathbf{x}_1, \ldots, \mathbf{x}_n$ by a single vector $\mathbf{x}_0$. To be more specific, suppose that we want to find a vector $\mathbf{x}_0$ such that the sum of the squared distances between $\mathbf{x}_0$ and the various $\mathbf{x}_k$ is as small as possible. We define the squared-error criterion function $J_0(\mathbf{x}_0)$ by

$$J_0(\mathbf{x}_0) = \sum_{k=1}^{n} ||\mathbf{x}_0 - \mathbf{x}_k||^2, \tag{78}$$

and seek the value of $\mathbf{x}_0$ that minimizes J_0. It is simple to show that the solution to this problem is given by $\mathbf{x}_0 = \mathbf{m}$, where $\mathbf{m}$ is the sample mean,

$$\mathbf{m} = \frac{1}{n} \sum_{k=1}^{n} \mathbf{x}_k. \tag{79}$$

This can be easily verified by writing

$$J_0(\mathbf{x}_0) = \sum_{k=1}^{n} ||(\mathbf{x}_0 - \mathbf{m}) - (\mathbf{x}_k - \mathbf{m})||^2$$

$$= \sum_{k=1}^{n} ||\mathbf{x}_0 - \mathbf{m}||^2 - 2 \sum_{k=1}^{n} (\mathbf{x}_0 - \mathbf{m})^t (\mathbf{x}_k - \mathbf{m}) + \sum_{k=1}^{n} ||\mathbf{x}_k - \mathbf{m}||^2$$

$$= \sum_{k=1}^{n} ||\mathbf{x}_0 - \mathbf{m}||^2 - 2(\mathbf{x}_0 - \mathbf{m})^t \sum_{k=1}^{n} (\mathbf{x}_k - \mathbf{m}) + \sum_{k=1}^{n} ||\mathbf{x}_k - \mathbf{m}||^2$$

$$= \sum_{k=1}^{n} ||\mathbf{x}_0 - \mathbf{m}||^2 + \underbrace{\sum_{k=1}^{n} ||\mathbf{x}_k - \mathbf{m}||^2}_{independent\ of\ \mathbf{x}_0}. \tag{80}$$

Since the second sum is independent of $\mathbf{x}_0$, this expression is obviously minimized by the choice $\mathbf{x}_0 = \mathbf{m}$.

The sample mean is a zero-dimensional representation of the data set. It is simple, but it does not reveal any of the variability in the data. We can obtain a more interesting, one-dimensional representation by projecting the data onto a line running through the sample mean. Let $\mathbf{e}$ be a unit vector in the direction of the line. Then the equation of the line can be written as

$$\mathbf{x} = \mathbf{m} + a\mathbf{e}, \tag{81}$$

where the scalar a (which takes on any real value) corresponds to the distance of any point $\mathbf{x}$ from the mean $\mathbf{m}$. If we represent $\mathbf{x}_k$ by $\mathbf{m} + a_k\mathbf{e}$, we can find an "optimal" set of coefficients a_k by minimizing the squared-error criterion function

$$J_1(a_1, \ldots, a_n, \mathbf{e}) = \sum_{k=1}^{n} ||(\mathbf{m} + a_k\mathbf{e}) - \mathbf{x}_k||^2 = \sum_{k=1}^{n} ||a_k\mathbf{e} - (\mathbf{x}_k - \mathbf{m})||^2$$

$$= \sum_{k=1}^{n} a_k^2 ||\mathbf{e}||^2 - 2 \sum_{k=1}^{n} a_k \mathbf{e}^t (\mathbf{x}_k - \mathbf{m}) + \sum_{k=1}^{n} ||\mathbf{x}_k - \mathbf{m}||^2. \tag{82}$$

Recognizing that $||\mathbf{e}|| = 1$, partially differentiating with respect to a_k, and setting the derivative to zero, we obtain

$$a_k = \mathbf{e}^t(\mathbf{x}_k - \mathbf{m}). \tag{83}$$

Geometrically, this result merely says that we obtain a least-squares solution by projecting the vector $\mathbf{x}_k$ onto the line in the direction of $\mathbf{e}$ that passes through the sample mean.

SCATTER MATRIX

This brings us to the more interesting problem of finding the *best* direction $\mathbf{e}$ for the line. The solution to this problem involves the so-called *scatter matrix* $\mathbf{S}$ defined by

$$\mathbf{S} = \sum_{k=1}^{n}(\mathbf{x}_k - \mathbf{m})(\mathbf{x}_k - \mathbf{m})^t. \tag{84}$$

The scatter matrix should look familiar—it is merely $n - 1$ times the sample covariance matrix. It arises here when we substitute a_k found in Eq. 83 into Eq. 82 to obtain

$$\begin{aligned}
J_1(\mathbf{e}) &= \sum_{k=1}^{n} a_k^2 - 2\sum_{k=1}^{n} a_k^2 + \sum_{k=1}^{n} ||\mathbf{x}_k - \mathbf{m}||^2 \\
&= -\sum_{k=1}^{n}[\mathbf{e}^t(\mathbf{x}_k - \mathbf{m})]^2 + \sum_{k=1}^{n} ||\mathbf{x}_k - \mathbf{m}||^2 \\
&= -\sum_{k=1}^{n}\mathbf{e}^t(\mathbf{x}_k - \mathbf{m})(\mathbf{x}_k - \mathbf{m})^t\mathbf{e} + \sum_{k=1}^{n} ||\mathbf{x}_k - \mathbf{m}||^2 \\
&= -\mathbf{e}^t\mathbf{S}\mathbf{e} + \sum_{k=1}^{n} ||\mathbf{x}_k - \mathbf{m}||^2. \tag{85}
\end{aligned}$$

Clearly, the vector $\mathbf{e}$ that minimizes J_1 also maximizes $\mathbf{e}^t\mathbf{S}\mathbf{e}$. We use the method of Lagrange multipliers (described in Section A.3 of the Appendix) to maximize $\mathbf{e}^t\mathbf{S}\mathbf{e}$ subject to the constraint that $||\mathbf{e}|| = 1$. Letting λ be the undetermined multiplier, we differentiate

$$u = \mathbf{e}^t\mathbf{S}\mathbf{e} - \lambda(\mathbf{e}^t\mathbf{e} - 1) \tag{86}$$

with respect to $\mathbf{e}$ to obtain

$$\frac{\partial u}{\partial \mathbf{e}} = 2\mathbf{S}\mathbf{e} - 2\lambda\mathbf{e}. \tag{87}$$

Setting this gradient vector equal to zero, we see that $\mathbf{e}$ must be an eigenvector of the scatter matrix:

$$\mathbf{S}\mathbf{e} = \lambda\mathbf{e}. \tag{88}$$

In particular, because $\mathbf{e}^t\mathbf{S}\mathbf{e} = \lambda\mathbf{e}^t\mathbf{e} = \lambda$, it follows that to maximize $\mathbf{e}^t\mathbf{S}\mathbf{e}$, we want to select the eigenvector corresponding to the largest eigenvalue of the scatter matrix. In other words, to find the best one-dimensional projection of the data (best in the least-sum-of-squared-error sense), we project the data onto a line through the sample mean in the direction of the eigenvector of the scatter matrix having the largest eigenvalue.

This result can be readily extended from a one-dimensional projection to a d'-dimensional projection. In place of Eq. 81, we write

$$\mathbf{x} = \mathbf{m} + \sum_{i=1}^{d'} a_i \mathbf{e}_i, \tag{89}$$

where $d' \leq d$. It is not difficult to show that the criterion function

$$J_{d'} = \sum_{k=1}^{n} \left\| \left(\mathbf{m} + \sum_{i=1}^{d'} a_{ki} \mathbf{e}_i \right) - \mathbf{x}_k \right\|^2 \tag{90}$$

is minimized when the vectors $\mathbf{e}_1, \ldots, \mathbf{e}_{d'}$ are the d' eigenvectors of the scatter matrix having the largest eigenvalues. Because the scatter matrix is real and symmetric, these eigenvectors are orthogonal. They form a natural set of basis vectors for representing any feature vector $\mathbf{x}$. The coefficients a_i in Eq. 89 are the components of $\mathbf{x}$ in that basis, and are called the *principal components*. Geometrically, if we picture the data points $\mathbf{x}_1, \ldots, \mathbf{x}_n$ as forming a d-dimensional, hyperellipsoidally shaped cloud, then the eigenvectors of the scatter matrix are the principal axes of that hyperellipsoid. Principal component analysis reduces the dimensionality of feature space by restricting attention to those directions along which the scatter of the cloud is greatest.

3.8.2 Fisher Linear Discriminant

Although PCA finds components that are useful for representing data, there is no reason to assume that these components must be useful for discriminating between data in different classes. If we pool all of the samples, the directions that are discarded by PCA might be exactly the directions that are needed for distinguishing between classes. For example, if we had data for the printed uppercase letters O and Q, PCA might discover the gross features that characterize Os and Qs, but might ignore the tail that distinguishes an O from a Q. Where PCA seeks directions that are efficient for representation, *discriminant analysis* seeks directions that are efficient for discrimination.

We begin by considering the problem of projecting data from d dimensions onto a line. Of course, even if the samples formed well-separated, compact clusters in d-space, projection onto an arbitrary line will usually produce a confused mixture of samples from all of the classes and thus produce poor recognition performance. However, by moving the line around, we might be able to find an orientation for which the projected samples are well separated. This is exactly the goal of classical discriminant analysis.

Suppose that we have a set of n d-dimensional samples $\mathbf{x}_1, \ldots, \mathbf{x}_n, n_1$ in the subset $\mathcal{D}_1$ labeled ω_1 and n_2 in the subset $\mathcal{D}_2$ labeled ω_2. If we form a linear combination of the components of $\mathbf{x}$, we obtain the scalar dot product

$$y = \mathbf{w}^t \mathbf{x} \tag{91}$$

and a corresponding set of n samples $y_1, \ldots, y_n$ divided into the subsets $\mathcal{Y}_1$ and $\mathcal{Y}_2$. Geometrically, if $\|\mathbf{w}\| = 1$, each y_i is the projection of the corresponding $\mathbf{x}_i$ onto

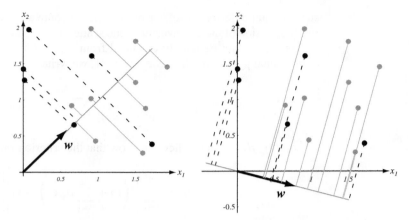

FIGURE 3.5. Projection of the same set of samples onto two different lines in the directions marked **w**. The figure on the right shows greater separation between the red and black projected points.

a line in the direction of **w**. Actually, the magnitude of **w** is of no real significance, because it merely scales y. The direction of **w** is important, however. If we imagine that the samples labeled ω_1 fall more or less into one cluster while those labeled ω_2 fall in another, we want the projections falling onto the line to be well separated, not thoroughly intermingled. Figure 3.5 illustrates the effect of choosing two different values for **w** for a two-dimensional example. It should be abundantly clear that if the original distributions are multimodal and highly overlapping, even the "best" **w** is unlikely to provide adequate separation, and thus this method will be of little use.

We now turn to the matter of finding the best such direction **w**, one we hope will enable accurate classification. A measure of the separation between the projected points is the difference of the sample means. If $\mathbf{m}_i$ is the d-dimensional sample mean given by

$$\mathbf{m}_i = \frac{1}{n_i} \sum_{\mathbf{x} \in \mathcal{D}_i} \mathbf{x}, \tag{92}$$

then the sample mean for the projected points is given by

$$\tilde{m}_i = \frac{1}{n_i} \sum_{y \in \mathcal{Y}_i} y$$
$$= \frac{1}{n_i} \sum_{\mathbf{x} \in \mathcal{D}_i} \mathbf{w}^t \mathbf{x} = \mathbf{w}^t \mathbf{m}_i \tag{93}$$

and is simply the projection of $\mathbf{m}_i$.

It follows that the distance between the projected means is

$$|\tilde{m}_1 - \tilde{m}_2| = |\mathbf{w}^t (\mathbf{m}_1 - \mathbf{m}_2)| \tag{94}$$

and that we can make this difference as large as we wish merely by scaling **w**. Of course, to obtain good separation of the projected data we really want the difference between the means to be large relative to some measure of the standard deviations for

SCATTER

each class. Rather than forming sample variances, we define the *scatter* for projected samples labeled ω_i by

$$\tilde{s}_i^2 = \sum_{y \in \mathcal{Y}_i} (y - \tilde{m}_i)^2. \tag{95}$$

Thus, $(1/n)(\tilde{s}_1^2 + \tilde{s}_2^2)$ is an estimate of the variance of the pooled data, and $\tilde{s}_1^2 + \tilde{s}_2^2$

WITHIN-CLASS
SCATTER

is called the total *within-class scatter* of the projected samples. The *Fisher linear discriminant* employs that linear function $\mathbf{w}^t\mathbf{x}$ for which the criterion function

$$J(\mathbf{w}) = \frac{|\tilde{m}_1 - \tilde{m}_2|^2}{\tilde{s}_1^2 + \tilde{s}_2^2} \tag{96}$$

is maximum (and independent of $\|\mathbf{w}\|$). While the $\mathbf{w}$ maximizing $J(\cdot)$ leads to the best separation between the two projected sets (in the sense just described), we will also need a threshold criterion before we have a true classifier. We first consider how to find the optimal $\mathbf{w}$, and later turn to the issue of thresholds.

SCATTER
MATRICES

To obtain $J(\cdot)$ as an explicit function of $\mathbf{w}$, we define the *scatter matrices* $\mathbf{S}_i$ and $\mathbf{S}_W$ by

$$\mathbf{S}_i = \sum_{\mathbf{x} \in \mathcal{D}_i} (\mathbf{x} - \mathbf{m}_i)(\mathbf{x} - \mathbf{m}_i)^t \tag{97}$$

and

$$\mathbf{S}_W = \mathbf{S}_1 + \mathbf{S}_2. \tag{98}$$

Then we can write

$$\tilde{s}_i^2 = \sum_{\mathbf{x} \in \mathcal{D}_i} (\mathbf{w}^t\mathbf{x} - \mathbf{w}^t\mathbf{m}_i)^2$$

$$= \sum_{\mathbf{x} \in \mathcal{D}_i} \mathbf{w}^t(\mathbf{x} - \mathbf{m}_i)(\mathbf{x} - \mathbf{m}_i)^t\mathbf{w}$$

$$= \mathbf{w}^t\mathbf{S}_i\mathbf{w}; \tag{99}$$

therefore the sum of these scatters can be written

$$\tilde{s}_1^2 + \tilde{s}_2^2 = \mathbf{w}^t\mathbf{S}_W\mathbf{w}. \tag{100}$$

Similarly, the separations of the projected means obeys

$$(\tilde{m}_1 - \tilde{m}_2)^2 = (\mathbf{w}^t\mathbf{m}_1 - \mathbf{w}^t\mathbf{m}_2)^2$$

$$= \mathbf{w}^t(\mathbf{m}_1 - \mathbf{m}_2)(\mathbf{m}_1 - \mathbf{m}_2)^t\mathbf{w}$$

$$= \mathbf{w}^t\mathbf{S}_B\mathbf{w}, \tag{101}$$

where

$$\mathbf{S}_B = (\mathbf{m}_1 - \mathbf{m}_2)(\mathbf{m}_1 - \mathbf{m}_2)^t. \tag{102}$$

WITHIN-CLASS
SCATTER

We call $\mathbf{S}_W$ the *within-class scatter matrix*. It is proportional to the sample covariance matrix for the pooled d-dimensional data. It is symmetric and positive semidefinite, and it is usually nonsingular if $n > d$. Likewise, $\mathbf{S}_B$ is called the *between-class scatter matrix*. It is also symmetric and positive semidefinite, but because it is the outer product of two vectors, its rank is at most one. In particular, for any $\mathbf{w}$, $\mathbf{S}_B\mathbf{w}$ is in the direction of $\mathbf{m}_1 - \mathbf{m}_2$, and $\mathbf{S}_B$ is quite singular.

BETWEEN-CLASS
SCATTER

In terms of $\mathbf{S}_B$ and $\mathbf{S}_W$, the criterion function $J(\cdot)$ can be written as

$$J(\mathbf{w}) = \frac{\mathbf{w}^t \mathbf{S}_B \mathbf{w}}{\mathbf{w}^t \mathbf{S}_W \mathbf{w}}. \tag{103}$$

This expression is well known in mathematical physics as the generalized Rayleigh quotient. It is easy to show that a vector $\mathbf{w}$ that maximizes $J(\cdot)$ must satisfy

$$\mathbf{S}_B\mathbf{w} = \lambda \mathbf{S}_W \mathbf{w}, \tag{104}$$

for some constant λ, which is a generalized eigenvalue problem (Problem 42). This can also be seen informally by noting that at an extremum of $J(\mathbf{w})$ a small change in $\mathbf{w}$ in Eq. 103 should leave unchanged the ratio of the numerator to the denominator. If $\mathbf{S}_W$ is nonsingular we can obtain a conventional eigenvalue problem by writing

$$\mathbf{S}_W^{-1}\mathbf{S}_B\mathbf{w} = \lambda \mathbf{w}. \tag{105}$$

In our particular case, it is unnecessary to solve for the eigenvalues and eigenvectors of $\mathbf{S}_W^{-1}\mathbf{S}_B$ due to the fact that $\mathbf{S}_B\mathbf{w}$ is always in the direction of $\mathbf{m}_1 - \mathbf{m}_2$. Because the scale factor for $\mathbf{w}$ is immaterial, we can immediately write the solution for the $\mathbf{w}$ that optimizes $J(\cdot)$:

$$\mathbf{w} = \mathbf{S}_W^{-1}(\mathbf{m}_1 - \mathbf{m}_2). \tag{106}$$

Thus, we have obtained $\mathbf{w}$ for Fisher's linear discriminant—the linear function yielding the maximum ratio of between-class scatter to within-class scatter. (The solution $\mathbf{w}$ given by Eq. 106 is sometimes called the *canonical variate*.) Thus the classification has been converted from a d-dimensional problem to a hopefully more manageable one-dimensional one. This mapping is many-to-one, and in theory it cannot possibly reduce the minimum achievable error rate if we have a very large training set. In general, one is willing to sacrifice some of the theoretically attainable performance for the advantages of working in one dimension. All that remains is to find the threshold, that is, the point along the one-dimensional subspace separating the projected points.

When the conditional densities $p(\mathbf{x}|\omega_i)$ are multivariate normal with equal covariance matrices $\mathbf{\Sigma}$, we can calculate the threshold directly. In that case we recall from Chapter 2 that the optimal decision boundary has the equation

$$\mathbf{w}^t\mathbf{x} + w_0 = 0 \tag{107}$$

where

$$\mathbf{w} = \mathbf{\Sigma}^{-1}(\boldsymbol{\mu}_1 - \boldsymbol{\mu}_2), \tag{108}$$

and where w_0 is a constant involving $\mathbf{w}$ and the prior probabilities. If we use sample means and the sample covariance matrix to estimate $\boldsymbol{\mu}_i$ and $\mathbf{\Sigma}$, we obtain a vector in the same direction as the $\mathbf{w}$ of Eq. 108 that maximized $J(\cdot)$. Thus, for the normal, equal-covariance case, the optimal decision rule is merely to decide ω_1 if Fisher's linear discriminant exceeds some threshold, and to decide ω_2 otherwise. More generally, if we smooth the projected data, or fit it with a univariate Gaussian, we then should choose w_0 where the posteriors in the one dimensional distributions are equal.

The computational complexity of finding the optimal $\mathbf{w}$ for the Fisher linear discriminant (Eq. 106) is dominated by the calculation of the within-category total scatter and its inverse, an $O(d^2 n)$ calculation.

3.8.3 Multiple Discriminant Analysis

For the c-class problem, the natural generalization of Fisher's linear discriminant involves $c - 1$ discriminant functions. Thus, the projection is from a d-dimensional space to a $(c - 1)$-dimensional space, and it is tacitly assumed that $d \geq c$. The generalization for the within-class scatter matrix is obvious:

$$\mathbf{S}_W = \sum_{i=1}^{c} \mathbf{S}_i \tag{109}$$

where, as before,

$$\mathbf{S}_i = \sum_{\mathbf{x} \in \mathcal{D}_i} (\mathbf{x} - \mathbf{m}_i)(\mathbf{x} - \mathbf{m}_i)^t \tag{110}$$

and

$$\mathbf{m}_i = \frac{1}{n_i} \sum_{\mathbf{x} \in \mathcal{D}_i} \mathbf{x}. \tag{111}$$

TOTAL MEAN VECTOR
TOTAL SCATTER MATRIX

The proper generalization for $\mathbf{S}_B$ is not quite so obvious. Suppose that we define a *total mean vector* $\mathbf{m}$ and a *total scatter matrix* $\mathbf{S}_T$ by

$$\mathbf{m} = \frac{1}{n} \sum_{\mathbf{x}} \mathbf{x} = \frac{1}{n} \sum_{i=1}^{c} n_i \mathbf{m}_i \tag{112}$$

and

$$\mathbf{S}_T = \sum_{\mathbf{x}} (\mathbf{x} - \mathbf{m})(\mathbf{x} - \mathbf{m})^t. \tag{113}$$

Then it follows that

$$\mathbf{S}_T = \sum_{i=1}^{c} \sum_{\mathbf{x} \in \mathcal{D}_i} (\mathbf{x} - \mathbf{m}_i + \mathbf{m}_i - \mathbf{m})(\mathbf{x} - \mathbf{m}_i + \mathbf{m}_i - \mathbf{m})^t$$

$$= \sum_{i=1}^{c} \sum_{\mathbf{x} \in \mathcal{D}_i} (\mathbf{x} - \mathbf{m}_i)(\mathbf{x} - \mathbf{m}_i)^t + \sum_{i=1}^{c} \sum_{\mathbf{x} \in \mathcal{D}_i} (\mathbf{m}_i - \mathbf{m})(\mathbf{m}_i - \mathbf{m})^t$$

$$= \mathbf{S}_W + \sum_{i=1}^{c} n_i (\mathbf{m}_i - \mathbf{m})(\mathbf{m}_i - \mathbf{m})^t. \tag{114}$$

It is natural to define this second term as a general between-class scatter matrix, so that the total scatter is the sum of the within-class scatter and the between-class scatter:

$$\mathbf{S}_B = \sum_{i=1}^{c} n_i (\mathbf{m}_i - \mathbf{m})(\mathbf{m}_i - \mathbf{m})^t \tag{115}$$

and

$$\mathbf{S}_T = \mathbf{S}_W + \mathbf{S}_B. \tag{116}$$

If we check the two-class case, we find that the resulting between-class scatter matrix is $n_1 n_2 / n$ times our previous definition.*

The projection from a d-dimensional space to a $(c - 1)$-dimensional space is accomplished by $c - 1$ discriminant functions

$$y_i = \mathbf{w}_i^t \mathbf{x} \qquad i = 1, \ldots, c - 1. \tag{117}$$

If the y_i are viewed as components of a vector $\mathbf{y}$ and the weight vectors $\mathbf{w}_i$ are viewed as the columns of a d-by-$(c - 1)$ matrix $\mathbf{W}$, then the projection can be written as a single matrix equation

$$\mathbf{y} = \mathbf{W}^t \mathbf{x}. \tag{118}$$

The samples $\mathbf{x}_1, \ldots, \mathbf{x}_n$ project to a corresponding set of samples $\mathbf{y}_1, \ldots, \mathbf{y}_n$, which can be described by their own mean vectors and scatter matrices. Thus, if we define

$$\tilde{\mathbf{m}}_i = \frac{1}{n_i} \sum_{\mathbf{y} \in \mathcal{Y}_i} \mathbf{y} \tag{119}$$

$$\tilde{\mathbf{m}} = \frac{1}{n} \sum_{i=1}^{c} n_i \tilde{\mathbf{m}}_i \tag{120}$$

$$\tilde{\mathbf{S}}_W = \sum_{i=1}^{c} \sum_{\mathbf{y} \in \mathcal{Y}_i} (\mathbf{y} - \tilde{\mathbf{m}}_i)(\mathbf{y} - \tilde{\mathbf{m}}_i)^t \tag{121}$$

and

$$\tilde{\mathbf{S}}_B = \sum_{i=1}^{c} n_i (\tilde{\mathbf{m}}_i - \tilde{\mathbf{m}})(\tilde{\mathbf{m}}_i - \tilde{\mathbf{m}})^t, \tag{122}$$

it is a straightforward matter to show that

$$\tilde{\mathbf{S}}_W = \mathbf{W}^t \mathbf{S}_W \mathbf{W} \tag{123}$$

and

$$\tilde{\mathbf{S}}_B = \mathbf{W}^t \mathbf{S}_B \mathbf{W}. \tag{124}$$

_____

*We could redefine $\mathbf{S}_B$ for the two-class case to obtain complete consistency, but there should be no misunderstanding of our usage.

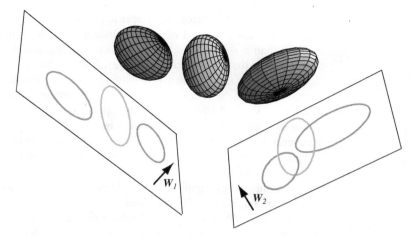

FIGURE 3.6. Three three-dimensional distributions are projected onto two-dimensional subspaces, described by a normal vectors $\mathbf{W}_1$ and $\mathbf{W}_2$. Informally, multiple discriminant methods seek the optimum such subspace, that is, the one with the greatest separation of the projected distributions for a given total within-scatter matrix, here as associated with $\mathbf{W}_1$.

These equations show how the within-class and between-class scatter matrices are transformed by the projection to the lower dimensional space (Fig. 3.6). What we seek is a transformation matrix $\mathbf{W}$ that in some sense maximizes the ratio of the between-class scatter to the within-class scatter. A simple scalar measure of scatter is the determinant of the scatter matrix. The determinant is the product of the eigenvalues, and hence is the product of the "variances" in the principal directions, thereby measuring the square of the hyperellipsoidal scattering volume (see also Eq. 46 in Chapter 2). Using this measure, we obtain the criterion function

$$J(\mathbf{W}) = \frac{|\tilde{\mathbf{S}}_B|}{|\tilde{\mathbf{S}}_W|} = \frac{|\mathbf{W}^t \mathbf{S}_B \mathbf{W}|}{|\mathbf{W}^t \mathbf{S}_W \mathbf{W}|}. \tag{125}$$

The problem of finding a rectangular matrix $\mathbf{W}$ that maximizes $J(\cdot)$ is tricky, though fortunately it turns out that the solution is relatively simple. The columns of an optimal $\mathbf{W}$ are the generalized eigenvectors that correspond to the largest eigenvalues in

$$\mathbf{S}_B \mathbf{w}_i = \lambda_i \mathbf{S}_W \mathbf{w}_i. \tag{126}$$

A few observations about this solution are in order. First, if $\mathbf{S}_W$ is nonsingular, this can be converted to a conventional eigenvalue problem as before. However, this is actually undesirable, since it requires an unnecessary computation of the inverse of $\mathbf{S}_W$. Instead, one can find the eigenvalues as the roots of the characteristic polynomial

$$|\mathbf{S}_B - \lambda_i \mathbf{S}_W| = 0 \tag{127}$$

and then solve

$$(\mathbf{S}_B - \lambda_i \mathbf{S}_W)\mathbf{w}_i = 0 \tag{128}$$

directly for the eigenvectors $\mathbf{w}_i$. Because $\mathbf{S}_B$ is the sum of c matrices of rank one or less, and because only $c-1$ of these are independent, $\mathbf{S}_B$ is of rank $c-1$ or less. Thus, no more than $c-1$ of the eigenvalues are nonzero, and the desired weight vectors correspond to these nonzero eigenvalues. If the within-class scatter is isotropic, the eigenvectors are merely the eigenvectors of $\mathbf{S}_B$, and the eigenvectors with nonzero eigenvalues span the space spanned by the vectors $\mathbf{m}_i - \mathbf{m}$. In this special case the columns of $\mathbf{W}$ can be found simply by applying the Gram-Schmidt orthonormalization procedure to the $c-1$ vectors $\mathbf{m}_i - \mathbf{m}$, $i = 1, \ldots, c-1$. Finally, we observe that in general the solution for $\mathbf{W}$ is not unique; the allowable transformations include rotating and scaling the axes in various ways. These are all linear transformations from a $(c-1)$-dimensional space to a $(c-1)$-dimensional space, however, and do not change things in any significant way; in particular, they leave the criterion function $J(\mathbf{W})$ invariant and the classifier unchanged.

If we have very little data, we would tend to project to a subspace of low dimension, while if there are more data, we can use a higher dimension, as we shall explore in Chapter 9. Once we have projected the distributions onto the optimal subspace (defined as above), we can use the methods of Chapter 2 to create our full classifier.

As in the two-class case, multiple discriminant analysis primarily provides a reasonable way of reducing the dimensionality of the problem. Parametric or nonparametric techniques that might not have been feasible in the original space may work well in the lower-dimensional space. In particular, it may be possible to estimate separate covariance matrices for each class and use the general multivariate normal assumption after the transformation where this could not be done with the original data. In general, if the transformation causes some unnecessary overlapping of the data and increases the theoretically achievable error rate, then the problem of classifying the data still remains. However, there are other ways to reduce the dimensionality of data, and we shall encounter this subject again in Chapter 10. We note that there are also alternative methods of discriminant analysis—such as the selection of features based on statistical significance—some of which are given in the references for this chapter. Of these, Fisher's method remains a fundamental and widely used technique.

⋆3.9 EXPECTATION-MAXIMIZATION (EM)

We saw in Chapter 2, Section 2.10 how we could classify a test point even when it has missing features. We can now extend our application of maximum-likelihood techniques to permit the *learning* of parameters governing a distribution from training points, some of which have missing features. If we had uncorrupted data, we could use maximum-likelihood, i.e., find $\hat{\boldsymbol{\theta}}$ that maximized the log-likelihood $l(\boldsymbol{\theta})$. The basic idea in the expectation-maximization or EM algorithm is to iteratively estimate the likelihood given the data that is present. The method has precursors in the Baum-Welch algorithm we will consider in Section 3.10.6.

Consider a full sample $\mathcal{D} = \{\mathbf{x}_1, \ldots, \mathbf{x}_n\}$ of points taken from a single distribution. Suppose, though, that here some features are missing; thus any sample point can be written as $\mathbf{x}_k = \{\mathbf{x}_{kg}, \mathbf{x}_{kb}\}$, i.e., comprising the "good" features and the missing, or "bad" ones (Chapter 2, Section 2.10). For notational convenience we separate these individual *features* (not samples) into two sets, $\mathcal{D}_g$ and $\mathcal{D}_b$ with $\mathcal{D} = \mathcal{D}_g \cup \mathcal{D}_b$ being the union of such features.

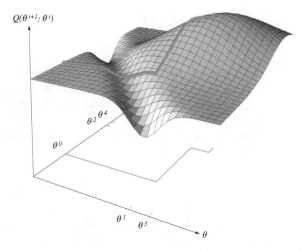

$Q(\theta^{i+1}; \theta^i)$

$\theta^2\,\theta^4$

θ^0

θ^1 θ^3

θ

FIGURE 3.7. The search for the best model via the EM algorithm starts with some initial value of the model parameters, θ^0. Then, via the **M step**, the optimal θ^1 is found. Next, θ^1 is held constant and the value θ^2 is found that optimizes $Q(\cdot;\ \cdot)$. This process iterates until no value of θ can be found that will increase $Q(\cdot;\ \cdot)$. Note in particular that this is different from a gradient search. For example here θ^1 is the global optimum (given fixed θ^0), and would not necessarily have been found via gradient search. (In this illustration, $Q(\cdot;\ \cdot)$ is shown symmetric in its arguments; this need not be the case in general, however.)

Next we form the function

$$Q(\boldsymbol{\theta};\ \boldsymbol{\theta}^i) = \mathcal{E}_{\mathcal{D}_b}[\ln p(\mathcal{D}_g, \mathcal{D}_b;\ \boldsymbol{\theta})|\mathcal{D}_g;\ \boldsymbol{\theta}^i], \tag{129}$$

where the use of the semicolon denotes, for instance on the left-hand side, that $Q(\boldsymbol{\theta};\ \boldsymbol{\theta}^i)$ is a function of $\boldsymbol{\theta}$ with $\boldsymbol{\theta}^i$ assumed fixed; on the right-hand side it denotes that the expected value is over the missing features assuming $\boldsymbol{\theta}^i$ are the true parameters describing the (full) distribution. The simplest way to interpret this, the central equation in expectation maximization, is the following. The parameter vector $\boldsymbol{\theta}^i$ is the current (best) estimate for the full distribution; $\boldsymbol{\theta}$ is a candidate vector for an improved estimate. Given such a candidate $\boldsymbol{\theta}$, the right-hand side of Eq. 129 calculates the likelihood of the data, including the unknown feature $\mathcal{D}_b$ *marginalized* with respect to the current best distribution, which is described by $\boldsymbol{\theta}^i$. Different candidate $\boldsymbol{\theta}$s will of course lead to different such likelihoods. Our algorithm will select the best such candidate $\boldsymbol{\theta}$ and call it $\boldsymbol{\theta}^{i+1}$—the one corresponding to the greatest $Q(\boldsymbol{\theta};\ \boldsymbol{\theta}^i)$.

If we continue to let i be an iteration counter and now let T be a preset convergence criterion, our algorithm (illustrated in Fig. 3.7) is as follows:

■ **Algorithm 1. (Expectation-Maximization)**

1 **begin initialize** $\boldsymbol{\theta}^0, T, i \leftarrow 0$
2   **do** $i \leftarrow i + 1$
3   **E step**: compute $Q(\boldsymbol{\theta};\ \boldsymbol{\theta}^i)$
4   **M step**: $\boldsymbol{\theta}^{i+1} \leftarrow \underset{\boldsymbol{\theta}}{\arg\max}\, Q(\boldsymbol{\theta};\ \boldsymbol{\theta}^i)$
5   **until** $Q(\boldsymbol{\theta}^{i+1};\ \boldsymbol{\theta}^i) - Q(\boldsymbol{\theta}^i;\ \boldsymbol{\theta}^{i-1}) \leq T$

6   **return** $\hat{\boldsymbol{\theta}} \leftarrow \boldsymbol{\theta}^{i+1}$
7   **end**

This so-called expectation-maximization (EM) algorithm is most useful when the optimization of $Q(\cdot;\ \cdot)$ is simpler than that of $l(\cdot)$. Most importantly, the algorithm guarantees that the log-likelihood of the good data (with the bad data marginalized) will increase monotonically, as explored in Problem 44. This is not the same as finding the particular value of the bad data that gives the maximum-likelihood of the full (completed) data, as can be seen in Example 2.

EXAMPLE 2   **Expectation-Maximization for a 2D Normal Model**

Suppose our data consists of four points in two dimensions, one point of which is missing a feature: $\mathcal{D} = \{\mathbf{x}_1, \mathbf{x}_2, \mathbf{x}_3, \mathbf{x}_4\} = \{\binom{0}{2}, \binom{1}{0}, \binom{2}{2}, \binom{*}{4}\}$, where $*$ represents the unknown value of the first feature of point $\mathbf{x}_4$. Thus our bad data $\mathcal{D}_b$ consists of the single feature x_{41}, and the good data $\mathcal{D}_g$ consists of all the rest. We assume our model is a Gaussian with diagonal covariance and arbitrary mean, and thus can be described by the parameter vector

$$\boldsymbol{\theta} = \begin{pmatrix} \mu_1 \\ \mu_2 \\ \sigma_1^2 \\ \sigma_2^2 \end{pmatrix}.$$

We take our initial guess to be a Gaussian centered on the origin having $\boldsymbol{\Sigma} = \mathbf{I}$, that is,

$$\boldsymbol{\theta}^0 = \begin{pmatrix} 0 \\ 0 \\ 1 \\ 1 \end{pmatrix}.$$

In finding our first improved estimate, $\boldsymbol{\theta}^1$, we must calculate $Q(\boldsymbol{\theta};\ \boldsymbol{\theta}^0)$ or, by Eq. 129,

$$Q(\boldsymbol{\theta};\ \boldsymbol{\theta}^0) = \mathcal{E}_{x_{41}}[\ln p(\mathbf{x}_g, \mathbf{x}_b;\ \boldsymbol{\theta})|\mathcal{D}_g;\ \boldsymbol{\theta}^0]$$

$$= \int_{-\infty}^{\infty} \left[\sum_{k=1}^{3} \ln p(\mathbf{x}_k|\boldsymbol{\theta}) + \ln p(\mathbf{x}_4|\boldsymbol{\theta}) \right] p(x_{41}|\boldsymbol{\theta}^0;\ x_{42} = 4)\ dx_{41}$$

$$= \sum_{k=1}^{3} [\ln p(\mathbf{x}_k|\boldsymbol{\theta})] + \int_{-\infty}^{\infty} \ln p\left(\binom{x_{41}}{4}\middle| \boldsymbol{\theta} \right) \underbrace{\frac{p\left(\binom{x_{41}}{4}|\boldsymbol{\theta}^0 \right)}{\left(\int_{-\infty}^{\infty} p\left(\binom{x'_{41}}{4}\middle| \boldsymbol{\theta}^0 \right)\ dx'_{41} \right)}}_{\equiv K}\ dx_{41},$$

where x_{41} is the unknown first feature of point $\mathbf{x}_4$, and K is a constant that can be brought out of the integral. We focus on the integral, substitute the equation for a general Gaussian, and find

$$Q(\boldsymbol{\theta};\ \boldsymbol{\theta}^0) = \sum_{k=1}^{3} [\ln p(\mathbf{x}_k|\boldsymbol{\theta})] + \frac{1}{K} \int_{-\infty}^{\infty} \ln p\left(\binom{x_{41}}{4}\middle| \boldsymbol{\theta} \right) \frac{1}{2\pi |\binom{1\ 0}{0\ 1}|} \exp\left[-\frac{1}{2}(x_{41}^2 + 4^2) \right]\ dx_{41}$$

$$= \sum_{k=1}^{3} [\ln p(\mathbf{x}_k|\boldsymbol{\theta})] - \frac{1 + \mu_1^2}{2\sigma_1^2} - \frac{(4 - \mu_2)^2}{2\sigma_2^2} - \ln(2\pi\sigma_1\sigma_2).$$

This completes the expectation or **E step**. Through a straightforward calculation, we find the values of $\boldsymbol{\theta}$ (i.e., μ_1, μ_2, σ_1, and σ_2) that maximize $Q(\cdot; \cdot)$, to get the next estimate:

$$\boldsymbol{\theta}^1 = \begin{pmatrix} 0.75 \\ 2.0 \\ 0.938 \\ 2.0 \end{pmatrix}.$$

This new mean and the $1/e$ ellipse of the new covariance matrix are shown in the figure. Subsequent iterations are conceptually the same, but require a bit more extensive calculation. The mean will remain at $\mu_2 = 2$. After three iterations the algorithm converges at the solution $\boldsymbol{\mu} = \begin{pmatrix} 1.0 \\ 2.0 \end{pmatrix}$ and $\boldsymbol{\Sigma} = \begin{pmatrix} 0.667 & 0 \\ 0 & 2.0 \end{pmatrix}$.

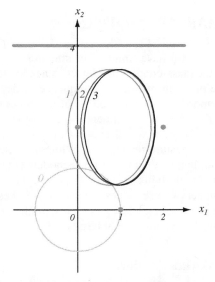

The four data points, one of which is missing the value of x_1 component, are shown in red. The initial estimate is a circularly symmetric Gaussian, centered on the origin (gray). (A better initial estimate could have been derived from the three known points.) Each iteration leads to an improved estimate, labeled by the iteration number i; here, after three iterations the algorithm has converged.

We must be careful and note that the EM algorithm leads to the greatest log-likelihood of the *good* data, with the bad data marginalized. There may be particular values of the bad data that give a different solution and an even greater log-likelihood. For instance, in this Example if the missing feature had value $x_{41} = 1$, so that $\mathbf{x}_4 = \begin{pmatrix} 1 \\ 4 \end{pmatrix}$, we would have a solution

$$\boldsymbol{\theta} = \begin{pmatrix} 1.0 \\ 2.0 \\ 0.5 \\ 2.0 \end{pmatrix}$$

and a log-likelihood for the *full* data (good plus bad) that is greater than for the good alone. Such an optimization, however, is not the goal of the canonical EM algorithm. Note too that if no data are missing, the calculation of $Q(\boldsymbol{\theta}; \boldsymbol{\theta}^i)$ is simple because no integrals are involved.

GENERALIZED EXPECTATION-MAXIMIZATION

Generalized expectation-maximization or GEM algorithms are a bit more lax than the EM algorithm as they require merely that an *improved* $\boldsymbol{\theta}^{i+1}$ be set in the **M step** (line 4) of the algorithm—not necessarily the *optimal*. Naturally, convergence will not be as rapid as for a proper EM algorithm, but GEM algorithms afford greater freedom to choose computationally simpler steps. One version of GEM finds the maximum-likelihood value of unknown features at each iteration step, then recalculates $\boldsymbol{\theta}$ in light of these new values—if indeed they lead to a greater likelihood.

In practice, the term expectation-maximization has come to mean loosely any iterative scheme in which the likelihood of *some* data increases with each step, even if such methods are not, technically speaking, the true EM algorithm as presented here.

3.10 HIDDEN MARKOV MODELS

So far, we have limited our attention to the problem of estimating the parameters in the class-conditional densities needed to make a single decision. Now we turn to the problem of making a sequence of decisions. In problems that have an inherent temporality—that is, consist of a process that unfolds in time—we may have states at time t that are influenced directly by a state at $t - 1$. Hidden Markov models (HMMs) have found greatest use in such problems—for instance, speech recognition or gesture recognition. While the notation and description is unavoidably more complicated than the simpler models considered up to this point, we stress that the same underlying ideas are exploited. Hidden Markov models have a number of parameters whose values are set so as to best explain training patterns for the known category. Later, a test pattern is classified by the model that has the highest posterior probability, that is, that best "explains" the test pattern.

3.10.1 First-Order Markov Models

We consider a sequence of states at successive times; the state at any time t is denoted $\omega(t)$. A particular sequence of length T is denoted by $\boldsymbol{\omega}^T = \{\omega(1), \omega(2), \ldots, \omega(T)\}$, as for instance we might have $\boldsymbol{\omega}^6 = \{\omega_1, \omega_4, \omega_2, \omega_2, \omega_1, \omega_4\}$. Note that the system can revisit a state at different steps, and not every state need be visited.

TRANSITION PROBABILITY

Our model for the production of any sequence is described by *transition probabilities* $P(\omega_j(t + 1)|\omega_i(t)) = a_{ij}$—the time-independent probability of having state ω_j at step $t + 1$ given that the state at time t was ω_i. There is no requirement that the transition probabilities be symmetric ($a_{ij} \neq a_{ji}$, in general) and a particular state may be visited in succession ($a_{ii} \neq 0$, in general), as illustrated in Fig. 3.8.

Suppose we are given a particular model $\boldsymbol{\theta}$—that is, the full set of a_{ij}—as well as a particular sequence $\boldsymbol{\omega}^T$. In order to calculate the probability that the model generated the particular sequence we simply multiply the successive probabilities. For instance, to find the probability that a particular model generated the sequence described above, we would have $P(\boldsymbol{\omega}^T|\boldsymbol{\theta}) = a_{14}a_{42}a_{22}a_{21}a_{14}$. If there is a prior probability on the first state $P(\omega(1) = \omega_i)$, we could include such a factor as well; for simplicity, we will ignore that detail for now.

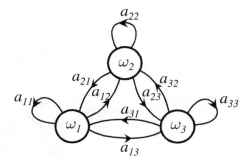

FIGURE 3.8. The discrete states, ω_i, in a basic Markov model are represented by nodes, and the transition probabilities, a_{ij}, are represented by links. In a first-order discrete-time Markov model, at any step t the full system is in a particular state $\omega(t)$. The state at step $t + 1$ is a random function that depends solely on the state at step t and the transition probabilities.

Up to here we have been discussing a Markov model, or technically speaking, a *first-order* discrete time Markov model, because the probability at $t + 1$ depends only on the states at t. For instance, in a Markov model for the production of spoken words, we might have states representing phonemes, and a Markov model for the production of a spoken work might have states representing phonemes. Such a Markov model for the word "cat" would have states for /k/, /a/ and /t/, with transitions from /k/ to /a/; transitions from /a/ to /t/; and transitions from /t/ to a final silent state.

Note, however, that in speech recognition the perceiver does not have access to the states $\omega(t)$. Instead, we measure some properties of the emitted sound. Thus we will have to augment our Markov model to allow for *visible states*—which are directly accessible to external measurement—as separate from the ω states, which are not.

3.10.2 First-Order Hidden Markov Models

We continue to assume that at every time step t the system is in a state $\omega(t)$ but now we also assume that it emits some (visible) symbol $v(t)$. While sophisticated Markov models allow for the emission of continuous functions (e.g., spectra), we will restrict ourselves to the case where a discrete symbol is emitted. As with the states, we define a particular sequence of such visible states as $\mathbf{V}^T = \{v(1), v(2), \ldots, v(T)\}$ and thus we might have $\mathbf{V}^6 = \{v_5, v_1, v_1, v_5, v_2, v_3\}$.

Our model is then that in any state $\omega(t)$ we have a probability of emitting a particular visible state $v_k(t)$. We denote this probability $P(v_k(t)|\omega_j(t)) = b_{jk}$. Because we have access only to the visible states, while the ω_j are unobservable, such a full model is called a *hidden Markov model* (Fig. 3.9).

3.10.3 Hidden Markov Model Computation

Now we define some new terms and clarify our notation. In general, networks such as those in Fig. 3.9 are called finite-state machines, and when they have associated transition probabilities, they are called Markov networks. They are strictly *causal*: The probabilities depend only upon *previous* states. A Markov model is called *ergodic* if every one of the states has a nonzero probability of occurring given some starting state. A *final* or *absorbing state* ω_0 is one which, if entered, is never left (i.e., $a_{00} = 1$).

ABSORBING
STATE

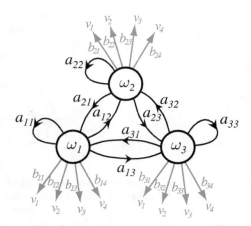

FIGURE 3.9. Three hidden units in an HMM and the transitions between them are shown in black while the visible states and the emission probabilities of visible states are shown in red. This model shows all transitions as being possible; in other HMMs, some such candidate transitions are not allowed.

As mentioned, we denote the transition probabilities a_{ij} among hidden states and b_{jk} for the probability of the emission of a visible state:

$$a_{ij} = P(\omega_j(t+1)|\omega_i(t))$$
$$b_{jk} = P(v_k(t)|\omega_j(t)). \tag{130}$$

We demand that some transition occur from step $t \rightarrow t+1$ (even if it is to the same state) and that some visible symbol be emitted after every step. Thus we have the normalization conditions:

$$\sum_j a_{ij} = 1 \quad \text{for all } i \tag{131}$$

and

$$\sum_k b_{jk} = 1 \quad \text{for all } j,$$

where the limits on the summations are over all hidden states and all visible symbols, respectively.

With these preliminaries behind us, we can now focus on the three central issues in hidden Markov models:

The Evaluation Problem. Suppose we have an HMM, complete with transition probabilities a_{ij} and b_{jk}. Determine the probability that a particular sequence of visible states $\mathbf{V}^T$ was generated by that model.

The Decoding Problem. Suppose we have an HMM as well as a set of observations $\mathbf{V}^T$. Determine the most likely sequence of *hidden* states $\boldsymbol{\omega}^T$ that led to those observations.

The Learning Problem. Suppose we are given the coarse structure of a model (the number of states and the number of visible states) but *not* the probabilities a_{ij} and b_{jk}. Given a set of training observations of visible symbols, determine these parameters.

We consider each of these problems in turn.

3.10.4 Evaluation

The probability that the model produces a sequence $\mathbf{V}^T$ of visible states is

$$P(\mathbf{V}^T) = \sum_{r=1}^{r_{max}} P(\mathbf{V}^T|\boldsymbol{\omega}_r^T)P(\boldsymbol{\omega}_r^T), \tag{132}$$

where each r indexes a particular sequence $\boldsymbol{\omega}_r^T = \{\omega(1), \omega(2), \ldots, \omega(T)\}$ of T hidden states. In the general case of c hidden states, there will be $r_{max} = c^T$ possible terms in the sum of Eq. 132, corresponding to all possible sequences of length T. Thus, according to Eq. 132, in order to compute the probability that the model generated the particular sequence of T visible states $\mathbf{V}^T$, we should take each conceivable sequence of hidden states, calculate the probability they produce $\mathbf{V}^T$, and then add up these probabilities. The probability of a particular visible sequence is merely the product of the corresponding (hidden) transition probabilities a_{ij} and the (visible) output probabilities b_{jk} of each step.

Because we are dealing here with a first-order Markov process, the second factor in Eq. 132, which describes the transition probability for the hidden states, can be rewritten as:

$$P(\boldsymbol{\omega}_r^T) = \prod_{t=1}^{T} P(\omega(t)|\omega(t-1)), \tag{133}$$

that is, a product of the a_{ij}'s according to the hidden sequence in question. In Eq. 133, $\omega(T) = \omega_0$ is some final absorbing state, which uniquely emits the visible state v_0. In speech recognition applications, ω_0 typically represents a null state or lack of utterance, and v_0 is some symbol representing silence. Because of our assumption that the output probabilities depend only upon the hidden state, we can write the first factor in Eq. 132 as

$$P(\mathbf{V}^T|\boldsymbol{\omega}_r^T) = \prod_{t=1}^{T} P(v(t)|\omega(t)), \tag{134}$$

that is, a product of b_{jk}'s according to the hidden state and the corresponding visible state. We can now use Eqs. 133 and 134 to express Eq. 132 as

$$P(\mathbf{V}^T) = \sum_{r=1}^{r_{max}} \prod_{t=1}^{T} P(v(t)|\omega(t))P(\omega(t)|\omega(t-1)). \tag{135}$$

Despite its formal complexity, Eq. 135 has a straightforward interpretation. The probability that we observe the particular sequence of T visible states $\mathbf{V}^T$ is equal to the sum over all r_{max} possible sequences of hidden states of the conditional probability that the system has made a particular transition multiplied by the probability that it then emitted the visible symbol in our target sequence. All these are captured in our parameters a_{ij} and b_{jk}, and thus Eq. 135 can be evaluated directly. Alas, this is an $O(c^T T)$ calculation, which is quite prohibitive in practice. For instance, if $c = 10$ and $T = 20$, we must perform on the order of 10^{21} calculations.

A computationally simpler algorithm for the same goal is as follows. We can calculate $P(\mathbf{V}^T)$ recursively, because each term $P(v(t)|\omega(t))P(\omega(t)|\omega(t-1))$ involves only $v(t)$, $\omega(t)$ and $\omega(t-1)$. We do this by defining

$$\alpha_j(t) = \begin{cases} 0 & t = 0 \text{ and } j \neq \text{initial state} \\ 1 & t = 0 \text{ and } j = \text{initial state} \\ \left[\sum_i \alpha_i(t-1)a_{ij}\right] b_{jk}v(t) & \text{otherwise,} \end{cases} \qquad (136)$$

where the notation $b_{jk}v(t)$ means the transition probability b_{jk} selected by the visible state emitted at time t. Thus the only nonzero contribution to the sum is for the index k which matches the visible state $v(t)$. Consequently, $\alpha_j(t)$ represents the probability that our HMM is in hidden state ω_j at step t having generated the first t elements of $\mathbf{V}^T$. This calculation is implemented in the *Forward algorithm* in the following way:

■ **Algorithm 2. (HMM Forward)**

1 **initialize** $t \leftarrow 0, a_{ij}, b_{jk}$, visible sequence $\mathbf{V}^T, \alpha_j(0)$
2   **for** $t \leftarrow t + 1$
3          $\alpha_j(t) \leftarrow b_{jk}v(t) \sum_{i=1}^c \alpha_i(t-1)a_{ij}$
4   **until** $t = T$
5 **return** $P(\mathbf{V}^T) \leftarrow \alpha_0(T)$ for the final state
6 **end**

---

where in line 5, α_0 denotes the probability of the associated sequence ending to the known final state. The *Forward algorithm* has, thus, a computational complexity of $O(c^2 T)$—*far* more efficient than the complexity associated with exhaustive enumeration of paths of Eq. 135 (Fig. 3.10). For the illustration of $c = 10$, $T = 20$ above, we would need only on the order of 2000 calculations—more than 17 orders of magnitude faster than that to examine each path individually.

We shall have cause to use the *Backward algorithm*, which is the time-reversed version of the *Forward algorithm*.

■ **Algorithm 3. (HMM Backward)**

1 **initialize** $\beta_j(T), t \leftarrow T, a_{ij}, b_{jk}$, visible sequence $\mathbf{V}^T$
2   **for** $t \leftarrow t - 1$;
3          $\beta_i(t) \leftarrow \sum_{j=1}^c \beta_j(t+1)a_{ij}b_{jk}v(t+1)$
4   **until** $t = 1$
5 **return** $P(\mathbf{V}^T) \leftarrow \beta_i(0)$ for the known initial state
6 **end**

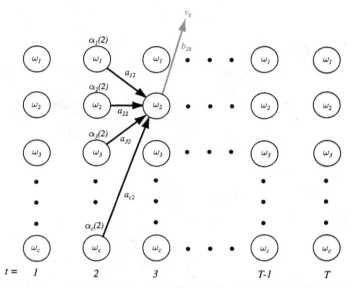

FIGURE 3.10. The computation of probabilities by the Forward algorithm can be visualized by means of a trellis—a sort of "unfolding" of the HMM through time. Suppose we seek the probability that the HMM was in state ω_2 at $t = 3$ and generated the observed visible symbol up through that step (including the observed visible symbol v_k). The probability the HMM was in state $\omega_i(t = 2)$ and generated the observed sequence through $t = 2$ is $\alpha_i(t)$ for $i = 1, 2, \ldots, c$. To find $\alpha_2(3)$ we must sum these and multiply the probability that state ω_2 emitted the observed symbol v_k. Formally, for this particular illustration we have $\alpha_2(3) = b_{2k} \sum_{i=1}^{c} \alpha_i(2) a_{i2}$.

EXAMPLE 3   Hidden Markov Model

To clarify the evaluation problem, consider an HMM such as shown in Fig. 3.9, but with an explicit absorber state and unique null visible symbol v_0 with the following transition probabilities (where the matrix indexes begin at 0):

$$a_{ij} = \begin{pmatrix} 1 & 0 & 0 & 0 \\ 0.2 & 0.3 & 0.1 & 0.4 \\ 0.2 & 0.5 & 0.2 & 0.1 \\ 0.8 & 0.1 & 0.0 & 0.1 \end{pmatrix}$$

and

$$b_{jk} = \begin{pmatrix} 1 & 0 & 0 & 0 & 0 \\ 0 & 0.3 & 0.4 & 0.1 & 0.2 \\ 0 & 0.1 & 0.1 & 0.7 & 0.1 \\ 0 & 0.5 & 0.2 & 0.1 & 0.2 \end{pmatrix}$$

What is the probability it generates the particular sequence $\mathbf{V}^4 = \{v_1, v_3, v_2, v_0\}$? Suppose we know the initial hidden state at $t = 0$ to be ω_1. The visible symbol at each step is shown at the top of the figure, and the $\alpha_i(t)$ is shown in each unit. The circles show the value for $\alpha_i(t)$ as we progress left to right. The product $a_{ij}b_{jk}$ is shown along each transition link for the step $t = 1$ to $t = 2$. The final probability, $P(\mathbf{V}^T|\boldsymbol{\theta})$, is hence 0.0011.

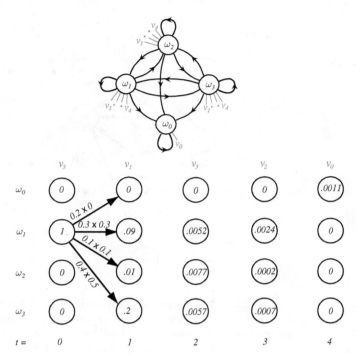

The HMM (above) consists of four hidden states (one of which is an absorber state, ω_0), each emitting one of five visible states; only the allowable transitions to visible states are shown. The trellis for this HMM is shown below. The numbers shown in the nodes are $\alpha_j(t)$—the probability the model generated the observed visible sequence up to t. For instance, we know that the system was in hidden state ω_1 at $t = 0$, and thus $\alpha_1(0) = 1$ and $\alpha_j(0) = 0$ for $j \neq 1$. The arrows show the calculation of $\alpha_j(1)$. For instance, because visible state v_1 was emitted at $t = 1$, we have $\alpha_0(1) = \alpha_1(0)a_{10}b_{01} = 1[0.2 \times 0] = 0$, as shown by the top arrow. Likewise the next highest arrow corresponds to the calculation $\alpha_1(1) = \alpha_1(0)a_{11}b_{11} = 1[0.3 \times 0.3] = 0.09$. In this example, the calculation of $\alpha_j(1)$ is particularly simple, because only transitions from the known initial hidden state need be considered; all other transitions have zero contribution to $\alpha_j(1)$. For subsequent times, however, the calculation requires a *sum* over all hidden states at the previous time, as given by line 3 in the Forward algorithm. The probability shown in the final (absorbing) state gives the probability of the full sequence observed, $P(\mathbf{V}^T|\boldsymbol{\theta}) = 0.0011$.

If we denote our model—the a's and b's—by $\boldsymbol{\theta}$, we have by Bayes formula that the probability of the model given the observed sequence is

$$P(\boldsymbol{\theta}|\mathbf{V}^T) = \frac{P(\mathbf{V}^T|\boldsymbol{\theta})P(\boldsymbol{\theta})}{P(\mathbf{V}^T)}. \tag{137}$$

In HMM pattern recognition we would have a number of HMMs, one for each category, and classify a test sequence according to the model with the highest probability. Thus in HMM speech recognition we could have a model for "cat" and another one for "dog" and for a test utterance determine which model has the highest probability. In practice, nearly all HMMs for speech are *left-to-right* models (Fig. 3.11).

LEFT-TO-RIGHT
MODEL

The Forward algorithm gives us $P(\mathbf{V}^T|\boldsymbol{\theta})$. The prior probability of the model, $P(\boldsymbol{\theta})$, is given by some external source, such as a *language model* in the case of speech. This prior probability might depend upon the semantic context, or the previ-

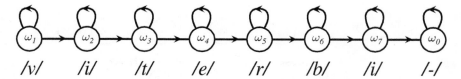

ω_1 /v/ ω_2 /i/ ω_3 /t/ ω_4 /e/ ω_5 /r/ ω_6 /b/ ω_7 /i/ ω_0 /-/

FIGURE 3.11. A left-to-right HMM commonly used in speech recognition. For instance, such a model could describe the utterance "viterbi," where ω_1 represents the phoneme /v/, ω_2 represents /i/,..., and ω_0 a final silent state. Such a left-to-right model is more restrictive than the general HMM in Fig. 3.9 because it precludes transitions "back" in time.

ous words, or yet other information. In the absence of such information, it is traditional to assume a uniform prior density on $\boldsymbol{\theta}$, and hence ignore it in any classification problem. (This is an example of a "noninformative" prior.)

3.10.5 Decoding

Given a sequence of visible states $\mathbf{V}^T$, the decoding problem is to find the most probable sequence of hidden states. While we might consider enumerating every possible path and calculating the probability of the visible sequence observed, this is an $O(c^T T)$ calculation and prohibitive. Instead, we use perhaps the simplest decoding algorithm:

■ **Algorithm 4. (HMM decoding)**

1 **begin initialize** Path $\leftarrow \{\}, t \leftarrow 0$
2   **for** $t \leftarrow t + 1$
3          $j \leftarrow j + 1$
4   **for** $j \leftarrow j + 1$
5              $\alpha_j(t) \leftarrow b_{jk}v(t) \sum_{i=1}^{c} \alpha_i(t-1)a_{ij}$
6   **until** $j = c$
7          $j' \leftarrow \arg\max_j \alpha_j(t)$
8   *Append* $\omega_{j'}$ *to* Path
9   **until** $t = T$
10   **return** Path
11 **end**

A closely related algorithm uses logarithms of the probabilities and calculates total probabilities by addition of such logarithms; this method has complexity $O(c^2 T)$ (Problem 52).

The black line in Fig. 3.12 corresponds to Path, and it connects the hidden states with the highest value of α_j at each step t. There is a difficulty, however. Note that there is no guarantee that the path is in fact a *valid* one—it might not be consistent with the underlying models. For instance, it is possible that the path actually implies a transition that is forbidden by the model, as illustrated in Example 4.

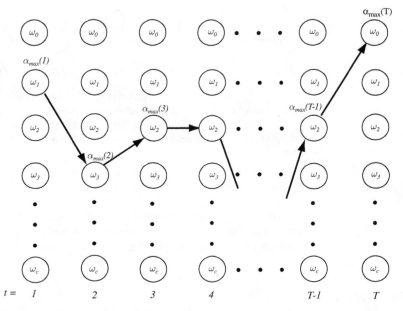

FIGURE 3.12. The decoding algorithm finds at each time step t the state that has the highest probability of having come from the previous step and generated the observed visible state v_k. The full path is the sequence of such states. Because this is a local optimization (dependent only upon the single previous time step, not the full sequence), the algorithm does not guarantee that the path is indeed allowable. For instance, it might be possible that the maximum at $t = 5$ is ω_1 and at $t = 6$ is ω_2, and thus these would appear in the path. This can even occur if $a_{12} = P(\omega_2(t+1)|\omega_1(t)) = 0$, precluding that transition.

EXAMPLE 4   **HMM Decoding**

We find the path for the data of Example 3 for the sequence $\{\omega_1, \omega_3, \omega_2, \omega_1, \omega_0\}$. Note especially that the transition from ω_3 to ω_2 is not allowed according to the transition probabilities a_{ij} given in Example 3. The path *locally* optimizes the probability through the trellis.

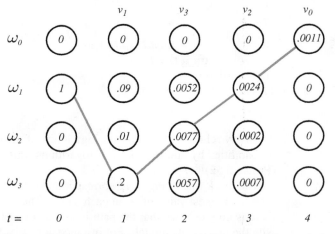

The locally optimal path through the HMM trellis of Example 3.

HMMs address the problem of rate invariance in the following two ways. The first is that the transition probabilities themselves incorporate probabilistic structure of the durations. Moreover, using postprocessing, we can delete repeated states and just get the sequence somewhat independent of variations in rate. Thus in postprocessing we can convert the sequence $\{\omega_1, \omega_1, \omega_3, \omega_2, \omega_2, \omega_2\}$ to $\{\omega_1, \omega_3, \omega_2\}$, which would be appropriate for speech recognition, where the fundamental phonetic units are not repeated in natural speech.

3.10.6 Learning

The goal in HMM learning is to determine model parameters—the transition probabilities a_{ij} and b_{jk}—from an ensemble of training samples. There is no known method for obtaining the optimal or most likely set of parameters from the data, but we can nearly always determine a good solution by a straightforward technique.

The Forward-Backward Algorithm

The forward-backward algorithm is an instance of a generalized expectation-maximization algorithm. The general approach will be to iteratively update the weights in order to better explain the observed training sequences.

Above, we defined $\alpha_i(t)$ as the probability that the model is in state $\omega_i(t)$ and has generated the target sequence up to step t. We can analogously define $\beta_i(t)$ to be the probability that the model is in state $\omega_i(t)$ and *will generate* the remainder of the given target sequence, that is, from $t + 1 \rightarrow T$. We express $\beta_i(t)$ as

$$
\beta_i(t) = \begin{cases} 0 & \omega_i(t) \neq \omega_0 \text{ and } t = T \\ 1 & \omega_i(t) = \omega_0 \text{ and } t = T \\ \sum_j \beta_j(t+1)a_{ij}b_{jk}v(t+1) & \text{otherwise.} \end{cases} \tag{138}
$$

To understand Eq. 138, imagine we knew $\alpha_i(t)$ up to step $T - 1$, and we wanted to calculate the probability that the model would generate the remaining single visible symbol. This probability, $\beta_i(T)$, is just the probability we make a transition to state $\omega_i(T)$ multiplied by the probability that this hidden state emitted the correct final visible symbol. By the definition of $\beta_i(T)$ in Eq. 138, this will be either 0 (if $\omega_i(T)$ is not the final hidden state) or 1 (if it is). Thus it is clear that $\beta_i(T - 1) = \sum_j a_{ij}b_{jk}v(T)\beta_j(T)$. Now that we have determined $\beta_i(T - 1)$, we can repeat the process, to determine $\beta_i(T - 2)$, and so on, *backward* through the trellis of Fig. 3.12.

But the $\alpha_i(t)$ and $\beta_i(t)$ we determined are merely *estimates* of their true values, because we don't know the actual value of the transition probabilities a_{ij} and b_{jk} in Eq. 138. We can calculate an improved value by first defining $\gamma_{ij}(t)$—the probability of transition between $\omega_i(t - 1)$ and $\omega_j(t)$, given the model generated the entire training sequence $\mathbf{V}^T$ by *any* path. We do this by defining $\gamma_{ij}(t)$, as follows:

$$
\gamma_{ij}(t) = \frac{\alpha_i(t-1)a_{ij}b_{jk}\beta_j(t)}{P(\mathbf{V}^T|\boldsymbol{\theta})}, \tag{139}
$$

where $P(\mathbf{V}^T|\boldsymbol{\theta})$ is the probability that the model generated sequence $\mathbf{V}^T$ by any path. Thus $\gamma_{ij}(t)$ is the probability of a transition from state $\omega_i(t - 1)$ to $\omega_j(t)$ given that the model generated the complete visible sequence $\mathbf{V}^T$.

We can now calculate an improved estimate for a_{ij}. The expected number of transitions between state $\omega_i(t-1)$ and $\omega_j(t)$ at *any* time in the sequence is simply $\sum_{t=1}^{T} \gamma_{ij}(t)$, whereas the total expected number of any transitions from ω_i is $\sum_{t=1}^{T} \sum_k \gamma_{ik}(t)$. Thus $\hat{a}_{ij}$ (the estimate of the probability of a transition from $\omega_i(t-1)$ to $\omega_j(t)$) can be found by taking the ratio between the expected number of transitions from ω_i to ω_j and the total expected number of *any* transitions from ω_i. That is:

$$\hat{a}_{ij} = \frac{\sum_{t=1}^{T} \gamma_{ij}(t)}{\sum_{t=1}^{T} \sum_k \gamma_{ik}(t)}. \tag{140}$$

In the same way, we can obtain an improved estimate $\hat{b}_{jk}$ by calculating the ratio between the frequency that any particular symbol v_k is emitted and that for any symbol. Thus we have

$$\hat{b}_{jk} = \frac{\sum_{\substack{t=1 \\ v(t)=v_k}}^{T} \sum_l \gamma_{jl}(t)}{\sum_{t=1}^{T} \sum_l \gamma_{jl}(t)}. \tag{141}$$

In short, then, we start with rough or arbitrary estimates of a_{ij} and b_{jk}, calculate improved estimates by Eqs. 140 and 141, and repeat until some convergence criterion is met (e.g., sufficiently small change in the estimated values of the parameters on subsequent iterations). This is the *Baum-Welch* or *forward-backward algorithm*—an example of a generalized expectation-maximization algorithm (Section 3.9):

■ **Algorithm 5. (Forward-Backward)**

1 **begin initialize** a_{ij}, b_{jk}, training sequence $\mathbf{V}^T$, convergence criterion θ, $z \leftarrow 0$
2   **do** $z \leftarrow z+1$
3   compute $\hat{a}(z)$ from $a(z-1)$ and $b(z-1)$ by Eq. 140
4   compute $\hat{b}(z)$ from $a(z-1)$ and $b(z-1)$ by Eq. 141
5         $a_{ij}(z) \leftarrow \hat{a}_{ij}(z)$
6         $b_{jk}(z) \leftarrow \hat{b}_{jk}(z)$
7   **until** $\max_{i,j,k}[a_{ij}(z) - a_{ij}(z-1), b_{jk}(z) - b_{jk}(z-1)] < \theta;$
                                                     (convergence achieved)
8   **return** $a_{ij} \leftarrow a_{ij}(z)$; $b_{jk} \leftarrow b_{jk}(z)$
9 **end**

The stopping or convergence criterion in line 7 halts learning when no estimated transition probability changes more than a predetermined amount, θ. In typical speech recognition applications, convergence requires several presentations of each training sequence (fewer than five is common). Other popular stopping criteria are based on overall probability that the learned model could have generated the full training data.

SUMMARY

If we know a parametric form of the class-conditional probability densities, we can reduce our learning task from one of finding the distribution itself to that of finding the *parameters* (represented by a vector $\boldsymbol{\theta}_i$ for each category ω_i) and use the resulting distributions for classification. The maximum-likelihood method seeks to find the parameter value that is best supported by the training data—that is, maximizes the probability of obtaining the samples actually observed. (In practice, for computational simplicity one typically uses log-likelihood.) In Bayesian estimation the parameters are considered random variables having a known prior density; the training data convert this to an *a posteriori* density. The recursive Bayes method updates the Bayesian parameter estimate incrementally, that is, as each training point is sampled. While Bayesian estimation is, in principle, to be preferred, maximum-likelihood methods are generally easier to implement and in the limit of large training sets give classifiers nearly as accurate.

A sufficient statistic **s** for $\boldsymbol{\theta}$ is a function of the samples that contains all information needed to determine $\boldsymbol{\theta}$. Once we know the sufficient statistic for models of a given form (e.g., exponential family), we need only estimate their value from data to create our classifier—no other functions of the data are relevant.

The Fisher linear discriminant finds a good subspace in which categories are best separated; other, general classification techniques can then be applied in the subspace. Fisher's method can be extended to cases with multiple categories projected onto subspaces of higher dimension than a line.

Expectation maximization is an iterative scheme to maximize model parameters, even when some data are missing. Each iteration employs two steps: the expectation or **E step**, which requires marginalizing over the missing variables given the current model, and the maximization or **M step**, in which the optimum parameters of a new model are chosen. Generalized expectation-maximization algorithms demand merely that parameters be *improved*—not optimized—on each iteration and these have been applied to the training of a large range of models.

Hidden Markov models consist of nodes representing hidden states, interconnected by links describing the conditional probabilities of a transition between the states. Each hidden state also has an associated set of probabilities of emitting a particular visible states. HMMs can be useful in modeling sequences, particularly context dependent ones, such as phonemes in speech. All the transition probabilities can be learned (estimated) iteratively from sample sequences by means of the *forward-backward* or *Baum-Welch* algorithm, an example of a generalized EM algorithm. Classification proceeds by finding the single model among candidates that is most likely to have produced a given observed sequence.

BIBLIOGRAPHICAL AND HISTORICAL REMARKS

Maximum-likelihood and Bayes estimation have a long history. The Bayesian approach to learning in pattern recognition began by the suggestion that the proper way to use samples when the conditional densities are unknown is the calculation of $P(\omega_i|\mathbf{x}, \mathcal{D})$, as described in reference [7]. Bayes himself appreciated the role of noninformative priors. An analysis of different priors from statistics appears in references [17], [24], and [5] has an extensive list of references. The Gibbs algorithm was described in reference [28] and analyzed in reference [16].

Principal component analysis is a classical statistical technique [19]; it is very useful in a wide range of engineering applications. Fisher's early work on linear discriminants [13] is well described in reference [26] and a number of standard textbooks such as references [9, 12, 15, 26] and [35].

The expectation-maximization algorithm is due to Dempster et al. [11], and a thorough overview and history appears in reference [27]. On-line or incremental versions of EM are described in references [20] and [36]. The definitive compendium of work on missing data, including much beyond our discussion here, is [31].

Markov developed what later became called the Markov framework [25] in order to analyze the the text of his fellow Russian Pushkin's masterpiece **Eugene Onegin**. Hidden Markov models were introduced by Baum and collaborators [3, 4] and have had their greatest applications in speech recognition [29, 30] and, to a lesser extent, statistical language learning [8, 18], as well as in sequence identification (such as in DNA sequences) [2, 23]. Hidden Markov methods have been extended to two-dimensions and applied to recognizing characters in optical document images [22]. The decoding algorithm is related to pioneering work of Viterbi and followers [14, 37]. The relationship between hidden Markov models and graphical models such as Bayesian belief nets is explored in reference [34].

Knuth's classic [21] was the earliest compendium of the central results on computational complexity, the majority due to himself. The standard books [10], which inspired several homework problems below, are a bit more accessible for those without deep backgrounds in computer science. Finally, several other pattern recognition textbooks—such as references [6, 32] and [35], which take a somewhat different approach to the field—can be recommended.

PROBLEMS

Section 3.2

1. Let x have an exponential density

$$p(x|\theta) = \begin{cases} \theta e^{-\theta x} & x \geq 0 \\ 0 & \text{otherwise.} \end{cases}$$

(a) Plot $p(x|\theta)$ versus x for $\theta = 1$. Plot $p(x|\theta)$ versus θ, $(0 \leq \theta \leq 5)$, for $x = 2$.

(b) Suppose that n samples $x_1, \ldots, x_n$ are drawn independently according to $p(x|\theta)$. Show that the maximum-likelihood estimate for θ is given by

$$\hat{\theta} = \frac{1}{\frac{1}{n}\sum_{k=1}^{n} x_k}.$$

(c) On your graph generated with $\theta = 1$ in part (a), mark the maximum-likelihood estimate $\hat{\theta}$ for large n.

2. Let x have a uniform density

$$p(x|\theta) \sim U(0, \theta) = \begin{cases} 1/\theta & 0 \leq x \leq \theta \\ 0 & \text{otherwise.} \end{cases}$$

(a) Suppose that n samples $\mathcal{D} = \{x_1, \ldots, x_n\}$ are drawn independently according to $p(x|\theta)$. Show that the maximum-likelihood estimate for θ is $\max[\mathcal{D}]$—that is, the value of the maximum element in $\mathcal{D}$.

(b) Suppose that $n = 5$ points are drawn from the distribution and the maximum value of which happens to be $\max_k x_k = 0.6$. Plot the likelihood $p(\mathcal{D}|\theta)$ in the range $0 \le \theta \le 1$. Explain in words why you do not need to know the values of the other four points.

3. Maximum-likelihood methods apply to estimates of prior probabilities as well. Let samples be drawn by successive, independent selections of a state of nature ω_i with unknown probability $P(\omega_i)$. Let $z_{ik} = 1$ if the state of nature for the kth sample is ω_i and $z_{ik} = 0$ otherwise.

(a) Show that

$$P(z_{i1}, \ldots, z_{in}|P(\omega_i)) = \prod_{k=1}^{n} P(\omega_i)^{z_{ik}}(1 - P(\omega_i))^{1-z_{ik}}.$$

(b) Show that the maximum-likelihood estimate for $P(\omega_i)$ is

$$\hat{P}(\omega_i) = \frac{1}{n}\sum_{k=1}^{n} z_{ik}.$$

Interpret your result in words.

4. Let $\mathbf{x}$ be a d-dimensional binary (0 or 1) vector with a multivariate Bernoulli distribution

$$P(\mathbf{x}|\boldsymbol{\theta}) = \prod_{i=1}^{d} \theta_i^{x_i}(1 - \theta_i)^{1-x_i},$$

where $\boldsymbol{\theta} = (\theta_1, \ldots, \theta_d)^t$ is an unknown parameter vector, θ_i being the probability that $x_i = 1$. Show that the maximum-likelihood estimate for $\boldsymbol{\theta}$ is

$$\hat{\boldsymbol{\theta}} = \frac{1}{n}\sum_{k=1}^{n} \mathbf{x}_k.$$

5. Let each component x_i of $\mathbf{x}$ be binary valued (0 or 1) in a two-category problem with $P(\omega_1) = P(\omega_2) = 0.5$. Suppose that the probability of obtaining a 1 in any component is

$$p_{i1} = p$$
$$p_{i2} = 1 - p,$$

and we assume for definiteness $p > 1/2$. The probability of error is known to approach zero as the dimensionality d approaches infinity. This problem asks you to explore the behavior as we increase the number of *features* in a *single* sample—a complementary situation.

(a) Suppose that a single sample $\mathbf{x} = (x_1, \ldots, x_d)^t$ is drawn from category ω_1. Show that the maximum-likelihood estimate for p is given by

$$\hat{p} = \frac{1}{d} \sum_{i=1}^{d} x_i.$$

(b) Describe the behavior of $\hat{p}$ as d approaches infinity. Indicate why such behavior means that by letting the number of features increase without limit we can obtain an error-free classifier even though we have only one sample from each class.

(c) Let $T = 1/d \sum_{j=1}^{d} x_j$ represent the proportion of 1's in a single sample. Plot $P(T|\omega_i)$ vs. T for the case $P = 0.6$, for small d and for large d (e.g., $d = 11$ and $d = 111$, respectively). Explain your answer in words.

6. Derive Eqs. 18 and 19 for the maximum-likelihood estimation of the mean and covariance of a multidimensional Gaussian. State clearly any assumptions you need to invoke.

7. Show that if our model is poor, the maximum-likelihood classifier we derive is not the best—even among our (poor) model set—by exploring the following example. Suppose we have two equally probable categories (i.e., $P(\omega_1) = P(\omega_2) = 0.5$). Furthermore, we know that $p(x|\omega_1) \sim N(0, 1)$ but *assume* that $p(x|\omega_2) \sim N(\mu, 1)$. (That is, the parameter θ we seek by maximum-likelihood techniques is the mean of the second distribution.) Imagine, however, that the *true* underlying distribution is $p(x|\omega_2) \sim N(1, 10^6)$.

(a) What is the value of our maximum-likelihood estimate $\hat{\mu}$ in our poor model, given a large amount of data?

(b) What is the decision boundary arising from this maximum-likelihood estimate in the poor model?

(c) Ignore for the moment the maximum-likelihood approach, and use the methods from Chapter 2 to derive the Bayes optimal decision boundary given the *true* underlying distributions: $p(x|\omega_1) \sim N(0, 1)$ and $p(x|\omega_2) \sim N(1, 10^6)$. Be careful to include all portions of the decision boundary.

(d) Now consider again classifiers based on the (poor) model assumption of $p(x|\omega_2) \sim N(\mu, 1)$. Using your result immediately above, find a *new* value of μ that will give lower error than the maximum-likelihood classifier.

(e) Discuss these results, with particular attention to the role of knowledge of the underlying model.

8. Consider an extreme case of the general issue discussed in Problem 7, one in which it is possible that the maximum-likelihood solution leads to the *worst* possible classifier, that is, one with an error that approaches 100% (in probability). Suppose our data in fact comes from two one-dimensional distributions of the forms

$$p(x|\omega_1) \sim [(1 - k)\delta(x - 1) + k\delta(x + X)]$$

and

$$p(x|\omega_2) \sim [(1 - k)\delta(x + 1) + k\delta(x - X)],$$

where X is positive, $0 \leq k < 0.5$ represents the portion of the total probability mass concentrated at the point $\pm X$, and $\delta(\cdot)$ is the Dirac delta function. Suppose our poor models are of the form $p(x|\omega_1, \mu_1) \sim N(\mu_1, \sigma_1^2)$ and $p(x|\omega_2, \mu_2) \sim N(\mu_2, \sigma_2^2)$ and we form a maximum-likelihood classifier.

(a) Consider the symmetries in the problem and show that in the infinite data case the decision boundary will always be at $x = 0$, regardless of k and X.

(b) Recall that the maximum-likelihood estimate of either mean, $\hat{\mu}_i$, is the mean of its distribution. For a fixed k, find the value of X such that the maximum-likelihood estimates of the means "switch," that is, where $\hat{\mu}_1 \geq \hat{\mu}_2$.

(c) Plot the true distributions and the Gaussian estimates for the particular case $k = .2$ and $X = 5$. What is the classification error in this case?

(d) Find a dependence $X(k)$ which will guarantee that the estimated mean $\hat{\mu}_1$ of $p(x|\omega_1)$ is less than zero. (By symmetry, this will also ensure $\hat{\mu}_2 > 0$.)

(e) Given your $X(k)$ just derived, state the classification error in terms of k.

(f) Suppose we constrained our model space such that $\sigma_1^2 = \sigma_2^2 = 1$ (or indeed any other constant). Would that change the above results?

(g) Discuss how if our model is wrong (here, does not include the delta functions), the error can approaches 100% (in probability). Does this surprising answer arise because we have found some local minimum in parameter space?

9. Prove the invariance property of maximum-likelihood estimators—that is, that if $\hat{\theta}$ is the maximum-likelihood estimate of θ, then for any differentiable function $\tau(\cdot)$, the maximum-likelihood estimate of $\tau(\theta)$ is $\tau(\hat{\theta})$.

10. Suppose we employ a novel method for estimating the mean of a data set $\mathcal{D} = \{\mathbf{x}_1, \mathbf{x}_2, \dots, \mathbf{x}_n\}$: We assign the mean to be the value of the first point in the set, that is, $\mathbf{x}_1$.

(a) Show that this method is unbiased.

(b) State why this method is nevertheless highly undesirable.

11. One measure of the difference between two distributions in the same space is the *Kullback-Leibler divergence* of Kullback-Leibler "distance":

$$D_{KL}(p_1(\mathbf{x}), p_2(\mathbf{x})) = \int p_1(\mathbf{x}) \ln \frac{p_1(\mathbf{x})}{p_2(\mathbf{x})} \, d\mathbf{x}.$$

(This "distance" does not obey the requisite symmetry and triangle inequalities for a metric.) Suppose we seek to approximate an arbitrary distribution $p_1(\mathbf{x})$ by a normal $p_2(\mathbf{x}) \sim N(\mu, \Sigma)$. Show that the values that lead to the smallest Kullback-Leibler divergence are the obvious ones:

$$\mu = \mathcal{E}_1[\mathbf{x}]$$
$$\Sigma = \mathcal{E}_1[(\mathbf{x} - \mu)(\mathbf{x} - \mu)^t],$$

where the expectation taken is over the density $p_1(\mathbf{x})$.

Section 3.3

12. Justify all the statements in the text leading from Eq. 24 to Eq. 25.

Section 3.4

13. Let $p(\mathbf{x}|\mathbf{\Sigma}) \sim N(\mathbf{\mu}, \mathbf{\Sigma})$ where $\mathbf{\mu}$ is known and $\mathbf{\Sigma}$ is unknown. Show that the maximum-likelihood estimate for $\mathbf{\Sigma}$ is given by

$$\widehat{\mathbf{\Sigma}} = \frac{1}{n} \sum_{k=1}^{n} (\mathbf{x}_k - \mathbf{\mu})(\mathbf{x}_k - \mathbf{\mu})^t$$

by carrying out the following argument:

(a) Prove the matrix identity $\mathbf{a}^t \mathbf{A} \mathbf{a} = \text{tr}[\mathbf{A} \mathbf{a} \mathbf{a}^t]$, where the trace, $\text{tr}[\mathbf{A}]$, is the sum of the diagonal elements of $\mathbf{A}$.

(b) Show that the likelihood function can be written in the form

$$p(\mathbf{x}_1, \ldots, \mathbf{x}_n|\mathbf{\Sigma}) = \frac{1}{(2\pi)^{nd/2}} |\mathbf{\Sigma}^{-1}|^{n/2} \exp\left[-\frac{1}{2}\text{tr}\left[\mathbf{\Sigma}^{-1}\sum_{k=1}^{n}(\mathbf{x}_k - \mathbf{\mu})(\mathbf{x}_k - \mathbf{\mu})^t\right]\right].$$

(c) Let $\mathbf{A} = \mathbf{\Sigma}^{-1}\widehat{\mathbf{\Sigma}}$ and $\lambda_1, \ldots, \lambda_n$ be the eigenvalues of $\mathbf{A}$; show that your result above leads to

$$p(\mathbf{x}_1, \ldots, \mathbf{x}_n|\mathbf{\Sigma}) = \frac{1}{(2\pi)^{nd/2}|\widehat{\mathbf{\Sigma}}|^{n/2}} (\lambda_1 \cdots \lambda_d)^{n/2} \exp\left[-\frac{n}{2}(\lambda_1 + \cdots + \lambda_d)\right].$$

(d) Complete the proof by showing that the likelihood is maximized by the choice $\lambda_1 = \cdots = \lambda_d = 1$. Explain your reasoning.

14. Suppose that $p(\mathbf{x}|\mathbf{\mu}_i, \mathbf{\Sigma}, \omega_i) \sim N(\mathbf{\mu}_i, \mathbf{\Sigma})$, where $\mathbf{\Sigma}$ is a common covariance matrix for all c classes. Let n samples $\mathbf{x}_1, \ldots, \mathbf{x}_n$ be drawn as usual, and let $l_1, \ldots, l_n$ be their labels, so that $l_k = i$ if the state of nature for $\mathbf{x}_k$ was ω_i.

(a) Show that

$$p(\mathbf{x}_1, \ldots, \mathbf{x}_n, l_1, \ldots, l_n|\mathbf{\mu}_1, \ldots, \mathbf{\mu}_c, \mathbf{\Sigma})$$
$$= \frac{\prod_{k=1}^{n} P(\omega_{l_k})}{(2\pi)^{nd/2}|\mathbf{\Sigma}|^{n/2}} \exp\left[-\frac{1}{2}\sum_{k=1}^{n}(\mathbf{x}_k - \mathbf{\mu}_{l_k})^t \mathbf{\Sigma}^{-1}(\mathbf{x}_k - \mathbf{\mu}_{l_k})\right].$$

(b) Using the results for samples drawn from a single normal population, show that the maximum-likelihood estimates for $\mathbf{\mu}_i$ and $\mathbf{\Sigma}$ are given by

$$\hat{\mathbf{\mu}} = \frac{\sum_{l_k=i} \mathbf{x}_k}{\sum_{l_k=1} 1}$$

and

$$\widehat{\mathbf{\Sigma}} = \frac{1}{n} \sum_{k=1}^{n} (\mathbf{x}_k - \hat{\mathbf{\mu}}_{l_k})(\mathbf{x}_k - \hat{\mathbf{\mu}}_{l_k})^t.$$

Interpret your answer in words.

15. Consider the problem of learning the mean of a univariate normal distribution. Let $n_0 = \sigma^2 / \sigma_0^2$ be the dogmatism, and imagine that μ_0 is formed by averaging n_0 fictitious samples x_k, $k = -n_0 + 1, -n_0 + 2, \ldots, 0$.

    (a) Show that Eqs. 31 and 32 for μ_n and σ_n^2 yield

    $$\mu_n = \frac{1}{n + n_0} \sum_{k=-n_0+1}^{n} x_k$$

    and

    $$\sigma_n^2 = \frac{\sigma^2}{n + n_0}.$$

    (b) Use this result to give an interpretation of the prior density $p(\mu) \sim N(\mu_0, \sigma_0^2)$.

16. Suppose that $\mathbf{A}$ and $\mathbf{B}$ are nonsingular matrices of the same order.

    (a) Prove the matrix identity

    $$(\mathbf{A}^{-1} + \mathbf{B}^{-1})^{-1} = \mathbf{A}(\mathbf{A} + \mathbf{B})^{-1}\mathbf{B} = \mathbf{B}(\mathbf{A} + \mathbf{B})^{-1}\mathbf{A}.$$

    (b) Must these matrices be square for this identity to hold?

    (c) Use this result in showing that Eqs. 45 and 46 do indeed follow from Eqs. 41 and 42.

Section 3.5

17. The purpose of this problem is to derive the Bayesian classifier for the d-dimensional multivariate Bernoulli case. As usual, work with each class separately, interpreting $P(\mathbf{x}|\mathcal{D})$ to mean $P(\mathbf{x}|\mathcal{D}_i, \omega_i)$. Let the conditional probability for a given category be given by

$$P(\mathbf{x}|\boldsymbol{\theta}) = \prod_{i=1}^{d} \theta_i^{x_i} (1 - \theta_i)^{1 - x_i},$$

and let $\mathcal{D} = \{\mathbf{x}_1, \ldots, \mathbf{x}_n\}$ be a set of n samples independently drawn according to this probability density.

    (a) If $\mathbf{s} = (s_1, \ldots, s_d)^t$ is the sum of the n samples, show that

    $$P(\mathcal{D}|\boldsymbol{\theta}) = \prod_{i=1}^{d} \theta_i^{s_i} (1 - \theta_i)^{n - s_i}.$$

    (b) Assuming a uniform prior distribution for $\boldsymbol{\theta}$ and using the identity

    $$\int_0^1 \theta^m (1 - \theta)^n \, d\theta = \frac{m! n!}{(m + n + 1)!},$$

    show that

$$p(\boldsymbol{\theta}|\mathcal{D}) = \prod_{i=1}^{d} \frac{(n+1)!}{s_i!(n-s_i)!} \theta_i^{s_i} (1-\theta_i)^{n-s_i}.$$

(c) Plot this density for the case $d = 1, n = 1$ and for the two resulting possibilities for s_1.

(d) Integrate the product $P(\mathbf{x}|\boldsymbol{\theta})p(\boldsymbol{\theta}|\mathcal{D})$ over $\boldsymbol{\theta}$ to obtain the desired conditional probability

$$P(\mathbf{x}|\mathcal{D}) = \prod_{i=1}^{d} \left(\frac{s_i+1}{n+2}\right)^{x_i} \left(1 - \frac{s_i+1}{n+2}\right)^{1-x_i}.$$

(e) If we think of obtaining $P(\mathbf{x}|\mathcal{D})$ by substituting an estimate $\hat{\boldsymbol{\theta}}$ for $\boldsymbol{\theta}$ in $P(\mathbf{x}|\boldsymbol{\theta})$, what is the effective Bayesian estimate for $\boldsymbol{\theta}$?

18. Consider how knowledge of an invariance can guide our creation of priors in the following case. Suppose we have a binary (0 or 1) variable x, chosen independently with a probability $p(\theta) = p(x = 1)$. Imagine we have observed $\mathcal{D}^n = \{x_1, x_2, \ldots, x_n\}$ and now wish to evaluate the probability that $x_{n+1} = 1$, which we express as a ratio:

$$\frac{P(x_{n+1} = 1|\mathcal{D}^n)}{P(x_{n+1} = 0|\mathcal{D}^n)}.$$

(a) Define $s = x_1 + \cdots + x_n$ and $p(t) = P(x_1 + \cdots + x_{n+1} = t)$. Assume now invariance of exchangeability, i.e., that the samples in any set $\mathcal{D}^n$ could have been selected in an arbitrary order and it would not affect any probabilities. Show how this assumption of exchangeability implies the ratio in question can be written

$$\frac{p(s+1)/\binom{n+1}{s+1}}{p(s)/\binom{n+1}{s}},$$

where $\binom{n+1}{s} = \frac{(n+1)!}{s!(n+1-s)!}$ is the binomial coefficient.

(b) Evaluate this ratio given the assumption $p(s) \simeq p(s+1)$, when n and $n - s$ and s are not too small. Interpret your answer in words.

(c) In the binomial framework, we now seek a prior $p(\theta)$ such that $p(s)$ does not depend upon s, where

$$p(s) = \int_0^1 \binom{n}{s} \theta^s (1-\theta)^{n-s} p(\theta) \, d\theta.$$

Show that this requirement is satisfied if $p(\theta)$ is uniform, i.e., $p(\theta) \sim U(0, 1)$.

19. Assume we have training data from a Gaussian distribution of known covariance $\boldsymbol{\Sigma}$ but unknown mean $\boldsymbol{\mu}$. Suppose further that this mean itself is random, and characterized by a Gaussian density having mean $\mathbf{m}_0$ and covariance $\boldsymbol{\Sigma}_0$.

(a) What is the MAP estimator for $\boldsymbol{\mu}$?

    (b) Suppose we transform our coordinates by a linear transform $\mathbf{x}' = \mathbf{Ax}$, for non-singular matrix $\mathbf{A}$, and accordingly for other terms. Determine whether your MAP estimator gives the appropriate estimate for the transformed mean $\boldsymbol{\mu}'$. Explain.

20. Consider the problem of noninformative priors.

    (a) Fill in the details leading to Eq. 55.

    (b) Suppose a density is defined on the unit circle $0 \leq 2\pi$ and has an orientation parameter θ_0 and a spread σ_θ. What are noninformative priors on these parameters?

21. State the conditions on $p(\mathbf{x}|\boldsymbol{\theta})$, on $p(\boldsymbol{\theta})$, and on $\mathcal{D}^n$ that ensure that the estimate $p(\boldsymbol{\theta}|\mathcal{D}^n)$ in Eq. 53 converges in the limit $n \to \infty$.

22. Demonstrate that the classification error arising from the Gibbs algorithm is bounded by twice the expected error of the Bayes optimal classifier in the following two-category, one-dimensional example. Suppose it is known that the distribution for the first category has a triangle distribution centered on $x = 0$ and "halfwidth" 1.0, according to:

$$p(x|\omega_1) \equiv T(0, 1) = \begin{cases} 1 - |x| & \text{for } |x| < 1 \\ 0 & \text{otherwise,} \end{cases}$$

and the second category has a uniform distribution

$$p(x|\omega_2) \equiv U(\mu, 1) = \begin{cases} 1/2 & \text{for } |x - \mu| < 1 \\ 0 & \text{otherwise.} \end{cases}$$

Suppose too that there is a prior on the unknown mean of the second category, $p(\mu)$, that is uniform in the range $0 \leq \mu \leq 2$.

    (a) Integrate over the unknown parameter to find $p(x|\omega_2)$.

    (b) Use this as a basis to find the decision point x^* for a Bayesian classifier.

    (c) Calculate the expected error of your Bayesian classifier.

    (d) Now consider the Gibbs algorithm, in which a single value of μ is chosen randomly, and used as if it is the true one. Integrate over test points and over μ to calculate the expected error of a classifier using the Gibbs algorithm.

    (e) Compare your results in parts (c) and (d) and verify that for this case the expected error from the Gibbs algorithm is bounded by twice that of your Bayes classifier.

Section 3.6

23. Employ the notation of the chapter and suppose $\mathbf{s}$ is a sufficient statistic for which $p(\boldsymbol{\theta}|\mathbf{s}, \mathcal{D}) = p(\boldsymbol{\theta}|\mathbf{s})$. Assume $p(\boldsymbol{\theta}|\mathbf{s}) \neq 0$ and prove that $p(\mathcal{D}|\mathbf{s}, \boldsymbol{\theta})$ is independent of $\boldsymbol{\theta}$.

24. Using the results given in Table 3.1, show that the maximum-likelihood estimate for the parameter θ of a Rayleigh distribution is given by

$$\hat{\theta} = \frac{1}{\frac{1}{n}\sum_{k=1}^{n} x_k^2}.$$

25. Using the results given in Table 3.1, show that the maximum-likelihood estimate for the parameter θ of a Maxwell distribution is given by

$$\hat{\theta} = \frac{3/2}{\frac{1}{n}\sum_{k=1}^{n} x_k^2}.$$

26. Using the results given in Table 3.1, show that the maximum-likelihood estimate for the parameter θ of a multinomial distribution is given by

$$\hat{\theta}_i = \frac{s_i}{\sum_{j=1}^{d} s_j}.$$

where the vector $\mathbf{s} = (s_1, \ldots, s_d)^t$ is the average of the n samples $\mathbf{x}_1, \ldots, \mathbf{x}_n$.

27. Demonstrate that sufficiency is an integral concept, i.e., that if $\mathbf{s}$ is sufficient for $\boldsymbol{\theta}$, then corresponding components of $\mathbf{s}$ and $\boldsymbol{\theta}$ need not be sufficient. Do this for the case of a univariate Gaussian $p(x) \sim N(\mu, \sigma^2)$ where $\boldsymbol{\theta} = \binom{\mu}{\sigma^2}$ is the full vector of parameters.

    (a) Verify that the statistic

    $$\mathbf{s} = \binom{s_1}{s_2} = \left[\begin{array}{c} \frac{1}{n}\sum_{k=1}^{n} x_k \\ \frac{1}{n}\sum_{k=1}^{n} x_k^2 \end{array} \right]$$

    is indeed sufficient for $\boldsymbol{\theta}$, as in Table 3.1.1.

    (b) Show that s_1 taken alone is not sufficient for μ. Does your answer depend upon whether σ^2 is known?

    (c) Show that s_2 taken alone is not sufficient for σ^2. Does your answer depend upon whether μ is known?

28. Suppose $\mathbf{s}$ is a statistic for which $p(\boldsymbol{\theta}|\mathbf{x}, \mathcal{D}) = p(\boldsymbol{\theta}|\mathbf{s})$.

    (a) Assume $p(\boldsymbol{\theta}|\mathbf{s}) \neq 0$, and prove that $p(\mathcal{D}|\mathbf{s}, \boldsymbol{\theta})$ is independent of $\boldsymbol{\theta}$.

    (b) Create an example to show that the inequality $p(\boldsymbol{\theta}|\mathbf{s}) \neq 0$ is required for your proof above.

29. Consider the Cauchy distribution,

$$p(x) = \frac{1}{\pi b} \cdot \frac{1}{1 + \left(\frac{x-a}{b}\right)^2},$$

    for $b > 0$ and arbitrary real a.

    (a) Confirm that the distribution is indeed normalized.

    (b) For a fixed a and b, try to calculate the mean and the standard deviation of the distribution. Explain your results.

    (c) Prove that this distribution has no sufficient statistics for the mean and standard deviation.

30. Show that the estimator of Eq. 21 is indeed unbiased for:
    (a) Normal distributions.
    (b) Cauchy distributions.
    (c) Binomial distributions.
    (d) Prove that the estimator of Eq. 20 is asymptotically unbiased.

Section 3.7

31. In the following, suppose a and b are positive constants greater than 1 and n a variable parameter.
    (a) Is $a^{n+1} = O(a^n)$?
    (b) Is $a^{bn} = O(a^n)$?
    (c) Is $a^{n+b} = O(a^n)$?
    (d) Prove $f(n) = O(f(n))$.

32. Consider the evaluation of a polynomial function $f(x) = \sum_{i=0}^{n-1} a_i x^i$ at a point x, where the n coefficients a_i are given.
    (a) Write pseudocode for a simple $\Theta(n^2)$-time algorithm for evaluating $f(x)$.
    (b) Show that such a polynomial can be rewritten as

    $$f(x) = \sum_{i=0}^{n-1} a_i x^i = (\cdots(a_{n-1}x + a_{n-2})x + \cdots + a_1)x + a_0,$$

    and so forth—a method known as *Horner's rule*. Use the rule to write pseudocode for a $\Theta(n)$-time algorithm for evaluating $f(x)$.

33. For each of the short procedures, state the computational complexity in terms of the variables N, M, P, and K, as appropriate. Assume that all data structures are defined, that those without indexes are scalars and that those with indexes have the number of dimensions shown.
    (a) 1 **begin for** $i \leftarrow i + 1$
        2              $s \leftarrow s + i^3$
        3   **until** $i = N$
        4 **return** s
        5 **end**
    (b) 1 **begin for** $i \leftarrow i + 1$
        2              $s \leftarrow s + x_i \times x_i$
        3   **until** $i = N$
        4 **return** $\sqrt{s}$
        5 **end**
    (c) 1 **begin for** $j \leftarrow j + 1$
        2   **for** $i \leftarrow i + 1$
        3             $s_j \leftarrow s_j + w_{ij} x_i$
        4   **until** $i = I$
        5   **until** $j = J$
        6   **for** $k \leftarrow k + 1$
        7   **for** $j \leftarrow j + 1$
        8           $r_k \leftarrow r_k + w_{jk} s_j$
        9   **until** $j = J$
       10   **until** $k = K$
       11 **end**

34. Consider a computer having a uniprocessor that can perform one operation per nanosecond (10^{-9} sec). The left column of the table shows the functional dependence of such operations in different hypothetical algorithms. For each such function, fill in the maximum size n that can be performed in the total time listed along the top.

$f(n)$	1 sec	1 hour	1 day	1 year
$\log_2 n$				
$\sqrt{n}$				
n				
$n \log_2 n$				
n^2				
n^3				
2^n				
e^n				
$n!$				

35. Let the sample mean $\hat{\boldsymbol{\mu}}_n$ and the sample covariance matrix $\mathbf{C}_n$ for a set of n samples $\mathbf{x}_1, \ldots, \mathbf{x}_n$ (each of which is d-dimensional) be defined by

$$\hat{\boldsymbol{\mu}}_n = \frac{1}{n} \sum_{k=1}^{n} \mathbf{x}_k$$

and

$$\mathbf{C}_n = \frac{1}{n-1} \sum_{k=1}^{n} (\mathbf{x}_k - \hat{\boldsymbol{\mu}}_n)(\mathbf{x}_k - \hat{\boldsymbol{\mu}}_n)^t.$$

We call these the "nonrecursive" formulas.

(a) What is the computational complexity of calculating $\hat{\boldsymbol{\mu}}_n$ and $\mathbf{C}_n$ by these formulas?

(b) Show that alternative, "recursive" techniques for calculating $\hat{\boldsymbol{\mu}}_n$ and $\mathbf{C}_n$ based on the successive addition of new samples $\mathbf{x}_{n+1}$ can be derived using the recursion relations

$$\hat{\boldsymbol{\mu}}_{n+1} = \hat{\boldsymbol{\mu}}_n + \frac{1}{n+1}(\mathbf{x}_{n+1} - \hat{\boldsymbol{\mu}}_n)$$

and

$$\mathbf{C}_{n+1} = \frac{n-1}{n}\mathbf{C}_n + \frac{1}{n+1}(\mathbf{x}_{n+1} - \hat{\boldsymbol{\mu}}_n)(\mathbf{x}_{n+1} - \hat{\boldsymbol{\mu}}_n)^t.$$

   (c) What is the computational complexity of finding $\hat{\boldsymbol{\mu}}_n$ and $\mathbf{C}_n$ by these recursive methods?

   (d) Describe situations where you might prefer to use the recursive method for computing $\hat{\boldsymbol{\mu}}_n$ and $\mathbf{C}_n$, and ones where you might prefer the nonrecursive method.

36. In pattern classification, one is often interested in the inverse of the covariance matrix—for instance, when designing a Bayes classifier for Gaussian distributions. Note that the nonrecursive calculation of $\mathbf{C}_n^{-1}$ (the inverse of the covariance matrix based on n samples, cf. Problem 35) might require the $O(n^3)$ inversion of $\mathbf{C}_n$ by standard matrix methods. We now explore an alternative, "recursive" method for computing $\mathbf{C}_n^{-1}$.

   (a) Prove the so-called Sherman-Morrison-Woodbury matrix identity

$$(\mathbf{A} + \mathbf{x}\mathbf{y}^t)^{-1} = \mathbf{A}^{-1} - \frac{\mathbf{A}^{-1}\mathbf{x}\mathbf{y}^t\mathbf{A}^{-1}}{1 + \mathbf{y}^t\mathbf{A}^{-1}\mathbf{x}}.$$

   (b) Use this and the results of Problem 35 to show that

$$\mathbf{C}_{n+1}^{-1} = \frac{n}{n-1}\left[\mathbf{C}_n^{-1} - \frac{\mathbf{C}_n^{-1}(\mathbf{x}_{n+1} - \hat{\boldsymbol{\mu}}_n)(\mathbf{x}_{n+1} - \hat{\boldsymbol{\mu}}_n)^t\mathbf{C}_n^{-1}}{\frac{n^2-1}{n} + (\mathbf{x}_{n+1} - \hat{\boldsymbol{\mu}}_n)^t\mathbf{C}_n^{-1}(\mathbf{x}_{n+1} - \hat{\boldsymbol{\mu}}_n)}\right].$$

   (c) What is the computational complexity of this calculation?

   (d) Describe situations where you would use the recursive method, and ones where you would use instead the nonrecursive method.

37. Suppose we wish to simplify (or regularize) a Gaussian classifier for two categories by means of shrinkage. Suppose that the estimated distributions are $N(\boldsymbol{\mu}_1, \boldsymbol{\Sigma}_1)$ and $N(\boldsymbol{\mu}_2, \boldsymbol{\Sigma}_2)$. In order to employ shrinkage of an assumed common covariance toward the identity matrix as given in Eq. 77, show that one must first normalize the data to have unit variance.

Section 3.8

38. Let $p_\mathbf{x}(\mathbf{x}|\omega_i)$ be arbitrary densities with means $\boldsymbol{\mu}_i$ and covariance matrices $\boldsymbol{\Sigma}_i$—not necessarily normal—for $i = 1, 2$. Let $y = \mathbf{w}^t\mathbf{x}$ be a projection, and let the induced one-dimensional densities $p(y|\omega_i)$ have means μ_i and variances σ_i^2.

   (a) Show that the criterion function

$$J_1(\mathbf{w}) = \frac{(\mu_1 - \mu_2)^2}{\sigma_1^2 + \sigma_2^2}$$

   is maximized by

$$\mathbf{w} = (\boldsymbol{\Sigma}_1 + \boldsymbol{\Sigma}_2)^{-1}(\boldsymbol{\mu}_1 - \boldsymbol{\mu}_2).$$

   (b) If $P(\omega_i)$ is the prior probability for ω_i, show that

$$J_2(\mathbf{w}) = \frac{(\mu_1 - \mu_2)^2}{P(\omega_1)\sigma_1^2 + P(\omega_2)\sigma_2^2}$$

is maximized by

$$\mathbf{w} = [P(\omega_1)\mathbf{\Sigma}_1 + P(\omega_2)\mathbf{\Sigma}_2]^{-1}(\boldsymbol{\mu}_1 - \boldsymbol{\mu}_2).$$

(c) To which of these criterion functions is the $J(\mathbf{w})$ of Eq. 96 more closely related? Explain.

39. The expression

$$J_1 = \frac{1}{n_1 n_2} \sum_{y_i \in \mathcal{Y}_1} \sum_{y_j \in \mathcal{Y}_2} (y_i - y_j)^2$$

clearly measures the total within-group scatter.

(a) Show that this within-group scatter can be written as

$$J_1 = (m_1 - m_2)^2 + \frac{1}{n_1}s_1^2 + \frac{1}{n_2}s_2^2.$$

(b) Show that the total scatter is

$$J_2 = \frac{1}{n_1}s_1^2 + \frac{1}{n_2}s_2^2.$$

(c) If $y = \mathbf{w}^t\mathbf{x}$, show that the $\mathbf{w}$ optimizing J_1 subject to the constraint that $J_2 = 1$ is given by

$$\mathbf{w} = \lambda \left(\frac{1}{n_1}\mathbf{S}_1 + \frac{1}{n_2}\mathbf{S}_2\right)^{-1}(\mathbf{m}_1 - \mathbf{m}_2),$$

where

$$\lambda = \left[(\mathbf{m}_1 - \mathbf{m}_2)^t \left(\frac{1}{n_1}\mathbf{S}_1 + \frac{1}{n_2}\mathbf{S}_2\right)(\mathbf{m}_1 - \mathbf{m}_2)\right]^{1/2},$$

$$\mathbf{m}_i = \frac{1}{n_i} \sum_{\mathbf{x} \in \mathcal{D}_i} \mathbf{x},$$

and

$$\mathbf{S}_i = \sum_{\mathbf{x} \in \mathcal{D}_i} n_i(\mathbf{m}_i - \mathbf{m})(\mathbf{m}_i - \mathbf{m})^t.$$

40. If $\mathbf{S}_B$ and $\mathbf{S}_W$ are two real, symmetric, d-by-d matrices, it is well known that there exists a set of n eigenvalues $\lambda_1, \ldots, \lambda_n$ satisfying $|\mathbf{S}_B - \lambda\mathbf{S}_W| = 0$, with a corresponding set of n eigenvectors $\mathbf{e}_1, \ldots, \mathbf{e}_n$ satisfying $\mathbf{S}_B\mathbf{e}_i = \lambda_i\mathbf{S}_W\mathbf{e}_i$. Furthermore, if $\mathbf{S}_W$ is positive definite, the eigenvectors can always be normalized so that $\mathbf{e}_i^t\mathbf{S}_W\mathbf{e}_j = \delta_{ij}$ and $\mathbf{e}_i^t\mathbf{S}_B\mathbf{e}_j = \lambda_i\delta_{ij}$. Let $\tilde{\mathbf{S}}_W = \mathbf{W}^t\mathbf{S}_W\mathbf{W}$ and $\tilde{\mathbf{S}}_B = \mathbf{W}^t\mathbf{S}_B\mathbf{W}$, where $\mathbf{W}$ is a d-by-n matrix whose columns correspond to n distinct eigenvectors.

(a) Show that $\tilde{\mathbf{S}}_W$ is the n-by-n identity matrix $\mathbf{I}$ and that $\tilde{\mathbf{S}}_B$ is a diagonal matrix whose elements are the corresponding eigenvalues. (This shows that the discriminant functions in multiple discriminant analysis are uncorrelated.)

(b) What is the value of $J = |\tilde{\mathbf{S}}_B|/|\tilde{\mathbf{S}}_W|$?

(c) Let $\mathbf{y} = \mathbf{W}^t \mathbf{x}$ be transformed by scaling the axes with a nonsingular n-by-n diagonal matrix $\mathbf{D}$ and by rotating this result with an orthogonal matrix $\mathbf{Q}$ where $\mathbf{y}' = \mathbf{Q}\mathbf{D}\mathbf{y}$. Show that J is invariant to this transformation.

41. Consider two normal distributions with arbitrary but equal covariances. Prove that the Fisher linear discriminant, for suitable threshold, can be derived from the negative of the log-likelihood ratio.

42. Consider the criterion function $J(\mathbf{w})$ required for the Fisher linear discriminant.

(a) Fill in the steps leading from Eqs. 96, 98, and 102 to Eq. 103.

(b) Use matrix methods to show that the solution to Eq. 103 is indeed given by Eq. 104.

(c) At the extreme of $J(\mathbf{w})$, a small change in $\mathbf{w}$ must leave $J(\mathbf{w})$ unchanged. Consider a small perturbation away from the optimal, $\mathbf{w} + \Delta \mathbf{w}$, and derive the solution condition of Eq. 104.

43. Consider multidiscriminant versions of Fisher's method for the case of c Gaussian distributions in d dimensions, each having the same covariance $\mathbf{\Sigma}$ (otherwise arbitrary) but different means. Solve for the optimal subspace in terms of $\mathbf{\Sigma}$ and the c mean vectors.

Section 3.9

44. Consider the convergence of the expectation-maximization algorithm, that is, that if $l(\boldsymbol{\theta}, \mathcal{D}_g) = \ln p(\mathcal{D}_g; \boldsymbol{\theta})$ is not already optimum, then the EM algorithm increases it. Prove this as follows:

(a) First note that

$$l(\boldsymbol{\theta}; \mathcal{D}_g) = \ln p(\mathcal{D}_g, \mathcal{D}_b; \boldsymbol{\theta}) - \ln p(\mathcal{D}_b | \mathcal{D}_g; \boldsymbol{\theta}).$$

Let $\mathcal{E}'[\cdot]$ denote the expectation with respect to the distribution $p(\mathcal{D}_b | \mathcal{D}_g; \boldsymbol{\theta}')$. Take such an expectation of $l(\boldsymbol{\theta}; \mathcal{D}_g)$, and express your answer in terms of $Q(\boldsymbol{\theta}; \boldsymbol{\theta}')$ of Eq. 129.

(b) Define $\phi(\mathcal{D}_b) = p(\mathcal{D}_b | \mathcal{D}_g; \boldsymbol{\theta}) / p(\mathcal{D}_b | \mathcal{D}_g; \boldsymbol{\theta}')$ to be the ratio of expectations assuming the two distributions. Show that $\mathcal{E}'[\ln \phi(\mathcal{D}_b)] \leq \mathcal{E}'[\phi(\mathcal{D}_b)] - 1 = 0$.

(c) Use this result to show that if $Q(\boldsymbol{\theta}^{t+1}; \boldsymbol{\theta}^t) > Q(\boldsymbol{\theta}^t; \boldsymbol{\theta}^t)$, achieved by the **M step** in Algorithm 1, then $l(\boldsymbol{\theta}^{t+1}; \mathcal{D}_g) > l(\boldsymbol{\theta}^t; \mathcal{D}_g)$.

45. Suppose we seek to estimate $\boldsymbol{\theta}$ describing a multidimensional distribution from data $\mathcal{D}$, some of whose points are missing features. Consider an iterative algorithm in which the maximum-likelihood value of the missing values is calculated, then assumed to be correct for the purposes of re-estimating $\boldsymbol{\theta}$ and the process iterated.

(a) Is this always equivalent to an expectation-maximization algorithm, or just a generalized expectation-maximization algorithm?

(b) If it is an expectation-maximization algorithm, what is $Q(\boldsymbol{\theta}; \boldsymbol{\theta}^t)$, as described by Eq. 129?

46. Consider data $\mathcal{D} = \left\{ \binom{2}{3}, \binom{3}{1}, \binom{5}{4}, \binom{4}{*}, \binom{*}{6} \right\}$, sampled from a two-dimensional uniform distribution

$$p(\mathbf{x}) \sim U(\mathbf{x}_l, \mathbf{x}_u) = \begin{cases} \frac{1}{|x_{u1}-x_{l1}||x_{u2}-x_{l2}|} & \text{if } x_{l1} \le x_1 \le x_{u1} \\ & \text{and } x_{l2} \le x_2 \le x_{u2} \\ \epsilon & \text{otherwise,} \end{cases}$$

where $*$ represents missing feature values and ϵ is a very small positive constant that can be neglected when normalizing the density within the above bounds.

(a) Start with an initial estimate

$$\boldsymbol{\theta}^0 = \begin{pmatrix} \mathbf{x}_l \\ \mathbf{x}_u \end{pmatrix} = \begin{pmatrix} 0 \\ 0 \\ 10 \\ 10 \end{pmatrix},$$

and analytically calculate $Q(\boldsymbol{\theta}; \boldsymbol{\theta}^0)$—the **E step** in the EM algorithm.

(b) Find the $\boldsymbol{\theta}$ that maximizes your $Q(\boldsymbol{\theta}; \boldsymbol{\theta}^0)$—the **M step**. You may make some simplifying assumptions.

(c) Plot your data and the bounding rectangle.

(d) Without having to iterate further, state the estimate of $\boldsymbol{\theta}$ that would result after convergence of the EM algorithm.

47. Consider data $\mathcal{D} = \left\{ \binom{1}{1}, \binom{3}{3}, \binom{2}{*} \right\}$, sampled from a two-dimensional (separable) distribution $p(x_1, x_2) = p(x_1)p(x_2)$, with

$$p(x_1) \sim \begin{cases} \frac{1}{\theta_1} e^{-x_1/\theta_1} & \text{if } x_1 \ge 0 \\ 0 & \text{otherwise,} \end{cases}$$

and

$$p(x_2) \sim U(0, \theta_2) = \begin{cases} \frac{1}{\theta_2} & \text{if } 0 \le x_2 \le \theta \\ 0 & \text{otherwise.} \end{cases}$$

As usual, $*$ represents a missing feature value.

(a) Start with an initial estimate $\boldsymbol{\theta}^0 = \binom{2}{4}$ and analytically calculate $Q(\boldsymbol{\theta}; \boldsymbol{\theta}^0)$—the **E step** in the EM algorithm. Be sure to consider the normalization of your distribution.

(b) Find the $\boldsymbol{\theta}$ that maximizes your $Q(\boldsymbol{\theta}; \boldsymbol{\theta}^0)$—the **M step**.

(c) Plot your data on a two-dimensional graph and indicate the new parameter estimates.

48. Repeat Problem 47 but with data $\mathcal{D} = \left\{ \binom{1}{1}, \binom{3}{3}, \binom{*}{2} \right\}$.

Section 3.10

49. Consider training an HMM by the forward-backward algorithm, for a single sequence of length T where each symbol could be one of c values. What is the computational complexity of a single revision of all values $\hat{a}_{ij}$ and $\hat{b}_{jk}$?

50. The standard method for calculating the probability of a sequence in a given HMM is to use the forward probabilities $\alpha_i(t)$.

(a) Show by a simple substitution that a symmetric method can be derived using the backward probabilities $\beta_i(t)$.

(b) Prove that one can get the probability by combining the forward and the backward probabilities at any place in the middle of the sequence. That is, show that

$$P(\boldsymbol{\omega}^{T'}) = \sum_{i=1}^{T'} \alpha_i(t)\beta_i(t),$$

where $\boldsymbol{\omega}^{T'}$ is a particular sequence of length $T' < T$.

(c) Show that your formula reduces to the known values at the beginning and end of the sequence.

51. Suppose we have a large number of symbol sequences emitted from an HMM that has a particular transition probability $a_{i'j'} = 0$ for some single value of i' and j'. We use such sequences to train a new HMM, one that happens also to start with its $a_{i'j'} = 0$. Prove that this parameter will remain 0 throughout training by the forward-backward algorithm. In other words, if the topology of the trained model (pattern of non-zero connections) matches that of the generating HMM, it will remain so after training.

52. Consider the decoding algorithm (*Algorithm 4*) in the text.

(a) Take logarithms of HMM model parameters and write pseudocode for an equivalent algorithm.

(b) Explain why taking logarithms is an $O(n)$ calculation, and thus the complexity of your algorithm in (a) is $O(c^2 T)$.

COMPUTER EXERCISES

Several exercises will make use of the following three-dimensional data sampled from three categories, denoted ω_i.

	ω_1			ω_2			ω_3		
sample	x_1	x_2	x_3	x_1	x_2	x_3	x_1	x_2	x_3
1	0.42	−0.087	0.58	−0.4	0.58	0.089	0.83	1.6	−0.014
2	−0.2	−3.3	−3.4	−0.31	0.27	−0.04	1.1	1.6	0.48
3	1.3	−0.32	1.7	0.38	0.055	−0.035	−0.44	−0.41	0.32
4	0.39	0.71	0.23	−0.15	0.53	0.011	0.047	−0.45	1.4
5	−1.6	−5.3	−0.15	−0.35	0.47	0.034	0.28	0.35	3.1
6	−0.029	0.89	−4.7	0.17	0.69	0.1	−0.39	−0.48	0.11
7	−0.23	1.9	2.2	−0.011	0.55	−0.18	0.34	−0.079	0.14
8	0.27	−0.3	−0.87	−0.27	0.61	0.12	−0.3	−0.22	2.2
9	−1.9	0.76	−2.1	−0.065	0.49	0.0012	1.1	1.2	−0.46
10	0.87	−1.0	−2.6	−0.12	0.054	−0.063	0.18	−0.11	−0.49

Section 3.2

1. Consider Gaussian density models in different dimensions.

   (a) Write a program to find the maximum-likelihood values $\hat{\mu}$ and $\hat{\sigma}^2$. Apply your program individually to each of the three features x_i of category ω_1 in the table above.

   (b) Modify your program to apply to two-dimensional Gaussian data $p(\mathbf{x}) \sim N(\boldsymbol{\mu}, \boldsymbol{\Sigma})$. Apply your data to each of the three possible pairings of two features for ω_1.

   (c) Modify your program to apply to three-dimensional Gaussian data. Apply your data to the full three-dimensional data for ω_1.

   (d) Assume your three-dimensional model is separable, so that

   $$\boldsymbol{\Sigma} = diag(\sigma_1^2, \sigma_2^2, \sigma_3^2).$$

   Write a program to estimate the mean and the diagonal components of $\boldsymbol{\Sigma}$. Apply your program to the data in ω_2.

   (e) Compare your results for the mean of each feature μ_i calculated in the above ways. Explain why they are the same or different.

   (f) Compare your results for the variance of each feature σ_i^2 calculated in the above ways. Explain why they are the same or different.

Section 3.3

2. Consider a one-dimensional model of a triangular density governed by two scalar parameters:

   $$p(x|\boldsymbol{\theta}) \equiv T(\mu, \delta) = \begin{cases} (\delta - |x - \mu|)/\delta^2 & \text{for } |x - \mu| < \delta \\ 0 & \text{otherwise,} \end{cases}$$

   where $\boldsymbol{\theta} = \binom{\mu}{\delta}$. Write a program to calculate the density $p(x|\mathcal{D})$ via Bayesian methods (Eq. 25) and apply it to the x_2 feature of category ω_2. Assume your priors on the parameters are uniform throughout the range of the data. Plot your resulting posterior density $p(x|\mathcal{D})$.

Section 3.4

3. Consider Bayesian estimation of the mean of a one-dimensional Gaussian. Suppose you are given the prior for the mean is $p(\mu) \sim N(\mu_0, \sigma_0)$.

   (a) Write a program that plots the density $p(x|\mathcal{D})$ given μ_0, σ_0, σ and training set $\mathcal{D} = \{x_1, x_2, \dots, x_n\}$.

   (b) Estimate σ for the x_2 component of ω_3 in the table above. Now assume $\mu_0 = -1$ and plot your estimated densities $p(x|\mathcal{D})$ for each of the following values of the dogmatism, σ^2/σ_0^2: 0.1, 1.0, 10, and 100.

Section 3.5

4. Suppose we have reason to believe that our data is sampled from a two-dimensional uniform density

   $$p(\mathbf{x}|\boldsymbol{\theta}) \sim U(\mathbf{x}_l, \mathbf{x}_u) = \begin{cases} \frac{1}{|x_{u1} - x_{l1}||x_{u2} - x_{l2}|} & \text{for } x_{l1} \le x_1 \le x_{u1} \text{ and } x_{l2} \le x_2 \le x_{u2} \\ 0 & \text{otherwise,} \end{cases}$$

where x_{l1} is the x_1 component of the "lower" bounding point $\mathbf{x}_l$, and analogously for the x_2 component and for the upper point. Suppose we have reliable prior information that the density is zero outside the box defined by $\mathbf{x}_l = \binom{-6}{-6}$ and $\mathbf{x}_u = \binom{+6}{+6}$. Write a program that calculates $p(\mathbf{x}|\mathcal{D})$ via recursive Bayesian estimation and apply it to the x_1–x_2 components of ω_1, in sequence, from the table above. For each expanding data set $\mathcal{D}^n$ ($2 \leq n \leq 10$) plot your posterior density.

Section 3.6

5. Write a single program to calculate sufficient statistics for any members of the exponential family (Eq. 69). Assume that the x_3 data from ω_3 in the table come from an exponential density, and use your program to calculate the sufficient statistics for each of the following exponential forms: Gaussian, Rayleigh, and Maxwell.

Section 3.7

6. Consider error rates in different dimensions.

   (a) Use maximum-likelihood to train a dichotomizer using the three-dimensional data for categories ω_1 and ω_2 in the table above. Numerically integrate to estimate the classification error rate.

   (b) Now consider the data projected into a two-dimensional subspace. For each of the three subspaces—defined by $x_1 = 0$ or $x_2 = 0$ or $x_3 = 0$—train a Gaussian dichotomizer. Numerically integrate to estimate the error rate.

   (c) Now consider the data projected onto one-dimensional subspaces, defined by each of the three axes. Train a Gaussian classifier, and numerically integrate to estimate the error rate.

   (d) Discuss the rank order of the error rates you find.

   (e) Assuming that you re-estimate the distribution in the different dimensions, logically must the Bayes error be higher in the projected spaces?

7. Repeat the steps in Exercise 6 but for categories ω_1 and ω_3.

8. Consider the classification of Gaussian data employing shrinkage of covariance matrices to a common one.

   (a) Generate 20 training points from each of three equally probable three-dimensional Gaussian distributions $N(\boldsymbol{\mu}_i, \boldsymbol{\Sigma}_i)$ with the following parameters:

$$\boldsymbol{\mu}_1 = (0,0,0)^t, \qquad \boldsymbol{\Sigma}_1 = diag[3,5,2]$$

$$\boldsymbol{\mu}_2 = (1,5,-3)^t, \qquad \boldsymbol{\Sigma}_2 = \begin{pmatrix} 1 & 0 & 0 \\ 0 & 4 & 1 \\ 0 & 1 & 6 \end{pmatrix}$$

$$\boldsymbol{\mu}_3 = (0,0,0)^t, \qquad \boldsymbol{\Sigma}_3 = 10\,\mathbf{I}.$$

   (b) Write a program to estimate the means and covariances of your data.

   (c) Write a program that takes α and shrinks these estimated covariance matrices according to Eq. 76.

   (d) Plot the training error as a function of α, where $0 < \alpha < 1$.

(e) Use your program from part (a) to generate 50 test points from each category. Plot the test error as a function of α.

Section 3.8

9. Consider the Fisher linear discriminant method.

(a) Write a general program to calculate the optimal direction $\mathbf{w}$ for a Fisher linear discriminant based on three-dimensional data.

(b) Find the optimal $\mathbf{w}$ for categories ω_2 and ω_3 in the table above.

(c) Plot a line representing your optimal direction $\mathbf{w}$ and mark on it the positions of the projected points.

(d) In this subspace, fit each distribution with a (univariate) Gaussian, and find the resulting decision boundary.

(e) What is the training error (the error on the training points themselves) in the optimal subspace you found in (b)?

(f) For comparison, repeat parts (d) and (e) using instead the nonoptimal direction $\mathbf{w} = (1.0, 2.0, -1.5)^t$. What is the training error in this nonoptimal subspace?

10. Consider the multicategory generalization of the Fisher linear discriminant, applied to the data in the table above.

(a) Write a general program to calculate the optimal $\mathbf{w}$ for multiple discriminant analysis. Use your program to find the optimal two-dimensional plane (described by normal vector $\mathbf{w}$) for the three-dimensional data in the table.

(b) In the subspace, fit a circularly symmetric Gaussian to the data, and use a simple linear classifier in each to find the decision boundaries in the subspace.

(c) What is the error on the training set?

(d) Classify following points : $(1.40, -0.36, -0.41)^t$, $(0.62, 1.30, 1.11)^t$ and $(-0.11, 1.60, 1.51)^t$.

(e) For comparison, repeat parts (b) and (c) for the nonoptimal direction $\mathbf{w} = (-0.5, -0.5, 1.0)^t$. Explain the difference between your training errors in the two cases.

Section 3.9

11. Suppose we know that the ten data points in category ω_1 in the table above come from a three-dimensional Gaussian. Suppose, however, that we do not have access to the x_3 components for the even-numbered data points.

(a) Write an EM program to estimate the mean and covariance of the distribution. Start your estimate with $\boldsymbol{\mu}^0 = \mathbf{0}$ and $\boldsymbol{\Sigma}^0 = \mathbf{I}$, the three-dimensional identity matrix.

(b) Compare your final estimate with that for the case when there is no missing data.

12. Suppose we know that the ten data points in category ω_2 in the table above come from a three-dimensional uniform distribution $p(\mathbf{x}|\omega_2) \sim U(\mathbf{x}_l, \mathbf{x}_u)$. Suppose, however, that we do not have access to the x_3 components for the even-numbered data points.

(a) Write an EM program to estimate the six scalars comprising $\mathbf{x}_l$ and $\mathbf{x}_u$ of the distribution. Start your estimate with $\mathbf{x}_l = (-2, -2, -2)^t$ and $\mathbf{x}_u = (+2, +2, +2)^t$.

(b) Compare your final estimate with that for the case when there is no missing data.

Section 3.10

13. Consider the use of hidden Markov models for classifying sequences of four visible states, A–D. Train two hidden Markov models, each consisting of three hidden states (plus a null initial state and a null final state), fully connected, with the following data. Assume that each sequence starts with a null symbol and ends with an end null symbol (not listed).

sample	ω_1	ω_2
1	AABBCCDD	DDCCBBAA
2	ABBCBBDD	DDABCBA
3	ACBCBCD	CDCDCBABA
4	ADB	DDBBA
5	ACBCBABCDD	DADACBBAA
6	BABAADDD	CDDCCBA
7	BABCDCC	BDDBCAAAA
8	ABDBBCCDD	BBABBDDDCD
9	ABAAACDCCD	DDADDBCAA
10	ABD	DDCAAA

(a) Print out the full transition matrices for each of the models.

(b) Assume equal prior probabilities for the two models and classify each of the following sequences: ABBBCDDD, DADBCBAA, CDCBABA, and ADBBBCD.

(c) As above, classify the test pattern BADBDCBA. Find the prior probabilities for your two trained models that would lead to equal posteriors for your two categories when applied to this pattern.

BIBLIOGRAPHY

[1] Russell G. Almond. *Graphical Belief Modelling*. Chapman & Hall, New York, 1995.

[2] Pierre Baldi, Søren Brunak, Yves Chauvin, Jacob Engelbrecht, and Anders Krogh. Hidden Markov models for human genes. In Stephen J. Hanson, Jack D. Cowan, and C. Lee Giles, editors, *Advances in Neural Information Processing Systems*, volume 6, pages 761–768, Morgan Kaufmann, San Mateo, CA, 1994.

[3] Leonard E. Baum and Ted Petrie. Statistical inference for probabilistic functions of finite state Markov chains. *Annals of Mathematical Statistics*, 37:1554–1563, 1966.

[4] Leonard E. Baum, Ted Petrie, George Soules, and Norman Weiss. A maximization technique occurring in the statistical analysis of probabilistic functions of Markov chains. *Annals of Mathematical Statistics*, 41(1):164–171, 1970.

[5] José M. Bernardo and Adrian F. M. Smith. *Bayesian Theory*. Wiley, New York, 1996.

[6] Christopher M. Bishop. *Neural Networks for Pattern Recognition*. Oxford University Press, Oxford, UK, 1995.

[7] David Braverman. Learning filters for optimum pattern recognition. *IRE Transactions on Information Theory*, IT-8:280–285, 1962.

[8] Eugene Charniak. *Statistical Language Learning*. MIT Press, Cambridge, MA, 1993.

[9] Herman Chernoff and Lincoln E. Moses. *Elementary Decision Theory*. Wiley, New York, 1959.

[10] Thomas H. Cormen, Charles E. Leiserson, and Ronald L. Rivest. *Introduction to Algorithms*. MIT Press, Cambridge, MA, 1990.

[11] Arthur P. Dempster, Nan M. Laird, and Donald B. Rubin. Maximum-likelihood from incomplete data via the EM algorithm (with discussion). *Journal of the Royal Statistical Society, Series B*, 39:1–38, 1977.

[12] Pierre A. Devijver and Josef Kittler. *Pattern Recognition: A Statistical Approach*. Prentice-Hall, London, 1982.

[13] Ronald A. Fisher. The use of multiple measurements in taxonomic problems. *Annals of Eugenics*, 7 Part II:179–188, 1936.

[14] G. David Forney, Jr. The Viterbi algorithm. *Proceedings of the IEEE*, 61:268–278, 1973.

[15] Keinosuke Fukunaga. *Introduction to Statistical Pattern Recognition*. Academic Press, New York, second edition, 1990.

[16] David Haussler, Michael Kearns, and Robert Schapire. Bounds on the sample complexity of Bayesian learning using information theory and the VC dimension. *Machine Learning*, 14:84–114, 1994.

[17] Harold Jeffreys. *Theory of Probability*. Oxford University Press, Oxford, UK, 1961 reprint edition, 1939.

[18] Frederick Jelinek. *Statistical Methods for Speech Recognition*. MIT Press, Cambridge, MA, 1997.

[19] Ian T. Jolliffe. *Principal Component Analysis*. Springer-Verlag, New York, 1986.

[20] Michael I. Jordan and Robert A. Jacobs. Hierarchical mixtures of experts and the EM algorithm. *Neural Computation*, 6(2):181–214, 1994.

[21] Donald E. Knuth. *The Art of Computer Programming*, volume 1. Addison-Wesley, Reading, MA, first edition, 1973.

[22] Gary E. Kopec and Phil A. Chou. Document image decoding using Markov source models. *IEEE Transactions on Pattern Analysis and Machine Intelligence*, 16(6):602–617, 1994.

[23] Anders Krogh, Michael Brown, I. Saira Mian, Kimmen Sjölander, and David Haussler. Hidden Markov models in computational biology: Applications to protein modelling. *Journal of Molecular Biology*, 235:1501–1531, 1994.

[24] Dennis Victor Lindley. The use of prior probability distributions in statistical inference and decision. In Jerzy Neyman and Elizabeth L. Scott, editors, *Proceedings Fourth Berkeley Symposium on Mathematical Statistics and Probability*, pages 453–468, University of California Press, Berkeley, CA, 1961.

[25] Andrei Andreivich Markov. Issledovanie zamechatelnogo sluchaya zavisimykh ispytanii (investigation of a remarkable case of dependant trials). *Izvestiya Petersburgskoi akademii nauk, 6th ser.*, 1(3):61–80, 1907.

[26] Geoffrey J. McLachlan. *Discriminant Analysis and Statistical Pattern Recognition*. Wiley, New York, 1992.

[27] Geoffrey J. McLachlan and Thiriyambakam Krishnan. *The EM Algorithm and Extensions*. Wiley, New York, 1996.

[28] Manfred Opper and David Haussler. Generalization performance of Bayes optimal prediction algorithm for learning a perceptron. *Physical Review Letters*, 66(20):2677–2681, 1991.

[29] Lawrence Rabiner and Biing-Hwang Juang. *Fundamentals of Speech Recognition*. Prentice-Hall, Englewood Cliffs, NJ, 1993.

[30] Lawrence R. Rabiner. A tutorial on hidden Markov models and selected applications in speech recognition. *Proceedings of IEEE*, 77(2):257–286, 1989.

[31] Donald B. Rubin and Roderick J. A. Little. *Statistical Analysis with Missing Data*. Wiley, New York, 1987.

[32] Jürgen Schürmann. *Pattern Classification: A Unified View of Statistical and Neural Approaches*. Wiley, New York, 1996.

[33] Ross D. Shachter. Evaluating influence diagrams. *Operations Research*, 34(6):871–882, 1986.

[34] Padhraic Smyth, David Heckerman, and Michael Jordan. Probabilistic independence networks for hidden Markov probability models. *Neural Computation*, 9(2):227–269, 1997.

[35] Charles W. Therrien. *Decision Estimation and Classification: An Introduction to Pattern Recognition and Related Topics*. Wiley, New York, 1989.

[36] D. Michael Titterington. Recursive parameter estimation using incomplete data. *Journal of the Royal Statistical Society, Series B*, 46(2):257–267, 1984.

[37] Andrew J. Viterbi. Error bounds for convolutional codes and an asymptotically optimal decoding algorithm. *IEEE Transactions on Information Theory*, IT-13(2):260–269, 1967.

NONPARAMETRIC TECHNIQUES

4.1 INTRODUCTION

In Chapter 3 we treated supervised learning under the assumption that the forms of the underlying density functions were known. Alas, in most pattern recognition applications this assumption is suspect; the common parametric forms rarely fit the densities actually encountered in practice. In particular, all of the classical parametric densities are unimodal (have a single local maximum), whereas many practical problems involve multimodal densities. Furthermore, our hopes are rarely fulfilled that a high-dimensional density might be accurately represented as the product of one-dimensional functions. In this chapter we shall examine *nonparametric* procedures that can be used with arbitrary distributions and without the assumption that the forms of the underlying densities are known.

There are several types of nonparametric methods of interest in pattern recognition. One consists of procedures for estimating the density functions $p(\mathbf{x}|\omega_j)$ from sample patterns. If these estimates are satisfactory, they can be substituted for the true densities when designing the classifier. Another consists of procedures for directly estimating the *a posteriori* probabilities $P(\omega_j|\mathbf{x})$. This is closely related to nonparametric design procedures such as the nearest-neighbor rule, which bypass probability estimation and go directly to decision functions.

4.2 DENSITY ESTIMATION

The basic ideas behind many of the methods of estimating an unknown probability density function are very simple, although rigorous demonstrations that the estimates converge require considerable care. The most fundamental techniques rely on the fact that the probability P that a vector $\mathbf{x}$ will fall in a region $\mathcal{R}$ is given by

$$P = \int_{\mathcal{R}} p(\mathbf{x}') \, d\mathbf{x}'. \tag{1}$$

Thus P is a smoothed or averaged version of the density function $p(\mathbf{x})$, and we can estimate this smoothed value of p by estimating the probability P. Suppose that n samples $\mathbf{x}_1, \ldots, \mathbf{x}_n$ are drawn independently and identically distributed (i.i.d.) according to the probability law $p(\mathbf{x})$. Clearly, the probability that k of these n fall in $\mathcal{R}$ is given by the binomial law

$$P_k = \binom{n}{k} P^k (1 - P)^{n-k}, \tag{2}$$

and the expected value for k is

$$\mathcal{E}[k] = nP. \tag{3}$$

Moreover, this binomial distribution for k peaks very sharply about the mean, so that we expect that the ratio k/n will be a very good estimate for the probability P, and hence for the smoothed density function. This estimate is especially accurate when n is very large. If we now assume that $p(\mathbf{x})$ is continuous and that the region $\mathcal{R}$ is so small that p does not vary appreciably within it, we can write

$$\int_{\mathcal{R}} p(\mathbf{x}') \, d\mathbf{x}' \simeq p(\mathbf{x})V, \tag{4}$$

where $\mathbf{x}$ is a point within $\mathcal{R}$ and V is the volume enclosed by $\mathcal{R}$. Combining Eqs. 1, 3, and 4, we arrive at the following obvious estimate for $p(\mathbf{x})$,

$$p(\mathbf{x}) \simeq \frac{k/n}{V}, \tag{5}$$

as illustrated in Fig. 4.1.

There are several problems that remain—some practical and some theoretical. If we fix the volume V and take more and more training samples, the ratio k/n will

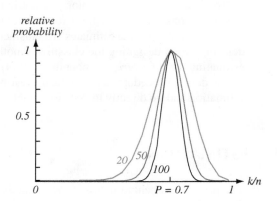

FIGURE 4.1. The relative probability an estimate given by Eq. 4 will yield a particular value for the probability density, here where the true probability was chosen to be 0.7. Each curve is labeled by the total number of patterns n sampled, and is scaled to give the same maximum (at the true probability). The form of each curve is binomial, as given by Eq. 2. For large n, such binomials peak strongly at the true probability. In the limit $n \to \infty$, the curve approaches a delta function, and we are guaranteed that our estimate will give the true probability.

converge (in probability) as desired, but then we have only obtained an estimate of the space-averaged value of $p(\mathbf{x})$,

$$\frac{P}{V} = \frac{\int_{\mathcal{R}} p(\mathbf{x}') \, d\mathbf{x}'}{\int_{\mathcal{R}} d\mathbf{x}'}. \tag{6}$$

If we want to obtain $p(\mathbf{x})$ rather than just an averaged version of it, we must be prepared to let V approach zero. However, if we fix the number n of samples and let V approach zero, the region will eventually become so small that it will enclose no samples, and our estimate $p(\mathbf{x}) \simeq 0$ will be useless. Or if by chance one or more of the training samples coincide at $\mathbf{x}$, the estimate diverges to infinity, which is equally useless.

From a practical standpoint, we note that the number of samples is always limited. Thus, the volume V cannot be allowed to become arbitrarily small. If this kind of estimate is to be used, one will have to accept a certain amount of variance in the ratio k/n and a certain amount of averaging of the density $p(\mathbf{x})$.

From a theoretical standpoint, it is interesting to ask how these limitations can be circumvented if an unlimited number of samples is available. Suppose we use the following procedure. To estimate the density at $\mathbf{x}$, we form a sequence of regions $\mathcal{R}_1, \mathcal{R}_2, \ldots$, containing $\mathbf{x}$—the first region to be used with one sample, the second with two, and so on. Let V_n be the volume of $\mathcal{R}_n$, k_n be the number of samples falling in $\mathcal{R}_n$, and $p_n(\mathbf{x})$ be the nth estimate for $p(\mathbf{x})$:

$$p_n(\mathbf{x}) = \frac{k_n/n}{V_n}. \tag{7}$$

If $p_n(\mathbf{x})$ is to converge to $p(\mathbf{x})$, three conditions appear to be required:

- $\lim_{n \to \infty} V_n = 0$
- $\lim_{n \to \infty} k_n = \infty$
- $\lim_{n \to \infty} k_n/n = 0$.

The first condition assures us that the space averaged P/V will converge to $p(\mathbf{x})$, provided that the regions shrink uniformly and that $p(\cdot)$ is continuous at $\mathbf{x}$. The second condition, which only makes sense if $p(\mathbf{x}) \neq 0$, assures us that the frequency ratio will converge (in probability) to the probability P. The third condition is clearly necessary if $p_n(\mathbf{x})$ given by Eq. 7 is to converge at all. It also says that although a huge number of samples will eventually fall within the small region $\mathcal{R}_n$, they will form a negligibly small fraction of the total number of samples.

There are two common ways of obtaining sequences of regions that satisfy these conditions (Fig. 4.2). One is to shrink an initial region by specifying the volume V_n as some function of n, such as $V_n = 1/\sqrt{n}$. It then must be shown that the random variables k_n and k_n/n behave properly or, more to the point, that $p_n(\mathbf{x})$ converges to $p(\mathbf{x})$. This is basically the Parzen-window method that will be examined in Section 4.3. The second method is to specify k_n as some function of n, such as $k_n = \sqrt{n}$. Here the volume V_n is grown until it encloses k_n neighbors of $\mathbf{x}$. This is the k_n-nearest-neighbor estimation method. Both of these methods do in fact converge, although it is difficult to make meaningful statements about their finite-sample behavior.

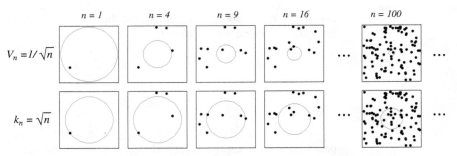

FIGURE 4.2. There are two leading methods for estimating the density at a point, here at the center of each square. The one shown in the top row is to start with a large volume centered on the test point and shrink it according to a function such as $V_n = 1/\sqrt{n}$. The other method, shown in the bottom row, is to decrease the volume in a data-dependent way, for instance letting the volume enclose some number $k_n = \sqrt{n}$ of sample points. The sequences in both cases represent random variables that generally converge and allow the true density at the test point to be calculated.

4.3 PARZEN WINDOWS

The Parzen-window approach to estimating densities can be introduced by temporarily assuming that the region $\mathcal{R}_n$ is a d-dimensional hypercube. If h_n is the length of an edge of that hypercube, then its volume is given by

$$V_n = h_n^d. \tag{8}$$

WINDOW FUNCTION

We can obtain an analytic expression for k_n, the number of samples falling in the hypercube, by defining the following *window function*:

$$\varphi(\mathbf{u}) = \begin{cases} 1 & |u_j| \leq 1/2; \quad j = 1, \ldots, d \\ 0 & \text{otherwise.} \end{cases} \tag{9}$$

Thus, $\varphi(\mathbf{u})$ defines a unit hypercube centered at the origin. It follows that $\varphi((\mathbf{x} - \mathbf{x}_i)/h_n)$ is equal to unity if $\mathbf{x}_i$ falls within the hypercube of volume V_n centered at $\mathbf{x}$, and is zero otherwise. The number of samples in this hypercube is therefore given by

$$k_n = \sum_{i=1}^{n} \varphi \left(\frac{\mathbf{x} - \mathbf{x}_i}{h_n} \right), \tag{10}$$

and when we substitute this into Eq. 7 we obtain the estimate

$$p_n(\mathbf{x}) = \frac{1}{n} \sum_{i=1}^{n} \frac{1}{V_n} \varphi \left(\frac{\mathbf{x} - \mathbf{x}_i}{h_n} \right). \tag{11}$$

This equation suggests a more general approach to estimating density functions. Rather than limiting ourselves to the hypercube window function of Eq. 9, suppose we allow a more general class of window functions. In such a case, Eq. 11 expresses our estimate for $p(\mathbf{x})$ as an average of functions of $\mathbf{x}$ and the samples $\mathbf{x}_i$. In essence, the window function is being used for *interpolation*—each sample contributing to the estimate in accordance with its distance from $\mathbf{x}$.

It is natural to ask that the estimate $p_n(\mathbf{x})$ be a legitimate density function, that is, that it be nonnegative and integrate to one. This can be assured by requiring the window function itself be a density function. To be more precise, if we require that

$$\varphi(\mathbf{x}) \geq 0 \tag{12}$$

and

$$\int \varphi(\mathbf{u})d\mathbf{u} = 1, \tag{13}$$

and if we maintain the relation $V_n = h_n^d$, then it follows at once that $p_n(\mathbf{x})$ also satisfies these conditions.

Let us examine the effect that the *window width* h_n has on $p_n(\mathbf{x})$. If we define the function $\delta_n(\mathbf{x})$ by

$$\delta_n(\mathbf{x}) = \frac{1}{V_n} \varphi\left(\frac{\mathbf{x}}{h_n}\right), \tag{14}$$

then we can write $p_n(\mathbf{x})$ as the average

$$p_n(\mathbf{x}) = \frac{1}{n} \sum_{i=1}^{n} \delta_n(\mathbf{x} - \mathbf{x}_i). \tag{15}$$

Because $V_n = h_n^d$, h_n clearly affects both the amplitude and the width of $\delta_n(\mathbf{x})$ (Fig. 4.3). If h_n is very large, then the amplitude of δ_n is small, and $\mathbf{x}$ must be far from $\mathbf{x}_i$ before $\delta_n(\mathbf{x} - \mathbf{x}_i)$ changes much from $\delta_n(\mathbf{0})$. In this case, $p_n(\mathbf{x})$ is the superposition of n broad, slowly changing functions and is a very smooth "out-of-focus" estimate of $p(\mathbf{x})$. On the other hand, if h_n is very small, then the peak value of $\delta_n(\mathbf{x} - \mathbf{x}_i)$ is large and occurs near $\mathbf{x} = \mathbf{x}_i$. In this case $p(\mathbf{x})$ is the superposition of n sharp pulses centered at the samples—an erratic, "noisy" estimate (Fig. 4.4). For any value of h_n, the distribution is normalized, that is,

$$\int \delta_n(\mathbf{x} - \mathbf{x}_i)\, d\mathbf{x} = \int \frac{1}{V_n} \varphi\left(\frac{\mathbf{x} - \mathbf{x}_i}{h_n}\right) d\mathbf{x} = \int \varphi(\mathbf{u})d\mathbf{u} = 1. \tag{16}$$

Thus, as h_n approaches zero, $\delta_n(\mathbf{x} - \mathbf{x}_i)$ approaches a Dirac delta function centered at $\mathbf{x}_i$, and $p_n(\mathbf{x})$ approaches a superposition of delta functions centered at the samples.

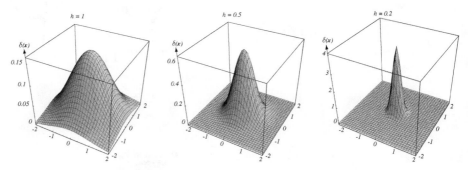

FIGURE 4.3. Examples of two-dimensional circularly symmetric normal Parzen windows for three different values of h. Note that because the $\delta(\mathbf{x})$ are normalized, different vertical scales must be used to show their structure.

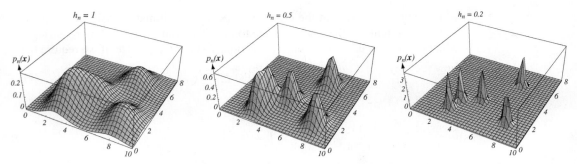

FIGURE 4.4. Three Parzen-window density estimates based on the same set of five samples, using the window functions in Fig. 4.3. As before, the vertical axes have been scaled to show the structure of each distribution.

Clearly, the choice of h_n (or V_n) has an important effect on $p_n(\mathbf{x})$. If V_n is too large, the estimate will suffer from too little resolution; if V_n is too small, the estimate will suffer from too much statistical variability. With a limited number of samples, the best we can do is to seek some acceptable compromise. However, with an unlimited number of samples, it is possible to let V_n slowly approach zero as n increases and have $p_n(\mathbf{x})$ converge to the unknown density $p(\mathbf{x})$.

In discussing convergence, we must recognize that we are talking about the convergence of a sequence of random variables, because for any fixed $\mathbf{x}$ the value of $p_n(\mathbf{x})$ depends on the random samples $\mathbf{x}_1, \ldots, \mathbf{x}_n$. Thus, $p_n(\mathbf{x})$ has some mean $\bar{p}_n(\mathbf{x})$ and variance $\sigma_n^2(\mathbf{x})$. We shall say that the estimate $p_n(\mathbf{x})$ converges to $p(\mathbf{x})$ if*

$$\lim_{n \to \infty} \bar{p}_n(\mathbf{x}) = p(\mathbf{x}) \tag{17}$$

and

$$\lim_{n \to \infty} \sigma_n^2(\mathbf{x}) = 0. \tag{18}$$

To prove convergence we must place conditions on the unknown density $p(\mathbf{x})$, on the window function $\varphi(\mathbf{u})$, and on the window width h_n. In general, continuity of $p(\cdot)$ at $\mathbf{x}$ is required, and the conditions imposed by Eqs. 12 and 13 are customarily invoked. Below we show that the following additional conditions assure convergence:

$$\sup_{\mathbf{u}} \varphi(\mathbf{u}) < \infty \tag{19}$$

$$\lim_{\|\mathbf{u}\| \to \infty} \varphi(\mathbf{u}) \prod_{i=1}^{d} u_i = 0 \tag{20}$$

$$\lim_{n \to \infty} V_n = 0 \tag{21}$$

and

$$\lim_{n \to \infty} n V_n = \infty. \tag{22}$$

---

*This type of convergence is called *convergence in mean square*. Other ways to define the modes of convergence of sequences of random variables are presented in advanced books on probability theory.

Equations 19 and 20 keep $\varphi(\cdot)$ well-behaved, and they are satisfied by most density functions that one might think of using for window functions. Equations 21 and 22 state that the volume V_n must approach zero, but at a rate slower than $1/n$. We shall now see why these are the basic conditions for convergence.

4.3.1 Convergence of the Mean

Consider first $\bar{p}_n(\mathbf{x})$, the mean of $p_n(\mathbf{x})$. Because the samples $\mathbf{x}_i$ are i.i.d. according to the (unknown) density $p(\mathbf{x})$, we have

$$\bar{p}_n(\mathbf{x}) = \mathcal{E}[p_n(\mathbf{x})]$$

$$= \frac{1}{n}\sum_{i=1}^{n}\mathcal{E}\left[\frac{1}{V_n}\varphi\left(\frac{\mathbf{x}-\mathbf{x}_i}{h_n}\right)\right]$$

$$= \int \frac{1}{V_n}\varphi\left(\frac{\mathbf{x}-\mathbf{v}}{h_n}\right)p(\mathbf{v})\,d\mathbf{v}$$

$$= \int \delta_n(\mathbf{x}-\mathbf{v})p(\mathbf{v})\,d\mathbf{v}. \tag{23}$$

CONVOLUTION

This equation shows that the expected value of the estimate is an averaged value of the unknown density—a *convolution* of the unknown density and the window function (see Section A.4.11 of the Appendix). Thus, $\bar{p}_n(\mathbf{x})$ is a blurred version of $p(\mathbf{x})$ as seen through the averaging window. But as V_n approaches zero, $\delta_n(\mathbf{x}-\mathbf{v})$ approaches a delta function centered at $\mathbf{x}$. Thus, if p is continuous at $\mathbf{x}$, Eq. 21 ensures that $\bar{p}_n(\mathbf{x})$ will approach $p(\mathbf{x})$ as n approaches infinity.

4.3.2 Convergence of the Variance

Equation 23 shows that there is no need for an infinite number of samples to make $\bar{p}_n(\mathbf{x})$ approach $p(\mathbf{x})$; one can achieve this for any n merely by letting V_n approach zero. Of course, for a particular set of n samples the resulting "spiky" estimate is useless; this fact highlights the need for us to consider the variance of the estimate. Because $p_n(\mathbf{x})$ is the sum of functions of statistically independent random variables, its variance is the sum of the variances of the separate terms, and hence

$$\sigma_n^2(\mathbf{x}) = \sum_{i=1}^{n}\mathcal{E}\left[\left(\frac{1}{nV_n}\varphi\left(\frac{\mathbf{x}-\mathbf{x}_i}{h_n}\right)-\frac{1}{n}\bar{p}_n(\mathbf{x})\right)^2\right]$$

$$= n\,\mathcal{E}\left[\frac{1}{n^2V_n^2}\varphi^2\left(\frac{\mathbf{x}-\mathbf{x}_i}{h_n}\right)\right]-\frac{1}{n}\bar{p}_n^2(\mathbf{x})$$

$$= \frac{1}{nV_n}\int \frac{1}{V_n}\varphi^2\left(\frac{\mathbf{x}-\mathbf{v}}{h_n}\right)p(\mathbf{v})\,d\mathbf{v}-\frac{1}{n}\bar{p}_n^2(\mathbf{x}). \tag{24}$$

By dropping the second term, bounding $\varphi(\cdot)$ and using Eq. 23, we obtain

$$\sigma_n^2(\mathbf{x}) \le \frac{\sup(\varphi(\cdot))\,\bar{p}_n(\mathbf{x})}{nV_n}. \tag{25}$$

Clearly, to obtain a small variance we want a large value for V_n, not a small one—a large V_n smoothes out the local variations in density. However, because the numerator stays finite as n approaches infinity, we can let V_n approach zero and still obtain zero variance, provided that nV_n approaches infinity. For example, we can let $V_n = V_1/\sqrt{n}$ or $V_1/\ln n$ or any other function satisfying Eqs. 21 and 22.

This is the principal theoretical result. Unfortunately, it does not tell us how to choose $\varphi(\cdot)$ and V_n to obtain good results in the finite sample case. Indeed, unless we have more knowledge about $p(\mathbf{x})$ than the mere fact that it is continuous, we have no direct basis for optimizing finite sample results.

4.3.3 Illustrations

It is interesting to see how the Parzen window method behaves on some simple examples, and particularly to see the effect of the window function. Consider first the case where $p(\mathbf{x})$ is a zero-mean, unit-variance, univariate normal density. Let the window function be of the same form:

$$\varphi(u) = \frac{1}{\sqrt{2\pi}} e^{-u^2/2}. \tag{26}$$

Finally, let $h_n = h_1/\sqrt{n}$, where h_1 is a parameter at our disposal. Thus $p_n(x)$ is an average of normal densities centered at the samples:

$$p_n(x) = \frac{1}{n} \sum_{i=1}^{n} \frac{1}{h_n} \varphi\left(\frac{x - x_i}{h_n}\right). \tag{27}$$

While it is not hard to evaluate Eqs. 23 and 24 to find the mean and variance of $p_n(x)$, it is even more interesting to see numerical results. When a particular set of normally distributed random samples was generated and used to compute $p_n(x)$, the results shown in Fig. 4.5 were obtained. These results depend both on n and h_1. For $n = 1$, $p_n(x)$ is merely a single Gaussian centered about the first sample, which of course has neither the mean nor the variance of the true distribution. For $n = 10$ and $h_1 = 0.1$ the contributions of the individual samples are clearly discernible; this is not the case for $h_1 = 1$ and $h_1 = 0.5$. As n gets larger, the ability of $p_n(x)$ to resolve variations in $p(x)$ increases. Concomitantly, $p_n(x)$ appears to be more sensitive to local sampling irregularities when n is large, although we are assured that $p_n(x)$ will converge to the smooth normal curve as n goes to infinity. While one should not judge on visual appearance alone, it is clear that many samples are required to obtain an accurate estimate. Figure 4.6 shows analogous results in two dimensions.

As a second one-dimensional example, we let $\varphi(x)$ and h_n be the same as in Fig. 4.5, but let the unknown density be a mixture of a uniform and a triangle density. Figure 4.7 shows the behavior of Parzen-window estimates for this density. As before, the case $n = 1$ tells more about the window function than it tells about the unknown density. For $n = 16$, none of the estimates is particularly good, but results for $n = 256$ and $h_1 = 1$ are beginning to appear acceptable.

4.3.4 Classification Example

In classifiers based on Parzen-window estimation, we estimate the densities for each category and classify a test point by the label corresponding to the maximum posterior. If there are multiple categories with unequal priors we can easily include these too (Problem 4). The decision regions for a Parzen-window classifier depend upon the choice of window function, of course, as illustrated in Fig. 4.8. In general, the training error—the empirical error on the training points themselves—can be made

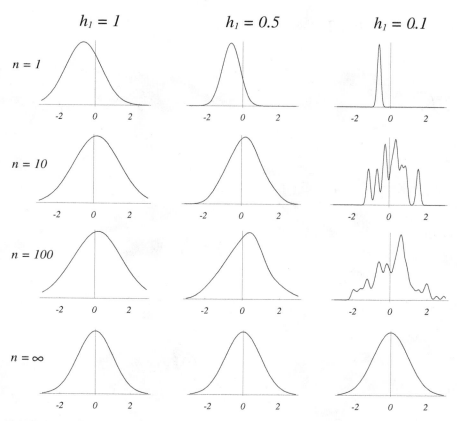

FIGURE 4.5. Parzen-window estimates of a univariate normal density using different window widths and numbers of samples. The vertical axes have been scaled to best show the structure in each graph. Note particularly that the $n = \infty$ estimates are the same (and match the true density function), regardless of window width.

arbitrarily low by making the window width sufficiently small.* However, the goal of creating a classifier is to classify novel patterns, and alas a low training error does not guarantee a small *test* error, as we shall explore in Chapter 9. Although a generic Gaussian window *shape* is plausible, in the absence of other information about the underlying distributions there is little theoretical justification of one window *width* over another.

These density estimation and classification examples illustrate some of the power and some of the limitations of nonparametric methods. Their power resides in their generality. Exactly the same procedure was used for the unimodal normal case and the bimodal mixture case, and we did not need to make any assumptions about the distributions ahead of time. With enough samples, we are essentially assured of convergence to an arbitrarily complicated target density. On the other hand, the number of samples needed may be very large indeed—much greater than would be required if we knew the form of the unknown density. Little or nothing in the way of data reduction is provided, which leads to severe requirements for computation time and storage. Moreover, the demand for a large number of samples grows exponentially

*We ignore cases in which the same feature vector has been assigned to multiple categories.

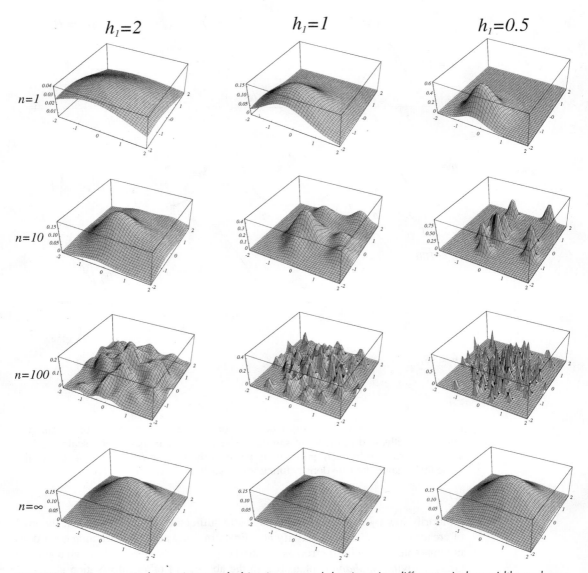

FIGURE 4.6. Parzen-window estimates of a bivariate normal density using different window widths and numbers of samples. The vertical axes have been scaled to best show the structure in each graph. Note particularly that the $n = \infty$ estimates are the same (and match the true distribution), regardless of window width.

with the dimensionality of the feature space. This limitation is called the "curse of dimensionality," and severely restricts the practical application of such nonparametric procedures (Problem 11). The fundamental reason for the curse of dimensionality is that high-dimensional functions have the potential to be much more complicated than low-dimensional ones, and that those complications are harder to discern. The only way to beat the curse is to incorporate knowledge about the data that is *correct*.

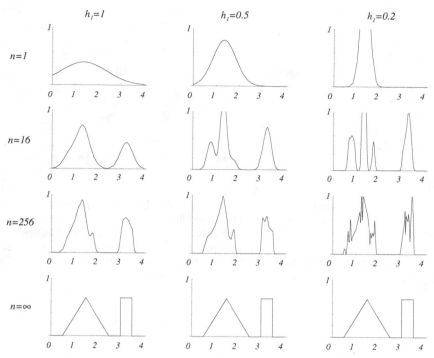

FIGURE 4.7. Parzen-window estimates of a bimodal distribution using different window widths and numbers of samples. Note particularly that the $n = \infty$ estimates are the same (and match the true distribution), regardless of window width.

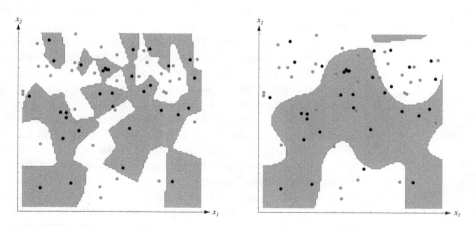

FIGURE 4.8. The decision boundaries in a two-dimensional Parzen-window dichotomizer depend on the window width h. At the left a small h leads to boundaries that are more complicated than for large h on same data set, shown at the right. Apparently, for these data a small h would be appropriate for the upper region, while a large h would be appropriate for the lower region; no single window width is ideal overall.

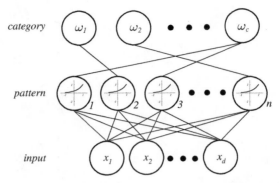

FIGURE 4.9. A probabilistic neural network (PNN) consists of d input units, n pattern units, and c category units. Each pattern unit forms the inner product of its weight vector and the normalized pattern vector $\mathbf{x}$ to form $z = \mathbf{w}^t\mathbf{x}$, and then it emits $\exp[(z-1)/\sigma^2]$. Each category unit sums such contributions from the pattern unit connected to it. This ensures that the activity in each of the category units represents the Parzen-window density estimate using a circularly symmetric Gaussian window of covariance $\sigma^2\mathbf{I}$, where $\mathbf{I}$ is the d-by-d identity matrix.

4.3.5 Probabilistic Neural Networks (PNNs)

Most pattern recognition methods can be implemented in a parallel fashion that trades space complexity for time complexity. These implementations are naturally represented as artificial neural networks, a topic that we treat in detail in Chapter 6. As a preview, we can take this opportunity to show how the Parzen-window method can be implemented as a neural network known as a Probabilistic Neural Network (Fig. 4.9). Suppose we wish to form a Parzen estimate based on n patterns, each of which is d-dimensional, randomly sampled from c classes. The PNN for this case

INPUT UNIT

PATTERN UNIT

CATEGORY UNIT

WEIGHT

consists of d *input units* comprising the input layer, where each unit is connected to each of the n *pattern units*; each pattern unit is, in turn, connected to one and only one of the c category units. The connections from the input to pattern units represent modifiable *weights*, which will be trained. (While these weights are merely parameters and could be represented by a vector $\boldsymbol{\theta}$, in keeping with the established terminology in neural networks we shall use the symbol $\mathbf{w}$.) Each category unit computes the sum of the pattern units connected to it.

The PNN is trained in the following way. First, each pattern $\mathbf{x}$ of the training set is normalized to have unit length—that is, is scaled so that $\sum_{i=1}^{d} x_i^2 = 1$. The first normalized training pattern is placed on the input units. The modifiable weights linking the input units and the first pattern unit are set such that $\mathbf{w}_1 = \mathbf{x}_1$. (Note that because of the normalization of $\mathbf{x}_1$, $\mathbf{w}_1$ is normalized too.) Then, a single connection from the first pattern unit is made to the category unit corresponding to the known class of that pattern. The process is repeated with each of the remaining training patterns, setting the weights to the successive pattern units such that $\mathbf{w}_k = \mathbf{x}_k$ for $k = 1, 2, \ldots, n$. After such training we have a network that is fully connected between input and pattern units, and sparsely connected from pattern to category units. If we denote the components of the jth pattern as x_{jk} and the weights to the jth pattern unit w_{jk}, for $j = 1, 2, \ldots, n$ and $k = 1, 2, \ldots, d$, then our algorithm is as follows:

▪ **Algorithm 1. (PNN training)**

1 **begin initialize** $j \leftarrow 0, n, a_{ji} \leftarrow 0$ for $j = 1, \ldots, n;\ i = 1, \ldots, c$
2 $\quad$ **do** $j \leftarrow j + 1$
3 $\qquad x_{jk} \leftarrow x_{jk} \Big/ \left(\sum_i^d x_{ji}^2 \right)^{1/2}$ $\quad$ (normalize)
4 $\qquad w_{jk} \leftarrow x_{jk}$ $\qquad\qquad$ (train)
5 $\qquad$ **if** $\mathbf{x}_j \in \omega_i$ **then** $a_{ji} \leftarrow 1$
6 $\quad$ **until** $j = n$
7 **end**

The trained network is then used for classification in the following way. A nor-malized test pattern $\mathbf{x}$ is placed at the input units. Each pattern unit computes the inner product to yield the *net activation* or simply *net*,

NET ACTIVATION

$$net_k = \mathbf{w}_k^t \mathbf{x}, \tag{28}$$

and emits a nonlinear function of net_k; each category unit sums the contributions from all pattern units connected to it. The nonlinear function is $e^{(net_k - 1)/\sigma^2}$, where σ is a parameter set by the user and determines the width of the effective Gaussian window. This *activation function* or transfer function, here must be an exponential to implement the Parzen windows algorithm. To see this, consider an (unnormalized) Gaussian window centered on the position of one of the training patterns $\mathbf{w}_k$. We work backwards from the desired Gaussian window function to infer the nonlinear activation function function that should be employed by the pattern units. That is, if we let our effective width h_n be a constant, the window function is

ACTIVATION FUNCTION

$$\varphi\left(\frac{\mathbf{x} - \mathbf{w}_k}{h_n}\right) \propto \overbrace{e^{-(\mathbf{x}-\mathbf{w}_k)^t(\mathbf{x}-\mathbf{w}_k)/2\sigma^2}}^{\text{desired Gaussian}}$$

$$= e^{-(\mathbf{x}^t\mathbf{x} + \mathbf{w}_k^t\mathbf{w}_k - 2\mathbf{x}^t\mathbf{w}_k)/2\sigma^2} = \underbrace{e^{(net_k - 1)/\sigma^2}}_{\substack{\text{activation} \\ \text{function}}}, \tag{29}$$

where we have used our normalization conditions $\mathbf{x}^t\mathbf{x} = \mathbf{w}_k^t\mathbf{w}_k = 1$. Thus each pattern unit contributes to its associated category unit a signal equal to the proba-bility the test point was generated by a Gaussian centered on the associated training point. The sum of these local estimates (computed at the corresponding category unit) gives the discriminant function $g_i(\mathbf{x})$—the Parzen-window estimate of the un-derlying distribution. The $\max_i g_i(\mathbf{x})$ operation gives the desired category for the test point (Algorithm 2).

▪ **Algorithm 2. (PNN Classification)**

1 **begin initialize** $k \leftarrow 0, \mathbf{x} \leftarrow$ test pattern
2 $\quad$ **do** $k \leftarrow k + 1$
3 $\qquad net_k \leftarrow \mathbf{w}_k^t \mathbf{x}$

4   **if** $a_{ki} = 1$ **then** $g_i \leftarrow g_i + \exp[(net_k - 1)/\sigma^2]$
5   **until** $k = n$
6   **return** $class \leftarrow \arg\max_i g_i(\mathbf{x})$

7 **end**

One of the benefits of PNNs is their speed of learning, because the learning rule (i.e., setting $\mathbf{w}_k = \mathbf{x}_k$) is simple and requires only a single pass through the training data. The space complexity (amount of memory) for the PNN is easy to determine by counting the number of connections in Fig. 4.9—$O((n + 1)d)$. This can be quite severe for instance in a hardware application, because both n and d can be quite large. The time complexity for classification by the parallel implementation of Fig. 4.9 is $O(1)$, since the n inner products of Eq. 28 can be done in parallel. Thus this PNN architecture could find uses where recognition speed is important and storage is not a severe limitation. Another benefit is that new training patterns can be incorporated into a previously trained classifier quite easily; this might be important for a particular on-line application.

4.3.6 Choosing the Window Function

As we have seen, one of the problems encountered in the Parzen-window/PNN approach concerns the choice of the sequence of cell-volume sizes V_1, V_2, ... or overall window size (or indeed other window parameters, such as shape or orientation). For example, if we take $V_n = V_1/\sqrt{n}$, the results for any finite n will be very sensitive to the choice for the initial volume V_1. If V_1 is too small, most of the volumes will be empty, and the estimate $p_n(\mathbf{x})$ will be very erratic (Fig. 4.7). On the other hand, if V_1 is too large, important spatial variations in $p(\mathbf{x})$ may be lost due to averaging over the cell volume. Furthermore, it may well be the case that a cell volume appropriate for one region of the feature space might be entirely unsuitable in a different region (Fig. 4.8). In Chapter 9 we shall consider general methods, including cross-validation, which are often used in conjunction with Parzen windows. In brief, cross-validation requires taking some small portion of the data to form a *validation set*. The classifier is trained on the remaining patterns in the training set, but the window width is adjusted to give the smallest error on the validation set.

4.4 k_n–NEAREST-NEIGHBOR ESTIMATION

PROTOTYPES

A potential remedy for the problem of the unknown "best" window function is to let the cell volume be a function of the *training data*, rather than some arbitrary function of the overall number of samples. For example, to estimate $p(\mathbf{x})$ from n training samples or *prototypes* we can center a cell about $\mathbf{x}$ and let it grow until it captures k_n samples, where k_n is some specified function of n. These samples are the k_n *nearest-neighbors* of $\mathbf{x}$. If the density is high near $\mathbf{x}$, the cell will be relatively small, which leads to good resolution. If the density is low, it is true that the cell will grow large, but it will stop soon after it enters regions of higher density. In either case, if we take

$$p_n(\mathbf{x}) = \frac{k_n/n}{V_n} \tag{30}$$

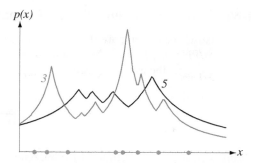

FIGURE 4.10. Eight points in one dimension and the k-nearest-neighbor density estimates, for $k = 3$ and 5. Note especially that the discontinuities in the slopes in the estimates generally lie *away* from the positions of the prototype points.

we want k_n to go to infinity as n goes to infinity, since this assures us that k_n/n will be a good estimate of the probability that a point will fall in the cell of volume V_n. However, we also want k_n to grow sufficiently slowly that the size of the cell needed to capture k_n training samples will shrink to zero. Thus, it is clear from Eq. 30 that the ratio k_n/n must go to zero. Although we shall not supply a proof, it can be shown that the conditions $\lim_{n\to\infty} k_n = \infty$ and $\lim_{n\to\infty} k_n/n = 0$ are necessary and sufficient for $p_n(\mathbf{x})$ to converge to $p(\mathbf{x})$ in probability at all points where $p(\mathbf{x})$ is continuous (Problem 5). If we take $k_n = \sqrt{n}$ and assume that $p_n(\mathbf{x})$ is a reasonably good approximation to $p(\mathbf{x})$, we then see from Eq. 30 that $V_n \simeq 1/(\sqrt{n}\,p(\mathbf{x}))$. Thus, V_n again has the form $V_1/\sqrt{n}$, but the initial volume V_1 is determined by the nature of the data rather than by some arbitrary choice on our part. It is interesting to note that although $p_n(\mathbf{x})$ is continuous, its slope (or gradient) is not. Furthermore, the points of discontinuity are rarely the same as the prototype points (Figs. 4.10 and 4.11).

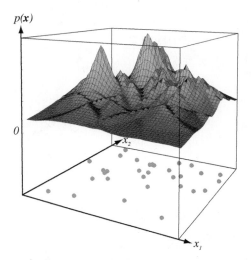

FIGURE 4.11. The k-nearest-neighbor estimate of a two-dimensional density for $k = 5$. Notice how such a finite n estimate can be quite "jagged," and notice that discontinuities in the slopes generally occur along lines away from the positions of the points themselves.

4.4.1 k_n–Nearest-Neighbor and Parzen-Window Estimation

It is instructive to compare the performance of this method with that of the Parzen-window/PNN method on the data used in the previous examples. With $n = 1$ and $k_n = \sqrt{n} = 1$, the estimate becomes

$$p_n(x) = \frac{1}{2|x - x_1|}. \tag{31}$$

This is clearly a poor estimate of $p(x)$, with its integral embarrassing us by diverging to infinity. As shown in Fig. 4.12, the estimate becomes considerably better as n gets larger, even though the integral of the estimate remains infinite. This unfortunate fact is compensated by the fact that $p_n(x)$ never plunges to zero just because no samples fall within some arbitrary cell or window. While this might seem to be a meager compensation, it can be of considerable value in higher-dimensional spaces.

As with the Parzen-window approach, we could obtain a family of estimates by taking $k_n = k_1\sqrt{n}$ and choosing different values for k_1. However, in the absence of any additional information, one choice is as good as another, and we can be confident only that the results will be correct in the infinite data case. For classification, one

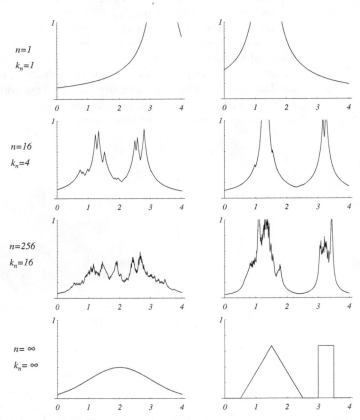

FIGURE 4.12. Several k-nearest-neighbor estimates of two unidimensional densities: a Gaussian and a bimodal distribution. Notice how the finite n estimates can be quite "spiky."

popular method is to adjust the window width until the classifier has the lowest error on a separate set of samples, also drawn from the target distributions, a technique we shall explore in Chapter 9.

4.4.2 Estimation of *A Posteriori* Probabilities

The techniques discussed in the previous sections can be used to estimate the *a posteriori* probabilities $P(\omega_i|\mathbf{x})$ from a set of n labeled samples by using the samples to estimate the densities involved. Suppose that we place a cell of volume V around $\mathbf{x}$ and capture k samples, k_i of which turn out to be labeled ω_i. Then the obvious estimate for the joint probability $p(\mathbf{x}, \omega_i)$ is

$$p_n(\mathbf{x}, \omega_i) = \frac{k_i/n}{V},\tag{32}$$

and thus a reasonable estimate for $P(\omega_i|\mathbf{x})$ is

$$P_n(\omega_i|\mathbf{x}) = \frac{p_n(\mathbf{x}, \omega_i)}{\sum_{j=1}^{c} p_n(\mathbf{x}, \omega_j)} = \frac{k_i}{k}.\tag{33}$$

That is, the estimate of the *a posteriori* probability that ω_i is the state of nature is merely the fraction of the samples within the cell that are labeled ω_i. Consequently, for minimum error rate we select the category most frequently represented within the cell. If there are enough samples and if the cell is sufficiently small, it can be shown that this will yield performance approaching the best possible.

When it comes to choosing the size of the cell, it is clear that we can use either the Parzen-window approach or the k_n-nearest-neighbor approach. In the first case, V_n would be some specified function of n, such as $V_n = 1/\sqrt{n}$. In the second case, V_n would be expanded until some specified number of samples were captured, such as $k = \sqrt{n}$. In either case, as n goes to infinity an infinite number of samples will fall within the infinitely small cell. The fact that the cell volume could become arbitrarily small and yet contain an arbitrarily large number of samples would allow us to learn the unknown probabilities with virtual certainty and thus eventually obtain optimum performance. Interestingly enough, we shall now see that we can obtain comparable performance if we base our decision solely on the label of the *single* nearest neighbor of $\mathbf{x}$.

4.5 THE NEAREST-NEIGHBOR RULE

We begin by letting $\mathcal{D}^n = \{\mathbf{x}_1, \ldots, \mathbf{x}_n\}$ denote a set of n labeled prototypes and letting $\mathbf{x}' \in \mathcal{D}^n$ be the prototype nearest to a test point $\mathbf{x}$. Then the *nearest-neighbor rule* for classifying $\mathbf{x}$ is to assign it the label associated with $\mathbf{x}'$. The nearest-neighbor rule is a suboptimal procedure; its use will usually lead to an error rate greater than the minimum possible, the Bayes rate. We shall see, however, that with an unlimited number of prototypes the error rate is never worse than twice the Bayes rate.

Before we get immersed in details, let us try to gain a heuristic understanding of why the nearest-neighbor rule should work so well. To begin with, note that the label θ' associated with the nearest neighbor is a random variable, and the probability that $\theta' = \omega_i$ is merely the *a posteriori* probability $P(\omega_i|\mathbf{x}')$. When the number of samples is very large, it is reasonable to assume that $\mathbf{x}'$ is sufficiently close to $\mathbf{x}$ that $P(\omega_i|\mathbf{x}') \simeq P(\omega_i|\mathbf{x})$. Because this is exactly the probability that nature will be in state ω_i, the nearest-neighbor rule is effectively matching probabilities with nature.

If we define $\omega_m(\mathbf{x})$ by

$$P(\omega_m|\mathbf{x}) = \max_i P(\omega_i|\mathbf{x}), \tag{34}$$

VORONOI TESSELATION

then the Bayes decision rule always selects ω_m. This rule allows us to partition the feature space into cells consisting of all points closer to a given training point $\mathbf{x}'$ than to any other training points. All points in such a cell are thus labeled by the category of the training point—a so-called *Voronoi tesselation* of the space (Fig. 4.13).

When $P(\omega_m|\mathbf{x})$ is close to unity, the nearest-neighbor selection is almost always the same as the Bayes selection. That is, when the minimum probability of error is small, the nearest-neighbor probability of error is also small. When $P(\omega_m|\mathbf{x})$ is close to $1/c$, so that all classes are essentially equally likely, the selections made by the nearest-neighbor rule and the Bayes decision rule are rarely the same, but the probability of error is approximately $1 - 1/c$ for both. While more careful analysis is clearly necessary, these observations should make the good performance of the nearest-neighbor rule less surprising.

Our analysis of the behavior of the nearest-neighbor rule will be directed at obtaining the infinite-sample conditional average probability of error $P(e|\mathbf{x})$, where the averaging is with respect to the training samples. The unconditional average probability of error will then be found by averaging $P(e|\mathbf{x})$ over all $\mathbf{x}$:

$$P(e) = \int P(e|\mathbf{x})\, p(\mathbf{x})\, d\mathbf{x}. \tag{35}$$

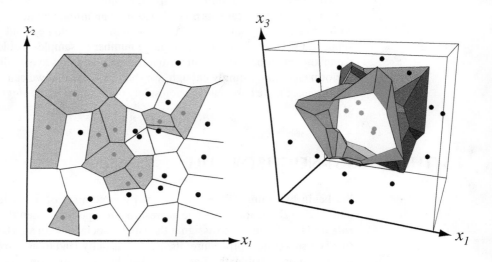

FIGURE 4.13. In two dimensions, the nearest-neighbor algorithm leads to a partitioning of the input space into Voronoi cells, each labeled by the category of the training point it contains. In three dimensions, the cells are three-dimensional, and the decision boundary resembles the surface of a crystal.

In passing we should recall that the Bayes decision rule minimizes $P(e)$ by minimizing $P(e|\mathbf{x})$ for every $\mathbf{x}$. Recall from Chapter 2 that if we let $P^*(e|\mathbf{x})$ be the minimum possible value of $P(e|\mathbf{x})$, and P^* be the minimum possible value of $P(e)$, then

$$P^*(e|\mathbf{x}) = 1 - P(\omega_m|\mathbf{x}) \tag{36}$$

and

$$P^* = \int P^*(e|\mathbf{x}) p(\mathbf{x}) \, d\mathbf{x}. \tag{37}$$

4.5.1 Convergence of the Nearest Neighbor

We now wish to evaluate the average probability of error for the nearest-neighbor rule. In particular, if $P_n(e)$ is the n-sample error rate, and if

$$P = \lim_{n \to \infty} P_n(e), \tag{38}$$

then we want to show that

$$P^* \le P \le P^*\left(2 - \frac{c}{c-1} P^*\right). \tag{39}$$

We begin by observing that when the nearest-neighbor rule is used with a particular set of n samples, the resulting error rate will depend on the accidental characteristics of the samples. In particular, if different sets of n samples are used to classify $\mathbf{x}$, different vectors $\mathbf{x}'$ will be obtained for the nearest neighbor of $\mathbf{x}$. Because the decision rule depends on this nearest neighbor, we have a conditional probability of error $P(e|\mathbf{x}, \mathbf{x}')$ that depends on both $\mathbf{x}$ and $\mathbf{x}'$. By averaging over $\mathbf{x}'$, we obtain

$$P(e|\mathbf{x}) = \int P(e|\mathbf{x}, \mathbf{x}') p(\mathbf{x}'|\mathbf{x}) \, d\mathbf{x}'. \tag{40}$$

It is usually very difficult to obtain an exact expression for the conditional density $p(\mathbf{x}'|\mathbf{x})$. However, because $\mathbf{x}'$ is by definition the nearest neighbor of $\mathbf{x}$, we expect this density to be very peaked in the immediate vicinity of $\mathbf{x}$, and very small elsewhere. Furthermore, as n goes to infinity we expect $p(\mathbf{x}'|\mathbf{x})$ to approach a delta function centered at $\mathbf{x}$, making the evaluation of Eq. 40 trivial. To show that this is indeed the case, we must assume that at the given $\mathbf{x}$, $p(\cdot)$ is continuous and not equal to zero. Under these conditions, the probability that any sample falls within a hypersphere $\mathcal{S}$ centered about $\mathbf{x}$ is some positive number P_s:

$$P_s = \int_{\mathbf{x}' \in \mathcal{S}} p(\mathbf{x}') \, d\mathbf{x}'. \tag{41}$$

Thus, the probability that all n of the independently drawn samples fall outside this hypersphere is $(1 - P_s)^n$, which approaches zero as n goes to infinity. Thus $\mathbf{x}'$ converges to $\mathbf{x}$ in probability, and $p(\mathbf{x}'|\mathbf{x})$ approaches a delta function, as expected. In fact, by using measure theoretic methods one can make even stronger (as well as more rigorous) statements about the convergence of $\mathbf{x}'$ to $\mathbf{x}$, but this result is sufficient for our purposes.

4.5.2 Error Rate for the Nearest-Neighbor Rule

We now turn to the calculation of the conditional probability of error $P_n(e|\mathbf{x}, \mathbf{x}')$. To avoid a potential source of confusion, we must state the problem with somewhat greater care than has been exercised so far. For example, to show explicitly that $\mathbf{x}'$, the nearest neighbor of $\mathbf{x}$, can change as we increase the number n of samples, we now denote the nearest neighbor by $\mathbf{x}'_n$. When we say that we have n independently drawn labeled samples, we are talking about n pairs of random variables $(\mathbf{x}_1, \theta_1), (\mathbf{x}_2, \theta_2), \ldots, (\mathbf{x}_n, \theta_n)$, where θ_j may be any of the c states of nature $\omega_1, \ldots, \omega_c$. We assume that these pairs were generated by selecting a state of nature ω_j for θ_j with probability $P(\omega_j)$ and then selecting an $\mathbf{x}_j$ according to the probability law $p(\mathbf{x}|\omega_j)$, with each pair being selected independently. Suppose that during classification, nature selects a pair $(\mathbf{x}, \theta)$, and also suppose that $\mathbf{x}'_n$, labeled θ'_n, is the training sample nearest $\mathbf{x}$. Because the state of nature when $\mathbf{x}'_n$ was drawn is independent of the state of nature when $\mathbf{x}$ is drawn, we have

$$P(\theta, \theta'_n | \mathbf{x}, \mathbf{x}'_n) = P(\theta|\mathbf{x})P(\theta'_n|\mathbf{x}'_n). \tag{42}$$

Now if we use the nearest-neighbor decision rule, we commit an error whenever $\theta \neq \theta'_n$. Thus, the conditional probability of error $P_n(e|\mathbf{x}, \mathbf{x}'_n)$ is given by

$$P_n(e|\mathbf{x}, \mathbf{x}'_n) = 1 - \sum_{i=1}^{c} P(\theta = \omega_i, \theta'_n = \omega_i|\mathbf{x}, \mathbf{x}'_n)$$

$$= 1 - \sum_{i=1}^{c} P(\omega_i|\mathbf{x})P(\omega_i|\mathbf{x}'_n). \tag{43}$$

To obtain $P_n(e)$ we must substitute this expression into Eq. 40 for $P_n(e|\mathbf{x})$ and then average the result over $\mathbf{x}$. This is very difficult, in general, but as we remarked earlier the integration called for in Eq. 40 becomes trivial as n goes to infinity and $p(\mathbf{x}'_n|\mathbf{x})$ approaches a delta function. If $P(\omega_i|\mathbf{x})$ is continuous at $\mathbf{x}$, we thus obtain

$$\lim_{n \to \infty} P_n(e|\mathbf{x}) = \int \left[1 - \sum_{i=1}^{c} P(\omega_i|\mathbf{x})P(\omega_i|\mathbf{x}'_n)\right]\delta(\mathbf{x}'_n - \mathbf{x})\, d\mathbf{x}'_n$$

$$= 1 - \sum_{i=1}^{c} P^2(\omega_i|\mathbf{x}). \tag{44}$$

Therefore, provided that we can exchange some limits and integrals, the asymptotic nearest-neighbor error rate is given by

$$P = \lim_{n \to \infty} P_n(e)$$

$$= \lim_{n \to \infty} \int P_n(e|\mathbf{x})p(\mathbf{x})\, d\mathbf{x}$$

$$= \int \left[1 - \sum_{i=1}^{c} P^2(\omega_i|\mathbf{x})\right] p(\mathbf{x})\, d\mathbf{x}. \tag{45}$$

4.5.3 Error Bounds

While Eq. 45 presents an exact result, it is more illuminating to obtain bounds on P in terms of the Bayes rate P^*. An obvious lower bound on P is P^* itself. Furthermore, it can be shown that for any P^* there is a set of conditional and prior probabilities for which the bound is achieved, so in this sense it is a tight lower bound.

The problem of establishing a tight upper bound is more interesting. The basis for hoping for a low upper bound comes from observing that if the Bayes rate is low, $P(\omega_i|\mathbf{x})$ is near 1.0 for some i, say $i = m$. Thus the integrand in Eq. 45 is approximately $1 - P^2(\omega_m|\mathbf{x}) \simeq 2(1 - P(\omega_m|\mathbf{x}))$, and since

$$P^*(e|\mathbf{x}) = 1 - P(\omega_m|\mathbf{x}), \tag{46}$$

integration over $\mathbf{x}$ might yield about twice the Bayes rate, which is still low and acceptable for some applications. To obtain an exact upper bound, we must find out how large the nearest-neighbor error rate P can become for a given Bayes rate P^*. Thus, Eq. 45 leads us to ask how small $\sum_{i=1}^{c} P^2(\omega_i|\mathbf{x})$ can be for a given $P(\omega_m|\mathbf{x})$. First we write

$$\sum_{i=1}^{c} P^2(\omega_i|\mathbf{x}) = P^2(\omega_m|\mathbf{x}) + \sum_{i \neq m} P^2(\omega_i|\mathbf{x}), \tag{47}$$

and then seek to bound this sum by minimizing the second term subject to the following constraints:

- $P(\omega_i|\mathbf{x}) \geq 0$
- $\sum_{i \neq m} P(\omega_i|\mathbf{x}) = 1 - P(\omega_m|\mathbf{x}) = P^*(e|\mathbf{x})$.

With a little thought we see that $\sum_{i=1}^{c} P^2(\omega_i|\mathbf{x})$ is minimized if all of the *a posteriori* probabilities except the mth are equal. The second constraint yields

$$P(\omega_i|\mathbf{x}) = \begin{cases} \dfrac{P^*(e|\mathbf{x})}{c-1} & i \neq m \\ 1 - P^*(e|\mathbf{x}) & i = m. \end{cases} \tag{48}$$

Thus we have the inequalities

$$\sum_{i=1}^{c} P^2(\omega_i|\mathbf{x}) \geq (1 - P^*(e|\mathbf{x}))^2 + \frac{P^{*2}(e|\mathbf{x})}{c-1} \tag{49}$$

and

$$1 - \sum_{i=1}^{c} P^2(\omega_i|\mathbf{x}) \leq 2P^*(e|\mathbf{x}) - \frac{c}{c-1} P^{*2}(e|\mathbf{x}). \tag{50}$$

This immediately shows that $P \leq 2P^*$, since we can substitute this result in Eq. 45 and merely drop the second term. However, a tighter bound can be obtained by observing that the variance is:

$$\text{Var}[P^*(e|\mathbf{x})] = \int [P^*(e|\mathbf{x}) - P^*]^2 p(\mathbf{x}) \, d\mathbf{x}$$

$$= \int P^{*2}(e|\mathbf{x}) p(\mathbf{x}) \, d\mathbf{x} - P^{*2} \geq 0,$$

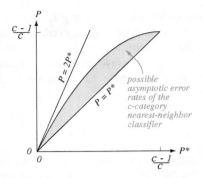

FIGURE 4.14. Bounds on the nearest-neighbor error rate P in a c-category problem given infinite training data, where P^* is the Bayes error (Eq. 52). At low error rates, the nearest-neighbor error rate is bounded above by twice the Bayes rate.

so that

$$\int P^{*2}(e|\mathbf{x})\,p(\mathbf{x})\,d\mathbf{x} \geq P^{*2}, \tag{51}$$

with equality holding if and only if the variance of $P^*(e|\mathbf{x})$ is zero. Using this result and substituting Eq. 50 into Eq. 45, we obtain the desired bounds on the nearest-neighbor error P in the case of an infinite number of samples:

$$P^* \leq P \leq P^*\left(2 - \frac{c}{c-1}P^*\right). \tag{52}$$

It is easy to show that this upper bound is achieved in the so-called zero-information case in which the densities $p(\mathbf{x}|\omega_i)$ are identical, so that $P(\omega_i|\mathbf{x}) = P(\omega_i)$ and furthermore $P^*(e|\mathbf{x})$ is independent of $\mathbf{x}$ (Problem 17). Thus the bounds given by Eq. 52 are as tight as possible, in the sense that for any P^* there exist conditional and prior probabilities for which the bounds are achieved. In particular, the Bayes rate P^* can be anywhere between 0 and $(c-1)/c$ and the bounds meet at the two extreme values for the probabilities. When the Bayes rate is small, the upper bound is approximately twice the Bayes rate (Fig. 4.14).

Because P is always less than or equal to $2P^*$, if one had an infinite collection of data and used an arbitrarily complicated decision rule, one could at most cut the error rate in half. In this sense, at least half of the classification information in an infinite data set resides in the nearest neighbor.

It is natural to ask how well the nearest-neighbor rule works in the finite-sample case, and how rapidly the performance converges to the asymptotic value. Unfortunately, despite prolonged effort on such problems, the only statements that can be made in the general case are negative. It can be shown that convergence can be arbitrarily slow, and the error rate $P_n(e)$ need not even decrease monotonically with n. As with other nonparametric methods, it is difficult to obtain anything other than asymptotic results without making further assumptions about the underlying probability structure (Problems 13 and 14).

4.5.4 The k-Nearest-Neighbor Rule

An obvious extension of the nearest-neighbor rule is the *k-nearest-neighbor rule*. As one would expect from the name, this rule classifies $\mathbf{x}$ by assigning it the label

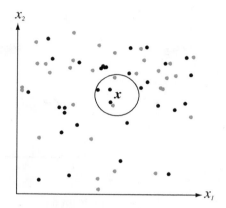

FIGURE 4.15. The k-nearest-neighbor query starts at the test point **x** and grows a spherical region until it encloses k training samples, and it labels the test point by a majority vote of these samples. In this $k = 5$ case, the test point **x** would be labeled the category of the black points.

most frequently represented among the k nearest samples; in other words, a decision is made by examining the labels on the k nearest neighbors and taking a vote (Fig. 4.15). We shall not go into a thorough analysis of the k-nearest-neighbor rule. However, by considering the two-class case with k odd (to avoid ties), we can gain some additional insight into these procedures.

The basic motivation for considering the k-nearest-neighbor rule rests on our earlier observation about matching probabilities with nature. We notice first that if k is fixed and the number n of samples is allowed to approach infinity, then all of the k nearest neighbors will converge to **x**. Hence, as in the single-nearest-neighbor cases, the labels on each of the k-nearest-neighbors are random variables, which independently assume the values ω_i with probabilities $P(\omega_i|\mathbf{x})$, $i = 1, 2$. If $P(\omega_m|\mathbf{x})$ is the larger *a posteriori* probability, then the Bayes decision rule always selects ω_m. The single-nearest-neighbor rule selects ω_m with probability $P(\omega_m|\mathbf{x})$. The k-nearest-neighbor rule selects ω_m if a majority of the k nearest neighbors are labeled ω_m, an event of probability

$$\sum_{i=(k+1)/2}^{k} \binom{k}{i} P(\omega_m|\mathbf{x})^i [1 - P(\omega_m|\mathbf{x})]^{k-i}. \tag{53}$$

In general, the larger the value of k, the greater the probability that ω_m will be selected.

We could analyze the k-nearest-neighbor rule in much the same way that we analyzed the single-nearest-neighbor rule. However, since the arguments become more involved and supply little additional insight, we shall content ourselves with stating the results. It can be shown that if k is odd, the large-sample two-class error rate for the k-nearest-neighbor rule is bounded above by the function $C_k(P^*)$, where $C_k(P^*)$ is defined to be the smallest concave function of P^* greater than

$$\sum_{i=0}^{(k-1)/2} \binom{k}{i} \left[(P^*)^{i+1}(1 - P^*)^{k-i} + (P^*)^{k-i}(1 - P^*)^{i+1} \right]. \tag{54}$$

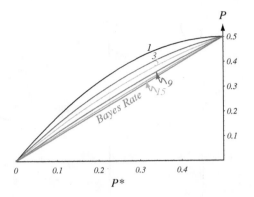

FIGURE 4.16. The error rate for the k-nearest-neighbor rule for a two-category problem is bounded by $C_k(P^*)$ in Eq. 54. Each curve is labeled by k; when $k = \infty$, the estimated probabilities match the true probabilities and thus the error rate is equal to the Bayes rate, that is, $P = P^*$.

Here the summation over the first bracketed term represents the probability of error due to i points coming from the category having the minimum probability and $k - i > i$ points from the other category. The summation over the second term in the brackets is the probability that $k - i$ points are from the minimum-probability category and $i + 1 < k - i$ from the higher probability category. Both of these cases constitute errors under the k-nearest-neighbor decision rule, and thus we must add them to find the full probability of error (Problem 18).

Figure 4.16 shows the bounds on the k-nearest-neighbor error rates for several values of k. As k increases, the upper bounds get progressively closer to the lower bound—the Bayes rate. In the limit as k goes to infinity, the two bounds meet and the k-nearest-neighbor rule becomes optimal.

At the risk of sounding repetitive, we conclude by commenting once again on the finite-sample situation encountered in practice. The k-nearest-neighbor rule can be viewed as another attempt to estimate the *a posteriori* probabilities $P(\omega_i | \mathbf{x})$ from samples. We want to use a large value of k to obtain a reliable estimate. On the other hand, we want all of the k nearest neighbors $\mathbf{x}'$ to be very near $\mathbf{x}$ to be sure that $P(\omega_i | \mathbf{x}')$ is approximately the same as $P(\omega_i | \mathbf{x})$. This forces us to choose a compromise k that is a small fraction of the number of samples. It is only in the limit as n goes to infinity that we can be assured of the nearly optimal behavior of the k-nearest-neighbor rule.

4.5.5 Computational Complexity of the k–Nearest-Neighbor Rule

The computational complexity of the nearest-neighbor algorithm—both in space (storage of prototypes) and time (search)—has received a great deal of analysis. There are a number of elegant theorems from computational geometry on the construction of Voronoi tesselations and nearest-neighbor searches in one- and two-dimensional spaces. However, because the greatest use of nearest-neighbor techniques is for problems with many features, we concentrate on the more general d-dimensional case.

Suppose we have n labeled training samples in d dimensions and we seek the (single) closest to a test point $\mathbf{x}$ ($k = 1$). In the most naive approach we inspect each stored point in turn, calculate its Euclidean distance to $\mathbf{x}$, retaining the identity only of the current closest one. Each distance calculation is $O(d)$, and thus this search

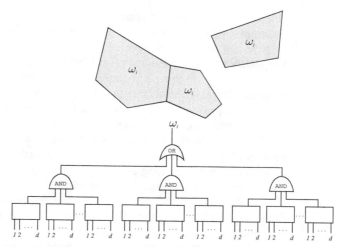

FIGURE 4.17. A parallel nearest-neighbor circuit can perform search in constant—that is, $O(1)$—time. The d-dimensional test pattern **x** is presented to each box, which calculates which side of a cell's face **x** lies on. If it is on the "close" side of every face of a cell, it lies in the Voronoi cell of the stored pattern, and receives its label. In the case shown, each of the three AND gates corresponds to a single Voronoi cell.

is $O(dn)$. An alternative but straightforward parallel implementation is shown in Fig. 4.17, which is $O(1)$ in time and $O(n)$ in space.

There are three general algorithmic techniques for reducing the computational burden in nearest-neighbor searches: computing partial distances, prestructuring, and editing the stored prototypes. In *partial distance*, we calculate the distance using some subset r of the full d dimensions, and if this partial distance is too great, we do not compute further. The partial distance based on r selected dimensions is

PARTIAL DISTANCE

$$D_r(\mathbf{a}, \mathbf{b}) = \left(\sum_{k=1}^{r} (a_k - b_k)^2 \right)^{1/2} \tag{55}$$

where $r < d$. Intuitively speaking, partial distance methods assume that what we know about the distance in a subspace is indicative of the full space. Of course, the partial distance is strictly nondecreasing as we add the contributions from more and more dimensions. Consequently, we can confidently terminate a distance calculation to any prototype once its partial distance is greater than the full $r = d$ Euclidean distance to the current closest prototype.

SEARCH TREE   In prestructuring we create some form of *search tree* in which prototypes are selectively linked. During classification, we compute the distance of the test point to one or a few stored "entry" or "root" prototypes and then consider only the prototypes linked to it. Of these, we find the one that is closest to the test point, and we recursively consider only subsequent linked prototypes. If the tree is properly structured, we will reduce the total number of prototypes that need to be searched.

Consider a trivial illustration of prestructuring in which we store a large number of prototypes that happen to be distributed uniformly in the unit square—that is, $p(\mathbf{x}) \sim U\left(\binom{0}{0}, \binom{1}{1}\right)$. Imagine we prestructure this set using four entry or root prototypes—at $\binom{1/4}{1/4}$, $\binom{1/4}{3/4}$, $\binom{3/4}{1/4}$ and $\binom{3/4}{3/4}$—each fully linked only to points in its corresponding quadrant. When a test pattern **x** is presented, the closest of these four prototypes is determined, and then the search is limited to the prototypes in the corresponding quadrant. In this way, 3/4 of the prototypes need never be queried.

Note that in this method we are no longer guaranteed to find the closest prototype. For instance, suppose the test point is near a boundary of the quadrants—for example, $\mathbf{x} = \begin{pmatrix} 0.499 \\ 0.499 \end{pmatrix}$. In this particular case, only prototypes in the first quadrant will be searched. Note, however, that the closest prototype might actually be in one of the other three quadrants, somewhere near $\begin{pmatrix} 0.5 \\ 0.5 \end{pmatrix}$. This illustrates a very general property in pattern recognition: the tradeoff of search complexity against accuracy.

More sophisticated search trees will have each stored prototype linked to a small number of others, and a full analysis of these methods would take us far afield. Nevertheless, here too so long as we do not query all training prototypes, we are not guaranteed that the nearest prototype will be found.

The third method for reducing the complexity of nearest-neighbor search is to eliminate "useless" prototypes during training, a technique known variously as *editing*, *pruning* or *condensing*. A simple method to reduce the $O(n)$ space complexity is to eliminate prototypes that are surrounded by training points of the same category label. This leaves the decision boundaries—and hence the error—unchanged, while reducing recall times. A simple editing algorithm is as follows.

EDITING

■ **Algorithm 3. (Nearest-Neighbor Editing)**

1 **begin initialize** $j \leftarrow 0$, $\mathcal{D} \leftarrow$ data set, $n \leftarrow$ # prototypes
2   construct the full Voronoi diagram of $\mathcal{D}$
3   **do** $j \leftarrow j + 1$; for each prototype $\mathbf{x}'_j$
4   find the Voronoi neighbors of $\mathbf{x}'_j$
5   **if** any neighbor is not from the same class as $\mathbf{x}'_j$, **then** mark $\mathbf{x}'_j$
6   **until** $j = n$
7   discard all points that are not marked
8   construct the Voronoi diagram of the remaining (marked) prototypes
9 **end**

The complexity of this editing algorithm is $O(d^3 n^{\lfloor d/2 \rfloor} \ln n)$, where the "floor" operation ($\lfloor \cdot \rfloor$) implies $\lfloor d/2 \rfloor = k$ if d is even and $2k - 1$ if d is odd (Problem 10).

According to Algorithm 3, if a prototype contributes to a decision boundary (i.e., at least one of its neighbors is from a different category), then it remains in the set; otherwise it is edited away (Problem 15). This algorithm does not guarantee that the *minimal* set of points is found (Problem 16), nevertheless, it is one of the examples in pattern recognition in which the computational complexity can be reduced—sometimes significantly—without affecting the accuracy. One drawback of such pruned nearest-neighbor systems is that one generally cannot add training data later, because the pruning step requires knowledge of *all* the training data ahead of time (Computer exercise 5). We conclude this section by noting the obvious—that is, that we can combine these three complexity reduction methods. We might first edit the prototypes, then form a search tree during training, and finally compute partial distances during classification.

4.6 METRICS AND NEAREST-NEIGHBOR CLASSIFICATION

The nearest-neighbor classifier relies on a metric or "distance" function between patterns. While so far we have assumed the Euclidean metric in d dimensions, the notion of a metric is far more general, and we now turn to the use alternative measures of distance to address key problems in classification. First let us review the properties of a metric. A metric $D(\cdot, \cdot)$ is merely a function that gives a generalized scalar distance between two argument patterns.

4.6.1 Properties of Metrics

A metric must have four properties: For all vectors **a**, **b**, and **c**, these properties are as follows:

nonnegativity: $D(\mathbf{a}, \mathbf{b}) \geq 0$

reflexivity: $D(\mathbf{a}, \mathbf{b}) = 0$ if and only if $\mathbf{a} = \mathbf{b}$

symmetry: $D(\mathbf{a}, \mathbf{b}) = D(\mathbf{b}, \mathbf{a})$

triangle inequality: $D(\mathbf{a}, \mathbf{b}) + D(\mathbf{b}, \mathbf{c}) \geq D(\mathbf{a}, \mathbf{c})$.

It is easy to verify that the Euclidean formula for distance in d dimensions,

$$D(\mathbf{a}, \mathbf{b}) = \left(\sum_{k=1}^{d} (a_k - b_k)^2 \right)^{1/2}, \tag{56}$$

possesses the properties of metric. Although one can always compute the Euclidean distance between two vectors, the results may or may not be meaningful. For example, if the space is transformed by multiplying each coordinate by an arbitrary constant, the Euclidean distance relationships in the transformed space can be very different from the original distance relationships, even though the transformation merely amounts to a different choice of units for the features (Problem 19). Such scale changes can have a major impact on nearest-neighbor classifiers (Fig. 4.18).

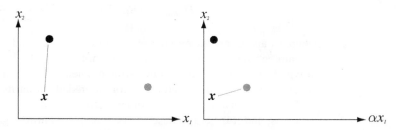

FIGURE 4.18. Scaling the coordinates of a feature space can change the distance relationships computed by the Euclidean metric. Here we see how such scaling can change the behavior of a nearest-neighbor classifer. Consider the test point **x** and its nearest neighbor. In the original space (left), the black prototype is closest. In the figure at the right, the x_1 axis has been rescaled by a factor $\alpha = 1/3$; now the nearest prototype is the red one. If there is a large disparity in the ranges of the full data in each dimension, a common procedure is to rescale all the data to equalize such ranges, and this is equivalent to changing the metric in the original space.

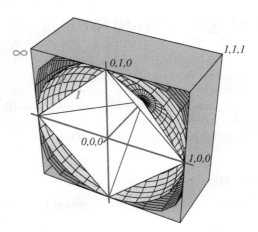

FIGURE 4.19. Each colored surface consists of points a distance 1.0 from the origin, measured using different values for k in the Minkowski metric (k is printed in red). Thus the white surfaces correspond to the L_1 norm (Manhattan distance), the light gray sphere corresponds to the L_2 norm (Euclidean distance), the dark gray ones correspond to the L_4 norm, and the pink box corresponds to the L_∞ norm.

One general class of metrics for d-dimensional patterns is the Minkowski metric

$$L_k(\mathbf{a}, \mathbf{b}) = \left(\sum_{i=1}^{d} |a_i - b_i|^k \right)^{1/k}, \tag{57}$$

also referred to as the L_k norm (Problem 20); thus, the Euclidean distance is the L_2 norm. The L_1 norm is sometimes called the *Manhattan* or *city block* distance, the shortest path between $\mathbf{a}$ and $\mathbf{b}$, each segment of which is parallel to a coordinate axis. (The name derives from the fact that the streets of Manhattan run north-south and east-west.) Suppose we compute the distances between the projections of $\mathbf{a}$ and $\mathbf{b}$ onto each of the d coordinate axes. The L_∞ distance between $\mathbf{a}$ and $\mathbf{b}$ corresponds to the maximum of these projected distances (Fig. 4.19).

The *Tanimoto metric* finds most use in taxonomy, where the distance between two sets is defined as

$$D_{Tanimoto}(\mathcal{S}_1, \mathcal{S}_2) = \frac{n_1 + n_2 - 2n_{12}}{n_1 + n_2 - n_{12}}, \tag{58}$$

where n_1 and n_2 are the number of elements in sets $\mathcal{S}_1$ and $\mathcal{S}_2$, respectively, and n_{12} is the number that is in both sets. The Tanimoto metric finds greatest use for problems in which two patterns or features—the elements in the set—are either the same or different, and there is no natural notion of graded similarity (Problem 27).

The selection among these or other metrics is generally dictated by computational concerns, and it is hard to base a choice on prior knowledge about the distributions. One exception is when there is great difference in the range of the data along different axes in a multidimensional data space. Here, we should scale the data—or equivalently alter the metric—as suggested in Fig. 4.18.

4.6.2 Tangent Distance

There may be drawbacks inherent in the uncritical use of a particular metric in nearest-neighbor classifiers, and these drawbacks can be overcome by the careful

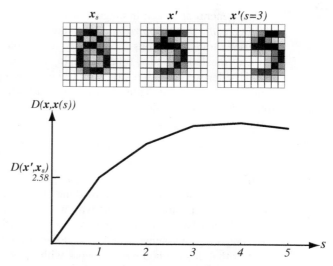

FIGURE 4.20. The uncritical use of Euclidean metric cannot address the problem of translation invariance. Pattern $\mathbf{x}'$ represents a handwritten 5, and $\mathbf{x}'(s=3)$ represents the same shape but shifted three pixels to the right. The Euclidean distance $D(\mathbf{x}', \mathbf{x}'(s=3))$ is much larger than $D(\mathbf{x}', \mathbf{x}_8)$, where $\mathbf{x}_8$ represents the handwritten 8. Nearest-neighbor classification based on the Euclidean distance in this way leads to very large errors. Instead, we seek a distance measure that would be insensitive to such translations, or indeed other known invariances, such as scale or rotation.

use of more general measures of distance. One crucial such problem is that of invariance. Consider a 100-dimensional pattern $\mathbf{x}'$ representing a 10×10 pixel grayscale image of a handwritten 5. Consider too the Euclidean distance from $\mathbf{x}'$ to the pattern representing an image that is shifted horizontally but otherwise identical (Fig. 4.20). Even if the relative shift s is a mere three pixels, the Euclidean distance grows very large—much greater than the distance to an unshifted 8. Clearly the Euclidean metric is of little use in a nearest-neighbor classifier that must be insensitive to such translations.

Likewise, other transformations, such as overall rotation or scale of the image, would not be well accommodated by Euclidean distance in this manner. Such drawbacks are especially pronounced if we demand that our classifier be simultaneously invariant to *several* transformations, such as horizontal translation, vertical translation, overall scale, rotation, line thickness, shear, and so on (Computer exercises 7 and 8). While we could preprocess the images by shifting their centers to coalign, then have the same bounding box, and so forth, such an approach has its own difficulties, such as sensitivity to outlying pixels or to noise. We explore here alternatives to such preprocessing.

Ideally, we would not compute the distance between two patterns until we had transformed them to be as similar to one another as possible. Alas, the computational complexity of such transformations is often quite high. Merely rotating a $k \times k$ image by a known amount and interpolating to a new grid is $O(k^2)$. But of course we do not know the proper rotation angle ahead of time and must search through several values, each value requiring a distance calculation to test whether or not the optimal setting has been found. If we must search for the optimal set of parameters for *several* transformations for each stored prototype during classification, the computational burden is prohibitive (Problem 25).

The general approach in tangent distance classifiers is to use a novel measure of distance and a linear *approximation* to the arbitrary transforms. Suppose we believe there are r transformations applicable to our problem, such as horizontal translation, vertical translation, shear, rotation, scale, and line thinning. During construction of the classifier we take each stored prototype $\mathbf{x}'$ and perform each of the transformations $\mathcal{F}_i(\mathbf{x}'; \alpha_i)$ on it. Thus $\mathcal{F}_i(\mathbf{x}'; \alpha_i)$ could represent the image described by $\mathbf{x}'$, rotated by a small angle α_i. We then construct a *tangent vector* $\mathbf{TV}_i$ for each transformation:

TANGENT
VECTOR

$$\mathbf{TV}_i = \mathcal{F}_i(\mathbf{x}'; \alpha_i) - \mathbf{x}'. \tag{59}$$

While such a transformation may be computationally intensive—as, for instance, the line thinning transform—it need be done only once, during training when computational constraints are lax. In this way we construct for each prototype $\mathbf{x}'$ an $r \times d$ matrix $\mathbf{T}$, consisting of the tangent vectors at $\mathbf{x}'$. (Such vectors can be orthonormalized, but we need assume here only that they are linearly independent. It should also be clear that this method will not work with binary images, since they lack a proper

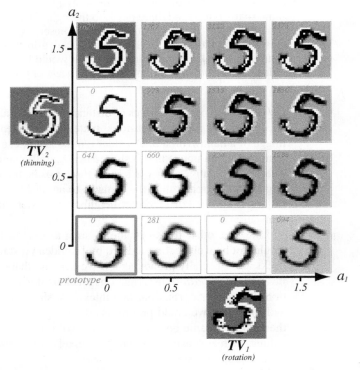

FIGURE 4.21. The pixel image of the handwritten 5 prototype at the lower left was subjected to two transformations, rotation, and line thinning, to obtain the tangent vectors $\mathbf{TV}_1$ and $\mathbf{TV}_2$; images corresponding to these tangent vectors are shown outside the axes. Each of the 16 images within the axes represents the prototype plus linear combination of the two tangent vectors with coefficients a_1 and a_2. The small red number in each image is the Euclidean distance between the tangent approximation and the image generated by the unapproximated transformations. Of course, this Euclidean distance is 0 for the prototype and for the cases $a_1 = 1, a_2 = 0$ and $a_1 = 0, a_2 = 1$. (The patterns generated with $a_1 + a_2 > 1$ have a gray background because of automatic grayscale conversion of images with negative pixel values.)

notion of derivative. If the data are binary, then, it is traditional to blur the images before creating a tangent-distance-based classifier.)

Each point in the subspace spanned by the r tangent vectors passing through $\mathbf{x}'$ represents the linearized approximation to the full combination of transforms, as shown in Fig. 4.22. During classification we search for the point in the tangent space that is closest to a test point $\mathbf{x}$—the linear approximation to our ideal. As we shall see, this search can be quite fast.

Now we turn to computing the tangent distance from a test point $\mathbf{x}$ to a particular stored prototype $\mathbf{x}'$. Formally, given a matrix $\mathbf{T}$ consisting of the r tangent vectors at $\mathbf{x}'$, the tangent distance from $\mathbf{x}'$ to $\mathbf{x}$ is

$$D_{tan}(\mathbf{x}', \mathbf{x}) = \min_{\mathbf{a}}[\|(\mathbf{x}' + \mathbf{T}\mathbf{a}) - \mathbf{x}\|], \tag{60}$$

that is, the Euclidean distance from $\mathbf{x}$ to the tangent space of $\mathbf{x}'$. Equation 60 describes the so-called "one-sided" tangent distance, because only one pattern, $\mathbf{x}'$, is transformed. The two-sided tangent distance allows both $\mathbf{x}$ and $\mathbf{x}'$ to be transformed but improves the accuracy only slightly at a large added computational burden (Problem 23); for this reason we shall concentrate on the one-sided version.

During classification of $\mathbf{x}$ we will find its tangent distance to $\mathbf{x}'$ by finding the optimizing value of $\mathbf{a}$ required by Eq. 60. This minimization is actually quite simple,

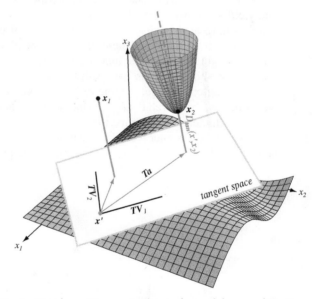

FIGURE 4.22. A stored prototype $\mathbf{x}'$, if transformed by combinations of two basic transformations, would fall somewhere on a complicated curved surface in the full d-dimensional space (gray). The tangent space at $\mathbf{x}'$ is an r-dimensional Euclidean space, spanned by the tangent vectors (here $\mathbf{TV}_1$ and $\mathbf{TV}_2$). The tangent distance $D_{tan}(\mathbf{x}', \mathbf{x})$ is the smallest Euclidean distance from $\mathbf{x}$ to the tangent space of $\mathbf{x}'$, shown in the solid red lines for two points, $\mathbf{x}_1$ and $\mathbf{x}_2$. Thus although the Euclidean distance from $\mathbf{x}'$ to $\mathbf{x}_1$ is less than that to $\mathbf{x}_2$, for the tangent distance the situation is reversed. The Euclidean distance from $\mathbf{x}_2$ to the tangent space of $\mathbf{x}'$ is a quadratic function of the parameter vector $\mathbf{a}$, as shown by the pink paraboloid. Thus simple gradient descent methods can find the optimal vector $\mathbf{a}$ and hence the tangent distance $D_{tan}(\mathbf{x}', \mathbf{x}_2)$.

because the squared distance we want to minimize is a quadratic function of **a**, as shown in pink in Fig. 4.22. We find the optimal **a** via simple search technique, such as iterative gradient descent or matrix methods, described more fully in Chapter 5.

*4.7 FUZZY CLASSIFICATION

Occasionally we may have informal knowledge about a problem domain where we seek to build a classifier. For instance, we might feel, generally speaking, that an adult salmon is oblong and light in color, while a sea bass is stouter and dark. The approach taken in *fuzzy classification* is to create so-called "fuzzy category memberships functions," which convert an objectively measurable parameter into a subjective "category memberships," which are then used for classification. We must stress immediately that the term "categories" used in this context refers not to the final class as we have been discussing, but instead to overlapping ranges of feature values. For instance, if we consider the feature value of lightness, we might split this into five "categories"—dark, medium-dark, medium, medium-light and light. In order to avoid misunderstandings, we shall use quotations when discussing such "categories."

For example we might have the lightness and shape of a fish be judged as in Fig. 4.23. Next we need a way to convert an objective measurement in several features into a category decision about the fish, and for this we need a *merging* or *conjunction rule*—a way to take the "category memberships" (e.g., lightness and shape) and yield a number to be used for making the final decision. Here fuzzy logic practitioners have at their disposal a large number of possible functions. Indeed, most functions can be used and there are few principled criteria to prefer one over another. One guiding principle that is often invoked is that in the extreme cases when the membership functions have value 0 or 1, the conjunction reduces to standard predicate logic; likewise, symmetry in the arguments is virtually always assumed. Nevertheless, there are no strong principled reasons to impose these conditions, nor are they sufficient to determine the "categories."

CONJUNCTION RULE

Suppose the designer feels that the final category based on lightness and shape can be described as medium-light *and* oblong. While the heuristic category membership

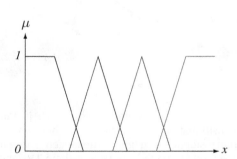

FIGURE 4.23. "Category membership" functions, derived from the designer's prior knowledge, together with a conjunction rule lead to discriminants. In this figure, *x* might represent an objectively measurable value such as the reflectivity of a fish's skin. The designer believes there are four relevant ranges, which might be called dark, medium-dark, medium-light, and light. The categories for the feature, of course, are not the same as the true categories or classes for the patterns.

function, $\mu(\cdot)$, converts the objective measurements to two "category memberships," we now need a *conjunction rule* to transform the component "membership values" into a discriminant function. There are many ways to do this, but the most popular is

$$\mu_x(x) \cdot \mu_y(y) \tag{61}$$

and the obvious extension if there are more than two features.

The "fuzzy" method just described appears similar, at base, to Parzen windows, probabilistic neural networks, and as we shall see in Chapter 6, radial basis functions. This similarity leads to a question frequently debated: Are fuzzy category memberships just probabilities (or proportial to probabilities)? It must be emphasized first that the classical notion of probability applies to more than just the relative frequency of an event; in particular, it quantifies our uncertainty of a proposition. Even before the introduction of fuzzy methods and category membership functions, the statistics, pattern recognition, and even mathematical philosophy communities argued a great deal over the fundamental nature of probability. Some questioned the applicability of the concept to single, nonrepeatable events, feeling that statements about a single event (What was the probability of rain on Tuesday?) were meaningless. Such discussion made it quite clear that "probability" need not apply only to repeatable events. Instead, since the first half of the 20th century, probability has been used as the logic of reasonable inference—work that highlighted the notion of *subjective probability*. Moreover, pattern recognition practitioners had happily used discriminant functions without concern over whether they represented probabilities,

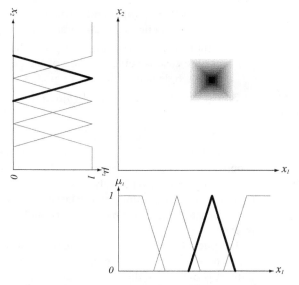

FIGURE 4.24. "Category membership" functions and a conjunction rule based on the designer's prior knowledge lead to discriminant functions. Here x_1 and x_2 are objectively measurable feature values. The designer believes that a particular class can be described as the conjunction of two "category memberships," here shown bold. Here the conjunction rule of Eq. 61 is used to give the discriminant function. The resulting discriminant function for the final category is indicated by the grayscale in the middle: the greater the discriminant, the darker. The designer constructs discriminant functions for other categories in a similar way (possibly also using disjunctions or other logical relations). During classification, the maximum discriminant function is chosen.

subjective probabilities, approximations to frequencies, or other fundamental entities. As such, the above discussion of fuzzy techniques can be subsumed by classical notions of probability, where probability—understood in its stronger broad sense—subsumes category memberships (Problem 29).

While a full analysis of these conceptual foundations would lead us away from our development of pattern recognition techniques, it pays to consider the claims of fuzzy logic proponents, because in order to be a good pattern recognition practitioner, we must understand what is or is not afforded by any technique. Proponents of fuzzy logic contend that category membership functions do not represent probabilities—subjective or not. Fuzzy logic practitioners point to examples such as when a half teaspoon of sugar is placed in a cup of tea, and they conclude that the "membership" in the category *sweet* is 0.5 and that it would be incorrect to state that the *probability* the tea was sweet was 50%. But this situation can be viewed simply as some sweetness feature having value 0.5, and there is some discriminant function, whose arguments include this feature value.

Rather than debate the fundamental nature of probability, we should really be concerned with the nature of inference, that is, how we take measurements and infer a category. Let us assume that it is possible to compute a meaningful degree of belief about memberships in categories a, b or c given data d by mathematical functions (not necessarily probabilities) of the form $P(a|d)$, $P(b|d)$, and $P(c|d)$. What are the minimum characteristics that we should demand of such functions? One very reasonable set of required characteristics is provided by Cox's axioms (also called Cox-Jaynes axioms):

1. If $P(a|d) > P(b|d)$ and $P(b|d) > P(c|d)$ then $P(a|d) > P(c|d)$. That is, degrees of belief have a natural ordering, given by real numbers.
2. $P(\text{not } a|d) = F_1[P(a|d)]$. That is, the degree of belief that a proposition is *not* the case is some function of the degree of belief that it *is* the case. Note that such degrees of belief are graded values.
3. $P(a, b|d) = F_2[P(a|d), P(b|a, d)]$.

These three axioms determine how to calculate degrees of belief—that is, how to do logical inference. We can choose the scaling such that the smallest value of any proposition is 0 and the largest is 1.0. In that case, it can be shown (Problem 30) that the function $F_1(a)$ equals $1 - a$, and the function $F_2(a, b)$ equals $a \times b$. From these two functions, along with classical inference, we get the laws of probability. Any consistent inference method is formally equivalent to standard probabilistic inference.

In spite of the arguments on such foundational issues, many practitioners are happy to use fuzzy logic, feeling that "whatever works" should be part of their repertoire. It is important, therefore, to understand the methodological strengths and limitations of the method. The limitations are formidable:

- Fuzzy methods are cumbersome to use in high dimensions or on complex problems or in problems with dozens or hundreds of features.
- The amount of information the designer can be expected to bring to a problem is quite limited—the number, positions, and widths of "category memberships."
- Because of their lack of normalization, pure fuzzy methods are poorly suited to problems in which there is a changing cost matrix λ_{ij} (Computer exercise 9).

- Pure fuzzy methods do not make use of training data. When such pure fuzzy methods (as outlined above) have unacceptable performance, it has been traditional to try to graft on adaptive (e.g., "neuro-fuzzy") methods.

If there is a contribution of fuzzy approaches to pattern recognition, it would lie in guiding the steps by which one takes knowledge in a linguistic form and casts it into discriminant functions. A severe limitation of pure fuzzy methods is that they do not rely on data, and when unsatisfactory results on problems of moderate size, it has been traditional to try to use neural or other adaptive techniques to compensate.

*4.8 REDUCED COULOMB ENERGY NETWORKS

We have seen how the Parzen-window method uses a fixed window throughout the feature space. However, in some regions a small window width would be appropriate, while elsewhere a large width would be appropriate. The k-nearest-neighbor method addressed this problem by adjusting the region based on the density of the points. Informally speaking, an approach that is intermediate between these two is to adjust the size of the window during training according to the distance to the nearest point of a *different* category. Such region adjustment can be implemented in a neural network.

One representative method—called the *reduced Coulomb energy* or RCE network—has the form shown in Fig. 4.25, which has the same topology as a probabilistic neural network (Fig. 4.9). (The method got its name because some of its equations resemble those in electrostatics describing the energy associated with a collection of charged particles.) In an RCE network, each pattern unit has an adjustable parameter that corresponds to the radius of a d-dimensional sphere in the input space. During training, each radius is adjusted so that each pattern unit covers a region as large as possible without containing a training point from another category.

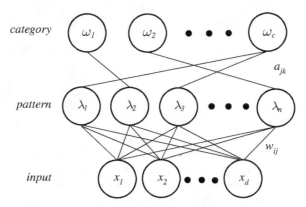

FIGURE 4.25. An RCE network is topologically equivalent to the PNN of Fig. 4.9. During training, normalized weights are adjusted to have the same values as the normalized pattern presented, just as in a PNN. In this way, distances can be calculated by an inner product. Pattern units in an RCE network also have a modifiable threshold corresponding to a "radius" λ. During training, each threshold is adjusted so that its radius is as large as possible without containing training patterns from a different category.

◼ **Algorithm 4.   (RCE Training)**

1 <u>**begin**</u> <u>**initialize**</u> $j \leftarrow 0$, $n \leftarrow$ # patterns, $\epsilon \leftarrow$ small param, $\lambda_m \leftarrow$ max radius
2   <u>**do**</u> $j \leftarrow j + 1$
3            $w_{ij} \leftarrow x_i$ (train weight)
4            $\hat{\mathbf{x}} \leftarrow \arg \min_{\mathbf{x}' \notin \omega_k} D(\mathbf{x}, \mathbf{x}')$ (find nearest point not in ω_i)
5            $\lambda_j \leftarrow \min[\max[D(\hat{\mathbf{x}}, \mathbf{x}'), \epsilon], \lambda_m]$ (set radius)
6   <u>**if**</u> $\mathbf{x} \in \omega_k$ <u>**then**</u> $a_{jk} \leftarrow 1$
7   <u>**until**</u> $j = n$
8 <u>**end**</u>

There are several subtleties that we need not consider right here. For instance, if the radius of a pattern unit becomes too small (i.e., less than some threshold λ_{min}), then it indicates that different categories are highly overlapping. In that case, the pattern unit is called a "probabilistic" unit and is marked accordingly.

During classification, a normalized test point is classified by the associated label obtained through training. Any region that is overlapped is considered ambiguous

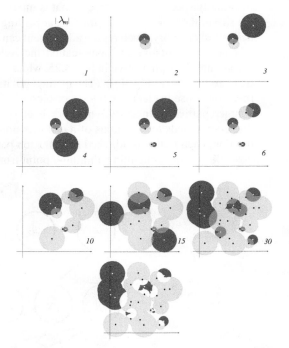

FIGURE 4.26. During training of an RCE network, each pattern has a parameter—equivalent to a radius in the d-dimensional space—that is adjusted to be as large as possible without enclosing any points from a different category (up to a maximum λ_m). As new patterns are presented, each such radius is decreased so that no sphere encloses a pattern of a different category. In this way, each sphere can enclose only patterns having the same category label. In this figure, the regions corresponding to one category are pink, and the other category are gray. Ambiguous regions (those enclosed by spheres of both categories) are shown in dark red. The number of points is shown in each component figure. The figure at the bottom shows the final decision regions, colored by category.

(Fig. 4.26). Such ambiguous regions can be useful, since the teacher can be queried as to the identity of points in that region. If we continue to let λ_j be the radius around stored prototype $\mathbf{x}'_j$ and now let $\mathcal{D}_t$ be the set of stored prototypes in whose hyperspheres the normalized test point $\mathbf{x}$ lies, then our classification algorithm is written as follows:

■ **Algorithm 5.   (RCE Classification)**

```
1  begin initialize j ← 0, k ← 0, x ← test pattern, Dt ← {}
2          do j ← j + 1
3              if D(x, x'j) < λj then  Dt ← Dt ∪ x'j
4          until j = n
5              if label of all x'j ∈ Dt is the same then return label of all xk ∈ Dt
6                                                  else return "ambiguous" label
7  end
```

4.9 APPROXIMATIONS BY SERIES EXPANSIONS

The nonparametric methods described thus far suffer from the requirement that in general all of the samples must be stored or that the designer have extensive knowledge of the problem. Because a large number of samples is needed to obtain good estimates, the memory requirements can be severe. In addition, considerable computation time may be required each time one of the methods is used to estimate $p(\mathbf{x})$ or classify a new $\mathbf{x}$.

In certain circumstances the Parzen-window procedure can be modified to reduce these problems considerably. The basic idea is to approximate the window function by a finite series expansion that is acceptably accurate in the region of interest. If we are fortunate and can find two sets of functions $\psi_j(\mathbf{x})$ and $\chi_j(\mathbf{x})$ that allow the expansion

$$\varphi\left(\frac{\mathbf{x} - \mathbf{x}_i}{h_n}\right) = \sum_{j=1}^{m} a_j \psi_j(\mathbf{x}) \chi_j(\mathbf{x}_i), \tag{62}$$

then we can split the dependence upon $\mathbf{x}$ and $\mathbf{x}_i$ as

$$\sum_{i=1}^{n} \varphi\left(\frac{\mathbf{x} - \mathbf{x}_i}{h_n}\right) = \sum_{j=1}^{m} a_j \psi_j(\mathbf{x}) \sum_{i=1}^{n} \chi_j(\mathbf{x}_i). \tag{63}$$

Then from Eq. 11 we have

$$p_n(\mathbf{x}) = \sum_{j=1}^{m} b_j \psi_j(\mathbf{x}), \tag{64}$$

where

$$b_j = \frac{a_j}{n V_n} \sum_{i=1}^{n} \chi_j(\mathbf{x}_i). \tag{65}$$

If a sufficiently accurate expansion can be obtained with a reasonable value for m, this approach has some obvious advantages. The information in the n samples is reduced to the m coefficients b_j. If additional samples are obtained, Eq. 65 for b_j can be updated easily, and the number of coefficients remains unchanged. If the functions $\psi_j(\cdot)$ and $\chi_j(\cdot)$ are polynomial functions of the components of $\mathbf{x}$ and $\mathbf{x}_i$, the expression for the estimate $p_n(\mathbf{x})$ is also a polynomial, which can be computed relatively efficiently. Furthermore, use of this estimate $p(\mathbf{x}|\omega_i)P(\omega_i)$ leads to a simple way of obtaining *polynomial discriminant functions*.

POLYNOMIAL
DISCRIMINANT

Before becoming too enthusiastic, however, we should note one of the problems with this approach. A key property of a useful window function is its tendency to peak at the origin and fade away elsewhere. Thus $\varphi((\mathbf{x}-\mathbf{x}_i)/h_n)$ should peak sharply at $\mathbf{x} = \mathbf{x}_i$ and should contribute little to the approximation of $p_n(\mathbf{x})$ for $\mathbf{x}$ far from $\mathbf{x}_i$. Unfortunately, polynomials have the annoying property of becoming unbounded. Thus, in a polynomial expansion we might find the terms associated with an $\mathbf{x}_i$ far from $\mathbf{x}$ contributing most (rather than least) to the expansion. It is quite important, therefore, to be sure that the expansion of each window function is in fact accurate in the region of interest, and this may well require a large number of terms.

There are many types of series expansions one might consider. Readers familiar with integral equations will naturally interpret Eq. 62 as an expansion of the kernel $\varphi(\mathbf{x}, \mathbf{x}_i)$ in a series of eigenfunctions. (In analogy with eigenvectors and eigenvalues, eigenfunctions are solutions to certain differential equations with fixed real-number coefficients.) Rather than computing eigenfunctions, one might choose any reasonable set of functions orthogonal over the region of interest and obtain a least-squares fit to the window function. We shall take an even more straightforward approach and expand the window function in a Taylor series. For simplicity, we confine our attention to a one-dimensional example using a Gaussian window function:

EIGENFUNCTION

$$\sqrt{\pi}\,\varphi(u) = e^{-u^2}$$

$$\simeq \sum_{j=0}^{m-1}(-1)^j \frac{u^{2j}}{j!}.$$

This expansion is most accurate near $u = 0$, and is in error by less than $u^{2m}/m!$. If we substitute $u = (x - x_i)/h$, we obtain a polynomial of degree $2(m - 1)$ in x and x_i. For example, if $m = 2$ the window function can be approximated as

$$\sqrt{\pi}\varphi\left(\frac{x - x_i}{h}\right) \simeq 1 - \left(\frac{x - x_i}{h}\right)^2$$

$$= 1 + \frac{2}{h^2}x\,x_i - \frac{1}{h^2}x^2 - \frac{1}{h^2}\,x_i^2,$$

and thus

$$\sqrt{\pi}\,p_n(x) = \frac{1}{nh}\sum_{i=1}^{n}\sqrt{\pi}\varphi\left(\frac{x - x_i}{h}\right) \simeq b_0 + b_1 x + b_2 x^2, \tag{66}$$

where the coefficients are

$$b_0 = \frac{1}{h} - \frac{1}{h^3}\frac{1}{n}\sum_{i=1}^{n}x_i^2$$

$$b_1 = \frac{2}{h^3} \frac{1}{n} \sum_{i=1}^{n} x_i$$

$$b_2 = -\frac{1}{h^3}.$$

This simple expansion condenses the information in n samples into the values b_0, b_1, and b_2. It is accurate if the largest value of $|x - x_i|$ is not greater than h. Unfortunately, this restricts us to a very wide window that is not capable of much resolution. By taking more terms we can use a narrower window. If we let r be the largest value of $|x - x_i|$ and use the fact that the error in the m-term expansion of $\sqrt{\pi}\, \varphi((x - x_i)/h)$ is less than $(r/h)^{2m}/m!$, then using Stirling's approximation for $m!$ we find that the error in approximating $p_n(x)$ is less than

$$\frac{1}{\sqrt{\pi}h} \frac{(r/h)^{2m}}{m!} \simeq \frac{1}{\sqrt{\pi}h\sqrt{2\pi m}} \left[\left(\frac{e}{m}\right)\left(\frac{r}{h}\right)^2 \right]^m . \tag{67}$$

Thus, the error becomes small only when $m > e(r/h)^2$. This implies the need for many terms if the window size h is small relative to the distance r from x to the most distant sample. Although this example is rudimentary, similar considerations arise in the multidimensional case even when more sophisticated expansions are used, and the procedure is most attractive when the window size is relatively large.

SUMMARY

There are two overarching approaches to nonparametric estimation for pattern classification: In one the densities are estimated (and then used for classification), while in the other the category is chosen directly. The former approach is exemplified by Parzen windows and their hardware implementation, probabilistic neural networks. The latter is exemplified by k-nearest-neighbor and several forms of relaxation networks. In the limit of infinite training data, the $k = 1$ nearest-neighbor error rate is bounded from above by twice the Bayes error rate. The extremely high space complexity of the nominal nearest-neighbor method can be reduced by editing (e.g., removing those prototypes that are surrounded by prototypes of the same category), prestructuring the data set for efficient search, or partial distance calculations. Novel distance measures, such as the tangent distance, can be used in the nearest-neighbor algorithm for incorporating known transformation invariances.

Fuzzy classification methods employ heuristic choices of "category membership" functions and heuristic conjunction rules to obtain discriminant functions. Any benefit of such techniques is limited to cases where there is very little (or no) training data and small numbers of features, as well as when the knowledge can be gleaned from the designer's prior beliefs.

Relaxation methods such as potential functions create "basins of attraction" surrounding training prototypes; when a test pattern lies in such a basin, the corresponding prototype can be easily identified along with its category label. Reduced Coulomb energy networks are one in the class of such relaxation networks, and the basins are adjusted to be as large as possible yet not include prototypes from other categories.

BIBLIOGRAPHICAL AND HISTORICAL REMARKS

Parzen introduced his window method for estimating density functions [31], and its use in regression was pioneered by Nadaraya [30] and Watson [45]. Its natural application to classification problems stems from the work of Specht [41], including its PNN hardware implementation [42]. Nearest-neighbor methods were introduced in references [13] and [14], but it was over 15 years later that computer power had increased, thereby making it practical and renewing interest in its theoretical foundations. Cover and Hart's foundational work on asymptotic bounds [9] were further expanded through the analysis of Devroye [12]. The first pruning or editing work in reference [20] was followed by a number of related algorithms, such as that described in references [4] and [44]. The k-nearest neighbor was explored in reference [32]. The computational complexity of nearest neighbor (Voronoi) is described in reference [35]; work on search, as described in Knuth's classic [26], has proven to be of greater use, in general. Much of the research on reducing the computational complexity of nearest-neighbor search comes from the vector quantization and compression community; for instance partial distance calculations are described in reference [18]. Friedman has an excellent analysis of some of the unintuitive properties of high-dimensional spaces, and indirectly nearest-neighbor classifiers, an inspiration for several problems here [16]. He and his colleagues have explored the use of tree data structures that speed search during classification [17]. The definitive collection of seminal papers in nearest-neighbor classification is found in reference [11].

The notion of tangent distance was introduced by Simard and colleagues [40] and was explored by a number of others [21]. Sperduti and Stork introduced a prestructuring and novel search criterion that speeds search in tangent-based classifiers [43]. The greatest successes of tangent methods have been in optical character recognition, but the method can be applied in other domains, so long as the invariances are known. The study of general invariance has been most profitable when limited to a particular domain, and readers seeking further background should consult reference [29] for computer vision and [34] for speech. Background on image transformations is covered in reference [15].

Early treatments of the use of potential functions for pattern classification are references [2] and [6]. Such work is closely allied with later efforts including those leading to the RCE network described in references [37] and [36].

The philosophical debate concerning frequency, probability, graded category membership, and so on, has a long history [28]. Keynes espoused a theory of probability as the logic of probable inference, and he did not need to rely on the notion of repeatability, frequency, and so on. We subscribe to the traditional view that probability is a conceptual and formal relation between hypotheses and conclusions—here, specifically between data and category. The limiting cases of such rational belief are certainty (on the one hand) and impossibility (on the other). Classical theory of probability cannot be based solely on classical logic, which has no formal notions for the probability of an event. While the rules in Keynes' probability [25] were taken as axiomatic, Cox [10] and later Jaynes [23] sought to place a formal underpinning. Many years after these debates, "fuzzy" methods were proposed from the computer science [46]. A formal equivalence of fuzzy category membership functions and probability is given in reference [19], which in turn is based on Cox [10]. Cheeseman has made some remarkably clear and forceful rebuttals to the assertions that fuzzy methods represent something beyond the notion of subjective probability [7, 8]; representative expositions to the contrary include references [5] and [27]. The many connectives for fuzzy logic have been listed in reference [3], though with little concern for the fundamental justification of one over the others. Readers unconcerned with foundational issues, and whether fuzzy methods provide any representational power or other benefits above standard probability (including subjective probability), can consult reference [24], which is loaded with over 3000 references.

PROBLEMS

Section 4.3

1. Show that Eqs. 19–22 are sufficient to ensure convergence in Eqs. 17 and 18.

2. Consider a normal $p(x) \sim N(\mu, \sigma^2)$ and Parzen-window function $\varphi(x) \sim N(0, 1)$. Show that the Parzen-window estimate

$$p_n(x) = \frac{1}{nh_n} \sum_{i=1}^{n} \varphi\left(\frac{x - x_i}{h_n}\right)$$

has the following properties:

(a) $\bar{p}_n(x) \sim N(\mu, \sigma^2 + h_n^2)$

(b) $\text{Var}[p_n(x)] \simeq \frac{1}{2nh_n\sqrt{\pi}} p(x)$

(c) $p(x) - \bar{p}_n(x) \simeq \frac{1}{2}\left(\frac{h_n}{\sigma}\right)^2 \left[1 - \left(\frac{x-\mu}{\sigma}\right)^2\right] p(x)$

for small h_n. (Note that if $h_n = h_1/\sqrt{n}$, this result implies that the error due to bias goes to zero as $1/n$, whereas the standard deviation of the noise only goes to zero as $\sqrt[4]{n}$.)

3. Let $p(x) \sim U(0, a)$ be uniform from 0 to a, and let a Parzen window be defined as $\varphi(x) = e^{-x}$ for $x > 0$ and 0 for $x \leq 0$.

   (a) Show that the mean of such a Parzen-window estimate is given by

$$\bar{p}_n(x) = \begin{cases} 0 & x < 0 \\ \frac{1}{a}(1 - e^{-x/h_n}) & 0 \leq x \leq a \\ \frac{1}{a}(e^{a/h_n} - 1)e^{-x/h_n} & a \leq x. \end{cases}$$

   (b) Plot $\bar{p}_n(x)$ versus x for $a = 1$ and $h_n = 1, 1/4$, and $1/16$.

   (c) How small does h_n have to be to have less than 1% bias over 99% of the range $0 < x < a$?

   (d) Find h_n for this condition if $a = 1$, and plot $\bar{p}_n(x)$ in the range $0 \leq x \leq 0.05$.

4. Suppose in a c-category supervised learning environment we sample the full distribution $p(\mathbf{x})$ and subsequently train a PNN classifier according to *Algorithm* 1.

   (a) Show that even if there are unequal category priors and hence unequal numbers of points in each category, the recognition method properly accounts for such priors.

   (b) Suppose we have trained a PNN with the assumption of equal category priors, but later wish to use it for a problem having the cost matrix λ_{ij}, representing the cost of choosing category ω_i when in fact the pattern came from ω_j. How should we do this?

(c) Suppose instead we know a cost matrix λ_{ij} *before* training. How shall we train a PNN for minimum risk?

Section 4.4

5. Show that Eq. 30 converges in probability to $p(\mathbf{x})$ given the conditions $\lim_{n\to\infty} k_n \to \infty$ and $\lim_{n\to\infty} k_n/n \to 0$.

6. Let $\mathcal{D} = \{\mathbf{x}_1, \ldots, \mathbf{x}_n\}$ be a set of n independent labeled samples and let $\mathcal{D}_k(\mathbf{x}) = \{\mathbf{x}'_1, \ldots, \mathbf{x}'_k\}$ be the k nearest neighbors of $\mathbf{x}$. Recall that the k-nearest-neighbor rule for classifying $\mathbf{x}$ is to give $\mathbf{x}$ the label most frequently represented in $\mathcal{D}_k(\mathbf{x})$. Consider a two-category problem with $P(\omega_1) = P(\omega_2) = 1/2$. Assume further that the conditional densities $p(\mathbf{x}|\omega_i)$ are uniform within unit hyperspheres a distance of ten units apart.

(a) Show that if k is odd the average probability of error is given by

$$P_n(e) = \frac{1}{2^n} \sum_{j=0}^{(k-1)/2} \binom{n}{j}.$$

(b) Show that for this case the single-nearest neighbor rule has a lower error rate than the k-nearest-neighbor error rate for $k > 1$.

(c) If k is allowed to increase with n but is restricted by $k < a\sqrt{n}$, show that $P_n(e) \to 0$ as $n \to \infty$.

Section 4.5

7. Prove that the Voronoi cells induced by the single-nearest neighbor algorithm must always be convex. That is, for any two points $\mathbf{x}_1$ and $\mathbf{x}_2$ in a cell, all points on the line linking $\mathbf{x}_1$ and $\mathbf{x}_2$ must also lie in the cell.

8. It is easy to see that the nearest-neighbor error rate P can equal the Bayes rate P^* if $P^* = 0$ (the best possibility) or if $P^* = (c - 1)/c$ (the worst possibility). One might ask whether or not there are problems for which $P = P^*$ when P^* is between these extremes.

(a) Show that the Bayes rate for the one-dimensional case where $P(\omega_i) = 1/c$ and

$$P(x|\omega_i) = \begin{cases} 1 & 0 \le x \le \frac{cr}{c-1} \\ 1 & i \le x \le i+1 - \frac{cr}{c-1} \\ 0 & \text{elsewhere} \end{cases}$$

is $P^* = r$.

(b) Show that for this case the nearest-neighbor rate is $P = P^*$.

9. Consider the following set of two-dimensional vectors from three categories:

ω_1		ω_2		ω_3	
x_1	x_2	x_1	x_2	x_1	x_2
10	0	5	10	2	8
0	−10	0	5	−5	2
5	−2	5	5	10	−4

(a) Plot the decision boundary resulting from the nearest-neighbor rule just for categorizing ω_1 and ω_2. Find the sample means $\mathbf{m}_1$ and $\mathbf{m}_2$ and on the same figure sketch the decision boundary corresponding to classifying $\mathbf{x}$ by assigning it to the category of the nearest sample mean.

(b) Repeat part (a) for categorizing only ω_1 and ω_3.

(c) Repeat part (a) for categorizing only ω_2 and ω_3.

(d) Repeat part (a) for a three-category classifier, classifying ω_1, ω_2 and ω_3.

10. Prove that the computational complexity of the basic nearest-neighbor editing algorithm (*Algorithm* 3) for n points in d dimension is $O(d^3 n^{\lfloor d/2 \rfloor} \ln n)$.

11. To understand the "curse of dimensionality" in greater depth, consider the effects of high dimensions on the simple nearest-neighbor algorithm. Suppose we need to estimate a density function $p(\mathbf{x})$ in the unit hypercube in $\mathbf{R}^d$ based on n samples. If $p(\mathbf{x})$ is complicated, we need dense samples to learn it well.

(a) Let n_1 denote the number of samples in a "dense" sample in $\mathbf{R}^1$. What is the sample size for the "same density" in $\mathbf{R}^d$? If $n_1 = 100$, what sample size is needed in a 20-dimensional space?

(b) Show that the interpoint distances are all large and roughly equal in $\mathbf{R}^d$, and that neighborhoods that have even just a few points must have large radii.

(c) Find $l_d(p)$, the length of a hypercube edge in d dimensions that contains the fraction p of points ($0 \leq p \leq 1$). To better appreciate the implications of your result, calculate: $l_5(0.01)$, $l_5(0.1)$, $l_{20}(0.01)$, and $l_{20}(0.1)$.

(d) Show that nearly all points are close to a face of the full space (e.g., the unit hypercube in d dimensions). Do this by calculating the L_∞ distance from one point to the closest other point. This shows that nearly all points are closer to a face than to another training point. (Argue that L_∞ is more favorable than L_2 distance, and happens to be easier to calculate.) The result shows that most points are on or near the convex hull of training samples and that nearly every point is an "outlier" with respects to all the others.

12. Show how the "curse of dimensionality" (Problem 11) can be "overcome" by choosing or assuming that your model is of a particular sort. Suppose that we are estimating a function of the form $y = f(\mathbf{x}) + N(0, \sigma^2)$.

(a) Suppose the true function is linear, $f(\mathbf{x}) = \sum_{j=1}^d a_j x_j$, and that the approximation is $\hat{f}(\mathbf{x}) = \sum_{j=1}^d \hat{a}_j x_j$. Of course, the fit coefficients are

$$\hat{a}_j = \arg \min_{a_j} \sum_{i=1}^n \left[y_i - \sum_{j=1}^d a_j x_{ij} \right]^2,$$

for $j = 1, \ldots, d$. Prove that $\mathcal{E}[f(\mathbf{x}) - \hat{f}(\mathbf{x})]^2 = d\sigma^2/n$, that is, that it increases linearly with d, and not exponentially as the curse of dimensionality might otherwise suggest.

(b) Generalize your result from part (a) to the case where a function is expressed in a different basis set, that is, $f(\mathbf{x}) = \sum_{i=1}^{M} a_i B_i(\mathbf{x})$ for some well-behaved basis set $B_i(x)$, and hence that the result does not depend on the fact that we have used a *linear* basis.

13. Consider classifiers based on samples with priors $P(\omega_1) = P(\omega_2) = 0.5$ and the distributions

$$p(x|\omega_1) = \begin{cases} 2x & \text{for } 0 \leq x \leq 1 \\ 0 & \text{otherwise,} \end{cases}$$

and

$$p(x|\omega_2) = \begin{cases} 2 - 2x & \text{for } 0 \leq x \leq 1 \\ 0 & \text{otherwise.} \end{cases}$$

(a) What is the Bayes decision rule and the Bayes classification error?

(b) Suppose we randomly select a single point from ω_1 and a single point from ω_2, and create a nearest-neighbor classifier. Suppose too we select a test point from one of the categories (ω_1 for definiteness). Integrate to find the expected error rate $P_1(e)$.

(c) Repeat with two training samples from each category and a single test point in order to find $P_2(e)$.

(d) Generalize to show that in general,

$$P_n(e) = \frac{1}{3} + \frac{1}{(n+1)(n+3)} + \frac{1}{2(n+2)(n+3)}.$$

Confirm this formula makes sense in the $n = 1$ case.

(e) Compare $\lim_{n \to \infty} P_n(e)$ with the Bayes error.

14. Repeat Problem 13 but with

$$p(x|\omega_1) = \begin{cases} 3/2 & \text{for } 0 \leq x \leq 2/3 \\ 0 & \text{otherwise,} \end{cases}$$

and

$$p(x|\omega_2) = \begin{cases} 3/2 & \text{for } 1/3 \leq x \leq 1 \\ 0 & \text{otherwise.} \end{cases}$$

15. Expand in greater detail Algorithm 3 and add a conditional branch that will speed it. Assuming the data points come from c categories and there are, on average, k Voronoi neighbors of any point $\mathbf{x}$, on average how much faster will your improved algorithm be?

16. Consider the simple nearest-neighbor editing algorithm (Algorithm 3).

(a) Show by counterexample that this algorithm does not yield the minimum set of points. (Consider a problem where the points from each of two categories are constrained to be on the intersections of a two-dimensional Cartesian grid.)

(b) Create a sequential editing algorithm in which each point is considered in turn, and retained or rejected before the next point is considered. Prove whether the final classifier produced from the remaining points does or does not depend upon the sequence the points are considered.

17. Consider classification problem where each of the c categories possesses the same distribution as well as prior $P(\omega_i) = 1/c$. Prove that the upper bound in Eq. 52, that is,

$$P \leq P^* \left(2 - \frac{c}{c-1} P^* \right),$$

is achieved in this "zero-information" case.

18. Derive Eq. 54, being sure to state all assumptions you invoke.

Section 4.6

19. Consider the Euclidean metric in d dimensions:

$$D(\mathbf{a}, \mathbf{b}) = \sqrt{\sum_{k=1}^{d} (a_k - b_k)^2}.$$

Suppose we rescale each axis by a fixed factor; that is, let $x_k' = \alpha_k x_k$ for real, nonzero constants $\alpha_k, k = 1, 2, \ldots, d$. Prove that the resulting space is a metric space. Discuss the import of this fact for standard nearest-neighbor classification methods.

20. Prove that the Minkowski metric indeed possesses the four properties required of all metrics.

21. Consider a noniterative method for finding the tangent distance between $\mathbf{x}'$ and $\mathbf{x}$, given the matrix $\mathbf{T}$ consisting of the r (column) tangent vectors $\mathbf{TV}_i$ at $\mathbf{x}'$.

    (a) Follow the treatment in the text and take the gradient of the squared Euclidean distance in the $\mathbf{a}$ parameter space to find an equation that must be solved for the optimal $\mathbf{a}$.

    (b) Solve your first derivative equation to find the optimizing $\mathbf{a}$.

    (c) Compute the second derivative of $D^2(\cdot, \cdot)$ to prove that your solution must be a *minimum* squared distance, and not a maximum or inflection point.

    (d) If there are r tangent vectors (invariances) in a d-dimensional space, what is the computational complexity of your method?

    (e) In practice, the iterative method described in the text requires only a few (roughly 5) iterations for problems in handwritten OCR. Compare the complexities of your analytic solution to that of the iterative scheme.

22. Consider a tangent-distance based classifier based on n prototypes, each representing a $k \times k$ pixel pattern of a handwritten character. Suppose there are r invariances we believe characterize the problem. What is the storage requirements (space complexity) of such a tangent-based classifier?

23. The two-sided tangent distance allows *both* the stored prototype $\mathbf{x}'$ and the test point $\mathbf{x}$ to be transformed. Thus if $\mathbf{T}$ is the matrix of the r tangent vectors for $\mathbf{x}'$ and $\mathbf{S}$ likewise at $\mathbf{x}$, the two-sided tangent distance is

$$D_{2tan}(\mathbf{x}', \mathbf{x}) = \min_{\mathbf{a}, \mathbf{b}} [\|(\mathbf{x}' + \mathbf{Ta}) - (\mathbf{x} + \mathbf{Sb})\|].$$

    (a) Follow the logic in Problem 21 and calculate the gradient with respect to the $\mathbf{a}$ parameter vector and to the $\mathbf{b}$ parameter vector.

(b) State two update rules for an iterative scheme for **a** and **b**.

(c) Prove that there is a unique minimum as a function of **a** and **b**. Describe this geometrically.

(d) In an iterative scheme, we would alternatively take steps in the **a** parameter space, then in the **b** parameter space. What is the computational complexity to this approach to the two-sided tangent distance classifier?

(e) Why is the actual complexity for classification in a two-sided tangent distance classifier even more severe than your result in (d) would suggest?

24. Consider the two-sided tangent distance described in Problem 23. Suppose we restrict ourselves to n prototypes $\mathbf{x}'$ in d dimensions, each with an associated matrix $\mathbf{T}$ of r tangent vectors, which we assume are linearly independent. Determine whether the two-sided tangent distance does or does not satisfy each of the requirements of a metric: nonnegativity, reflexivity, symmetry, and the triangle inequality.

25. Consider the computational complexity of nearest-neighbor classifier for $k \times k$ pixel grayscale images of handwritten digits. Instead of using tangent distance, we will search for the parameters of full nonlinear transforms before computing a Euclidean distance. Suppose the number of operations needed to perform each of the r transformations (e.g., rotation, line thinning, shear, and so forth) is $a_i k^2$, where for the sake of simplicity we assume $a_i \simeq 10$. Suppose too that for the test of each prototype we must search though $A \simeq 5$ such values, and judge it by the Euclidean distance.

(a) Given a transformed image, how many operations are required to calculate the Euclidean distance to a stored prototype?

(b) Find the number of operations required per search.

(c) Suppose there are n prototypes. How many operations are required to find the nearest neighbor, given such transforms?

(d) Assume for simplicity that no complexity reduction methods have been used (such as editing, partial distance, graph creation). If the number of prototypes is $n = 10^6$ points and there are $r = 6$ transformations, and basic operations on our computer require 10^{-9} secs., how long does it take to classify a single point?

26. Explore the effect of r on the accuracy of nearest-neighbor search based on partial distance. Assume we have a large number n of points randomly placed in a d-dimensional hypercube. Suppose we have a test point $\mathbf{x}$, also selected randomly in the hypercube, and seek its nearest neighbor. By definition, if we use the full d-dimensional Euclidean distance, we are guaranteed to find its nearest neighbor. Suppose, however, we use the partial distance

$$D_r(\mathbf{x}, \mathbf{x}') = \left(\sum_{i=1}^{r} (x_i - x_i')^2 \right)^{1/2}.$$

(a) Plot the probability that a partial distance search finds the true closest neighbor of an arbitrary point $\mathbf{x}$ as a function of r for fixed n ($1 \leq r \leq d$) for $d = 10$.

(b) Consider the effect of r on the accuracy of a nearest-neighbor classifier. Assume we have have $n/2$ prototypes from each two categories in a hypercube of length 1 on a side. The density for each category is separable into the product of (linear) ramp functions, highest at one side of the range, and zero at the other side. Thus the density for category ω_1 is highest at $(0, 0, \ldots, 0)^t$ and zero at $(1, 1, \ldots, 1)^t$, while the density for ω_2 is highest at $(1, 1, \ldots, 1)^t$ and zero at $(0, 0, \ldots, 0)^t$. State by inspection the Bayesian decision boundary.

(c) Calculate the Bayes error rate.

(d) Estimate through simulation the probability of correct classification of a point $\mathbf{x}$, randomly selected from one of the category densities, as a function of r in a partial distance metric.

27. Consider the Tanimoto metric applied to sets having discrete elements.

(a) Determine whether the four properties of a metric are obeyed by

$$D_{Tanimoto}(\mathcal{S}_1, \mathcal{S}_2) = \frac{n_1 + n_2 - 2n_{12}}{n_1 + n_2 - n_{12}},$$

as given in Eq. 58.

(b) Consider the following six words as mere sets of unordered letters: pattern, pat, pots, stop, taxonomy, and elementary. Use the Tanimoto metric to rank order all $\binom{6}{2} = 30$ possible pairings of these sets.

(c) Is the triangle inequality obeyed for these six patterns?

Section 4.7

28. Suppose someone asks you whether a cup of water is hot or cold, and you respond that it is warm. Explain why this exchange need not indicate that the membership of the cup in some "hot" class is a graded value less than 1.0.

29. Consider the design a fuzzy classifier for three types of fish based on two features: length and lightness. The designer feels that there are five ranges of length: short, medium-short, medium, medium-large, and large. Similarly, lightness falls into three ranges: dark, medium, and light. The designer uses the triangle function

$$\hat{T}(x; \mu_i, \delta_1) = \begin{cases} 1 - \frac{|x-\mu_i|}{\delta_1} & x \le |\mu_i - \delta_1| \\ 0 & \text{otherwise} \end{cases}$$

for the intermediate values, and he or she uses an open triangle function for the extremes, that is,

$$\hat{C}(x; \mu_i, \delta_2) = \begin{cases} 1 & x > \mu_i \\ 1 - \frac{x-\mu_i}{\delta_2} & \mu_i - \delta_2 \le x \le \mu_i \\ 0 & \text{otherwise,} \end{cases}$$

and its symmetric version.

Suppose we have for the length $\delta_1 = 5$, $\mu_1 = 5$, $\mu_2 = 7$, $\mu_3 = 9$, $\mu_4 = 11$ and $\mu_5 = 13$, and for lightness $\delta_2 = 30$, $\mu_1 = 30$, $\mu_2 = 50$, and $\mu_3 = 70$. Suppose the designer feels that $\omega_1 = $ *medium-light and long*, $\omega_2 = $ *dark and short*, and $\omega_3 = $ *medium dark and long*, where the conjunction rule "and" is defined in Eq. 61.

(a) Write the algebraic form of the discriminant functions.

(b) If every "category membership function" were rescaled by a constant, would classification change?

(c) Classify the pattern $\mathbf{x} = (7.5, 60)^t$.

(d) Suppose that instead we knew that your classification in part (c) was wrong. Would we have any principled way to know whether the error was due to the number of category membership functions? their functional form? the conjunction rule?

30. Given Cox's axioms and the notation in the text, prove that if the scaling is between 0 and 1, the functions are $F_1(a) = 1 - a$ and $F_2(a, b) = a \times b$. Explain how these two particular functional forms imply that any system of inference obeying Cox's axioms is formally equivalent to the laws of probability.

Section 4.8

31. Suppose that through standard training of an RCE network (*Algorithm* 4), all the radii have been reduced to values less than λ_m. Prove that there is no subset of the training data that will yield the same category decision boundary.

Section 4.9

32. Consider a window function $\varphi(x) \sim N(0, 1)$ and a density estimate

$$p_n(x) = \frac{1}{nh_n} \sum_{i=1}^{n} \varphi\left(\frac{x - x_i}{h_n}\right).$$

Approximate this estimate by factoring the window function and expanding the factor $e^{x - x_i / h_n^2}$ in a Taylor series about the origin as follows:

(a) Show that in terms of the normalized variable $u = x/h_n$ the m-term approximation is given by

$$p_{nm}(x) = \frac{1}{\sqrt{2\pi} h_n} e^{-u^2/2} \sum_{j=0}^{m-1} b_j u^j$$

where

$$b_j = \frac{1}{n} \sum_{i=1}^{n} \frac{1}{j!} u_i^j e^{-u_i^2/2}.$$

(b) Suppose that the n samples happen to be extremely tightly clustered about $u = u_0$. Show that the two-term approximation peaks at the two points where $u^2 + u/u_0 - 1 = 0$.

(c) Show that one peak occurs approximately at $u = u_0$, as desired, if $u_0 \ll 1$, but that it moves only to $u = 1$ for $u_0 \gg 1$.

(d) Confirm your answer to part (c) by plotting $p_{n2}(u)$ versus u for $u_0 = 0.01, 1$, and 10. (You may need to rescale the graphs vertically.)

COMPUTER EXERCISES

Several exercises will make use of the following three-dimensional data sampled from three categories.

sample	ω_1			ω_2			ω_3		
	x_1	x_2	x_3	x_1	x_2	x_3	x_1	x_2	x_3
1	0.28	1.31	−6.2	0.011	1.03	−0.21	1.36	2.17	0.14
2	0.07	0.58	−0.78	1.27	1.28	0.08	1.41	1.45	−0.38
3	1.54	2.01	−1.63	0.13	3.12	0.16	1.22	0.99	0.69
4	−0.44	1.18	−4.32	−0.21	1.23	−0.11	2.46	2.19	1.31
5	−0.81	0.21	5.73	−2.18	1.39	−0.19	0.68	0.79	0.87
6	1.52	3.16	2.77	0.34	1.96	−0.16	2.51	3.22	1.35
7	2.20	2.42	−0.19	−1.38	0.94	0.45	0.60	2.44	0.92
8	0.91	1.94	6.21	−0.12	0.82	0.17	0.64	0.13	0.97
9	0.65	1.93	4.38	−1.44	2.31	0.14	0.85	0.58	0.99
10	−0.26	0.82	−0.96	0.26	1.94	0.08	0.66	0.51	0.88

Section 4.2

1. Explore some of the properties of density estimation in the following way.

  (a) Write a program to generate points according to a uniform distribution in a unit cube, $−1/2 \le x_i \le 1/2$ for $i = 1, 2, 3$. Generate 10^4 such points.

  (b) Write a program to estimate the density at the origin based on your 10^4 points as a function of the size of a cubical window function of size h. Plot your estimate as a function of h, for $0 < h \le 1$.

  (c) Evaluate the density at the origin using n of your points and the volume of a cube window which just encloses n points. Plot your estimate as a function of $n = 1, \ldots, 10^4$.

  (d) Write a program to generate 10^4 points from a spherical Gaussian density (with $\mathbf{\Sigma} = \mathbf{I}$) centered on the origin. Repeat parts (b) and (c) using your Gaussian data.

  (e) Discuss any qualitative differences between the functional dependencies of your estimation results for the uniform and Gaussian densities.

Section 4.3

2. Consider Parzen-window estimates and classifiers for points in the table above. Let your window function be a spherical Gaussian, i.e.,

$$\varphi((\mathbf{x} − \mathbf{x}_i)/h) \propto \exp[−(\mathbf{x} − \mathbf{x}_i)^t(\mathbf{x} − \mathbf{x}_i)/(2h^2)].$$

  (a) Write a program to classify an arbitrary test point $\mathbf{x}$ based on the Parzen window estimates. Train your classifier using the three-dimensional data from your three categories in the table above. Set $h = 1$ and classify the following three points: $(0.50, 1.0, 0.0)^t$, $(0.31, 1.51, −0.50)^t$, and $(−0.3, 0.44, −0.1)^t$.

  (b) Repeat with $h = 0.1$.

Section 4.4

3. Consider k-nearest-neighbor density estimations in different numbers of dimensions.

(a) Write a program to find the k-nearest-neighbor density for n (unordered) points in one dimension. Use your program to plot such a density estimate for the x_1 values in category ω_3 in the table above for $k = 1, 3$, and 5.

(b) Write a program to find the k-nearest-neighbor density estimate for n points in two dimensions. Use your program to plot such a density estimate for the $(x_1, x_2)^t$ values in ω_2 for $k = 1, 3$, and 5.

(c) Write a program to form a k-nearest-neighbor classifier for the three-dimensional data from the three categories in the table above. Use your program with $k = 1, 3$, and 5 to estimate the relative densities at the following points: $(-0.41, 0.82, 0.88)^t$, $(0.14, 0.72, 4.1)^t$, and $(-0.81, 0.61, -0.38)^t$.

Section 4.5

4. Write a program to create a Voronoi tesselation in two dimensions as follows.

(a) First derive analytically the equation of a line separating two arbitrary points.

(b) Given the full data set $\mathcal{D}$ of prototypes and a particular point $\mathbf{x} \in \mathcal{D}$, write a program to create a list of line segments comprising the Voronoi cell of $\mathbf{x}$.

(c) Use your program to form the Voronoi tesselation of the x_1 and x_2 features from the data of ω_1 and ω_3 in the table above. Plot your Voronoi diagram.

(d) Write a program to find the category decision boundary based on this full set $\mathcal{D}$.

(e) Implement a version of the pruning method described in *Algorithm* 3. Prune your data set from (c) to form a condensed set.

(f) Apply your programs from (c) and (d) to form the Voronoi tesselation and boundary for your condensed data set. Compare the decision boundaries you found for the full and the condensed sets.

5. Explore the tradeoff between computational complexity (as it relates to partial distance calculations) and search accuracy in nearest-neighbor classifiers in the following exercise.

(a) Write a program to generate n prototypes from a uniform distributions in a six-dimensional hypercube centered on the origin. Use your program to generate 10^6 points for category ω_1, 10^6 different points for category ω_2, and likewise for ω_3 and ω_4. Denote this full set $\mathcal{D}$.

(b) Use your program to generate a test set $\mathcal{D}_t$ of $n = 100$ points, also uniformly distributed in the 6-dimensional unit hypercube.

(c) Write a program to implement the nearest-neighbor neighbor algorithm. Use this program to label each of your points in $\mathcal{D}_t$ by the category of its nearest neighbor in $\mathcal{D}$. From now on we will assume that the labels you find are in fact the true ones, and thus the "test error" is zero.

(d) Write a program to perform nearest-neighbor classification using partial distance, based on just the first r features of each vector. Suppose we define the search accuracy as the percentage of points in $\mathcal{D}_t$ that are associated with their particular closest prototype in $\mathcal{D}$. (Thus for $r = 6$, this accuracy is 100%, by construction.) For $1 \leq r \leq 6$ in your partial distance classifier, estimate the search accuracy. Plot a curve of this search accuracy versus r. What value of r would give a 90% search accuracy? (Round r to the nearest integer.)

(e) Estimate the "wall clock time"—the overall time required by your computer to perform the search—as a function of r. If T is the time for a full search in six dimensions, what value of r requires roughly $T/2$? What is the search accuracy in that case?

(f) Suppose instead we define search accuracy as the *classification* accuracy. Estimate this classification accuracy for a partial distance nearest-neighbor classifier using your points of $\mathcal{D}_t$. Plot this accuracy for $1 \leq r \leq 6$. Explain your result.

(g) Repeat (e) for this classification accuracy. If T is the time for full search in d dimensions, what value of r requires roughly $T/2$? What is the classification search accuracy in this case?

Section 4.6

6. Consider nearest-neighbor classifiers employing different values of k in the L_k norm or Minkowski metric.

(a) Write a program to implement a nearest-neighbor classifier for c categories, using the Minkowski metric or L_k norm, where k can be selected at classification time.

(b) Use the three dimensional data in the table above to classify the following points using the L_k norm for $k = 1, 2, 4$ and ∞: $(2.21, 1.9, 0.43)^t$, $(-0.15, 1.17, 6.19)^t$, and $(0.01, 1.34, 2.60)^t$.

7. Create a 10×10 pixel grayscale pattern $\mathbf{x}'$ of a handwritten 4.

(a) Plot the Euclidean distance between the 100-dimensional vectors corresponding to $\mathbf{x}'$ and a horizontally shifted version of it as a function of the horizontal offset.

(b) Shift $\mathbf{x}'$ by two pixels to the right to form the tangent vector $\mathbf{TV}_1$. Write a program to calculate the tangent distance for shifted patterns using your $\mathbf{TV}_1$. Plot the tangent distance as a function of the displacement of the test pattern. Compare your graphs and explain the implications.

8. Repeat Computer exercise 7, but for a handwritten 7 and vertical translations.

Section 4.7

9. Assume that size, color, and shape are appropriate descriptions of fruit, and use fuzzy methods to classify fruit. In particular, assume that all "category membership" functions are either triangular (with center μ and full half-width δ) or, at the extremes, are left- or right-open triangular functions.

Suppose the size features (measured in centimeters) are: small ($\mu = 2$), medium ($\mu = 4$), large ($\mu = 6$), and extra-large ($\mu = 8$). In all cases we assume the category membership functions have $\delta = 3$. Suppose shape is described by the eccentricity, here the ratio of the major axis to minor axis lengths: thin ($\mu = 2, \delta = .6$), oblong ($\mu = 1.6, \delta = .3$), oval ($\mu = 1.4, \delta = .2$) and spherical ($\mu = 1.1, \delta = .2$). Suppose color here is represented by some measure of the mixture of red to yellow: yellow ($\mu = .1, \delta = .1$), yellow-orange ($\mu = 0.3, \delta = 0.3$), orange ($\mu = 0.5, \delta = 0.3$), orange-red ($\mu = 0.7, \delta = 0.3$) and red ($\mu = 0.9, \delta = 0.3$). The fuzzy practitioner believes the following are good descriptions of some common fruit:

- ω_1 = cherry = {small and spherical and red}
- ω_2 = orange = {medium and spherical and orange}
- ω_3 = banana = {large and thin and yellow}
- ω_4 = peach = {medium and spherical and orange-red}
- ω_5 = plum = {medium and spherical and red}
- ω_6 = lemon = {medium and oblong and yellow}
- ω_7 = grapefruit = {medium and spherical and yellow}

(a) Write a program to take any objective pattern and classify it.

(b) Classify each of these {size, shape, color}: $\{2.5, 1.0, 0.95\}$, $\{7.5, 1.9, 0.2\}$ and $\{5.0, 0.5, 0.4\}$.

(c) Suppose there is a cost associated with classification, as described by a cost matrix λ_{ij}—the cost of selecting ω_i given that the true category is ω_j. Suppose the cost matrix is

$$\lambda_{ij} = \begin{pmatrix} 0 & 1 & 1 & 0 & 2 & 2 & 1 \\ 1 & 0 & 2 & 2 & 0 & 0 & 1 \\ 1 & 2 & 0 & 1 & 0 & 0 & 2 \\ 0 & 2 & 1 & 0 & 2 & 2 & 2 \\ 2 & 0 & 0 & 2 & 0 & 1 & 1 \\ 2 & 0 & 0 & 2 & 1 & 0 & 2 \\ 1 & 1 & 2 & 2 & 1 & 2 & 0 \end{pmatrix}.$$

Reclassify the patterns in (b) for minimum cost.

Section 4.8

10. Explore relaxation networks in the following way.

(a) Write a program to implement an RCE classifier in three dimensions. Let the maximum radius be $\lambda_m = 0.5$. Train your classifier with the data from the three categories in the table above. For these data, how many times was any sphere reduced in size? (If the same sphere is reduced two times, count that as twice.)

(b) Use your classifier to classify the following:

$$(0.53, -0.44, 1.1)^t, (-0.49, 0.44, 1.11)^t, \quad \text{and} \quad (0.51, -0.21, 2.15)^t.$$

If the classification of any point is ambiguous, state which are the candidate categories.

Section 4.9

11. Consider a classifier based on a Taylor series expansion of a Gaussian window function. Let k be the highest power of x_i in a Taylor series expansion of each of the independent features of a two-dimensional Gaussian. Below, consider just the x_1 and x_2 features of categories ω_2 and ω_3 in the table above. For each value $k = 2, 4$, and 6, classify the following three points: $(0.56, 2.3, 0.10)^t$, $(0.60, 5.1, 0.86)^t$, and $(-0.95, 1.3, 0.16)^t$.

BIBLIOGRAPHY

[1] David W. Aha, editor. *Lazy Learning*. Kluwer, Boston, MA, 1997.

[2] Mark A. Aizerman, Emmanuil M. Braverman, and Leo I. Rozonoer. The Robbins-Monro process and the method of potential functions. *Automation and Remote Control*, 26:1882–1885, 1965.

[3] Claudi Alsina, Enric Trillas, and Llorenc Valverde. On some logical connectives for fuzzy set theory. *Journal of Mathematical Analysis and Applications*, 93(1):15–26, 1983.

[4] David Avis and Binay K. Bhattacharya. Algorithms for computing d-dimensional Voronoi diagrams and their duals. In Franco P. Preparata, editor, *Advances in Computing Research: Computational Geometry*, pages 159–180, JAI Press, Greenwich, CT, 1983.

[5] James C. Bezdek and Sankar K. Pal, editors. *Fuzzy Models for Pattern Recognition: Methods that Search for Structures in Data*. IEEE Press, New York, 1992.

[6] Emmanuil M. Braverman. On the potential function method. *Automation and Remote Control*, 26:2130–2138, 1965.

[7] Peter Cheeseman. In defense of probability. In *Proceedings of the Ninth International Joint Conference on Artificial Intelligence*, pages 1002–1009, Morgan Kaufmann, San Mateo, CA, 1985.

[8] Peter Cheeseman. Probabilistic versus fuzzy reasoning. In Laveen N. Kanal and John F. Lemmer, editors, *Uncertainty in Artificial Intelligence*, pages 85–102. Elsevier Science Publishers, Amsterdam, 1986.

[9] Thomas M. Cover and Peter E. Hart. Nearest neighbor pattern classification. *IEEE Transactions on Information Theory*, IT-13(1):21–27, 1967.

[10] Richard T. Cox. Probability, frequency, and reasonable expectation. *American Journal of Physics*, 14(1):1–13, 1946.

[11] Belur V. Dasarathy, editor. *Nearest Neighbor (NN) Norms: NN Pattern Classification Techniques*. IEEE Computer Society, Washington, DC, 1991.

[12] Luc P. Devroye. On the inequality of Cover and Hart in nearest neighbor discrimination. *IEEE Transactions on Pattern Analysis and Machine Intelligence*, PAMI-3(1):75–78, 1981.

[13] Evelyn Fix and Joseph L. Hodges, Jr. Discriminatory analysis: Nonparametric discrimination: Consistency properties. *USAF School of Aviation Medicine*, 4:261–279, 1951.

[14] Evelyn Fix and Joseph L. Hodges, Jr. Discriminatory analysis: Nonparametric discrimination: Small sample performance. *USAF School of Aviation Medicine*, 11:280–322, 1952.

[15] James D. Foley, Andries Van Dam, Steven K. Feiner, and John F. Hughes. *Fundamentals of Interactive Computer Graphics: Principles and Practice*. Addison-Wesley, Reading, MA, second edition, 1990.

[16] Jerome H. Friedman. An overview of predictive learning and function approximation. In Vladimir Cherkassky, Jerome H. Friedman, and Harry Wechsler, editors, *From Statistics to Neural Networks: Theory and Pattern Recognition Applications*, pages 1–61, Springer-Verlag, NATO ASI, New York, 1994.

[17] Jerome H. Friedman, Jon Louis Bentley, and Raphael Ari Finkel. An algorithm for finding best matches in logarithmic expected time. *ACM Transactions on Mathematical Software*, 3(3):209–226, 1977.

[18] Allen Gersho and Robert M. Gray. *Vector Quantization and Signal Processing*. Kluwer Academic Publishers, Boston, MA, 1992.

[19] Richard M. Golden. *Mathematical Methods for Neural Network Analysis and Design*. MIT Press, Cambridge, MA, 1996.

[20] Peter Hart. The condensed nearest neighbor rule. *IEEE Transactions on Information Theory*, IT-14(3):515–516, 1968.

[21] Trevor Hastie, Patrice Simard, and Eduard Säckinger. Learning prototype models for tangent distance. In Gerald Tesauro, David S. Touretzky, and Todd K. Leen, editors, *Advances in Neural Information Processing Systems*, volume 7, pages 999–1006, Cambridge, MA, 1995. MIT Press.

[22] Anil K. Jain and Madras D. Ramaswami. Classifier design with Parzen windows. In Edzard S. Gelsema and Laveen N. Kanal, editors, *Pattern Recognition and Artificial Intelligence*, pages 211–227. Elsevier Science Publishers, New York, 1988.

[23] Edwin T. Jaynes and G. Larry Bretthorst. *Probability Theory: The Logic of Science*. Cambridge U. Press, 2003.

[24] Abraham Kandel. *Fuzzy Techniques in Pattern Recognition*. Wiley, New York, 1982.

[25] John Maynard Keynes. *A Treatise on Probability*. Macmillan, New York, 1929.

[26] Donald E. Knuth. *The Art of Computer Programming*, volume 1. Addison-Wesley, Reading, MA, first edition, 1973.

[27] Bart Kosko. Fuzziness vs. probability. *International Journal of General Systems*, 17(2):211–240, 1990.

[28] Jan Lukasiewicz. Logical foundations of probability theory. In Ludwik Borkowski, editor, *Jan Lukasiewicz: Selected Works*, pages 16–43. North-Holland, Amsterdam, 1970.

[29] Joseph L. Mundy and Andrews Zisserman, editors. *Geometric Invariance in Computer Vision*. MIT Press, Cambridge, MA, 1992.

[30] Elizbar A. Nadaraya. On estimating regression. *Theory of Probability and Its Applications*, 9(1):141–142, 1964.

[31] Emanuel Parzen. On estimation of a probability density function and mode. *Annals of Mathematical Statistics*, 33(3):1065–1076, 1962.

[32] Edward A. Patrick and Frederick P. Fischer, III. A generalized k-nearest neighbor rule. *Information and Control*, 16(2):128–152, 1970.

[33] Witold Pedrycz and Fernando Gomide. *An Introduction to Fuzzy Sets*. MIT Press, Cambridge, MA, 1998.

[34] Joseph S. Perkell and Dennis H. Klatt, editors. *Invariance and Variability in Speech Processes*. Lawrence Erlbaum Associates, Hillsdale, NJ, 1986.

[35] Franco P. Preparata and Michael Ian Shamos. *Computational Geometry: An Introduction*. Springer-Verlag, New York, 1985.

[36] Douglas L. Reilly and Leon N Cooper. An overview of neural networks: Early models to real world systems. In Steven F. Zornetzer, Joel L. Davis, Clifford Lau, and Thomas McKenna, editors, *An Introduction to Neural and Electronic Networks*, pages 229–250. Academic Press, New York, second edition, 1995.

[37] Douglas L. Reilly, Leon N Cooper, and Charles Elbaum. A neural model for category learning. *Biological Cybernetics*, 45(1):35–41, 1982.

[38] Bernhard Schölkopf, Christopher J. C. Burges, and Alexander J. Smola, editors. *Advances in Kernel Methods: Support Vector Learning*. MIT Press, Cambridge, MA, 1999.

[39] Bernard W. Silverman and M. Christopher Jones. E. Fix and J. L. Hodges (1951): An important contribution to nonparametric discriminant analysis and density estimation. *International Statistical Review*, 57(3):233–247, 1989.

[40] Patrice Simard, Yann Le Cun, and John Denker. Efficient pattern recognition using a new transformation distance. In Stephen J. Hanson, Jack D. Cowan, and C. Lee Giles, editors, *Advances in Neural Information Processing Systems*, volume 5, pages 50–58, Morgan Kaufmann, San Mateo, CA, 1993.

[41] Donald F. Specht. Generation of polynomial discriminant functions for pattern recognition. *IEEE Transactions on Electronic Computers*, EC-16(3):308–319, 1967.

[42] Donald F. Specht. Probabilistic neural networks. *Neural Networks*, 3(1):109–118, 1990.

[43] Alessandro Sperduti and David G. Stork. A rapid graph-based method for arbitrary transformation-invariant pattern classification. In Gerald Tesauro, David S. Touretzky, and Todd K. Leen, editors, *Advances in Neural Information Processing Systems*, volume 7, pages 665–672, MIT Press, Cambridge, MA, 1995.

[44] Godfried T. Toussaint, Binay K. Bhattacharya, and Ronald S. Poulsen. Application of Voronoi diagrams to nonparametric decision rules. In *Proceedings of Computer Science and Statistics: The 16th Symposium on the Interface*, pages 97–108, North-Holland, Amsterdam, 1984.

[45] Geoffrey S. Watson. Smooth regression analysis. *Sankhyā: The Indian Journal of Statistics, Series A*, 26:359–372, 1964.

[46] Lotfi Zadeh. Fuzzy sets. *Information and Control*, 8(3):338–353, 1965.

LINEAR DISCRIMINANT FUNCTIONS

5.1 INTRODUCTION

We assumed in Chapter 3 that the forms for the underlying *probability densities* were known, and that we will use the training samples to estimate the values of their parameters. In this chapter we shall instead assume we know the proper forms for the *discriminant functions*, and use the samples to estimate the values of parameters of the classifier. We shall examine various procedures for determining discriminant functions, some of which are statistical and some of which are not. None of them, however, requires knowledge of the forms of underlying probability distributions, and in this limited sense they can be said to be nonparametric.

Throughout this chapter we shall be concerned with discriminant functions that are either linear in the components of **x**, or linear in some given set of functions of **x**. Linear discriminant functions have a variety of pleasant analytical properties. As we have seen in Chapter 2, they can be optimal if the underlying distributions are cooperative, such as Gaussians having equal covariance, as might be obtained through an intelligent choice of feature detectors. Even when they are not optimal, we might be willing to sacrifice some performance in order to gain the advantage of their simplicity. Linear discriminant functions are relatively easy to compute and in the absence of information suggesting otherwise, linear classifiers are attractive candidates for initial, trial classifiers. They also illustrate a number of very important principles that will be used more fully in neural networks (Chapter 6).

TRAINING ERROR The problem of finding a linear discriminant function will be formulated as a problem of minimizing a criterion function. The obvious criterion function for classification purposes is the *sample risk*, or *training error*—the average loss incurred in classifying the set of training samples. We must emphasize right away, however, that despite the attractiveness of this criterion, it is fraught with problems. While our goal will be to classify novel test patterns, a small training error does not guarantee a small test error—a fascinating and subtle problem that will command our attention in Chapter 9. As we shall see here, it is difficult to derive the minimum-risk linear discriminant anyway, and for that reason we investigate several related criterion functions that are analytically more tractable.

215

Much of our attention will be devoted to studying the convergence properties and computational complexities of various gradient descent procedures for minimizing criterion functions. The similarities between many of the procedures sometimes makes it difficult to keep the differences between them clear, and for this reason we have included a summary of the principal results in Table 5.1 at the end of Section 5.10.

5.2 LINEAR DISCRIMINANT FUNCTIONS AND DECISION SURFACES

A discriminant function that is a linear combination of the components of $\mathbf{x}$ can be written as

$$g(\mathbf{x}) = \mathbf{w}^t\mathbf{x} + w_0, \tag{1}$$

THRESHOLD WEIGHT

where $\mathbf{w}$ is the *weight vector* and w_0 the *bias* or *threshold weight*. As we saw in Chapter 2, for the general case there will be c such discriminant functions, one for each of c categories. We shall return to this case, but first consider a simpler situation, that is, when there are only two categories.

5.2.1 The Two-Category Case

For a discriminant function of the form of Eq. 1, a two-category classifier implements the following decision rule: Decide ω_1 if $g(\mathbf{x}) > 0$ and ω_2 if $g(\mathbf{x}) < 0$. Thus, $\mathbf{x}$ is assigned to ω_1 if the inner product $\mathbf{w}^t\mathbf{x}$ exceeds the threshold $-w_0$ and to ω_2 otherwise. If $g(\mathbf{x}) = 0$, $\mathbf{x}$ can ordinarily be assigned to either class, but in this chapter we shall leave the assignment undefined. Figure 5.1 shows a typical implementation, a clear example of the general structure of a pattern recognition system we saw in Chapter 2.

The equation $g(\mathbf{x}) = 0$ defines the decision surface that separates points assigned to ω_1 from points assigned to ω_2. When g(**x**) is linear, this decision surface is a

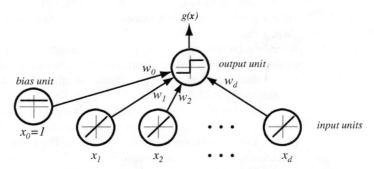

FIGURE 5.1. A simple linear classifier having d input units, each corresponding to the values of the components of an input vector. Each input feature value x_i is multiplied by its corresponding weight w_i; the effective input at the output unit is the sum all these products, $\sum w_i x_i$. We show in each unit its effective input-output function. Thus each of the d input units is linear, emitting exactly the value of its corresponding feature value. The single bias unit unit always emits the constant value 1.0. The single output unit emits a $+1$ if $\mathbf{w}^t\mathbf{x} + w_0 > 0$ or a -1 otherwise.

hyperplane. If $\mathbf{x}_1$ and $\mathbf{x}_2$ are both on the decision surface, then

$$\mathbf{w}^t\mathbf{x}_1 + w_0 = \mathbf{w}^t\mathbf{x}_2 + w_0$$

or

$$\mathbf{w}^t(\mathbf{x}_1 - \mathbf{x}_2) = 0,$$

and this shows that $\mathbf{w}$ is normal to any vector lying in the hyperplane. In general, the hyperplane H divides the feature space into two half-spaces: decision region $\mathcal{R}_1$ for ω_1 and region $\mathcal{R}_2$ for ω_2. Because $g(\mathbf{x}) > 0$ if $\mathbf{x}$ is in $\mathcal{R}_1$, it follows that the normal vector $\mathbf{w}$ points into $\mathcal{R}_1$. It is sometimes said that any $\mathbf{x}$ in $\mathcal{R}_1$ is on the *positive* side of H, and any $\mathbf{x}$ in $\mathcal{R}_2$ is on the *negative* side.

The discriminant function $g(\mathbf{x})$ gives an algebraic measure of the distance from $\mathbf{x}$ to the hyperplane. Perhaps the easiest way to see this is to express $\mathbf{x}$ as

$$\mathbf{x} = \mathbf{x}_p + r\frac{\mathbf{w}}{\|\mathbf{w}\|},$$

where $\mathbf{x}_p$ is the normal projection of $\mathbf{x}$ onto H, and r is the desired algebraic distance—positive if $\mathbf{x}$ is on the positive side and negative if $\mathbf{x}$ is on the negative side. Then, because $g(\mathbf{x}_p) = 0$,

$$g(\mathbf{x}) = \mathbf{w}^t\mathbf{x} + w_0 = r\|\mathbf{w}\|,$$

or

$$r = \frac{g(\mathbf{x})}{\|\mathbf{w}\|}.$$

In particular, the distance from the origin to H is given by $w_0/\|\mathbf{w}\|$. If $w_0 > 0$, the origin is on the positive side of H, and if $w_0 < 0$, it is on the negative side. If $w_0 = 0$, then $g(\mathbf{x})$ has the homogeneous form $\mathbf{w}^t\mathbf{x}$, and the hyperplane passes through the origin. A geometric illustration of these algebraic results is given in Fig. 5.2.

To summarize, a linear discriminant function divides the feature space by a hyperplane decision surface. The orientation of the surface is determined by the normal

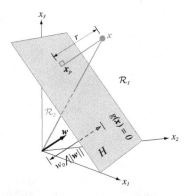

FIGURE 5.2. The linear decision boundary H, where $g(\mathbf{x}) = \mathbf{w}^t\mathbf{x} + w_0 = 0$, separates the feature space into two half-spaces $\mathcal{R}_1$ (where $g(\mathbf{x}) > 0$) and $\mathcal{R}_2$ (where $g(\mathbf{x}) < 0$).

vector $\mathbf{w}$, and the location of the surface is determined by the bias w_0. The discriminant function $g(\mathbf{x})$ is proportional to the signed distance from $\mathbf{x}$ to the hyperplane, with $g(\mathbf{x}) > 0$ when $\mathbf{x}$ is on the positive side, and $g(\mathbf{x}) < 0$ when $\mathbf{x}$ is on the negative side.

5.2.2   The Multicategory Case

There is more than one way to devise multicategory classifiers employing linear discriminant functions. For example, we might reduce the problem to c two-class problems, where the ith problem is solved by a linear discriminant function that separates points assigned to ω_i from those not assigned to ω_i. A more extravagant approach would be to use $c(c - 1)/2$ linear discriminants, one for every pair of classes. As illustrated in Fig. 5.3, both of these approaches can lead to regions in which the classification is undefined. We shall avoid this problem by adopting the approach taken in Chapter 2, defining c linear discriminant functions

$$g_i(\mathbf{x}) = \mathbf{w}_i^t \mathbf{x} + w_{i0} \qquad i = 1, \ldots, c, \qquad (2)$$

LINEAR MACHINE

and assigning $\mathbf{x}$ to ω_i if $g_i(\mathbf{x}) > g_j(\mathbf{x})$ for all $j \neq i$; in case of ties, the classification is left undefined. The resulting classifier is called a *linear machine*. A linear machine

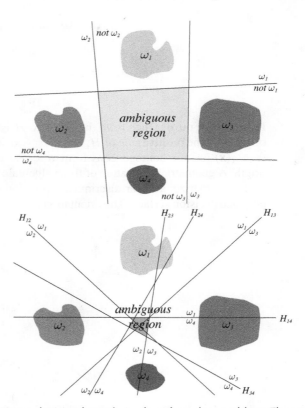

FIGURE 5.3. Linear decision boundaries for a four-class problem. The top figure shows ω_i/not ω_i dichotomies while the bottom figure shows ω_i/ω_j dichotomies and the corresponding decision boundaries H_{ij}. The pink regions have ambiguous category assignments.

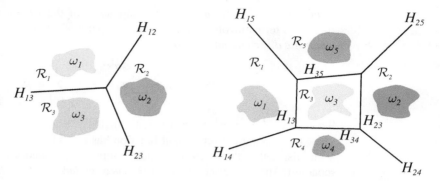

FIGURE 5.4. Decision boundaries produced by a linear machine for a three-class problem and a five-class problem.

divides the feature space into c decision regions, with $g_i(\mathbf{x})$ being the largest discriminant if $\mathbf{x}$ is in region $\mathcal{R}_i$. If $\mathcal{R}_i$ and $\mathcal{R}_j$ are contiguous, the boundary between them is a portion of the hyperplane H_{ij} defined by

$$g_i(\mathbf{x}) = g_j(\mathbf{x})$$

or

$$(\mathbf{w}_i - \mathbf{w}_j)^t \mathbf{x} + (w_{i0} - w_{j0}) = 0.$$

It follows at once that $\mathbf{w}_i - \mathbf{w}_j$ is normal to H_{ij}, and the signed distance from $\mathbf{x}$ to H_{ij} is given by $(g_i(\mathbf{x}) - g_j(\mathbf{x}))/\|\mathbf{w}_i - \mathbf{w}_j\|$. Thus, with the linear machine it is not the weight vectors themselves but their *differences* that are important. While there are $c(c - 1)/2$ pairs of regions, they need not all be contiguous, and the total number of hyperplane segments appearing in the decision surfaces is often fewer than $c(c - 1)/2$, as shown in Fig. 5.4.

It is easy to show that the decision regions for a linear machine are convex, and this restriction surely limits the flexibility and accuracy of the classifier (Problems 2 and 3). In particular, every decision region is singly connected, and this tends to make the linear machine most suitable for problems for which the conditional densities $p(\mathbf{x}|\omega_i)$ are unimodal. Nevertheless, we must be careful: There are multimodal distributions for which linear discriminants give excellent results and unimodal distributions for which they give poor classification results (Problem 1).

5.3 GENERALIZED LINEAR DISCRIMINANT FUNCTIONS

The linear discriminant function $g(\mathbf{x})$ can be written as

$$g(\mathbf{x}) = w_0 + \sum_{i=1}^{d} w_i x_i, \tag{3}$$

QUADRATIC
DISCRIMINANT

where the coefficients w_i are the components of the weight vector **w**. By adding additional terms involving the products of pairs of components of **x**, we obtain the *quadratic discriminant function*

$$g(\mathbf{x}) = w_0 + \sum_{i=1}^{d} w_i x_i + \sum_{i=1}^{d} \sum_{j=1}^{d} w_{ij} x_i x_j. \tag{4}$$

Because $x_i x_j = x_j x_i$, we can assume that $w_{ij} = w_{ji}$ with no loss in generality. Thus, the quadratic discriminant function has an additional $d(d+1)/2$ coefficients at its disposal with which to produce more complicated separating surfaces. The separating surface defined by $g(\mathbf{x}) = 0$ is a second-degree or *hyperquadric* surface. If the symmetric matrix $\mathbf{W} = [w_{ij}]$ is nonsingular the linear terms in $g(\mathbf{x})$ can be eliminated by translating the axes. The basic character of the separating surface can be described in terms of the scaled matrix $\overline{\mathbf{W}} = \mathbf{W}/(\mathbf{w}^t \mathbf{W}^{-1} \mathbf{w} - 4w_0)$. If $\overline{\mathbf{W}}$ is a positive multiple of the identity matrix, the separating surface is a *hypersphere*. If $\overline{\mathbf{W}}$ is positive definite, the separating surfaces is a *hyperellipsoid*. If some of the eigenvalues of $\overline{\mathbf{W}}$ are positive and others are negative, the surface is one of the variety of types of *hyperhyperboloids* (Problem 12). As we observed in Chapter 2, these are the kinds of separating surfaces that arise in the general multivariate Gaussian case.

POLYNOMIAL
DISCRIMINANT

By continuing to add terms such as $w_{ijk} x_i x_j x_k$, we can obtain the class of *polynomial discriminant functions*. These can be thought of as truncated series expansions of some arbitrary $g(\mathbf{x})$, and this in turn suggests the *generalized linear discriminant function*

$$g(\mathbf{x}) = \sum_{i=1}^{\hat{d}} a_i y_i(\mathbf{x}) \tag{5}$$

or

$$g(\mathbf{x}) = \mathbf{a}^t \mathbf{y}, \tag{6}$$

PHI FUNCTION

where **a** is now a $\hat{d}$-dimensional weight vector and where the $\hat{d}$ functions $y_i(\mathbf{x})$—sometimes called φ functions—can be arbitrary functions of **x**. Such functions might be computed by a feature detecting subsystem. By selecting these functions judiciously and letting $\hat{d}$ be sufficiently large, one can approximate any desired discriminant function by such an expansion. The resulting discriminant function is not linear in **x**, but it *is* linear in **y**. The $\hat{d}$ functions $y_i(\mathbf{x})$ merely map points in d-dimensional **x**-space to points in $\hat{d}$-dimensional **y**-space. The homogeneous discriminant $\mathbf{a}^t \mathbf{y}$ separates points in this transformed space by a hyperplane passing through the origin. Thus, the mapping from **x** to **y** reduces the problem to one of finding a homogeneous linear discriminant function.

Some of the advantages and disadvantages of this approach can be clarified by considering a simple example. Let the quadratic discriminant function be

$$g(x) = a_1 + a_2 x + a_3 x^2, \tag{7}$$

so that the three-dimensional vector **y** is given by

$$\mathbf{y} = \begin{pmatrix} 1 \\ x \\ x^2 \end{pmatrix}. \tag{8}$$

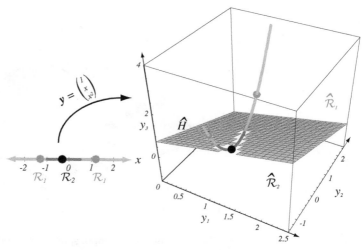

FIGURE 5.5. The mapping $\mathbf{y} = (1, x, x^2)^t$ takes a line and transforms it to a parabola in three dimensions. A plane splits the resulting $\mathbf{y}$-space into regions corresponding to two categories, and this in turn gives a nonsimply connected decision region in the one-dimensional x-space.

The mapping from x to $\mathbf{y}$ is illustrated in Fig. 5.5. The data remain inherently one-dimensional, because varying x causes $\mathbf{y}$ to trace out a curve in three dimensions. Thus, one thing to notice immediately is that if x is governed by a probability law $p(x)$, the induced density $\hat{p}(\mathbf{y})$ will be degenerate, being zero everywhere except on the curve, where it is infinite. This is a common problem whenever $\hat{d} > d$, and the mapping takes points from a lower-dimensional space to a higher-dimensional space.

The plane $\hat{H}$ defined by $\mathbf{a}^t \mathbf{y} = 0$ divides the $\mathbf{y}$-space into two decision regions $\hat{\mathcal{R}}_1$ and $\hat{\mathcal{R}}_2$. Figure 5.5 shows the separating plane corresponding to $\mathbf{a} = (-1, 1, 2)^t$, the decision regions $\hat{\mathcal{R}}_1$ and $\hat{\mathcal{R}}_2$, and their corresponding decision regions $\mathcal{R}_1$ and $\mathcal{R}_2$ in the original x-space. The quadratic discriminant function $g(x) = -1 + x + 2x^2$ is positive if $x < -1$ or if $x > 0.5$, and hence $\mathcal{R}_1$ is multiply connected. Thus although the decision regions in $\mathbf{y}$-space are convex, this is by no means the case in x-space. More generally speaking, even with relatively simple functions $y_i(\mathbf{x})$, decision surfaces induced in an $\mathbf{x}$-space can be fairly complex.

Unfortunately, the curse of dimensionality often makes it hard to capitalize on this flexibility in practice. A complete quadratic discriminant function involves $\hat{d} = (d + 1)(d + 2)/2$ terms. If d is modestly large, say $d = 50$, this requires the computation of a great many terms; inclusion of cubic and in general kth-order components in the polynomial leads to $O(d^k)$ terms. Furthermore, the $\hat{d}$ components of the weight vector $\mathbf{a}$ must be determined from training samples. If we think of $\hat{d}$ as specifying the number of degrees of freedom for the discriminant function, it is natural to require that the number of samples be not less than the number of degrees of freedom (cf. Chapter 9). Clearly, a general series expansion of $g(\mathbf{x})$ can easily lead to completely unrealistic requirements for computation and data. We shall see in Section 5.11 that this drawback can be accommodated by imposing a constraint of large margins, or bands between the training patterns, however. In this case, we are not technically speaking fitting all the free parameters; instead, we are relying on the assumption that the mapping to a high-dimensional space does not impose any spurious structure or relationships among the training points. Alternatively,

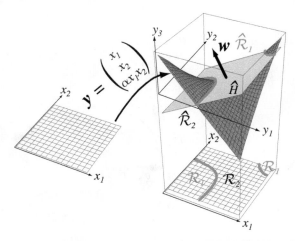

FIGURE 5.6. The two-dimensional input space **x** is mapped through a polynomial function f to **y**. Here the mapping is $y_1 = x_1$, $y_2 = x_2$ and $y_3 \propto x_1 x_2$. A linear discriminant in this transformed space is a hyperplane, which cuts the surface. Points to the positive side of the hyperplane $\hat{H}$ correspond to category ω_1, and those beneath it correspond to category ω_2. Here, in terms of the **x** space, $\mathcal{R}_1$ is a not simply connected.

multilayer neural networks approach this problem by employing multiple copies of a single nonlinear function of the input features, as we shall see in Chapter 6.

While it may be hard to realize the potential benefits of a generalized linear discriminant function, we can at least exploit the convenience of being able to write $g(\mathbf{x})$ in the homogeneous form $\mathbf{a}^t \mathbf{y}$. In the particular case of the linear discriminant function we have

$$g(\mathbf{x}) = w_0 + \sum_{i=1}^{d} w_i x_i = \sum_{i=0}^{d} w_i x_i \tag{9}$$

where we set $x_0 = 1$. Thus we can write

$$\mathbf{y} = \begin{bmatrix} 1 \\ x_1 \\ \vdots \\ x_d \end{bmatrix} = \begin{bmatrix} 1 \\ \mathbf{x} \end{bmatrix}, \tag{10}$$

AUGMENTED
VECTOR
and **y** is sometimes called an *augmented feature vector*. Likewise, an *augmented weight vector* can be written as:

$$\mathbf{a} = \begin{bmatrix} w_0 \\ w_1 \\ \vdots \\ w_d \end{bmatrix} = \begin{bmatrix} w_0 \\ \mathbf{w} \end{bmatrix}. \tag{11}$$

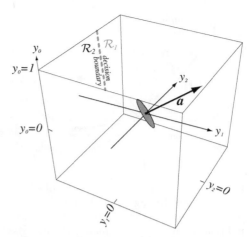

FIGURE 5.7. A three-dimensional augmented feature space $\mathbf{y}$ and augmented weight vector $\mathbf{a}$ (at the origin). The set of points for which $\mathbf{a}^t\mathbf{y} = 0$ is a plane (or more generally, a hyperplane) perpendicular to $\mathbf{a}$ and passing through the origin of $\mathbf{y}$-space, as indicated by the red disk. Such a plane need not pass through the origin of the two-dimensional feature space of the problem, as illustrated by the dashed decision boundary shown at the top of the box. Thus there exists an augmented weight vector $\mathbf{a}$ that will lead to any straight decision line in $\mathbf{x}$-space.

This mapping from d-dimensional $\mathbf{x}$-space to $(d + 1)$-dimensional $\mathbf{y}$-space is mathematically trivial but nonetheless quite convenient. The addition of a constant component to $\mathbf{x}$ preserves all distance relationships among samples. The resulting $\mathbf{y}$ vectors all lie in a d-dimensional subspace, which is the $\mathbf{x}$-space itself. The hyperplane decision surface $\hat{H}$ defined by $\mathbf{a}^t\mathbf{y} = 0$ passes through the origin in $\mathbf{y}$-space, even though the corresponding hyperplane H can be in any position in $\mathbf{x}$-space. The distance from $\mathbf{y}$ to $\hat{H}$ is given by $|\mathbf{a}^t\mathbf{y}|/\|\mathbf{a}\|$, or $|g(\mathbf{x})|/\|\mathbf{a}\|$. Because $\|\mathbf{a}\| \geq \|\mathbf{w}\|$, this distance is less than, or at most equal to, the distance from $\mathbf{x}$ to H. By using this mapping we reduce the problem of finding a weight vector $\mathbf{w}$ and a threshold weight w_0 to the problem of finding a single weight vector $\mathbf{a}$ (Fig. 5.7).

5.4 THE TWO-CATEGORY LINEARLY SEPARABLE CASE

Suppose now that we have a set of n samples $\mathbf{y}_1, \ldots, \mathbf{y}_n$, some labeled ω_1 and some labeled ω_2. We want to use these samples to determine the weights $\mathbf{a}$ in a linear discriminant function $g(\mathbf{x}) = \mathbf{a}^t\mathbf{y}$. Suppose we have reason to believe that there exists a solution for which the probability of error is very low. Then a reasonable approach is to look for a weight vector that classifies all of the samples correctly. If such a weight vector exists, the samples are said to be *linearly separable*.

A sample $\mathbf{y}_i$ is classified correctly if $\mathbf{a}^t\mathbf{y}_i > 0$ and $\mathbf{y}_i$ is labeled ω_1 or if $\mathbf{a}^t\mathbf{y}_i < 0$ and $\mathbf{y}_i$ is labeled ω_2. This suggests a "normalization" that simplifies the treatment of the two-category case, namely, the replacement of all samples labeled ω_2 by their negatives. With this "normalization" we can forget the labels and look for a weight vector $\mathbf{a}$ such that $\mathbf{a}^t\mathbf{y}_i > 0$ for *all* of the samples. Such a weight vector is called a *separating vector* or more generally a *solution vector*.

LINEARLY
SEPARABLE

SEPARATING
VECTOR

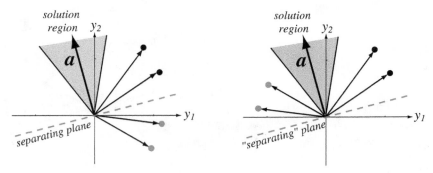

FIGURE 5.8. Four training samples (black for ω_1, red for ω_2) and the solution region in feature space. The figure on the left shows the raw data; the solution vectors leads to a plane that separates the patterns from the two categories. In the figure on the right, the red points have been "normalized"—that is, changed in sign. Now the solution vector leads to a plane that places all "normalized" points on the same side.

5.4.1 Geometry and Terminology

WEIGHT SPACE

The weight vector **a** can be thought of as specifying a point in *weight space*. Each sample $\mathbf{y}_i$ places a constraint on the possible location of a solution vector. The equation $\mathbf{a}^t\mathbf{y}_i = 0$ defines a hyperplane through the origin of weight space having $\mathbf{y}_i$ as a normal vector. The solution vector—if it exists—must be on the positive side of every hyperplane. Thus, a solution vector must lie in the intersection of n half-spaces; indeed any vector in this region is a solution vector. The corresponding region is called the *solution region*, and it should not be confused with the decision region in feature space corresponding to any particular category. A two-dimensional example illustrating the solution region for both the normalized and the unnormalized case is shown in Fig. 5.8.

SOLUTION
REGION

From this discussion, it should be clear that the solution vector—again, if it exists—is not unique. There are several ways to impose additional requirements to constrain the solution vector. One possibility is to seek a unit-length weight vector that maximizes the minimum distance from the samples to the separating plane. Another possibility is to seek the minimum-length weight vector satisfying $\mathbf{a}^t\mathbf{y}_i \geq b$ for all i, where b is a positive constant called the *margin*. As shown in Fig. 5.9, the solution region resulting form the intersections of the halfspaces for which $\mathbf{a}^t\mathbf{y}_i \geq b > 0$ lies within the previous solution region, being insulated from the old boundaries by the distance $b/\|\mathbf{y}_i\|$.

MARGIN

The motivation behind these attempts to find a solution vector closer to the "middle" of the solution region is the natural belief that the resulting solution is more likely to classify new test samples correctly. In most of the cases we shall treat, however, we shall be satisfied with any solution strictly within the solution region. Our chief concern will be to see that any iterative procedure used does not converge to a limit point on the boundary. This problem can always be avoided by the introduction of a margin, that is, by requiring that $\mathbf{a}^t\mathbf{y}_i \geq b > 0$ for all i.

5.4.2 Gradient Descent Procedures

The approach we shall take to finding a solution to the set of linear inequalities $\mathbf{a}^t\mathbf{y}_i > 0$ will be to define a criterion function $J(\mathbf{a})$ that is minimized if **a** is a solution vector. This reduces our problem to one of minimizing a scalar function—a

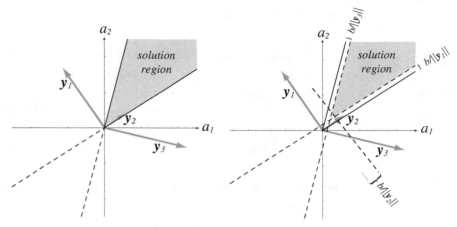

FIGURE 5.9. The effect of the margin on the solution region. At the left is the case of no margin ($b = 0$) equivalent to a case such as shown at the left in Fig. 5.8. At the right is the case $b > 0$, shrinking the solution region by margins $b/\|\mathbf{y}_i\|$.

problem that can often be solved by a gradient descent procedure. Basic gradient descent is very simple. We start with some arbitrarily chosen weight vector $\mathbf{a}(1)$ and compute the gradient vector $\nabla J(\mathbf{a}(1))$. The next value $\mathbf{a}(2)$ is obtained by moving some distance from $\mathbf{a}(1)$ in the direction of steepest descent, i.e., along the negative of the gradient. In general, $\mathbf{a}(k + 1)$ is obtained from $\mathbf{a}(k)$ by the equation

$$\mathbf{a}(k + 1) = \mathbf{a}(k) - \eta(k)\nabla J(\mathbf{a}(k)), \qquad (12)$$

LEARNING RATE where η is a positive scale factor or *learning rate* that sets the step size. We hope that such a sequence of weight vectors will converge to a solution minimizing $J(\mathbf{a})$. In algorithmic form we have:

■ **Algorithm 1. (Basic Gradient Descent)**

1 **begin initialize a**, threshold θ, $\eta(\cdot)$, $k \leftarrow 0$
2 $\qquad$ **do** $k \leftarrow k + 1$
3 $\qquad\qquad$ $\mathbf{a} \leftarrow \mathbf{a} - \eta(k)\nabla J(\mathbf{a})$
4 $\qquad$ **until** $|\eta(k)\nabla J(\mathbf{a})| < \theta$
5 $\qquad$ **return a**
6 **end**

The many problems associated with gradient descent procedures are well known. Fortunately, we shall be constructing the functions we want to minimize, and shall be able to avoid the most serious of these problems. One that will confront us repeatedly, however, is the choice of the learning rate $\eta(k)$. If $\eta(k)$ is too small, convergence is needlessly slow, whereas if $\eta(k)$ is too large, the correction process will overshoot and can even diverge (Section 5.6.1).

We now consider a principled method for setting the learning rate. Suppose that the criterion function can be well-approximated by the second-order expansion around a value $\mathbf{a}(k)$ as

$$J(\mathbf{a}) \simeq J(\mathbf{a}(k)) + \nabla J^t(\mathbf{a} - \mathbf{a}(k)) + \frac{1}{2}(\mathbf{a} - \mathbf{a}(k))^t \mathbf{H}\,(\mathbf{a} - \mathbf{a}(k)), \qquad (13)$$

HESSIAN MATRIX

where $\mathbf{H}$ is the *Hessian matrix* of second partial derivatives $\partial^2 J/\partial a_i \partial a_j$ evaluated at $\mathbf{a}(k)$. Then, substituting $\mathbf{a}(k+1)$ from Eq. 12 into Eq. 13 we find

$$J(\mathbf{a}(k+1)) \simeq J(\mathbf{a}(k)) - \eta(k)\|\nabla J\|^2 + \frac{1}{2}\eta^2(k)\nabla J^t \mathbf{H} \nabla J.$$

From this it follows (Problem 13) that $J(\mathbf{a}(k+1))$ can be minimized by the choice

$$\eta(k) = \frac{\|\nabla J\|^2}{\nabla J^t \mathbf{H} \nabla J}, \qquad (14)$$

where $\mathbf{H}$ depends on $\mathbf{a}$, and thus indirectly on k. This then is the optimal choice of $\eta(k)$ given the assumptions mentioned. Note that if the criterion function $J(\mathbf{a})$ is quadratic throughout the region of interest, then $\mathbf{H}$ is constant and η is a constant independent of k.

NEWTON'S
ALGORITHM

An alternative approach, obtained by ignoring Eq. 12 and choosing $\mathbf{a}(k+1)$ to minimize the second-order expansion, is *Newton's algorithm* where line 3 in Algorithm 1 is replaced by

$$\mathbf{a}(k+1) = \mathbf{a}(k) - \mathbf{H}^{-1}\nabla J, \qquad (15)$$

leading to the following algorithm:

■ **Algorithm 2. (Newton Descent)**

1 **begin initialize a**, threshold θ
2   **do**
3              $\mathbf{a} \leftarrow \mathbf{a} - \mathbf{H}^{-1}\nabla J(\mathbf{a})$
4   **until** $|\mathbf{H}^{-1}\nabla J(\mathbf{a})| < \theta$
5   **return a**
6 **end**

(We should point out that Newton's algorithm works for the quadratic error functions we have been considering, but not in the nonquadratic error functions of multilayer neural networks we shall meet in Chapter 6.) Simple gradient descent and Newton's algorithm are compared in Fig. 5.10.

Generally speaking, Newton's algorithm will usually give a greater improvement *per step* than the simple gradient descent algorithm, even with the optimal value of $\eta(k)$. However, Newton's algorithm is not applicable if the Hessian matrix $\mathbf{H}$ is singular. Furthermore, even when $\mathbf{H}$ is nonsingular, the $O(d^3)$ time required for matrix inversion on each iteration can easily offset the descent advantage. In fact, it often takes less time to set $\eta(k)$ to a constant η that is smaller than necessary and make a few more corrections than it is to compute the optimal $\eta(k)$ at each step (Computer exercise 1).

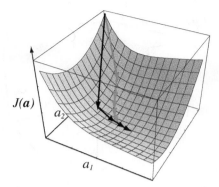

FIGURE 5.10. The sequence of weight vectors given by a simple gradient descent method (red) and by Newton's (second order) algorithm (black). Newton's method typically leads to greater improvement per step, even when using optimal learning rates for both methods. However the added computational burden of inverting the Hessian matrix used in Newton's method is not always justified, and simple gradient descent may suffice.

5.5 MINIMIZING THE PERCEPTRON CRITERION FUNCTION

5.5.1 The Perceptron Criterion Function

Consider now the problem of constructing a criterion function for solving the linear inequalities $\mathbf{a}^t \mathbf{y}_i > 0$. The most obvious choice is to let $J(\mathbf{a};\ \mathbf{y}_1, \ldots, \mathbf{y}_n)$ be the number of samples misclassified by $\mathbf{a}$. However, because this function is piecewise constant, it is obviously a poor candidate for a gradient search. A better choice is the *Perceptron criterion function*

$$J_p(\mathbf{a}) = \sum_{\mathbf{y} \in \mathcal{Y}} (-\mathbf{a}^t \mathbf{y}), \tag{16}$$

where $\mathcal{Y}(\mathbf{a})$ is the set of samples *misclassified* by $\mathbf{a}$. (If no samples are misclassified, $\mathcal{Y}$ is empty and we define J_p to be zero.) Because $\mathbf{a}^t \mathbf{y} \leq 0$ if $\mathbf{y}$ is misclassified, $J_p(\mathbf{a})$ is never negative, being zero only if $\mathbf{a}$ is a solution vector or if $\mathbf{a}$ is on the decision boundary. Geometrically, $J_p(\mathbf{a})$ is proportional to the sum of the distances from the misclassified samples to the decision boundary. Figure 5.11 illustrates J_p for a simple two-dimensional example.

Because the jth component of the gradient of J_p is $\partial J_p / \partial a_j$, we see from Eq. 16 that

$$\boldsymbol{\nabla} J_p = \sum_{\mathbf{y} \in \mathcal{Y}} (-\mathbf{y}), \tag{17}$$

and hence the update rule becomes

$$\mathbf{a}(k+1) = \mathbf{a}(k) + \eta(k) \sum_{\mathbf{y} \in \mathcal{Y}_k} \mathbf{y}, \tag{18}$$

where $\mathcal{Y}_k$ is the set of samples misclassified by $\mathbf{a}(k)$. Thus the Perceptron algorithm is as follows:

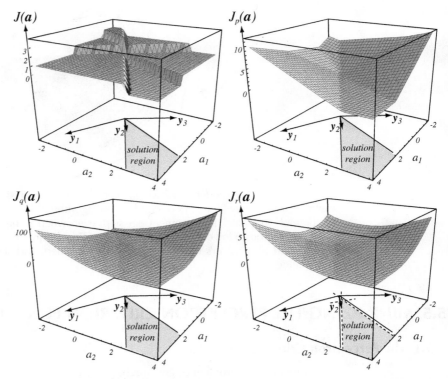

FIGURE 5.11. Four learning criteria as a function of weights in a linear classifier. At the upper left is the total number of patterns misclassified, which is piecewise constant and hence unacceptable for gradient descent procedures. At the upper right is the Perceptron criterion (Eq. 16), which is piecewise linear and acceptable for gradient descent. The lower left is squared error (Eq. 32), which has nice analytic properties and is useful even when the patterns are not linearly separable. The lower right is the square error with margin (Eq. 33). A designer may adjust the margin b in order to force the solution vector to lie toward the middle of the $b = 0$ solution region in hopes of improving generalization of the resulting classifier.

■ **Algorithm 3. (Batch Perceptron)**

1   <u>**begin initialize**</u> $\mathbf{a}$, $\eta(\cdot)$, criterion θ, $k \leftarrow 0$
2   <u>**do**</u> $k \leftarrow k + 1$
3            $\mathbf{a} \leftarrow \mathbf{a} + \eta(k)\sum_{\mathbf{y} \in \mathcal{Y}_k} \mathbf{y}$
4   <u>**until**</u> $|\eta(k)\sum_{\mathbf{y} \in \mathcal{Y}_k} \mathbf{y}| < \theta$
5   <u>**return a**</u>
6   <u>**end**</u>

Thus, the batch Perceptron algorithm for finding a solution vector can be stated very simply: The next weight vector is obtained by adding some multiple of the sum BATCH TRAINING of the misclassified samples to the present weight vector. We use the term "batch" to refer to the fact that (in general) a large group of samples is used when computing

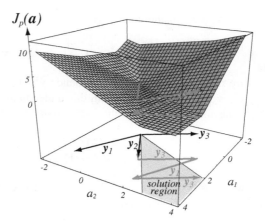

FIGURE 5.12. The Perceptron criterion, $J_p(\mathbf{a})$, is plotted as a function of the weights a_1 and a_2 for a three-pattern problem. The weight vector begins at **0**, and the algorithm sequentially adds to it vectors equal to the "normalized" misclassified patterns themselves. In the example shown, this sequence is $\mathbf{y}_1 + \mathbf{y}_2 + \mathbf{y}_3$, $\mathbf{y}_2$, $\mathbf{y}_3$, $\mathbf{y}_1$, $\mathbf{y}_3$, at which time the vector lies in the solution region and iteration terminates. Note that the second update (by $\mathbf{y}_3$) takes the candidate vector *farther* from the solution region than after the first update (cf. Theorem 5.1).

each weight update. (We shall soon see alternative methods based on single samples.) Figure 5.12 shows how this algorithm yields a solution vector for a simple two-dimensional example with $\mathbf{a}(1) = \mathbf{0}$, and $\eta(k) = 1$. We shall now show that it will yield a solution for any linearly separable problem.

5.5.2 Convergence Proof for Single-Sample Correction

We begin our examination of convergence properties of the Perceptron algorithm with a variant that is easier to analyze. Rather than testing $\mathbf{a}(k)$ on all of the samples and basing our correction of the set $\mathcal{Y}_k$ of misclassified training samples, we shall consider the samples in a sequence and shall modify the weight vector whenever it misclassifies a *single* sample. For the purposes of the convergence proof, the detailed nature of the sequence is unimportant as long as every sample appears in the sequence infinitely often. The simplest way to assure this is to repeat the samples cyclically, though from a practical point of view random selection is often to be preferred (Section 5.8.5). Clearly neither the batch nor this single-sample version of the Perceptron algorithm are on-line because we must store and potentially revisit all of the training patterns.

Two further simplifications help to clarify the exposition. First, we shall temporarily restrict our attention to the case in which $\eta(k)$ is constant—the so-called

Fixed Increment

fixed-increment case. It is clear from Eq. 18 that if $\eta(t)$ is constant, it merely serves to scale the samples; thus, in the fixed-increment case we can take $\eta(t) = 1$ with no loss in generality. The second simplification merely involves notation. When the samples are considered sequentially, some will be misclassified. Because we shall only change the weight vector when there is an error, we really need only pay attention to the misclassified samples. Thus we denote the sequence of samples using superscripts—that is, by $\mathbf{y}^1$, $\mathbf{y}^2$, ..., $\mathbf{y}^k$, ..., where each $\mathbf{y}^k$ is one of the n samples $\mathbf{y}_1, \ldots, \mathbf{y}_n$ and where each $\mathbf{y}^k$ is misclassified. For example, if the samples $\mathbf{y}_1$, $\mathbf{y}_2$, and $\mathbf{y}_3$ are considered cyclically and if the marked samples

$$\overset{\downarrow}{\mathbf{y}_1}, \ \mathbf{y}_2, \ \overset{\downarrow}{\mathbf{y}_3}, \ \overset{\downarrow}{\mathbf{y}_1}, \ \overset{\downarrow}{\mathbf{y}_2}, \ \mathbf{y}_3, \ \mathbf{y}_1, \ \overset{\downarrow}{\mathbf{y}_2}, \ \dots \tag{19}$$

FIXED-
INCREMENT
RULE
are misclassified, then the sequence $\mathbf{y}^1, \ \mathbf{y}^2, \ \mathbf{y}^3, \ \mathbf{y}^4, \ \mathbf{y}^5, \dots$ denotes the sequence $\mathbf{y}_1, \ \mathbf{y}_3, \ \mathbf{y}_1, \ \mathbf{y}_2, \ \mathbf{y}_2, \dots$. With this understanding, the *fixed-increment rule* for generating a sequence of weight vectors can be written as

$$\begin{aligned} \mathbf{a}(1) & \qquad \text{arbitrary} \\ \mathbf{a}(k+1) &= \mathbf{a}(k) + \mathbf{y}^k \quad k \geq 1 \end{aligned} \tag{20}$$

where $\mathbf{a}^t(k)\mathbf{y}^k \leq 0$ for all k. If we let n denote the total number of patterns, the algorithm is as follows:

■ **Algorithm 4. (Fixed-Increment Single-Sample Perceptron)**

1 **begin initialize** $\mathbf{a}, k \leftarrow 0$
2   **do** $k \leftarrow (k+1) \bmod n$
3   **if** $\mathbf{y}^k$ is misclassified by $\mathbf{a}$ **then** $\mathbf{a} \leftarrow \mathbf{a} + \mathbf{y}^k$
4   **until** all patterns properly classified
5   **return** $\mathbf{a}$
6 **end**

The fixed-increment Perceptron rule is the simplest of many algorithms that have been proposed for solving systems of linear inequalities. Geometrically, its interpretation in weight space is particularly clear. Because $\mathbf{a}(k)$ misclassifies $\mathbf{y}^k$, $\mathbf{a}(k)$ is not on the positive side of the $\mathbf{y}^k$ hyperplane $\mathbf{a}^t\mathbf{y}^k = 0$. The addition of $\mathbf{y}^k$ to $\mathbf{a}(k)$ moves the weight vector directly toward and perhaps across this hyperplane. Whether the hyperplane is crossed or not, the new inner product $\mathbf{a}^t(k+1)\mathbf{y}^k$ is larger than the old inner product $\mathbf{a}^t(k)\mathbf{y}^k$ by the amount $\|\mathbf{y}^k\|^2$, and the correction is hence moving the weight vector in a good direction (Fig. 5.13).

Clearly this algorithm can only terminate if the samples are linearly separable; we now prove that indeed it terminates so long as the samples are linearly separable.

■ **Theorem 5.1. (Perceptron Convergence)** If training samples are linearly separable, then the sequence of weight vectors given by Algorithm 4 will terminate at a solution vector.

Proof. In seeking a proof, it is natural to try to show that each correction brings the weight vector closer to the solution region. That is, one might try to show that if $\hat{\mathbf{a}}$ is any solution vector, then $\|\mathbf{a}(k+1) - \hat{\mathbf{a}}\|$ is smaller than $\|\mathbf{a}(k) - \hat{\mathbf{a}}\|$. While this turns out not to be true in general (cf. steps 6 and 7 in Fig. 5.13), we shall see that it is true for solution vectors that are sufficiently long.

Let $\hat{\mathbf{a}}$ be any solution vector, so that $\hat{\mathbf{a}}^t\mathbf{y}_i$ is strictly positive for all i, and let α be a positive scale factor. From Eq. 20 we have

$$\mathbf{a}(k+1) - \alpha\hat{\mathbf{a}} = (\mathbf{a}(k) - \alpha\hat{\mathbf{a}}) + \mathbf{y}^k$$

and hence

$$\|\mathbf{a}(k+1) - \alpha\hat{\mathbf{a}}\|^2 = \|\mathbf{a}(k) - \alpha\hat{\mathbf{a}}\|^2 + 2(\mathbf{a}(k) - \alpha\hat{\mathbf{a}})^t\mathbf{y}^k + \|\mathbf{y}^k\|^2.$$

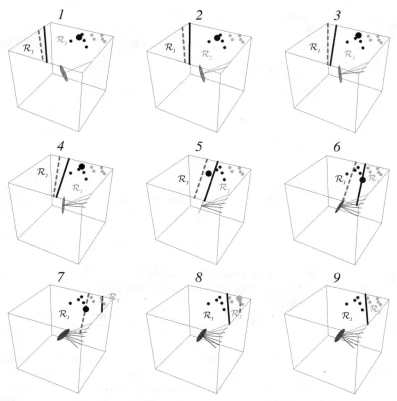

FIGURE 5.13. Samples from two categories, ω_1 (black) and ω_2 (red) are shown in augmented feature space, along with an augmented weight vector **a**. At each step in a fixed-increment rule, one of the misclassified patterns, $\mathbf{y}^k$, is shown by the large dot. A correction $\Delta\mathbf{a}$ (proportional to the pattern vector $\mathbf{y}^k$) is added to the weight vector—toward an ω_1 point or away from an ω_2 point. This changes the decision boundary from the dashed position (from the previous update) to the solid position. The sequence of resulting **a** vectors is shown, where later values are shown darker. In this example, by step 9 a solution vector has been found and the categories are successfully separated by the decision boundary shown.

Because $\mathbf{y}^k$ was misclassified, $\mathbf{a}^t(k)\mathbf{y}^k \leq 0$ and thus

$$\|\mathbf{a}(k+1) - \alpha\hat{\mathbf{a}}\|^2 \leq \|\mathbf{a}(k) - \alpha\hat{\mathbf{a}}\|^2 - 2\alpha\hat{\mathbf{a}}^t\mathbf{y}^k + \|\mathbf{y}^k\|^2.$$

Because $\hat{\mathbf{a}}^t\mathbf{y}^k$ is strictly positive, the second term will dominate the third if α is sufficiently large. In particular, if we let β be the maximum length of a pattern vector,

$$\beta^2 = \max_i \|\mathbf{y}_i\|^2, \tag{21}$$

and let γ be the smallest inner product of the solution vector with any pattern vector, that is,

$$\gamma = \min_i \left[\hat{\mathbf{a}}^t\mathbf{y}_i\right] > 0, \tag{22}$$

then we have the inequality

$$\|\mathbf{a}(k+1) - \alpha\hat{\mathbf{a}}\|^2 \leq \|\mathbf{a}(k) - \alpha\hat{\mathbf{a}}\|^2 - 2\alpha\gamma + \beta^2.$$

If we choose

$$\alpha = \frac{\beta^2}{\gamma}, \tag{23}$$

we obtain

$$\|\mathbf{a}(k+1) - \alpha\hat{\mathbf{a}}\|^2 \leq \|\mathbf{a}(k) - \alpha\hat{\mathbf{a}}\|^2 - \beta^2.$$

Thus, the squared distance from $\mathbf{a}(k)$ to $\alpha\hat{\mathbf{a}}$ is reduced by at least β^2 at each correction, and after k corrections we obtain

$$\|\mathbf{a}(k+1) - \alpha\hat{\mathbf{a}}\|^2 \leq \|\mathbf{a}(1) - \alpha\hat{\mathbf{a}}\|^2 - k\beta^2. \tag{24}$$

Because this squared distance cannot become negative, it follows that the sequence of corrections must terminate after no more than k_0 corrections, where

$$k_0 = \frac{\|\mathbf{a}(1) - \alpha\hat{\mathbf{a}}\|^2}{\beta^2}. \tag{25}$$

Because a correction occurs whenever a sample is misclassified and because each sample appears infinitely often in the sequence, it follows that when corrections cease the resulting weight vector must classify all of the samples correctly.

The number k_0 gives us a bound on the number of corrections. If $\mathbf{a}(1) = \mathbf{0}$, we get the following particularly simple expression for k_0:

$$k_0 = \frac{\alpha^2\|\hat{\mathbf{a}}\|^2}{\beta^2} = \frac{\beta^2\|\hat{\mathbf{a}}\|^2}{\gamma^2} = \frac{\max_i \|\mathbf{y}_i\|^2 \|\hat{\mathbf{a}}\|^2}{\min_i [\mathbf{y}_i^t\hat{\mathbf{a}}]^2}. \tag{26}$$

The denominator in Eq. 26 shows that the difficulty of the problem is essentially determined by the samples most nearly orthogonal to the solution vector. Unfortunately, it provides no help when we face an unsolved problem, because the bound is expressed in terms of a solution vector that is unknown. In general, it is clear that linearly separable problems can be made arbitrarily difficult to solve by making the samples almost coplanar (Computer exercise 2). Nevertheless, if the training samples are linearly separable, the fixed-increment rule will yield a solution after a finite number of corrections.

5.5.3 Some Direct Generalizations

The fixed increment rule can be generalized to provide a variety of related algorithms. We shall briefly consider two variants of particular interest. The first variant introduces a *variable increment* $\eta(k)$ and a margin b, and it calls for a correction whenever $\mathbf{a}^t(k)\mathbf{y}^k$ fails to exceed the margin. The update is given by

VARIABLE
INCREMENT

$$\begin{aligned} \mathbf{a}(1) & \qquad \text{arbitrary} \\ \mathbf{a}(k+1) &= \mathbf{a}(k) + \eta(k)\mathbf{y}^k \quad k \geq 1, \end{aligned} \tag{27}$$

where now $\mathbf{a}^t(k)\mathbf{y}^k \leq b$ for all k. Thus for n patterns, our algorithm is:

■ **Algorithm 5. (Variable-Increment Perceptron with Margin)**

1 **begin initialize** $\mathbf{a}$, threshold θ, margin b, $\eta(\cdot)$, $k \leftarrow 0$
2   **do** $k \leftarrow (k+1) \bmod n$
3   **if** $\mathbf{a}^t\mathbf{y}^k \leq b$ **then** $\mathbf{a} \leftarrow \mathbf{a} + \eta(k)\mathbf{y}^k$
4   **until** $\mathbf{a}^t\mathbf{y}^k > b$ for all k
5   **return** $\mathbf{a}$
6 **end**

It can be shown that if the samples are linearly separable and if

$$\eta(k) \geq 0, \tag{28}$$

$$\lim_{m \to \infty} \sum_{k=1}^{m} \eta(k) = \infty \tag{29}$$

and

$$\lim_{m \to \infty} \frac{\sum_{k=1}^{m} \eta^2(k)}{\left(\sum_{k=1}^{m} \eta(k)\right)^2} = 0, \tag{30}$$

then $\mathbf{a}(k)$ converges to a solution vector $\mathbf{a}$ satisfying $\mathbf{a}^t\mathbf{y}_i > b$ for all i (Problem 19). In particular, these conditions on $\eta(k)$ are satisfied if $\eta(k)$ is a positive constant or if it decreases as $1/k$.

Another variant of interest is our original gradient descent algorithm for J_p,

$$\begin{aligned} \mathbf{a}(1) \quad & \text{arbitrary} \\ \mathbf{a}(k+1) = \mathbf{a}(k) &+ \eta(k) \sum_{\mathbf{y} \in \mathcal{Y}_k} \mathbf{y}, \end{aligned} \tag{31}$$

where $\mathcal{Y}_k$ is the set of training samples misclassified by $\mathbf{a}(k)$. It is easy to see that this algorithm will also yield a solution once one recognizes that if $\hat{\mathbf{a}}$ is a solution vector for $\mathbf{y}_1, \ldots, \mathbf{y}_n$, then it correctly classifies the correction vector

$$\mathbf{y}^k = \sum_{\mathbf{y} \in \mathcal{Y}_k} \mathbf{y}.$$

In greater detail, then, the algorithm is as follows:

■ **Algorithm 6. (Batch Variable Increment Perceptron)**

1 **begin initialize** $\mathbf{a}$, $\eta(\cdot)$, $k \leftarrow 0$
2   **do** $k \leftarrow (k+1) \bmod n$
3         $\mathcal{Y}_k = \{\}$
4         $j = 0$
5   **do** $j \leftarrow j+1$

6   **if** $\mathbf{y}_j$ is misclassified **then** Append $\mathbf{y}_j$ to $\mathcal{Y}_k$
7   **until** $j = n$
8                              $\mathbf{a} \leftarrow \mathbf{a} + \eta(k) \sum_{\mathbf{y} \in \mathcal{Y}_k} \mathbf{y}$
9   **until** $\mathcal{Y}_k = \{\}$
10   **return a**
11 **end**

The benefit of batch gradient descent is that the trajectory of the weight vector is smoothed, compared to that in corresponding single-sample algorithms (e.g., Algorithm 5), because at each update the full set of misclassified patterns is used—the local statistical variations in the misclassified patterns tend to cancel while the large-scale trend does not. Thus, if the samples are linearly separable, all of the possible correction vectors form a linearly separable set, and if $\eta(k)$ satisfies Eqs. 28–30, the sequence of weight vectors produced by the gradient descent algorithm for $J_p(\cdot)$ will always converge to a solution vector.

It is interesting to note that the conditions on $\eta(k)$ are satisfied if $\eta(k)$ is a positive constant, if it decreases as $1/k$, or even if it increases as k. Generally speaking, one would prefer to have $\eta(k)$ become smaller as time goes on. This is particularly true if there is reason to believe that the set of samples is not linearly separable, because it reduces the disruptive effects of a few "bad" samples. However, in the separable case it is a curious fact that one can allow $\eta(k)$ to become larger and still obtain a solution.

This observation brings out one of the differences between theory and practice. From a theoretical viewpoint, it is interesting that we can obtain a solution in a finite number of steps for any finite set of separable samples, for any initial weight vector $\mathbf{a}(1)$, for any nonnegative margin b, and for any scale factor $\eta(k)$ satisfying Eqs. 28–30. From a practical viewpoint, we want to make wise choices for these quantities. Consider the margin b, for example. If b is much smaller than $\eta(k)\|\mathbf{y}^k\|^2$, the amount by which a correction increases $\mathbf{a}^t(k)\mathbf{y}^k$, it is clear that it will have little if any effect. If it is much larger than $\eta(k)\|\mathbf{y}^k\|^2$, many corrections will be needed to satisfy the conditions $\mathbf{a}^t(k)\mathbf{y}^k > b$. A value close to $\eta(k)\|\mathbf{y}^k\|^2$ is often a useful compromise. In addition to these choices for $\eta(k)$ and b, the scaling of the components of $\mathbf{y}^k$ can also have a great effect on the results. The possession of a convergence theorem does not remove the need for thought in applying these techniques.

A close descendant of the Perceptron algorithm is the Winnow algorithm, which has applicability to separable training data. The key difference is that while the weight vector returned by the Perceptron algorithm has components a_i ($i = 0, \ldots, d$), in Winnow they are scaled according to $2\sinh[a_i]$. In one version, the balanced Winnow algorithm, there are separate "positive" and "negative" weight vectors, $\mathbf{a}^+$ and $\mathbf{a}^-$, each associated with one of the two categories to be learned. Corrections on the positive weight are made if and only if a training pattern in ω_1 is misclassified; conversely, corrections on the negative weight are made if and only if a training pattern in ω_2 is misclassified.

WINNOW
ALGORITHM

■ **Algorithm 7.   (Balanced Winnow)**

1 **begin initialize** $\mathbf{a}^+, \mathbf{a}^-, \eta(\cdot), k \leftarrow 0, \alpha > 1$
2   **if** $\mathrm{Sgn}[\mathbf{a}^{+t}\mathbf{y}_k - \mathbf{a}^{-t}\mathbf{y}_k] \neq z_k$ (pattern misclassified)
3   **then if** $z_k = +1$ **then** $a_i^+ \leftarrow \alpha^{+y_i} a_i^+;\ a_i^- \leftarrow \alpha^{-y_i} a_i^-$ for all i

4   **if** $z_k = -1$ **then** $a_i^+ \leftarrow \alpha^{-y_i} a_i^+$; $a_i^- \leftarrow \alpha^{+y_i} a_i^-$ for all i

5   **return** $\mathbf{a}^+, \mathbf{a}^-$

6 **end**

There are two main benefits of such a version of the Winnow algorithm. The first is that during training each of the two constituent weight vectors moves in a uniform direction and this means that for separable data the "gap," determined by these two vectors, can never increase in size. This leads to a convergence proof that, while somewhat more complicated, is nevertheless more general than the Perceptron convergence theorem (cf. Bibliography). The second benefit is that convergence is generally faster than in a Perceptron, because for proper setting of learning rate, each constituent weight does not overshoot its final value. This benefit is especially pronounced whenever a large number of irrelevant or redundant features are present (Computer exercise 6).

5.6 RELAXATION PROCEDURES

We have seen how a linear classifier is trained through the minimization of the Perceptron criterion of Eq. 16. We can generalize this approach, in so-called "relaxation procedures," to include a broader class of criterion functions and methods for minimizing them.

5.6.1 The Descent Algorithm

The criterion function $J_p(\cdot)$ is by no means the only function we can construct that is minimized when $\mathbf{a}$ is a solution vector. A close but distinct relative is

$$J_q(\mathbf{a}) = \sum_{y \in \mathcal{Y}} (\mathbf{a}^t \mathbf{y})^2, \tag{32}$$

where $\mathcal{Y}(\mathbf{a})$ again denotes the set of training samples misclassified by $\mathbf{a}$. Both J_p and J_q focus attention on the misclassified samples. The chief difference is that the gradient of J_q is continuous, whereas the gradient of J_p is not. Thus, J_q presents a smoother surface to search (Fig. 5.11). Unfortunately, J_q is so smooth near the boundary of the solution region that the sequence of weight vectors can converge to a point on the boundary. It is particularly embarrassing to spend some time following the gradient merely to reach the boundary point $\mathbf{a} = \mathbf{0}$. Another problem with J_q is that its value can be dominated by the longest sample vectors. Both of these problems are avoided by the criterion function

$$J_r(\mathbf{a}) = \frac{1}{2} \sum_{y \in \mathcal{Y}} \frac{(\mathbf{a}^t \mathbf{y} - b)^2}{\|\mathbf{y}\|^2}, \tag{33}$$

where now $\mathcal{Y}(\mathbf{a})$ is the set of samples for which $\mathbf{a}^t \mathbf{y} \leq b$. (If $\mathcal{Y}(\mathbf{a})$ is empty, we define J_r to be zero.) Thus, $J_r(\mathbf{a})$ is never negative, and is zero if and only if $\mathbf{a}^t \mathbf{y} \geq b$ for all of the training samples. The gradient of J_r is given by

$$\nabla J_r = \sum_{y \in \mathcal{Y}} \frac{\mathbf{a}^t \mathbf{y} - b}{\|\mathbf{y}\|^2} \mathbf{y},$$

and the update rule is

$$\mathbf{a}(1) \quad \text{arbitrary}$$
$$\mathbf{a}(k+1) = \mathbf{a}(k) + \eta(k) \sum_{\mathbf{y} \in \mathcal{Y}} \frac{b - \mathbf{a}^t \mathbf{y}}{\|\mathbf{y}\|^2} \mathbf{y}. \tag{34}$$

Thus the relaxation algorithm becomes:

■ **Algorithm 8. (Batch Relaxation with Margin)**

```
1  begin initialize a, η(·), b, k ← 0
2      do k ← (k + 1) mod n
3          𝒴_k = {}
4          j ← 0
5          do j ← j + 1
6              if a^t y^j ≤ b  then Append y^j to 𝒴_k
7          until j = n
8          a ← a + η(k) Σ_{y∈𝒴} (b−a^t y)/‖y‖² y
9      until 𝒴_k = {}
10     return a
11 end
```

As before, we find it easier to prove convergence when the samples are considered one at a time rather than jointly—that is, single-sample rather than batch. We also limit our attention to the fixed-increment case, $\eta(k) = \eta$. Thus, we are again led to consider a sequence $\mathbf{y}^1, \mathbf{y}^2, \ldots$ formed from those samples that call for the weight vector to be corrected. The single-sample correction rule analogous to Eq. 34 is

$$\mathbf{a}(1) \quad \text{arbitrary}$$
$$\mathbf{a}(k+1) = \mathbf{a}(k) + \eta \frac{b - \mathbf{a}^t(k)\mathbf{y}^k}{\|\mathbf{y}^k\|^2} \mathbf{y}^k, \tag{35}$$

where $\mathbf{a}^t(k)\mathbf{y}^k \leq b$ for all k. The algorithm is as follows:

■ **Algorithm 9. (Single-Sample Relaxation with Margin)**

```
1  begin initialize a, η(·), k ← 0
2          do k ← (k + 1) mod n
3              if a^t y^k ≤ b  then a ← a + η(k) (b−a^t y^k)/‖y^k‖² y^k
4          until a^t y^k > b for all y^k
5      return a
6  end
```

This algorithm is known as the *single-sample relaxation rule with margin*, and it has a simple geometrical interpretation. The quantity

$$r(k) = \frac{b - \mathbf{a}^t(k)\mathbf{y}^k}{\|\mathbf{y}^k\|} \tag{36}$$

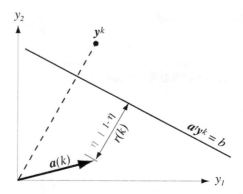

FIGURE 5.14. In each step of a basic relaxation algorithm, the weight vector is moved a proportion η of the way toward the hyperplane defined by $\mathbf{a}^t\mathbf{y}^k = b$.

is the distance from $\mathbf{a}(k)$ to the hyperplane $\mathbf{a}^t\mathbf{y}^k = b$. Because $\mathbf{y}^k/\|\mathbf{y}^k\|$ is the unit normal vector for the hyperplane, Eq. 35 calls for $\mathbf{a}(k)$ to be moved a certain fraction η of the distance from $\mathbf{a}(k)$ to the hyperplane. If $\eta = 1$, $\mathbf{a}(k)$ is moved exactly to the hyperplane, so that the "tension" created by the inequality $\mathbf{a}^t(k)\mathbf{y}^k \leq b$ is "relaxed" (Fig. 5.14). From Eq. 35, after a correction, we obtain

$$\mathbf{a}^t(k+1)\mathbf{y}^k - b = (1 - \eta)(\mathbf{a}^t(k)\mathbf{y}^k - b). \tag{37}$$

UNDER-
RELAXATION

OVER-
RELAXATION

If $\eta < 1$, then $\mathbf{a}^t(k+1)\mathbf{y}^k$ is still less than b, while if $\eta > 1$, then $\mathbf{a}^t(k+1)\mathbf{y}^k$ is greater than b. These conditions are referred to as *underrelaxation* and *overrelaxation*, respectively. In general, we shall restrict η to the range $0 < \eta < 2$ (Figs. 5.14 and 5.15).

5.6.2 Convergence Proof

When the relaxation rule is applied to a set of linearly separable samples, the number of corrections may or may not be finite. If it is finite, then of course we have obtained a solution vector. If it is not finite, we shall see that $\mathbf{a}(k)$ converges to a limit vector on the boundary of the solution region. Because the region in which $\mathbf{a}^t\mathbf{y} \geq b$ is contained in a larger region where $\mathbf{a}^t\mathbf{y} > 0$ if $b > 0$, this implies that $\mathbf{a}(k)$ will enter

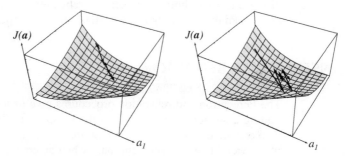

FIGURE 5.15. At the left, underrelaxation ($\eta < 1$) leads to needlessly slow descent, or even failure to converge. Overrelaxation ($1 < \eta < 2$, shown at the right) describes overshooting; nevertheless, convergence will ultimately be achieved.

this larger region at least once, eventually remaining there for all k greater than some finite k_0.

The proof depends upon the fact that if $\hat{\mathbf{a}}$ is *any* vector in the solution region—that is, any vector satisfying $\hat{\mathbf{a}}^t \mathbf{y}_i > b$ for all i—then at each step, $\mathbf{a}(k)$ gets closer to $\hat{\mathbf{a}}$. This fact follows at once from Eq. 35, because

$$\|\mathbf{a}(k+1) - \hat{\mathbf{a}}\|^2 = \|\mathbf{a}(k) - \hat{\mathbf{a}}\|^2 - 2\eta \frac{(b - \mathbf{a}^t(k)\mathbf{y}^k)}{\|\mathbf{y}^k\|^2}(\hat{\mathbf{a}} - \mathbf{a}(k))^t \mathbf{y}^k$$

$$+ \eta^2 \frac{(b - \mathbf{a}^t(k)\mathbf{y}^k)^2}{\|\mathbf{y}^k\|^2} \tag{38}$$

and

$$(\hat{\mathbf{a}} - \mathbf{a}(k))^t \mathbf{y}^k > b - \mathbf{a}^t(k)\mathbf{y}^k \geq 0, \tag{39}$$

so that

$$\|\mathbf{a}(k+1) - \hat{\mathbf{a}}\|^2 \leq \|\mathbf{a}(k) - \hat{\mathbf{a}}\|^2 - \eta(2 - \eta)\frac{(b - \mathbf{a}^t(k)\mathbf{y}^k)^2}{\|\mathbf{y}^k\|^2}. \tag{40}$$

Because we restrict η to the range $0 < \eta < 2$, it follows that $\|\mathbf{a}(k+1) - \hat{\mathbf{a}}\| \leq \|\mathbf{a}(k) - \hat{\mathbf{a}}\|$. Thus, the vectors in the sequence $\mathbf{a}(1), \mathbf{a}(2), \ldots$ get closer and closer to $\hat{\mathbf{a}}$, and in the limit as k goes to infinity the distance $\|\mathbf{a}(k) - \hat{\mathbf{a}}\|$ approaches some limiting distance $r(\hat{\mathbf{a}})$. This means that as k goes to infinity, $\mathbf{a}(k)$ is confined to the surface of a hypersphere with center $\hat{\mathbf{a}}$ and radius $r(\hat{\mathbf{a}})$. Because this is true for any $\hat{\mathbf{a}}$ in the solution region, the limiting $\mathbf{a}(k)$ is confined to the intersection of the hyperspheres centered about all of the possible solution vectors.

We now show that the common intersection of these hyperspheres is a single point on the boundary of the solution region. Suppose first that there are at least two points $\mathbf{a}'$ and $\mathbf{a}''$ on the common intersection. Then $\|\mathbf{a}' - \hat{\mathbf{a}}\| = \|\mathbf{a}'' - \hat{\mathbf{a}}\|$ for every $\hat{\mathbf{a}}$ in the solution region. But this implies that the solution region is contained in the $(\hat{d} - 1)$-dimensional hyperplane of points equidistant from $\mathbf{a}'$ to $\mathbf{a}''$, whereas we know that the solution region is $\hat{d}$-dimensional. (Stated formally, if $\hat{\mathbf{a}}^t \mathbf{y}_i > 0$ for $i = 1, \ldots, n$, then for any $\hat{d}$-dimensional vector $\mathbf{v}$, we have $(\hat{\mathbf{a}} + \epsilon \mathbf{v})^t \mathbf{y} > 0$ for $i = 1, \ldots, n$ if ϵ is sufficiently small.) Thus, $\mathbf{a}(k)$ converges to a single point $\mathbf{a}$. This point is certainly not inside the solution region, because then the sequence would be finite. It is not outside either, because each correction causes the weight vector to move η times its distance from the boundary plane, thereby preventing the vector from being bounded away from the boundary forever. Hence the limit point must be on the boundary.

5.7 NONSEPARABLE BEHAVIOR

ERROR-
CORRECTING
PROCEDURE

The Perceptron and relaxation procedures give us a number of simple methods for finding a separating vector when the samples are linearly separable. All of these methods are called *error-correcting procedures* because they call for a modification of the weight vector when and only when an error is encountered. Their success on separable problems is largely due to this relentless search for an error-free solution. In practice, one would only consider the use of these methods if there was reason to believe that the error rate for the optimal linear discriminant function is low.

Of course, even if a separating vector is found for the training samples, it does not follow that the resulting classifier will perform well on independent test data. A moment's reflection will show that *any* set of fewer than $2\hat{d}$ samples is likely to be linearly separable—a matter we shall return to in Chapter 9. Thus, one should use several times that many design samples to overdetermine the classifier, thereby ensuring that the performance on training and test data will be similar. Unfortunately, sufficiently large design sets are almost certainly *not* linearly separable. This makes it important to know how the error-correction procedures will behave when the samples are nonseparable.

Because no weight vector can correctly classify every sample in a nonseparable set (by definition), it is clear that the corrections in an error-correction procedure can never cease. Each algorithm produces an infinite sequence of weight vectors, any member of which may or may not yield a useful "solution." The exact nonseparable behavior of these rules has been studied thoroughly in a few special cases. It is known, for example, that the length of the weight vectors produced by the fixed-increment rule are bounded. Empirical rules for terminating the correction procedure are often based on this tendency for the length of the weight vector to fluctuate near some limiting value. From a theoretical viewpoint, if the components of the samples are integer-valued, the fixed-increment procedure yields a finite-state process. If the correction process is terminated at some arbitrary point, the weight vector may or may not be in a good state. By averaging the weight vectors produced by the correction rule, one can reduce the risk of obtaining a bad solution by accidentally choosing an unfortunate termination time.

A number of similar heuristic modifications to the error-correction rules have been suggested and studied empirically. The goal of these modifications is to obtain acceptable performance on nonseparable problems while preserving the ability to find a separating vector on separable problems. A common suggestion is the use of a variable increment $\eta(k)$, with $\eta(k)$ approaching zero as k approaches infinity. The rate at which $\eta(k)$ approaches zero is quite important. If it is too slow, the results will still be sensitive to those training samples that render the set nonseparable. If it is too fast, the weight vector may converge prematurely with less than optimal results. One way to choose $\eta(k)$ is to make it a function of recent performance, decreasing it as performance improves. Another way is to program $\eta(k)$ by a choice such as $\eta(k) = \eta(1)/k$. When we examine stochastic approximation techniques, we shall see that this latter choice is the theoretical solution to an analogous problem. Before we take up this topic, however, we shall consider an approach that sacrifices the ability to obtain a separating vector for good compromise performance on both separable and nonseparable problems.

5.8 MINIMUM SQUARED-ERROR PROCEDURES

The criterion functions we have considered thus far have focused their attention on the misclassified samples. We shall now consider a criterion function that involves *all* of the samples. Where previously we have sought a weight vector **a** making all of the inner products $\mathbf{a}^t \mathbf{y}_i$ positive, now we shall try to make $\mathbf{a}^t \mathbf{y}_i = b_i$, where the b_i are some arbitrarily specified positive constants. Thus, we have replaced the problem of finding the solution to a set of linear inequalities with the more stringent but better understood problem of finding the solution to a set of linear equations.

5.8.1 Minimum Squared-Error and the Pseudoinverse

The treatment of simultaneous linear equations is simplified by introducing matrix notation. Let $\mathbf{Y}$ be the n-by-$\hat{d}$ matrix ($\hat{d} = d + 1$) whose ith row is the vector $\mathbf{y}_i^t$, and let $\mathbf{b}$ be the column vector $\mathbf{b} = (b_1, \ldots, b_n)^t$. Then our problem is to find a weight vector $\mathbf{a}$ satisfying

$$
\begin{pmatrix}
y_{10} & y_{11} & \cdots & y_{1d} \\
y_{20} & y_{21} & \cdots & y_{2d} \\
\vdots & \vdots & & \vdots \\
\vdots & \vdots & & \vdots \\
\vdots & \vdots & & \vdots \\
y_{n0} & y_{n1} & \cdots & y_{nd}
\end{pmatrix}
\begin{pmatrix}
a_0 \\
a_1 \\
\vdots \\
a_d
\end{pmatrix}
=
\begin{pmatrix}
b_1 \\
b_2 \\
\vdots \\
\vdots \\
\vdots \\
b_n
\end{pmatrix}
\quad \text{or} \quad \mathbf{Ya} = \mathbf{b}. \tag{41}
$$

If $\mathbf{Y}$ were nonsingular, we could write $\mathbf{a} = \mathbf{Y}^{-1}\mathbf{b}$ and obtain a formal solution at once. However, $\mathbf{Y}$ is rectangular, usually with more rows than columns. When there are more equations than unknowns, $\mathbf{a}$ is overdetermined, and ordinarily no exact solution exists. However, we can seek a weight vector $\mathbf{a}$ that minimizes some function of the error between $\mathbf{Ya}$ and $\mathbf{b}$. If we define the error vector $\mathbf{e}$ by

$$
\mathbf{e} = \mathbf{Ya} - \mathbf{b} \tag{42}
$$

then one approach is to try to minimize the squared length of the error vector. This is equivalent to minimizing the sum-of-squared-error criterion function

$$
J_s(\mathbf{a}) = \|\mathbf{Ya} - \mathbf{b}\|^2 = \sum_{i=1}^{n}(\mathbf{a}^t\mathbf{y}_i - b_i)^2. \tag{43}
$$

The problem of minimizing the sum of squared error is a classical one. It can be solved by a gradient search procedure, as we shall see below. A simple closed-form solution can also be found by forming the gradient

$$
\nabla J_s = \sum_{i=1}^{n} 2(\mathbf{a}^t\mathbf{y}_i - b_i)\mathbf{y}_i = 2\mathbf{Y}^t(\mathbf{Ya} - \mathbf{b}) \tag{44}
$$

and setting it equal to zero. This yields the necessary condition

$$
\mathbf{Y}^t\mathbf{Ya} = \mathbf{Y}^t\mathbf{b}, \tag{45}
$$

and in this way we have converted the problem of solving $\mathbf{Ya} = \mathbf{b}$ to that of solving $\mathbf{Y}^t\mathbf{Ya} = \mathbf{Y}^t\mathbf{b}$. This celebrated equation has the great advantage that the $\hat{d}$-by-$\hat{d}$ matrix $\mathbf{Y}^t\mathbf{Y}$ is square and often nonsingular. If it is nonsingular, we can solve for $\mathbf{a}$ uniquely as

$$
\mathbf{a} = (\mathbf{Y}^t\mathbf{Y})^{-1}\mathbf{Y}^t\mathbf{b}
$$
$$
= \mathbf{Y}^{\dagger}\mathbf{b}, \tag{46}
$$

where the $\hat{d}$-by-n matrix

$$
\mathbf{Y}^{\dagger} \equiv (\mathbf{Y}^t\mathbf{Y})^{-1}\mathbf{Y}^t \tag{47}
$$

PSEUDOINVERSE is called the *pseudoinverse* of $\mathbf{Y}$. Note that if $\mathbf{Y}$ is square and nonsingular, the pseudoinverse coincides with the regular inverse. Note also that $\mathbf{Y}^{\dagger}\mathbf{Y} = \mathbf{I}$, but $\mathbf{YY}^{\dagger} \neq \mathbf{I}$ in general. However, a minimum-squared-error (MSE) solution always exists. In particular, if $\mathbf{Y}^{\dagger}$ is defined more generally by

$$\mathbf{Y}^{\dagger} \equiv \lim_{\epsilon \to 0}(\mathbf{Y}^{t}\mathbf{Y} + \epsilon\mathbf{I})^{-1}\mathbf{Y}^{t}, \tag{48}$$

it can be shown that this limit always exists, and that $\mathbf{a} = \mathbf{Y}^{\dagger}\mathbf{b}$ is an MSE solution to $\mathbf{Ya} = \mathbf{b}$.

The MSE solution depends on the margin vector $\mathbf{b}$, and we shall see that different choices for $\mathbf{b}$ give the solution different properties. If $\mathbf{b}$ is fixed arbitrarily, there is no reason to believe that the MSE solution yields a separating vector in the linearly separable case. However, it is reasonable to hope that by minimizing the squared-error criterion function we might obtain a useful discriminant function in both the separable and the nonseparable cases. Below we examine two properties of the solution that support this hope.

EXAMPLE 1   Constructing a Linear Classifier by Matrix Pseudoinverse

---

Suppose we have the following two-dimensional points for two categories: ω_1: $(1, 2)^t$ and $(2, 0)^t$, and ω_2: $(3, 1)^t$ and $(2, 3)^t$, as shown in black and red, respectively, in the accompanying figure.

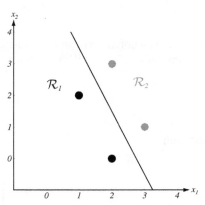

Four training points and the decision boundary $\mathbf{a}^t \begin{pmatrix} 1 \\ x_1 \\ x_2 \end{pmatrix} = 0$, where $\mathbf{a}$ was found by means of a pseudoinverse technique.

Our matrix $\mathbf{Y}$ is therefore

$$\mathbf{Y} = \begin{pmatrix} 1 & 1 & 2 \\ 1 & 2 & 0 \\ -1 & -3 & -1 \\ -1 & -2 & -3 \end{pmatrix}$$

and after a few simple calculations we find that its pseudoinverse is

$$\mathbf{Y}^\dagger = (\mathbf{Y}^t\mathbf{Y})^{-1}\mathbf{Y}^t = \begin{pmatrix} 5/4 & 13/12 & 3/4 & 7/12 \\ -1/2 & -1/6 & -1/2 & -1/6 \\ 0 & -1/3 & 0 & -1/3 \end{pmatrix}$$

We arbitrarily let all the margins be equal—that is, $\mathbf{b} = (1, 1, 1, 1)^t$. Our solution is $\mathbf{a} = \mathbf{Y}^\dagger\mathbf{b} = (11/3, -4/3, -2/3)^t$, and leads to the decision boundary shown in the figure. Other choices for $\mathbf{b}$ would typically lead to different decision boundaries, of course.

5.8.2 Relation to Fisher's Linear Discriminant

In this section we shall show that with the proper choice of the vector $\mathbf{b}$, the MSE discriminant function $\mathbf{a}^t\mathbf{y}$ is directly related to Fisher's linear discriminant (Chapter 3 Section 3.8.2). To do this, we must return to the use of linear rather than generalized linear discriminant functions. We assume that we have a set of n d-dimensional samples $\mathbf{x}_1, \ldots, \mathbf{x}_n$, n_1 of which are in the subset $\mathcal{D}_1$ labeled ω_1, and n_2 of which are in the subset $\mathcal{D}_2$ labeled ω_2. Furthermore, we assume that a sample $\mathbf{y}_i$ is formed from $\mathbf{x}_i$ by adding a threshold component $x_0 = 1$ to make an *augmented pattern vector*.

AUGMENTED PATTERN VECTOR

Further, if the sample is labeled ω_2, then the entire pattern vector is multiplied by -1—the "normalization" we saw in Section 5.4.1. With no loss in generality, we can assume that the first n_1 samples are labeled ω_1 and the second n_2 are labeled ω_2. Then the matrix $\mathbf{Y}$ can be partitioned as follows:

$$\mathbf{Y} = \begin{bmatrix} \mathbf{1}_1 & \mathbf{X}_1 \\ -\mathbf{1}_2 & -\mathbf{X}_2 \end{bmatrix},$$

where $\mathbf{1}_i$ is a column vector of n_i ones, and $\mathbf{X}_i$ is an n_i-by-d matrix whose rows are the samples labeled ω_i. We partition $\mathbf{a}$ and $\mathbf{b}$ correspondingly, with

$$\mathbf{a} = \begin{bmatrix} w_0 \\ \mathbf{w} \end{bmatrix}$$

and with

$$\mathbf{b} = \begin{bmatrix} \dfrac{n}{n_1}\mathbf{1}_1 \\ \dfrac{n}{n_2}\mathbf{1}_2 \end{bmatrix}.$$

We shall now show that this special choice for $\mathbf{b}$ links the MSE solution to Fisher's linear discriminant.

We begin by writing Eq. 45 for $\mathbf{a}$ in terms of the partitioned matrices:

$$\begin{bmatrix} \mathbf{1}_1^t & -\mathbf{1}_2^t \\ \mathbf{X}_1^t & -\mathbf{X}_2^t \end{bmatrix}\begin{bmatrix} \mathbf{1}_1 & \mathbf{X}_1 \\ -\mathbf{1}_2 & -\mathbf{X}_2 \end{bmatrix}\begin{bmatrix} w_0 \\ \mathbf{w} \end{bmatrix} = \begin{bmatrix} \mathbf{1}_1^t & -\mathbf{1}_2^t \\ \mathbf{X}_1^t & -\mathbf{X}_2^t \end{bmatrix}\begin{bmatrix} \dfrac{n}{n_1}\mathbf{1}_1 \\ \dfrac{n}{n_2}\mathbf{1}_2 \end{bmatrix}. \quad (49)$$

By defining the sample means $\mathbf{m}_i$ and the pooled sample scatter matrix $\mathbf{S}_W$ as

$$\mathbf{m}_i = \frac{1}{n_i} \sum_{\mathbf{x} \in \mathcal{D}_i} \mathbf{x} \qquad i = 1, 2 \tag{50}$$

and

$$\mathbf{S}_W = \sum_{i=1}^{2} \sum_{\mathbf{x} \in \mathcal{D}_i} (\mathbf{x} - \mathbf{m}_i)(\mathbf{x} - \mathbf{m}_i)^t, \tag{51}$$

we can multiply the matrices of Eq. 49 and obtain

$$\begin{bmatrix} n & (n_1\mathbf{m}_1 + n_2\mathbf{m}_2)^t \\ (n_1\mathbf{m}_1 + n_2\mathbf{m}_2) & \mathbf{S}_W + n_1\mathbf{m}_1\mathbf{m}_1^t + n_2\mathbf{m}_2\mathbf{m}_2^t \end{bmatrix} \begin{bmatrix} w_0 \\ \mathbf{w} \end{bmatrix} = \begin{bmatrix} 0 \\ n(\mathbf{m}_1 - \mathbf{m}_2) \end{bmatrix}.$$

This can be viewed as a pair of equations, the first of which can be solved for w_0 in terms of $\mathbf{w}$:

$$w_0 = -\mathbf{m}^t\mathbf{w}, \tag{52}$$

where $\mathbf{m}$ is the mean of all of the samples. Substituting this in the second equation and performing a few algebraic manipulations, we obtain

$$\left[\frac{1}{n}\mathbf{S}_W + \frac{n_1 n_2}{n^2}(\mathbf{m}_1 - \mathbf{m}_2)(\mathbf{m}_1 - \mathbf{m}_2)^t\right]\mathbf{w} = \mathbf{m}_1 - \mathbf{m}_2. \tag{53}$$

Because the vector $(\mathbf{m}_1 - \mathbf{m}_2)(\mathbf{m}_1 - \mathbf{m}_2)^t\mathbf{w}$ is in the direction of $\mathbf{m}_1 - \mathbf{m}_2$ for any value of $\mathbf{w}$, we can write

$$\frac{n_1 n_2}{n^2}(\mathbf{m}_1 - \mathbf{m}_2)(\mathbf{m}_1 - \mathbf{m}_2)^t\mathbf{w} = (1 - \alpha)(\mathbf{m}_1 - \mathbf{m}_2),$$

where α is some scalar. Then Eq. 53 yields

$$\mathbf{w} = \alpha n \mathbf{S}_W^{-1}(\mathbf{m}_1 - \mathbf{m}_2), \tag{54}$$

which, except for an unimportant scale factor, is identical to the solution for Fisher's linear discriminant. In addition, we obtain the threshold weight w_0 and the following decision rule: Decide ω_1 if $\mathbf{w}^t(\mathbf{x} - \mathbf{m}) > 0$; otherwise decide ω_2.

5.8.3 Asymptotic Approximation to an Optimal Discriminant

Another property of the MSE solution that recommends its use is that if $\mathbf{b} = \mathbf{1}_n$, it approaches a minimum mean-squared-error approximation to the Bayes discriminant function

$$g_0(\mathbf{x}) = P(\omega_1|\mathbf{x}) - P(\omega_2|\mathbf{x}) \tag{55}$$

in the limit as the number of samples approaches infinity. To demonstrate this fact, we must assume that the samples are drawn independently, identically distributed (i.i.d.) according to the probability law

$$p(\mathbf{x}) = p(\mathbf{x}|\omega_1)P(\omega_1) + p(\mathbf{x}|\omega_2)P(\omega_2). \tag{56}$$

In terms of the augmented vector $\mathbf{y}$, the MSE solution yields the series expansion $g(\mathbf{x}) = \mathbf{a}^t\mathbf{y}$, where $\mathbf{y} = \mathbf{y}(\mathbf{x})$. If we define the mean-squared approximation error by

$$\epsilon^2 = \int [\mathbf{a}^t\mathbf{y} - g_0(\mathbf{x})]^2 p(\mathbf{x})\, d\mathbf{x}, \tag{57}$$

then our goal is to show that ϵ^2 is minimized by the solution $\mathbf{a} = \mathbf{Y}^\dagger \mathbf{1}_n$.

The proof is simplified if we preserve the distinction between category ω_1 and category ω_2 samples. In terms of the unnormalized data, the criterion function J_s becomes

$$J_s(\mathbf{a}) = \sum_{\mathbf{y}\in\mathcal{Y}_1} (\mathbf{a}^t\mathbf{y} - 1)^2 + \sum_{\mathbf{y}\in\mathcal{Y}_2} (\mathbf{a}^t\mathbf{y} + 1)^2$$

$$= n\left[\frac{n_1}{n} \frac{1}{n_1} \sum_{\mathbf{y}\in\mathcal{Y}_1} (\mathbf{a}^t\mathbf{y} - 1)^2 + \frac{n_2}{n} \frac{1}{n_2} \sum_{\mathbf{y}\in\mathcal{Y}_2} (\mathbf{a}^t\mathbf{y} + 1)^2 \right]. \tag{58}$$

Thus, by the law of large numbers, as n approaches infinity $(1/n)J_s(\mathbf{a})$ approaches

$$\bar{J}(\mathbf{a}) = P(\omega_1)\mathcal{E}_1[(\mathbf{a}^t\mathbf{y} - 1)^2] + P(\omega_2)\mathcal{E}_2[(\mathbf{a}^t\mathbf{y} + 1)^2], \tag{59}$$

with probability one, where

$$\mathcal{E}_1[(\mathbf{a}^t\mathbf{y} - 1)^2] = \int (\mathbf{a}^t\mathbf{y} - 1)^2 p(\mathbf{x}|\omega_1)\, d\mathbf{x}$$

and

$$\mathcal{E}_2[(\mathbf{a}^t\mathbf{y} + 1)^2] = \int (\mathbf{a}^t\mathbf{y} + 1)^2 p(\mathbf{x}|\omega_2)\, d\mathbf{x}.$$

Now, if we recognize from Eq. 55 that

$$g_0(\mathbf{x}) = \frac{p(\mathbf{x}, \omega_1) - p(\mathbf{x}, \omega_2)}{p(\mathbf{x})},$$

we see that

$$\bar{J}(\mathbf{a}) = \int (\mathbf{a}^t\mathbf{y} - 1)^2 p(\mathbf{x}, \omega_1)\, d\mathbf{x} + \int (\mathbf{a}^t\mathbf{y} + 1)^2 p(\mathbf{x}, \omega_2)\, d\mathbf{x}$$

$$= \int (\mathbf{a}^t\mathbf{y})^2 p(\mathbf{x})\, d\mathbf{x} - 2\int \mathbf{a}^t\mathbf{y} g_0(\mathbf{x}) p(\mathbf{x})\, d\mathbf{x} + 1$$

$$= \underbrace{\int [\mathbf{a}^t\mathbf{y} - g_0(\mathbf{x})]^2 p(\mathbf{x})\, d\mathbf{x}}_{\epsilon^2} + \underbrace{\left[1 - \int g_0^2(\mathbf{x}) p(\mathbf{x})\, d\mathbf{x}\right]}_{\text{independent of } \mathbf{a}}. \tag{60}$$

The second term in this sum is independent of the weight vector $\mathbf{a}$. Hence, the $\mathbf{a}$ that minimizes J_s also minimizes ϵ^2—the mean-squared-error between $\mathbf{a}^t\mathbf{y}$ and $g(\mathbf{x})$ (Fig. 5.16). In Chapter 6 we shall see that analogous properties also hold for many multilayer neural networks.

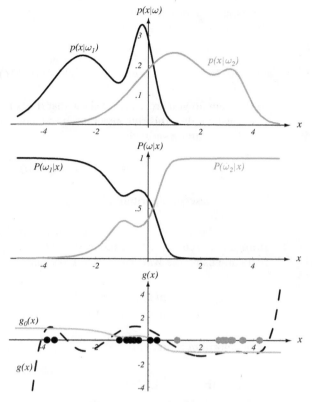

FIGURE 5.16. The top figure shows two class-conditional densities, and the middle figure the posteriors, assuming equal priors. Minimizing the MSE error also minimizes the mean-squared-error between $\mathbf{a}^t\mathbf{y}$ and the discriminant function $g(\mathbf{x})$ (here a seventh-order polynomial) measured over the data distribution, as shown at the bottom. Note that the resulting $g(x)$ best approximates $g_0(x)$ in the regions where the data points lie.

This result gives considerable insight into the MSE procedure. By approximating $g_0(\mathbf{x})$, the discriminant function $\mathbf{a}^t\mathbf{y}$ gives direct information about the posterior probabilities $P(\omega_1|\mathbf{x}) = (1 + g_0)/2$ and $P(\omega_2|\mathbf{x}) = (1 - g_0)/2$. The quality of the approximation depends on the functions $y_i(\mathbf{x})$ and the number of terms in the expansion $\mathbf{a}^t\mathbf{y}$. Unfortunately, the mean-square-error criterion places emphasis on points where $p(\mathbf{x})$ is larger, rather than on points near the decision surface $g_0(\mathbf{x}) = 0$. Thus, the discriminant function that "best" approximates the Bayes discriminant does not necessarily minimize the probability of error. Despite this property, the MSE solution has interesting properties and has received considerable attention in the literature. We shall encounter the mean-square approximation of $g_0(\mathbf{x})$ again when we consider stochastic approximation methods and multilayer neural networks.

5.8.4 The Widrow-Hoff or LMS Procedure

We remarked earlier that $J_s(\mathbf{a}) = \|\mathbf{Ya} - \mathbf{b}\|^2$ could be minimized by a gradient descent procedure. Such an approach has two advantages over merely computing the pseudoinverse: (1) It avoids the problems that arise when $\mathbf{Y}^t\mathbf{Y}$ is singular, and (2) it avoids the need for working with large matrices. In addition, the computation involved is effectively a feedback scheme which automatically copes with some of the computational problems due to roundoff or truncation. Because $\nabla J_s = 2\mathbf{Y}^t(\mathbf{Ya} - \mathbf{b})$, the obvious update rule is

$$\begin{aligned} \mathbf{a}(1) \quad &\text{arbitrary} \\ \mathbf{a}(k + 1) = \mathbf{a}(k) &+ \eta(k)\mathbf{Y}^t(\mathbf{Ya}(k) - \mathbf{b}). \end{aligned}$$

In Problem 26 you are asked to show that if $\eta(k) = \eta(1)/k$, where $\eta(1)$ is any positive constant, then this rule generates a sequence of weight vectors that converges to a limiting vector $\mathbf{a}$ satisfying

$$\mathbf{Y}^t(\mathbf{Ya}(\mathbf{k}) - \mathbf{b}) = 0.$$

Thus, the descent algorithm always yields a solution regardless of whether or not $\mathbf{Y}^t\mathbf{Y}$ is singular.

While the $\hat{d}$-by-$\hat{d}$ matrix $\mathbf{Y}^t\mathbf{Y}$ is usually smaller than the $\hat{d}$-by-n matrix $\mathbf{Y}^\dagger$, the storage requirements can be reduced still further by considering the samples sequentially and using the *Widrow-Hoff* or *LMS rule* (least-mean-squared):

LMS RULE

$$\left.\begin{aligned} \mathbf{a}(1) \quad &\text{arbitrary} \\ \mathbf{a}(k + 1) = \mathbf{a}(k) &+ \eta(k)(b(k) - \mathbf{a}^t(k)\mathbf{y}^k)\mathbf{y}^k, \end{aligned}\right\} \tag{61}$$

or in algorithm form:

■ **Algorithm 10. (LMS)**

1 **begin initialize** $\mathbf{a}$, $\mathbf{b}$, threshold θ, $\eta(\cdot)$, $k \leftarrow 0$
2   **do** $k \leftarrow (k + 1) \bmod n$
3          $\mathbf{a} \leftarrow \mathbf{a} + \eta(k)(b_k - \mathbf{a}^t\mathbf{y}^k)\mathbf{y}^k$
4   **until** $|\eta(k)(b_k - \mathbf{a}^t\mathbf{y}^k)\mathbf{y}^k| < \theta$
5   **return** $\mathbf{a}$
6 **end**

At first glance this descent algorithm appears to be essentially the same as the relaxation rule. The primary difference is that the relaxation rule is an error-correction rule, so that $\mathbf{a}^t(k)\mathbf{y}^k$ does not equal b_k, and thus the corrections never cease. Therefore, $\eta(k)$ must decrease with k to obtain convergence, the choice $\eta(k) = \eta(1)/k$ being common. Exact analysis of the behavior of the Widrow-Hoff rule in the deterministic case is rather complicated, and merely indicates that the sequence of weight vectors tends to converge to the desired solution. Instead of pursuing this topic further, we shall turn to a very similar rule that arises from a stochastic descent procedure. We note, however, that the solution need not give a separating vector, even if one exists, as shown in Fig. 5.17 (Computer exercise 10).

5.8.5 Stochastic Approximation Methods

All of the iterative descent procedures we have considered thus far have been described in deterministic terms. We are given a particular set of samples, and we generate a particular sequence of weight vectors. In this section we digress briefly to consider an MSE procedure in which the samples are drawn randomly, resulting in

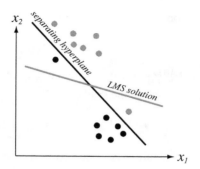

FIGURE 5.17. The LMS algorithm need not converge to a separating hyperplane, even if one exists. Because the LMS solution minimizes the sum of the squares of the distances of the training points to the hyperplane, for this example the plane is rotated clockwise compared to a separating hyperplane.

a random sequence of weight vectors. We will return in Chapter 7 to the theory of stochastic approximation though here some of the main ideas will be presented without proof.

Suppose that samples are drawn independently by selecting a state of nature with probability $P(\omega_i)$ and then selecting an $\mathbf{x}$ according to the probability law $p(\mathbf{x}|\omega_i)$. For each $\mathbf{x}$ we let θ be its *label*, with $\theta = +1$ if $\mathbf{x}$ is labeled ω_1 and $\theta = -1$ if $\mathbf{x}$ is labeled ω_2. Then the data consist of an infinite sequence of independent pairs $(\mathbf{x}, \theta_1)$, $(\mathbf{x}_2, \theta_2)$, ..., $(\mathbf{x}_k, \theta_k)$, ... Even though the label variable θ is binary-valued, it can be thought of as a noisy version of the Bayes discriminant function $g_0(\mathbf{x})$. This follows from the observation that

$$P(\theta = 1|\mathbf{x}) = P(\omega_1|\mathbf{x}),$$

and

$$P(\theta = -1|\mathbf{x}) = P(\omega_2|\mathbf{x}),$$

so that the conditional mean of θ is given by

$$\mathcal{E}_{\theta|\mathbf{x}}[\theta] = \sum_{\theta} \theta P(\theta|\mathbf{x}) = P(\omega_1|\mathbf{x}) - P(\omega_2|\mathbf{x}) = g_0(\mathbf{x}). \tag{62}$$

Suppose that we wish to approximate $g_0(\mathbf{x})$ by the finite series expansion

$$g(\mathbf{x}) = \mathbf{a}^t \mathbf{y} = \sum_{i=1}^{\hat{d}} a_i y_i(\mathbf{x}),$$

where both the basis functions $y_i(\mathbf{x})$ and the number of terms $\hat{d}$ are known. Then we can seek a weight vector $\hat{\mathbf{a}}$ that minimizes the mean-squared approximation error

$$\epsilon^2 = \mathcal{E}[(\mathbf{a}^t \mathbf{y} - g_0(\mathbf{x}))^2]. \tag{63}$$

Minimization of ϵ^2 would appear to require knowledge of Bayes discriminant $g_0(\mathbf{x})$. However, as one might have guessed from the analogous situation in Section 5.8.3, it can be shown that the weight vector $\hat{\mathbf{a}}$ that minimizes ϵ^2 also minimizes the criterion function

$$J_m(\mathbf{a}) = \mathcal{E}[(\mathbf{a}^t\mathbf{y} - \theta)^2]. \tag{64}$$

This should also be plausible from the fact that θ is essentially a noisy version of $g_0(\mathbf{x})$. Because the gradient is

$$\nabla J_m = 2\mathcal{E}[(\mathbf{a}^t\mathbf{y} - \theta)\mathbf{y}], \tag{65}$$

we can obtain the closed-form solution

$$\hat{\mathbf{a}} = \mathcal{E}[\mathbf{y}\mathbf{y}^t]^{-1}\mathcal{E}[\theta\mathbf{y}]. \tag{66}$$

Thus, one way to use the samples is to estimate $\mathcal{E}[\mathbf{y}\mathbf{y}^t]$ and $\mathcal{E}[\theta\mathbf{y}]$, and use Eq. 66 to obtain the MSE-optimal linear discriminant. An alternative is to minimize $J_m(\mathbf{a})$ by a gradient descent procedure. Suppose that in place of the true gradient we substitute the noisy version $2(\mathbf{a}^t\mathbf{y}_k - \theta_k)\mathbf{y}_k$. This leads to the update rule

$$\mathbf{a}(k+1) = \mathbf{a}(k) + \eta(\theta_k - \mathbf{a}^t(k)\mathbf{y}_k)\mathbf{y}_k, \tag{67}$$

which is basically just the Widrow-Hoff rule. It can be shown (Problem 23) that if $\mathcal{E}[\mathbf{y}\mathbf{y}^t]$ is nonsingular and if the coefficients $\eta(k)$ satisfy

$$\lim_{m\to\infty} \sum_{k=1}^{m} \eta(k) = +\infty \tag{68}$$

and

$$\lim_{m\to\infty} \sum_{k=1}^{m} \eta^2(k) < \infty \tag{69}$$

then $\mathbf{a}(k)$ converges to $\hat{\mathbf{a}}$ in mean square:

$$\lim_{k\to\infty} \mathcal{E}[\|\mathbf{a}(k) - \hat{\mathbf{a}}\|^2] = 0. \tag{70}$$

The reasons we need these conditions on $\eta(k)$ are simple. The first condition keeps the weight vector from converging so fast that a systematic error will remain forever uncorrected. The second condition ensures that random fluctuations are eventually suppressed. Both conditions are satisfied by the conventional choice $\eta(k) = 1/k$. Unfortunately, this kind of programmed decrease of $\eta(k)$, independent of the problem at hand, often leads to very slow convergence.

Of course, this is neither the only nor the best descent algorithm for minimizing J_m. For example, if we note that the matrix of second partial derivatives for J_m is given by

$$D = 2\mathcal{E}[\mathbf{y}\mathbf{y}^t],$$

we see that Newton's rule for minimizing J_m (Eq. 15) is

$$\mathbf{a}(k+1) = \mathbf{a}(k) + \mathcal{E}[\mathbf{y}\mathbf{y}^t]^{-1}\mathcal{E}[(\theta - \mathbf{a}^t\mathbf{y})\mathbf{y}].$$

A stochastic analog of this rule is

$$\mathbf{a}(k+1) = \mathbf{a}(k) + \mathbf{R}_{k+1}(\theta_k - \mathbf{a}^t(k)\mathbf{y}_k)\mathbf{y}_k. \tag{71}$$

with

$$\mathbf{R}_{k+1}^{-1} = \mathbf{R}_k^{-1} + \mathbf{y}_k\mathbf{y}_k^t, \tag{72}$$

or, equivalently,*

$$\mathbf{R}_{k+1} = \mathbf{R}_k - \frac{\mathbf{R}_k\mathbf{y}_k(\mathbf{R}_k\mathbf{y}_k)^t}{1 + \mathbf{y}_k^t\mathbf{R}_k\mathbf{y}_k}. \tag{73}$$

This rule also produces a sequence of weight vectors that converges to the optimal solution in mean square. Its convergence is faster, but it requires more computation per step (Computer exercise 8).

REGRESSION
FUNCTION
STOCHASTIC
APPROXIMATION

These gradient procedures can be viewed as methods for minimizing a criterion function, or finding the zero of its gradient, in the presence of noise. In the statistical literature, functions such as J_m and ∇J_m that have the form $\mathcal{E}[f(\mathbf{a}, \mathbf{x})]$ are called *regression functions*, and the iterative algorithms are called *stochastic approximation procedures*. Two well-known ones are (1) the Kiefer-Wolfowitz procedure for minimizing a regression function, and (2) the Robbins-Monro procedure for finding a root of a regression function. Often the easiest way to obtain a convergence proof for a particular descent or approximation procedure is to show that it satisfies the convergence conditions for these more general procedures. Unfortunately, an exposition of these methods in their full generality would lead us rather far afield, and we must close this digression by referring the interested reader to the references in the Bibliography.

5.9 THE HO-KASHYAP PROCEDURES

The procedures we have considered thus far differ in several ways. The Perceptron and relaxation procedures find separating vectors if the samples are linearly separable, but do not converge on nonseparable problems. The MSE procedures yield a weight vector whether the samples are linearly separable or not, but there is no guarantee that this vector is a separating vector in the separable case (Fig. 5.17). If the margin vector $\mathbf{b}$ is chosen arbitrarily, all we can say is that the MSE procedures minimize $\|\mathbf{Ya} - \mathbf{b}\|^2$. Now if the training samples happen to be linearly separable, then there exists an $\hat{\mathbf{a}}$ and a $\hat{\mathbf{b}}$ such that

$$\mathbf{Y}\hat{\mathbf{a}} = \hat{\mathbf{b}} > 0,$$

where by $\hat{\mathbf{b}} > 0$, we mean that every component of $\hat{\mathbf{b}}$ is positive. Clearly, were we to take $\mathbf{b} = \hat{\mathbf{b}}$ and apply the MSE procedure, we would obtain a separating vector. Of course, we usually do not know $\hat{\mathbf{b}}$ beforehand. However, we shall now see how the MSE procedure can be modified to obtain both a separating vector $\mathbf{a}$ and a margin

---

*This recursive formula for computing R_k, which is roughly $(1/k)\mathcal{E}[\mathbf{yy}^t]^{-1}$, cannot be used if R_k is singular.

vector **b**. The underlying idea comes from the observation that if the samples are separable, and if both **a** and **b** in the criterion function

$$J_s(\mathbf{a}, \mathbf{b}) = \|\mathbf{Ya} - \mathbf{b}\|^2 \tag{74}$$

are allowed to vary (subject to the constraint $\mathbf{b} > \mathbf{0}$), then the minimum value of J_s is zero, and the **a** that achieves that minimum is a separating vector.

5.9.1 The Descent Procedure

To minimize J_s in Eq. 74 we shall use a modified gradient descent procedure. The gradient of J_s with respect to **a** is given by

$$\nabla_{\mathbf{a}} J_s = 2\mathbf{Y}^t(\mathbf{Ya} - \mathbf{b}), \tag{75}$$

and the gradient of J_s with respect to **b** is given by

$$\nabla_{\mathbf{b}} J_s = -2(\mathbf{Ya} - \mathbf{b}). \tag{76}$$

For any value of **b**, we can always take

$$\mathbf{a} = \mathbf{Y}^\dagger \mathbf{b}, \tag{77}$$

thereby obtaining $\nabla_{\mathbf{a}} J_s = \mathbf{0}$ and minimizing J_s with respect to **a** in one step. We are not so free to modify **b**, however, because we must respect the constraint $\mathbf{b} > \mathbf{0}$, and we must avoid a descent procedure that converges to $\mathbf{b} = \mathbf{0}$. One way to prevent **b** from converging to zero is to start with $\mathbf{b} > \mathbf{0}$ and to refuse to reduce any of its components. We can do this and still try to follow the negative gradient if we first set all positive components of $\nabla_{\mathbf{b}} J_s$ to zero. Thus, if we let $|\mathbf{v}|$ denote the vector whose components are the magnitudes of the corresponding components of **v**, we are led to consider an update rule for the margin of the form

$$\mathbf{b}(k + 1) = \mathbf{b}(k) - \eta \frac{1}{2}[\nabla_{\mathbf{b}} J_s - |\nabla_{\mathbf{b}} J_s|]. \tag{78}$$

Using Eqs. 76 and 77, and being a bit more specific, we obtain the *Ho-Kashyap rule* for minimizing $J_s(\mathbf{a}, \mathbf{b})$:

$$\begin{aligned} \mathbf{b}(1) &> \mathbf{0} \quad \text{but otherwise arbitrary} \\ \mathbf{b}(k + 1) &= \mathbf{b}(k) + 2\eta(k)\mathbf{e}^+(k), \end{aligned} \tag{79}$$

where $\mathbf{e}(k)$ is the error vector

$$\mathbf{e}(k) = \mathbf{Ya}(k) - \mathbf{b}(k), \tag{80}$$

$\mathbf{e}^+(k)$ is the positive part of the error vector

$$\mathbf{e}^+(k) = \frac{1}{2}(\mathbf{e}(k) + |\mathbf{e}(k)|), \tag{81}$$

and

$$\mathbf{a}(k) = \mathbf{Y}^\dagger \mathbf{b}(k), \qquad k = 1, 2, \ldots. \tag{82}$$

Thus if we let b_{min} be a small convergence criterion and Abs[**e**] denote the positive part of **e**, our algorithm is as follows:

■ **Algorithm 11. (Ho-Kashyap)**

1 **begin initialize a**, **b**, $\eta(\cdot) < 1$, threshold b_{min}, k_{max}
2   **do** $k \leftarrow (k + 1)$ mod n
3   **e** $\leftarrow$ **Ya** − **b**
4         $\mathbf{e}^+ \leftarrow 1/2(\mathbf{e} + \text{Abs}[\mathbf{e}])$
5   **b** $\leftarrow$ **b** $+ 2\eta(k)\mathbf{e}^+$
6   **a** $\leftarrow$ **Y**†**b**
7   **if** Abs[**e**] $\leq b_{min}$ **then return a**, **b** and **exit**
8   **until** $k = k_{max}$
9   print "NO SOLUTION FOUND"
10 **end**

Because the weight vector **a**(k) is completely determined by the margin vector **b**(k), this is basically an algorithm for producing a sequence of margin vectors. The initial vector **b**(1) is positive to begin with, and if $\eta > 0$, all subsequent vectors **b**(k) are positive. We might worry that if none of the components of **e**(k) is positive, so that **b**(k) stops changing, we might fail to find a solution. However, we shall see that in that case either **e**(k) = **0** and we have a solution, or **e**(k) $\leq$ 0 and we have proof that the samples are not linearly separable.

5.9.2 Convergence Proof

We shall now show that if the samples are linearly separable, and if $0 < \eta < 1$, then the Ho-Kashyap algorithm will yield a solution vector in a finite number of steps. To make the algorithm terminate, we should add a terminating condition stating that corrections cease once a solution vector is obtained or some large criterion number of iterations have occurred. However, it is mathematically more convenient to let the corrections continue and show that the error vector **e**(k) either becomes zero for some finite k, or converges to zero as k goes to infinity.

It is clear that either **e**(k) = **0** for some k—say k_0—or there are no zero vectors in the sequence **e**(1), **e**(2), ... In the first case, once a zero vector is obtained, no further changes occur to **a**(k), **b**(k), or **e**(k), and **Ya**(k) = **b**(k) > **0** for all $k \geq k_0$. Thus, if we happen to obtain a zero error vector, the algorithm automatically terminates with a solution vector.

Suppose now that **e**(k) is never zero for finite k. To see that **e**(k) must nevertheless converge to zero, we begin by asking whether or not we might possibly obtain an **e**(k) with no positive components. This would be most unfortunate, because we would have **Ya**(k) $\leq$ **b**(k), and since $\mathbf{e}^+(k)$ would be zero, we would obtain no further changes in **a**(k), **b**(k), or **e**(k). Fortunately, this can never happen if the samples are linearly separable. A proof is simple and is based on the fact that if $\mathbf{Y}^t\mathbf{Ya}(k) = \mathbf{Y}^t\mathbf{b}$, then $\mathbf{Y}^t\mathbf{e}(k) = \mathbf{0}$. But if the samples are linearly separable, there exists an $\hat{\mathbf{a}}$ and a $\hat{\mathbf{b}} > \mathbf{0}$ such that

$$\mathbf{Y}\hat{\mathbf{a}} = \hat{\mathbf{b}}.$$

Thus,

$$\mathbf{e}^t(k)\mathbf{Y}\hat{\mathbf{a}} = 0 = \mathbf{e}^t(k)\hat{\mathbf{b}},$$

and because all the components of $\hat{\mathbf{b}}$ are positive, either $\mathbf{e}(k) = \mathbf{0}$ or at least one of the components of $\mathbf{e}(k)$ must be positive. Because we have excluded the case $\mathbf{e}(k) = \mathbf{0}$, it follows that $\mathbf{e}^+(k)$ cannot be zero for finite k.

The proof that the error vector always converges to zero exploits the fact that the matrix $\mathbf{Y}\mathbf{Y}^\dagger$ is symmetric, positive semidefinite, and satisfies

$$(\mathbf{Y}\mathbf{Y}^\dagger)^t(\mathbf{Y}\mathbf{Y}^\dagger) = \mathbf{Y}\mathbf{Y}^\dagger. \tag{83}$$

Although these results are true in general, for simplicity we demonstrate them only for the case where $\mathbf{Y}^t\mathbf{Y}$ is nonsingular. In this case $\mathbf{Y}\mathbf{Y}^\dagger = \mathbf{Y}(\mathbf{Y}^t\mathbf{Y})^{-1}\mathbf{Y}^t$, and the symmetry is evident. Because $\mathbf{Y}^t\mathbf{Y}$ is positive definite, so is $(\mathbf{Y}^t\mathbf{Y})^{-1}$; thus, $\mathbf{b}\mathbf{Y}(\mathbf{Y}^t\mathbf{Y})^{-1}\mathbf{Y}^t\mathbf{b} \geq \mathbf{0}$ for any $\mathbf{b}$, and $\mathbf{Y}\mathbf{Y}^\dagger$ is at least positive semidefinite. Finally, Eq. 83 follows from

$$(\mathbf{Y}\mathbf{Y}^\dagger)^t(\mathbf{Y}\mathbf{Y}^\dagger) = [\mathbf{Y}(\mathbf{Y}^t\mathbf{Y})^{-1}\mathbf{Y}^t][\mathbf{Y}(\mathbf{Y}^t\mathbf{Y})^{-1}\mathbf{Y}^t].$$

To see that $\mathbf{e}(k)$ must converge to zero, we eliminate $\mathbf{a}(k)$ between Eqs. 80 and 82 and obtain

$$\mathbf{e}(k) = (\mathbf{Y}\mathbf{Y}^\dagger - \mathbf{I})\mathbf{b}(k).$$

Then, using a constant learning rate and Eq. 79 we obtain the recursion relation

$$\mathbf{e}(k + 1) = (\mathbf{Y}\mathbf{Y}^\dagger - \mathbf{I})(\mathbf{b}(k) + 2\eta\mathbf{e}^+(k))$$
$$= \mathbf{e}(k) + 2\eta(\mathbf{Y}\mathbf{Y}^\dagger - \mathbf{I})\mathbf{e}^+(k), \tag{84}$$

so that

$$\frac{1}{4}\|\mathbf{e}(k+1)\|^2 = \frac{1}{4}\|\mathbf{e}(k)\|^2 + \eta\mathbf{e}^t(k)(\mathbf{Y}\mathbf{Y}^\dagger - \mathbf{I})\mathbf{e}^+(k) + \|\eta(\mathbf{Y}\mathbf{Y}^\dagger - \mathbf{I})\mathbf{e}^+(k)\|^2.$$

Both the second and the third terms simplify considerably. Because $\mathbf{e}^t(k)\mathbf{Y} = 0$, the second term becomes

$$\eta\mathbf{e}^t(k)(\mathbf{Y}\mathbf{Y}^\dagger - \mathbf{I})\mathbf{e}^+(k) = -\eta\mathbf{e}^t(k)\mathbf{e}^{+t}(k) = -\eta\|\mathbf{e}^+(k)\|^2,$$

the nonzero components of $\mathbf{e}^+(k)$ being the positive components of $\mathbf{e}(k)$. Because $\mathbf{Y}\mathbf{Y}^\dagger$ is symmetric and is equal to $(\mathbf{Y}\mathbf{Y}^\dagger)^t(\mathbf{Y}\mathbf{Y}^\dagger)$, the third term simplifies to

$$\|\eta(\mathbf{Y}\mathbf{Y}^\dagger - \mathbf{I})\mathbf{e}^+(k)\|^2 = \eta^2\mathbf{e}^{+t}(k)(\mathbf{Y}\mathbf{Y}^\dagger - \mathbf{I})^t(\mathbf{Y}\mathbf{Y}^\dagger - \mathbf{I})\mathbf{e}^+(k)$$
$$= \eta^2\|\mathbf{e}^+(k)\|^2 - \eta^2\mathbf{e}^+(k)\mathbf{Y}\mathbf{Y}^\dagger\mathbf{e}^+(k),$$

and thus we have

$$\frac{1}{4}(\|\mathbf{e}(k)\|^2 - \|\mathbf{e}(k+1)\|^2) = \eta(1 - \eta)\|\mathbf{e}^+(k)\|^2 + \eta^2\mathbf{e}^{+t}(k)\mathbf{Y}\mathbf{Y}^\dagger\mathbf{e}^+(k). \tag{85}$$

Because $\mathbf{e}^+(k)$ is nonzero by assumption and because $\mathbf{YY}^\dagger$ is a positive semi-definite, $\|\mathbf{e}(k)\|^2 > \|\mathbf{e}(k+1)\|^2$ if $0 < \eta < 1$. Thus the sequence $\|\mathbf{e}(1)\|^2$, $\|\mathbf{e}(2)\|^2, \ldots$ is monotonically decreasing and must converge to some limiting value $\|\mathbf{e}\|^2$. But for convergence to take place, $\mathbf{e}^+(k)$ must converge to zero, so that all the positive components of $\mathbf{e}(k)$ must converge to zero. Because $\mathbf{e}^t(k)\hat{\mathbf{b}} = 0$ for all k, it follows that all of the components of $\mathbf{e}(k)$ must converge to zero. Thus, if $0 < \eta < 1$ and if the samples are linearly separable, $\mathbf{a}(k)$ will converge to a solution vector as k goes to infinity.

If we test the signs of the components of $\mathbf{Ya}(k)$ at each step and terminate the algorithm when they are all positive, we will in fact obtain a separating vector in a finite number of steps. This follows from the fact that $\mathbf{Ya}(k) = \mathbf{b}(k) + \mathbf{e}(k)$, and that the components of $\mathbf{b}(k)$ never decrease. Thus, if b_{min} is the smallest component of $\mathbf{b}(1)$ and if $\mathbf{e}(k)$ converges to zero, then $\mathbf{e}(k)$ must enter the hypersphere $\|\mathbf{e}(k)\| = b_{min}$ after a finite number of steps, at which point $\mathbf{Ya}(k) > 0$. Although we ignored terminating conditions to simplify the proof, such a terminating condition would always be used in practice.

5.9.3 Nonseparable Behavior

If the convergence proof just given is examined to see how the assumption of separability was employed, it will be seen that it was needed twice. First, the fact that $\mathbf{e}^t(k)\hat{\mathbf{b}} = 0$ was used to show that either $\mathbf{e}(k) = \mathbf{0}$ for some finite k, or $\mathbf{e}^+(k)$ is never zero and corrections go on forever. Second, this same constraint was used to show that if $\mathbf{e}^+(k)$ converges to zero, $\mathbf{e}(k)$ must also converge to zero.

If the samples are not linearly separable, it no longer follows that if $\mathbf{e}^+(k)$ is zero then $\mathbf{e}(k)$ must be zero. Indeed, on a nonseparable problem one may well obtain a nonzero error vector having no positive components. If this occurs, the algorithm automatically terminates and we have proof that the samples are not separable.

What happens if the patterns are not separable, but $\mathbf{e}^+(k)$ is never zero? In this case it still follows that

$$\mathbf{e}(k+1) = \mathbf{e}(k) + 2\eta(\mathbf{YY}^\dagger - \mathbf{I})\mathbf{e}^+(k) \tag{86}$$

and

$$\frac{1}{4}(\|\mathbf{e}(k)\|^2 - \|\mathbf{e}(k+1)\|^2) = \eta(1-\eta)\|\mathbf{e}^+(k)\|^2 + \eta^2\mathbf{e}^{+t}(k)\mathbf{YY}^\dagger\mathbf{e}^+(k). \tag{87}$$

Thus, the sequence $\|\mathbf{e}(1)\|^2$, $\|\mathbf{e}(2)\|^2$, $\ldots$ must still converge, though the limiting value $\|\mathbf{e}\|^2$ cannot be zero. Convergence requires that $\mathbf{e}^+(k) = \mathbf{0}$ for some finite k, or $\mathbf{e}^+(k)$ converges to zero while $\|\mathbf{e}(k)\|$ is bounded away from zero. Thus, the Ho-Kashyap algorithm provides us with a separating vector in the separable case, and with evidence of nonseparability in the nonseparable case. However, there is no bound on the number of steps needed to disclose nonseparability.

5.9.4 Some Related Procedures

If we write $\mathbf{Y}^\dagger = (\mathbf{Y}^t\mathbf{Y})^{-1}\mathbf{Y}^t$ and make use of the fact that $\mathbf{Y}^t\mathbf{e}(k) = 0$, we can modify the Ho-Kashyap rule as follows

$$
\begin{aligned}
\mathbf{b}(1) \quad &> \quad 0 \qquad \text{but otherwise arbitrary} \\
\mathbf{a}(1) \quad &= \quad \mathbf{Y}^\dagger \mathbf{b}(1) \\
\mathbf{b}(k+1) \quad &= \quad \mathbf{b}(k) + \eta(\mathbf{e}(k) + |\mathbf{e}(k)|) \\
\mathbf{a}(k+1) \quad &= \quad \mathbf{a}(k) + \eta \mathbf{Y}^\dagger |\mathbf{e}(k)|,
\end{aligned}
\tag{88}
$$

where, as usual,

$$
\mathbf{e}(k) = \mathbf{Y}\mathbf{a}(k) - \mathbf{b}(k).
\tag{89}
$$

This then gives the algorithm for fixed learning rate:

■ **Algorithm 12. (Modified Ho-Kashyap)**

---

1 **begin initialize a, b**, $\eta < 1$, threshold b_{min}, k_{max}
2 $\quad$ **do** $k \leftarrow (k+1) \bmod n$
3 $\quad\quad$ $\mathbf{e} \leftarrow \mathbf{Y}\mathbf{a} - \mathbf{b}$
4 $\quad\quad$ $\mathbf{e}^+ \leftarrow 1/2(\mathbf{e} + \text{Abs}[\mathbf{e}])$
5 $\quad\quad$ $\mathbf{b} \leftarrow \mathbf{b} + 2\eta(k)(\mathbf{e} + \text{Abs}[\mathbf{e}])$
6 $\quad\quad$ $\mathbf{a} \leftarrow \mathbf{Y}^\dagger \mathbf{b}$
7 $\quad\quad$ **if** $\text{Abs}[\mathbf{e}] \le b_{min}$ **then return a, b** *and* **exit**
8 $\quad$ **until** $k = k_{max}$
9 $\quad$ print "NO SOLUTION FOUND"
10 **end**

This algorithm differs from the Perceptron and relaxation algorithms for solving linear inequalities in at least three ways: (1) It varies both the weight vector **a** and the margin vector **b**, (2) it provides evidence of nonseparability, but (3) it requires the computation of the pseudoinverse of **Y**. Even though this last computation need be done only once, it can be time consuming, and it requires special treatment if $\mathbf{Y}^t\mathbf{Y}$ is singular. An interesting alternative algorithm that resembles Eq. 88 but avoids the need for computing $\mathbf{Y}^\dagger$ is

$$
\begin{aligned}
\mathbf{b}(1) &> 0 \quad \text{but otherwise arbitrary,} \\
\mathbf{a}(1) &\qquad \text{arbitrary,} \\
\mathbf{b}(k+1) &= \mathbf{b}(k) + (\mathbf{e}(k) + |\mathbf{e}(k)|), \\
\mathbf{a}(k+1) &= \mathbf{a}(k) + \eta \mathbf{R}\mathbf{Y}^t |\mathbf{e}(k)|,
\end{aligned}
\tag{90}
$$

where **R** is an arbitrary, constant, positive-definite $\hat{d}$-by-$\hat{d}$ matrix. We shall show that if η is properly chosen, this algorithm also yields a solution vector in a finite number of steps, provided that a solution exists. Furthermore, if no solution exists, the vector $\mathbf{Y}^t |\mathbf{e}(k)|$ either vanishes, exposing the nonseparability, or converges to zero.

The proof is fairly straightforward. Whether the samples are linearly separable or not, Eqs. 89 and 90 show that

$$
\begin{aligned}
\mathbf{e}(k+1) &= \mathbf{Y}\mathbf{a}(k+1) - \mathbf{b}(k+1) \\
&= (\eta \mathbf{Y}\mathbf{R}\mathbf{Y}^t - \mathbf{I})|\mathbf{e}(k)|.
\end{aligned}
$$

We can find, then, that the squared magnitude is

$$\|\mathbf{e}(k+1)\|^2 = |\mathbf{e}(k)|^t (\eta^2 \mathbf{YRY}^t \mathbf{YRY} - 2\eta \mathbf{YRY}^t + \mathbf{I})|\mathbf{e}(k)|,$$

and, furthermore,

$$\|\mathbf{e}(k)\|^2 - \|\mathbf{e}(k+1)\|^2 = (\mathbf{Y}^t|\mathbf{e}(k)|)^t \mathbf{A}(\mathbf{Y}^t|\mathbf{e}(k)|), \tag{91}$$

where

$$\mathbf{A} = 2\eta \mathbf{R} - \eta^2 \mathbf{RY}^t \mathbf{R}. \tag{92}$$

Clearly, if η is positive but sufficiently small, $\mathbf{A}$ will be approximately $2\eta \mathbf{R}$ and hence positive definite. Thus, if $\mathbf{Y}^t|\mathbf{e}(k)| \neq 0$ we will have $\|\mathbf{e}(k)\|^2 > \|\mathbf{e}(k+1)\|^2$.

At this point we must distinguish between the separable and the nonseparable case. In the separable case there exists an $\hat{\mathbf{a}}$ and a $\hat{\mathbf{b}} > 0$ satisfying $\mathbf{Y}\hat{\mathbf{a}} = \hat{\mathbf{b}}$. Thus, if $|\mathbf{e}(k)| \neq 0$,

$$|\mathbf{e}(k)|^t \mathbf{Y}\hat{\mathbf{a}} = |\mathbf{e}(k)|^t \hat{\mathbf{b}} > 0,$$

so that $\mathbf{Y}^t|\mathbf{e}(k)|$ cannot be zero unless $\mathbf{e}(k)$ is zero. Thus, the sequence $\|\mathbf{e}(1)\|^2$, $\|\mathbf{e}(2)\|^2, \ldots$ is monotonically decreasing and must converge to some limiting value $\|\mathbf{e}\|^2$. But for convergence to take place, $\mathbf{Y}^t|\mathbf{e}(k)|$ must converge to zero, which implies that $|\mathbf{e}(k)|$ and hence $\mathbf{e}(k)$ must converge to zero. Because $\mathbf{e}(k)$ starts out positive and never decreases, it follows that $\mathbf{a}(k)$ must converge to a separating vector. Moreover, by the same argument used before, a solution must actually be obtained after a finite number of steps.

In the nonseparable case, $\mathbf{e}(k)$ can neither be zero nor converge to zero. It may happen that $\mathbf{Y}^t|\mathbf{e}(k)| = 0$ at some step, which would provide proof of nonseparability. However, it is also possible for the sequence of corrections to go on forever. In this case, it again follows that the sequence $\|\mathbf{e}(1)\|^2$, $\|\mathbf{e}(2)\|^2, \ldots$ must converge to a limiting value $\|\mathbf{e}\|^2 \neq 0$, and that $\mathbf{Y}^t|\mathbf{e}(k)|$ must converge to zero. Thus, we again obtain evidence of nonseparability in the nonseparable case.

Before closing this discussion, let us look briefly at the question of choosing η and $\mathbf{R}$. The simplest choice for $\mathbf{R}$ is the identity matrix, in which case $\mathbf{A} = 2\eta \mathbf{I} - \eta^2 \mathbf{Y}^t \mathbf{Y}$. This matrix will be positive definite, thereby ensuring convergence, if $0 < \eta < 2/\lambda_{max}$, where λ_{max} is the largest eigenvalue of $\mathbf{Y}^t \mathbf{Y}$. Because the trace of $\mathbf{Y}^t \mathbf{Y}$ is both the sum of the eigenvalues of $\mathbf{Y}^t \mathbf{Y}$ and the sum of the squares of the elements of $\mathbf{Y}$, one can use the pessimistic bound $\hat{d}\lambda_{max} \leq \sum_i \|\mathbf{y}_i\|^2$ in selecting a value for η.

A more interesting approach is to change η at each step, selecting that value that maximizes $\|\mathbf{e}(k)\|^2 - \|\mathbf{e}(k+1)\|^2$. Equations 91 and 92 give

$$\|\mathbf{e}(k)\|^2 - \|\mathbf{e}(k+1)\|^2 = |\mathbf{e}(k)|^t \mathbf{Y}(2\eta \mathbf{R} - \eta^2 \mathbf{RY}^t \mathbf{YR})\mathbf{Y}^t|\mathbf{e}(k)|. \tag{93}$$

By differentiating with respect to η, we obtain the optimal value

$$\eta(k) = \frac{|\mathbf{e}(k)|^t \mathbf{YRY}^t|\mathbf{e}(k)|}{|\mathbf{e}(k)|^t \mathbf{YRY}^t \mathbf{YRY}^t|\mathbf{e}(k)|}, \tag{94}$$

which, for $\mathbf{R} = \mathbf{I}$, simplifies to

$$\eta(k) = \frac{\|\mathbf{Y}^t|\mathbf{e}(k)|\ \|^2}{\|\mathbf{Y}\mathbf{Y}^t|\mathbf{e}(k)|\ \|^2}. \tag{95}$$

This same approach can also be used to select the matrix $\mathbf{R}$. By replacing $\mathbf{R}$ in Eq. 93 by the symmetric matrix $\mathbf{R} + \delta\mathbf{R}$ and neglecting second-order terms, we obtain

$$\delta(\|\mathbf{e}(k)\|^2 - \|\mathbf{e}(k+1)\|^2) = |\mathbf{e}(k)|\mathbf{Y}[\delta\mathbf{R}^t(\mathbf{I} - \eta\mathbf{Y}^t\mathbf{Y}\mathbf{R}) + (\mathbf{I} - \eta\mathbf{R}\mathbf{Y}^t\mathbf{Y})\delta\mathbf{R}]\mathbf{Y}^t|\mathbf{e}(k)|.$$

Thus, the decrease in the squared error vector is maximized by the choice

$$\mathbf{R} = \frac{1}{\eta}(\mathbf{Y}^t\mathbf{Y})^{-1} \tag{96}$$

and because $\eta\mathbf{R}\mathbf{Y}^t = \mathbf{Y}^\dagger$, the descent algorithm becomes virtually identical with the original Ho-Kashyap algorithm.

*5.10 LINEAR PROGRAMMING ALGORITHMS

The Perceptron, relaxation and Ho-Kashyap procedures are basically gradient descent procedures for solving simultaneous linear inequalities. Linear programming techniques are procedures for maximizing or minimizing linear functions subject to linear equality or inequality constraints. This at once suggests that one might be able to solve linear inequalities by using them as the constraints in a suitable linear programming problem. In this section we shall consider two of several ways that this can be done. The reader need have no knowledge of linear programming to understand these formulations, though such knowledge would certainly be useful in applying the techniques.

5.10.1 Linear Programming

OBJECTIVE FUNCTION

A classical linear programming problem can be stated as follows: Find a vector $\mathbf{u} = (u_1, \ldots, u_m)^t$ that minimizes the linear (scalar) *objective function*

$$z = \boldsymbol{\alpha}^t\mathbf{u} \tag{97}$$

subject to the constraint

$$\mathbf{A}\mathbf{u} \geq \boldsymbol{\beta}, \tag{98}$$

SIMPLEX ALGORITHM

where $\boldsymbol{\alpha}$ is an m-by-1 *cost vector*, $\boldsymbol{\beta}$ is an l-by-1 vector, and $\mathbf{A}$ is an l-by-m matrix. The *simplex algorithm* is the classical iterative procedure for solving this problem (Fig. 5.18). For technical reasons, it requires the imposition of one more constraint, namely, $\mathbf{u} \geq 0$.

If we think of $\mathbf{u}$ as being the weight vector $\mathbf{a}$, this constraint is unacceptable, because in most cases the solution vector will have both positive and negative components. However, suppose that we write

$$\mathbf{a} \equiv \mathbf{a}^+ - \mathbf{a}^-, \tag{99}$$

where

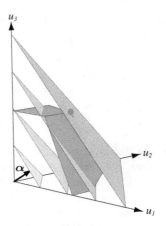

FIGURE 5.18. Surfaces of constant $z = \boldsymbol{\alpha}^t \mathbf{u}$ are shown in gray, while constraints of the form $\mathbf{A}\mathbf{u} = \boldsymbol{\beta}$ are shown in red. The simplex algorithm finds an extremum of z given the constraints, that is, where the gray plane intersects the red at a single point.

$$\mathbf{a}^+ \equiv \frac{1}{2}(|\mathbf{a}| + \mathbf{a}) \tag{100}$$

and

$$\mathbf{a}^- \equiv \frac{1}{2}(|\mathbf{a}| - \mathbf{a}). \tag{101}$$

Then both $\mathbf{a}^+$ and $\mathbf{a}^-$ are nonnegative, and by identifying the components of $\mathbf{u}$ with the components of $\mathbf{a}^+$ and $\mathbf{a}^-$, for example, we can accept the constraint $\mathbf{u} \geq 0$.

5.10.2 The Linearly Separable Case

Suppose that we have a set of n samples $\mathbf{y}_1, \ldots, \mathbf{y}_n$ and we want a weight vector $\mathbf{a}$ that satisfies $\mathbf{a}^t \mathbf{y}_i \geq b_i > 0$ for all i. How can we formulate this as a linear programming problem? One approach is to introduce what is called an *artificial variable* $\tau \geq 0$ by writing

$$\mathbf{a}^t \mathbf{y}_i + \tau \geq b_i.$$

If τ is sufficiently large, there is no problem in satisfying these constraints; for example, they are satisfied if $\mathbf{a} = \mathbf{0}$ and $\tau = \max_i b_i$.* However, this hardly solves our original problem. What we want is a solution with $\tau = 0$, which is the smallest value τ can have and still satisfy $\tau \geq 0$. Thus, we are led to consider the following problem: Minimize τ over all values of τ and $\mathbf{a}$ that satisfy the conditions $\mathbf{a}^t \mathbf{y}_i \geq b_i$ and $\tau \geq 0$. If the answer is zero, the samples are linearly separable, and we have a solution. If the answer is positive, there is no separating vector, but we have proof that the samples are nonseparable.

---

*In the terminology of linear programming, any solution satisfying the constraints is called a *feasible solution*. A feasible solution for which the number of nonzero variables does not exceed the number of constraints (not counting the simplex requirement for nonnegative variables) is called a *basic feasible solution*. Thus, the solution $\mathbf{a} = \mathbf{0}$ and $\tau = \max_i b_i$ is a basic feasible solution. Possession of such a solution simplifies the application of the simplex algorithm.

Formally, our problem is to find a vector $\mathbf{u}$ that minimizes the objective function $z = \boldsymbol{\alpha}^t \mathbf{u}$ subject to the constraints $\mathbf{A}\mathbf{u} \geq \boldsymbol{\beta}$ and $\mathbf{u} \geq 0$, where

$$
\mathbf{A} = \begin{bmatrix} \mathbf{y}_1^t & -\mathbf{y}_1^t & 1 \\ \mathbf{y}_2^t & -\mathbf{y}_2^t & 1 \\ \vdots & \vdots & \vdots \\ \mathbf{y}_n^t & -\mathbf{y}_n^t & 1 \end{bmatrix}, \quad \mathbf{u} = \begin{bmatrix} \mathbf{a}^+ \\ \mathbf{a}^- \\ \tau \end{bmatrix}, \quad \boldsymbol{\alpha} = \begin{bmatrix} \mathbf{0} \\ \mathbf{0} \\ 1 \end{bmatrix}, \quad \boldsymbol{\beta} = \begin{bmatrix} b_1 \\ b_2 \\ \vdots \\ b_n \end{bmatrix}.
$$

Thus, the linear programming problem involves $m = 2\hat{d} + 1$ variables and $l = n$ constraints, plus the simplex algorithm constraints $\mathbf{u} \geq 0$. The simplex algorithm will find the minimum value of the objective function $z = \boldsymbol{\alpha}^t \mathbf{u} = \tau$ in a finite number of steps, and it will exhibit a vector $\hat{\mathbf{u}}$ yielding that value. If the samples are linearly separable, the minimum value of τ will be zero, and a solution vector $\hat{\mathbf{a}}$ can be obtained from $\hat{\mathbf{u}}$. If the samples are not separable, the minimum value of τ will be positive. The resulting $\hat{\mathbf{u}}$ is usually not very useful as an approximate solution, but at least one obtains proof of nonseparability.

5.10.3 Minimizing the Perceptron Criterion Function

In the vast majority of pattern classification applications we cannot assume that the samples are linearly separable. In particular, when the patterns are not separable, one still wants to obtain a weight vector that classifies as many samples correctly as possible. Unfortunately, the number of errors is not a linear function of the components of the weight vector, and its minimization is not a linear programming problem. However, it turns out that the problem of minimizing the Perceptron criterion function can be posed as a problem in linear programming. Because minimization of this criterion function yields a separating vector in the separable case and a reasonable solution in the nonseparable case, this approach is quite attractive.

Recall from Section 5.5 that the basic Perceptron criterion function is given by

$$
J_p(\mathbf{a}) = \sum_{\mathbf{y} \in \mathcal{Y}} (-\mathbf{a}^t \mathbf{y}), \tag{102}
$$

where $\mathcal{Y}(\mathbf{a})$ is the set of training samples misclassified by $\mathbf{a}$. To avoid the useless solution $\mathbf{a} = \mathbf{0}$, we introduce a positive margin vector $\mathbf{b}$ and write

$$
J_p'(\mathbf{a}) = \sum_{\mathbf{y} \in \mathcal{Y}'} (b_i - \mathbf{a}^t \mathbf{y}), \tag{103}
$$

where $\mathbf{y}_i \in \mathcal{Y}'$ if $\mathbf{a}^t \mathbf{y}_i \leq b_i$. Clearly, J_p' is a piecewise-linear function of $\mathbf{a}$, not a linear function, and linear programming techniques are not immediately applicable. However, by introducing n artificial variables and their constraints, we can construct an equivalent linear objective function. Consider the problem of finding vectors $\mathbf{a}$ and $\boldsymbol{\tau}$ that minimize the linear function

$$
z = \sum_{i=1}^{n} \tau_i
$$

subject to the constraints

$$
\tau_k \geq 0 \quad \text{and} \quad \tau_i \geq b_i - \mathbf{a}^t \mathbf{y}_i.
$$

Of course for any fixed value of $\mathbf{a}$, the minimum value of z is exactly equal to $J_p'(\mathbf{a})$, since under these constraints the best we can do is to take $\tau_i = max[0, b_i - \mathbf{a}^t \mathbf{y}_i]$. If we minimize z over $\mathbf{t}$ *and* $\mathbf{a}$, we shall obtain the minimum possible value of $J_p'(\mathbf{a})$. Thus, we have converted the problem of minimizing $J_p'(\mathbf{a})$ to one of minimizing a linear function z subject to linear inequality constraints. Letting $\mathbf{u}_n$ denote an n-dimensional unit vector, we obtain the following problem with $m = 2\hat{d} + n$ variables and $l = n$ constraints: Minimize $\boldsymbol{\alpha}^t \mathbf{u}$ subject to $\mathbf{Au} \geq \boldsymbol{\beta}$ and $\mathbf{u} \geq 0$, where

$$\mathbf{A} = \begin{bmatrix} \mathbf{y}_1^t & -\mathbf{y}_1^t & 1 & 0 & \cdots & 0 \\ \mathbf{y}_2^t & -\mathbf{y}_2^t & 0 & 1 & \cdots & 0 \\ \vdots & \vdots & \vdots & \vdots & \ddots & \vdots \\ \mathbf{y}_n^t & -\mathbf{y}_n^t & 0 & 0 & \cdots & 1 \end{bmatrix}, \quad \mathbf{u} = \begin{bmatrix} \mathbf{a}^+ \\ \mathbf{a}^- \\ \boldsymbol{\tau} \end{bmatrix},$$

$$\boldsymbol{\alpha} = \begin{bmatrix} \mathbf{0} \\ \mathbf{0} \\ \mathbf{1}_n \end{bmatrix}, \quad \boldsymbol{\beta} = \begin{bmatrix} b_1 \\ b_2 \\ \vdots \\ b_n \end{bmatrix}.$$

The choice $\mathbf{a} = 0$ and $\tau_i = b_i$ provides a basic feasible solution to start the simplex algorithm, and the simplex algorithm will provide an $\hat{\mathbf{a}}$ minimizing $J_p'(\mathbf{a})$ in a finite number of steps.

We have shown two ways to formulate the problem of finding a linear discriminant function as a problem in linear programming. There are other possible formulations, the ones involving the so-called *dual problem* being of particular interest from a computational standpoint. Generally speaking, methods such as the simplex method are merely sophisticated gradient descent methods for extremizing linear functions subject to linear constraints. The coding of a linear programming algorithm is usually more complicated than the coding of the simpler descent procedures we described earlier, and these descent procedures generalize naturally to multilayer neural networks. However, general purpose linear programming packages can often be used directly or modified appropriately with relatively little effort. When this can be done, one can secure the advantage of guaranteed convergence on both separable and nonseparable problems.

The various algorithms for finding linear discriminant functions presented in this chapter are summarized in Table 5.1. It is natural to ask which one is best, but none uniformly dominates or is uniformly dominated by all others. The choice depends upon such considerations as desired characteristics, ease of programming, the number of samples, and the dimensionality of the samples. If a linear discriminant function can yield a low error rate, any of these procedures, intelligently applied, can provide good performance.

*5.11 SUPPORT VECTOR MACHINES

We have seen how to train linear machines with margins. *Support vector machines* (SVMs) are motivated by many of the same considerations, but rely on preprocessing the data to represent patterns in a high dimension—typically much higher than the original feature space. With an appropriate nonlinear mapping $\varphi(\cdot)$ to a sufficiently high dimension, data from two categories can always be separated by a hyperplane

Table 5.1. Descent Procedures for Obtaining Linear Discriminant Functions

Name	Criterion	Algorithm	Conditions	Remarks
Fixed increment	$J_p = \sum_{a^t y \leq 0} (-a^t y)$	$a(k+1) = a(k) + y^k$ $(a^t(k)y^k \leq 0)$	—	Finite convergence if linearly separable to solution with $a^t y > 0$; $a(k)$ always bounded.
Variable Increment	$J'_p = \sum_{a^t y \leq b} -(a^t y - b)$	$a(k+1) = a(k) + \eta(k)y^k$ $(a^t(k)y^k \leq b)$	$\eta(k) \geq 0$ $\sum \eta(k) \to \infty$ $\dfrac{\sum \eta^2(k)}{\left(\sum \eta(k)\right)^2} \to 0$	Converges if linearly separable to solution with $a^t y > b$; Finite convergence if $0 < \alpha \leq \eta(k) \leq \beta < \infty$.
Relaxation	$J_r = \frac{1}{2} \sum_{a^t y \leq b} \frac{(a^t y - b)^2}{\|y\|^2}$	$a(k+1) = a(k) + \eta \frac{b - a^t(k)y^k}{\|y^k\|^2} y^k$ $(a^t(k)y^k \leq b)$	$0 < \eta < 2$	Converges if linearly separable to solution with $a^t y \geq b$. If $b > 0$, finite convergence to solution with $a^t y > 0$.
Widrow-Hoff (LMS)	$J_s = \sum_i (a^t y_i - b_i)^2$	$a(k+1) = a(k) + \eta(k)(b_k - a^t(k)y^k)y^k$	$\eta(k) > 0$ $\eta(k) \to 0$	Tends toward solution minimizing J_s.
Stochastic approximation	$J_m = \mathcal{E}\left[(a^t y - z)^2\right]$	$a(k+1) = $ $a(k) + \eta(k)(z_k - a^t(k)y^k)y^k$ $a(k+1) = $ $a(k) + R(k)(z(k) - a(k)^t y^k)y^k$ $R^{-1}(k+1) = R^{-1}(k) + y_k y_k^t$	$\sum \eta(k) \to \infty$ $\sum \eta^2(k) \to L < \infty$	Involves an infinite number of randomly drawn samples; converges in mean square to a solution minimizing J_m and provides an MSE approximation to Bayes discriminant.
Pseudoinverse	$J_s = \|Ya - b\|^2$	$a = Y^\dagger b$	—	Classical MSE solution; special choices for b yield Fisher's linear discriminant and MSE approximation to Bayes discriminant.

	Criterion	Algorithm	Parameters	Comments
Ho-Kashyap	$J_s = \|\mathbf{Ya} - \mathbf{b}\|^2$	$\mathbf{b}(k+1) = \mathbf{b}(k) + \eta(\mathbf{e}(k) + |\mathbf{e}(k)|)$ $\mathbf{e}(k) = \mathbf{Ya}(k) - \mathbf{b}(k)$ $\mathbf{a}(k) = \mathbf{Y}^\dagger \mathbf{b}(k)$	$0 < \eta < 1$ $\mathbf{b}(1) > 0$	$\mathbf{a}(k)$ is MSE solution for each $\mathbf{b}(k)$; finite convergence if linearly separable; if $\mathbf{e}(k) \leq 0$ but $\mathbf{e}(k) \neq 0$, the samples are nonseparable.
		$\mathbf{b}(k+1) = \mathbf{b}(k) + \eta(\mathbf{e}(k) + (|\mathbf{e}(k)|)$ $\mathbf{a}(k+1) = \mathbf{a}(k) + \eta \mathbf{R}\mathbf{Y}^t |\mathbf{e}(k)|$	$\eta(k) = \dfrac{|\mathbf{e}(k)|^t \mathbf{YRY}^t |\mathbf{e}(k)|}{|\mathbf{e}(k)|^t \mathbf{YRY}^t \mathbf{YRY}^t |\mathbf{e}(k)|}$ is optimum; $\mathbf{R}$ symmetric positive definite; $\mathbf{b}(1) > 0$	Finite convergence if linearly separable; if $\mathbf{Y}^t |\mathbf{e}(k)| = 0$, but $\mathbf{e}(k) \neq 0$, the samples are nonseparable.
Linear Programming	$\tau = \max_{\mathbf{a}^t \mathbf{y}_i \leq b_i}\left[-(\mathbf{a}^t \mathbf{y}_i - b_i)\right]$	Simplex algorithm	$\mathbf{a}^t \mathbf{y}_i + \tau \geq b_i$ $\tau \geq 0$	Finite convergence in both separable and nonseparable cases; useful solution only if separable.
	$J_p' = \sum_{i=1}^n \tau_i$ $= \sum_{\mathbf{a}^t \mathbf{y}_i \leq b_i} -(\mathbf{a}^t \mathbf{y}_i - b_i)$	Simplex algorithm	$\mathbf{a}^t \mathbf{y}_i + \tau \geq b_i$ $\tau_i \geq 0$	Finite convergence in both separable and nonseparable cases; useful solution only if separable.

(Problem 29). Here we assume each pattern $\mathbf{x}_k$ has been transformed to $\mathbf{y}_k = \varphi(\mathbf{x}_k)$; we return to the choice of $\varphi(\cdot)$ below. For each of the n patterns, $k = 1, 2, \ldots, n$, we let $z_k = \pm 1$, according to whether pattern k is in ω_1 or ω_2. A linear discriminant in an augmented $\mathbf{y}$ space is

$$g(\mathbf{y}) = \mathbf{a}^t \mathbf{y}, \tag{104}$$

where both the weight vector and the transformed pattern vector are augmented (by $a_0 = w_0$ and $y_0 = 1$, respectively). Thus a separating hyperplane ensures

$$z_k g(\mathbf{y}_k) \geq 1, \qquad k = 1, \ldots, n, \tag{105}$$

much as was shown in Fig. 5.8.

In Section 5.9, the margin was any positive distance from the decision hyperplane. The goal in training a Support Vector Machine is to find the separating hyperplane with the *largest* margin; we expect that the larger the margin, the better generalization of the classifier. As illustrated in Fig. 5.2, the distance from any hyperplane to a (transformed) pattern $\mathbf{y}$ is $|g(\mathbf{y})|/||\mathbf{a}||$, and assuming that a positive margin b exists, Eq. 105 implies

$$\frac{z_k g(\mathbf{y}_k)}{||\mathbf{a}||} \geq b, \qquad k = 1, \ldots, n; \tag{106}$$

the goal is to find the weight vector $\mathbf{a}$ that maximizes b. Of course, the solution vector can be scaled arbitrarily and still preserve the hyperplane, and thus to ensure uniqueness we impose the constraint $b \, ||\mathbf{a}|| = 1$; that is, we demand the solution to Eqs. 104 and 105 also minimize $||\mathbf{a}||^2$.

SUPPORT
VECTOR

The *support vectors* are the (transformed) training patterns for which Eq. 105 represents an equality—that is, the support vectors are (equally) close to the hyperplane (Fig. 5.19). The support vectors are the training samples that define the optimal separating hyperplane and are the most difficult patterns to classify. Informally speaking, they are the patterns most informative for the classification task.

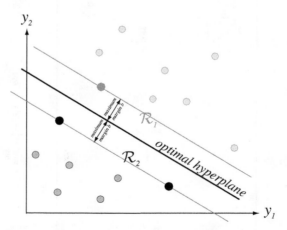

FIGURE 5.19. Training a support vector machine consists of finding the optimal hyperplane, that is, the one with the maximum distance from the nearest training patterns. The support vectors are those (nearest) patterns, a distance b from the hyperplane. The three support vectors are shown as solid dots.

A conceptually simple method of training a support vector machine is based on a small modification to the familiar Perceptron training rule (Algorithm 4). Recall that the Perceptron learning rule required merely updating the weight vector by an amount proportional to *any* randomly chosen misclassified pattern. A support vector machine can, instead, be trained by choosing the current *worst*-classified pattern. During most of the training period, such a pattern is one on the wrong side of the current decision boundary, farthest from that boundary. At the end of the training period, such a pattern will be one of the support vectors (Problem 31).

Alas, in practice, finding the worst-case pattern is computationally expensive, since for each update we must search through the entire training set to find the worst-classified pattern. As such, this simple method is used only for small problems. Before we return to the more common method of training an SVM having many patterns, we consider the error of such a classifier.

If N_s denotes the total number of support vectors, then for n training patterns the expected value of the generalization error rate is bounded, according to

$$\mathcal{E}_n[error] \leq \frac{\mathcal{E}_n[N_s]}{n}, \tag{107}$$

LEAVE-ONE-OUT
BOUND

where the expectation is over all training sets of size n drawn from the (stationary) distributions describing the categories. This bound is independent of the dimensionality of the space of transformed vectors, determined by $\varphi(\cdot)$. We will return to this topic in Chapter 9, but for now we can understand this informally by means of the *leave-one-out bound*. Suppose we have n points in the training set, we train an SVM on $n-1$ of them, and we test on the single remaining point. If that remaining point happens to be a support vector for the full n sample case, then there will be an error; otherwise, there will not. Note that if we can find a transformation $\varphi(\cdot)$ that well separates the data—so the expected number of support vectors is small—then Eq. 107 shows that the expected error rate will be lower.

5.11.1 SVM Training

We now turn to the problem of training an SVM. The first step is, of course, to choose the nonlinear φ-functions that map the input to a higher-dimensional space. Often this choice will be informed by the designer's knowledge of the problem domain. In the absence of such information, one might choose to use polynomials, Gaussians, or yet other basis functions. The dimensionality of the mapped space can be arbitrarily high (though in practice it may be limited by computational resources).

We begin by recasting the problem of minimizing the magnitude of the weight vector constrained by the separation into an unconstrained problem by the method of Lagrange undetermined multipliers. Thus from Eq. 106 and our goal of minimizing $||\mathbf{a}||$, we construct the functional

$$L(\mathbf{a}, \boldsymbol{\alpha}) = \frac{1}{2}||\mathbf{a}||^2 - \sum_{k=1}^{n} \alpha_k[z_k \mathbf{a}^t \mathbf{y}_k - 1] \tag{108}$$

and seek to minimize $L()$ with respect to the weight vector $\mathbf{a}$, and maximize it with respect to the undetermined multipliers $\alpha_k \geq 0$. The last term in Eq. 108 expresses the goal of classifying the points correctly. It can be shown using the so-called Kuhn-Tucker construction (Problem 33) that this optimization can be reformulated as maximizing

$$L(\boldsymbol{\alpha}) = \sum_{k=1}^{n} \alpha_i - \frac{1}{2} \sum_{k,j}^{n} \alpha_k \alpha_j z_k z_j \mathbf{y}_j^t \mathbf{y}_k, \tag{109}$$

subject to the constraints

$$\sum_{k=1}^{n} z_k \alpha_k = 0 \qquad \alpha_k \geq 0, \qquad k = 1, \dots, n, \tag{110}$$

given the training data. While these equations can be solved using quadratic programming, a number of alternate schemes have been devised (cf. Bibliography).

EXAMPLE 2   SVM for the XOR Problem

The exclusive-OR is the simplest problem that cannot be solved using a linear discriminant operating directly on the features. The points $k = 1, 3$ at $\mathbf{x} = (1, 1)^t$ and $(-1, -1)^t$ are in category ω_1 (red in the figure), while $k = 2, 4$ at $\mathbf{x} = (1, -1)^t$ and $(-1, 1)^t$ are in ω_2 (black in the figure). Following the approach of SVMs, we preprocess the features to map them to a higher dimension space where they can be linearly separated. While many φ-functions could be used, here we use the simplest expansion up to second order: 1, $\sqrt{2}x_1$, $\sqrt{2}x_2$, $\sqrt{2}x_1 x_2$, x_1^2 and x_2^2, where the $\sqrt{2}$ is convenient for normalization.

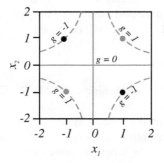 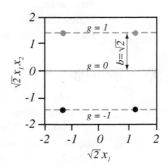

The XOR problem in the original $x_1 x_2$-space is shown at the left; the two red patterns are in category ω_1 and the two black ones in ω_2. These four training patterns $\mathbf{x}$ are mapped to a six-dimensional space by 1, $\sqrt{2}x_1$, $\sqrt{2}x_2$, $\sqrt{2}x_1 x_2$, x_1^2 and x_2^2. In this space, the optimal hyperplane is found to be $g(x_1, x_2) = x_1 x_2 = 0$ and the margin is $b = \sqrt{2}$. A two-dimensional projection of this space is shown at the right. The hyperplanes through the support vectors are $\sqrt{2}x_1 x_2 = \pm 1$, and correspond to the hyperbolas $x_1 x_2 = \pm 1$ in the original feature space, as shown.

We seek to maximize Eq. 109,

$$\sum_{k=1}^{4} \alpha_k - \frac{1}{2} \sum_{k,j}^{n} \alpha_k \alpha_j z_k z_j \mathbf{y}_j^t \mathbf{y}_k$$

subject to the constraints (Eq. 110)

$$\alpha_1 - \alpha_2 + \alpha_3 - \alpha_4 = 0$$

$$0 \le \alpha_k \qquad k = 1, 2, 3, 4.$$

It is clear from the symmetry of the problem that $\alpha_1 = \alpha_3$ and that $\alpha_2 = \alpha_4$ at the solution. While we could use iterative gradient descent as described in Section 5.9, for this small problem we can use analytic techniques instead. The solution is $a_k^* = 1/8$, for $k = 1, 2, 3, 4$, and from the last term in Eq. 108 this implies that all four training patterns are support vectors—an unusual case due to the highly symmetric nature of the XOR problem.

The final discriminant function is $g(\mathbf{x}) = g(x_1, x_2) = x_1 x_2$, and the decision hyperplane is defined by $g = 0$, which properly classifies all training patterns. The margin is easily computed from the solution $||\mathbf{a}||$ and is found to be $b = 1/||\mathbf{a}|| = \sqrt{2}$. The figure at the right shows the margin projected into two dimensions of the five-dimensional transformed space. Problem 30 asks you to consider this margin as viewed in other two-dimensional projected subspaces.

An important benefit of the support vector machine approach is that the complexity of the resulting classifier is characterized by the number of support vectors rather than the dimensionality of the transformed space. As a result, SVMs tend to be less prone to problems of overfitting than some other methods.

5.12 MULTICATEGORY GENERALIZATIONS

There is no uniform way to extend all of the two-category procedures we have discussed to the multicategory case. In Section 5.2.2 we defined a multicategory classifier called a linear machine which classifies a pattern by computing c linear discriminant functions

$$g_i(\mathbf{x}) = \mathbf{w}^t \mathbf{x} + w_{i0} \quad i = 1, \dots, c,$$

and assigning $\mathbf{x}$ to the category corresponding to the largest discriminant. This is a natural generalization for the multiclass case, particularly in view of the results of Chapter 2 for the multivariate normal problem. It can be extended simply to generalized linear discriminant functions by letting $\mathbf{y}(\mathbf{x})$ be a $\hat{d}$-dimensional vector of functions of $\mathbf{x}$, and by writing

$$g_i(\mathbf{x}) = \mathbf{a}_i^t \mathbf{y} \quad i = 1, \dots, c, \tag{111}$$

where again $\mathbf{x}$ is assigned to ω_i if $g_i(\mathbf{x}) > g_j(\mathbf{x})$ for all $j \ne i$.

The generalization of our procedures from a two-category linear classifier to a multicategory linear machine is simplest in the linearly-separable case. Suppose that we have a set of labeled samples $\mathbf{y}_1, \mathbf{y}_2, \dots, \mathbf{y}_n$, with n_1 in the subset $\mathcal{Y}_1$ labeled ω_1, n_2 in the subset $\mathcal{Y}_2$ labeled $\omega_2, \dots$, and n_c in the subset $\mathcal{Y}_c$ labeled ω_c. We say that this set is linearly separable if there exists a linear machine that classifies all of them correctly. That is, if these samples are linearly separable, then there exists a set of weight vectors $\hat{\mathbf{a}}_1, \dots, \hat{\mathbf{a}}_c$ such that if $\mathbf{y}_k \in \mathcal{Y}_i$, then

$$\hat{\mathbf{a}}_i^t \mathbf{y}_k > \hat{\mathbf{a}}_j^t \mathbf{y}_k \tag{112}$$

for all $j \ne i$.

5.12.1 Kesler's Construction

One of the pleasant things about this definition in Eq. 112 is that it is possible to manipulate these inequalities and reduce the multicategory problem to the two-category case. Suppose for the moment that $\mathbf{y}_k \in \mathcal{Y}_1$, so that Eq. 112 becomes

$$\hat{\mathbf{a}}_1^t \mathbf{y}_k - \hat{\mathbf{a}}_j^t \mathbf{y}_k > 0, \quad j = 2, \ldots, c. \tag{113}$$

This set of $c - 1$ inequalities can be thought of as requiring that the $c\hat{d}$-dimensional weight vector

$$\hat{\boldsymbol{\alpha}} = \begin{bmatrix} \mathbf{a}_1 \\ \mathbf{a}_2 \\ \vdots \\ \mathbf{a}_c \end{bmatrix}$$

correctly classifies all $c - 1$ of the $c\hat{d}$-dimensional sample sets

$$\boldsymbol{\eta}_{12} = \begin{bmatrix} \mathbf{y} \\ -\mathbf{y} \\ \mathbf{0} \\ \vdots \\ \mathbf{0} \end{bmatrix}, \quad \boldsymbol{\eta}_{13} = \begin{bmatrix} \mathbf{y} \\ \mathbf{0} \\ -\mathbf{y} \\ \vdots \\ \mathbf{0} \end{bmatrix}, \quad \cdots, \quad \boldsymbol{\eta}_{1c} = \begin{bmatrix} \mathbf{y} \\ \mathbf{0} \\ \mathbf{0} \\ \vdots \\ -\mathbf{y} \end{bmatrix}.$$

In other words, each $\boldsymbol{\eta}_{1j}$ corresponds to "normalizing" the patterns in ω_1 and ω_j. More generally, if $\mathbf{y} \in \mathcal{Y}_i$, we construct $(c - 1)c$ $\hat{d}$-dimensional training samples $\boldsymbol{\eta}_{ij}$ by partitioning $\boldsymbol{\eta}_{ij}$ into c $\hat{d}$-dimensional subvectors, with the ith subvector being $\mathbf{y}$, the jth being $-\mathbf{y}$, and all others being zero. Clearly, if $\hat{\boldsymbol{\alpha}}^t \boldsymbol{\eta}_{ij} > 0$ for $j \neq i$, then the linear machine corresponding to the components of $\hat{\boldsymbol{\alpha}}$ classifies $\mathbf{y}$ correctly.

This so-called Kesler construction multiplies the dimensionality of the data by c and the number of samples by $c - 1$, which does not make its direct use attractive. Its importance resides in the fact that it allows us to convert many multicategory error-correction procedures to two-category procedures for the purpose of obtaining a convergence proof.

5.12.2 Convergence of the Fixed-Increment Rule

We now use use Kesler's construction to prove convergence for a generalization of the fixed-increment rule for a linear machine. Suppose that we have a set of n linearly-separable samples $\mathbf{y}_1, \ldots, \mathbf{y}_n$, and we use them to form an infinite sequence in which every sample appears infinitely often. Let L_k denote the linear machine whose weight vectors are $\mathbf{a}_1(k), \ldots, \mathbf{a}_c(k)$. Starting with an arbitrary initial linear machine L_1, we want to use the sequence of samples to construct a sequence of linear machines that converges to a solution machine, one that classifies all of the samples correctly. We shall propose an error-correction rule in which weight changes are made if and only if the present linear machine misclassifies a sample. Let $\mathbf{y}^k$ denote the kth sample requiring correction, and suppose that $\mathbf{y}^k \in \mathcal{Y}_i$. Because $\mathbf{y}^k$ requires correction, there must be at least one $j \neq i$ for which

$$\mathbf{a}_i^t(k)\mathbf{y}^k \leq \mathbf{a}_j^t(k)\mathbf{y}^k. \tag{114}$$

Then the fixed-increment rule for correcting L_k is

$$\mathbf{a}_i(k+1) = \mathbf{a}_i(k) + \mathbf{y}^k$$
$$\mathbf{a}_j(k+1) = \mathbf{a}_j(k) - \mathbf{y}^k \qquad (115)$$
$$\mathbf{a}_l(k+1) = \mathbf{a}_l(k), \; l \neq i \text{ and } l \neq j.$$

That is, the weight vector for the desired category is incremented by the pattern, the weight vector for the incorrectly chosen category is decremented, and all other weights are left unchanged (Problem 36, Computer exercise 12).

We shall now show that this rule must lead to a solution machine after a finite number of corrections. The proof is simple. For each linear machine L_k there corresponds a weight vector

$$\boldsymbol{\alpha}_k = \begin{bmatrix} \mathbf{a}_1(k) \\ \vdots \\ \mathbf{a}_c(k) \end{bmatrix}.$$

For each sample $\mathbf{y} \in \mathcal{Y}_i$ there are $c - 1$ samples $\boldsymbol{\eta}_{ij}$ formed as described in Section 5.12.1. In particular, corresponding to the vector $\mathbf{y}^k$ satisfying Eq. 114 there is a vector

$$\boldsymbol{\eta}_{ij}^k = \begin{bmatrix} \vdots \\ \mathbf{y}^k \\ \vdots \\ -\mathbf{y}^k \\ \vdots \end{bmatrix} \begin{matrix} \\ \leftarrow i \\ \\ \leftarrow j \\ \\ \end{matrix}$$

satisfying

$$\boldsymbol{\alpha}^t(k)\boldsymbol{\eta}_{ij}^k \leq 0.$$

Furthermore, the fixed-increment rule for correcting L_k is the fixed-increment rule for correcting $\boldsymbol{\alpha}(k)$, namely,

$$\boldsymbol{\alpha}(k+1) = \boldsymbol{\alpha}(k) + \boldsymbol{\eta}_{ij}^k.$$

Thus, we have obtained a complete correspondence between the multicategory case and the two-category case, in which the multicategory procedure produces a sequence of samples $\boldsymbol{\eta}^1, \boldsymbol{\eta}^2, \ldots, \boldsymbol{\eta}^k, \ldots$ and a sequence of weight vectors $\boldsymbol{\alpha}_1, \boldsymbol{\alpha}_2, \ldots, \boldsymbol{\alpha}_k, \ldots$ By our results for the the two-category case, this latter sequence cannot be infinite, but must terminate in a solution vector. Hence, the sequence $L_1, L_2, \ldots, L_k, \ldots$ must terminate in a solution machine after a finite number of corrections.

This use of Kesler's construction to establish equivalences between multicategory and two-category procedures is a powerful theoretical tool. It can be used to extend all of our results for the Perceptron and relaxation procedures to the multicategory case, and it applies as well to the error-correction rules for the method of potential functions (Problem 38). Unfortunately, it is not as directly useful in generalizing the MSE or the linear programming approaches.

5.12.3 Generalizations for MSE Procedures

Perhaps the simplest way to obtain a natural generalization of the MSE procedures to the multiclass case is to consider the problem as a set of c two-class problems. The ith problem is to obtain a weight vector $\mathbf{a}_i$ that is minimum-squared-error solution to the equations

$$\mathbf{a}_i^t \mathbf{y} = 1 \qquad \text{for all } \mathbf{y} \in \mathcal{Y}_i$$
$$\mathbf{a}_i^t \mathbf{y} = -1 \qquad \text{for all } \mathbf{y} \notin \mathcal{Y}_i.$$

In light of the results of Section 5.8.3, we see that if the number of samples is very large we will obtain a minimum mean-squared-error approximation to the Bayes discriminant function

$$P(\omega_i | \mathbf{x}) - P(\text{not } \omega_i | \mathbf{x}) = 2P(\omega_i | \mathbf{x}) - 1.$$

This observation has two immediate consequences. First, it suggests a modification in which we seek a weight vector $\mathbf{a}_i$ that is a minimum-squared-error solution to the equations

$$\mathbf{a}_i^t \mathbf{y} = 1 \qquad \text{for all } \mathbf{y} \in \mathcal{Y}_i$$
$$\mathbf{a}_i^t \mathbf{y} = 0 \qquad \text{for all } \mathbf{y} \notin \mathcal{Y}_i \tag{116}$$

so that $\mathbf{a}^t \mathbf{y}$ will be a minimum mean-squared-error approximation to $P(\omega_i | \mathbf{x})$. Second, it justifies the use of the resulting discriminant functions in a linear machine, in which we assign $\mathbf{y}$ to ω_i if $\mathbf{a}_i^t \mathbf{y} > \mathbf{a}_j^t \mathbf{y}$ for all $j \neq i$.

The pseudoinverse solution to the multiclass MSE problem can be written in a form analogous to the form for the two-class case. Let $\mathbf{Y}$ be the n-by-$\hat{d}$ matrix of training samples, which we assume to be partitioned as

$$\mathbf{Y} = \begin{bmatrix} \mathbf{Y}_1 \\ \mathbf{Y}_2 \\ \vdots \\ \mathbf{Y}_c \end{bmatrix}, \tag{117}$$

with the samples labeled ω_i comprising the rows of $\mathbf{Y}_i$. Similarly, let $\mathbf{A}$ be the $\hat{d}$-by-c matrix of weight vectors

$$\mathbf{A} = [\mathbf{a}_1 \ \mathbf{a}_2 \ \cdots \ \mathbf{a}_c], \tag{118}$$

and let $\mathbf{B}$ be the n-by-c matrix

$$\mathbf{B} = \begin{bmatrix} \mathbf{B}_1 \\ \mathbf{B}_2 \\ \vdots \\ \mathbf{B}_c \end{bmatrix}, \tag{119}$$

where all of the elements of $\mathbf{B}_i$ are zero except for those in the ith column, which are unity. Then the trace of the "squared" error matrix $(\mathbf{YA} - \mathbf{B})^t (\mathbf{YA} - \mathbf{B})$ is minimized by the solution*

$$\mathbf{A} = \mathbf{Y}^\dagger \mathbf{B}, \tag{120}$$

where, as usual, $\mathbf{Y}^\dagger$ is the pseudoinverse of $\mathbf{Y}$.

This result can be generalized in a theoretically interesting fashion. Let λ_{ij} be the loss incurred for deciding ω_i when the true state of nature is ω_j, and let the jth submatrix of $\mathbf{B}$ be given by

$$\mathbf{B}_j = - \begin{bmatrix} \lambda_{1j} & \lambda_{2j} & \cdots & \lambda_{cj} \\ \lambda_{1j} & \lambda_{2j} & \cdots & \lambda_{cj} \\ \vdots & & & \vdots \\ \lambda_{1j} & \lambda_{2j} & \cdots & \lambda_{cj} \end{bmatrix} \updownarrow n_j \qquad j = 1, \ldots, c. \tag{121}$$

Then, as the number of samples approaches infinity, the solution $\mathbf{A} = \mathbf{Y}^\dagger \mathbf{B}$ yields discriminant functions $\mathbf{a}_i^t \mathbf{y}$ that provide a minimum-mean-square-error approximation to the Bayes discriminant function

$$g_{0i} = - \sum_{j=1}^{c} \lambda_{ij} P(\omega_i | \mathbf{x}). \tag{122}$$

The proof of this is a direct extension of the proof given in Section 5.8.3 (Problem 37).

SUMMARY

This chapter considers discriminant functions that are a linear function of a set of parameters, generally called weights. In all two-category cases, such discriminants lead to hyperplane decision boundaries, either in the feature space itself or in a space where the features have been mapped by a nonlinear function (general linear discriminants).

In broad overview, techniques such as the Perceptron algorithm adjust the parameters to increase the inner product with patterns in category ω_1 and decrease the inner product with those in ω_2. A very general approach is to form some criterion function and perform gradient descent. Different criterion functions have different strengths and weaknesses related to computation and convergence, none uniformly dominates the others. One can use linear algebra to solve for the weights (parameters) directly, by means of pseudoinverse matrices for small problems.

In support vector machines, the input is mapped by a nonlinear function to a high-dimensional space, and the optimal hyperplane found the one that has the largest margin. The support vectors are those (transformed) patterns that determine the margin; they are informally the hardest patterns to classify, and the most informative ones for designing the classifier. An upper bound on expected error rate of the classifier depends linearly upon the expected number of support vectors.

---

*If we let $\mathbf{b}_i$ denote the ith column of $\mathbf{B}$, the trace of $(\mathbf{YA} - \mathbf{B})^t (\mathbf{YA} - \mathbf{B})$ is equal to the sum of the squared lengths of the error vectors $\mathbf{Ya}_i - \mathbf{b}_i$. The solution $\mathbf{A} = \mathbf{Y}^\dagger \mathbf{B}$ not only minimizes this sum, but also minimizes each term in the sum.

For multicategory problems, the linear machines create decision boundaries consisting of sections of such hyperplanes. One can prove convergence of multicategory algorithms by first converting them to two-category algorithms and using the two-category proofs. The simplex algorithm finds the optimum of a linear function subject to (inequality) constraints, and it can be used for training linear classifiers.

Linear discriminants, while useful, are not sufficiently general for arbitrary challenging pattern recognition problems (such as those involving multimodal or nonconvex densities) unless an appropriate nonlinear mapping (φ function) can be found. In this chapter we have not considered any principled approaches to choosing such functions, but we turn to that topic in Chapter 6.

BIBLIOGRAPHICAL AND HISTORICAL REMARKS

Because linear discriminant functions are so amenable to analysis, far more papers have been written about them than the subject deserves. Historically, all of this work begins with the classic paper by Ronald A. Fisher [5]. The application of linear discriminant function to pattern classification was well described in reference [9], which posed the problem of optimal (minimum-risk) linear discriminant and proposed plausible gradient descent procedures to determine a solution from samples. Unfortunately, little can be said about such procedures without knowing the underlying distributions, and even then the situation is analytically complex. The design of multicategory classifiers using two-category procedures stems from reference [16]. Minsky and Papert's **Perceptrons** [13] was influential in pointing out the weaknesses of linear classifiers—weaknesses that were overcome by the methods we shall study in Chapter 6. The Winnow algorithms [10] in the error-free case [7, 11] and subsequent work in the general case have been useful in the computational learning community, as they allow one to derive convergence bounds.

While this work was statistically oriented, many of the pattern recognition papers that appeared in the late 1950s and early 1960s adopted other viewpoints. One viewpoint was that of neural networks, in which individual neurons were modeled as threshold elements, two-category linear machines—work that had its origins in the famous paper by McCulloch and Pitts [12].

As linear machines have been applied to larger and larger data sets in higher and higher dimensions, the computational burden of linear programming [1] has made this approach less popular. Stochastic approximation methods have been applied to this problem, as described in reference [21].

Early papers on the key ideas in Support Vector Machines are references [2] and [15]. A more extensive treatment, including complexity control, can be found in references [18] and [14]—material we shall visit in Chapter 9. A clear presentation of the method is [4], which provided the inspiration behind our Example 2. Guyon and Stork provide an overview of the relationships among linear classifiers, including Support Vector Machines [8]. The Kuhn-Tucker construction, used in the SVM training method described in the text and explored in Problem 33, is from reference [6] and is used in reference [17]. The fundamental result is that exactly one of the following three cases holds. (1) The original (primal) conditions have an optimal solution; in that case the dual cases do too, and their objective values are equal, or (2) the primal conditions are infeasible; in that case the dual is either unbounded or itself infeasible, or (3) the primal conditions are unbounded; in that case the dual is infeasible.

PROBLEMS

Section 5.2

1. Explore the applicability of linear discriminants for unimodal and multimodal problems in two dimensions through the following.

   (a) Sketch two multimodal distributions for which a linear discriminant could give excellent or possibly even the optimal classification accurcy.

   (b) Sketch two unimodal distributions for which even the best linear discriminant would give poor classification accuracy.

   (c) Consider two circular Gaussian distributions $p(\mathbf{x}|\omega_i) \sim N(\boldsymbol{\mu}_i, a_i \mathbf{I})$ and $P(\omega_i)$ for $i = 1, 2$ where $\mathbf{I}$ is the identity matrix and the other parameters can take on arbitrary values. Without performing any explicit calculations, explain whether the optimal decision boundary for this two-category problem must be a line. If not, sketch an example where the optimal discriminant is not a line.

2. Consider a linear machine with discriminant functions $g_i(\mathbf{x}) = \mathbf{w}^t \mathbf{x} + w_{i0}$, $i = 1, \ldots, c$. Show that the decision regions are convex by showing that if $\mathbf{x}_1 \in \mathcal{R}_i$ and $\mathbf{x}_2 \in \mathcal{R}_i$ then $\lambda \mathbf{x}_1 + (1 - \lambda)\mathbf{x}_2 \in \mathcal{R}_i$ for $0 \leq \lambda \leq 1$.

3. Figure 5.3 illustrates the two most popular methods for designing a c-category classifier from linear boundary segments. Another method is to save the full $\binom{c}{2}$ linear ω_i/ω_j boundaries, and classify any point by taking a *vote* based on all these boundaries. Prove whether the resulting decision regions must be convex. If they need not be convex, construct a nonpathological example yielding at least one nonconvex decision region.

4. Consider the hyperplane used in discrimination.

   (a) Show that the distance from the hyperplane $g(\mathbf{x}) = \mathbf{w}^t \mathbf{x} + w_0 = 0$ to the point $\mathbf{x}_a$ is $|g(\mathbf{x}_a)|/\|\mathbf{w}\|$ by minimizing $\|\mathbf{x} - \mathbf{x}_a\|^2$ subject to the constraint $g(\mathbf{x}) = 0$.

   (b) Show that the projection of $\mathbf{x}_a$ onto the hyperplane is given by

$$\mathbf{x}_p = \mathbf{x}_a - \frac{g(\mathbf{x}_a)}{\|\mathbf{w}\|^2} \mathbf{w}.$$

5. Consider the three-category linear machine with discriminant functions $g_i(\mathbf{x}) = \mathbf{w}_i^t \mathbf{x} + w_{i0}$, $i = 1, 2, 3$.

   (a) For the special case where $\mathbf{x}$ is two-dimensional and the threshold weights w_{i0} are zero, sketch the weight vectors with their tails at the origin, the three lines joining their heads, and the decision boundaries.

   (b) How does this sketch change when a constant vector $\mathbf{c}$ is added to each of the three weight vectors?

6. In the multicategory case, a set of samples is said to be linearly separable if there exists a linear machine that can classify them all correctly. If any samples labeled ω_i can be separated from all others by a single hyperplane, we shall say TOTAL LINEAR SEPARABILITY the samples are *totally linearly separable*. Show that totally linearly separable samples must be linearly separable, but that the converse need not be true. (For the converse, you may wish to consider a case in which a linear machine like the one in Problem 5 separates the samples.)

7. A set of samples is said to be *pairwise linearly separable* if there exist $c(c-1)/2$ hyperplanes H_{ij} such that H_{ij} separates samples labeled ω_i from samples ω_j. Show that a pairwise-linearly separable set of patterns may not be linearly separable.

8. Let $\{\mathbf{y}_1, \ldots, \mathbf{y}_n\}$ be a finite set of linearly separable training samples, and let $\mathbf{a}$ be called a solution vector if $\mathbf{a}^t \mathbf{y}_i \geq b$ for all i. Show that the minimum-length solution vector is unique. (You may wish to consider the effect of averaging two solution vectors.)

9. The *convex hull* of a set of vectors $\mathbf{x}_i$, $i = 1, \ldots, n$ is the set of all vectors of the form

$$\mathbf{x} = \sum_{i=1}^{n} \alpha_i \mathbf{x}_i,$$

where the coefficients α_i are nonnegative and sum to one. Given two sets of vectors, show that either they are linearly separable or their convex hulls intersect. (To answer this, suppose that both statements are true, and consider the classification of a point in the intersection of the convex hulls.)

10. A classifier is said to be a *piecewise linear machine* if its discriminant functions have the form

$$g_i(\mathbf{x}) = \max_{j=1,\ldots,n_i} g_{ij}(\mathbf{x}),$$

where

$$g_{ij}(\mathbf{x}) = \mathbf{w}_{ij}^t \mathbf{x} + w_{ij0}, \qquad \begin{aligned} i &= 1, \ldots, c \\ j &= 1, \ldots, n_i. \end{aligned}$$

(a) Indicate how a piecewise linear machine can be viewed in terms of a linear machine for classifying subclasses of patterns.

(b) Show that the decision regions of a piecewise linear machine can be non-convex and even multiply connected.

(c) Sketch a plot of $g_{ij}(x)$ for a one-dimensional example in which $n_1 = 2$ and $n_2 = 1$ to illustrate your answer to part (b).

11. Let the d components of $\mathbf{x}$ be either 0 or 1. Suppose we assign $\mathbf{x}$ to ω_1 if the number of nonzero components of $\mathbf{x}$ is odd, and to ω_2 otherwise. (This is called the *d-bit parity problem*.)

(a) Show that this dichotomy is not linearly separable if $d > 1$.

(b) Show that this problem can be solved by a piecewise linear machine with $d + 1$ weight vectors $\mathbf{w}_{ij}$ (see Problem 10). (To answer this, you may wish to consider vectors of the form $\mathbf{w}_{ij} = \alpha_{ij}(1, 1, \ldots, 1)^t$.)

Section 5.3

12. Consider the quadratic discriminant function (Eq. 4)

$$g(\mathbf{x}) = w_0 + \sum_{i=1}^{d} w_i x_i + \sum_{i=1}^{d} \sum_{j=1}^{d} w_{ij} x_i x_j,$$

and define the symmetric, nonsingular matrix $\mathbf{W} = [w_{ij}]$. Show that the basic character of the decision boundary can be described in terms of the scaled matrix $\overline{\mathbf{W}} = \mathbf{W}/(\mathbf{w}^t \mathbf{W}^{-1}\mathbf{w} - 4w_0)$ as follows:

(a) If $\overline{\mathbf{W}} \propto \mathbf{I}$ (the identity matrix), then the decision boundary is a hypersphere.

(b) If $\overline{\mathbf{W}}$ is positive definite, then the decision boundary is a hyperellipsoid.

(c) If some eigenvalues of $\overline{\mathbf{W}}$ are positive and some negative, then the decision boundary is a hyperhyperboloid.

(d) Suppose $\mathbf{w} = \begin{pmatrix} 5 \\ 2 \\ -3 \end{pmatrix}$ and $\mathbf{W} = \begin{pmatrix} 1 & 2 & 0 \\ 2 & 5 & 1 \\ 0 & 1 & -3 \end{pmatrix}$. What is the character of the solution?

(e) Repeat part (d) for $\mathbf{w} = \begin{pmatrix} 2 \\ -1 \\ 3 \end{pmatrix}$ and $\mathbf{W} = \begin{pmatrix} 1 & 2 & 3 \\ 2 & 0 & 4 \\ 3 & 4 & -5 \end{pmatrix}$.

Section 5.4

13. Derive Eq. 14, where $J(\cdot)$ depends on the iteration step k.

14. Consider the sum-of-squared-error criterion function (Eq. 43),

$$J_s(\mathbf{a}) = \sum_{i=1}^{n} (\mathbf{a}^t \mathbf{y}_i - b_i)^2.$$

Let $b_i = b$ and consider the following six training points:

$$\omega_1 : (1, 5)^t \qquad (2, 9)^t \qquad (-5, -3)^t$$
$$\omega_2 : (2, -3)^t \qquad (-1, -4)^t \qquad (0, 2)^t$$

(a) Calculate the Hessian matrix for this problem.

(b) Assuming the quadratic criterion function calculate the optimal learning rate η.

Section 5.5

15. In the convergence proof for the Perceptron algorithm (Theorem 5.1) the scale factor α was taken to be β^2/γ.

(a) Using the notation of Section 5.5, show that if α is greater than $\beta^2/(2\gamma)$, the maximum number of corrections is given by

$$k_0 = \frac{\|\mathbf{a}_1 - \alpha\mathbf{a}\|^2}{2\alpha\gamma - \beta^2}.$$

(b) If $\mathbf{a}_1 = \mathbf{0}$, what value of α minimizes k_0?

16. Modify the convergence proof given in Section 5.5.2 (Theorem 5.1) to prove the convergence of the following correction procedure: starting with an arbitrary initial weight vector $\mathbf{a}_1$, correct $\mathbf{a}(k)$ according to

$$\mathbf{a}(k + 1) = \mathbf{a}(k) + \eta(k)\mathbf{y}^k,$$

if and only if $\mathbf{a}^t(k)\mathbf{y}^k$ fails to exceed the margin b, where $\eta(k)$ is bounded by $0 < \eta_a \leq \eta(k) \leq \eta_b < \infty$. What happens if b is negative?

17. Let $\{\mathbf{y}_1, \ldots, \mathbf{y}_n\}$ be a finite set of linearly separable samples in d dimensions.

    (a) Suggest an exhaustive procedure that will find a separating vector in a finite number of steps. (You might wish to consider weight vectors whose components are integer-valued.)

    (b) What is the computational complexity of your procedure?

18. Consider the criterion function

$$J_q(\mathbf{a}) = \sum_{\mathbf{y} \in \mathcal{Y}(\mathbf{a})} (\mathbf{a}^t \mathbf{y} - b)^2$$

where $\mathcal{Y}(\mathbf{a})$ is the set of samples for which $\mathbf{a}^t \mathbf{y} \le b$. Suppose that $\mathbf{y}_1$ is the only sample in $\mathcal{Y}(\mathbf{a}(k))$. Show that $\nabla J_q(\mathbf{a}(k)) = 2(\mathbf{a}^t(k)\mathbf{y}_1 - b)\mathbf{y}_1$ and that the matrix of second partial derivatives is given by $\mathbf{D} = 2\mathbf{y}_1\mathbf{y}_1^t$. Use this to show that when the optimal $\eta(k)$ is used in Eq. 12 the gradient descent algorithm yields

$$\mathbf{a}(k + 1) = \mathbf{a}(k) + \frac{b - \mathbf{a}^t \mathbf{y}_1}{\|\mathbf{y}_1\|^2} \mathbf{y}_1.$$

19. Given the conditions in Eqs. 28–30, show that $\mathbf{a}(k)$ in the variable increment descent rule indeed converges for $\mathbf{a}^t \mathbf{y}_i > b$ for all i.

Section 5.6

20. Sketch a figure to illustrate the proof in Section 5.6.2. Be sure to take a general case, and label all variables.

Section 5.8

21. Show that the scale factor α in the MSE solution corresponding to Fisher's linear discriminant (Section 5.8.2) is given by

$$\alpha = \left[1 + \frac{n_1 n_2}{n} (\mathbf{m}_1 - \mathbf{m}_2)^t \mathbf{S}_W^{-1} (\mathbf{m}_1 - \mathbf{m}_2) \right]^{-1}$$

22. Generalize the results of Section 5.8.3 to show that the vector $\mathbf{a}$ that minimizes the criterion function

$$J_s'(\mathbf{a}) = \sum_{\mathbf{y} \in \mathcal{Y}_1} (\mathbf{a}^t \mathbf{y} - (\lambda_{21} - \lambda_{11}))^2 + \sum_{\mathbf{y} \in \mathcal{Y}_2} (\mathbf{a}^t \mathbf{y} - (\lambda_{12} - \lambda_{22}))^2$$

provides asymptotically a minimum-mean-squared-error approximation to the Bayes discriminant function $(\lambda_{21} - \lambda_{11})P(\omega_1|\mathbf{x}) - (\lambda_{12} - \lambda_{22})P(\omega_2|\mathbf{x})$.

23. Using the notation in the text, show that if $\mathcal{E}[\mathbf{y}\mathbf{y}^t]$ is nonsingular, and if the rate coefficients $\eta(k)$ for iteration index k satisfy Eqs. 68 and 69, then the weight vector $\mathbf{a}(k)$ convergest to $\hat{\mathbf{a}}$ in mean square, that is, $\lim_{k \to \infty} \mathcal{E}[\|\mathbf{a}(k) - \hat{\mathbf{a}}\|^2] = 0$, as stated in Eq. 70.

24. Consider the criterion function $J_m(\mathbf{a}) = \mathcal{E}[(\mathbf{a}^t \mathbf{y}(\mathbf{x}) - z)^2]$ and the Bayes discriminant function $g_0(\mathbf{x})$.

    (a) Show that

$$J_m = \mathcal{E}[(\mathbf{a}^t \mathbf{y} - g_0)^2] - 2\mathcal{E}[(\mathbf{a}^t \mathbf{y} - g_0)(z - g_0)] + \mathcal{E}[(z - g_0)^2].$$

(b) Use the fact that the conditional mean of z is $g_0(\mathbf{x})$ in showing that the $\hat{\mathbf{a}}$ that minimizes J_m also minimizes $\mathcal{E}[(\mathbf{a}^t\mathbf{y} - g_0)^2]$.

25. A scalar analog of the relation $\mathbf{R}_{k+1}^{-1} = \mathbf{R}_k^{-1} + \mathbf{y}_k\mathbf{y}_k^t$ used in stochastic approximation is $\eta^{-1}(k+1) = \eta^{-1}(k) + y_k^2$.

(a) Show that this has the closed-form solution

$$\eta(k) = \frac{\eta(1)}{1 + \eta(1)\sum_{i=1}^{k-1} y_i^2}.$$

(b) Assume that $\eta(1) > 0$ and $0 < a \leq y_i^2 \leq b < \infty$, and indicate why this sequence of coefficients will satisfy $\sum \eta(k) \to \infty$ and $\sum \eta(k)^2 \to L < \infty$.

26. Show for the Widrow-Hoff or LMS rule that if $\eta(k) = \eta(1)/k$, then the sequence of weight vectors converges to a limiting vector $\mathbf{a}$ satisfying $\mathbf{Y}^\dagger(\mathbf{Ya} - \mathbf{b}) = 0$ (Eq. 61).

Section 5.9

27. Consider the following six data points:

$$\omega_1:(1, 2)^t \quad (2, -4)^t \quad (-3, -1)^t$$
$$\omega_2:(2, 4)^t \quad (-1, -5)^t \quad (5, 0)^t$$

(a) Are they linearly separable?
(b) Using the approach in the text, assume $\mathbf{R} = \mathbf{I}$, the identity matrix, and calculate the optimal learning rate η by Eq. 95.

Section 5.10

28. The linear programming problem formulated in Section 5.10.2 involved minimizing a single artificial variable τ under the constraints $\mathbf{a}^t\mathbf{y}_i + \tau > b_i$ and $\tau \geq 0$. Show that the resulting weight vector minimizes the criterion function

$$J_\tau(\mathbf{a}) = \max_{\mathbf{a}^t\mathbf{y}_i \leq b_i} [b_i - \mathbf{a}^t\mathbf{y}_i].$$

Section 5.11

29. Discuss qualitatively why if samples from two categories are distinct (i.e., no feature point is labeled by both categories), there always exists a nonlinear mapping to a higher dimension that leaves the points linearly separable.

30. The figure in Example 2 shows the maximum margin for a Support Vector Machine applied to the exclusive-OR problem mapped to a five-dimensional space. That figure shows the training patterns and contours of the discriminant function, as projected in the two-dimensional subspace defined by the features $\sqrt{2}x_1$ and $\sqrt{2}x_1x_2$. Ignore the constant feature, and consider the other four features. For each of the $\binom{4}{2} - 1 = 5$ pairs of features other than the one shown in the Example, show the patterns and the lines corresponding to the discriminant $g = \pm 1$. Are the margins in your figures the same? Explain why or why not?

31. Write pseudocode to implement a simple learning method for support vector machines by modifying the Perceptron learning algorithm (Algorithm 4). Be

sure to write a detailed step, including mathematical form, for describing the current worst-classified training pattern. Explain why at the end of the training period, weight updates will involve only the support vectors.

32. Consider a Support Vector Machine and the following training data from two categories:

$$\omega_1:(1, 1)^t \quad (2, 2)^t \quad (2, 0)^t$$
$$\omega_2:(0, 0)^t \quad (1, 0)^t \quad (0, 1)^t$$

(a) Plot these six training points, and construct by inspection the weight vector for the optimal hyperplane, and the optimal margin.

(b) What are the support vectors?

(c) Construct the solution in the dual space by finding the Lagrange undetermined multipliers, α_i. Compare your result to that in part (a).

33. This problem asks you to follow the Kuhn-Tucker theorem to convert the constrained optimization problem in support vector machines into a dual, unconstrained one. For SVMs, the goal is to find the minimum-length weight vector $\mathbf{a}$ subject to the (classification) constraints

$$z_k \mathbf{a}^t \mathbf{y}_k \geq 1 \quad k = 1, \ldots, n,$$

where $z_k = \pm 1$ indicates the target getegory of each of the n patterns $\mathbf{y}_k$. Note that $\mathbf{a}$ and $\mathbf{y}$ are augmented (by a_0 and $y_0 = 1$, respectively).

(a) Consider the unconstrained optimization associated with SVMs:

$$L(\mathbf{a}, \boldsymbol{\alpha}) = \frac{1}{2}||\mathbf{a}||^2 - \sum_{k=1}^{n} \alpha_k[z_k \mathbf{a}^t \mathbf{y}_k - 1].$$

In the space determined by the components of $\mathbf{a}$, and the n (scalar) undetermined multipliers α_k, the desired solution is a *saddle point*, rather than a global maximum or minimum. Explain.

(b) Next eliminate the dependency of this ("primal") functional upon $\mathbf{a}$ (i.e., reformulated the optimization in a dual form) by the following steps. Note that at the saddle point of the primal functional, we have

$$\frac{\partial L(\mathbf{a}^*, \boldsymbol{\alpha}^*)}{\partial \mathbf{a}} = \mathbf{0}.$$

Solve for the partial derivatives and conclude

$$\sum_{k=1}^{n} \alpha_k^* z_k = 0 \quad \alpha_k^* \geq 0, \; k = 1, \ldots, n.$$

(c) Prove that at this inflection point, the optimal hyperplane is a linear combination of the training vectors:

$$\mathbf{a}^* = \sum_{k=1}^{n} \alpha_k^* z_k \mathbf{y}_k.$$

(d) According to the Kuhn-Tucker theorem (cf. Bibliography), an undetermined multiplier α_k^* is non-zero only if the corresponding sample $\mathbf{y}_k$ satisfies $z_k \mathbf{a}^t \mathbf{y}_k = 1$. Show that this can be expressed as

$$\alpha_k^*[z_k \mathbf{a}^{*t} \mathbf{y}_k - 1] = 0 \qquad k = 1, \ldots, n.$$

Recall that the samples where α_k^* are nonzero (i.e., $z_k \mathbf{a}^t \mathbf{y}_k = 1$), are the support vectors.

(e) Use the results from parts (b) and (c) to eliminate the weight vector in the functional, and thereby construct the dual functional

$$\tilde{L}(\mathbf{a}, \boldsymbol{\alpha}) = \frac{1}{2}||\mathbf{a}||^2 - \sum_{k=1}^{n} \alpha_k z_k \mathbf{a}^t \mathbf{y}_k + \sum_{k=1}^{n} \alpha_k.$$

(f) Substitute the solution $\mathbf{a}^*$ from part (c) to find the dual functional

$$\tilde{L}(\boldsymbol{\alpha}) = -\frac{1}{2} \sum_{j,k=1}^{n} \alpha_j \alpha_k z_j z_k (\mathbf{y}_j^t \mathbf{y}_k) + \sum_{j=1}^{n} \alpha_j.$$

34. Repeat Example 2 using the same $\varphi(\cdot)$ but with the following four points:

$$\begin{aligned}
\omega_1 &: (1, 5)^t \quad (-2, -4)^t \\
\omega_2 &: (2, 3)^t \quad (-1, 5)^t.
\end{aligned}$$

Section 5.12

35. Suppose that for each two-dimensional training point $\mathbf{y}_i$ in category ω_1 there is a corresponding (symmetric) point in ω_2 at $-\mathbf{y}_i$.

(a) Prove that a separating hyperplane (should one exist) or LMS solution must go through the origin.

(b) Consider such a symmetric, six-point problem with the following points:

$$\begin{aligned}
\omega_1 &: \quad (1, 2)^t \quad \quad (2, -4)^t \quad \quad \mathbf{y} \\
\omega_2 &: (-1, -2)^t \quad (-2, 4)^t \quad \quad -\mathbf{y}
\end{aligned}$$

Find the mathematical conditions on $\mathbf{y}$ such that the LMS solution for this problem *not* give a separating hyperplane.

(c) Generalize your result as follows. Suppose ω_1 consists of $\mathbf{y}_1$ and $\mathbf{y}_2$ (known) and the symmetric versions in ω_2. What is the condition on $\mathbf{y}_3$ such that the LMS solution does not separate the points.

36. Write pseudocode for a fixed increment multicategory algorithm based on Eq. 115. Discuss the strengths and weakness of such an implementation.

37. Generalize the discussion in Section 5.8.3 in order to prove that the solution derived from Eq. 120 provides a minimum-mean-square-error approximation to the Bayes discriminant function given in Eq. 122.

38. Use Kesler's construction to establish equivalences between multicategory and two-category procedures. Generalize the Perceptron and relaxation procedures to the multicategory case.

COMPUTER EXERCISES

Several of the exercises use the data in the following table.

sample	ω_1 x_1	x_2	ω_2 x_1	x_2	ω_3 x_1	x_2	ω_4 x_1	x_2
1	0.1	1.1	7.1	4.2	−3.0	−2.9	−2.0	−8.4
2	6.8	7.1	−1.4	−4.3	0.5	8.7	−8.9	0.2
3	−3.5	−4.1	4.5	0.0	2.9	2.1	−4.2	−7.7
4	2.0	2.7	6.3	1.6	−0.1	5.2	−8.5	−3.2
5	4.1	2.8	4.2	1.9	−4.0	2.2	−6.7	−4.0
6	3.1	5.0	1.4	−3.2	−1.3	3.7	−0.5	−9.2
7	−0.8	−1.3	2.4	−4.0	−3.4	6.2	−5.3	−6.7
8	0.9	1.2	2.5	−6.1	−4.1	3.4	−8.7	−6.4
9	5.0	6.4	8.4	3.7	−5.1	1.6	−7.1	−9.7
10	3.9	4.0	4.1	−2.2	1.9	5.1	−8.0	−6.3

Section 5.4

1. Consider basic gradient descent (Algorithm 1) and the Perceptron criterion (Eq. 16) applied to the data in the tables.

   (a) Apply both to the two-dimensional data in order to discriminate categories ω_1 and ω_3. For the gradient descent, use $\eta(k) = 0.01$. Plot the criterion function as function of the iteration number.

   (b) Estimate the total number of mathematical operations in the two algorithms.

   (c) Plot the convergence time versus learning rate. What is the minimum learning rate that fails to lead to convergence?

2. Write a program to implement the batch Perceptron algorithm (Algorithm 3).

   (a) Starting with $\mathbf{a} = \mathbf{0}$, apply your program to the training data from ω_1 and ω_2. Note the number of iterations required for convergence.

   (b) Apply your program to ω_3 and ω_2. Again, note the number of iterations required for convergence.

   (c) Explain the difference between the iterations required in the two cases.

POCKET
ALGORITHM

3. The *Pocket algorithm* uses the criterion of longest sequence of correctly classified points, and can be used in conjunction with a number of basic learning algorithms. For instance, the Pocket version of the Perceptron algorithm works in a sort of ratchet scheme as follows. There are two sets of weights, one for the normal Perceptron algorithm, and a separate one (not directly used for training) which is kept "in your pocket." Both are randomly chosen at the start. The "pocket" weights are tested on the full data set to find the longest run of patterns properly classified. (At the beginning, this run will be short.) The Perceptron weights are trained as usual, but after every weight update (or after some finite number of such weight updates), the Perceptron weight is tested on data points, randomly selected, to determine the longest run of properly classified points.

If this length is greater than with the pocket weights, the Perceptron weights replace the pocket weights, and Perceptron training continues. In this way, the pocket weights continually improve, classifying longer and longer runs of randomly selected points.

(a) Write a pocket version of the single-sample Perceptron algorithm (Algorithm 4).

(b) Apply it to the data from ω_1 and ω_3. How often are the pocket weights updated?

4. Explore through simulation, several equations governing the rate of convergence of the Perceptron algorithm.

(a) Generate 25 three-dimensional points for each of two categories according to Gaussian distributions, $p(\mathbf{x}|\omega_i) \sim N(\boldsymbol{\mu}_i, \mathbf{I})$, and $\boldsymbol{\mu}_1 = \mathbf{0}$, $\boldsymbol{\mu}_2 = (4, 4, 4)^t$.

(b) Randomly choose an initial weight vector $\mathbf{a}$ with the constraint $\|\mathbf{a}\| = 0.1$ (i.e., $\mathbf{a}$ lies on a three-dimensional sphere of radius 0.1). Implement the single-sample Perceptron algorithm (Algorithm 4) using your initial weight vector and the data from part (a).

(c) Calculate β^2 according to Eq. 21.

(d) After training is complete, calculate γ according to Eq. 22 and verify that the value of k_0 from Eq. 25 is consistent with your simulation results.

5. Show that the first five points of categories ω_3 and ω_4 are not linearly separable. Construct by hand a nonlinear mapping of the feature space to make them linearly separable and train a Perceptron classifier on them. What is the classification error on the remaining (transformed) points in the table?

6. Consider a version of the Balanced Winnow training algorithm (Algorithm 7). Classification of test data is given by line 2. Compare the converge rate of Balanced Winnow with the fixed-increment, single-sample Perceptron (Algorithm 4) on a problem with large number of redundant features, as follows.

(a) Generate a training set of 2000 100-dimensional patterns (1000 from each of two categories) in which only the first ten features are informative, in the following way. For patterns in category ω_1, each of the first ten features are chosen randomly and uniformly from the range $+1 \le x_i \le 2$, for $i = 1, \ldots, 10$. Conversely, for patterns in ω_2, each of the first ten features are chosen randomly and uniformly from the range $-2 \le x_i \le -1$. All other features from both categories are chosen from the range $-2 \le x_i \le +2$.

(b) Construct by hand the obvious separating hyperplane.

(c) Adjust the learning rates so that your two algorithms have roughly the same convergence rate on the full training set when only the first ten features are considered. That is, assume each of the 2000 training patterns consists of just the first ten features.

(d) Now apply your two algorithms to 2000 50-dimensional patterns, in which the first ten features are informative and the remaining 40 are not. Plot the total number of errors versus iteration.

(e) Now apply your two algorithms to the full training set of 2000 100-dimensional patterns.

(f) Summarize your answers to parts (c)–(e).

Section 5.6

7. Consider relaxation methods as described in the text.

(a) Implement batch relaxation with margin (Algorithm 8), set $b = 0.1$ and $\mathbf{a}(1) = \mathbf{0}$ and apply it to the data in ω_1 and ω_3. Plot the criterion function as a function of the number of passes through the training set.

(b) Repeat for $b = 0.5$ and $\mathbf{a}(1) = \mathbf{0}$. Explain qualitatively any difference you find in the convergence rates.

(c) Modify your program to use single sample learning. Again, plot the criterion as a function of the number of passes through the training set.

(d) Discuss any differences, being sure to consider the learning rate.

Section 5.8

8. Write a single-sample relaxation algorithm using Eq. 72 to update the matrix $\mathbf{R}$, as described in the text. Apply your program to the ω_2 and ω_3 data in the table above.

Section 5.9

9. Implement the Ho-Kashyap algorithm (Algorithm 11) and apply to the data in categories ω_1 and ω_3. Repeat for categories ω_4 and ω_2.

Section 5.10

10. Write a program to implement the LMS algorithm (Algorithm 10) for two categories and two-dimensional data. Suppose ω_1 consists of 21 points: 10 copies of $(0, 0)^t$, 10 copies of $(0, 1)^t$ and a single point at $(1, -2)^t$. The other category, ω_2 consists of 10 copies of $(1, 0)^t$, 10 copies of $(1, 1)^t$ and a single point at $(0, 3)^t$. Find the LMS solution for the weight vector and the equation of the corresponding hyperplane. Does it separate the classes? Are the classes linearly separable? Explain.

Section 5.11

11. Write a program to implement the Support Vector Machine algorithm. Train an SVM classifer with data from ω_3 and ω_4 in the following way. Preprocess each training pattern to form a new vector having components $1, x_1, x_2, x_1^2, x_1 x_2,$ and x_2^2.

(a) Train your classifier with just the first patterns in ω_3 and ω_4 and find the separating hyperplane and the margin.

(b) Repeat part (a) using the first two points in the two categories (four points total). What is the equation of the separating hyperplane, the margin, and the support vectors?

(c) Repeat part (b) with the first three points in each category (six points total), the first four points, and so on, until the transformed patterns cannot be linearly separated in the transformed space.

Section 5.12

12. Write a program based on Kesler's construction to implement a multicategory generalization of basic LMS algorithm (Algorithm 10).

(a) Apply it to the data in all four categories in the table.

(b) Use your algorithm in a two-category mode to form ω_i /not ω_i boundaries for $i = 1, 2, 3, 4$. Find any regions whose categorization by your system is ambiguous. That is, give representative patterns **x** where your classifier cannot return a category label unambiguously.

BIBLIOGRAPHY

[1] Henry D. Block and Simon A. Levin. On the bounded-ness of an iterative procedure for solving a system of linear inequalities. *Proceedings of the American Mathematical Society*, 26:229–235, 1970.

[2] Bernhard E. Boser, Isabelle Guyon, and Vladimir Vapnik. A training algorithm for optimal margin classifiers. In David Haussler, editor, *Proceedings of the 4th Workshop on Computational Learning Theory*, pages 144–152, ACM Press, San Mateo, CA, 1992.

[3] Hervé Bourlard and Yves Kamp. Auto-association by multilayer perceptrons and singular value decomposition. *Biological Cybernetics*, 59:291–294, 1988.

[4] Vladimir Cherkassky and Filip Mulier. *Learning from Data: Concepts, Theory, and Methods*. Wiley, New York, 1998.

[5] Ronald A. Fisher. The use of multiple measurements in taxonomic problems. *Annals of Eugenics*, 7 Part II:179–188, 1936.

[6] David Gale, Harold W. Kuhn, and Albert W. Tucker. Linear programming and the theory of games. In Tjalling C. Koopmans, editor, *Activity Analysis of Production and Allocation*, pages 317–329. Wiley, New York, 1951.

[7] Adam J. Grove, Nicholas Littlestone, and Dale Schuurmans. General convergence results for linear discriminant updates. In *Proceedings of the COLT 97*, pages 171–183. ACM Press, 1997.

[8] Isabelle Guyon and David G. Stork. Linear discriminant and support vector classifiers. In Alex Smola, Peter Bartlett, Bernhard Schölkopf, and Dale Schuurmans, editors, *Advances in large margin classifiers*. MIT Press, Cambridge, MA, 1999.

[9] Wilbur H. Highleyman. Linear decision functions, with application to pattern recognition. *Proceedings of the IRE*, 50:1501–1514, 1962.

[10] Nicholas Littlestone. Learning quickly when irrelevant attributes abound: A new linear-threshold algorithm. *Machine Learning*, 2(4):285–318, 1988.

[11] Nicholas Littlestone. Redundant noisy attributes, attribute errors, and linear-threshold learning using Win-now. In Manfred K. Warmuth and Leslie G. Valiant, editors, *Proceedings of COLT 91*, pages 147–156, Morgan Kaufmann, San Mateo, CA, 1991.

[12] Warren S. McCulloch and Walter Pitts. A logical calculus of ideas imminent in nervous activity. *Bulletin of Mathematical Biophysics*, 5:115–133, 1943.

[13] Marvin L. Minsky and Seymour A. Papert. *Perceptrons: An Introduction to Computational Geometry*. MIT Press, Cambridge, MA, 1969.

[14] Bernhard Schölkopf, Christopher J. C. Burges, and Alexander J. Smola, editors. *Advances in Kernel Methods: Support Vector Learning*. MIT Press, Cambridge, MA, 1999.

[15] Bernhard Schölkopf, Christopher J. C. Burges, and Vladimir Vapnik. Extracting support data for a given task. In Usama M. Fayyad and Ramasamy Uthurasamy, editors, *Proceedings of the First International Conference on Knowledge Discovery and Data Mining*, pages 252–257. AAAI Press, Menlo Park, CA, 1995.

[16] Fred W. Smith. Design of multicategory pattern classifiers with two-category classifier design procedures. *IEEE Transactions on Computers*, C-18(6):548–551, 1969.

[17] Gilbert Strang. *Introduction to Applied Mathematics*. Wellesley-Cambridge Press, Wellesley, MA, 1986.

[18] Vladimir Vapnik. *The Nature of Statistical Learning Theory*. Springer-Verlag, New York, 1995.

[19] Šarūnas Raudys. Evolution and generalization of a single neurone: I. Single-layer perceptron as seven statistical classifiers. *Neural Networks*, 11(2):283–296, 1998.

[20] Šarūnas Raudys. Evolution and generalization of a single neurone: II. Complexity of statistical classifiers and sample size considerations. *Neural Networks*, 11(2):297–313, 1998.

[21] Stephen S. Yau and John M. Schumpert. Design of pattern classifiers with the updating property using stochastic approximation techniques. *IEEE Transactions on Computers*, C-17(9):861–872, 1968.

MULTILAYER NEURAL NETWORKS

6.1 INTRODUCTION

We saw in Chapter 5 a number of methods for training classifiers consisting of input units connected by modifiable weights to output units. The LMS algorithm, in particular, provided a powerful gradient descent method for reducing the error, even when the patterns are not linearly separable. Unfortunately, the class of solutions that can be obtained from such networks—comprising hyperplane decision boundaries—while surprisingly good on a range of real-world problems, is simply not general enough in demanding applications: there are many problems for which linear discriminants are insufficient for minimum error.

With a clever choice of nonlinear φ functions, however, we can obtain arbitrary decision regions, in particular those leading to minimum error. The central difficulty is, naturally, choosing the appropriate nonlinear functions. One brute force approach might be to choose a complete basis set such as all polynomials, but this will not work; such a classifier would have too many free parameters to be determined from a limited number of training patterns (Chapter 9). Alternatively, we may have prior knowledge relevant to the classification problem and this might guide our choice of nonlinearity. However, we have seen no principled or automatic method for finding the nonlinearities in the absence of such information. What we seek, then, is a way to *learn* the nonlinearity at the same time as the linear discriminant. This is the approach of multilayer neural networks or multilayer Perceptrons: The parameters governing the nonlinear mapping are learned at the same time as those governing the linear discriminant.

We shall revisit the limitations of the two-layer networks of the previous chapter,* and see how three- and four-layer nets overcome those drawbacks—indeed

---

*Some authors describe such networks as *single* layer networks because they have only one layer of modifiable weights, but we shall instead refer to them based on the number of layers of *units*.

how such multilayer networks can, at least in principle, provide the optimal solution to an arbitrary classification problem. There is nothing particularly magical about multilayer neural networks; they implement *linear* discriminants, but in a space where the inputs have been mapped nonlinearly. The key power provided by such networks is that they admit fairly simple algorithms where the form of the nonlinearity can be learned from training data. The models are thus extremely powerful, have nice theoretical properties, and apply well to a vast array of real-world applications.

BACK-
PROPAGATION

One of the most popular methods for training such multilayer networks is based on gradient descent in error—the *backpropagation algorithm* or generalized delta rule, a natural extension of the LMS algorithm. We shall study backpropagation in depth, first of all because it is powerful, useful, and relatively easy to understand, but also because many other training methods can be seen as modifications of it. The backpropagation training method is simple even for complex models having hundreds or thousands of parameters. In part because of the intuitive graphical representation and the simplicity of design of these models, designers can test different models quickly and easily; neural networks are thus a flexible heuristic technique for doing statistical pattern recognition with complicated models. The conceptual and algorithmic simplicity of backpropagation, along with its manifest success on many real-world problems, make it a mainstay in adaptive pattern recognition.

While the basic theory of backpropagation is simple, a number of heuristics—some a bit subtle—are often used to improve performance and increase training speed. Guided by an analysis of networks and their function we can make informed choices of the scaling of input values and initial weights, desired output values, and more. We shall also discuss alternative training schemes, for instance ones that are faster, or adjust their complexity automatically in response to training data.

Network architecture or topology plays an important role for neural net classification, and the optimal topology will depend upon the problem at hand. It is here that another great benefit of networks becomes apparent: often knowledge of the problem domain which might be of an informal or heuristic nature can be easily incorporated into network architectures through choices in the number of hidden layers, units, feedback connections, and so on. Thus setting the topology of the network is heuristic model selection. The practical ease in selecting models by setting the network topology and estimating parameters via backpropagation enables classifier designers to try out alternative models fairly simply.

REGULAR-
IZATION

A deep problem in the use of neural network techniques involves regularization, that is, selecting or adjusting the complexity of the network. Whereas the number of inputs and outputs is given by the feature space and number of categories, the total number of weights or parameters in the network is not—or at least not directly. If too many free parameters are used, generalization will be poor; conversely if too few parameters are used, the training data cannot be learned adequately. How shall we adjust the complexity to achieve the best generalization? We shall explore a number of methods for complexity adjustment, and we will return in Chapter 9 to consider their theoretical foundations.

It is crucial to remember that neural networks do not exempt designers from intimate and detailed knowledge of the data and problem domain. Networks provide a powerful and speedy tool for building classifiers, and as with any tool or technique, one gains intuition and expertise through analysis and repeated experimentation over a broad range of problems.

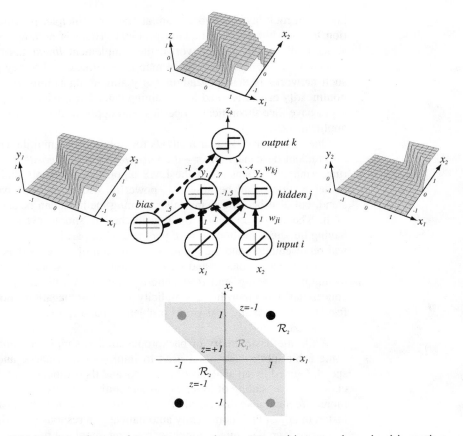

FIGURE 6.1. The two-bit parity or exclusive-OR problem can be solved by a three-layer network. At the bottom is the two-dimensional feature $x_1 x_2$-space, along with the four patterns to be classified. The three-layer network is shown in the middle. The input units are linear and merely distribute their feature values through multiplicative weights to the hidden units. The hidden and output units here are linear threshold units, each of which forms the linear sum of its inputs times their associated weight to yield *net*, and emits a +1 if this *net* is greater than or equal to 0, and −1 otherwise, as shown by the graphs. Positive or "excitatory" weights are denoted by solid lines, negative or "inhibitory" weights by dashed lines; each weight magnitude is indicated by the line's thickness, and is labeled. The single output unit sums the weighted signals from the hidden units and bias to form its *net*, and emits a +1 if its *net* is greater than or equal to 0 and emits a −1 otherwise. Within each unit we show a graph of its input-output or activation function—$f(net)$ versus *net*. This function is linear for the input units, a constant for the bias, and a step or sign function elsewhere. We say that this network has a 2-2-1 fully connected topology, describing the number of units (other than the bias) in successive layers.

6.2 FEEDFORWARD OPERATION AND CLASSIFICATION

HIDDEN LAYER

Figure 6.1 shows a simple three-layer neural network. This one consists of an input layer, a *hidden layer*,* and an output layer, interconnected by modifiable weights,

---

*We call any units that are neither input nor output units "hidden" because their activations are not directly "seen" by the external environment—that is, the input or output.

BIAS UNIT

NEURON

represented by links between layers. There is, furthermore, a single *bias unit* that is connected to each unit other than the input units. Cleary, such a network is an extension of the two-layer networks we studied in Chapter 5. The function of units is loosely based on properties of biological neurons, and hence they are sometimes called "neurons." We are interested in the use of such networks for pattern recognition, where the input units represent the components of a feature vector and where signals emitted by output units will be the values of the discriminant functions used for classification.

We can clarify our notation and describe the feedforward operation of such a network on what is perhaps the simplest nonlinear problem: the exclusive-OR (XOR) problem (Fig. 6.1); a three-layer network can indeed solve this problem whereas a linear machine operating directly on the features cannot.

NET ACTIVATION

Each two-dimensional input vector is presented to the input layer, and the output of each input unit equals the corresponding component in the vector. Each hidden unit computes the weighted sum of its inputs to form its scalar *net activation* which we denote simply as *net*. That is, the net activation is the inner product of the inputs with the weights at the hidden unit. As in Chapter 5, for simplicity we augment both the input vector, by appending a feature value $x_0 = 1$, as well as the weight vector, by appending a value w_0, and we can then write

$$net_j = \sum_{i=1}^{d} x_i w_{ji} + w_{j0} = \sum_{i=0}^{d} x_i w_{ji} \equiv \mathbf{w}_j^t \mathbf{x}, \tag{1}$$

where the subscript i indexes units in the input layer, j in the hidden; w_{ji} denotes the input-to-hidden layer weights at the hidden unit j. In analogy with neurobiology, such weights or connections are sometimes called "synapses" and the values of the connections the "synaptic weights." Each hidden unit emits an output that is a nonlinear function of its activation, $f(net)$, that is,

SYNAPSE

$$y_j = f(net_j). \tag{2}$$

Figure 6.1 shows a simple threshold or *sign* (read "signum") function,

$$f(net) = \text{Sgn}(net) \equiv \begin{cases} 1 & \text{if } net \geq 0 \\ -1 & \text{if } net < 0, \end{cases} \tag{3}$$

ACTIVATION FUNCTION

but as we shall see, other functions have more desirable properties and are hence more commonly used. This $f(\cdot)$ is sometimes called the *activation function* or merely "nonlinearity" of a unit, and it serves as a φ function discussed in Chapter 5. We have assumed the *same* nonlinearity is used at the various hidden and output units, though this is not crucial.

Each output unit similarly computes its net activation based on the hidden unit signals as

$$net_k = \sum_{j=1}^{n_H} y_j w_{kj} + w_{k0} = \sum_{j=0}^{n_H} y_j w_{kj} = \mathbf{w}_k^t \mathbf{y}, \tag{4}$$

where the subscript k indexes units in the output layer and n_H denotes the number of hidden units. We have mathematically treated the bias unit as equivalent to one of the hidden units whose output is always $y_0 = 1$. In this example, there is only one output unit. However, anticipating a more general case, we shall refer to its output as

z_k. An output unit computes the nonlinear function of its *net*, emitting

$$z_k = f(net_k), \tag{5}$$

where in the figure we assume that this nonlinearity is also a sign function. Clearly, the output z_k can also be thought of as a function of the input feature vector **x**. When there are c output units, we can think of the network as computing c discriminant functions $z_k = g_k(\mathbf{x})$, and can classify the input according to which discriminant function is largest. In a two-category case, it is traditional to use a single output unit and label a pattern by the sign of the output z.

It is easy to verify that the three-layer network with the weight values listed indeed solves the XOR problem. The hidden unit computing y_1 acts like a two-layer Perceptron, and it computes the boundary $x_1 + x_2 + 0.5 = 0$; input vectors for which $x_1 + x_2 + 0.5 \geq 0$ lead to $y_1 = 1$, and all other inputs lead to $y_1 = -1$. Likewise the other hidden unit computes the boundary $x_1 + x_2 - 1.5 = 0$. The final output unit emits $z_1 = +1$ if and only if $y_1 = +1$ and $y_2 = +1$. Using the terminology of computer logic, the units are behaving like gates, where the first hidden unit is an OR gate, the second hidden unit is an AND gate, and the output unit implements

$$z_k = y_1 \text{ AND NOT } y_2 = (x_1 \text{ OR } x_2) \text{ AND NOT } (x_1 \text{ AND } x_2) \tag{6}$$

$$= x_1 \text{ XOR } x_2,$$

giving rise to the appropriate nonlinear decision region shown in the figure—the XOR problem is solved.

6.2.1 General Feedforward Operation

From the above example, it should be clear that nonlinear multilayer networks—that is, ones with input units, hidden units, and output units—have greater computational or *expressive power* than similar networks that otherwise lack hidden units. That is, they can implement more functions. Indeed, we shall see in Section 6.2.2 that given sufficient number of hidden units of a general type *any* function can be so represented.

EXPRESSIVE
POWER

Clearly, we can generalize the above discussion to more inputs, other nonlinearities, and arbitrary number of output units. For classification, we will have c output units, one for each of the categories, and the signal from each output unit is the discriminant function $g_k(\mathbf{x})$. We gather the results from Eqs. 1, 2, 4, and 5, to express such discriminant functions as

$$g_k(\mathbf{x}) \equiv z_k = f\left(\sum_{j=1}^{n_H} w_{kj} \, f\left(\sum_{i=1}^{d} w_{ji} x_i + w_{j0}\right) + w_{k0}\right). \tag{7}$$

This, then, is the class of functions that can be implemented by a three-layer neural network. In general, as we elaborate in Section 6.8.1, the activation function does not have to be a sign function. Indeed, we shall often require the activation functions to be continuous and differentiable. We can even allow the activation functions in the output layer to be different from the activation functions in the hidden layer, or indeed have different activation functions for each individual unit. Although we will have cause to use such networks later, to simplify the mathematical analysis and reveal the essential concepts, we will temporarily assume that the activation functions are all the same.

6.2.2 Expressive Power of Multilayer Networks

It is natural to ask if *every* decision can be implemented by such a three-layer network as described by Eq. 7. The answer, due ultimately to Kolmogorov but refined by others, is "yes"—any continuous function from input to output can be implemented in a three-layer net, given sufficient number of hidden units n_H, proper nonlinearities, and weights. In particular, any posterior probabilities can be represented by a three-layer net. Just as we did in Chapter 2, for this c-category classification case we can merely apply a max[·] function to the set of network outputs, and thereby obtain any decision boundary.

Specifically, Kolmogorov proved that any continuous function $g(\mathbf{x})$ defined on the unit hypercube I^n ($I = [0, 1]$ and $n \geq 2$) can be represented in the form

$$g(\mathbf{x}) = \sum_{j=1}^{2n+1} \Xi_j \left(\sum_{i=1}^{d} \psi_{ij}(x_i) \right) \tag{8}$$

for properly chosen functions Ξ_j and ψ_{ij}. In any particular probem, we can always scale the input region of interest to lie in a hypercube, and thus this condition on the feature space is not limiting. Equation 8 can be expressed in neural network terminology as follows: Each of $2n + 1$ hidden units takes as input a sum of d nonlinear functions, one for each input feature x_i. Each hidden unit emits a nonlinear function Ξ of its total input; the output unit merely emits the sum of the contributions of the hidden units.

Unfortunately, the relationship of Kolmogorov's theorem to practical neural networks is a bit tenuous, for several reasons. In particular, the functions Ξ_j and ψ_{ij} are not the simple weighted sums passed through nonlinearities favored in neural networks. In fact those functions can be extremely complex; they are not smooth, and indeed for subtle mathematical reasons they cannot be smooth. As we shall soon see, smoothness is important for gradient descent learning. Most importantly, Kolmogorov's theorem tells us very little about how to find the nonlinear functions based on data—the central problem in network-based pattern recognition.

A more intuitive proof of the universal expressive power of three-layer nets is inspired by Fourier's theorem that any continuous function $g(\mathbf{x})$ can be approximated arbitrarily closely by a possibly infinite sum of harmonic functions (Problem 2). One can imagine a network whose hidden units implement such harmonic functions. Proper hidden-to-output weights related to the coefficients in a Fourier synthesis would then enable the full network to implement the desired function. Informally speaking, we need not build up harmonic functions for Fourier-like synthesis of a desired function. Instead a sufficiently large number of "bumps" at different input locations, of different amplitude and sign, can be put together to give our desired function. Such localized bumps might be implemented in a number of ways, for instance by S-shaped activation functions grouped appropriately (Fig. 6.2). The Fourier analogy and bump constructions are conceptual tools, and they do not explain the way networks in fact function. In short, this is not how neural networks "work": We never find that through training (Section 6.3) simple networks build a Fourier-like representation, nor do they learn to group S-shaped functions to get component bumps. However, these analogies help to explain the reasons for the expressive power of multilayer networks.

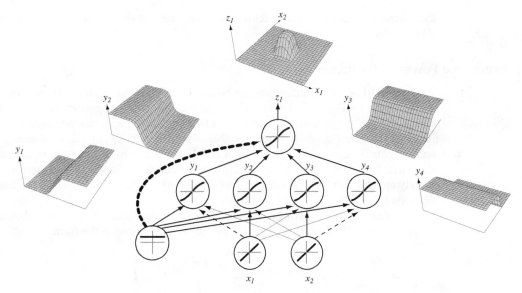

FIGURE 6.2. A 2-4-1 network (with bias) along with the response functions at different units; each hidden output unit has sigmoidal activation function $f(\cdot)$. In the case shown, the hidden unit outputs are paired in opposition thereby producing a "bump" at the output unit. Given a sufficiently large number of hidden units, any continuous function from input to output can be approximated arbitrarily well by such a network.

These latter constructions, showing that any desired function can be implemented by a three-layer network, are of greater theoretical than practical interest because the constructions give neither the number of hidden units required nor the proper weight values. Even if there *were* a constructive proof, it would be of little use in pattern recognition because we do not know the desired function anyway—it is related to the training patterns in a very complicated way. All in all, then, these results on the expressive power of networks give us confidence we are on the right track, but shed little practical light on the problems of designing and *training* neural networks—their main benefit for pattern recognition (Fig. 6.3).

6.3 BACKPROPAGATION ALGORITHM

We have just seen that any function from input to output can be implemented as a three-layer neural network. We now turn to the crucial problem of setting the weights based on training patterns and the desired output.

Backpropagation is one of the simplest and most general methods for supervised training of multilayer neural networks—it is the natural extension of the LMS algorithm for linear systems we saw in Chapter 5. Other methods may be faster or have other desirable properties, but few are more instructive. The LMS algorithm worked for two-layer systems because the error, proportional to the square of the difference between the actual output and the desired output, could be evaluated for each output unit. Similarly, in a three-layer net it is a straightforward matter to find how the output, and thus the error, depends on the hidden-to-output layer weights. In fact, this dependency is the same as in the analogous two-layer case, and thus the learning rule is the same.

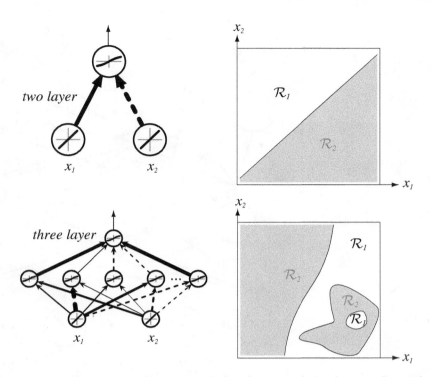

FIGURE 6.3. Whereas a two-layer network classifier can only implement a linear decision boundary, given an adequate number of hidden units, three-, four- and higher-layer networks can implement arbitrary decision boundaries. The decision regions need not be convex or simply connected.

But how should the input-to-hidden weights be learned, the ones governing the nonlinear transformation of the input vectors? If the "proper" outputs for a hidden unit were known for any pattern, the input-to-hidden weights could be adjusted to approximate it. However, there is no explicit teacher to state what the hidden unit's output should be. This is called the *credit assignment* problem. The power of backpropagation is that it allows us to calculate an effective error for each hidden unit, and thus derive a learning rule for the input-to-hidden weights.

CREDIT ASSIGNMENT

Networks have two primary modes of operation: feedforward and learning. Feedforward operation, such as illustrated in our XOR example above, consists of presenting a pattern to the input units and passing the signals through the network in order to yield outputs from the output units. Supervised learning consists of presenting an input pattern and changing the network parameters to bring the actual outputs closer to the desired teaching or *target* values. Figure 6.4 shows a three-layer network and the notation we shall use.

TARGET PATTERN

6.3.1 Network Learning

The basic approach in learning is to start with an untrained network, present a training pattern to the input layer, pass the signals through the net and determine the output at the output layer. Here these outputs are compared to the target values; any difference corresponds to an error. This error or criterion function is some scalar function of the weights and is minimized when the network outputs match the de-

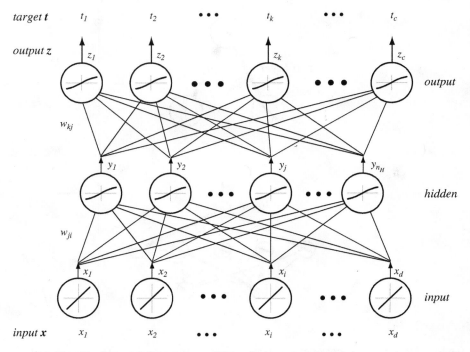

FIGURE 6.4. A d-n_H-c fully connected three-layer network and the notation we shall use. During feedforward operation, a d-dimensional input pattern $\mathbf{x}$ is presented to the input layer; each input unit then emits its corresponding component x_i. Each of the n_H hidden units computes its net activation, net_j, as the inner product of the input layer signals with weights w_{ji} at the hidden unit. The hidden unit emits $y_j = f(net_j)$, where $f(\cdot)$ is the nonlinear activation function, shown here as a sigmoid. Each of the c output units functions in the same manner as the hidden units do, computing net_k as the inner product of the hidden unit signals and weights at the output unit. The final signals emitted by the network, $z_k = f(net_k)$, are used as discriminant functions for classification. During network training, these output signals are compared with a teaching or target vector $\mathbf{t}$, and any difference is used in training the weights throughout the network.

sired outputs. Thus the weights are adjusted to reduce this measure of error. Here we present the learning rule on a per pattern basis, and we return to other protocols later.

TRAINING ERROR   We consider the *training error* on a pattern to be the sum over output units of the squared difference between the desired output t_k given by a teacher and the actual output z_k, much as we had in the LMS algorithm for two-layer nets:

$$J(\mathbf{w}) \equiv \frac{1}{2} \sum_{k=1}^{c} (t_k - z_k)^2 = \frac{1}{2} \|\mathbf{t} - \mathbf{z}\|^2, \tag{9}$$

where $\mathbf{t}$ and $\mathbf{z}$ are the target and the network output vectors of length c and $\mathbf{w}$ represents all the weights in the network.

The backpropagation learning rule is based on gradient descent. The weights are initialized with random values, and then they are changed in a direction that will reduce the error:

$$\Delta \mathbf{w} = -\eta \frac{\partial J}{\partial \mathbf{w}}, \tag{10}$$

or in component form

$$\Delta w_{pq} = -\eta \frac{\partial J}{\partial w_{pq}}, \tag{11}$$

LEARNING RATE

where η is the *learning rate*, and merely indicates the relative size of the change in weights. The power of Eqs. 10 and 11 is in their simplicity: They merely demand that we take a step in weight space that lowers the criterion function. It is clear from Eq. 9 that the criterion function can never be negative; furthermore, the learning rule guarantees that learning will stop, except in pathological cases. This iterative algorithm requires taking a weight vector at iteration m and updating it as

$$\mathbf{w}(m+1) = \mathbf{w}(m) + \Delta \mathbf{w}(m), \tag{12}$$

where m indexes the particular pattern presentation.

We now turn to the problem of evaluating Eq. 11 for a three-layer net. Consider first the hidden-to-output weights, w_{kj}. Because the error is not explicitly dependent upon w_{jk}, we must use the chain rule for differentiation:

$$\frac{\partial J}{\partial w_{kj}} = \frac{\partial J}{\partial net_k} \frac{\partial net_k}{\partial w_{kj}} = -\delta_k \frac{\partial net_k}{\partial w_{kj}}, \tag{13}$$

SENSITIVITY

where the *sensitivity* of unit k is defined to be

$$\delta_k = -\partial J / \partial net_k, \tag{14}$$

and describes how the overall error changes with the unit's net activation. Assuming that the activation function $f(\cdot)$ is differentiable, we differentiate Eq. 9 and find that for such an output unit, δ_k is simply

$$\delta_k = -\frac{\partial J}{\partial net_k} = -\frac{\partial J}{\partial z_k} \frac{\partial z_k}{\partial net_k} = (t_k - z_k) f'(net_k). \tag{15}$$

The last derivative in Eq. 13 is found using Eq. 4:

$$\frac{\partial net_k}{\partial w_{kj}} = y_j. \tag{16}$$

Taken together, these results give the weight update or learning rule for the hidden-to-output weights:

$$\Delta w_{kj} = \eta \delta_k y_j = \eta (t_k - z_k) f'(net_k) y_j. \tag{17}$$

In the case that the output unit is linear—that is, $f(net_k) = net_k$ and $f'(net_k) = 1$—then Eq. 17 is simply the LMS rule we saw in Chapter 5.

The learning rule for the input-to-hidden units is more subtle, indeed, it is the crux of the solution to the credit assignment problem. From Eq. 11, and again using the chain rule, we calculate

$$\frac{\partial J}{\partial w_{ji}} = \frac{\partial J}{\partial y_j} \frac{\partial y_j}{\partial net_j} \frac{\partial net_j}{\partial w_{ji}}. \tag{18}$$

The first term on the right-hand side involves all of the weights w_{kj}, and requires just a bit of care:

$$\frac{\partial J}{\partial y_j} = \frac{\partial}{\partial y_j} \left[\frac{1}{2} \sum_{k=1}^{c} (t_k - z_k)^2 \right]$$

$$= - \sum_{k=1}^{c} (t_k - z_k) \frac{\partial z_k}{\partial y_j}$$

$$= - \sum_{k=1}^{c} (t_k - z_k) \frac{\partial z_k}{\partial net_k} \frac{\partial net_k}{\partial y_j}$$

$$= - \sum_{k=1}^{c} (t_k - z_k) f'(net_k) w_{kj}. \tag{19}$$

For the second step above we had to use the chain rule yet again. The final sum over output units in Eq. 19 expresses how the hidden unit output, y_j, affects the error at each output unit. This will allow us to compute an effective target activation for each hidden unit. In analogy with Eq. 15 we use Eq. 19 to define the sensitivity for a hidden unit as

$$\delta_j \equiv f'(net_j) \sum_{k=1}^{c} w_{kj} \delta_k. \tag{20}$$

Equation 20 is the core of the solution to the credit assigment problem: The sensitivity at a hidden unit is simply the sum of the individual sensitivities at the output units weighted by the hidden-to-output weights w_{kj}, all multiplied by $f'(net_j)$. Thus the learning rule for the input-to-hidden weights is

$$\Delta w_{ji} = \eta x_i \delta_j = \eta \underbrace{\left[\sum_{k=1}^{c} w_{kj} \delta_k \right] f'(net_j)}_{\delta_j} x_i. \tag{21}$$

Equations 17 and 21, together with training protocols described below, give the backpropagation algorithm, or more specifically the "backpropagation of errors" algorithm. It is so-called because during training an error must be propagated from the output layer *back* to the hidden layer in order to perform the learning of the input-to-hidden weights by Eq. 21 (Fig. 6.5). At base then, backpropagation is just gradient descent in layered models where application of the chain rule through continuous functions allows the computation of derivatives of the criterion function with respect to all model weights.

As with all gradient-descent procedures, the exact behavior of the backpropagation algorithm depends on the starting point. While it might seem natural to start by setting the weight values to 0, Eq. 21 reveals that that would have very undesirable consequences. If the weights w_{kj} to the output layer were ever all zero, the back-propagated error would also be zero and the input-to-hidden weights would never change! This is the reason we start the process with random values for weights, as discussed further in Section 6.8.8.

These learning rules make intuitive sense. Consider first the rule for learning weights at the output units (Eq. 17). The weight update at unit k should indeed be

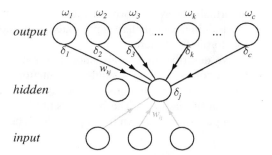

FIGURE 6.5. The sensitivity at a hidden unit is proportional to the weighted sum of the sensitivities at the output units: $\delta_j = f'(net_j) \sum_{k=1}^{c} w_{kj}\delta_k$. The output unit sensitivities are thus propagated "back" to the hidden units.

proportional to $(t_k - z_k)$: If we get the desired output ($z_k = t_k$), then there should be no weight change. For the typical sigmoidal $f(\cdot)$ we shall use most often, $f'(net_k)$ is always positive. Thus if y_j and $(t_k - z_k)$ are both positive, then the actual output is too small and the weight must be increased; indeed, the proper sign is given by the learning rule. Finally, the weight update should be proportional to the input value; if $y_j = 0$, then hidden unit j has no effect on the output and hence the error, and thus changing w_{ji} will not change the error on the pattern presented. A similar analysis of Eq. 21 yields insight of the input-to-hidden weights (Problem 5).

Although our analysis was done for a special case of a particularly simple three-layer network, it can readily be extended to much more general networks. With moderate notational and bookkeeping effort (see Problems 7 and 11), the backpropagation learning algorithm can be generalized directly to feed-forward networks in which

- Input units include a bias unit.
- Input units are connected directly to output units as well as to hidden units.
- There are more than three layers of units.
- There are different nonlinearities for different layers.
- Each unit has its own nonlinearity.
- Each unit has a different learning rate.

Because stability problems arise, it is a more subtle matter to incorporate learning into networks having connections *within* a layer, or feedback connections from units in higher layers back to those in lower layers. We shall consider such *recurrent networks* in Section 6.10.5. First, however, we take a closer look at the convergence of the backpropagation algorithm in simpler situations.

6.3.2 Training Protocols

TRAINING SET

STOCHASTIC
TRAINING

BATCH TRAINING

In broad overview, supervised training consists in presenting to the network those patterns whose category label we know—the *training set*—finding the output of the net and adjusting the weights so as to make the actual output more like the desired target values. The three most useful training protocols are: stochastic, batch, and on-line. In *stochastic training* patterns are chosen randomly from the training set, and the network weights are updated for each pattern presentation. This method is called stochastic because the training data can be considered a random variable. In *batch training*, all patterns are presented to the network before learning takes place.

ON-LINE
PROTOCOL

EPOCH

In virtually every case we must make several passes through the training data. In *on-line* training, each pattern is presented once and only once; there is no use of memory for storing the patterns.*

We describe the overall amount of pattern presentations by *epoch*—where one epoch corresponds to a single presentations of all patterns in the training set. For other variables being constant, the number of epochs is an indication of the relative amount of learning.[†] The basic stochastic and batch protocols of backpropagation for n patterns are shown in the procedures below.

■ **Algorithm 1. (Stochastic Backpropagation)**

1 **begin initialize** n_H, $\mathbf{w}$, criterion θ, η, $m \leftarrow 0$
2   **do** $m \leftarrow m + 1$
3         $\mathbf{x}^m \leftarrow$ randomly chosen pattern
4         $w_{ji} \leftarrow w_{ji} + \eta \delta_j x_i; \quad w_{kj} \leftarrow w_{kj} + \eta \delta_k y_j$
5   **until** $\|\nabla J(\mathbf{w})\| < \theta$
6   **return w**
7 **end**

In the on-line version of backpropagation, line 3 of Algorithm 1 is replaced by sequential selection of training patterns (Problem 9). Line 5 makes the algorithm end when the change in the criterion function $J(\mathbf{w})$ is smaller than some preset value θ. While this is perhaps the simplest meaningful *stopping criterion*, others generally lead to better performance, as we shall discuss in Section 6.8.14.

STOPPING
CRITERION

In the batch training protocol, all the training patterns are presented first and their corresponding weight updates summed; only then are the actual weights in the network updated. This process is iterated until some stopping criterion is met.

So far we have considered the error on a single pattern, but in fact we want to consider an error defined over the entirety of patterns in the training set. While we have to be careful about ambiguities in notation, we can nevertheless write this total training error as the sum over the errors on n individual patterns:

$$J = \sum_{p=1}^{n} J_p. \tag{22}$$

In stochastic training, a weight update may reduce the error on the single pattern being presented, yet *increase* the error on the full training set. Given a large number of such individual updates, however, the total error as given in Eq. 22 decreases.

■ **Algorithm 2. (Batch backpropagation)**

1 **begin initialize** n_H, $\mathbf{w}$, criterion θ, η, $r \leftarrow 0$
2   **do** $r \leftarrow r + 1$ (increment epoch)
3         $m \leftarrow 0; \; \Delta w_{ji} \leftarrow 0; \; \Delta w_{kj} \leftarrow$

---

*In Chapter 9 we shall discuss a fourth protocol, *learning with queries*, where the output of the network is used to *select* new training patterns.

[†]The notion of epoch does not apply to on-line training, where the number of pattern presentations is the more appropriate measure.

$$
\begin{array}{ll}
4 & \quad\quad\quad \underline{\textbf{do}}\ m \leftarrow m + 1 \\
5 & \quad\quad\quad\quad \mathbf{x}^m \leftarrow \text{select pattern} \\
6 & \quad\quad\quad\quad \Delta w_{ji} \leftarrow \Delta w_{ji} + \eta \delta_j x_i; \quad \Delta w_{kj} \leftarrow \Delta w_{kj} + \eta \delta_k y_j \\
7 & \quad\quad\quad \underline{\textbf{until}}\ m = n \\
8 & \quad\quad\quad\quad w_{ji} \leftarrow w_{ji} + \Delta w_{ji}; \quad w_{kj} \leftarrow w_{kj} + \Delta w_{kj} \\
9 & \quad\quad\quad \underline{\textbf{until}}\ \| \nabla J(\mathbf{w}) \| < \theta \\
10 & \quad\ \underline{\textbf{return}}\ \mathbf{w} \\
11 & \underline{\textbf{end}}
\end{array}
$$

In batch backpropagation, we need not select patterns randomly, because the weights are updated only after all patterns have been presented once. We shall consider the merits and drawbacks of each protocol in Section 6.8.

6.3.3 Learning Curves

Before training has begun, the error on the training set is typically high; through learning the error becomes lower, as shown in a *learning curve* (Fig. 6.6). The (per pattern) training error ultimately reaches an asymptotic value which depends upon the Bayes error, the amount of training data, and the expressive power (e.g., the number of weights) in the network: The higher the Bayes error and the fewer the number of such weights, the higher this asymptotic value is likely to be. Because batch backpropagation performs gradient descent in the criterion function, if the learning rate is not too high the training error tends to decrease monotonically. The average error on an independent test set is virtually always higher than on the training set, and while it generally decreases, it can increase or oscillate.

In addition to the use of the training set, there are two conceptual uses of independently selected patterns. One is to state the performance of the fielded network; for this we use the *test* set. Another use is to decide when to stop training; for this VALIDATION SET we use a *validation set*. As we shall discuss in Section 6.8.14, we stop training at a minimum of the error on the validation set.

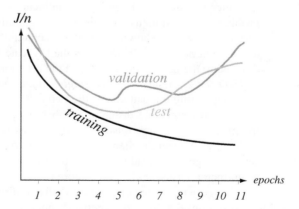

FIGURE 6.6. A learning curve shows the criterion function as a function of the amount of training, typically indicated by the number of epochs or presentations of the full training set. We plot the average error per pattern, that is, $1/n \sum_{p=1}^{n} J_p$. The validation error and the test or generalization error per pattern are virtually always higher than the training error. In some protocols, training is stopped at the first minimum of the validation set.

Figure 6.6 also shows the average error on a *validation set*—patterns not used directly for gradient descent training, and thus indirectly representative of novel patterns yet to be classified. The validation set can be used in a stopping criterion in both batch and stochastic protocols; gradient descent learning of the training set is stopped when a minimum is reached in the validation error (e.g., near epoch 5 in the figure). We shall return in Chapter 9 to understand in greater depth why this technique of *validation*, or more general *cross-validation*, often leads to networks having improved recognition accuracy.

6.4 ERROR SURFACES

Because backpropagation is based on gradient descent in a criterion function, we can gain understanding and intuition about the algorithm by studying error surfaces themselves—the function $J(\mathbf{w})$. Of course, such an error surface depends upon the classification task; nevertheless, there are some general properties of error surfaces that seem to hold over a broad range of real-world pattern recognition problems. One of the issues that concerns us is local minima; if many local minima plague the error landscape, then it is unlikely that the network will find the *global* minimum. Below we shall see whether this necessarily leads to poor performance. Another issue is the presence of plateaus—regions where the error varies only slightly as a function of weights. If such plateaus are plentiful, we can expect training according to Algorithms 1 and 2 to be slow. Because training typically begins with small weights, the error surface in the neighborhood of $\mathbf{w} \simeq \mathbf{0}$ will determine the general direction of descent. What can we say about the error in this region? Most interesting real-world problems are of high dimensionality. Are there any *general* properties of high-dimensional error functions?

We now explore these issues in some illustrative systems.

6.4.1 Some Small Networks

Consider the simplest three-layer nonlinear network, here solving a two-category problem in one dimension; this 1-1-1 sigmoidal network (and bias) is shown in Fig. 6.7. The data shown are linearly separable, and the optimal decision boundary, a point near $x_1 = 0$, separates the two categories. During learning, the weights descend to the global minimum, and the problem is solved.

Here the error surface has a *single* (global) minimum, which yields the decision point separating the patterns of the two categories. Different plateaus in the surface correspond roughly to different numbers of patterns properly classified; the maximum number of such misclassified patterns is four in this example. The plateau regions, where weight change does not lead to a change in error, here correspond to sets of weights that lead to roughly the same decision point in the input space. Thus as w_1 increases and w_2 becomes more negative, the surface shows that the error does not change, a result that can be informally confirmed by looking at the network itself.

Now consider the same network applied to another, harder, one-dimensional problem—one that is not linearly separable (Fig. 6.8). First, note that overall the error surface is slightly higher than in Fig. 6.7 because even the best solution attainable with this network leads to one pattern being misclassified. As before, the different plateaus in error correspond to different numbers of training patterns properly learned. However, one must not confuse the squared error measure with

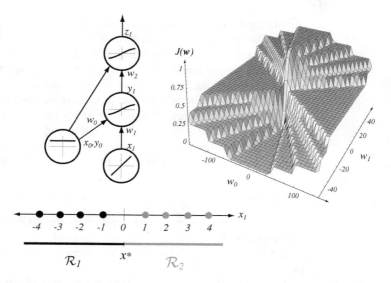

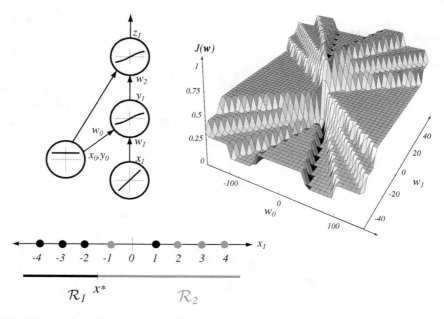

FIGURE 6.7. Eight one-dimensional patterns (four in each of two classes) are to be learned by a 1-1-1 network with steep sigmoidal hidden and output units with bias. The error surface as a function of w_0 and w_1 is also shown, where the bias weights are assigned their final values. The network starts with random weights; through stochastic training, it descends to a global minimum in error. Note especially that a low error solution exists, which indeed leads to a decision boundary separating the training points into their two categories.

FIGURE 6.8. As in Fig. 6.7, except here the patterns are not linearly separable; the error surface is slightly higher than in that figure. Note too from the error surface that there are two forms of minimum error solution; these correspond to $-2 < x^* < -1$ and $1 < x^* < 2$, in which one pattern is misclassified.

classification error. For instance, here there are two general ways to misclassify exactly two patterns, but these have different errors. Incidentally, a 1-3-1 network (but not a 1-2-1 network) can solve this problem (Computer exercise 3).

From these very simple examples, where the correspondences among weight values, decision boundary, and error are manifest, we can see how the error of the global minimum is lower when the problem can be solved and that there are plateaus corresponding to sets of weights that lead to nearly the same decision boundary. Furthermore, the surface near $\mathbf{w} \simeq \mathbf{0}$, the traditional region for starting learning, has high error and happens in this case to have a large slope; if the starting point had differed somewhat, the network would descend to the same final weight values.

6.4.2 The Exclusive-OR (XOR)

A somewhat more complicated problem is the XOR problem we have seen before. Figure 6.9 shows several two-dimensional slices through the nine-dimensional weight space of the 2-2-1 sigmoidal network with bias. The slices shown include a global minimum in the error.

Notice first that the error varies a bit more gradually as a function of a *single* weight than does the error in the networks solving the problems in Figs. 6.7 and 6.8. This is because in a large network any single weight has on average a smaller relative contribution to the output. Ridges, valleys, and a variety of other shapes can all be seen in the surface. Several local minima in the high-dimensional weight space exist, which here correspond to solutions that classify three (but not four) patterns. Although it is hard to show it graphically, the error surface is invariant with respect to certain discrete permutations. For instance, if the labels on the two hidden units are exchanged, and the weight values changed appropriately, the shape of the error surface is unaffected (Problem 13).

6.4.3 Larger Networks

Alas, the intuition we gain from considering error surfaces for small networks gives only hints of what is going on in large networks, and at times can be quite misleading. For a network with many weights solving a complicated high-dimensional classification problem, the error varies quite gradually as a single weight is changed. Nevertheless, the surfaces can have troughs, valleys, canyons, and a host of shapes.

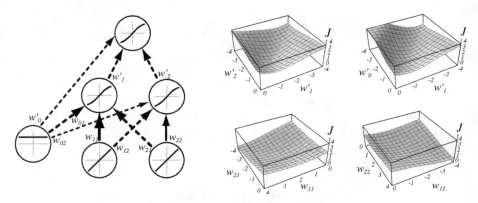

FIGURE 6.9. Two-dimensional slices through the nine-dimensional error surface after extensive training for a 2-2-1 network solving the XOR problem.

Whereas in low-dimensional spaces, local minima can be plentiful, in high dimension, the problem of local minima is different: The high-dimensional space may afford more ways (dimensions) for the system to "get around" a barrier or local maximum during learning. The more superfluous the weights, the less likely it is a network will get trapped in local minima. However, networks with an unnecessarily large number of weights are undesirable because of the dangers of overfitting, as we shall see in Section 6.11.

6.4.4 How Important Are Multiple Minima?

The possibility of the presence of multiple local minima is one reason that we resort to iterative gradient descent—analytic methods are highly unlikely to find a single global minimum, especially in high-dimensional weight spaces. In computational practice, we do not want our network to be caught in a local minimum having high training error because this usually indicates that key features of the problem have not been learned by the network. In such cases it is traditional to reinitialize the weights and train again, possibly also altering other parameters in the net (Section 6.8).

In many problems, convergence to a nonglobal minimum is acceptable, if the error is nevertheless fairly low. Furthermore, common stopping criteria demand that training terminate even before the minimum is reached, and thus it is not essential that the network be converging toward the *global* minimum for acceptable performance. In short, the presence of multiple minima does not necessarily present difficulties in training nets, and a few simple heuristics can often overcome such problems (Section 6.8).

6.5 BACKPROPAGATION AS FEATURE MAPPING

Because the hidden-to-output layer leads to a linear discriminant, the novel computational power provided by multilayer neural nets can be attributed to the nonlinear warping of the input to the representation at the hidden units. Let us consider this transformation, again with the help of the XOR problem.

Figure 6.10 shows a three-layer net addressing the XOR problem. For any input pattern in the $x_1 x_2$-space, we can show the corresponding output of the two hidden units in the $y_1 y_2$-space. With small initial weights, the net activation of each hidden unit is small, and thus the *linear* portion of their activation function is used. Such a linear transformation from **x** to **y** leaves the patterns linearly *in*separable (Problem 1). However, as learning progresses and the input-to-hidden weights increase in magnitude, the nonlinearities of the hidden units warp and distort the mapping from input to the hidden unit space. The linear decision boundary at the end of learning found by the hidden-to-output weights is shown by the straight dashed line; the nonlinearly separable problem at the inputs is transformed into a linearly separable at the hidden units.

We can illustrate such distortion in the three-bit parity problem, where the output is equal to $+1$ if the number of 1s in the input is odd, and -1 otherwise—a generalization of the XOR or two-bit parity problem (Fig. 6.11). As before, early in learning the hidden units operate in their linear range and thus the representation after the hidden units remains linearly *in*separable—the patterns from the two categories lie at alternating vertexes of a cube. After learning causes the weights to become larger, the nonlinearities of the hidden units are expressed and patterns have been moved and are linearly separable, as shown.

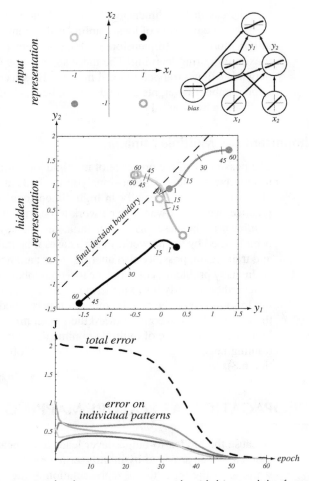

FIGURE 6.10. A 2-2-1 backpropagation network with bias and the four patterns of the XOR problem are shown at the top. The middle figure shows the outputs of the hidden units for each of the four patterns; these outputs move across the $y_1 y_2$-space as the network learns. In this space, early in training (epoch 1) the two categories are not linearly separable. As the input-to-hidden weights learn, as marked by the number of epochs, the categories become linearly separable. The dashed line is the linear decision boundary determined by the hidden-to-output weights at the end of learning; indeed the patterns of the two classes are separated by this boundary. The bottom graph shows the learning curves—the error on individual patterns and the total error as a function of epoch. Note that, as frequently happens, the total training error decreases monotonically, even though this is not the case for the error on each individual pattern.

Figure 6.12 shows a different two-dimensional two-category problem and the pattern representations in a 2-2-1 and in a 2-3-1 network of sigmoidal hidden units. Note that in the two-hidden-unit net, the categories are separated somewhat, but not enough for error-free classification; the expressive power of the net is not sufficiently high. In contrast, the three-hidden-unit net *can* separate the patterns. In general, given sufficiently many hidden units in a sigmoidal network, any set of different patterns can be learned in this way.

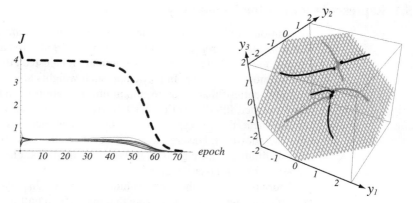

FIGURE 6.11. A 3-3-1 backpropagation network with bias can indeed solve the three-bit parity problem. The representation of the eight patterns at the hidden units ($y_1 y_2 y_3$-space) as the system learns and the planar decision boundary found by the hidden-to-output weights at the end of learning. The patterns of the two classes are indeed separated by this plane, as desired. The learning curve shows the error on individual patterns and the total error J as a function of epoch.

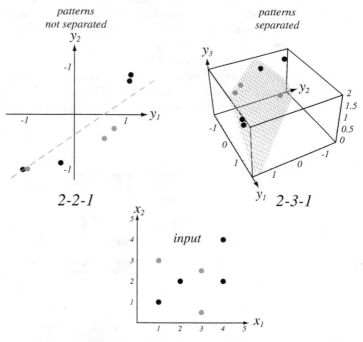

FIGURE 6.12. Seven patterns from a two-dimensional two-category nonlinearly separable classification problem are shown at the bottom. The figure at the top left shows the hidden unit representations of the patterns in a 2-2-1 sigmoidal network with bias fully trained to the global error minimum; the linear boundary implemented by the hidden-to-output weights is marked as a gray dashed line. Note that the categories are almost linearly separable in this $y_1 y_2$-space, but one training point is misclassified. At the top right is the analogous hidden unit representation for a fully trained 2-3-1 network with bias. Because of the higher dimension of the hidden layer representation, the categories are now linearly separable; indeed the learned hidden-to-output weights implement a plane that separates the categories.

6.5.1 Representations at the Hidden Layer—Weights

In addition to visualizing on the transformation of *patterns* in a network, we can also consider the representation of learned *weights*. Because the hidden-to-output weights merely lead to a linear discriminant, it is instead the input-to-hidden weights that are most instructive. In particular, such weights at a single hidden unit describe the input pattern that leads to maximum activation of that hidden unit, analogous to a "matched filter" (Section 6.10.3). Because the hidden unit activation functions are nonlinear, the correspondence with classical methods such as matched filters is not exact, however. Nevertheless, it is occasionally convenient to think of the hidden units as finding feature groupings useful for the linear classifier implemented by the hidden-to-output layer weights.

MATCHED FILTER

Figure 6.13 shows the input-to-hidden weights, displayed as images, for a simple task of character recognition. Note that one hidden unit seems "tuned" or "matched" to a pair of horizontal bars while the other is tuned to a single lower bar. Both of these feature groupings are useful building blocks for the patterns presented. In complex, high-dimensional problems, however, the pattern of learned weights may not appear to be simply related to the features we suspect are appropriate for the task. This could be because we may be mistaken about which are the true, relevant feature groupings; nonlinear interactions between features may be significant in a problem and such interactions are not manifest in the patterns of weights at a single hidden unit; or the network may have too many weights, and thus the feature selectivity is low. Thus, while analyses of learned weights may be suggestive, the whole endeavor must be approached with caution.

It is generally much harder to represent the hidden-to-output layer weights in terms of input features. Not only do the hidden units themselves already encode a somewhat abstract pattern, there is moreover no natural ordering or arrangement of the hidden units analogous to that of the input units in Fig. 6.13. Together with the

sample training patterns

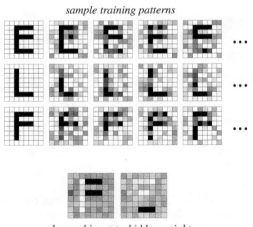

learned input-to-hidden weights

FIGURE 6.13. The top images represent patterns from a large training set used to train a 64-2-3 sigmoidal network for classifying three characters. The bottom figures show the input-to-hidden weights, represented as patterns, at the two hidden units after training. Note that these learned weights indeed describe feature groupings useful for the classification task. In large networks, such patterns of learned weights may be difficult to interpret in this way.

fact that the output of hidden units are nonlinearly related to the inputs, this makes analyzing hidden-to-output weights somewhat problematic. Often the best we can do is list the patterns of input weights for hidden units that have strong connections to the output unit in question (Computer exercise 9).

6.6 BACKPROPAGATION, BAYES THEORY AND PROBABILITY

While multilayer neural networks may appear to be somewhat ad hoc, we now show that when trained via backpropagation on a sum-squared error criterion, they provide a least squares fit to the Bayes discriminant functions.

6.6.1 Bayes Discriminants and Neural Networks

As we saw in Chapter 5, the LMS algorithm computed the approximation to the Bayes discriminant function for two-layer nets. We now generalize this result in two ways: to multiple categories and to nonlinear functions implemented by three-layer neural networks. We use the network of Fig. 6.4 and let $g_k(\mathbf{x}; \mathbf{w})$ be the output of the kth output unit—the discriminant function corresponding to category ω_k. Recall first Bayes formula,

$$P(\omega_k|\mathbf{x}) = \frac{p(\mathbf{x}|\omega_k)P(\omega_k)}{\sum_{i=1}^{c} p(\mathbf{x}|\omega_i)p(\omega_i)} = \frac{p(\mathbf{x}, \omega_k)}{p(\mathbf{x})}, \tag{23}$$

and the Bayes decision for any pattern $\mathbf{x}$: Choose the category ω_k having the largest discriminant function $g_k(\mathbf{x}) = P(\omega_k|\mathbf{x})$.

Suppose we train a network having c output units with a target signal according to

$$t_k(\mathbf{x}) = \begin{cases} 1 & \text{if } \mathbf{x} \in \omega_k \\ 0 & \text{otherwise.} \end{cases} \tag{24}$$

(In practice, teaching values of ± 1 are to be preferred, as we shall see in Section 6.8; we use the values 0–1 in this derivation for analytical simplicity.) The contribution to the criterion function based on a single output unit k for finite number of training samples $\mathbf{x}$ is

$$J(\mathbf{w}) = \sum_{\mathbf{x}} [g_k(\mathbf{x}; \mathbf{w}) - t_k]^2 \tag{25}$$

$$= \sum_{\mathbf{x} \in \omega_k} [g_k(\mathbf{x}; \mathbf{w}) - 1]^2 + \sum_{\mathbf{x} \notin \omega_k} [g_k(\mathbf{x}; \mathbf{w}) - 0]^2$$

$$= n \left\{ \frac{n_k}{n} \frac{1}{n_k} \sum_{\mathbf{x} \in \omega_k} [g_k(\mathbf{x}; \mathbf{w}) - 1]^2 + \frac{n - n_k}{n} \frac{1}{n - n_k} \sum_{\mathbf{x} \notin \omega_k} [g_k(\mathbf{x}; \mathbf{w}) - 0]^2 \right\},$$

where n is the total number of training patterns, n_k of which are in ω_k. In the limit of infinite data we can use Bayes formula to express Eq. 25 as (Problem 17)

$$\lim_{n\to\infty} \frac{1}{n} J(\mathbf{w}) \equiv \tilde{J}(\mathbf{w}) \tag{26}$$

$$= P(\omega_k) \int [g_k(\mathbf{x};\ \mathbf{w}) - 1]^2 \, p(\mathbf{x}|\omega_k) d\mathbf{x} + P(\omega_{i\neq k}) \int g_k^2(\mathbf{x};\ \mathbf{w}) p(\mathbf{x}|\omega_{i\neq k}) \, d\mathbf{x}$$

$$= \int g_k^2(\mathbf{x};\ \mathbf{w}) p(\mathbf{x}) d\mathbf{x} - 2 \int g_k(\mathbf{x};\ \mathbf{w}) p(\mathbf{x}, \omega_k) d\mathbf{x} + \int p(\mathbf{x}, \omega_k) \, d\mathbf{x}$$

$$= \int [g_k(\mathbf{x};\ \mathbf{w}) - P(\omega_k|\mathbf{x})]^2 \, p(\mathbf{x}) \, d\mathbf{x} + \underbrace{\int P(\omega_k|\mathbf{x}) P(\omega_{i\neq k}|\mathbf{x}) p(\mathbf{x}) d\mathbf{x}}_{\text{independent of } \mathbf{w}}.$$

The backpropagation rule changes weights to minimize the left-hand side of Eq. 26, and thus it minimizes

$$\int [g_k(\mathbf{x};\ \mathbf{w}) - P(\omega_k|\mathbf{x})]^2 p(\mathbf{x}) \, d\mathbf{x}. \tag{27}$$

Because this is true for each category ω_k ($k = 1, 2, \ldots, c$), backpropagation minimizes the sum (Problem 22):

$$\sum_{k=1}^{c} \int [g_k(\mathbf{x};\ \mathbf{w}) - P(\omega_k|\mathbf{x})]^2 \, p(\mathbf{x}) \, d\mathbf{x}. \tag{28}$$

Thus in the limit of infinite data the outputs of the trained network will approximate the true *a posteriori* probabilities in a least-squares sense; that is, the output units represent the *a posteriori* probabilities,

$$g_k(\mathbf{x};\ \mathbf{w}) \simeq P(\omega_k|\mathbf{x}). \tag{29}$$

We must be cautious in interpreting these results, however. A key assumption underlying the argument is that the network can indeed represent the functions $P(\omega_k|\mathbf{x})$; with insufficient hidden units, this will not be true. Nevertheless, the above results show that neural networks can have very desirable limiting properties.

6.6.2 Outputs as Probabilities

In the previous subsection we saw one way to make the c output units of a trained net represent probabilities by training with $0 - 1$ target values. While indeed given infinite amounts of training data, the outputs will represent probabilities. If, however, these conditions do not hold—in particular we have only a *finite* amount of training data—then the outputs will not represent probabilities; for instance, there is not even a guarantee that they sum to 1.0. In fact, if the sum of the network outputs differs significantly from 1.0 within some range of the input space, it is an indication that the network is not accurately modeling the posteriors. This, in turn, may suggest changing the network topology, number of hidden units, or other aspects of the net (Section 6.8).

One approach toward approximating probabilities is to choose the output unit nonlinearity to be exponential rather than sigmoidal—$f(net_k) \propto e^{net_k}$—and for each pattern to normalize the outputs to sum to 1.0,

$$z_k = \frac{e^{net_k}}{\sum_{m=1}^{c} e^{net_m}},$$ (30)

SOFTMAX
WINNER-TAKE-
ALL

while training using $0 - 1$ target signals. This is the *softmax* method—a smoothed or continuous version of a *winner-take-all* nonlinearity in which the maximum output is transformed to 1.0, and all others reduced to 0.0. The softmax output finds theoretical justification if for each category ω_k the hidden unit representations **y** can be assumed to come from an exponential distribution (Problem 20 and Computer exercise 10).

As we have seen, then, a neural network classifier trained in this manner approximates the posterior probabilities $P(\omega_i|\mathbf{x})$, and this depends upon the prior probabilities of the categories. If such a trained network is to be used on problems in which the priors have been changed, it is a simple matter to rescale each network output, $g_i(\mathbf{x}) = P(\omega_i|\mathbf{x})$, by the ratio of such priors, though this need not ensure minimal error. The use of softmax is appropriate when the network is to be used for estimating probabilities. While probabilities as network outputs can certainly be used for classification, other representations—for instance, where outputs can be positive or negative and need not sum to 1.0—are preferable, as we shall see in Section 6.8.4.

*6.7 RELATED STATISTICAL TECHNIQUES

While the graphical, topological representation of networks is useful and a guide to intuition, we must not forget that the underlying mathematics of the feedforward operation is governed by Eq. 7. A number of statistical methods bear similarities to that equation. For instance, *projection pursuit regression*, or simply projection pursuit, implements

PROJECTION
PURSUIT

$$z = \sum_{j=1}^{jmax} w_j f_j(\mathbf{v}_j^t \mathbf{x} + v_{j0}) + w_0.$$ (31)

Here each $\mathbf{v}_j$ and v_{j0} together define the projection of the input **x** onto one of j_{max} different d-dimensional hyperplanes. These projections are transformed by nonlinear functions $f_j(\cdot)$ whose values are then linearly combined at the output; traditionally, sigmoidal or Gaussian functions are used. The $f_j(\cdot)$ have been called *ridge functions* because for peaked $f_j(\cdot)$, the system output appears as a ridge in a two-dimensional input space. Equation 31 implements a mapping to a scalar function z; in a c-category classification problem there would be c such outputs. In computational practice, the parameters are learned in groups minimizing an LMS error—for example, first the components of $\mathbf{v}_1$ and v_{10}, then $\mathbf{v}_2$ and v_{20} up to $\mathbf{v}_{jmax}$ and v_{jmax0}; then the w_j and w_0, iterating until convergence.

RIDGE FUNCTION

Such models are related to the three-layer networks we have seen in that the $\mathbf{v}_j$ and v_{j0} are analogous to the input-to-hidden weights at a hidden unit and the effective output unit is linear. The class of functions $f_j(\cdot)$ at such hidden units are more general and have more free parameters than do sigmoids. Moreover, such a model can have an output much larger than 1.0, as might be needed in a general regression task. In the classification tasks we have considered, a saturating output such as a sigmoid is more appropriate.

GENERALIZED
ADDITIVE
MODEL

Another technique related to multilayer neural nets is *generalized additive models*, which implement

$$z = f\left(\sum_{i=1}^{d} f_i(x_i) + w_0\right),\tag{32}$$

where again $f(\cdot)$ is often chosen to be a sigmoid, and the functions $f_i(\cdot)$ operating on the input features are nonlinear, and sometimes chosen to be sigmoidal. Such models are trained by iteratively adjusting parameters of the component nonlinearities $f_i(\cdot)$. Indeed, the basic three-layer neural networks of Section 6.2 implement a special case of general additive models (Problem 24), though the training methods differ.

MULTIVARIATE
ADAPTIVE
REGRESSION
SPLINE

An extremely flexible technique having many adjustable parameters is *multivariate adaptive regression splines* (MARS). In this technique, localized spline functions (polynomials adjusted to ensure continuous derivative) are used in the initial processing. Here the output is the weighted sum of M products of splines:

$$z = \sum_{k=1}^{M} w_k \prod_{r=1}^{r_k} \phi_{kr}(x_{q(k,r)}) + w_0,\tag{33}$$

where the kth basis function is the product of r_k one-dimensional spline functions ϕ_{kr}, and w_0 is a scalar offset. The splines depend on the input values x_q, such as the feature component of an input, and this index is denoted $q(k, r)$. Naturally, in a c-category task, there would be one such output for each category.

In broad overview, training in MARS begins by fitting the data with a spline function along each feature dimension in turn. The spline that best fits the data, in a sum-squared-error sense, is retained. This is the $r = 1$ term in Eq. 33. Next, each of the other feature dimensions is considered, one by one. For each such dimension, candidate splines are selected based on the data fit using the *product* of that spline with the one previously selected, thereby giving the product $r = 1 \rightarrow 2$. The best such second spline is retained, thereby giving the $r = 2$ term. In this way, splines are added incrementally up to some value r_k, where some desired quality of fit is achieved. The weights w_k are learned via gradient descent on an LMS criterion.

For several reasons, multilayer neural nets have all but supplanted projection pursuit, MARS, and earlier related techniques in practical pattern recognition research. Backpropagation is simpler than learning in projection pursuit and MARS, especially when the number of training patterns and the dimension is large; heuristic information can be incorporated more simply into nets (Section 6.8.12); nets admit a variety of simplification or regularization methods (Section 6.11) that have no direct counterpart in those earlier methods. It is, moreover, usually simpler to refine a trained neural net using additional training data than it is to modify classifiers based on projection pursuit or MARS.

6.8 PRACTICAL TECHNIQUES FOR IMPROVING BACKPROPAGATION

Up to this point we have ignored a number of practical considerations for the sake of simplicity. Although the above analyses are correct, a naive application of the procedures can lead to very slow convergence, poor performance or other unsatisfactory results. Thus we now turn to a number of practical suggestions for training networks by backpropagation. While it is difficult to give mathematical proofs for them, these suggestions can be based on a number of plausible heuristics and have been found to be useful in many practical applications.

6.8.1 Activation Function

There are a number of properties we seek for $f(\cdot)$, but we must not lose sight of the fact that backpropagation will work with virtually any activation function, given that a few simple conditions such as continuity of $f(\cdot)$ and its derivative are met. In any given classification problem we may have a good reason for selecting a particular activation function. For instance, if we have prior information that the distributions arise from a mixture of Gaussians, then Gaussian activation functions are appropriate.

When not guided by such problem dependent information, what general properties might we seek in $f(\cdot)$? First, of course, $f(\cdot)$ must be nonlinear—otherwise the three-layer network provides no computational power above that of a two-layer net (Problem 1). A second desirable property is that $f(\cdot)$ saturate—that is, have some maximum and minimum output value. This will keep the weights and activations bounded, and thus keep the training time limited. Saturation is a particularly desirable property when the output is meant to represent a probability. It is also desirable for models of biological neural networks, where the output represents a neural firing rate. It may not be desirable in networks used for regression, where a wide dynamic range may be required.

A third property is continuity and smoothness—that is, that $f(\cdot)$ and $f'(\cdot)$ be defined throughout the range of their argument. Recall that the fact that we could take a derivative of $f(\cdot)$ was crucial in the derivation of the backpropagation learning rule. The rule would not, therefore, work with the threshold or sign function of Eq. 3. Backpropagation can be made to work with *piecewise* linear activation functions, but with added complexity and few benefits.

Monotonicity is another convenient, but nonessential, property for $f(\cdot)$—we might wish that the derivative have the same sign throughout the range of the argument, for example, $f'(\cdot) \geq 0$. If f is *not* monotonic and has multiple local maxima, additional and undesirable local extrema in the error surface may become introduced. Nonmonotonic activation functions such as radial basis functions can be used if proper care is taken (Section 6.10.1). Another desirable property is linearity for a small value of *net*, which will enable the system to implement a linear model if adequate for yielding low error.

SIGMOID
One class of functions that has all the above desired properties is the *sigmoid*, such as a hyperbolic tangent. The sigmoid is smooth, differentiable, nonlinear, and saturating. It also admits a linear model if the network weights are small. A minor benefit is that the derivative $f'(\cdot)$ can be easily expressed in terms of $f(\cdot)$ itself (Problem 10).

POLYNOMIAL CLASSIFIER
We mention in passing that *polynomial classifiers* use activation functions of the form $x_1, x_2, \ldots, x_d, x_1^2, x_2^2, \ldots, x_d^2, x_1 x_2, \ldots, x_1 x_d$, and so forth—all terms up to some limit; training is via gradient descent too. One drawback here is that the outputs of the hidden units (φ functions) can become extremely large even for realistic problems (Problem 29). By employing saturating sigmoidal activation functions, neural networks avoid this problem.

DISTRIBUTED REPRESENTA-TION
A hidden layer of sigmoidal units affords a *distributed* or *global* representation of the input. That is, any particular input **x** is likely to yield activity throughout *several* hidden units. In contrast, if the hidden units have activation functions that have

significant response only for inputs within a small range, then an input **x** generally leads to *fewer* hidden units being active—a *local representation*. (Nearest neighbor classifiers employ local representations, of course.) It is often found in practice that when there are few training points, distributed representations are superior because more of the data influences the posteriors at any given input region.

The sigmoid is the most widely used activation function for the above reasons, and in most of the following we shall employ sigmoids.

6.8.2 Parameters for the Sigmoid

Given that we will use the sigmoidal activation function, there remain a number of parameters to set. It is best to keep the function centered on zero and antisymmetric— or as an "odd" function, that is, $f(-net) = -f(net)$—rather than one whose value is always positive. Together with the data preprocessing described in Section 6.8.3, antisymmetric sigmoids lead to faster learning. Nonzero means in the input variables and activation functions make some of the eigenvalues of the Hessian matrix large (Section 6.9.1), and this slows the learning, as we shall see.

Sigmoid functions of the form

$$f(net) = a \tanh(b\ net) = a \left[\frac{e^{+b\ net} - e^{-b\ net}}{e^{+b\ net} + e^{-b\ net}} \right] \tag{34}$$

work well. The *overall* range and slope are not important, because it is their relationship to parameters such as the learning rate and magnitudes of the inputs and targets that affect learning (Problem 23). For convenience, though, we choose $a = 1.716$ and $b = 2/3$ in Eq. 34—values that ensure $f'(0) \simeq 0.5$, that the linear range is $-1 < net < +1$, and that the extrema of the second derivative occur roughly at $net \simeq \pm 2$ (Fig. 6.14).

6.8.3 Scaling Input

Suppose we were using a two-input network to classify fish based on the features of mass (measured in grams) and length (measured in meters). Such a representation will have serious drawbacks for a neural network classifier: The numerical value of the mass will be orders of magnitude larger than that for length. During training, the network will adjust weights from the "mass" input unit far more than for the "length" input—indeed the error will hardly depend upon the tiny length values. If, however, the same physical information were presented but with mass measured in kilograms and length in millimeters, the situation would be reversed. Naturally we do not want our classifier to prefer one of these features over the other, because they differ solely in the arbitrary representation. The difficulty arises even for features having the same units but differing in overall magnitude, of course—for instance, if a fish's length and its fin thickness were both measured in millimeters.

In order to avoid such difficulties, the input patterns should be shifted so that the average over the training set of each feature is zero. Moreover, the full data set should then be scaled to have the same variance in each feature component—here chosen to be 1.0 for reasons that will be clear in Section 6.8.8. That is, we *standardize* the training patterns. This data standardization plays a role similar to a whitening transformation applied to the training set (Chapter 2), and need be done once, before the start of network training. It represents a small one-time computational burden (Problem 27). Standardization can only be done for stochastic and batch learning

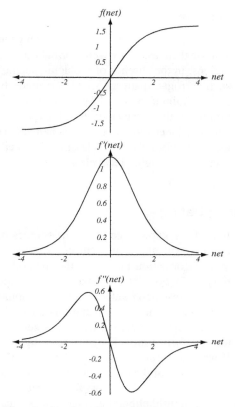

FIGURE 6.14. A useful activation function $f(net)$ is an anti-symmetric sigmoid. For the parameters given in the text, $f(net)$ is nearly linear in the range $-1 < net < +1$ and its second derivative, $f''(net)$, has extrema near $net \simeq \pm 2$.

protocols, but not on-line protocols where the full data set is never available at any one time. Naturally, any test pattern must be subjected to the same transformation before it is to be classified by the network.

6.8.4 Target Values

For pattern recognition, we typically train with the pattern and its category label, and thus we use a one-of-c representation for the target vector. Because the output units saturate at ± 1.716, we might naively feel that the target values should be those values; however, that would present a difficulty. For any finite value of net_k, the output could never reach these saturation values, and thus there would be error. Full training would never terminate because weights would become extremely large as net_k would be driven to $\pm \infty$.

This difficulty can be avoided by using teaching values of $+1$ for the target category and -1 for the non-target categories. For instance, in a four-category problem if the pattern is in category ω_3, the following target vector should be used: $\mathbf{t} = (-1, -1, +1, -1)^t$. Of course, this target representation yields efficient learning for categorization—the outputs here do not represent *a posterior* probabilities (Section 6.6.2).

6.8.5 Training with Noise

When the training set is small, one can generate virtual or surrogate training patterns and use them as if they were normal training patterns sampled from the source distributions. In the absence of problem-specific information, a natural assumption is that such surrogate patterns should be made by adding d-dimensional Gaussian noise to true training points. In particular, for the standardized inputs described in Section 6.8.3, the variance of the added noise should be less than 1.0 (e.g., 0.1) and the category label should be left unchanged. This method of training with noise can be used with virtually every classification method, though it generally does not improve accuracy for highly local classifiers such as ones based on the nearest neighbor.

6.8.6 Manufacturing Data

If we have knowledge about the sources of variation among patterns, for instance, due to geometrical invariances, we can "manufacture" training data that conveys more information than does the method of training with uncorrelated noise (Section 6.8.5). For instance, in an optical character recognition problem, an input image may be presented rotated by various amounts. Hence during training we can take any particular training pattern and rotate its image to "manufacture" a training point that may be representative of a much larger training set. Likewise, we might scale a pattern, perform simple image processing to simulate a bold face character, and so on. If we have information about the range of expected rotation angles, or the variation in thickness of the character strokes, we should manufacture the data accordingly.

While this method bears formal equivalence to incorporating prior information in a maximum-likelihood approach, it is usually much simpler to implement, because we need only the forward model for generating patterns. As with training with noise, manufacturing data can be used with a wide range of pattern recognition methods. A drawback is that the memory requirements may be large and overall training may be slow.

6.8.7 Number of Hidden Units

While the number of input units and output units are dictated by the dimensionality of the input vectors and the number of categories, respectively, the number of hidden units is not simply related to such obvious properties of the classification problem. The number of hidden units, n_H, governs the expressive power of the net—and thus the complexity of the decision boundary. If the patterns are well-separated or linearly separable, then few hidden units are needed; conversely, if the patterns are drawn from complicated densities that are highly interspersed, then more hidden units are needed. Without further information there is no foolproof method for setting the number of hidden units before training.

Figure 6.15 shows the training and test error on a two-category classification problem for networks that differ solely in their number of hidden units. For large n_H, the training error can become small because such networks have high expressive power and become tuned to the particular training set. Nevertheless, in this regime, the test error is unacceptably high, an example of overfitting we shall study again in Chapter 9. At the other extreme of too few hidden units, the net does not have enough free parameters to fit the training data well, and again the test error is high. We seek some intermediate number of hidden units that will give low test error.

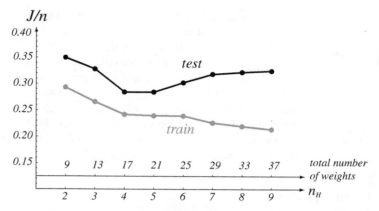

FIGURE 6.15. The error per pattern for networks fully trained but differing in the numbers of hidden units, n_H. Each $2 - n_H - 1$ network with bias was trained with 90 two-dimensional patterns from each of two categories, sampled from a mixture of three Gaussians, and thus $n = 180$. The minimum of the test error occurs for networks in the range $4 \leq n_H \leq 5$, i.e., the range of weights 17 to 21. This illustrates the rule of thumb that choosing networks with roughly $n/10$ weights often gives low test error.

The number of hidden units determines the total number of weights in the net—which we consider informally as the number of degrees of freedom—and thus it is plausible that we should not have more weights than the total number of training points, n. A convenient rule of thumb is to choose the number of hidden units such that the total number of weights in the net is roughly $n/10$. This seems to work well over a range of practical problems. It must be noted, however, that many successful systems employ more than this number. A more principled method is to adjust the complexity of the network in response to the training data, for instance, start with a "large" number of hidden units and "decay," prune, or eliminate weights—techniques we shall study in Section 6.11 and in Chapter 9.

6.8.8 Initializing Weights

UNIFORM
LEARNING

First, we can see from Eq. 21 that we cannot initialize the weights to 0, otherwise learning cannot take place. Thus we must confront the problem of choosing their starting values. Suppose we have fixed the network topology and thus have set the number of hidden units. We now seek to set the initial weight values in order to have fast and *uniform learning*, that is, all weights reach their final equilibrium values at about the same time. One form of nonuniform learning occurs when one category is learned well before others. In this undesirable case, the distribution of errors differs markedly from Bayes, and the overall error rate is typically higher than necessary. (The data standarization described above also helps to ensure uniform learning.)

In setting weights in a given layer, we choose weights randomly from a *single* distribution to help ensure uniform learning. Because data standardization gives positive and negative values equally, on average, we want positive and negative weights as well; thus we choose weights from a uniform distribution $-\tilde{w} < w < +\tilde{w}$, for some $\tilde{w}$ yet to be determined. If $\tilde{w}$ is chosen too small, the net activation of a hidden unit will be small and the linear model will be implemented. Alternatively, if $\tilde{w}$ is too large, the hidden unit may saturate even before learning begins. Because $net_j \simeq \pm 1$ are the limits to its linear range, we set $\tilde{w}$ such that the net activation at a hidden unit is in the range $-1 < net_j < +1$ (Fig. 6.14).

In order to calculate $\tilde{w}$, we consider a hidden unit accepting input from d input units. Suppose too that the same distribution is used to initialize all the weights, namely, a uniform distribution in the range $-\tilde{w} < w < +\tilde{w}$. On average, then, the net activation from d random variables of variance 1.0 from our standarized input through such weights will be $\tilde{w}\sqrt{d}$. As mentioned, we would like this net activation to be roughly in the range $-1 < net < +1$. This implies that $\tilde{w} = 1/\sqrt{d}$; thus input weights should be chosen in the range $-1/\sqrt{d} < w_{ji} < +1/\sqrt{d}$. The same argument holds for the hidden-to-output weights, where here the number of connected units is n_H; hidden-to-output weights should be initialized with values chosen in the range $-1/\sqrt{n_H} < w_{kj} < +1/\sqrt{n_H}$.

6.8.9 Learning Rates

NONUNIFORM
LEARNING

In principle, so long as the learning rate is small enough to ensure convergence, its value determines only the speed at which the network attains a minimum in the criterion function $J(\mathbf{w})$, not the final weight values themselves. In practice, however, because networks are rarely trained fully to a training error minimum (Section 6.8.14), the learning rate can indeed affect the quality of the final network. If some weights converge significantly earlier than others (nonuniform learning) then the network may not perform equally well throughout the full range of inputs, or equally well for the patterns in each category. Figure 6.16 shows the effect of different learning rates on convergence in a single dimension.

The optimal learning rate is the one that leads to the local error minimum in one learning step. A principled method of setting the learning rate comes from assuming the criterion function can be reasonably approximated by a quadratic, and this gives

$$\frac{\partial^2 J}{\partial w^2} \Delta w = \frac{\partial J}{\partial w}, \tag{35}$$

as illustrated in Fig. 6.17. The optimal rate is found directly to be

$$\eta_{opt} = \left(\frac{\partial^2 J}{\partial w^2} \right)^{-1}. \tag{36}$$

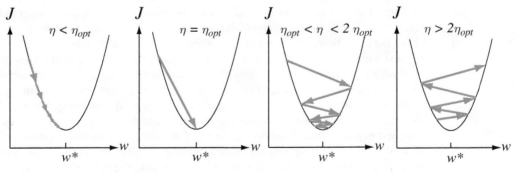

FIGURE 6.16. Gradient descent in a one-dimensional quadratic criterion with different learning rates. If $\eta < \eta_{opt}$, convergence is assured, but training can be needlessly slow. If $\eta = \eta_{opt}$, a single learning step suffices to find the error minimum. If $\eta_{opt} < \eta < 2\eta_{opt}$, the system will oscillate but nevertheless converge, but training is needlessly slow. If $\eta > 2\eta_{opt}$, the system diverges.

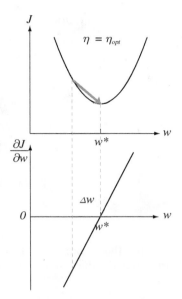

FIGURE 6.17. If the criterion function is quadratic (above), its derivative is linear (below). The optimal learning rate η_{opt} ensures that the weight value yielding minimum error, w^*, is found in a single learning step.

Of course the maximum learning rate that will give convergence is $\eta_{max} = 2\eta_{opt}$. It should be noted that a learning rate η in the range $\eta_{opt} < \eta < 2\eta_{opt}$ will lead to slower convergence (Computer exercise 8).

Thus, for rapid and uniform learning, we should calculate the second derivative of the criterion function with respect to *each* weight and set the optimal learning rate separately for each weight. We shall return in Section 6.9 to calculate second derivatives in networks, and to alternative descent and training methods. For typical problems addressed with sigmoidal networks and parameters discussed throughout this section, it is found that a learning rate of $\eta \simeq 0.1$ is often adequate as a first choice. The learning rate should be lowered if the criterion function diverges during learning, or instead should be raised if learning seems unduly slow.

6.8.10 Momentum

Error surfaces often have plateaus—regions in which the slope $dJ(\mathbf{w})/d\mathbf{w}$ is very small. These can arise when there are "too many" weights and thus the error depends only weakly upon any one of them. Momentum—loosely based on the notion from physics that moving objects tend to keep moving unless acted upon by outside forces—allows the network to learn more quickly when plateaus in the error surface exist. The approach is to alter the learning rule in stochastic backpropagation to include some fraction α of the previous weight update. Let $\Delta\mathbf{w}(m) = \mathbf{w}(m) - \mathbf{w}(m-1)$, and let $\Delta\mathbf{w}_{bp}(m)$ be the change in $\mathbf{w}(m)$ that would be called for by the backpropagation algorithm. Then

$$\mathbf{w}(m+1) = \mathbf{w}(m) + (1-\alpha)\Delta\mathbf{w}_{bp}(m) + \alpha\Delta\mathbf{w}(m-1) \qquad (37)$$

represents learning with momentum.

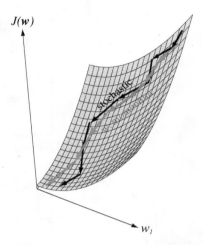

FIGURE 6.18. The incorporation of momentum into stochastic gradient descent by Eq. 37 (red arrows) reduces the variation in overall gradient directions and speeds learning.

Those who are familiar with digital signal processing will recognize this as a recursive or infinite-impulse-response low-pass filter that smooths the changes in $\mathbf{w}$. Obviously, α should not be negative, and for stability α must be less than 1.0. If $\alpha = 0$, the algorithm is the same as standard backpropagation. If $\alpha = 1$, the change suggested by backpropagation is ignored, and the weight vector moves with constant velocity. The weight changes are response to backpropagation if α is small, and sluggish if α is large. (Values typically used are $\alpha \simeq 0.9$.) Thus, the use of momentum "averages out" stochastic variations in weight updates during stochastic learning. By increasing stability, it can speed the learning process, even far from error plateaus (Fig. 6.18).

Algorithm 3 shows one way to incorporate momentum into gradient descent.

■ **Algorithm 3.   (Stochastic Backpropagation with Momentum)**

1 <u>**begin initialize**</u> $n_H, \mathbf{w}, \alpha(< 1), \theta, \eta, m \leftarrow 0, b_{ji} \leftarrow 0, b_{kj} \leftarrow 0$
2   <u>**do**</u> $m \leftarrow m + 1$
3             $\mathbf{x}^m \leftarrow$ randomly chosen pattern
4             $b_{ji} \leftarrow \eta(1 - \alpha)\delta_j x_i + \alpha b_{ji};\ b_{kj} \leftarrow \eta(1 - \alpha)\delta_k y_j + \alpha b_{kj}$
5             $w_{ji} \leftarrow w_{ji} + b_{ji};\ w_{kj} \leftarrow w_{kj} + b_{kj}$
6   <u>**until**</u> $\|\nabla J(\mathbf{w})\| < \theta$
7   <u>**return w**</u>
8 <u>**end**</u>

6.8.11 Weight Decay

One method of simplifying a network and avoiding overfitting is to impose a heuristic that the weights should be small. There is no principled reason why such a method of "weight decay" should always lead to improved network performance (indeed there are occasional cases where it leads to *degraded* performance), but it is found in most cases that it helps. The basic approach is to start with a network with "too many" weights and "decay" all weights during training. Small weights favor models that are more nearly linear (Problems 1 and 41). One of the reasons weight decay is so popular is its simplicity. After each weight update, every weight is simply "decayed" or shrunk according to

$$w^{new} = w^{old}(1 - \epsilon),$$ (38)

where $0 < \epsilon < 1$. In this way, weights that are not needed for reducing the error function become smaller and smaller, possibly to such a small value that they can be eliminated altogether. Those weights that *are* needed to solve the problem will not decay indefinitely. In weight decay, then, the system achieves a balance between pattern error (Eq. 67) and some measure of overall weight. It can be shown (Problem 42) that the weight decay is equivalent to gradient descent in a new effective error or criterion function:

$$J_{ef} = J(\mathbf{w}) + \frac{2\epsilon}{\eta} \mathbf{w}^t \mathbf{w}.$$ (39)

The second term on the right-hand side of Eq. 39 is sometimes called a regularization term (cf. Eq. 67) and in this case preferentially penalizes a single large weight. Another version of weight decay includes a decay parameter that depends upon the value of the weight itself, and this tends to distribute the penalty throughout the network:

$$J_{ef} = J(\mathbf{w}) + \frac{2\epsilon}{\eta} \sum_{i,j} \frac{w_{ij}^2/(\mathbf{w}^t\mathbf{w})}{1 + w_{ij}^2/(\mathbf{w}^t\mathbf{w})}.$$ (40)

6.8.12 Hints

Often we have insufficient training data for the desired classification accuracy and we would like to add information or constraints to improve the network. The approach of learning with *hints* is to add output units for addressing an ancillary problem, one different from but related to the specific classification problem at hand. The expanded network is trained on the classification problem of interest *and* the ancillary one, possibly simultaneously. For instance, suppose we seek to train a network to classify c phonemes based on some acoustic input. In a standard neural network we would have c output units. In learning with hints, we might add two ancillary output units, one which represents vowels and the other consonants. During training, the target vector must be lengthened to include components for the hint outputs. During classification, the hint units are not used; they and their hidden-to-output weights can be discarded (Fig. 6.19).

The benefit provided by hints is in improved feature selection. So long as the hints are related to the classification problem at hand, the feature groupings useful for the hint task are likely to aid category learning. For instance, the feature groupings useful for distinguishing vowel sounds from consonants in general are likely to be useful for distinguishing the /b/ from /oo/ or the /g/ from /ii/ categories in particular. Alternatively, one can train just the hint units in order to develop improved hidden unit representations.

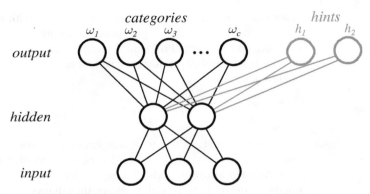

FIGURE 6.19. In learning with hints, the output layer of a standard network having c category units is augmented with hint units. During training, the target vectors are also augmented with signals for the hint units. In this way the input-to-hidden weights learn improved feature groupings. The hint units are not used during classification, and thus they and their hidden-to-output weights are removed from the trained network (red).

Learning with hints illustrates another benefit of neural networks: Hints are more easily incorporated into neural networks than into classifiers based on other algorithms, such as the nearest-neighbor or MARS.

6.8.13 On-Line, Stochastic or Batch Training?

Each of the three leading training protocols described in Section 6.3.2 has strengths and drawbacks. On-line learning is to be used when the amount of training data is so large, or when memory costs are so high, that storing the data is prohibitive. Most practical neural network classification problems are addressed instead with batch or stochastic protocols.

Batch learning is typically slower than stochastic learning. To see this, imagine a training set of 50 patterns that consists of 10 copies each of five patterns $(\mathbf{x}^1, \mathbf{x}^2, \dots, \mathbf{x}^5)$. In batch learning, the presentations of the duplicates of $\mathbf{x}^1$ provide as much information as a single presentation of $\mathbf{x}^1$ in the stochastic case. For example, suppose that in the batch case the learning rate is set optimally. The same weight change can be achieved with just a *single* presentation of each of the five different patterns in the batch case, so long as the learning rate is set correspondingly higher. Of course, true problems do not have exact duplicates of individual patterns; nevertheless, actual data sets are generally highly redundant, and the above analysis holds.

For most applications—especially ones employing large redundant training sets—stochastic training is hence to be preferred. Batch training admits some second-order techniques that cannot be easily incorporated into stochastic learning protocols and hence in some problems should be preferred, as we shall see in Section 6.9.

6.8.14 Stopped Training

In three-layer networks having many weights, excessive training can lead to poor generalization as the net implements a complex decision boundary "tuned" to the specific training data rather than the general properties of the underlying distributions. In training the two-layer networks of Chapter 5, we can usually train as long

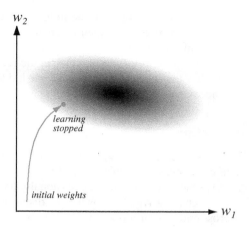

FIGURE 6.20. When weights are initialized with small magnitudes, stopped training leads to final weights that are smaller than they would be after extensive training. As such, stopped training behaves much like a form of weight decay.

as we like without fear that it would degrade final recognition accuracy because the *complexity* of the decision boundary is not changed—it is always simply a hyperplane. For this reason the general phenomenon should be called "overfitting," not "overtraining."

Because the network weights are initialized with small values, the units operate in their linear range and the full network implements linear discriminants. As training progresses, the nonlinearities of the units are expressed and the decision boundary warps. Qualitatively speaking, then, stopping the training before gradient descent is complete can help avoid overfitting. In practice, it is hard to know beforehand what stopping criterion θ should be set in line 5 of Algorithm 1. A far simpler method is to stop training when the error on a separate validation set reaches a minimum (Fig. 6.6). In Chapter 9 we shall explore the theory underlying this technique of validation and more generally cross-validation. We note in passing that weight decay behaves much like a form of stopped training (Fig. 6.20).

6.8.15 Number of Hidden Layers

The backpropagation algorithm applies equally well to networks with three, four, or more layers, so long as the units in such layers have differentiable activation functions. Because, as we have seen, three layers suffice to implement any arbitrary function, we would need special problem conditions or requirements to recommend the use of more than three layers.

One possible such requirement is translation, rotation or other distortion invariances. If the input layer represents the pixel image in an optical character recognition problem, we generally want such a recognizer to be invariant with respect to such transformations. It is easier for a four-layer net to learn translations than for a three-layer net. This is because each layer can generally easily learn an invariance within a limited range of parameters—for instance, a lateral shift of just two pixels. Stacking multiple layers, then, allows the full network to learn shifts of up to four pixels as the full invariance task is distributed throughout the net. Naturally, the weight initialization, learning rate, and data preprocessing arguments apply to these networks too. Some functions can be implemented more efficiently (i.e., with fewer total units) in networks with more than one hidden layer. It has been found empirically that networks with multiple hidden layers are more prone to getting caught in undesirable local minima, however.

In the absence of a problem-specific reason for multiple hidden layers, then, it is simplest to proceed using just a single hidden layer, but also to try two hidden layers if necessary.

6.8.16 Criterion Function

The squared error of Eq. 9 is the most common training criterion because it is simple to compute, is nonnegative, and simplifies the proofs of some theorems. Nevertheless, other training criteria occasionally have benefits. One popular alternative is the cross entropy which measures a "distance" between probability distributions. The cross entropy for n patterns is of the form:

$$J_{ce}(\mathbf{w}) = \sum_{m=1}^{n} \sum_{k=1}^{c} t_{mk} \ln(t_{mk}/z_{mk}), \tag{41}$$

where t_{mk} and z_{mk} are the target and the actual output of unit k for pattern m. Of course, to interpret J as an entropy, the target and output values must be interpreted as probabilities and must fall between 0 and 1.

MINKOWSKI ERROR

Yet another criterion function is based on the *Minkowski error*:

$$J_{Mink}(\mathbf{w}) = \sum_{m=1}^{n} \sum_{k=1}^{c} |z_{mk}(\mathbf{x}) - t_{mk}(\mathbf{x})|^{R}, \tag{42}$$

much as we saw in Chapter 4. It is a straightforward matter to derive the backpropagation rule for this error (Problem 30). While in general the rule is a bit more complex than for the ($R = 2$) sum-squared error we have considered, the Minkowski error for $1 \leq R < 2$ reduces the influence of long tails in the distributions—tails that may be quite far from the category decision boundaries. As such, the designer can adjust the "locality" of the classifier indirectly through choice of R; the smaller the R, the more local the classifier.

Most of the heuristics described in this section can be used alone or in combination. While they may interact in unexpected ways, all have found use in important pattern recognition problems and classifier designers should have experience with all of them.

*6.9 SECOND-ORDER METHODS

We have used a second-order analysis of the error in order to determine the optimal learning rate. We can use second-order information more fully in other ways, including the elimination of unneeded weights in a network.

6.9.1 Hessian Matrix

We derived the first-order derivatives of a sum-squared-error criterion function in three-layer networks, summarized in Eqs. 17 and 21. We now turn to second-order derivatives, which find use in rapid learning methods as well as in some pruning or regularization algorithms. We treat the familiar sum-squared-error criterion for a network with a single output,

$$J(\mathbf{w}) = \frac{1}{2} \sum_{m=1}^{n} (t_m - z_m)^2, \tag{43}$$

where t_m and z_m are the target and output signals, and n the total number of training patterns. The elements in the Hessian matrix are thus

$$\frac{\partial^2 J(\mathbf{w})}{\partial w_{ji} \partial w_{lk}} = \frac{1}{n} \left(\sum_{m=1}^{n} \frac{\partial J}{\partial w_{ji}} \frac{\partial J}{\partial w_{lk}} + \underbrace{\sum_{m=1}^{n} (z - t) \frac{\partial^2 J}{\partial w_{ji} \partial w_{lk}}}_{O(\|\mathbf{t}-\mathbf{z}\|)} \right), \tag{44}$$

where we have used the subscripts to refer to *any* weight in the network; hence i, j, l, and k could all take on values that describe input-to-hidden weights, or that describe hidden-to-output weights, or mixtures. Of course the Hessian matrix is symmetric. The second term in Eq. 44 is of order $O(\|\mathbf{t} - \mathbf{z}\|)$, which is generally small and thus often neglected. This approximation guarantees that the resulting approximation is positive definite and thus gradient descent will progress. In this *outer product approximation* our Hessian reduces to:

$$\mathbf{H} = \frac{1}{n} \sum_{m=1}^{n} \mathbf{X}^{[m]t} \mathbf{X}^{[m]} \tag{45}$$

where the superscript $[m]$ indexes the pattern and $\mathbf{X} = \partial J / \partial \mathbf{w}$ can be broken into two parts as

$$\mathbf{X} = \left(\begin{array}{c} \mathbf{X}_u \\ \mathbf{X}_v \end{array} \right). \tag{46}$$

Here $\mathbf{X}_v$ refers to derivatives with respect to the hidden-to-output weights, and $\mathbf{X}_u$ refers to the input-to-hidden weights. For a $d - n_H - 1$ neuron three-layer network, these vectors of derivatives can be written (Problem 31)

$$\mathbf{X}_v^t = (f'(net) y_1, \ldots, f'(net) y_{n_H}) \tag{47}$$

and

$$\mathbf{X}_u^t = (f'(net) f'(net_1) y_1 x_1, \ldots, f'(net) f'(net_{n_H} x_d)) \tag{48}$$

Equations 45, 47 and 48, enable the approximation to the Hessian be computed in a straightforward fashion.

6.9.2 Newton's Method

We can use a Taylor series to describe the change in the criterion function due to a change in weights $\Delta \mathbf{w}$ as

$$\Delta J(\mathbf{w}) = J(\mathbf{w} + \Delta \mathbf{w}) - J(\mathbf{w})$$

$$\simeq \left(\frac{\partial J(\mathbf{w})}{\partial \mathbf{w}}\right)^t \Delta \mathbf{w} + \frac{1}{2}\Delta \mathbf{w}^t \mathbf{H} \Delta \mathbf{w}, \tag{49}$$

where $\mathbf{H}$ is the Hessian matrix. We differentiate Eq. 49 with respect to $\Delta \mathbf{w}$ and find that $\Delta J(\mathbf{w})$ is minimized for

$$\left(\frac{\partial J(\mathbf{w})}{\partial \mathbf{w}}\right) + \mathbf{H}\Delta \mathbf{w} = \mathbf{0}. \tag{50}$$

Therefore, the optimum change in weights can be expressed as

$$\Delta \mathbf{w} = -\mathbf{H}^{-1}\left(\frac{\partial J(\mathbf{w})}{\partial \mathbf{w}}\right), \tag{51}$$

as we saw in Fig. 6.17. Thus, in gradient descent using Newton's algorithm, if we have an estimate for the weights at iteration m (that is, $\mathbf{w}(m)$), we can get an improved estimate using the weight change given by Eq. 51, that is,

$$\mathbf{w}(m+1) = \mathbf{w}(m) + \Delta \mathbf{w} \tag{52}$$

$$= \mathbf{w}(m) - \mathbf{H}^{-1}(m)\left(\frac{\partial J(\mathbf{w}(m))}{\partial \mathbf{w}}\right),$$

In Newton's algorithm we iteratively recompute $\mathbf{w}$ according to Eq. 52.

Alas, there are several drawbacks to this simple version of Newton's algorithm. The first is that for a network having N weights, the algorithm requires computing, storing and inverting the $N \times N$ Hessian matrix, which has complexity $O(N^3)$, making it impractical for all but small problems. Second, and more severe, is the fact that the algorithm need not converge in nonquadratic error surfaces, which often occur in practice. Nevertheless, an understanding of Newton's algorithm is excellent background for understanding more sophisticated methods, such as conjugate gradient descent (Section 6.9.4).

6.9.3 Quickprop

One of the simplest methods for using second-order information to increase training speed is the Quickprop algorithm. In this method, the weights are assumed to be independent, and the descent is optimized separately for each. The error surface is assumed to be quadratic and the coefficients for the particular parabola are determined by two successive evaluations of $J(w)$ and $dJ(w)/dw$. The single weight w is then moved to the computed minimum of the parabola (Fig. 6.21). It can be shown (Problem 35) that this approach leads to the following weight update rule,

$$\Delta w(m+1) = \frac{\frac{dJ}{dw}\big|_m}{\frac{dJ}{dw}\big|_{m-1} - \frac{dJ}{dw}\big|_m}\Delta w(m). \tag{53}$$

where the derivatives are evaluated at iterations m and $m - 1$, as indicated.

If the third- and higher-order terms in the error are nonnegligible, or if the assumption of weight independence does not hold, then the computed error minimum will not equal the true minimum, and further weight updates will be needed. When a

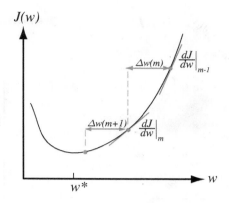

FIGURE 6.21. The quickprop weight update takes the error derivatives at two points separated by a known amount, and by Eq. 53 computes the next weight value. If the error can be fully expressed as a second-order function, then the weight update leads to the weight (w*) leading to minimum error.

number of obvious heuristics are imposed—to reduce the effects of estimation error when the surface is nearly flat, or the step actually increases the error—the method can be significantly faster than standard backpropagation. Another benefit is that each weight has, in effect, its own learning rate, and thus weights tend to converge at roughly the same time, thereby reducing problems due to nonuniform learning.

6.9.4 Conjugate Gradient Descent

Another fast learning method is conjugate gradient descent, which employs a series of line searches in weight or parameter space. One picks the first descent direction (for instance, the simple gradient) and moves along that direction until the local minimum in error is reached. The second descent direction is then computed: This direction—the "conjugate direction"—is the one along which the gradient does not change its *direction*, but merely its magnitude during the next descent. Descent along this direction will not "spoil" the contribution from the previous descent iterations (Fig. 6.22).

More specifically, we let $\Delta\mathbf{w}(m-1)$ represent the *direction* of a line search on step $m-1$. Note especially that this is not an overall *magnitude* of change, which will be determined by the line search. We demand that the subsequent direction, $\Delta\mathbf{w}(m)$, obey

$$\Delta\mathbf{w}^t(m-1)\mathbf{H}\Delta\mathbf{w}(m) = 0, \tag{54}$$

where $\mathbf{H}$ is the Hessian matrix. Pairs of descent directions that obey Eq. 54 are called "conjugate." If the Hessian is proportional to the identity matrix, then such directions are orthogonal in weight space. Conjugate gradient requires batch training, because the Hessian matrix is defined over the full training set.

The descent direction on iteration m is in the direction of the gradient plus a component along the previous descent direction:

$$\Delta\mathbf{w}(m) = -\nabla J(\mathbf{w}(m)) + \beta_m \Delta\mathbf{w}(m-1); \tag{55}$$

the relative proportions of these contributions is governed by β_m. This proportion can be derived by ensuring that the descent direction on iteration m does not spoil that

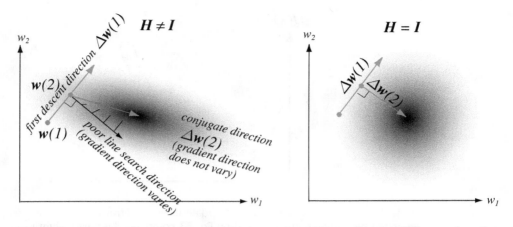

FIGURE 6.22. Conjugate gradient descent in weight space employs a sequence of line searches. If $\Delta\mathbf{w}(1)$ is the first descent direction, the second direction obeys $\Delta\mathbf{w}^t(1)\mathbf{H}\Delta\mathbf{w}(2) = 0$. Note especially that along this second descent, the gradient changes only in magnitude, not direction; as such, the second descent does not "spoil" the contribution due to the previous line search. In the case where the Hessian is diagonal (right), the directions of the line searches are orthogonal.

from direction $m - 1$, and indeed all earlier directions. It is generally calculated in one of two ways. The first formula (Fletcher-Reeves) is

$$\beta_m = \frac{\nabla J^t(\mathbf{w}(m))\nabla J(\mathbf{w}(m))}{\nabla J^t(\mathbf{w}(m-1))\nabla J(\mathbf{w}(m-1))}. \tag{56}$$

A slightly preferable formula (Polak-Ribiere) that is more robust in nonquadratic error functions is

$$\beta_m = \frac{\nabla J^t(\mathbf{w}(m))[\nabla J(\mathbf{w}(m)) - \nabla J(\mathbf{w}(m-1))]}{\nabla J^t(\mathbf{w}(m-1))\nabla J(\mathbf{w}(m-1))}. \tag{57}$$

Equations 55 and 37 show that conjugate gradient descent algorithm is analogous to calculating a "smart" momentum, where β_m plays the role of a momentum. If the error function is quadratic, then the convergence of conjugate gradient descent is guaranteed when the number of iterations equals the total number of weights.

EXAMPLE 1 Conjugate Gradient Descent

Consider finding the minimum of a simple quadratic criterion function centered on the origin of weight space, $J(\mathbf{w}) = 1/2(.2w_1^2 + w_2^2) = 1/2\mathbf{w}^t\mathbf{H}\mathbf{w}$, where by simple differentiation the Hessian is found to be $\mathbf{H} = \begin{pmatrix} .2 & 0 \\ 0 & 1 \end{pmatrix}$. We start descent at a randomly selected position, which for this figure happens to be $\mathbf{w}(0) = \begin{pmatrix} -8 \\ -4 \end{pmatrix}$. The first descent direction is determined by a simple gradient, which is easily found to be $-\nabla J(\mathbf{w}(0)) = -\begin{pmatrix} .4w_1(0) \\ 2w_2(0) \end{pmatrix} = \begin{pmatrix} 3.2 \\ 8 \end{pmatrix}$. In typical complex problems in high dimensions, the minimum along this direction is found using a line search; in this simple case the minimum can be found by calculus. We let s represent the distance along the first descent direction, and we find its value for the minimum of $J(\mathbf{w})$ according to

$$\frac{d}{ds}\left[\left[\begin{pmatrix}-8\\-4\end{pmatrix}+s\begin{pmatrix}3.2\\8\end{pmatrix}\right]^{t}\begin{pmatrix}.2 & 0\\0 & 1\end{pmatrix}\left[\begin{pmatrix}-8\\-4\end{pmatrix}+s\begin{pmatrix}3.2\\8\end{pmatrix}\right]\right]=0,$$

which has solution $s = 0.562$. Therefore the minimum along this direction is

$$\mathbf{w}(1) = \mathbf{w}(0) + 0.562(-\Delta J(\mathbf{w}(0)))$$

$$= \begin{pmatrix}-8\\-4\end{pmatrix}+0.562\begin{pmatrix}3.2\\8\end{pmatrix}=\begin{pmatrix}-6.202\\0.496\end{pmatrix}.$$

Now we turn to the use of conjugate gradients for the next descent. The simple gradient evaluated at $\mathbf{w}(1)$ is

$$-\nabla J(\mathbf{w}(1)) = -\begin{pmatrix}.4w_{1}(1)\\2w_{2}(1)\end{pmatrix}=\begin{pmatrix}2.48\\-0.99\end{pmatrix}.$$

(It is easy to verify that this direction, shown as a black arrow in the figure, does not point toward the global minimum at $\mathbf{w}^{*} = \mathbf{0}$.) We use the Fletcher-Reeves formula (Eq. 56) to construct the conjugate gradient direction:

$$\beta_{1} = \frac{\nabla J^{t}(\mathbf{w}(1))\nabla J(\mathbf{w}(1))}{\nabla J^{t}(\mathbf{w}(0))\nabla J(\mathbf{w}(0))} = \frac{(-2.48\ .99)\begin{pmatrix}-2.48\\.99\end{pmatrix}}{(-3.2\ 8)\begin{pmatrix}-3.2\\8\end{pmatrix}} = \frac{7.13}{74} = 0.096.$$

For this and all quadratic error surfaces, the Polak-Ribiere formula (Eq. 57) would give the same value as the Fletcher-Reeves formula. Thus the conjugate descent direction is

$$\nabla\mathbf{w}(1) = -\nabla J(\mathbf{w}(1)) + \beta_{1}\begin{pmatrix}1.6\\4\end{pmatrix}=\begin{pmatrix}2.788\\-.223\end{pmatrix}.$$

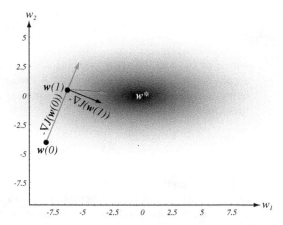

Conjugate gradient descent in a quadratic error landscape (indicated by the shading), shown in the density plot, starts at a random point $\mathbf{w}(0)$ and descends by a sequence of line searches. The first direction is given by the standard gradient and terminates at a minimum of the error—the point $\mathbf{w}(1)$. Standard gradient descent from $\mathbf{w}(1)$ would be along the black vector, "spoiling" some of the gains made by the first descent; it would, furthermore, miss the global minimum. Instead, the conjugate gradient (red vector) does not spoil the gains from the first descent, and it properly passes through the global error minimum at $\mathbf{w}^{*} = \begin{pmatrix}0\\0\end{pmatrix}$.

As above, rather than perform a traditional line search, we use calculus to find the error minimum along this second descent direction:

$$\frac{d}{ds}\left[[\mathbf{w}(1) + s\Delta\mathbf{w}(1)]^t\,\mathbf{H}\,[\mathbf{w}(1) + s\Delta\mathbf{w}(1)]\right] =$$

$$\frac{d}{ds}\left[\left[\begin{pmatrix} -6.202 \\ 0.496 \end{pmatrix} + s\begin{pmatrix} 2.788 \\ -.223 \end{pmatrix}\right]^t \begin{pmatrix} .2 & 0 \\ 0 & 1 \end{pmatrix} \left[\begin{pmatrix} -6.202 \\ 0.496 \end{pmatrix} + s\begin{pmatrix} 2.788 \\ -.223 \end{pmatrix}\right]\right] = 0,$$

which has solution $s = 2.23$. This yields the next minimum to be

$$\mathbf{w}(2) = \mathbf{w}(1) + s\Delta\mathbf{w}(1) = \begin{pmatrix} -6.202 \\ 0.496 \end{pmatrix} + 2.23\begin{pmatrix} 2.788 \\ -.223 \end{pmatrix} = \begin{pmatrix} 0 \\ 0 \end{pmatrix}.$$

Indeed, the conjugate gradient search finds the global minimum in this quadratic error function in two search steps—the number of dimensions of the space.

*6.10 ADDITIONAL NETWORKS AND TRAINING METHODS

We now consider some alternative networks and training methods that have proven effective in special classes of problems.

6.10.1 Radial Basis Function Networks (RBFs)

We have already considered several classifiers, such as Parzen windows, that employ densities estimated by localized basis functions such as Gaussians. In light of our discussion of gradient descent and backpropagation in particular, we now turn to a different method for training such networks. A radial basis function network with linear output units implements

$$z_k(\mathbf{x}) = \sum_{j=0}^{n_H} w_{kj}\phi_j(\mathbf{x}), \tag{58}$$

where we have included a $j = 0$ bias unit. If we define a vector $\boldsymbol{\phi}$ whose components are the hidden unit outputs, along with a matrix $\mathbf{W}$ whose entries are the hidden-to-output weights, then Eq. 58 can be rewritten as $\mathbf{z}(\mathbf{x}) = \mathbf{W}\boldsymbol{\phi}$. Minimizing the criterion function

$$J(\mathbf{w}) = \frac{1}{2}\sum_{m=1}^{n}\|\mathbf{y}(\mathbf{x}^m;\ \mathbf{w}) - \mathbf{t}^m\|^2 \tag{59}$$

is formally equivalent to the linear problem we saw in Chapter 5. We let $\mathbf{T}$ be the matrix consisting of target vectors and let $\boldsymbol{\phi}$ be the matrix whose columns are the vectors $\boldsymbol{\phi}$, then the solution weights obey

$$\boldsymbol{\phi}^t\boldsymbol{\phi}\mathbf{W}^t = \boldsymbol{\phi}^t\mathbf{T}, \tag{60}$$

and the solution can be written directly: $\mathbf{W}^t = \boldsymbol{\Phi}^\dagger\mathbf{T}$. Recall that $\boldsymbol{\phi}^\dagger$ is the pseudo-inverse of $\boldsymbol{\phi}$. One of the benefits of such radial basis function or RBF networks with linear output units is that the solution requires merely such standard linear techniques. Nevertheless, inverting large matrices can be computationally expensive, and thus the above method is generally confined to problems of moderate size.

If the output units are nonlinear, that is, if the network implements

$$z_k(\mathbf{x}) = f\left(\sum_{j=0}^{n_H} w_{kj}\phi_j(\mathbf{x})\right) \tag{61}$$

rather than Eq. 58, then standard backpropagation can be used. One need merely take derivatives of the localized activation functions. For classification problems it is traditional to use a sigmoid for the output units in order to keep the output values restricted to a fixed range. Some of the computational simplification afforded by sigmoidal activation functions in the hidden units functions is lost, but this presents no conceptual difficulties (Problem 38).

6.10.2 Special Bases

Occasionally we may have special information about the functional form of the distributions underlying categories, and then it makes sense to use corresponding hidden unit activation functions. In this way, fewer parameters need to be learned for a given quality of fit to the data. This is an example of increasing the bias of our model and thereby reducing the variance in the solution, a crucial topic we shall consider again in Chapter 9. For instance, if we know that each underlying distribution comes from a mixture of two Gaussians, naturally we would use Gaussian activation functions and use a learning rule that set the parameters, for instance as the mean and entries of the covariance matrix. This is very closely related to the model-dependent maximum-likelihood techniques we saw in Chapter 3.

6.10.3 Matched Filters

In Chapter 2 we treated the problem of designing a classifier in the ideal case that the full probability structure is known. Now we consider the design of a detector for a specific known pattern. This will lead us to the concept of a matched filter. While it comes as no surprise that the optimal detector for a pattern must "match" that pattern (in a way that will become clear), our treatment here lends deeper understanding of why this should be so. Matched filters appear in a wide range of detection problems, particularly those of time-varying signals. Although not a neural network, such filters have close relations to convolutional neural networks we will consider in Section 6.10.4. Furthermore, the maximum response in a hidden unit in a traditional three-layer network occurs when the input pattern "matches" the pattern of input weights leading to that hidden unit (Section 6.5.1).

IMPULSE
RESPONSE

Consider the problem of detecting a continuous signal $x(t)$ by a linear detector. We describe such a detector by its *impulse response*, $h(t)$, or, better yet, by its time-reversed impulse response $w(t) = h(-t)$. The output of a linear detector to an arbitrary input $x(t)$ is given by the integral

$$z(T) = \int_{-\infty}^{+\infty} x(t)w(t - T)\, dt. \tag{62}$$

where T is the relative offset of the signal.

Our goal in designing an optimal detector is thus to find the response function that gives the maximum output z; we denote this unknown filter function $\hat{w}(t)$. Of course the output can be made arbitrarily large by choosing a large $w(t)$. Such a solution is uninformative; we are interested in the *shape* of $w(t)$. For this reason, we impose a constraint

ENERGY

$$\int_{-\infty}^{+\infty} w^2(t)\,dt = const \tag{63}$$

where the constant is sometimes called the *energy* of the filter, in analogy with the total energy of a physical signal such as a sound wave or light wave.

The optimization problem, then, is to find the response that maximizes the output of Eq. 62 subject to the constraint of Eq. 63. Because we are searching for the *function* that leads to the maximum output, we employ calculus of variations. Specifically, we take the functional derivative of Eq. 63 and add Eq. 62 times an undetermined multiplier λ, and set it to zero. Without loss of generality, we can arbitrarily assign the offset to be $T = 0$, and thus we find

$$\delta z(0) = \delta \left(\int_{-\infty}^{+\infty} [w^2(t) + \lambda x(t)w(t)]\,dt \right) = 0 \tag{64}$$

or

$$\int_{-\infty}^{+\infty} \underbrace{[2w(t)\delta w(t) + \lambda x(t)\delta w(t)]}_{0 \text{ for extremum}}\,dt = 0. \tag{65}$$

Because Eq. 65 is true for all $\delta w(t)$, the integrand must vanish and this gives our solution: the optimal filter response is

$$\hat{w}(t) = \frac{-\lambda}{2}x(t). \tag{66}$$

In short, the optimal detector has a time-reversed impulse response proportional to the target signal (Fig. 6.23). The overall magnitude is determined by the energy constant, as conveyed by λ; it is easy to show that λ is negative. Finally, we should mention that technically speaking the above the derivation proves only that we have an *extremum*; nevertheless, it can be shown that indeed this solution gives a *maximum* (Problem 36).

6.10.4 Convolutional Networks

TIME DELAY
NEURAL
NETWORK

We can incorporate prior knowledge into the network architecture itself. For instance, if we demand that our classifier be insensitive to translations of the pattern, we can effectively replicate the recognizer at all such translations. This is the approach taken in time delay neural networks (or TDNNs).

Figure 6.24 shows a typical TDNN architecture; while the architecture consists of input, hidden and output layers, much as we have seen before, there is a crucial difference. Each hidden unit accepts input from a restricted spatial range of positions in the input layer. Hidden units at "delayed" locations (i.e., shifted to the right) accept

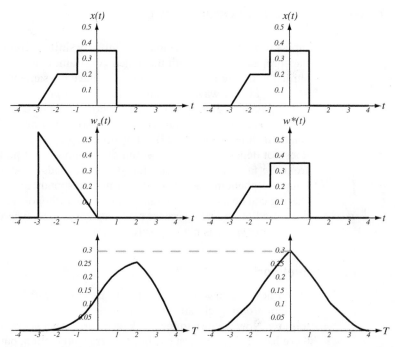

FIGURE 6.23. The left column shows a signal $x(t)$; beneath it is an arbitrary response function $w_a(t)$. At the bottom is the response of the filter as a function of the offset T, as given by Eq. 62. The right column shows the the case where the input and response function "match." The two response functions, $w_a(t)$ and $w^*(t)$ here have the same energy. Note particularly at the bottom that the maximum output in this case is greater than in the nonmatching case at the left.

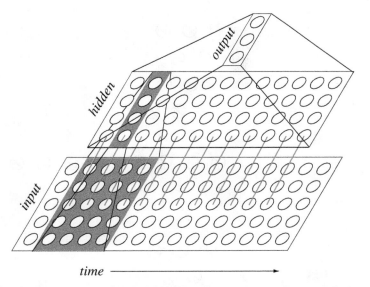

time ⟶

FIGURE 6.24. A time delay neural network (TDNN) uses weight sharing to ensure that patterns are recognized regardless of shift in one dimension; in practice, this dimension generally corresponds to time. Thus all weights shown in red are forced to have the same value. In this example, there are five input units at each time step. Because we hypothesize that the input patterns are of four time steps or less in duration, each of the hidden units at a given time step accepts inputs from only $4 \times 5 = 20$ input units, as highlighted in gray. An analogous translation constraint is also imposed between the hidden and output layer units.

inputs from the input layer that are similarly shifted. Training proceeds as in standard backpropagation, but with the added constraint that corresponding weights (that is, shifted to the right or left) are forced to have the same value—an example of *weight sharing*. In this way, the weights learned do not depend upon the position of the training pattern so long as the full pattern lies in the domain of the input layer.

WEIGHT
SHARING

The feedforward operation of the network during recognition is the same as in standard three-layer networks, but because of the weight sharing, the final output does not depend upon the position of the input pattern. The network gets its name from the fact that it was developed for and finds greatest use in speech and other temporal phenomena, where the shift corresponds to delays in time. Such weight sharing can be extended to translations in an orthogonal *spatial* dimensions, and has been used in optical character recognition systems, where the location of an image in the input space is not precisely known.

6.10.5 Recurrent Networks

Up to now we have considered only networks which use feedforward flow of information during classification; the only feedback flow was of error signals during training. Now we turn to feedback or *recurrent* networks. In their most general form, these have found greatest use in time series prediction, but we consider here just one specific type of recurrent net that has had some success in static classification tasks.

Figure 6.25 illustrates such a recurrent architecture, one in which the output unit values are fed back and duplicated as auxiliary inputs, augmenting the traditional

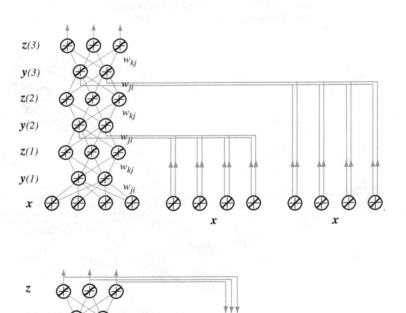

FIGURE 6.25. The form of recurrent network most useful for static classification has the architecture shown at the bottom, with the recurrent connections in red. It is functionally equivalent to a static network with many hidden layers and extensive weight sharing, as shown above.

feature values. During classification, a static pattern **x** is presented to the input units, the feedforward flow is computed, and the outputs are fed back as auxiliary inputs. This, in turn, leads to a different set of hidden unit activations, new output activations, and so on. Ultimately, the activations stabilize, and the final output values are used for classification. As such, this recurrent architecture, if "unfolded" in time, is equivalent to the static network shown at the top of the figure, where it must be understood that many sets of weights are constrained to be the same, as indicated.

Recurrent networks have proven effective in learning time-dependent signals whose structure varies over farily short periods. Such nets are less successful when applied to problems where the structure is longer term, since during training the error gets "diluted" when passed back through the layers many times.

6.10.6 Cascade-Correlation

The central notion underlying the training of networks by cascade-correlation is quite simple. We begin with a two-layer network and train to minimum of an LMS error. If the resulting training error is low enough, training is stopped. In the more common case in which the error is not low enough, we fix the weights but add a single hidden unit, fully connected from inputs and to output units. Then these new weights

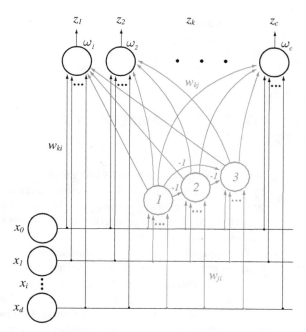

FIGURE 6.26. The training of a multilayer network via cascade-correlation begins with the input later fully connected to the output layer (black). Such weights, w_{ki}, are fully trained using an LMS criterion, as discussed in Chapter 5. If the resulting training error is not sufficiently low, a first hidden unit (labeled 1, in red) is introduced, fully interconnected from the input layer and to the output layer. These new red weights are fully trained, while the previous (black) ones are held fixed. If the resulting training error is still not sufficiently low, a second hidden unit (labeled 2) is likewise introduced, fully interconnected; it also receives a the output from each previous hidden unit, multiplied by −1. Training proceeds in this way, training successive hidden units until the training error is acceptably low.

are trained using an LMS criterion. If the resulting error is not sufficiently low, yet another hidden unit is added, fully connected from the input layer and to the output layer. Furthermore, the output of each previous hidden unit is multiplied by a fixed weight of -1 and presented to the new hidden unit; this prevents the new hidden unit from learning function already represented by the previous hidden units. Then the new weights are trained on an LMS criterion. Thus training proceeds by alternatively training weights, then (if needed) adding a new hidden unit, training the new modifiable weights, and so on. In this way the network grows to a size that depends upon the problem at hand (Fig. 6.26). The benefit of cascade-correlation is that it is often faster than traditional backpropagation since fewer weights are updated at any time.

■ **Algorithm 4. (Cascade-Correlation)**

$$
\begin{array}{ll}
1 & \underline{\textbf{begin initialize}} \ \mathbf{a}, \text{criterion } \theta, \eta, k \leftarrow 0 \\
2 & \quad \underline{\textbf{do}} \ m \leftarrow m + 1 \\
3 & \qquad w_{ki} \leftarrow w_{ki} - \eta \nabla J(\mathbf{w}) \\
4 & \quad \underline{\textbf{until}} \ \|\nabla J(\mathbf{w})\| < \theta \\
5 & \quad \underline{\textbf{if}} \ J(\mathbf{w}) > \theta \ \underline{\textbf{then}} \ \text{add hidden unit} \ \underline{\textbf{else}} \ \underline{\textbf{exit}} \\
6 & \quad \underline{\textbf{do}} \ m \leftarrow m + 1 \\
7 & \qquad w_{ji} \leftarrow w_{ji} - \eta \nabla J(\mathbf{w}); \ w_{kj} \leftarrow w_{kj} - \eta \nabla J(\mathbf{w}) \\
8 & \quad \underline{\textbf{until}} \ \|\nabla J(\mathbf{w})\| < \theta \\
9 & \quad \underline{\textbf{return}} \ \mathbf{w} \\
10 & \underline{\textbf{end}}
\end{array}
$$

6.11 REGULARIZATION, COMPLEXITY ADJUSTMENT AND PRUNING

Whereas the number of inputs and outputs of a three-layer network are determined by the problem itself, we do not know ahead of time the number of hidden units, or weights. If we have too many weights and thus degrees of freedom and train too long, there is a danger of overfitting. If we have too few weights, the training set cannot be learned.

One general approach of regularization is to make a new criterion function that depends not only on the classical training error we have seen, but also on classifier complexity. Specifically the new criterion function penalizes highly complex models; searching for the minimum in this criterion is to balance error on the training set with complexity. Formally, then, we can write the new error as the sum of the familiar error on the training set plus a regularization term, which expresses constraints or desirable properties of solutions:

$$ J = J_{pat} + \lambda J_{reg}. \tag{67} $$

The parameter λ is adjusted to impose the regularization more or less strongly. Clearly weight decay (Section 6.8.11) can be cast in this form, where J_{reg} is large for large weights.

Another approach is to eliminate—*prune*—weights that are least needed. It is natural to imagine that after training it is the weights with the smallest magnitude that can be eliminated. Such *magnitude-based pruning* can work, but is provably nonoptimal; sometimes weights with small magnitudes are important for learning the training data.

WALD STATISTIC

The fundamental idea in *Wald statistics* is that we can estimate the importance of a parameter in a model, and then eliminate the parameter having the least such importance. In a network, such a parameter would be a weight. The *Optimal Brain Damage* (OBD) algorithm and its descendent, *Optimal Brain Surgeon* (OBS), use a second-order approximation to predict how the training error depends upon a weight, and eliminate a weight that leads to the smallest increase in the training error.

OPTIMAL BRAIN DAMAGE
OPTIMAL BRAIN SURGEON

Optimal Brain Damage and Optimal Brain Surgeon share the same basic approach of training a network to local minimum in error at weight $\mathbf{w}^*$, and then pruning a weight that leads to the smallest increase in the training error. The predicted functional increase in the error for a change in full weight vector $\delta\mathbf{w}$ is

$$\delta J = \underbrace{\left(\frac{\partial J}{\partial \mathbf{w}}\right)^t \cdot \delta\mathbf{w}}_{\simeq 0} + \frac{1}{2}\,\delta\mathbf{w}^t \cdot \underbrace{\frac{\partial^2 J}{\partial \mathbf{w}^2}}_{\equiv \mathbf{H}} \cdot \delta\mathbf{w} + \underbrace{O(\|\delta\mathbf{w}\|^3)}_{\simeq 0} \tag{68}$$

where $\mathbf{H}$ is the Hessian matrix. The first term vanishes because we are at a local minimum in error; we ignore third- and higher-order terms. The general solution for minimizing this function given the constraint of deleting one weight is (Problem 44):

$$\delta\mathbf{w} = -\frac{w_q}{[\mathbf{H}^{-1}]_{qq}}\,\mathbf{H}^{-1} \cdot \mathbf{u}_q \qquad \text{and} \qquad L_q = \frac{1}{2}\,\frac{w_q^2}{[\mathbf{H}^{-1}]_{qq}}\,. \tag{69}$$

SALIENCY

Here, $\mathbf{u}_q$ is the unit vector along the qth direction in weight space and L_q is approximation to the *saliency* of weight q—the increase in training error if weight q is pruned and the other weights updated by the left equation in Eq. 69 (Problem 45).

We saw in Eq. 45 the outer product approximation for calculating the Hessian matrix. Note that Eq. 69 requires the *inverse* of $\mathbf{H}$. One method to calculating this inverse matrix is to start with a small value, $\mathbf{H}_0^{-1} = \alpha^{-1}\mathbf{I}$, where α is a small parameter—effectively a weight decay constant (Problem 43). Next we update the matrix with

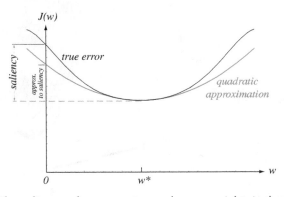

FIGURE 6.27. The saliency of a parameter, such as a weight, is the increase in the training error when that weight is set to zero. One can approximate the saliency by expanding the true error around a local minimum, w^*, and setting the weight to zero. In this example the approximated saliency is smaller than the true saliency; this is typically, but not always, the case.

each pattern according to

$$\mathbf{H}_{m+1}^{-1} = \mathbf{H}_m^{-1} - \frac{\mathbf{H}_m^{-1}\mathbf{X}_{m+1}\mathbf{X}_{m+1}^T\mathbf{H}_m^{-1}}{\frac{n}{a_m} + \mathbf{X}_{m+1}^T\mathbf{H}_m^{-1}\mathbf{X}_{m+1}}, \tag{70}$$

where the subscripts correspond to the pattern being presented and a_m decreases with m. After the full training set has been presented, the inverse Hessian matrix is given by $\mathbf{H}^{-1} = \mathbf{H}_n^{-1}$. Figure 6.28 illustrates the effects of pruning based on OBD, OBS, and the magnitude of a weight in a network.

In algorithmic form, the Optimal Brain Surgeon method is:

■ **Algorithm 5.   (Optimal Brain Surgeon)**

1 **begin initialize** $n_H, \mathbf{w}, \theta$
2   train a reasonably large network to minimum error
3   **do** compute $\mathbf{H}^{-1}$ by Eq. 70
4              $q^* \leftarrow \underset{q}{\arg\min}\, w_q^2/(2[\mathbf{H}^{-1}]_{qq})$ (saliency L_q)
5              $\mathbf{w} \leftarrow \mathbf{w} - \frac{w_{q^*}}{[\mathbf{H}^{-1}]_{q^*q^*}}\mathbf{H}^{-1}\mathbf{e}_{q^*}$
6   **until** $J(\mathbf{w}) > \theta$
7   **return w**
8 **end**

Optimal Brain Damage method is computationally simpler because the calculation of the inverse Hessian matrix in line 3 is particularly simple for a diagonal matrix. The above algorithm terminates when the error is greater than a criterion initialzed to be θ. Another approach is to change line 6 to terminate when the *change* in $J(\mathbf{w})$ due to the elimination of a weight is greater than some criterion value.

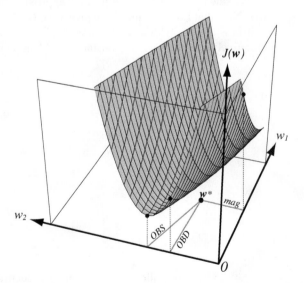

FIGURE 6.28. The figure shows a quadratic error surface as a function of weights, $J(\mathbf{w})$, and the global minimum at $\mathbf{w}^*$. In the second-order approximation to the criterion function, Optimal Brain Damage assumes the Hessian matrix is diagonal, while Optimal Brain Surgeon uses the full Hessian matrix.

SUMMARY

Multilayer nonlinear neural networks—nets with two or more layers of modifiable weights—trained by gradient descent methods such as backpropagation perform a maximum-likelihood estimation of the weight values in the model defined by the network topology. The nonlinear activation function $f(net)$ at the hidden units allows such networks to form an arbitrary decision boundary, so long as there are sufficiently many hidden units.

One of the great benefits of learning in such networks is the simplicity of the learning algorithm, the ease in model selection, and the incorporation of heuristic information and constraints. Such heuristics include weight decay (which penalizes large weights and thus effectively simplifies the classifier) and learning with hints (in which the network also learns an ancillary related task). The parameters of the nonlinear activation function $f(net)$, preprocessing of patterns, target values and weight initialization can all be founded on statistical principles that will ensure fast and uniform learning. The three primary learning protocols are stochastic, batch, and online. Discrete pruning algorithms such as Optimal Brain Surgeon and Optimal Brain Damage correspond to priors favoring *few* weights and can help avoid overfitting.

Alternative networks and training algorithms have benefits. For instance, radial basis functions are most useful when the data clusters. Cascade-correlation networks are generally faster than backpropagation.

BIBLIOGRAPHICAL AND HISTORICAL REMARKS

One of the earliest discussions of what we would now recognize as a neural network was due to the pioneer of modern computer science, Alan Turing, who in a 1948 paper described his "B-type unorganized machine," which consisted of networks of NAND gates [79]. (This forward-looking work was dismissed as a "schoolboy essay" by the head of the lab where Turing worked, Sir Charles Darwin, the grandson of the great English biologist and naturalist.) McCulloch and Pitts provided the first principled mathematical and logical treatment of the behavior of networks of simple neurons [51]. This early work addressed nonrecurrent as well as recurrent nets (those possessing "circles" in their terminology), but not learning. Its concentration on all-or-none or threshold function of neurons indirectly delayed the consideration of continuous valued neurons that would later dominate the field. These authors later wrote an extremely important paper on featural mapping, invariances, and learning in nervous systems and thereby advanced the conceptual development of pattern recognition [58].

Rosenblatt's work on the (two-layer) perceptron (cf. Chapter 5) [64, 65] was some of the earliest to address learning and was the first to include proofs about convergence. A number of stochastic methods, including Pandemonium [69, 70], were developed for training networks with several layers of processors, though in keeping with the preoccupation with threshold functions, such processors generally computed logical functions (AND or OR), rather than some continuous functions favored in later neural network research. The limitations of networks implementing linear discriminants—linear machines—were well known in the 1950s and 1960s and discussed by both their promoters [65] and their detractors [53].

A popular early method was to design by hand three-layer networks with fixed input-to-hidden weights, and then train the hidden-to-output weight; reference [86] provides a review. Much of the difficulty in finding learning algorithms for all layers in a multilayer neural network came from the prevalent use of linear threshold units. Because these do not have useful derivatives throughout their entire range, the current approach of applying the chain rule for derivatives and the resulting "back-propagation of errors" did not gain more adherents earlier.

The development of backpropagation was gradual, with several steps, not all of which were appreciated at the time they were introduced. The earliest application of adaptive methods that would ultimately become backpropagation came from the field of control [6]. Kalman filtering from electrical engineering used an analog error (difference between predicted and measured output) for adjusting gain parameters in predictors, as described in references [30] and [39]. Bryson, Denham, and Drey-fus showed how Lagrangian methods could train multilayer networks for control, as presented in reference [7]. We saw in the last chapter the work of Widrow, Hoff, and their colleagues [87, 88] in using analog signals and the LMS training criterion applied to pattern recognition in two-layer networks. Werbos [83, 84], too, discussed a method for calculating the derivatives of a function based on a sequence of samples, which, if interpreted carefully, carried the key ideas of backpropagation. Parker's early "Learning logic" [55, 56], developed independently, showed how layers of lin-ear units could be learned by a sufficient number of input-output pairs. This work lacked simulations on representative or challenging problems and terminology in the field, and alas was not appreciated adequately. Le Cun independently developed a learning algorithm for three-layer networks [10, in French] in which target values are propagated, rather than derivatives; the resulting learning algorithm is equivalent to standard backpropagation, as was pointed out shortly thereafter [11].

Without question, the paper by Rumelhart, Hinton and Williams [67], later ex-panded into a full and readable chapter [68], brought the backpropagation method to the attention of the widest audience. These authors clearly appreciated the power of the method, demonstrated it on key tasks (such as XOR), and applied it to pattern recognition more generally. An enormous number of papers and books of applications—speech perception, optical character recognition, data mining, finance, game playing, and much more—continues unabated. One view of the history of backpropagation is presented in reference [84]; two collections of key papers in the history of neural processing more generally, including many in pattern recognition, are presented in references [2] and [3]. One novel class of such networks includes generalization in the *production* of new patterns [22, 23].

Clear elementary papers on neural networks can be found in [37] and [48], and several good textbooks, which differ from the current one in their emphasis on neural networks over other pattern recognition techniques, can be recommended, particu-larly textbooks [4, 29, 31, 61] and [63]. An extensive treatment of the mathematical aspects of networks, much of which is beyond that needed for mastering the use of networks for pattern classification, can be found in reference [21]. There is continued exploration of the strong links between networks and more standard statistical meth-ods; White presents an overview [85], and books such as references [9] and [71] explore a number of close relationships. The important relation of multilayer per-ceptrons to Bayesian methods and probability estimation can be found in references [5, 14, 25, 45, 54, 62] and [66]. Original papers on projection pursuit and MARS can be found in references [16] and [35], respectively, and a good overview can be found in reference [63].

Shortly after its wide dissemination, the backpropagation algorithm was criticized for its lack of biological plausibility; in particular, Grossberg [24] discussed the non-local nature of the algorithm—that is, that synaptic weight values were transported without physical means. Somewhat later, Stork devised a local implementation of backpropagation [47, 75], and pointed out that it was nevertheless highly implausible as a biological model.

The discussions and debates over the relevance of Kolmogorov's theorem [40] to neural networks—such as references [13, 20, 34, 38, 41, 42] and [44]—have centered on their expressive power. The proof of the universal expressive power of three-layer nets based on bumps and Fourier ideas appears in reference [32]. The expressive power of networks having non-traditional activation functions was explored in references [76], [77], and elsewhere. The fact that three-layer networks can have local minima in the criterion function was explored in reference [52], and some of the properties of error surfaces were illustrated in reference [36].

The outer product approximation to the Hessian matrix and deeper analysis of second-order methods can be found in references [26, 46, 50] and [60]. Three-layer networks trained via cascade-correlation have been shown to perform well compared to standard three-layer nets trained via backpropagation [15]. Although there was little new from a learning theory presented in Fukushima's neocognitron [17, 18], its use of many layers and mixture of hand-crafted feature detectors and learning groupings showed how networks could address shift, rotation, and scale invariance. The method of matched filters predates neural networks; Stork and Levinson used matched filters to explore human visual response functions, as described in references [74] and [78].

A simple method of weight decay was introduced in reference [33], and it gained greater acceptance due to the work of Weigend and others [82]. The method of hints was introduced in reference [1]. While the Wald test [80, 81] has been used in traditional statistical research [72], its application to multilayer network pruning began with the work of Le Cun et al.'s Optimal Brain Damage method [12], later extended to include nondiagonal Hessian matrices [28, 26, 27], including some speedup methods [73]. An extensive review of the computation and use of second-order derivatives in networks can be found in reference [8], and a good review of pruning algorithms can be found in reference [60].

PROBLEMS

Section 6.2

1. Show that if the activation function of the hidden units is linear, a three-layer network is equivalent to a two-layer one. Use your result to explain why a three-layer network with linear hidden units cannot solve a nonlinearly separable problem such as XOR or d-bit parity.

2. Fourier's theorem can be used to show that a three-layer neural net with sigmoidal hidden units can approximate to arbitrary accuracy any posterior function. Consider two-dimensional input and a single output, $z(x_1, x_2)$. Recall that Fourier's theorem states that, given weak restrictions, any such function can be written as a possibly infinite sum of cosine functions, as

$$z(x_1, x_2) \approx \sum_{f_1} \sum_{f_2} A_{f_1 f_2} \cos(f_1 x_1) \, \cos(f_2 x_2),$$

with coefficients $A_{f_1 f_2}$.

(a) Use the trigonometric identity

$$\cos(\alpha)\,\cos(\beta) = \frac{1}{2}\cos(\alpha + \beta) + \frac{1}{2}\cos(\alpha - \beta)$$

to write $z(x_1, x_2)$ as a linear combination of terms $\cos(f_1 x_1 + f_2 x_2)$ and $\cos(f_1 x_1 - f_2 x_2)$.

(b) Show that $\cos(x)$ or indeed any continuous function $f(x)$ can be approximated to any accuracy by a linear combination of sign functions as:

$$f(x) \approx f(x_0) + \sum_{i=0}^{N} \left[f(x_{i+1}) - f(x_i) \right] \left[\frac{1 + \text{Sgn}(x - x_i)}{2} \right],$$

where the x_i are sequential values of x; the smaller $x_{i+1} - x_i$, the better the approximation.

(c) Put your results together to show that $z(x_1, x_2)$ can be expressed as a linear combination of step functions or sign functions whose arguments are themselves linear combinations of the input variables x_1 and x_2. Explain, in turn, why this implies that a three-layer network with sigmoidal hidden units and a linear output unit can approximate any function that can be expressed by a Fourier series.

(d) Does your construction guarantee that the derivative $df(x)/dx$ can be well approximated too?

Section 6.3

3. Consider a $d - n_H - c$ network trained with n patterns for m_e epochs.

(a) What is the space complexity in this problem? (Consider both the storage of network parameters as well as the storage of patterns, but not the program itself.)

(b) Suppose the network is trained in stochastic mode. What is the time complexity? Because this is dominated by the number of multiply-accumulates, use this as a measure of the time complexity.

(c) Suppose the network is trained in batch mode. What is the time complexity?

4. Prove that the formula for the sensitivity δ for a hidden unit in a three-layer net (Eq. 21) generalizes to a hidden unit in a four- (or higher-) layer network, where the sensitivity is the weighted sum of sensitivities of units in the next higher layer.

5. Explain in words why the backpropagation rule for training input-to-hidden weights makes intuitive sense by considering the dependency upon each of the terms in Eq. 21.

6. One might guess that the backpropagation learning rules should be *inversely* related to $f'(net)$—that is, that weight change should be *large* where the output does not vary. In fact, as shown in Eq. 17, the learning rule is linear in $f'(net)$. Explain intuitively why the learning rule should be linear in $f'(net)$.

7. Show that the learning rule described in Eqs. 17 and 21 works for bias, where $x_0 = y_0 = 1$ is treated as another input and hidden unit.

8. Consider a standard three-layer backpropagation net with d input units, n_H hidden units, c output units, and bias.

    (a) How many weights are in the net?

    (b) Consider the symmetry in the value of the weights. In particular, show that if the sign is flipped on every weight, the network function is unaltered.

    (c) Consider now the hidden unit exchange symmetry. There are no labels on the hidden units, and thus they can be exchanged (along with corresponding weights) and leave network function unaffected. Prove that the number of such equivalent labelings—the exchange symmetry factor—is thus $n_H 2^{n_H}$. Evaluate this factor for the case $n_H = 10$.

9. Write pseudocode for on-line version of backpropagation training, being careful to distinguish it from stochastic and batch procedures.

10. Express the derivative of a sigmoid in terms of the sigmoid itself in the following two cases (for positive constants a and b):

    (a) A sigmoid that is purely positive: $f(net) = \frac{1}{1+e^{a\,net}}$.

    (b) An antisymmetric sigmoid: $f(net) = a \tanh(b\,net)$.

11. Generalize the backpropagation to four layers, and individual (smooth, differentiable) activation functions at each unit. In particular, let x_i, y_j, v_l and z_k denote the activations on units in successive layers of a four-layer fully connected network, trained with target values t_k. Let f_{1i} be the activation function of unit i in the first layer, f_{2j} in the second layer, and so on. Write a program, with greater detail than that of Algorithm 1, showing the calculation of sensitivities, weight update, and so on, for your general four-layer network.

Section 6.4

12. Explain why the input-to-hidden weights must be different from each other (e.g., random) or else learning cannot proceed well (cf. Computer exercise 2). Specifically, what happens if the weights are initialized so as to have identical values?

13. If the labels on the two hidden units are exchanged (and the weight values changed appropriately), the shape of the error surface is unaffected. Consider a $d - n_H - c$ three-layer network. How many equivalent relabellings of the units (and their associated weights) are there?

14. Show that proper preprocessing of the data will lead to faster convergence, at least in a simple network 2-1 (two-layer) network with bias. Suppose the training data come from two Gaussians, $p(x|\omega_1) \sim N(-.5, 1)$ and $p(x|\omega_2) \sim N(+.5, 1)$. Let the teaching values for the two categories be $t = \pm 1$.

    (a) Write the error as a sum over the n patterns of a function of the weights, inputs, and other parameters.

    (b) Differentiate twice with respect to the weights to get the Hessian matrix $\mathbf{H}$.

    (c) Consider two data sets drawn from $p(\mathbf{x}|\omega_i) \sim N(\boldsymbol{\mu}_i, \mathbf{I})$ for $i = 1, 2$ and $\mathbf{I}$ is the 2-by-2 identity matrix. Calculate your Hessian matrix in terms of $\boldsymbol{\mu}_i$.

    (d) Calculate the maximum and minimum eigenvalues of the Hessian in terms of the components of $\boldsymbol{\mu}_i$.

    (e) Suppose $\boldsymbol{\mu}_1 = (1, 0)^t$ and $\boldsymbol{\mu}_2 = (0, 1)^t$. Calculate the ratio of the eigenvalues, and hence a measure of the convergence time.

(f) Now standardize your data, by subtracting means and scaling to have unit covariances in each of the two dimensions. That is, find two new distributions that have overall zero mean and the same covariance. Check your answer by calculating the ratio of the maximum to minimum eigenvalues.

(g) If T denotes the total training time for the unprocessed data, express the time required for the preprocessed data (cf. Computer exercise 12).

15. Consider the derivation of the bounds on the convergence time for simple gradient descent. Assume that the error function can be well described by a Hessian matrix $\mathbf{H}$ having d eigenvalues $\lambda_1, \lambda_2, \ldots, \lambda_d$, where λ_{max} and λ_{min} are the maximum and minimum values in this set. Assume that the learning rate has been set to the optimum for one of the dimensions, as based on Eq. 36.

(a) Express the optimum rate in terms of the appropriate eigenvalue, that is, either λ_{max} or λ_{min}.

(b) State a convergence criterion for learning.

(c) In terms of the variables given, compute the time for your system to meet your convergence criterion.

16. Assume that the criterion function $J(\mathbf{w})$ is well-described to second order by a Hessian matrix $\mathbf{H}$.

(a) Show that convergence of learning is ensured if the learning rate obeys $\eta < 2/\lambda_{max}$, where λ_{max} is the largest eigenvalue of $\mathbf{H}$.

(b) Show that the learning time is thus dependent upon the ratio of the largest to the smallest nonnegligible eigenvalue of $\mathbf{H}$.

(c) Explain why "standardizing" the training data can therefore reduce learning time.

(d) What is the relation between such standardizing and the whitening transform described in Chapter 2?

Section 6.6

17. Fill in the steps in the derivation leading to Eq. 26.

18. Confirm that one of the solutions to the minimum squared-error condition yields outputs that are indeed posterior probabilities. Do this as follows:

(a) To find the minimum of $\tilde{J}(\mathbf{w})$ in Eq. 28, calculate its derivative $\partial \tilde{J}(\mathbf{w})/\partial \mathbf{w}$; this will consist of the sum of two integrals. Set $\partial \tilde{J}(\mathbf{w})/\partial \mathbf{w} = 0$ and solve to obtain the natural solution.

(b) Apply Bayes rule and the normalization $P(\omega_k|\mathbf{x}) + P(\omega_{i \neq k}|\mathbf{x}) = 1$ to prove that the outputs $z_k = g_k(\mathbf{x}; \mathbf{w})$ are indeed equal to the posterior probabilities $P(\omega_k|\mathbf{x})$.

19. In the derivation that backpropagation finds a least squares fit to the posterior probabilities, it was implicitly assumed that the network could indeed represent the true underlying distribution. Explain where in the derivation this was assumed, and what in the subsequent steps may not hold if that assumption is violated.

20. Show that the softmax output (Eq. 30) indeed approximates posterior probabilities if the hidden unit outputs, $\mathbf{y}$, belong to the family of exponential distributions

as

$$p(\mathbf{y}|\omega_k) = \exp[A(\tilde{\mathbf{w}}_k) + B(\mathbf{y}, \phi) + \tilde{\mathbf{w}}_k^t \mathbf{y}]$$

for n_H-dimensional vectors $\tilde{\mathbf{w}}_k$ and $\mathbf{y}$, and scalar ϕ and scalar functions $A(\cdot)$ and $B(\cdot, \cdot)$. Proceed as follows:

(a) Given $p(\mathbf{y}|\omega_k)$, use Bayes theorem to write the posterior probability $P(\omega_k|\mathbf{y})$.

(b) Interpret the parameters $A(\cdot)$, $\tilde{\mathbf{w}}_k$, $B(\cdot, \cdot)$, and ϕ in light of your results.

21. Consider a three-layer network for classification with output units employing softmax (Eq. 30), trained with 0-1 signals.

(a) Derive the learning rule if the criterion function (per pattern) is sum squared error, that is,

$$J(\mathbf{w}) = \frac{1}{2} \sum_{k=1}^{c} (t_k - z_k)^2.$$

(b) Repeat for the criterion function is cross-entropy, that is,

$$J_{ce}(\mathbf{w}) = \sum_{k=1}^{c} t_k \ln \frac{t_k}{z_k}.$$

22. Clearly if the discriminant functions $g_{k_1}(\mathbf{x}; \mathbf{w})$ and $g_{k_2}(\mathbf{x}; \mathbf{w})$ were independent, the derivation of Eq. 28 would follow from Eq. 27. Show that the derivation is nevertheless valid despite the fact that these functions are implemented using the same input-to-hidden weights. Are the discriminant functions independent?

Section 6.7

23. Show that the slope of the sigmoid and the learning rates together determine the learning time.

(a) That is, show that if the slope of the sigmoid is increased by a factor of γ, and the learning rate decreased by a factor $1/\gamma$, the total learning time remains the same.

(b) Must the input be rescaled for this relationship to hold?

24. Show that the basic three-layer neural networks of Section 6.2 are special cases of general additive models by describing in detail the correspondences between Eqs. 7 and 32.

25. Show that the sigmoidal activation function acts to transmit the maximum information if its inputs are distributed normally. Recall that the entropy (a measure of information) is defined as $H = \int p(y)\ln[p(y)] \, dy$.

(a) Consider a continuous input variable x drawn from the density $p(x) \sim N(0, \sigma^2)$. What is entropy for this distribution?

(b) Suppose samples x are passed through an antisymmetric sigmoidal function to give $y = f(x)$, where the zero crossing of the sigmoid occurs at the peak of the Gaussian input, and the effective width of the linear region equal to the range $-\sigma < x < +\sigma$. What values of a and b in Eq. 34 ensure this?

(c) Calculate the entropy of the output distribution $p(y)$.

(d) Suppose instead that the activation function were a Dirac delta function $\delta(x - 0)$. What is the entropy of the resulting output distribution $p(y)$?

(e) Summarize your results of (c) and (d) in words.

Section 6.8

26. Consider the sigmoidal activation function:

$$f(net) = a \tanh(b\ net) = a \left[\frac{1 - e^{-2b\ net}}{1 + e^{-2b\ net}} \right] = \frac{2a}{1 + e^{-2b\ net}} - a.$$

(a) Show that its derivative $f'(net)$ can be written simply in terms of $f(net)$ itself.

(b) What are $f(net)$, $f'(net)$ and $f''(net)$ at $net = -\infty$? 0? $+\infty$?

(c) For which value of net is the second derivative $f''(net)$ extremal?

27. Consider the computational burden for standardizing data, as described in the text.

(a) What is the computational complexity of standardizing a training set of n d-dimensional patterns?

(b) Estimate the computational complexity of training. Use the heuristic for choosing the size of the network (i.e., number of weights) described in Section 6.8.7. Assume that the number of training epochs is nd.

(c) Use your results from (a) and (b) to express the computational burden of standardizing as a ratio. (Assume unknown constants have value 1.0.)

28. Derive the gradient descent learning rule for a three-layer network with linear input units and sigmoidal hidden and output units for the Minkowski error of Eq. 42 with arbitrary R. Confirm that your answer reduces to Eqs. 17 and 21 for $R = 2$.

29. Consider a $d - n_H - c$ three-layer neural network whose input units are linear and output units are sigmoidal but each hidden unit implements a particular polynomial function, trained on a sum-square error criterion. Specifically, let the output of hidden unit j be given by

$$o_j = w_{ji}x_i + w_{jm}x_m + q_j x_i x_m$$

for two prespecified inputs, i and $m \neq i$.

(a) Write gradient descent learning rule for the input-to-hidden weights and scalar parameter q_j.

(b) Does the learning rule for the hidden-to-output unit weights differ from that in the standard three-layer network described in the text?

(c) What might be some of the strengths and the weaknesses of such a network and its learning rule?

30. Derive the backpropagation learning rules for the Minkowski error defined in Eq. 42. Confirm that your result reduces to the standard learning rules for the case $R = 2$. Repeat your derivation for the Manhattan metric directly, that is, without referring to R.

Section 6.9

31. Show intermediate steps in the derivation of the full Hessian matrix for a sum-square-error criterion in a three-layer network, as given in Eqs. 47 and 48.

32. Repeat Problem 31 but for a cross-entropy error criterion.

33. Suppose a Hessian matrix for an error function is proportional to the identity matrix, that is, $\mathbf{H} \propto \mathbf{I}$.

    (a) Prove that in this case both the Polak-Ribiere and the Fletcher-Reeves formulas, Eqs. 57 and 56, used in conjugate gradient descent give $\beta = 0$.

    (b) Interpret your result. In particular state why for a Hessian matrix proportional to the identity matrix the two methods would give the same β, and moreover why the resulting value for β will be 0.

34. Consider an error surface that can be well approximated by a positive definite Hessian matrix associated with a sum-square error on the training set. Let $\mathbf{w}^*$ denote a network's optimal weight vector, that is, at the global error minimum, and let $\Delta\mathbf{w} = \mathbf{w} - \mathbf{w}^*$ denote the difference between an arbitrary weight vector and this global minimum. Prove that moving along the direction given in conjugate gradient descent initially shrinks $\|\Delta\mathbf{w}\|$, and thus the process converges.

35. Based on the discussion in the text, derive the quickprop learning rule in Eq. 53. Be sure to state any assumptions you need to make.

Section 6.10

36. Prove that the matched filter given in Eq. 66 gives a maximum (not merely an extremum) as follows. Throughout the fixed signal desired is $x(t)$.

    (a) Let $w(t) = w^*(t) + h(t)$ be a trial weight function, where $w^*(t)$ is the matched filter. We demand that the constraint of Eq. 63 be obeyed for this trial weight function, that is,

$$\int_{-\infty}^{+\infty} w^2(t)\, dt = \int_{-\infty}^{+\infty} w^{*2}(t)\, dt.$$

    Expand the left-hand side of this equation to prove

$$-2 \int_{-\infty}^{+\infty} h(t) w^*(t)\, dt = \int_{-\infty}^{+\infty} h^2(t)\, dt \geq 0. \tag{71}$$

    (b) Let z^* be the output of the matched filter to the target input, according to Eq. 62. Use your result in (a) to show that the output for the trial weights is

$$z = z^* - \frac{1}{-\lambda} \int_{-\infty}^{+\infty} h^2(t)\, dt.$$

    (c) The sign of λ is determined by the sign of z^*. Use Eq. 66 to show that λ has the opposite sign to that of z^*.

(d) Use all your above results to prove that $z^* = z$ if and only if $h(t) = 0$ for all t, i.e., that the matched filter $w^*(t)$ ensures the *maximum* output, and moreover $w^*(t)$ is unique.

37. Derive the central equations (i.e., analogous to Eq. 69) for OBD and OBS in a three-layer sigmoidal network for a cross-entropy error.

38. Derive the learning rule for a three-layer radial basis function neural network in which the hidden units are spherical Gaussians whose mean μ and amplitude are learned in response to data.

Section 6.11

39. Consider a general constant matrix $\mathbf{K}$ and variable vector parameter $\mathbf{x}$.

(a) Use summation notation with components explicit to derive the formula for the derivative:

$$\frac{d}{d\mathbf{x}}[\mathbf{x}^t \mathbf{K}\mathbf{x}] = (\mathbf{K} + \mathbf{K}^t)\mathbf{x}.$$

(b) Show simply that for the case where $\mathbf{K}$ is symmetric (as for instance the Hessian matrix $\mathbf{H} = \mathbf{H}^t$), we have

$$\frac{d}{d\mathbf{x}}[\mathbf{x}^t \mathbf{H}\mathbf{x}] = 2\mathbf{H}\mathbf{x}$$

as was used in the derivation of the Optimal Brain Damage and Optimal Brain Surgeon methods.

40. Weight decay is equivalent to doing gradient descent on an error that has a "complexity" term.

(a) Show that in the weight decay rule $w_{ij}^{new} = w_{ij}^{old}(1 - \epsilon)$ amounts to performing gradient descent in the error function $J_{ef} = J(\mathbf{w}) + \frac{2\epsilon}{\eta}\mathbf{w}^t\mathbf{w}$ (Eq. 39).

(b) Express γ in terms of the weight decay constant ϵ and learning rate η.

(c) Likewise, show that if $w_{mr}^{new} = w_{mr}^{old}(1 - \epsilon_{mr})$ where $\epsilon_{mr} = 1/(1 + w_{mr}^2)^2$, that the new effective error function is $J_{ef} = J(\mathbf{w}) + \gamma \sum_{mr} w_{mr}^2/(1 + w_{mr}^2)$. Find γ in terms of η and ϵ_{mr}.

(d) Consider a network with a wide range of magnitudes for weights. Describe qualitatively how the two different weight decay methods affect the network.

41. Show that the weight decay rule of Eq. 38 is equivalent to a prior on models that favors small weights.

42. Prove that the weight decay rule of Eq. 38 leads to the J_{reg} of Eq. 39.

43. Equation 69 for Optimal Brain Surgeon requires the inverse of $\mathbf{H}$. One method to calculating this inverse matrix is to start with a small value, $\mathbf{H}_0^{-1} = \alpha^{-1}\mathbf{I}$ and iteratively estimate $\mathbf{H}^{-1}$ by Eq. 70. In this context, show that α acts as a weight decay constant.

44. Derive the key equation in the Optimal Brain Surgeon algorithms, Eq. 69, from the Taylor expansion of the criterion function expressed in Eq. 68.

45. Consider the computational demands of the Optimal Brain Surgeon procedure as follows:

(a) Find the space and the time computational complexities for one step in the nominal OBS method.

(b) Find the space and the time computational complexities for pruning the *first* weight in OBS. What is it for pruning subsequent weights, if one uses Shur's decomposition method?

(c) Find the space and the time computational complexities for one step of OBD (without retraining).

(d) Find the saliency in Optimal Brain Damage, where one assumes $\mathbf{H}_{ij} = 0$ for $i \neq j$.

COMPUTER EXERCISES

Several exercises will make use of the following three-dimensional data sampled from three categories, denoted ω_i.

sample	ω_1			ω_2			ω_3		
	x_1	x_2	x_3	x_1	x_2	x_3	x_1	x_2	x_3
1	1.58	2.32	−5.8	0.21	0.03	−2.21	−1.54	1.17	0.64
2	0.67	1.58	−4.78	0.37	0.28	−1.8	5.41	3.45	−1.33
3	1.04	1.01	−3.63	0.18	1.22	0.16	1.55	0.99	2.69
4	−1.49	2.18	−3.39	−0.24	0.93	−1.01	1.86	3.19	1.51
5	−0.41	1.21	−4.73	−1.18	0.39	−0.39	1.68	1.79	−0.87
6	1.39	3.16	2.87	0.74	0.96	−1.16	3.51	−0.22	−1.39
7	1.20	1.40	−1.89	−0.38	1.94	−0.48	1.40	−0.44	0.92
8	−0.92	1.44	−3.22	0.02	0.72	−0.17	0.44	0.83	1.97
9	0.45	1.33	−4.38	0.44	1.31	−0.14	0.25	0.68	−0.99
10	−0.76	0.84	−1.96	0.46	1.49	0.68	−0.66	−0.45	0.08

Section 6.2

1. Consider a 2-2-1 network with bias, where the activation function at the hidden units and the output unit is a sigmoid $y_j = a \tanh(b \, net_j)$ for $a = 1.716$ and $b = 2/3$.

(a) Suppose the matrices describing the input-to-hidden weights (w_{ji} for $j = 1, 2$ and $i = 0, 1, 2$) and the hidden-to-output weights (w_{kj} for $k = 1$ and $j = 0, 1, 2$) are, respectively,

$$\begin{pmatrix} 0.5 & -0.5 \\ 0.3 & -0.4 \\ -0.1 & 1.0 \end{pmatrix} \quad \text{and} \quad \begin{pmatrix} 1.0 \\ -2.0 \\ 0.5 \end{pmatrix}.$$

The network is to be used to place patterns into one of two categories, based on the sign of the output unit signal. Shade a two-dimensional $x_1 x_2$ input space ($-5 \leq x_1, x_2 \leq +5$) black or white according to the category given by the network.

(b) Repeat part (a) with the following weight matrices:

$$\begin{pmatrix} -1.0 & 1.0 \\ -0.5 & 1.5 \\ 1.5 & -0.5 \end{pmatrix} \quad \text{and} \quad \begin{pmatrix} 0.5 \\ -1.0 \\ 1.0 \end{pmatrix}.$$

Section 6.3

2. Create a 3-1-1 sigmoidal network with bias to be trained to classify patterns from ω_1 and ω_2 in the table above. Use stochastic backpropagation (Algorithm 1) with learning rate $\eta = 0.1$ and sigmoid as described in Eq. 34 in Section 6.8.2.

   (a) Initialize all weights randomly in the range $-1 \leq w \leq +1$. Plot a learning curve—the training error as a function of epoch.

   (b) Now repeat (a) but with weights initialized to be the *same* throughout each level. In particular, let all input-to-hidden weights be initialized with $w_{ji} = 0.5$ and all hidden-to-output weights with $w_{kj} = -0.5$.

   (c) Explain the source of the differences between your learning curves (cf. Problem 12).

3. Consider the nonlinearly separable categorization problem shown in Fig. 6.8.

   (a) Train a 1-3-1 sigmoidal network with bias by means of batch backpropagation (Algorithm 2) to solve that problem.

   (b) Display your decision boundary by classifying points along the x-axis.

   (c) Repeat with a 1-2-1 network.

   (d) Inspect the decision boundary for your 1-3-1 network (or construct by hand an optimal one) and explain why no 1-2-1 network with sigmoidal hidden units can achieve it.

4. Write a backpropagation program for a 2-2-1 network with bias to solve the XOR problem (cf. Fig. 6.1).

   (a) Show the input-to-hidden weights and analyze the function of each hidden unit.

   (b) Plot the representation of each pattern as well as the final decision boundary in the $y_1 y_2$-space.

   (c) Although it was not used as a training pattern, show the representation of $\mathbf{x} = \mathbf{0}$ in your $y_1 y_2$-space.

5. Write a basic backpropagation program for a 3-3-1 network with bias to solve the three-bit parity problem where each input has value ± 1. That is, the output should be $+1$ if the number of inputs that have value $+1$ is even, and should be -1 if the number of such inputs is odd.

   (a) Show the input-to-hidden weights and analyze the function of each hidden unit.

   (b) Retrain several times from new random points until you get a local (but not global) minimum. Analyze the function of the hidden units now.

   (c) How many patterns are properly classified for your local minima? Explain.

6. Explore the effect of the number of hidden units on the accuracy of a $2 - n_H - 1$ neural network classifier (with bias) in a two-dimensional two-category problem with $p(\mathbf{x}|\omega_1) \sim N(\mathbf{0}, \mathbf{I})$ and $p(\mathbf{x}|\omega_2) \sim N\left(\binom{1}{.5}, \binom{3\ 1}{1\ 2}\right)$.

   (a) Generate a training set of 100 points, 50 from each category, as well as an independent set of 40 points (20 from each category).

   (b) Train your network fully with with different number of hidden units, $1 \leq n_H \leq 10$, and for each trained network plot the training and test error, as in Fig. 6.15. What number of hidden units gives the minimum training error?

What number of hidden units gives the minimum test error? (Call this last number n_H^*.)

(c) Re-initialize a $2 - n_H^* - 1$ network and train it. Plot learning curves, that is, the training error and the validation error as a function of the number of epochs. Suppose you were to stop training at the the minimum of this validation error. What is the value of the validation error at such a stopping point?

(d) Compare and explain the difference between the minimum of the validation error in part (c) with the corresponding validation error for n_H^* hidden units in part (b).

7. Train a 2-4-1 network having a different activation function at each hidden unit on a two-dimensional two-category problem with 2^k patterns chosen randomly from the unit square. Estimate k such that the expected error is 25%. Discuss your results.

Section 6.4

8. Construct a 3-1-3 network (with bias) having sigmoidal activation functions. Train your network on the data in the table above.

(a) Compute the Hessian matrix **H**.

(b) Find the eigenvalues and eigenvectors of **H**.

Section 6.5

9. Show that the hidden units of a network may find meaningful feature groupings in the following problem based on optical character recogntion.

(a) Let your input space consist of and 8×8 pixel grid. Generate a 100 training patterns for a B category in the following way. Start with a block-letter representation of the B where the "black" pixels have value -1.0 and the "white" pixels $+1.0$. Generate 100 different versions of this prototype B by adding independent random noise independently to each pixel. Let the distribution of noise be uniform between -0.5 and $+0.5$. Repeat the above for O and E. In this way you have created a set $\mathcal{D}$ of 300 training patterns.

(b) Train a 64-3-3 network (with bias) using your training set $\mathcal{D}$.

(c) Display the input-to-hidden weights as 8×8 images. separately for each hidden unit.

(d) Interpret the patterns of weights displayed in part (c).

Section 6.6

10. Consider training a three-layer network for estimating probabilities of four, equally probable three-dimensional distributions in the following.

(a) First generate a training set $\mathcal{D}$ of 4000 patterns, 1000 each from the following Gaussian distributions $p(\mathbf{x}|\omega_i) \sim N(\boldsymbol{\mu}_i, \boldsymbol{\Sigma}_i)$ for $i = 1, 2, 3, 4$:

i	$\boldsymbol{\mu}_i$	$\boldsymbol{\Sigma}_i$
1	**0**	**I**
2	$\begin{pmatrix} 0 \\ 1 \\ 0 \end{pmatrix}$	$\begin{pmatrix} 1 & 0 & 1 \\ 0 & 2 & 2 \\ 1 & 2 & 5 \end{pmatrix}$

i	μ_i	Σ_i
3	$\begin{pmatrix} -1 \\ 0 \\ 1 \end{pmatrix}$	Diag[2,6,1]
4	$\begin{pmatrix} 0 \\ 0.5 \\ 1 \end{pmatrix}$	Diag[2,1,3]

(b) Train a 3-3-4 network (with bias) using the 4000 patterns in $\mathcal{D}$. Use the softmax target values as given in Eq. 30.

(c) Use your trained network to estimate the posterior probabilities of each of the following five test patterns: $\mathbf{x}_1 = (0,0,0)^t$, $\mathbf{x}_2 = (-1,0,1)^t$, $\mathbf{x}_3 = (.5, -.5, 0)^t$, $\mathbf{x}_4 = (-1,0,0)^t$, and $\mathbf{x}_5 = (0,0,1)^t$.

(d) Use your network to classify each of the five test patterns.

(e) Use the techniques of Chapter 2 to calculate the *a posteriori* probabilities of the five test points. Compare these answer to your answers in part (c).

Section 6.7

11. Consider several gradient descent methods applied to a criterion function in one dimension: simple gradient descent with fixed learning rate η, optimized descent, Newton's method, and Quickprop. Consider first the criterion function $J(w) = w^2$ which of course has minimum $J = 0$ at $w = 0$. In all cases, start the descent at $w(0) = 1$. For definiteness, we consider convergence to be complete when $J(\mathbf{w}) < 0.001$.

(a) Plot the number of steps until convergence as a function of η for $\eta = 0.01, 0.03, 0.1, 0.3, 1, 3$.

(b) Calculate the optimum learning rate η_{opt} by Eq. 36, and confirm that this value is consistent with your results in part (a).

(c) Calculate the weight update by the Quickprop rule of Eq. 53.

Section 6.8

12. Demonstrate that preprocessing data can lead to significant reduction in time of learning. Consider a single linear output unit for a two-category classification task, with teaching values ± 1 and squared-error criterion.

(a) Write a program to train the three weights based on training samples.

(b) Generate 20 samples from each of two categories $P(\omega_1) = P(\omega_2) = .5$ and $p(\mathbf{x}|\omega_i) \sim N(\mu_i)$, $\mathbf{I}$, where $\mathbf{I}$ is the 2×2 identity matrix and $\mu_1 = (0,1)^t$ and $\mu_2 = (1,-1)^t$.

(c) Find the optimal learning rate empirically by trying a few values.

(d) Train to minimum error. Why is there no danger of overtraining in this case?

(e) Why can we be sure that it is at least possible that this network can achieve the minimum (Bayes) error?

(f) Generate 100 test samples, 50 from each category, and estimate the error rate.

(g) Now preprocess the data by subtracting off the mean and scaling standard deviation in each dimension.

(h) Repeat the above, and find the optimal learning rate.

(**i**) Find the error rate on the (transformed) test set.

(**j**) Verify that the accuracy is virtually the same in the two cases (any differences can be attributed to stochastic effects).

(**k**) Explain in words the underlying reasons for your results.

BIBLIOGRAPHY

[1] Yaser S. Abu-Mostafa. Learning from hints in neural networks. *Journal of Complexity*, 6(2):192–198, 1990.

[2] James A. Anderson, Andras Pellionisz, and Edward Rosenfeld, editors. *Neurocomputing 2: Directions for Research*. MIT Press, Cambridge, MA, 1990.

[3] James A. Anderson and Edward Rosenfeld, editors. *Neurocomputing: Foundations of Research*. MIT Press, Cambridge, MA, 1988.

[4] Christopher M. Bishop. *Neural Networks for Pattern Recognition*. Oxford University Press, Oxford, UK, 1995.

[5] John S. Bridle. Probabilistic interpretation of feedforward classification network outputs, with relationships to statistical pattern recognition. In Françoise Fogelman-Soulié and Jeanny Hérault, editors, *Neurocomputing: Algorithms, Architectures and Applications*, pages 227–236, Springer-Verlag, New York, 1990.

[6] Arthur E. Bryson, Jr., Walter Denham, and Stuart E. Dreyfus. Optimal programming problem with inequality constraints. I: Necessary conditions for extremal solutions. *American Institute of Aeronautics and Astronautics Journal*, 1(11):2544–2550, 1963.

[7] Arthur E. Bryson, Jr. and Yu-Chi Ho. *Applied Optimal Control*. Blaisdell, Waltham, MA, 1969.

[8] Wray L. Buntine and Andreas S. Weigend. Computing second derivatives in feed-forward networks: A review. *IEEE Transactions on Neural Networks*, 5(3):480–488, 1991.

[9] Vladimir Cherkassky, Jerome H. Friedman, and Harry Wechsler, editors. *From Statistics to Neural Networks: Theory and Pattern Recognition Applications*. NATO ASI. Springer, New York, 1994.

[10] Yann Le Cun. A learning scheme for asymmetric threshold networks. In *Proceedings of Cognitiva 85*, pages 599–604, Paris, France, 1985.

[11] Yann Le Cun. Learning processes in an asymmetric threshold network. In Elie Bienenstock, Françoise Fogelman-Soulié, and Gerard Weisbuch, editors, *Disordered Systems and Biological Organization*, pages 233–240, Les Houches, Springer-Verlag, France, 1986.

[12] Yann Le Cun, John S. Denker, and Sara A. Solla. Optimal Brain Damage. In David S. Touretzky, editor, *Advances in Neural Information Processing Systems*, volume 2, pages 598–605. Morgan Kaufmann, San Mateo, CA, 1990

[13] George Cybenko. Approximation by superpositions of a sigmoidal function. *Mathematical Control Signals Systems*, 2:303–314, 1989.

[14] John S. Denker and Yann Le Cun. Transforming neural-net output levels to probability distributions. In Richard Lippmann, John Moody, and David Touretzky, editors, *Advances in Neural Information Processing Systems*, volume 3, pages 853–859. Morgan Kaufmann, San Mateo, CA, 1991.

[15] Scott E. Fahlman and Christian Lebiere. The Cascade-Correlation learning architecture. In David S. Touretzky, editor, *Advances in Neural Information Processing Systems*, volume 2, pages 524–532. Morgan Kaufmann, San Mateo, CA, 1990.

[16] Jerome H. Friedman and Werner Stuetzle. Projection pursuit regression. *Journal of the American Statistical Association*, 76(376):817–823, 1981.

[17] Kunihiko Fukushima. Neocognitron: A self-organizing neural network model for a mechanism of pattern recognition unaffected by shift in position. *Biological Cybernetics*, 36:193–202, 1980.

[18] Kunihiko Fukushima, Sei Miyake, and Takayuki Ito. Neocognitron: A neural network model for a mechanism of visual pattern recognition. *IEEE Transactions on Systems, Man, and Cybernetics*, SMC-13(5):826–834, 1983.

[19] Federico Girosi, Michael Jones, and Tomaso Poggio. Regularization theory and neural networks architectures. *Neural Computation*, 7(2):219–269, 1995.

[20] Federico Girosi and Tomaso Poggio. Representation properties of networks: Kolmogorov's theorem is irrelevant. *Neural Computation*, 1(4):465–469, 1989.

[21] Richard M. Golden. *Mathematical Methods for Neural Network Analysis and Design*. MIT Press, Cambridge, MA, 1996.

[22] Igor Grebert, David G. Stork, Ron Keesing, and Steve Mims. Connectionist generalization for productions: An example from Gridfont. *Neural Networks*, 5(4):699–710, 1992.

[23] Igor Grebert, David G. Stork, Ron Keesing, and Steve Mims. Network generalization for production: Learning and producing styled letterforms. In John E. Moody, Stephen J. Hanson, and Richard P. Lippmann, editors, *Advances in Neural Information Processing Systems*, volume 4, pages 1118–1124. Morgan Kaufmann, San Mateo, CA, 1992.

[24] Stephen Grossberg. Competitive learning: From interactive activation to adaptive resonance. *Cognitive Science*, 11(1):23–63, 1987.

[25] John B. Hampshire, II and Barak A. Pearlmutter. Equivalence proofs for multi-layer Perceptron classifiers and

the Bayesian discriminant function. In David S. Touret-zky, Jeffrey L. Elman, Terrence J. Sejnowski, and Geoffrey E. Hinton, editors, *Proceedings of the 1990 Connectionst Models Summer School*, pages 159–172. Morgan Kaufmann, San Mateo, CA, 1990.

[26] Babak Hassibi and David G. Stork. Second-order derivatives for network pruning: Optimal Brain Surgeon. In Stephen J. Hanson, Jack D. Cowan, and C. Lee Giles, editors, *Advances in Neural Information Processing Systems*, volume 5, pages 164–171. Morgan Kaufmann, San Mateo, CA, 1993.

[27] Babak Hassibi, David G. Stork, and Greg Wolff. Optimal Brain Surgeon and general network pruning. In *Proceedings of the International Conference on Neural Networks*, volume 1, pages 293–299. IEEE, San Francisco, CA, 1993.

[28] Babak Hassibi, David G. Stork, Gregory Wolff, and Takahiro Watanabe. Optimal Brain Surgeon: Extensions and performance comparisons. In Jack D. Cowan, Gerald Tesauro, and Joshua Alspector, editors, *Advances in Neural Information Processing Systems*, volume 6, pages 263–270. Morgan Kaufmann, San Mateo, CA, 1994.

[29] Mohamad H. Hassoun. *Fundamentals of Artificial Neural Networks*. MIT Press, Cambridge, MA, 1995.

[30] Simon Haykin. *Adaptive Filter Theory*. Prentice-Hall, Englewood Cliffs, NJ, second edition, 1991.

[31] Simon Haykin. *Neural Networks: A Comprehensive Foundation*. Macmillan, New York, 1994.

[32] Robert Hecht-Nielsen. Theory of the backpropagation neural network. In *Proceeding of the International Joint Conference on Neural Networks (IJCNN)*, volume 1, pages 593–605. IEEE, New York, 1989.

[33] Geoffrey E. Hinton. Learning distributed representations of concepts. In *Proceedings of the Eighth Annual Conference of the Cognitive Science Society*, pages 1–12. Lawrence Erlbaum Associates, Hillsdale, NJ, 1986.

[34] Kurt Hornik, Maxwell Stinchcombe, and Halbert L. White, Jr. Multilayer feedforward networks are universal approximators. *Neural Networks*, 2(5):359–366, 1989.

[35] Peter J. Huber. Projection pursuit. *Annals of Statistics*, 13(2):435–475, 1985.

[36] Don R. Hush, John M. Salas, and Bill G. Horne. Error surfaces for multi-layer Perceptrons. In *Proceedings of International Joint Conference on Neural Networks (IJCNN)*, volume 1, pages 759–764. IEEE, New York, 1991.

[37] Anil K. Jain, Jianchang Mao, and K. Moidin Mohiuddin. Artificial neural networks: A tutorial. *Computer*, 29(3):31–44, 1996.

[38] Lee K. Jones. Constructive approximations for neural networks by sigmoidal functions. *Proceedings of the IEEE*, 78(10):1586–1589, 1990.

[39] Rudolf E. Kalman. A new approach to linear filtering and prediction problems. *Transactions of the ASME, Series D, Journal of Basic Engineering*, 82(1):34–45, 1960.

[40] Andreæi N. Kolmogorov. On the representation of continuous functions of several variables by superposition of continuous functions of one variable and addition. *Doklady Akademiia Nauk SSSR*, 114(5):953–956, 1957.

[41] Věra Kůrková. Kolmogorov's theorem is relevant. *Neural Computation*, 3(4):617–622, 1991.

[42] Věra Kůrková. Kolmogorov's theorem and multilayer neural networks. *Neural Computation*, 5(3):501–506, 1992.

[43] Chuck Lam and David G. Stork. Learning network topology. In Michael A. Arbib, editor, *The Handbook of Brain Theory and Neural Networks*. MIT Press, Cambridge, MA, second edition, 2001.

[44] Alan Lapedes and Ron Farber. How neural nets work. In Dana Z. Anderson, editor, *Advances in Neural Information Processing Systems*, pages 442–456. American Institute of Physics, New York, 1988.

[45] Dar-Shyang Lee, Sargur N. Srihari, and Roger Gaborski. Bayesian and neural network pattern recognition: A theoretical connection and empiricial results with handwritten characters. In Ishwar K. Sethi and Anil K. Jain, editors, *Artificial Neural Networks and Statistical Pattern Recognition: Old and New Connections*, chapter 5, pages 89–108. North-Holland, Amsterdam, 1991.

[46] Kenneth Levenberg. A method for the solution of certain non-linear problems in least squares. *Quarterly Journal of Applied Mathematics*, II(2):164–168, 1944.

[47] Daniel S. Levine. *Introduction to Neural and Cognitive Modeling*. Lawrence Erlbaum Associates, Hillsdale, NJ, 1991.

[48] Richard Lippmann. An introduction to computing with neural nets. *IEEE ASSP Magazine*, pages 4–22, April 1987.

[49] David Lowe and Andrew R. Webb. Optimized feature extraction and the Bayes decision in feed-forward classifier networks. *IEEE Transactions on Pattern Analysis and Machine Intelligence*, PAMI-13(4):355–364, 1991.

[50] Donald W. Marquardt. An algorithm for least-squares estimation of non-linear parameters. *Journal of the Society for Industrial and Applied Mathematics*, 11(2):431–441, 1963.

[51] Warren S. McCulloch and Walter Pitts. A logical calculus of ideas imminent in nervous activity. *Bulletin of Mathematical Biophysics*, 5:115–133, 1943.

[52] John M. McInerny, Karen G. Haines, Steve Biafore, and Robert Hecht-Nielsen. Back propagation error surfaces can have local minima. In *International Joint Conference on Neural Networks (IJCNN)*, volume 2, page 627. IEEE, New York, 1989.

[53] Marvin L. Minsky and Seymour A. Papert. *Perceptrons: An Introduction to Computational Geometry*. MIT Press, Cambridge, MA, 1969.

[54] Hermann Ney. On the probabilistic interpretation of neural network classifiers and discriminative training criteria. *IEEE Transactions on Pattern Analysis and Machine Intelligence*, PAMI-17(2):107–119, 1995.

[55] David B. Parker. Learning logic. Technical Report S81-64, File 1, Stanford University Office of Technology Licensing, 1982.

[56] David B. Parker. Learning logic. Technical Report TR-47, MIT Center for Research in Computational Economics and Management Science, 1985.

[57] Fernando Pineda. Recurrent backpropagation and the dynamical approach to adaptive neural computation. *Neural Computation*, 1(2):161–172, 1989.

[58] Walter Pitts and Warren S. McCulloch. How we know universals: The perception of auditory and visual forms. *Bulletin of Mathematical Biophysics*, 9:127–147, 1947.

[59] Tomaso Poggio and Federico Girosi. Regularization algorithms for learning that are equivalent to multilayer networks. *Science*, 247(4945):978–982, 1990.

[60] Russell Reed. Pruning algorithms—a survey. *IEEE Transactions on Neural Networks*, TNN-4(5):740–747, 1993.

[61] Russell D. Reed and Roberg J. Marks II. *Neural Smithing: Supervised Learning in Feedforward Artificial Neural Networks*. MIT Press, Cambridge, MA, 1999.

[62] Michael D. Richard and Richard P. Lippmann. Neural network classifiers estimate Bayesian *a-posteriori* probabilities. *Neural Computation*, 3(4):461–483, 1991.

[63] Brian D. Ripley. *Pattern Recognition and Neural Networks*. Cambridge University Press, Cambridge, UK, 1996.

[64] Frank Rosenblatt. The Perceptron: A probabilistic model for information storage and organization in the brain. *Psychological Review*, 65(6):386–408, 1958.

[65] Frank Rosenblatt. *Principles of Neurodynamics*. Spartan Books, Washington, DC, 1962.

[66] Dennis W. Ruck, Steven K. Rogers, Matthew Kabrisky, Mark E. Oxley, and Bruce W. Suter. The multilayer Perceptron as an approximation to a Bayes optimal discriminant function. *IEEE Transactions on Neural Networks*, TNN-1(4):296–298, 1990.

[67] David E. Rumelhart, Geoffrey E. Hinton, and Ronald J. Williams. Learning internal representations by back-propagating errors. *Nature*, 323(99):533–536, 1986.

[68] David E. Rumelhart, Geoffrey E. Hinton, and Ronald J. Williams. Learning internal representations by error propagation. In David E. Rumelhart, James L. McClelland, and the PDP Research Group, editors, *Parallel Distributed Processing*, volume 1, chapter 8, pages 318–362. MIT Press, Cambridge, MA, 1986.

[69] Oliver G. Selfridge. Pandemonium: A paradigm for learning. In *Mechanisation of Thought Processes: Proceedings of a Symposium held at the National Physical Laboratory*, pages 513–526, London, 1958. HMSO.

[70] Oliver G. Selfridge and Ulrich Neisser. Pattern recognition by machine. *Scientific American*, 203(2):60–68, 1960.

[71] Ishwar K. Sethi and Anil K. Jain, editors. *Artificial Neural Networks and Statistical Pattern Recognition: Old and New Connections*. North-Holland, Amsterdam, The Netherlands, 1991.

[72] Suzanne Sommer and Richard M. Huggins. Variables selection using the Wald test and a robust CP. *Applied Statistics*, 45(1):15–29, 1996.

[73] Achim Stahlberger and Martin Riedmiller. Fast network pruning and feature extraction using the unit-OBS algorithm. In Michael C. Mozer, Michael I. Jordan, and Thomas Petsche, editors, *Advances in Neural Information Processing Systems*, volume 9, pages 655–661. MIT Press, Cambridge, MA, 1997.

[74] David G. Stork. *Determination of symmetry and phase in human visual response functions: Theory and Experiment*. Ph.D. thesis, University of Maryland, College Park, MD, 1984.

[75] David G. Stork. Is backpropagation biologically plausible? In *Proceedings of the International Joint Conference on Neural Networks (IJCNN)*, pages II–241–246. IEEE, New York, 1989.

[76] David G. Stork and James D. Allen. How to solve the n-bit parity problem with two hidden units. *Neural Networks*, 5(6):923–926, 1992.

[77] David G. Stork and James D. Allen. How to solve the n-bit encoder problem with just one hidden unit. *Neurocomputing*, 5(3):141–143, 1993.

[78] David G. Stork and John Z. Levinson. Receptive fields and the optimal stimulus. *Science*, 216(4542):204–205, 1982.

[79] Alan M. Turing. Intelligent machinery. In Darrell C. Ince, editor, *Collected Works of A. M. Turing: Mechanical Intelligence*. Elsevier Science Publishers, Amsterdam, The Netherlands, 1992.

[80] Abraham Wald. Tests of statistical hypotheses concerning several parameters when the number of observations is large. *Transactions of the American Mathematical Society*, 54(3):426–482, 1943.

[81] Abraham Wald. *Statistical Decision Functions*. Wiley, New York, 1950.

[82] Andreas S. Weigend, David E. Rumelhart, and Bernardo A. Huberman. Generalization by weight-elimination with application to forecasting. In Richard P. Lippmann, John E. Moody, and David S. Touretzky, editors, *Advances in Neural Information Processing Systems*, volume 3, pages 875–882. Morgan Kaufmann, San Mateo, CA, 1991.

[83] Paul John Werbos. *Beyond Regression: New Tools for Prediction and Analysis in the Behavioral Sciences*. Ph.D. thesis, Harvard University, Cambridge, MA, 1974.

[84] Paul John Werbos. *The Roots of Backpropagation: From Ordered Derivatives to Neural Networks and Political Forecasting*. Wiley, New York, 1994.

[85] Halbert L. White, Jr. Learning in artifical neural networks: A statistical perspective. *Neural Computation*, 3(5):425–464, 1989.

[86] Bernard Widrow. 30 years of adaptive neural networks: Perceptron, Madaline, and Backpropagation. *Proceedings of IEEE*, 78(9):1415–1452, 1990.

[87] Bernard Widrow and Marcian E. Hoff, Jr. Adaptive switching circuits. *1960 IRE WESCON Convention Record*, pages 96–104, 1960.

[88] Bernard Widrow and Samuel D. Stearns, editors. *Adaptive Signal Processing*. Prentice-Hall, Englewood Cliffs, NJ, 1985.

STOCHASTIC METHODS

7.1 INTRODUCTION

Learning plays a central role in the construction of pattern classifiers. As we have seen, the general approach is to specify a model having one or more parameters and then estimate their values from training data. When the models are fairly simple and of low dimension, we can use analytic methods such as computing derivatives and solving equations explicitly to find optimal model parameters. If the models are somewhat more complicated, we may calculate local derivatives and use gradient methods, as in neural networks and some maximum-likelihood problems. In most high-dimensional and complicated models, there are multiple maxima and we must use a variety of tricks—such as performing the search multiple times from different starting conditions—to have any confidence that an acceptable local maximum has been found.

These methods become increasingly unsatisfactory as the models become more complex. A naive approach—exhaustive search through solution space—rapidly gets out of hand and is completely impractical for real-world problems. The more complicated the model, the less the prior knowledge, and the less the training data, the more we must rely on sophisticated search for finding acceptable model parameters. In this chapter we consider stochastic methods for finding parameters, where randomness plays a crucial role in search and learning. The general approach is to bias the search toward regions where we expect the solution to be and allow randomness—somehow—to help find good parameters.

We shall consider two general classes of such methods. The first, exemplified by Boltzmann learning, is based on concepts and techniques from physics, specifically statistical mechanics. The second, exemplified by genetic algorithms, is based on concepts from biology, specifically the mathematical theory of evolution. The former class has a highly developed and rigorous theory and has many successes in pattern recognition; hence it will command most of our effort. The latter class is more heuristic yet affords flexibility and can be attractive when adequate computational resources are available. We shall generally illustrate these techniques in cases that are simple, and which might also be addressed with standard gradient procedures; nevertheless we emphasize that these stochastic methods may be preferable in complex problems. The methods have high computational burden, and they would be of little use without computers.

7.2 STOCHASTIC SEARCH

We begin by discussing an important and general quadratic optimization problem. Even though there are analytical ways to approach quadratic optimization, they do not scale well to large problems. Thus, we focus here on methods of *search* through different candidate solutions. We then consider a form of stochastic search that finds use in learning for pattern recognition.

Suppose we have a large number of variables s_i, $i = 1, \ldots, N$, where each variable can take one of two discrete values, for simplicity chosen to be ± 1. The optimization problem is this: Find the values of the s_i so as to minimize the cost or ENERGY *energy*

$$E = -\frac{1}{2} \sum_{i,j=1}^{N} w_{ij} s_i s_j, \tag{1}$$

where the w_{ij} are symmetric, and can be positive or negative. We require the self-feedback terms to vanish (i.e., $w_{ii} = 0$), because the nonzero w_{ii} merely add an unimportant constant to E, independent of the s_i. This optimization problem can be visualized in terms of a network of nodes, where bi-directional links or interconnections correspond to the weights. Figure 7.1 shows such a network and the notation we shall use.

This network suggests a physical analogy which in turn will guide our choice of solution method. Imagine the network represents N physical magnets, each of which can have its north pole pointing up ($s_i = +1$) or pointing down ($s_i = -1$). The w_{ij} are functions of the physical separations between the magnets. Each pair of magnets has an associated interaction energy which depends upon their state, separation and other physical properties: $E_{ij} = -1/2\, w_{ij} s_i s_j$. The energy of the full system is the sum of all these interaction energies, as given in Eq. 1. The optimization task is to find the configuration of states of the magnets that is the most stable configuration, the one with lowest energy. This general optimization problem appears in a wide range of applications, in many of which the weights do not have a physical interpretation.* As mentioned, we shall be particularly interested in its application to learning methods.

Except for very small problems or few connections, it is infeasible to solve directly for the N values s_i that give the minimum energy—the space has 2^N possible configurations (Problem 4). It is tempting to propose a greedy algorithm to search for the optimum configuration: Begin by randomly assigning a state to each node. Next consider each node in turn and calculate the energy with it in the $s_i = +1$ state and then in the $s_i = -1$ state, and choose the one giving the lower energy. (Naturally, this decision needs to be based on only those nodes connected to node i with nonzero weight w_{ij}.) Alas, this greedy search is rarely successful because the system usually gets caught in local energy minima or never converges (Computer exercise 1). Another method is required.

7.2.1 Simulated Annealing

ANNEALING   In physics, the method for allowing a system such as many magnets or atoms in an alloy to find a low-energy configuration is based on *annealing*. In physical anneal-

---

*Similar generalized energy functions, called *Lyapunov functions* or *objective functions*, can be used for finding optimum states in many problem domains.

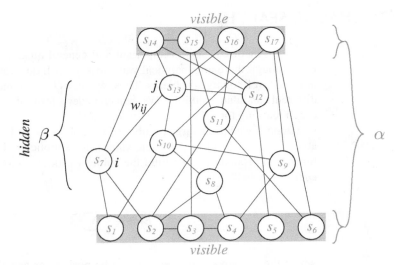

FIGURE 7.1. The class of optimization problems of Eq. 1 can be viewed in terms of a network of nodes or units, each of which can be in the $s_i = +1$ or $s_i = -1$ state. Every pair of nodes i and j is connected by bi-directional weights w_{ij}; if a weight between two nodes is zero, then no connection is drawn. (Because the networks we shall discuss can have an arbitrary interconnection, there is no notion of layers as in multilayer neural networks.) The optimization problem is to find a configuration (i.e., assignment of all s_i) that minimizes the energy described by Eq. 1. While for neural nets our convention was to show *functions* inside each node's circle, our convention in so-called Boltzmann networks here is to indicate the *state* of each node. The configuration of the full network is indexed by an integer γ, and because here there are 17 binary nodes, γ is bounded $0 \leq \gamma < 2^{17}$. When such a network is used for pattern recognition, the input and output nodes are said to be visible, and the remaining nodes are said to be hidden. The states of the visible nodes and hidden nodes are indexed by α and β, respectively, and in this case are bounded $0 \leq \alpha \leq 2^{10}$ and $0 \leq \beta < 2^7$.

ing the system is heated, thereby conferring randomness to each component (magnet). As a result, each variable can temporarily assume a value that is energetically *un*favorable and the full system explores configurations that have high energy. Annealing proceeds by gradually lowering the temperature of the system—ultimately toward zero and thus no randomness—so as to allow the system to relax into a low-energy configuration. Such annealing is effective because even at moderately high temperatures the system slightly favors regions in the configuration space that are overall lower in energy, and hence are more likely to contain the global minimum. As the temperature is lowered, the system has increased probability of finding the optimum configuration. This method is successful in a wide range of energy functions or energy "landscapes," though there are pathological cases such as the "golf course" landscape in Fig. 7.2 where it is unlikely to succeed. Fortunately, the problems in learning we shall consider rarely involve such pathological functions.

7.2.2 The Boltzmann Factor

The statistical properties of large number of interacting physical components at a temperature T, such as molecules in a gas or magnetic atoms in a solid, have been thoroughly analyzed. A key result, which relies on a few very natural assumptions, is the following: The probability the system is in a (discrete) configuration indexed

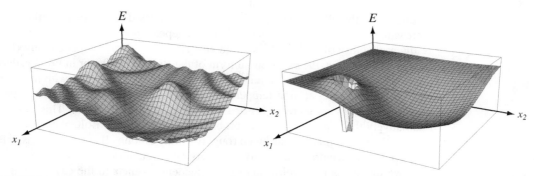

FIGURE 7.2. The energy function or energy "landscape" on the left is meant to suggest the types of optimization problems addressed by simulated annealing. The method uses randomness, governed by a control parameter or "temperature" T, to avoid getting stuck in local energy minima and thus to find the global minimum, like a small ball rolling in the landscape as it is shaken. The pathological "golf course" landscape at the right is, generally speaking, not amenable to solution via simulated annealing because the region of lowest energy is so small and is surrounded by energetically unfavorable configurations. The configuration spaces of the problems we shall address are discrete and are more accurately displayed in Fig. 7.6.

by γ having energy E_γ is given by

$$P(\gamma) = \frac{e^{-E_\gamma/T}}{Z(T)}, \tag{2}$$

BOLTZMANN FACTOR PARTITION FUNCTION

where Z is a normalization constant. The numerator is the *Boltzmann factor* and the denominator is the *partition function*, the sum over all possible configurations

$$Z(T) = \sum_{\gamma'} e^{-E_{\gamma'}/T}, \tag{3}$$

which guarantees that Eq. 2 represents a true probability.* The number of configurations is very high, 2^N, and in physical systems Z can be calculated only in simple cases. Fortunately, we need not calculate the partition function, as we shall see.

Because of the fundamental importance of the Boltzmann factor in our discussions, it pays to take a slight detour to understand it, at least in an informal way. Consider a different, but nontheless related system: one consisting of a large number of *noninteracting* magnets—that is, without interconnecting weights—in a uniform external magnetic field. If a magnet is pointing up, $s_i = +1$ (in the same direction as the field), it contributes a small positive energy to the total system; if the magnet is pointing down, it contributes a small negative energy. The total energy of the collection is thus proportional to the total number of magnets pointing up. The *probability* that the system has a particular total energy is related to the number of configurations that have that energy. Consider the highest-energy configuration, with all magnets pointing up. There is only $\binom{N}{N} = 1$ configuration that has this energy. The next-to-highest energy comes with just a single magnet pointing down; there are $\binom{N}{1} = N$ such configurations. The next-lower-energy configurations have two magnets pointing down; there are $\binom{N}{2} = N(N-1)/2$ of these configurations, and

---

*In the Boltzmann factor for physical systems there is a "Boltzmann constant," which converts a temperature into an energy; we can ignore this factor by scaling the temperature in our simulations.

so on. It can thus be shown that the number of states declines exponentially with increasing energy. Because of the statistical independence of the magnets, for large N the probability of finding the state in energy E also decays exponentially (Problem 7). In sum, then, the exponential form of the Boltzmann factor in Eq. 2 is due to the exponential decrease in the number of accessible configurations with increasing energy. Furthermore, at high temperature there is, roughly speaking, more energy available and thus an increased probability of higher-energy states. This describes qualitatively the dependence of the probability upon T in the Boltzmann factor: At high T, the probability is distributed roughly evenly among all configurations, while at low T it is concentrated at the lowest-energy configurations.

If we move from the collection of independent magnets to the case of magnets interconnected by weights, the situation is a bit more complicated. Now the energy associated with a magnet pointing up or down depends upon the states of others. Nonetheless, in the case of large N, the number of configurations decays exponentially with the energy of the configuration, as described by the Boltzmann factor of Eq. 2.

Simulated Annealing Algorithm

The above discussion and the physical analogy suggest the following *simulated annealing* method for finding the optimum configuration for our general optimization problem. Start with randomized states throughout the network and select a high initial "temperature" $T(1)$. (Of course in the simulation T is merely a control parameter that will control the randomness; it is not a true physical temperature.) Next, choose a node i randomly. Suppose its state is $s_i = +1$. Calculate the system energy in this configuration, E_a; next recalculate the energy, E_b, for a candidate new state $s_i = -1$. If this candidate state has a lower energy, accept this change in state. If, however, the energy is *higher*, accept this change with a probability equal to

$$e^{-\Delta E_{ab}/T}, \tag{4}$$

where $\Delta E_{ab} = E_b - E_a$. This occasional acceptance of a state that is energetically less favorable is crucial to the success of simulated annealing, and it is in marked distinction to naive gradient descent and the greedy approach. The key benefit is that it allows the system to jump out of unacceptable local energy minima. For example, at very high temperatures, every configuration has a Boltzmann factor with nearly the same value, namely $e^{-E/T} \approx e^0$. After normalization by the partition function, then, every configuration is roughly equally likely. This implies that every node is equally likely to be in either of its two states (Problem 6).

POLLING The algorithm continues *polling* (selecting and testing) the nodes randomly several times and setting their states in this way. Next the temperature is lowered and the polling repeated. Now, according to Eq. 4, there will be a slightly smaller probability that a candidate higher-energy state will be accepted. Next the algorithm polls all the nodes until each node has been visited several times. Then the temperature is lowered further, the polling is repeated, and so forth. At very low temperatures, the probability that an energetically less favorable state will be accepted is small, and thus the search becomes more like a greedy algorithm. Simulated annealing terminates when the temperature is very low (near zero). If this cooling has been sufficiently slow, the system then has a high probability of being in a low-energy state—hopefully the global energy minimum.

Because it is the *difference* in energies between the two states that determines the acceptance probabilities, we need only consider nodes connected to the one being polled: All the units *not* connected to the polled unit are in the same state and contribute the same total amount to the full energy.

We let $\mathcal{N}_i$ denote the set of nodes connected with nonzero weights to node i; a fully connected net would include the complete set of $N - 1$ remaining nodes. Furthermore, we let Rand[0, 1) denote a randomly selected positive real number less than 1. With this notation, then, the randomized or stochastic simulated annealing algorithm is as follows:

■ **Algorithm 1. (Stochastic Simulated Annealing)**

1 **begin** **initialize** $T(k), k_{max}, s_i(1), w_{ij}$ for $i, j = 1, \ldots, N$
2        $k \leftarrow 0$
3   **do** $k \leftarrow k + 1$
4   **do** select node i randomly; suppose its state is s_i
5                $E_a \leftarrow -1/2 \sum_{j}^{\mathcal{N}_i} w_{ij} s_i s_j$
6                $E_b \leftarrow -E_a$
7   **if** $E_b < E_a$
8   **then** $s_i \leftarrow -s_i$
9   **else if** $e^{-(E_b - E_a)/T(k)} > \text{Rand}[0, 1)$
10   **then** $s_i \leftarrow -s_i$
11   **until** all nodes polled several times
12   **until** $k = k_{max}$ or stopping criterion met
13   **return** E, s_i, for $i = 1, \ldots, N$
14 **end**

Because units are polled one at a time, the algorithm is occasionally called *sequential simulated annealing*. Note that in line 5, we define E_a based only on those units connected to the polled one—a slightly different convention than in Eq. 1. Changing the usage in this way has no effect, since in line 9 it is the *difference* in energies that determines transition probabilities.

There are several aspects of the algorithm that must be considered carefully—in particular the starting temperature, the rate at which the temperature is decreased, the ending temperature, and stopping criterion.

ANNEALING SCHEDULE

The function $T(k)$ (where k is an iteration index) is called the *cooling schedule* or the *annealing schedule*. We demand $T(1)$ to be sufficiently high that all configurations have roughly equal probability. This requires that the temperature be larger than the maximum difference in energy between any configurations. Such a high temperature allows the system to move to any configuration which may be needed, because the random initial configuration may be far from the optimal. The decrease in temperature must be both gradual and slow enough that the system can move to any part of the state space before being trapped in an unacceptable local minimum, points we shall consider below. At the very least, annealing must allow $N/2$ transitions, because a global optimum may differ from an arbitrary configuration by at most this number of steps. (In practice, annealing can require polling several orders of magnitude more times than this number.) The final temperature must be low enough (or equivalently k_{max} must be large enough or the stopping criterion must be good enough) that there is a negligible probability that if the system is in a global minimum it will move out. We may, furthermore, decide to record the current best configuration

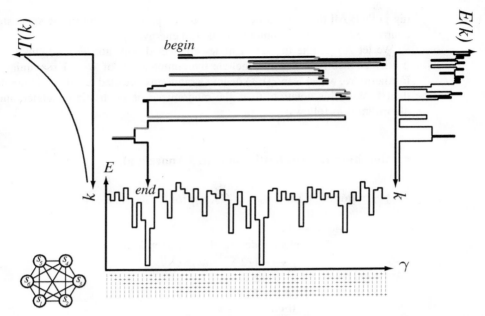

FIGURE 7.3. Stochastic simulated annealing (Algorithm 1) uses randomness, governed by a control parameter or "temperature" $T(k)$ to search through a discrete space for a minimum of an energy function. In this example there are $N = 6$ variables; the $2^6 = 64$ configurations are shown along the bottom as a column of $+$ and $-$ symbols. The plot of the associated energy of each configuration given by Eq. 1 for randomly chosen weights. Every transition corresponds to the change of just a single s_i. (The configurations have been arranged so that adjacent ones differ by the state of just a single node; nevertheless, most transitions corresponding to a single node appear far apart in this ordering.) Because the system energy is invariant with respect to a global interchange $s_i \leftrightarrow -s_i$, there are two "global" minima. The graph at the upper left shows the annealing schedule—the decreasing temperature versus iteration number k. The middle portion shows the configuration versus iteration number generated by Algorithm 1. The trajectory through the configuration space is colored red for transitions that increase the energy and black for those that decrease the energy. Such energetically unfavorable (red) transitions become rarer later in the anneal. The graph at the right shows the full energy $E(k)$, which decreases to the global minimum.

throughout the search, and use it if the final configuration returned by the stochastic search itself is worse—a method we shall mention again in Section 7.3.4.

Figure 7.3 shows that early in the annealing process when the temperature is high, the system explores a wide range of configurations. Later, as the temperature is lowered, only states "close" to the global minimum are tested. Throughout the process, each transition corresponds to the change in state of a single unit.

A typical choice of annealing schedule is $T(k + 1) = cT(k)$ with $0 < c < 1$. If computational resources are of no concern, a high initial temperature, large c, and large k_{max} are most desirable. Values in the range $0.8 < c < 0.99$ have been found to work well in many real-world problems. In practice the algorithm is slow, requiring many iterations and many passes through all the nodes, though for all but the smallest problems it is still faster than exhaustive search (Problem 5). We shall revisit the issue of parameter setting in the context of learning in Section 7.3.4.

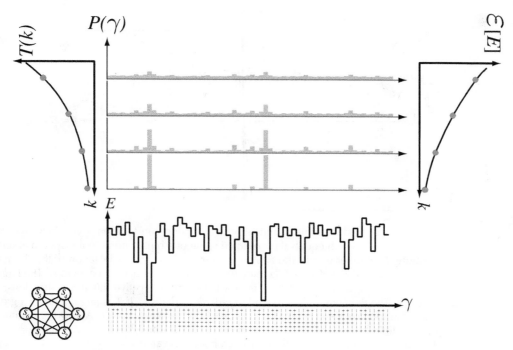

FIGURE 7.4. An estimate of the probability $P(\gamma)$ of being in a configuration denoted by γ is shown for four temperatures during a slow anneal. (These estimates, based on a large number of runs, are nearly the theoretical values $e^{-E_\gamma/T}$.) Early, at high T, each configuration is roughly equal in probability while late, at low T, the probability is strongly concentrated at the global minima. The expected value of the energy, $\mathcal{E}[E]$ (i.e., averaged at temperature T), decreases gradually during the anneal.

While Fig. 7.3 displayed a single trajectory through the configuration space, a more relevant property is the *probability* of being in a configuration as the system is annealed gradually. Figure 7.4 shows such probability distributions at four temperatures. Note especially that at the final, low temperature the probability is concentrated at the global minima, as desired. While this figure shows that for positive temperature all states have a nonzero probability of being visited, we must recognize that only a small fraction of configurations are in fact visited in any anneal. In short, in the vast majority of large problems, annealing does not require that all configurations be explored, and hence it is more efficient than exhaustive search.

7.2.3 Deterministic Simulated Annealing

Stochastic simulated annealing is slow, in part because of the discrete nature of the search through the space of all configurations—that is, an N-dimensional hypercube. Each trajectory is along a *single* edge, thereby missing full gradient information that would be provided by analog state values in the "interior" of the hypercube. A faster, alternative method is to allow each node to take on *analog* values during search; at the end of the search the values are forced to be $s_i = \pm 1$, as required by the optimization problem. Such a *deterministic* simulated annealing algorithm also follows from the physical analogy. Consider a single node (magnet) i connected to several others; each exerts a force tending to point node i up or down. In deterministic annealing

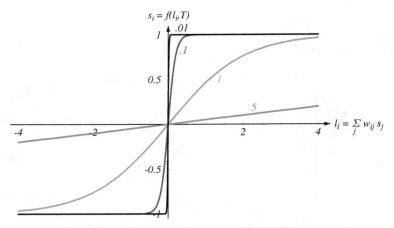

FIGURE 7.5. In deterministic annealing, each node can take on a continuous value $-1 \le s_i \le +1$, which equals the expected value of a binary node in the system at temperature T. In other words, the analog value s_i replaces the expectation of the discrete variable, $\mathcal{E}[s_i]$. We let l_i denote a force exerted by the nodes connected to s_i. The larger this force, the closer the analog s_i is to $+1$; the more negative this force, the closer is s_i to -1. The temperature T (marked in red) also affects s_i. If T is large, there is a great deal of randomness and even a large force will not ensure $s_i \approx +1$. At low temperature, there is little or no randomness and even a small positive force ensures that $s_i = +1$. Thus at the end of an anneal (low T), each node has value $s_i = +1$ or $s_i = -1$.

we sum the forces to find an *analog* value for s_i. If there is a large "positive" force, then $s_i \approx +1$; if a large negative force, then $s_i \approx -1$. In the general case s_i will lie between these limits.

The value of s_i also depends upon the temperature. At high T (large randomness) even a large upward force will not be enough to ensure $s_i = +1$, whereas at low temperature it will. If we let $l_i = \sum_j w_{ij} s_j$ be the force exerted on node i, then the updated value is

$$s_i = f(l_i, T) = \tanh[l_i/T], \tag{5}$$

RESPONSE
FUNCTION

where there is an implied scaling of the force and temperature in the *response function* $f(\cdot, \cdot)$, as shown in Fig. 7.5. In broad overview, deterministic annealing consists in setting an annealing schedule and then at each temperature finding an equilibrium analog value for every s_i. This analog value is merely the expected value of the discrete s_i in a system at temperature T (Problem 8). At low temperatures (i.e., at the end of the anneal), each variable will assume an extreme value ± 1, as can be seen in the low-T curve in Fig. 7.5.

It is instructive to consider the energy landscape for the continuous case. As can be seen from Eq. 1 and as Fig. 7.6 illustrates, the partial derivative of the energy E is linear in each variable. Thus, there are no local minima in any "cut" parallel to any axis. Note too that there are no stable local energy minima within the volume of the space; the energy minima always occur at the "corners"—that is, extreme $s_i = \pm 1$ for all i, as required by the optimization problem.

MEAN-FIELD
ANNEALING

This search method is sometimes called *mean-field annealing* because each node responds to the average or mean of the forces (fields) due to the nodes connected to it. In essence the method approximates the effects of all other magnets while ignoring

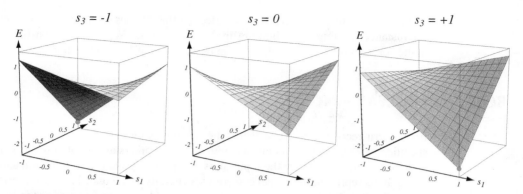

FIGURE 7.6. If the state variables s_i can assume analog values (as in mean-field annealing), the energy in Eq. 1 is a general quadratic form having minima at the extreme values $s_i = \pm 1$. In the case illustrated here $N = 3$ nodes are fully interconnected with arbitrary weights w_{ij}. While the total energy function is three-dimensional, we show two-dimensional surfaces for each of three values of s_3. The energy is linear in each variable so long as the other variables are held fixed. Furthermore, the energy is invariant with respect to the interchange of all variables $s_i \leftrightarrow -s_i$. In particular, here the global minimum occurs as $s_1 = -1$, $s_2 = +1$ and $s_3 = -1$ as well as the symmetric configuration $s_1 = +1$, $s_2 = -1$, and $s_3 = +1$ (red dots).

their mutual interactions and their response to the magnet in question, node i. Such annealing is also called *deterministic* because in principle we could deterministically solve the simultaneous equations governing the s_i as the temperature is lowered. The algorithm has a natural parallel mode of implementation, for instance where each value s_i is updated simultaneously and deterministically as the temperature is lowered. In an inherently serial simulation, however, the nodes are updated one at a time. Even though the nodes might be polled pseudorandomly, the algorithm is in principle deterministic—there need be no inherent randomness in the search. If we let $s_i(1)$ denote the initial state of unit i, the algorithm is as follows:

■ **Algorithm 2. (Deterministic Simulated Annealing)**

1 **begin initialize** $T(k)$, w_{ij}, $s_i(1)$, i, $j = 1, \ldots N$
2          $k \leftarrow 0$
3   **do** $k \leftarrow k + 1$
4   select node i randomly
5              $l_i \leftarrow \sum_j^{\mathcal{N}_i} w_{ij} s_j$
6              $s_i \leftarrow f(l_i, T(k))$
7   **until** $k = k_{max}$ or convergence criterion met
8   **return** E, s_i, $i = 1, \ldots, N$
9 **end**

In practice, deterministic and stochastic annealing give very similar solutions. In large real-world problems deterministic annealing is faster, sometimes by two or three orders of magnitude.

Simulated annealing can also be applied to other classes of optimization problem, for instance, finding the minimum in $\sum_{ijk} w_{ijk} s_i s_j s_k$. We will not consider such higher-order problems, though they can be the basis of learning methods as well.

7.3 BOLTZMANN LEARNING

CLAMP

For pattern recognition, we will use the network structure shown in Fig. 7.1 and specifically identify certain units as input units and others as output units. This is illustrated in Fig. 7.7, where the input units accept binary feature information and the output units represent the output categories, generally in the familiar 1-of-c representation. During classification the input units are held fixed or *clamped* to the feature values of the input pattern; the remaining units are annealed to find the lowest-energy, most probable configuration. The category information is then read from the final values of the output units. Of course, accurate recognition requires proper weights, and thus we now turn to a method for learning weights from training patterns. There are two closely related approaches to such learning, one based on stochastic and the other on deterministic simulated annealing.

7.3.1 Stochastic Boltzmann Learning of Visible States

Before we turn to our central concern—learning categories from training patterns—consider an alternative learning problem where we have a set of desired probabilities for *all* the visible units, $Q(\alpha)$, and seek weights so that the actual probability $P(\alpha)$,

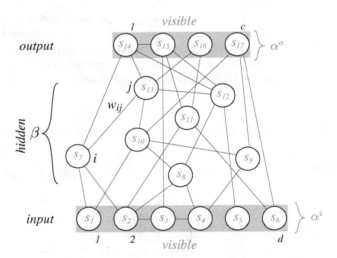

FIGURE 7.7. When a network such as shown in Fig. 7.1 is used for learning, it is important to distinguish between two types of visible units—the d input units and c output units, which receive external feature and category information—as well as the remaining, hidden units. The state of the full network is indexed by an integer γ, and because here there are 17 binary nodes, γ is bounded $0 \leq \gamma < 2^{17}$. The state of the input visible nodes is described by α^i and the output visible nodes by α^o. (The superscripts are not indexes, but merely refer to the input and output, respectively.) In the case shown, α^i is bound in the range $0 \leq \alpha^i < 2^d$ and α^o is bound in the range $0 \leq \alpha^0 < 2^c$. The state of the hidden nodes is indexed by β, which is bound in the range $0 \leq \beta < 2^{N-c-d}$.

achieved in random simulations, matches these probabilities over a given set of patterns. In this alternative learning problem the desired probabilities would be derived from training patterns containing both input (feature) and output (category) information. The actual probability describes the states of a network annealed with neither input nor output variables clamped.

We now make use of the distinction between configurations of "visible" units (the input and output), denoted α, and the hidden states, denoted β, shown in Fig. 7.1. For instance, whereas a and b (c.f. Eq. 4) refered to different configurations of the full system, α and β will specify visible and hidden configurations.

The probability of a visible configuration is the sum over all possible hidden configurations:

$$P(\alpha) = \sum_{\beta} P(\alpha, \beta)$$

$$= \frac{\sum_{\beta} e^{-E_{\alpha\beta}/T}}{Z}, \tag{6}$$

where $E_{\alpha\beta}$ is the system energy in the configuration defined by the visible and hidden parts, and Z is again the full partition function. Equation 6 is based on Eq. 3 and states simply that to find the probability of a given visible state α, we sum over all possible hidden states consistent with α. A natural measure of the difference between the actual and the desired probability distributions is the relative entropy, Kullback-Leibler distance or Kullback-Leibler divergence,

$$D_{KL}(Q(\alpha), P(\alpha)) = \sum_{\alpha} Q(\alpha) \log \frac{Q(\alpha)}{P(\alpha)}. \tag{7}$$

It is simple to show that D_{KL} is nonnegative and can be zero if and only if $P(\alpha) = Q(\alpha)$ for all α (see Appendix Section A.7.2). Note that Eq. 7 depends solely upon the visible units, not the hidden units.

Learning is based on gradient descent in the relative entropy. A set of training patterns defines $Q(\alpha)$, and we seek weights so that at some temperature T the actual distribution $P(\alpha)$ matches $Q(\alpha)$ as closely as possible. Thus we take an untrained network and update each weight according to:

$$\Delta w_{ij} = -\eta \frac{\partial D_{KL}}{\partial w_{ij}} = \eta \sum_{\alpha} \frac{Q(\alpha)}{P(\alpha)} \frac{\partial P(\alpha)}{\partial w_{ij}}, \tag{8}$$

where η is a learning rate. Note that while $P(\cdot)$ depends on the weights, $Q(\cdot)$ does not, and thus we used $\partial Q(\alpha)/\partial w_{ij} = 0$. From Eqs. 1 and 6 we find

$$\frac{\partial P(\alpha)}{\partial w_{ij}} = \frac{\sum_{\beta} e^{-E_{\alpha\beta}/T} s_i(\alpha\beta) s_j(\alpha\beta)}{TZ} - \frac{\left(\sum_{\beta} e^{E_{\alpha\beta}/T}\right) \sum_{\lambda\mu} e^{-E_{\lambda\mu}/T} s_i(\lambda\mu) s_j(\lambda\mu)}{TZ^2}$$

$$= \frac{1}{T} \left[\sum_{\beta} s_i(\alpha\beta) s_j(\alpha\beta) P(\alpha, \beta) - P(\alpha) \mathcal{E}[s_i s_j] \right]. \tag{9}$$

Here $s_i(\alpha\beta)$ is the state of node i in the full configuration specified by α and β. Of course, if node i is a visible one, then only the value of α is relevant; if the node is a hidden one, then only the value of β is relevant. (Our notation unifies these two cases.) The expectation value $\mathcal{E}[s_i s_j]$ is taken at temperature T. We gather terms and find from Eqs. 8 and 9 that the weight update should be

$$
\begin{aligned}
\Delta w_{ij} &= \frac{\eta}{T}\left[\sum_\alpha \frac{Q(\alpha)}{P(\alpha)}\sum_\beta s_i(\alpha\beta)s_j(\alpha\beta)P(\alpha,\beta) - \sum_\alpha Q(\alpha)\mathcal{E}[s_i s_j]\right] \\
&= \frac{\eta}{T}\left[\sum_{\alpha\beta} Q(\alpha)P(\beta|\alpha)s_i(\alpha\beta)s_j(\alpha\beta) - \mathcal{E}[s_i s_j]\right] \\
&= \frac{\eta}{T}\left[\underbrace{\mathcal{E}_Q[s_i s_j]_{\alpha\ clamped}}_{learning} - \underbrace{\mathcal{E}[s_i s_j]_{free}}_{unlearning}\right]
\end{aligned}
\tag{10}
$$

where $P(\alpha,\beta) = P(\beta|\alpha)P(\alpha)$. We have defined

$$
\mathcal{E}_Q[s_i s_j]_{\alpha\ clamped} = \sum_{\alpha\beta} Q(\alpha)P(\beta|\alpha)s_i(\alpha\beta)s_j(\alpha\beta)
\tag{11}
$$

to be the correlation of the variables s_i and s_j when the visible units are held fixed—clamped—in visible configuration α, averaged according to the probabilities of the training patterns, $Q(\alpha)$.

The first term on the right of Eq. 10 is informally referred to as the *learning component* or *teacher component* (as the visible units are held to values given by the teacher), and the second term is referred to as the *unlearning component* or *student component* (where the variables are free to vary). If $\mathcal{E}_Q[s_i s_j]_{\alpha\ clamped} = \mathcal{E}[s_i s_j]_{free}$, then $\Delta w_{ij} = 0$ and we have achieved the desired weights. The unlearning component reduces spurious correlations between units—spurious in that they are not due to the training patterns. A learning algorithm based on the above derivation would present each pattern in the full training set several times and adjust the weights by Eq. 10, just as we saw in numerous other training methods such as backpropagation (Chapter 6).

LEARNING
COMPONENT

UNLEARNING
COMPONENT

Stochastic Learning of Input-Output Associations

Now consider the problem of learning mappings from input to output—our real interest in pattern recognition. Here we want the network to learn associations between the (visible) states on the input units, denoted α^i, and states on the output units, denoted α^o, as shown in Fig. 7.7. Formally, we want $P(\alpha^o|\alpha^i)$ to match $Q(\alpha^o|\alpha^i)$ as closely as possible. The appropriate cost function here is the Kullback-Leibler divergence weighted by the probability of each input pattern:

$$
\overline{D}_{KL}(Q(\alpha^o|\alpha^i), P(\alpha^o|\alpha^i)) = \sum_{\alpha^i} P(\alpha^i)\sum_{\alpha^o} Q(\alpha^o|\alpha^i)\log\frac{Q(\alpha^o|\alpha^i)}{P(\alpha^o|\alpha^i)}.
\tag{12}
$$

Just as in Eq. 8, learning involves changing weights to reduce this weighted distance, that is,

$$
\Delta w_{ij} = -\eta\frac{\partial \overline{D}_{KL}}{\partial w_{ij}}.
\tag{13}
$$

The derivation of the full learning rule follows closely that leading to Eq. 11; the only difference is that the input units are clamped in both the learning and unlearning components (Problem 11). The result is that the weight update is

$$\Delta w_{ij} = \frac{\eta}{T}\left[\underbrace{\mathcal{E}_Q[s_i s_j]_{\alpha^i \alpha^o \ clamped}}_{learning} - \underbrace{\mathcal{E}[s_i s_j]_{\alpha^i \ clamped}}_{unlearning}\right]. \tag{14}$$

In Section 7.3.3 we shall present pseudocode for implementing the preferred, deterministic version of Boltzmann learning, but first we can gain intuition into the general method by considering the learning of a single pattern according to Eq. 14. Figure 7.8 shows a seven-unit network being trained with the input pattern $s_1 = +1$, $s_2 = +1$ and the output pattern $s_6 = -1$, $s_7 = +1$. In a typical 1-of-c representation, this desired output signal would represent category ω_2. Because during both training and classification, the input units s_1 and s_2 are clamped at the value $+1$, we have shown only the associated $2^5 = 32$ configurations at the right. The energy before training (Eq. 1), corresponding to randomly chosen weights, is shown in black. After the weights are trained by Eq. 14 using the pattern shown, the energy is changed (shown in red). Note that all states having the desired output pattern have their en-

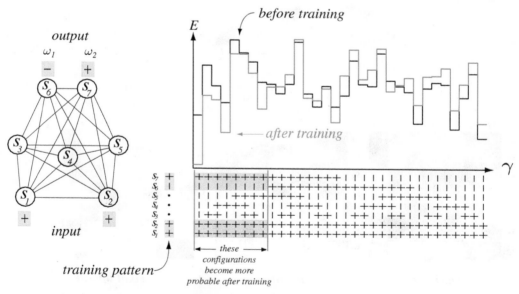

FIGURE 7.8. The fully connected seven-unit network on the left is being trained to assign the input pattern $s_1 = +1, s_2 = +1$ to class ω_2 by using the Boltzmann learning algorithm. During training, the output s_7 (corresponding to ω_2) is clamped at $+1$, and the output s_6 (corresponding to ω_1) is clamped at -1. All $2^5 = 32$ configurations with $s_1 = +1, s_2 = +1$ are shown at the right, along with their energy (Eq. 1). The black curve shows the energy before training; the red curve shows the energy after training. Note particularly that after training, all configurations that represent the full training pattern have been lowered in energy and hence are more probable; most other patterns become *less* probable after training. Thus, after training, if the input pattern $s_1 = +1, s_2 = +1$ is presented and the remaining network is annealed, there is an increased chance of yielding the outputs $s_6 = -1$ and $s_7 = +1$, as desired.

ergies lowered through training, just as we need. Thus when these input states are clamped and the remaining networked annealed, the desired output is more likely to be found.

Equation 14 appears a bit different from weight change equations we have encountered in Chapters 5 and 6, and it is worthwhile explaining it carefully. Figure 7.9 illustrates in greater detail the learning of the single training pattern in Fig. 7.8. Because s_1 and s_2 are clamped throughout, $\mathcal{E}_Q[s_1 s_2]_{\alpha^i \alpha^o\, clamped} = 1 = \mathcal{E}[s_1 s_2]_{\alpha^i\, clamped}$, and thus (from Eq. 14) the weight w_{12} is not changed. Consider a more general case, involving s_1 and s_7. During the learning phase, both units are clamped at $+1$ and thus the correlation is $\mathcal{E}_Q[s_1 s_7] = +1$. During the unlearning phase, the output s_7 is free to vary and the correlation is lower; in fact it happens to be negative. Thus, the learning rule seeks to increase the magnitude of w_{17} so that the input $s_1 = +1$ leads to $s_7 = +1$, as can be seen in the matrix on the right. Because hidden units are only weakly correlated (or anticorrelated), the weights linking hidden units are changed only slightly.

In learning a training set of many patterns, each pattern is presented in turn, and the weights are updated as just described. Learning ends when the actual output matches or nearly matches the desired output for all patterns (cf. Section 7.3.4). One benefit of such stochastic learning is that if the final error seems to be unacceptably high, we can merely increase the temperature and anneal—we do not need to re-initialize the weights and re-start the full anneal.

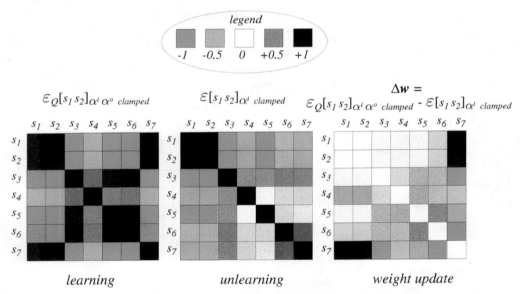

FIGURE 7.9. Boltzmann learning of a single pattern is illustrated for the seven-node network of Fig. 7.8. The (symmetric) matrix on the left shows the correlation of units for the learning component, where the input units are clamped to $s_1 = +1$, $s_2 = +1$ and the outputs are clamped to $s_6 = -1$, $s_7 = +1$. The middle matrix shows the unlearning component, where the inputs are clamped but outputs are free to vary. According to Eq. 14, the weight update should be proportional to the difference between those matrices, as indeed is shown on the right matrix. Notice, for instance, that because the correlation between s_1 and s_2 is large in both the learning and unlearning components (because those variables are clamped), there is no associated weight change, that is, $\Delta w_{12} = 0$. However, strong correlations between s_1 and s_7 in the learning but not in the unlearning component implies that the weight w_{17} should be increased, as can be seen in the weight update matrix.

7.3.2 Missing Features and Category Constraints

A key benefit of Boltzmann training (including its preferred implementation, described in Section 7.3.3) is its ability to deal with missing features, both during training and during classification. If a deficient binary pattern is used for training, input units corresponding to missing features are allowed to vary—they are temporarily treated as (unclamped) hidden units rather than clamped input units. As a result, during annealing such units assume values most consistent with the rest of the input pattern and the current state of the network (Problem 14). Likewise, when a deficient pattern is to be classified, any units corresponding to missing input features are not clamped and are allowed to assume any value.

Some subsidiary knowledge or constraints can be incorporated into a Boltzmann network during classification. Suppose in a five-category problem it is somehow known that a test pattern is neither in category ω_1 nor in category ω_4. (Such constraints could come from context or stages subsequent to the classifier itself.) During classification, then, the output units corresponding to ω_1 and ω_4 are clamped at $s_i = -1$ during the anneal, and the final category is read as usual. Of course in this example the possible categories are then limited to the unclamped output units, for ω_2, ω_3 and ω_5. Such constraint imposition may lead to an improved classification rate (Problem 15).

Pattern Completion

The problem of pattern completion is to estimate the full pattern given just a part of that pattern; as such, it is related to the problem of classification with missing features. Pattern completion is naturally addressed using Boltzmann networks. A fully interconnected network, with or without hidden units, is trained with a set of representative patterns; as before, the visible units correspond to the feature components. When a deficient pattern is presented, the units in a subset of the visible units are clamped to the components of the partial pattern, and the network annealed. The estimate of the unknown features appears on the remaining visible units, as illustrated in Fig. 7.10 (Computer exercise 4). Such pattern completion in Boltzmann networks can be more accurate when known category information is imposed at the output units.

HOPFIELD
NETWORK

Boltzmann networks without hidden or category units are closely related to so-called *Hopfield networks* or Hopfield autoassociation networks (Problem 12). Such networks store patterns but not their category labels. The learning rule for such networks does not require the full Boltzmann learning of Eq. 14. Instead, weights are set to be proportional to the correlation of the feature vectors, averaged over the training set,

$$w_{ij} \propto \mathcal{E}_Q[s_i s_j], \tag{15}$$

with $w_{ii} = 0$; furthermore, there is no need to consider temperature. Such learning is of course much faster than true Boltzmann learning using annealing. If a network fully trained by Eq. 15 is nevertheless annealed, as in full Boltzmann learning, there is no guarantee that the equilibrium correlations in the learning and unlearning phases are equal—that is, that $\Delta w_{ij} = 0$ (Problem 13).

The successes of such Hopfield networks in true pattern recognition have been modest, partly because the basic Hopfield network does not have as natural an output representation for categorization problems. Occasionally, however, they can be used in simple low-dimensional pattern completion or autoassociation problems. In

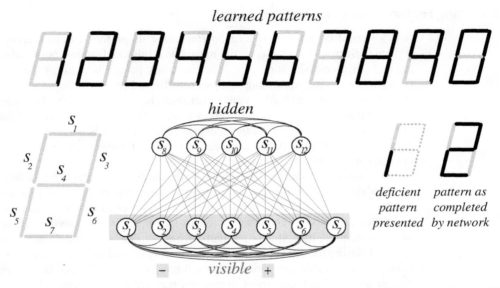

FIGURE 7.10. A Boltzmann network can be used for pattern completion—that is, filling in unknown features of a deficient pattern. Here, a 12-unit network with five hidden units has been trained with the 10 numeral patterns of a seven-segment digital display. The diagram at the lower left shows the correspondence between the display segments and nodes of the network. Along the top, a black segment is represented by a +1, and a light gray segment is represented as a −1. Consider the deficient pattern consisting of $s_2 = -1, s_5 = +1$, but with the other five inputs (shown as dotted lines in the pattern) unspecified. If these units are clamped and the full network is annealed, the remaining five visible units will assume values most probable given the clamped ones, as shown at the right.

particular, suppose that a network learns a set of patterns, which are represented as configurations in the net. Later, a noisy or deficient pattern is presented to the net, which then progresses to one of the stored configurations. Alas, it has been shown that the number of d-dimensional random patterns that can be stored in such a Hopfield network is roughly $0.14d$—very limited indeed. Furthermore, if the training patterns are *not* randomly distributed, then the number of patterns that can be stored is lower still. In a Boltzmann with hidden units such as we have discussed, however, the number of hidden units can be increased in order to allow more patterns to be stored.

Because Boltzmann networks include loops and feedback connections, the internal representations learned at the hidden units are often difficult to interpret. Occasionally, though, the pattern of weights from the input units suggests feature groupings that are important for the classification task.

7.3.3 Deterministic Boltzmann Learning

The computational complexity of stochastic Boltzmann learning in a network with hidden units is very high. Each pattern must be presented several times, and every anneal requires each unit to be polled several times. Just as mean-field annealing is usually preferable to stochastic annealing, so too a mean-field version of Boltzmann learning is preferable to the stochastic version. The basic approach in deterministic Boltzmann learning is to use Eq. 14 with mean-field annealing and analog values for the s_i. As mentioned, at the end of deterministic simulated annealing, the values of s_i are ± 1, as required by the problem.

Specifically, if we let $\mathcal{D}$ be the set of training patterns $\mathbf{x}$ containing feature and category information, the algorithm is as follows:

■ **Algorithm 3. (Deterministic Boltzmann Learning)**

1 **begin initialize** $\mathcal{D}, \eta, T(k), w_{ij}\ i, j = 1, \ldots, N$
2   **do** randomly select training pattern $\mathbf{x}$
3   randomize states s_i
4   anneal network with input and output clamped
5   at final, low T, calculate $[s_i s_j]_{\alpha^i \alpha^o clamped}$
6   randomize states s_i
7   anneal network with input clamped but output free
8   at final, low T, calculate $[s_i s_j]_{\alpha^i clamped}$
9              $w_{ij} \leftarrow w_{ij} + \eta/T \left[[s_i s_j]_{\alpha^i \alpha^o clamped} - [s_i s_j]_{\alpha^i clamped} \right]$
10   **until** $k = k_{max}$ or convergence criterion met
11 **return** w_{ij}
12 **end**

Using mean-field theory, it is possible to efficiently calculate *approximations* of the mean of correlations entering the gradient. The analog state s_i of each unit replaces its average value $\mathcal{E}[s_i]$ and could in theory be calculated by iteratively solving a set of nonlinear equations. The mean of correlations is then calculated by making the approximation $\mathcal{E}[s_i s_j] \approx \mathcal{E}[s_i]\mathcal{E}[s_j] \approx s_i s_j$, as shown in lines 5 and 8.

7.3.4 Initialization and Setting Parameters

As with virtually every classifier, there are several interrelated parameters that must be set in a Boltzmann network. The first are the network topology and number of hidden units. The number of visible units (input and output) is determined by the dimensions of the binary feature vectors and number of categories. In the absence of detailed information about the problem, we assume the network is fully interconnected, and thus merely the number of hidden units must be set. A popular alternative topology is obtained by eliminating interconnections among input units, as well as among output units. (Such a network is faster to train but will be somewhat less effective at pattern completion or classifying deficient patterns.) Of course, generally speaking, the harder the classification problem, the more hidden units will be needed. The question is then, How many hidden units should be used?

Suppose the training set $\mathcal{D}$ has n distinct patterns of input-output pairs. An *upper bound* on the minimum number of hidden units is n—one for each pattern—where for each pattern there is a corresponding unique hidden unit having value $s_i = +1$ while all others are -1. Such an internal representation can be ensured in the following way: For the particular hidden unit i, set w_{ij} to be positive for each input unit j corresponding to a $+1$ feature in its associated pattern; further set w_{ij} to be negative for input units corresponding to a -1 feature. For the remaining hidden units, the sign of the corresponding weights should be inverted. Next, the connection from hidden unit i to the output unit corresponding to the known category should be positive, but negative to all other output units. The resulting internal representation is closely related to that in the probabilistic neural network implementation of Parzen windows (Chapter 4). Naturally, this representation is undesirable as the number of weights grows exponentially with the number of patterns. Training becomes slow and generalization tends to be poor.

Because the states of the hidden units are binary-valued and because it takes $\lceil \log_2 n \rceil$ bits to specify n different items, there must be at least $\lceil \log_2 n \rceil$ hidden units if there is to be a distinct hidden configuration for each of the n patterns. Thus a *lower bound* on the number of hidden units is $\lceil \log_2 n \rceil$, which is necessary for a distinct hidden configuration for each pattern. Nevertheless, this bound need not be tight, because there may be no set of weights ensuring a unique representation (Problem 16). Aside from these bounds, it is hard to make firm statements about the number of hidden units needed—this number depends upon the inherent difficulty of the classification problem. It is traditional, then, to start with a somewhat large net and use weight decay. Much as we saw in backpropagation (Chapter 6), a Boltzmann network with "too many" hidden units and weights can be improved by means of weight decay. During training, a small increment ϵ is added to w_{ij} when s_i and s_j are both positive or both negative during learning phase, but subtracted in the unlearning phase. It is traditional to decrease ϵ throughout training. Such a version of weight decay tends to reduce the effects on the weights due to spurious random correlations in units and to eliminate unneeded weights, thereby improving generalization.

One of the benefits of Boltzmann networks over backpropagation networks is that "too many" hidden units in a backpropagation network tend to degrade performance more than "too many" in a Boltzmann network. This is because during learning, there is stochastic averaging over states in a Boltzmann network which tends to smooth decision boundaries; backpropagation networks have no such equivalent averaging. Of course, this averaging comes at a higher computational burden for Boltzmann networks.

The next matter to consider is weight initialization. Initializing all weights to zero is acceptable, but leads to unnecessarily slow learning. In the absence of information otherwise, we can expect that roughly half the weights will be positive and half negative. In a network with fully interconnected hidden units, there is nothing to differentiate the individual hidden units; thus we can arbitrarily initialize roughly half of the weights to have positive values and the rest negative. Learning speed is increased if weights are initialized with random values within a proper range. Assume a fully interconnected network having N units (and thus $N - 1 \approx N$ connections to each unit). Assume further that at any instant each unit has an equal chance of being in state $s_i = +1$ or $s_i = -1$. We seek initial weights that will make the net force on each unit a random variable with variance 1.0, roughly the useful range shown in Fig. 7.5. This implies that weights should be initialized randomly throughout the range $-\sqrt{3/N} < w_{ij} < +\sqrt{3/N}$, as shown in Problem 17.

If a very large number of iterations—several thousand—are needed, even $c = 0.99$ may be too small. In that case we can write $c = e^{-1/k_0}$, and thus $T(k) = T(1)e^{-k/k_0}$, and k_0 can be interpreted as a decay constant. The initial temperature $T(1)$ should be set high enough that virtually all candidate state transitions are accepted. While this condition can be ensured by choosing $T(1)$ extremely high, in order to reduce training time we seek the lowest adequate value of $T(1)$. A lower bound on the acceptable initial temperature depends upon the problem, but can be set empirically by monitoring state transitions in short simulations at candidate temperatures. Let m_1 be the number of energy-decreasing transitions (these are always accepted), and let m_2 be the number of energy-*increasing* queries according to the annealing algorithm; let $\mathcal{E}_+[\Delta E]$ denote the average increase in energy over such transitions. Then, from Eq. 4 we find that at the beginning of the anneal the acceptance ratio is

$$R = \frac{\text{number of accepted transitions}}{\text{number of proposed transitions}} \approx \frac{m_1 + m_2 \cdot \exp[-\mathcal{E}_+[\Delta E]/T(1)]}{m_1 + m_2}. \quad (16)$$

We rearrange terms and see that the initial temperature obeys

$$T(1) = \frac{\mathcal{E}_+[\Delta E]}{\ln[m_2] - \ln[m_2 R - m_1(1 - R)]}. \quad (17)$$

For any initial temperature set by the designer, the acceptance ratio may or may not be nearly the desired 1.0; nevertheless, Eq. 17 will be obeyed. The appropriate value for $T(1)$ is found through a simple iterative procedure. First, set $T(1)$ to zero and perform a sequence of m_0 trials (pollings of units); count empirically the number of energetically favorable (m_1) and energetically unfavorable (m_2) transitions. In general, $m_1 + m_2 < m_0$ because many candidate energy increasing transitions are rejected, according to Eq. 4. Next, use Eq. 17 to calculate a new, improved value of $T(1)$ from the observed m_1 and m_2. Perform another sequence of m_0 trials, observe new values for m_1 and m_2, recalculate $T(1)$, and so on. Repeat this procedure until $m_1 + m_2 \approx m_0$. The associated $T(1)$ gives an acceptance ratio $R \approx 1$, and thus it is to be used. In practice this method quickly yields a good starting temperature.

The next important parameter is the learning rate η in Eq. 14. Recall that the learning is based on gradient descent in the weighted Kullback-Leibler divergence between the actual and the desired distributions on the visible units. In Chapter 6 we derived bounds on the learning rate for multilayer neural networks by calculating the curvature of the error, and finding the maximum value of the learning rate that ensured stability. This curvature was based on a Hessian matrix, the matrix of second-order derivatives of the error with respect to the weights. In the case of an N-unit, fully connected Boltzmann network, whose $N(N - 1)/2$ weights are described by a vector $\mathbf{w}$, this curvature is proportional to $\mathbf{w}^t \mathbf{H} \mathbf{w}$, where

$$\mathbf{H} = \frac{\partial^2 \overline{D}_{KL}}{\partial \mathbf{w}^2} \quad (18)$$

is the appropriate Hessian matrix and the Kullback-Liebler divergence is given by Eq. 12. Given weak assumptions about the classification problem we can estimate this Hessian matrix; the stability requirement is then simply $\eta \leq T^2/N$ (Problem 18). Note that at large temperature T, a large learning rate is acceptable because the effective error surface is smoothed by high randomness.

While not technically parameter setting, one heuristic that provides modest computational speedup is to propose changing the states of *several* nodes simultaneously early in an anneal. The change in energy and acceptance probability are calculated as before. At the end of the anneal, however, polling should be of single units giving a detailed search for the optimum configuration. A method that occasionally improves the final solution is to update and store the current best configuration during an anneal. If the basic annealing converges to a local minimum that is worse than this stored configuration, this current optimal should be used.

There are two stopping criteria associated with Boltzmann learning. The first determines when to stop a single anneal (associated with either the learning or the unlearning components). Here, the final temperature should be so low that no energetically unfavorable transitions are accepted. Such information is readily apparent in the graph of the energy versus iteration number, such as shown at the right of Fig. 7.3. All N variables should be polled individually at the end of the anneal, to ensure that the final configuration is indeed a local (though perhaps not global) energy minimum.

The second stopping criterion controls the number of times each training pattern is presented to the network. Of course the proper criterion depends upon the inherent difficulty of the classification problem. In general, overtraining is less of a concern in Boltzmann networks than it is in multilayer neural networks trained via gradient descent. This is because the averaging over states in Boltzmann networks tends to smooth decision boundaries while overtraining in multilayer neural networks tunes the decision boundaries to the particular training set. A reasonable stopping criterion for Boltzmann networks is to monitor the error on a validation set (Chapter 9) and stop learning when this error no longer changes significantly.

*7.4 BOLTZMANN NETWORKS AND GRAPHICAL MODELS

While we have considered fully interconnected Boltzmann networks, the learning algorithm (Algorithm 3) applies equally well to networks with arbitrary connection topologies. Furthermore, it is easy to modify Boltzmann learning in order to impose constraints such as weight sharing. As a consequence, several popular recognition architectures—so-called graphical models such as Bayesian belief networks and Hidden Markov Models—have counterparts in structured Boltzmann networks, and this leads to new methods for training them.

Recall from Chapter 3 that Hidden Markov Models consist of several discrete hidden and visible states; at each discrete time step t, the system is in a single hidden state and emits a single visible state, denoted $\omega(t)$ and $v(t)$, respectively. The transition probabilities between hidden states at successive time steps are

$$a_{ij} = P(\omega_j(t+1)|\omega_i(t)) \tag{19}$$

and those between hidden and visible states at a given time are

$$b_{jk} = P(v_k(t)|\omega_j(t)). \tag{20}$$

The Forward-Backward or Baum-Welch algorithm (Algorithm 5 in Chapter 3) is traditionally used for learning these parameters from a pattern of T_f visible states[*] $\mathbf{V}^{T_f} = \{v(1), v(2), \ldots, v(T_f)\}$.

Recall that a Hidden Markov model can be "unfolded" in time to yield a trellis. A structured Boltzmann network with the same trellis topology—a *Boltzmann chain*—can be used to implement the same classification as the corresponding Hidden Markov Model (Fig. 7.11). Although it is often simpler to work in a representation where discrete states have multiple values, we temporarily work in a representation where the binary nodes take value $s_i = 0$ or $+1$, rather than ± 1 as in previous discussions. In this representation, a special case of the general energy (Eq. 1) includes terms for a particular sequence of visible, $\mathbf{V}^{T_f}$, and hidden states $\boldsymbol{\omega}^{T_f} = \{\omega(1), \omega(2), \ldots, \omega(T_f)\}$ and can be written as

BOLTZMANN
CHAIN

$$E_{\boldsymbol{\omega}\mathbf{V}} = E[\boldsymbol{\omega}^{T_f}, \mathbf{V}^{T_f}] = -\sum_{t=1}^{T_f-1} A_{ij} - \sum_{t=1}^{T_f} B_{jk} \tag{21}$$

---

[*]Here we use T_f to count the number of discrete time steps in order to avoid confusion with the temperature T in Boltzmann simulations.

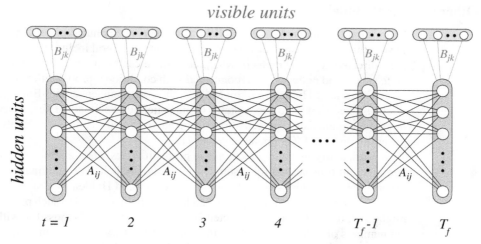

FIGURE 7.11. A Hidden Markov Model can be "unfolded" in time to show a trellis, which can be represented as a Boltzmann chain, as shown. The discrete hidden states are grouped into vertical sets, fully interconnected by weights A_{ij} (related to the Hidden Markov Model transition probabilities a_{ij}). The discrete visible states are grouped into horizontal sets, and are fully interconnected with the hidden states by weights B_{jk} (related to emission probabilities b_{jk}). Training the net with a single pattern, or list of T_f visible states, consists of clamping the visible states and performing Boltzmann learning throughout the full network, with the constraint that each of the time shifted weights labeled by a particular A_{ij} have the same numerical value.

where the particular values of the A_{ij} and B_{jk} terms depend implicitly upon the sequence. The choice of binary state representation implies that only the weights linking nodes that both have $s_i = +1$ appear in the energy. Each "legal" configuration—consisting of a single visible unit and a single hidden unit at each time—implies a set of A_{ij} and B_{jk} (Problem 20). The partition function is the sum over all legal states,

$$Z = \sum_{\boldsymbol{\omega} \times \mathbf{V}} e^{-E_{\boldsymbol{\omega}\mathbf{V}}/T},$$ (22)

which ensures normalization. The correspondence between the Boltzmann chain at temperature T and the unfolded Hidden Markov Model (trellis) implies

$$A_{ij} = T \ln a_{ij} \qquad \text{and} \qquad B_{jk} = T \ln b_{jk}.$$ (23)

(As in our discussion of Hidden Markov Models, we assume that the initial hidden state is known and thus there is no need to consider the correspondence of prior probabilities in the two approaches.) While the 0-1 binary representation of states in the structured network clarifies the relationship to Hidden Markov Models through Eq. 21, the more familiar representation $s_i = \pm 1$ works as well. Weights in the structured Boltzmann network are trained according to the method of Section 7.3, though the relation to transition probabilities in a Hidden Markov Model is no longer simple (Problem 21).

7.4.1 Other Graphical Models

In addition to Hidden Markov Models, a number of graphical models have analogs in structured Boltzmann networks. One of the most general includes Bayesian belief nets, directed acyclic graphs in which each node can be in one of a number of discrete states, and nodes are interconnected with conditional probabilities (Chapter 2). As in the case of Hidden Markov Models, the correspondence with Boltzmann networks is clearest if the discrete states in the belief net are binary states; nevertheless, in practice multistate representations more naturally enforce the constraints and are generally preferred (Computer exercise 7).

A particularly intriguing recognition problem arises when a temporal signal has two inherent time scales, for instance the rapid daily behavior in a financial market superimposed on slow seasonal variations. A standard Hidden Markov Model typically has a *single* inherent time scale and hence is poorly suited to such problems. We might seek to use two interconnected Hidden Markov Models, possibly with different numbers of hidden states. Alas, the Forward-Backward algorithm generally does not converge when applied to a model having closed loops, as when two Hidden Markov Models have cross-connections.

BOLTZMANN
ZIPPER

Here the correspondence with Boltzmann networks is particularly helpful. We can link two Boltzmann chains with cross-connections, as shown in Fig. 7.12, to form a *Boltzmann zipper*. The particular benefit of such an architecture is that it can learn both short-time structure (through the "fast" component chain) and long-time struc-

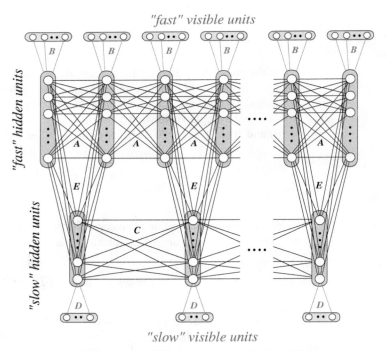

FIGURE 7.12. A Boltzmann zipper consists of two Boltzmann chains (cf. Fig. 7.11), whose hidden units are interconnected. The component chains differ in the rate at which visible features are sampled, and thus they capture structure at different temporal scales. Correlations are learned by the weights linking the hidden units, here labeled **E**. It is somewhat more difficult to train linked Hidden Markov Models to learn structure at different time scales.

ture (through the "slow" chain). The cross-connections, labeled by weight matrix **E** in the figure, learn correlations between the "fast" and "slow" internal representations. Unlike the case in Eq. 23, the **E** weights are not simply related to transition probabilities, however (Problem 22).

Boltzmann zippers can address problems such as acoustic speech recognition, where the fast chain learns the rapid transitions and structure of individual phonemes while the slow component chain learns larger structure associated with prosody and stress throughout a word or a full phrase. Related applications include speechreading (lipreading), where the fast chain learns the acoustic transitions and the slow chain learns the much slower transitions associated with the (visible) image of the talker's lips, jaw, and tongue.

*7.5 EVOLUTIONARY METHODS

POPULATION
SCORE

Inspired by the process of biological evolution, evolutionary methods of classifier design employ stochastic search for an optimal classifier. These methods admit a natural implementation on massively parallel computers. In broad overview, such methods proceed as follows. First, we create several classifiers—a *population*—each varying somewhat from the other. Next, we judge or *score* each classifier on a representative version of the classification task, such as accuracy on a set of labeled examples. In keeping with the analogy with biological evolution, the resulting (scalar) score is sometimes called the *fitness*. Then we rank these classifiers according to their score and retain the best classifiers, some portion of the total population. Again, in keeping with biological terminology, this is called *survival of the fittest*.

FITNESS

SURVIVAL
OF THE FITTEST

OFFSPRING
PARENT

We now stochastically alter the classifiers to produce the next generation—the children or *offspring*. Some offspring classifiers will have higher scores than their *parents* in the previous generation, some will have lower scores. The overall process is then repeated for the subsequent generation: The classifiers are scored and the best ones are retained, randomly altered to give yet another generation, and so on. In part because of the ranking, each generation has, on average, a slightly higher score than the previous one. The process is halted when the single best classifier in a generation has a score that exceeds a desired criterion value.

The method employs stochastic variations, and these in turn depend upon the fundamental representation of each classifier. There are two primary representations we shall consider: a string of binary bits (in basic genetic algorithms) and snippets of computer code (in genetic programming). In both cases a key property is that occasionally very large changes in the classifier are introduced. The presence of such large changes and random variations implies that evolutionary methods can find good classifiers even in extremely complex discontinuous spaces or "fitness landscapes" that are hard to address by techniques such as gradient descent.

7.5.1 Genetic Algorithms

CHROMOSOME

In basic genetic algorithms, the fundamental representation of each classifier is a binary string, called a *chromosome*. The mapping from the chromosome to the features and other aspects of the classifier depends upon the problem domain, and the designer has great latitude in specifying this mapping. In pattern classification, the score is usually chosen to be some monotonic function of the accuracy on a data set, possibly with penalty term to avoid overfitting. We use a desired fitness, θ, as the stopping criterion. Before we discuss these points in more depth, we first consider more specifically the structure of the basic genetic algorithm and then turn to the key notion of genetic operators, used in the algorithm; below we shall turn to the matter of the crossover and mutation rates, P_{co} and P_{mut}.

■ **Algorithm 4. (Basic Genetic Algorithm)**

1 **begin initialize** θ, P_{co}, P_{mut}, L N-bit chromosomes
2   **do** determine fitness of each chromosome, f_i, $i = 1, \ldots, L$
3   rank the chromosomes
4   **do** select two chromosomes with highest score
5   **if** Rand[0, 1) $< P_{co}$ **then** crossover the pair at a randomly
6   chosen bit
7   **else** change each bit with probability P_{mut};
8   remove the parent chromosomes
9   **until** N offspring have been created
10   **until** any chromosome's score f exceeds θ
11   **return** highest fitness chromosome (best classifier)
12 **end**

Figure 7.13 shows schematically the evolution of a population of classifiers given by Algorithm 4.

Genetic Operators

There are three primary genetic operators that govern reproduction—that is, producing offspring in the next generation described in lines 6 and 7 of Algorithm 4. The last two of these introduce variation into the chromosomes (Fig. 7.14):

Replication: A chromosome is merely reproduced, unchanged.

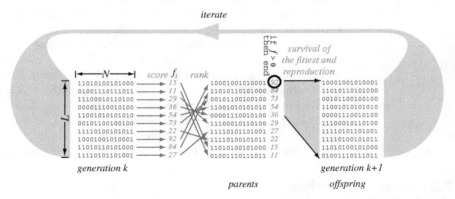

FIGURE 7.13. A basic genetic algorithm is a stochastic iterative search method. Each of the L classifiers in the population in generation k is represented by a string of bits of length N, called a chromosome (on the left). Each classifier is judged or scored according its performance on a classification task, giving L scalar values f_i. The chromosomes are then ranked according to these scores. The chromosomes are considered in descending order of score and are operated upon by the genetic operators of replication, crossover, and mutation to form the next generation of chromosomes—the offspring. The cycle repeats until a classifier exceeds the criterion score θ.

parents
(generation k)

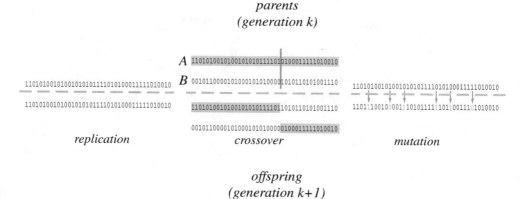

replication   *crossover*   *mutation*

offspring
(generation k+1)

FIGURE 7.14. Three basic genetic operations are used to transform a population of chromosomes at one generation to form a new generation. In replication, the chromosome is unchanged. Crossover involves the mixing or "mating" of two chromosomes to yield two new chromosomes. A position along the chromosomes is chosen randomly (red vertical line); then the first part of chromosome *A* is linked with the last part of chromosome *B*, and vice versa. In mutation, each bit is given a small chance of being changed from a 1 to a 0 or vice versa.

MATING

Crossover: Crossover involves the mixing—"mating"—of two chromosomes. A split point is chosen randomly along the length of either chromosome. The first part of chromosome *A* is spliced to the last part of chromosome *B*, and vice versa, thereby yielding two new chromosomes. The probability a given pair of chromosomes will undergo crossover is given by P_{co} in Algorithm 4.

Mutation: Each bit in a single chromosome is given a small chance, P_{mut}, of being changed from a 1 to a 0 or vice versa.

Other genetic operators may be employed, for instance *inversion*—where the chromosome is reversed front to back. This operator is used only rarely since inverting a chromosome with a high score nearly always leads to one with very low score. Below we shall briefly consider another operator, *insertions*.

Representation

When designing a classifier by means of genetic algorithms we must specify the mapping from a chromosome to properties of the classifier itself. Such mapping will depend upon the form of classifier and problem domain, of course. One of the earliest and simplest approaches is to let the bits specify features (such as pixels in a character recognition problem) in a two-layer perceptron with fixed weights (Chapter 5). The primary benefit of this particular mapping is that different segments of the chromosome, which generally remain undisturbed under the crossover operator, may evolve to recognize different *portions* of the input space such as the descender (lower) or the ascender (upper) portions of typed characters. As a result, occasionally the crossover operation will append a good segment for the ascender region in one chromosome to a good segment for the descender region in another, thereby yielding an excellent overall classifier.

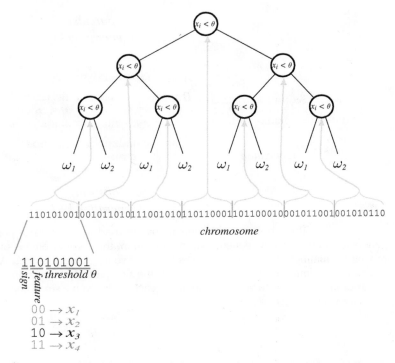

FIGURE 7.15. One natural mapping in genetic algorithms for pattern recognition is from a binary chromosome to a binary tree classifier, illustrated here for a simple binary decision tree. In this example, each of the nodes computes a query or question of the form "Is $\pm x_i < \theta$?" and is governed by nine bits in the chromosome. The first bit specifies a sign, and the next two bits specify the feature queried. The remaining six bits are a binary representation of the threshold θ. For instance, the leftmost node encodes the rule "Is $+x_3 < 41$?" (In practice, larger trees would be used for problems with four features.) During classification, a test pattern is presented to the top decision node, and depending upon the answer, passes to the right or to the left through the tree to a node at the next level. This process continues until the pattern comes to a category label (cf. Chapter 8).

Another mapping is to let different segments of the chromosome represent the weights in a multilayer neural net with a fixed topology. Likewise, a chromosome could represent a network topology itself, the presence of an individual bit implying two particular neurons are interconnected. One of the most natural representations is for the bits to specify properties of a decision tree classifier (Chapter 8), as shown in Fig. 7.15.

Scoring

For a c-category classification problem, it is generally most convenient to evolve c dichotomizers, each to distinguish a different ω_i from all other ω_j for $j \neq i$. During classification, the test pattern is presented to each of the c dichotomizers and assigned the label accordingly. The goal of classifier design is accuracy on future patterns, or if decisions have associated costs, then the goal is low expected cost. Such goals should be reflected in the method of scoring and selection in a genetic algorithm. It is natural to base the score on the classification accuracy measured on the data set. As we have seen numerous times, though, there is a danger that the classifier becomes "tuned" to the properties of the particular data set, however. (We can informally broaden our usage of the term "overfitting" from generic learning to apply to this search-based case as well.) One method for avoiding such overfitting is penalizing classifier complexity, and thus the score should have a term that penalizes overly large networks. Another method is to adjust the stopping criterion. Because the appropriate measure of classifier complexity and the stopping criterion depend strongly on the problem, it is hard to make specific guidelines in setting these parameters. Hence, designers should be prepared to explore these parameters in any practical application.

Selection

The process of selection specifies which chromosomes from one generation will be sources for chromosomes in the next generation. Up to here, we have assumed that the chromosomes would be ranked and selected in order of decreasing fitness until the next generation is complete. This has the benefit of generally pushing the population toward higher and higher scores. Nevertheless, the average improvement from one generation to the next depends upon the variance in the scores at a given generation, and because this standard fitness-based selection need not give high variance, other selection methods may prove superior.

FITNESS-
PROPORTIONAL
SELECTION

The principal alternative selection scheme is *fitness-proportional selection*, or fitness-proportional reproduction, in which the probability that each chromosome is selected is proportional to its fitness. While high-fitness chromosomes are preferentially selected, occasionally low-fitness chromosomes are selected, and this may preserve diversity and increase the variance of the population.

A minor modification of this method is to make the probability of selection proportional to some monotonically increasing function of the fitness. If the function has a positive second derivative, the probability that high-fitness chromosomes is enhanced. One version of this heuristic is inspired by the Boltzmann factor of Eq. 2; the probability that chromosome i with fitness f_i will be selected is

$$P(i) = \frac{e^{f_i/T}}{\mathcal{E}[e^{f_i/T}]},$$

(24)

where the expectation is over the current generation and T is a control parameter loosely referred to as a temperature. Early in the evolution the temperature is set high, giving all chromosomes roughly equal probability of being selected. Late in the evolution the temperature is set lower so as to find the chromosomes in the region of the optimal classifier. We can express such search by analogy to biology: Early in the search the population remains diverse and *explores* the fitness landscape in search of promising areas; later the population *exploits* the specific fitness opportunities in a small region of the space of possible classifiers.

7.5.2 Further Heuristics

There are many additional heuristics that can occasionally be of use. One of them concerns the adaptation of the crossover and mutation rates, P_{co} and P_{mut}. If these rates are too low, the average improvement from one generation to the next will be small, and the search is unacceptably long. Conversely, if these rates are too high, the evolution is undirected and similar to a highly inefficient random search. We can monitor the average improvement in fitness of each generation and adjust the mutation and crossover rates to ensure that such improvement is rapid. In practice, this is done by encoding the rates in the chromosomes themselves and allowing the genetic algorithm itself to select the proper values.

Another heuristic is to use ternary or n-ary chromosomes rather than the traditional binary ones. These representations provide little or no benefit at the algorithmic level, but may make the mapping to the classifier itself more natural and easier to compute. For instance, a ternary chromosome might be most appropriate if the classifier is a decision tree with three-way decisions, as discussed in Chapter 8.

Occasionally the mapping to the classifier will work for chromosomes of different length. For example, if the bits in the chromosome specify weights in a neural network, then longer chromosomes would describe networks with a larger number INSERTION of hidden units. In such a case we allow the *insertion* operator, which with a small probability inserts bits into the chromosome at a randomly chosen position. This so-called "messy" genetic algorithm method has a more appropriate counterpart in genetic programming, as we shall see in Section 7.6.

7.5.3 Why Do They Work?

Because there are many heuristics to choose as well as parameters to set, it is hard to make firm theoretical statements about building classifiers by means of evolutionary methods. The performance and search time depend upon the number of bits, the size of a population, the mutation and crossover rates, choice of features and mapping from chromosomes to the classifier itself, the inherent difficulty of the problem and possibly parameters associated with other heuristics.

A genetic algorithm restricted to mere replication and mutation is, at base, a version of stochastic random search. The incorporation of the crossover operator, which mates two chromosomes, provides a qualitatively different search, one that has no counterpart in stochastic grammars (Chapter 8). Crossover works by finding, rewarding, and recombining "good" segments of chromosomes, and the more faithfully the segments of the chromosomes represent such functional building blocks, the better we can expect genetic algorithms to perform. The only way to ensure this is with prior knowledge of the problem domain and the desired form of classifier.

*7.6 GENETIC PROGRAMMING

Genetic programming shares the same algorithmic structure of basic genetic algorithms, but differs in the representation of each classifier. Instead of chromosomes consisting of strings of bits, genetic programming uses snippets of computer programs made up of mathematical operators and variables. As a result, the genetic operators are somewhat different; moreover, the insertion operator now plays a significant role.

As illustrated in Fig. 7.16, the four principal operators in genetic programming are as follows:

Replication: A snippet is merely reproduced, unchanged.

Crossover: Crossover involves the mixing—"mating"—of two snippets. A split point is chosen from allowable locations in snippet A as well as from snippet B. The first part of snippet A is spliced to the back part of chromosome B, and vice versa, thereby yielding two new snippets.

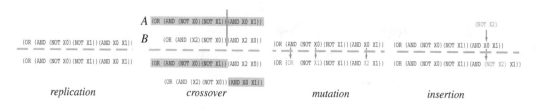

parents
(generation k)

replication   crossover   mutation   insertion

offspring
(generation k+1)

FIGURE 7.16. Four basic operators in genetic programming are used to transform a population of snippets of code at one generation to form a new generation. In replication, the snippet is unchanged. Crossover involves the mixing or "mating" of two snippets to yield two new snippets. A position along the snippet *A* is randomly chosen from the allowable locations (red vertical line); likewise one is chosen for snippet *B*. Then the front portion of *A* is spliced to the back portion of *B* and vice versa. In mutation, each element is given a small chance of being changed. There are several different types of elements, and replacements must be of the same type. For instance, only a number can replace another number; only a numerical operator that takes a single argument can replace a similar operator, and so on. In insertion, a randomly selected element is replaced by a compatible snippet, keeping the entire snippet grammatically well formed and meaningful.

Mutation: Each element in a single snippet is given a small chance of being changed to a different value. Such a change must be compatible with the syntax of the total snippet. For instance, a number can be replaced by another number; a mathematical operator that takes a single argument can be replaced by another such operator, and so forth.

INSERTION

Insertion: Insertion consists in replacing a single element in the snippet with another (short) snippet randomly chosen from a set.

In the *c*-category problem, it is simplest to form *c* dichotomizers just as in genetic algorithms. If the output of the classifier is positive, the test pattern belongs to category ω_i, if negative, then it is not in ω_i.

Representation

A program must be expressed in some language, and the choice affects the complexity of the procedure. Syntactically rich languages such as *C* or *C++* are complex and somewhat difficult to work with whereas here the syntactic simplicity of a language such as *Lisp* is advantageous. Many *Lisp* expressions can be written in the form (<operator> <operand> <operand>), where an <operand> can be a constant, a variable or another parenthesized expression. For example, the expressions (+ X 2) and (* 3 (+ Y 5)) are valid *Lisp* expressions for the arithmetic expressions $x + 2$ and $3(y + 5)$, respectively. These expressions are easily represented by a binary tree, with the operator being specified at the node and the operands appearing as the children (Fig. 7.17).

Whatever language is used, genetic programming operators used for mutation should replace variables and constants with variables and constants; likewise they should replace operators with functionally compatible operators. They should also

parents
(generation k)

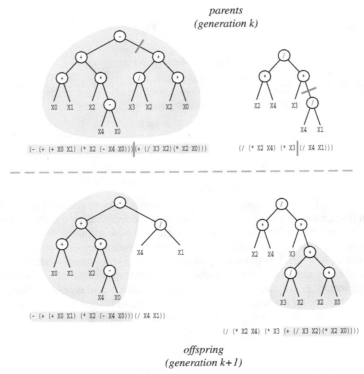

$(- (+ (+ X0 X1) (* X2 (- X4 X0)))) | (+ (/ X3 X2) (* X2 X0)))$ $(/ (* X2 X4) (* X3 | (/ X4 X1)))$

$(- (+ (+ X0 X1) (* X2 (- X4 X0))) (/ X4 X1))$

$(/ (* X2 X4) (* X3 (+ (/ X3 X2) (* X2 X0)))))$

offspring
(generation k+1)

FIGURE 7.17. Unlike the decision trees of Fig. 7.15 and Chapter 8, the trees shown here are merely a representation using the syntax of *Lisp* expressions that implement a single function. For instance, the upper-right (parent) tree implements $(x_2 x_4)/(x_3(x_4/x_1))$. Two parent snippets and break points chosen randomly from allowable ones are shown by the red line segments. The offspring are formed by the crossover operation, where the left side of the snippet for parent 1 is concatenated to the right side of the snippet from parent 2, and vice versa. The resulting functions have an implied threshold or sign function when used for classification. Thus the function will operate on the features of a test pattern and emit category ω_i if the function is positive, and NOT ω_i otherwise.

be required to produce syntactically valid results. Nevertheless, occasionally an ungrammatical code snippet may be produced. For that reason, it is traditional to employ a *wrapper*—a routine that decides whether the classifier is meaningful, and eliminates it if not.

WRAPPER

As with genetic algorithms, it is nearly impossible to make sound theoretical statements about genetic programming; even the rules of thumb learned from simulations in one domain, such as control or function optimization, are of little value in another domain, such as classification problems. Of course, the method works best in problems that are matched by the classifier representation, as simple operations such as multiplication, division, square roots, logical NOT, and so on.

Nevertheless, we can state that as computation continues to decrease in cost, an increasing share of the burden of solving classification problems will be assumed by computation rather than careful analysis, and here evolutionary techniques have promise for use in classification research.

SUMMARY

When a pattern recognition problem involves a model that is discrete or of such high complexity that analytic or gradient descent methods are unlikely to work, we may employ stochastic techniques—ones that at some level rely on randomness to find model parameters. Simulated annealing, based on physical annealing of metals, consists in randomly perturbing the system, and gradually decreasing the randomness to a low final level, in order to find an optimal solution. Boltzmann learning trains the weights in a network so that the probability of a desired final output is increased. Such learning relies on annealing and is based on gradient descent in the Kullback-Liebler divergence between two distributions of visible states at the output units: one distribution describes these units when clamped at the known category information, and the other describes them when they are free to assume values based on the activations throughout the network. Some graphical models, such as hidden Markov models and Bayes belief networks, have counterparts in structured Boltzmann networks, and this leads to new applications of Boltzmann learning.

Search methods based on evolution—genetic algorithms and genetic programming—perform highly parallel stochastic searches in a space set by the designer. The fundamental representation used in genetic algorithms is a string of bits, or chromosome; the representation in genetic programming is a snippet of computer code. Variation is introduced by means of crossover, mutation, and insertion. As with all classification methods, the better the features, the better the solution. There are many heuristics that can be employed, and there are parameters that must be set. As the cost of computation contiues to decline, computationally intensive methods, such as Boltzmann networks and evolutionary methods, should become increasingly popular.

BIBLIOGRAPHICAL AND HISTORICAL REMARKS

The general problem of search is of central interest in computer science and artificial intelligence, and is far too extensive to treat here. Nevertheless, techniques such as depth first, breadth first, branch-and-bound and A* [20], occasionally find use in fields touching upon pattern recognition, and practitioners should have at least a passing knowledge of them. Good overviews can be found in reference [34] and a number of textbooks on artificial intelligence, such as references [47, 55], and [67]. For rigor and completeness, Knuth's book on the subject is without peer [33].

The infinite monkey theorem, attributed to Sir Arthur Eddington, states that if there is a sufficiently large number of monkeys typing at typewriters, eventually one will bang out the script to *Hamlet*. It reflects one extreme of the tradeoff between prior knowledge about the location of a solution on the one hand and the effort of search required to find it on the other. Computers made available in the early 1950s permitted the first automated attempts at highly stochastic search, most notably the pioneering work of Metropolis and colleagues for simulating chemical processes [41]. One of the earliest and most influential applications of stochastic methods for pattern recognition was the Pandemonium learning method due to Selfridge, which used stochastic search for input weights in a feedforward network model [57]. Kirkpatrick, Gelatt and Vecchi [31], and independently Černý [64], introduced the Boltzmann factor to general stochastic search methods, the first example of simulated annealing. The statistical physics foundations of Boltzmann factors, at the present level of mathematical sophistication, can be found in reference [32]. The physical model of stochastic binary components was introduced by Wilhemlm Lenz in 1920, but it became associated with his doctoral student Ernst Ising several years thereafter and was first called the "Ising model" by Rudolf Peierls in 1936. It has spawned a great deal of theoretical and simulation research [21].

The use of simulated annealing for learning was proposed by Ackley, Hinton, and Sejnowski [2]. A good book on the method is reference [1], which described the procedure for initializing the temperature in simulated annealing and was the inspiration for Fig. 7.10. Peterson and Anderson introduced deterministic annealing and mean-field Boltzmann learning and described some of the (rare) conditions when the mean-field approximation might lead to nonoptimal solutions [51]. Hinton showed that the deterministic Boltzmann learning rule performs steepest descent in weight space [22].

A number of papers explore structured Boltzmann networks, including Hopfield's influential paper on networks for pattern completion or autoassociation [26]. The linear storage capacities of Hopfield networks quoted in the text, along with $n\log n$ relationships for partial storage, are derived in [40, 65] and [66]. The learning rule described in that work has roots in the *Learning matrix* of [59] and [60]. Harmonium [15, 58], another two-layer variant of a Boltzmann network, is primarily of historical interest. The relation of Boltzmann networks to graphical models such as Hidden Markov models has been explored in references [28, 38], and [56]; the last was the source for our discussion in Section 7.4. Implementation of constraints for Boltzmann machines was introduced in reference [43] and a second-order pruning algorithm was described in reference [50].

Boltzmann learning has been applied to a number of real-world pattern recognition problems, most notably speech recognition [8, 52] and stochastic restoration of images or pattern completion [17]. Because Boltzmann learning has high computational burden yet a natural VLSI implementation, a number of special-purpose chips have been fabricated, as described in references [24, 44], and [45]. The ordering of configurations in Fig. 7.3, in which neighboring configurations differ in just one bit, is a version of a Gray code; an elegant method for constructing such codes is described in [19].

Some of the earliest work inspired by evolution was put forth in references [13] and [14], but the computational power available was insufficient for anything but toy problems. Later, Rechenberg's "evolution strategies" were applied to optimization in aeronautical design problems [53]. His earliest work did not employ full *populations* of candidate solutions, nor the key operation of crossover. Evolutionary programming saves good parents while the method of evolutionary strategies generally does not. Neither employ mating, that is, crossover. Holland introduced genetic algorithms in 1975 [25], and like the algorithm itself, researchers have explored a very wide range of problems in search, optimization, and pattern recognition to find some that are best solved by such methods. A review appears in reference [6], and there is an increasing number of textbooks [18, 42], the latter with a more rigorous approach to the mathematics. Koza's extensive books on genetic programming provide a good introduction and include several illustrative simulations [35, 36], though relatively little on pattern recognition. There are several collections of papers on evolutionary techniques in pattern recognition, including reference [49], and though its title is very misleading, the collection of key papers in this history of evolutionary computation can nevertheless be recommended [12]. An intriguing effect due to the interaction of learning and evolution is the Baldwin effect, where learning can influence the rate of evolution [23]; it has been shown that too much learning (as well as too little learning) leads to slower evolution [29]. Evolutionary methods can lead to "nonoptimal" or inelegant solutions, and there is computational evidence that this occurs in nature [61, 62].

PROBLEMS

Section 7.1

1. One version of the infinite monkey theorem states that a single (immortal) monkey typing randomly will ultimately reproduce the script of *Hamlet*. Estimate the time needed for this, assuming that the monkey can type two characters per second and that the play has 50 pages, each containing roughly 80 lines and 40 characters per line. Assume there are 30 possible characters (a through z), space, period, exclamation point, and carriage return. Compare this time to the estimated age of the universe, 10^{10} years.

Section 7.2

2. Prove that for any optimization problem of the form of Eq. 1 having a nonsymmetric connection matrix, there is an equivalent optimization problem in which the matrix is replaced by its symmetric part.

3. The complicated energy landscape in the left of Fig. 7.2 is misleading for a number of reasons.

   (a) Discuss the difference between the continuous space shown in that figure and the discrete space for the true optimization problem of Eq. 1.

   (b) That figure shows a local minimum near the middle of the space. Given the nature of the discrete space, are any states closer to any "middle"?

   (c) Suppose the axes referred to *continuous* variables s_i (as in mean-field annealing). If each s_i obeyed a sigmoid (Fig. 7.5), could the energy landscape be nonmonotonic, as is shown in Fig. 7.2?

4. Consider exhaustive search for the minimum of the energy given in Eq. 1 for binary units and arbitrary connections w_{ij}. Suppose that on a uniprocessor it takes 10^{-8} sec to calculate the energy for each configuration. How long will it take to exhaustively search the space for $N = 100$ units? How long for $N = 1000$ units?

5. Suppose it takes a uniprocessor 10^{-10} sec to perform a single multiply-accumulate, $w_{ij}s_is_j$, in the calculation of the energy $E = -1/2 \sum_{ij} w_{ij}s_is_j$ given in Eq. 1.

   (a) Make some simplifying assumptions and write a formula for the total time required to search exhaustively for the minimum energy in a fully connected network of N nodes.

   (b) Plot your function using a log-log scale for $N = 1, \ldots, 10^5$.

   (c) What size network, N, could be searched exhaustively in a day? A year? A century?

6. Make and justify any necessary mathematical assumptions and show analytically that at high temperature, every configuration in a network of N units interconnected by weights is equally likely (cf. Fig. 7.1).

7. Derive the exponential form of the Boltzmann factor in the following way. Consider an isolated set of $M + N$ independent magnets, each of which can be in an

$s_i = +1$ or $s_i = -1$ state. There is a uniform magnetic field applied; this means that the energy of the $s_i = +1$ state has some positive energy, which we can arbitrarily set to 1; the $s_i = -1$ state has energy -1. The total energy of the system is therefore the sum of the number pointing up, k_u, minus the number pointing down, k_d; that is, $E_T = k_u - k_d$. (Of course, $k_u + k_d = M + N$ regardless of the total energy.)

The fundamental statistical assumptions describing this system are that the magnets are independent and that the probability a subsystem (that is, the N magnets), has a particular energy is proportional to the *number of configurations* that have this energy.

(a) Consider the subsystem of N magnets, which has energy E_N. Write an expression for the number of configurations $K(N, E_N)$ that have energy E_N.

(b) As in part (a), write a general expression for the number of configurations in the subsystem M magnets at energy E_M, that is, $K(M, E_M)$.

(c) Because the two subsystems consist of independent magnets, the total number of ways the full system can have total energy $E_T = E_N + E_M$ is the product $K(N, E_N)K(M, E_M)$. Write an analytic expression for this total number.

(d) In statistical physics, if $M \gg N$, the M-magnet subsystem is called the heat reservoir or heat bath. Assume that $M \gg N$, and write a series expansion for your answer to part (c).

(e) Use your answer in part (d) to show that the probability the N-unit system has energy E_N has the form of a Boltzmann factor, e^{-E_N}.

8. Prove that the analog value of s_i given by Eq. 5 is the expected value of a binary variable in temperature T in the following simple case. Consider a single binary magnet whose $s = +1$ state has energy $+E_0$ and whose $s = -1$ state has energy $-E_0$, as would occur if an external magnetic field has been applied.

(a) Construct the partition function Z by summing over the two possible states $\gamma' = 0$ and $\gamma' = 1$ according to Eq. 3.

(b) Recall that the probability of finding the system in state $s = +1$ is given by a Boltzmann factor divided by the partition function (Eq. 2). Define the (analog) expected value of the state to be

$$s = \mathcal{E}[s] = \Pr[s = +1](+1) + \Pr[s = -1](-1).$$

Show that this implies the analog state of a single magnet obeys Eq. 5.

(c) Argue that if the $N - 1$ other magnets in a large system can be assumed to give an average field (this is the mean-field approximation), then the analog value of a single magnet will obey a function of the form given in Eq. 5.

9. Consider Boltzmann networks applied to the exclusive-OR problem.

(a) A fully connected network consisting solely of two input units and a single output unit, whose sign gives the class, cannot solve the exclusive-OR problem. Prove this by writing a set of inequalities for the weights and show that they are inconsistent.

(b) As in part (a), prove that a fully connected Boltzmann network consisting solely of two input units and two output units representing the two categories cannot solve the exclusive-OR problem.

(c) Prove that a Boltzmann network of part (b) with a single hidden unit *can* implement the exclusive-OR problem.

10. Consider a fully connected Boltzmann network with two input units, a single hidden unit and a single (category) output unit. Construct by hand a set of weights w_{ij} for $i, j = 1, 2, 3, 4$ which allows the net to solve the exclusive-OR problem for a representation in which $s_i = \pm 1$.

Section 7.3

11. Show all intermediate steps in the derivation of Eq. 14 from Eq. 12. Be sure your notation distinguishes this case from that leading to Eq. 10.

12. Determine the weights in a six-unit Hopfield network trained with the following three patterns. Use the learning rule of Eq. 15:

$$\mathbf{x}^1 = \{+1, +1, +1, -1, -1, -1\}$$
$$\mathbf{x}^2 = \{+1, -1, +1, -1, +1, -1\}$$
$$\mathbf{x}^3 = \{-1, +1, -1, -1, +1, +1\}.$$

(a) Verify that each of the patterns gives a local minium in energy by perturbing each of the six units individually and monitoring the energy.
(b) Verify that the symmetric state $s_i \rightarrow -s_i$ for $i = 1, \ldots, 6$ also gives a local energy minimum of the same energy.

13. Repeat Problem 12 but with the eight-unit network and the following patterns:

$$\mathbf{x}^1 = \{+1, +1, +1, -1, -1, -1, -1, +1\}$$
$$\mathbf{x}^2 = \{+1, -1, +1, +1, +1, -1, +1, -1\}$$
$$\mathbf{x}^3 = \{-1, +1, -1, -1, +1, +1, -1, +1\}.$$

14. Show that a missing feature assumes the appropriate value when training a deficient pattern in a trained Boltzmann network. That is, show that under the conditions described in the text, a missing feature is filled in with its most probable value, given the state of the clamped units.

15. Show how the constraints that a pattern not be in a subset of categories improves the recognition for the others. That is, show by means of inequalities that clamping an incorrect category's output unit to -1 will increase the probability that the correct category's output unit will become $+1$ at the end of an anneal.

16. The text states that a lower bound on the number of hidden units needed in a Boltzmann network trained with n patterns is $\lceil \log_2 n \rceil$. This is, of course, the number of hiddens needed to allow a distinct hidden representation for each pattern. Show that this lower bound is not tight, because there may not be weights to ensure such a representation. Do this by considering a Boltzmann network with three input units, three hiddens, and a single output, addressing the 3-bit parity problem.

(a) Argue that the hidden representation must be equivalent to the input representation.
(b) Argue that there is no two-layer Boltzmann network (here, hidden to output) that can solve the three-bit parity problem. Explain why this implies that the $\lceil \log_2 n \rceil$ bound is not tight.

17. Consider the problem of initializing the N weights in a fully connected Boltzmann network. Let there be $N - 1 \approx N$ weights connected to each unit. Suppose too that the chance that any particular units will be in the $s_i = +1$ state is 0.5, and likewise for the $s_i = -1$ state. We seek weights such that the variance of the net activation of each unit is roughly 1.0, a reasonable measure of the end of the linear range of the sigmoid nonlinearity. The variance of l_i is

$$\text{Var}[l_i] = \sum_{j=1}^{N} \text{Var}[w_{ij} s_j] = N \text{Var}[w_{ij}] \text{Var}[s_j].$$

Set $\text{Var}[l_i] = 1$, and solve for $\text{Var}[w_{ij}]$ and thereby show that weights should be initialized randomly in the range $-\sqrt{3/N} < w_{ij} < +\sqrt{3/N}$.

18. Show that under reasonable conditions, the learning rate η in Eq. 14 for a Boltzmann network of N units should be bounded $\eta \leq T^2/N$ to ensure stability as follows:

(a) Take the derivative of Eq. 14 to prove that the Hessian is

$$\mathbf{H} = \frac{\partial^2 \overline{D}_{KL}}{\partial \mathbf{w}^2} = \frac{\partial^2 \overline{D}_{KL}}{\partial w_{ij} \partial w_{uv}}$$

$$= \frac{1}{T^2} \left[\mathcal{E}[s_i s_j s_u s_v] - \mathcal{E}[s_i s_j] \mathcal{E}[s_u s_v] \right].$$

(b) Use this to show that

$$\mathbf{w}^t \mathbf{H} \mathbf{w} \leq \frac{1}{T^2} \mathcal{E} \left[\left(\sum_{ij} |w_{ij}| \right)^2 \right].$$

(c) Suppose we normalize weights such that $\|\mathbf{w}\| = 1$ and thus

$$\sum_{ij} w_{ij} \leq \sqrt{N}.$$

Use this fact together with your answer to part (b) to show that the curvature of the $\overline{D}_{KL}$ obeys

$$\mathbf{w}^t \mathbf{H} \mathbf{w} \leq \frac{1}{T^2} \mathcal{E} \left[\left(\sqrt{N} \right)^2 \right] = \frac{N}{T^2}.$$

(d) Use the fact that stability demands the learning rate to be the inverse of the curvature, along with your answer in (c), to show that the learning rate should be bounded $\eta \leq T^2/N$.

Section 7.4

19. For any Hidden Markov Model, there exists a Boltzmann chain that implements the equivalent probability model. Show the converse is not true, that is, for every chain, there need not exist a functionally equivalent Hidden Markov Model. Use the fact that weights in a Boltzmann chain are bounded $-\infty < A_{ij}, B_{jk} < +\infty$, but probabilities in an Hidden Markov Model are positive and sum to 1.0.

20. How many legal paths are there through a Boltzmann chain as shown in Fig. 7.11 as a function of the number of time steps, hidden units and visible units (T_f, c, and k, respectively)?

21. The discussion of the relation between Boltzmann chains and Hidden Markov Models in the text assumed that the initial hidden state was known. Show that if this hidden state is not known, the energy of Eq. 21 has another term that describes the prior probability the system is in a particular hidden state.

22. Consider the Boltzmann zipper illustrated Fig. 7.12. Denote the cross-connections by weight matrix **E** which learn correlations between the "fast" and "slow" internal representations. Show that unlike the case in Eq. 23, the **E** weights are not simply related to transition probabilities. In particular, show that they need not be normalized. Must they always be positive?

Section 7.5

23. Consider the populations of size L of N-bit chromosomes.

 (a) Show the number of different populations is $\binom{L+2^N-1}{2^N-1}$.

 (b) Assume some number $1 \le L_s \le L$ are selected for reproduction in a given generation. Use your answer to part (a) to write an expression for the number of possible sets of parents as a function of L and L_s. (It is just the set, not their order, that is relevant here.)

 (c) Show that your answer to part (b) reduces to that in part (a) for the case $L_s = L$.

 (d) Show that your answer to part (b) gives L in the case $L_s = 1$.

Section 7.6

24. For each of the below snippets, mark suitable positions for breaks for the crossover operator, and the multiplication operator, *, and the addition operator, +, can take two or more operands.

 (a) (* (X0 (+ x4 x8)) x5 (SQRT 5))

 (b) (SQRT (X0 (+ x4 x8)))

 (c) (* (- (SIN X0) (* (TAN 3.4) (SQRT X4)))

 (d) (* (X0 (+ x4 x8)) x5 (SQRT 5))

 (e) Separate the following *Lisp* symbols into groups such that any member in a group can be replaced by another through the mutation operator in genetic programming:
 {+, X3, NOR, *, X0, 5.5, SQRT, /, X5, SIN, -, -4.5, NOT, OR, 2.7, TAN}

COMPUTER EXERCISES

Two of the exercises use the data in the following table where + denotes $+1$ and − denotes -1.

ω_1	ω_2	ω_3
--+-+-++	-++-+++-	+++++-++
--++-+++	-++-+---	+-+-++-+
+++++--+	-+++-++-	+-++-+++
-+-+-++-	-++--+-+	+-++-+--
----+++-	-+----++	+-++-++-
---+++-	-+--++-+	+--+-++-
-+-+++--	--+-++--	+++-+--+
-+-++--+	-+-+-+++	+-++-+-+
+-+-+-++	---++-++	++--+-++
+-+--+++	-++-++-+	+--+-++-

Section 7.2

1. Consider the problem of searching for a global minimum of the energy given in Eq. 1 for a system of N units, fully interconnected by weights randomly chosen in the range $-1/\sqrt{N} < w_{ij} < +1/\sqrt{N}$.

   (a) Let $N = 10$. Write a program to search through all 2^N configurations to find global minima, and apply it to your network. Verify that there are two "global" minima.

   (b) Write a program to perform the following version of gradient descent. Let the units be numbered and ordered $i = 1, \ldots, N$ for bookkeeping. For each configuration, find the unit with the lowest index i which can be changed to lower the total energy. Iteratively make this change until the system converges, or that it is clear that it will not converge. Plot the total system energy as a function of iteration, that is, $E(m)$ versus m.

   (c) Perform a search as in part (b) but with *random* polling of units. As before, plot $E(m)$ versus m.

   (d) Repeat parts (a – c) for $N = 100$ and $N = 1000$.

   (e) Discuss your results, paying particular attention to convergence and the problem of local minima.

2. Implement stochastic simulated annealing (Algorithm 1) to find the minimum of the energy of Eq. 1 for a six-unit system having weight connections described by the matrix

$$
\mathbf{w} = \begin{pmatrix}
0 & 5 & -3 & 4 & 4 & 1 \\
5 & 0 & -1 & 2 & -3 & 1 \\
-3 & -1 & 0 & 2 & 2 & 0 \\
4 & 2 & 2 & 0 & 3 & -3 \\
4 & -3 & 2 & 3 & 0 & 5 \\
1 & 1 & 0 & -3 & 5 & 0
\end{pmatrix}. \tag{25}
$$

(a) Let the initial temperature be $T(1) = 10$ and use an annealing schedule of the form $T(m + 1) = cT(m)$ where $c = 0.9$.

(b) Repeat part (a) but with $T(1) = 5$ and $c = 0.5$.

3. Repeat Computer exercise 2 but using deterministic simulated annealing, Algorithm 2.

Section 7.3

4. Train a Boltzmann network consisting of eight input units and ten category units with the characters of a seven-segment display shown in Fig. 7.10.

(a) Use the network to classify each of the ten patterns, and thus verify that all have been learned.

(b) Explore pattern completion in your network the following way. For each of the 2^8 possible patterns do pattern completion for several characters. Add hidden units and show that better performance results for ambiguous characters

5. Use Boltzmann learning in a network of eight input units, n_H hidden units and two output units to create a classifier to distinguish two categories.

(a) Let the number of hidden units be $n_H = 4$, and train your network using the patterns in ω_1 and ω_2, as listed in the table above. Use your trained network to classify the patterns ----+-++, -++-++++ and ++++++++.

(b) Repeat part (a) letting the number of hidden units be instead $n_H = 6$.

(c) Repeat parts (a) and (b) but for distinguishing ω_1 from ω_3.

(d) Repeat parts (a) and (b) but for distinguishing ω_2 from ω_3.

6. Explore the use of Boltzmann networks for pattern completion in the following way. Create a network with eight visible units and n_H hidden units. The visible units will encode the patterns presented; there are no category or output units.

(a) Use stochastic Boltzmann training to learn all ten points in ω_1 in the table above for the case of $n_H = 8$ hidden units. Now use your trained network to complete the following deficient patterns: 001010*, *1011001, and 0000*110, where * denotes a missing feature. In both learning and recall, let $T(1) = 50$ and $T(m + 1) = cT(m)$ for $c = 0.9$ describe your annealing schedule.

(b) Repeat part (a) but for the case of $n_H = 2$ hidden units.

(c) Explain the source of the difference in your results in (a) and (b).

Section 7.4

7. Consider a Bayes Belief network (Chapter 2) having two input nodes which are connected to a single output node but not to each other. The first input node can be in one of three discrete states, which we denote $\mathbf{x} = $ 100, 010 or 001. The other input node can be in one of four discrete states: $\mathbf{y} = $ 1000, 0100, 0010 or 0001. The output can likewise be in one of two states, $\mathbf{z} = $ 10 or 01.

(a) Create a structured Boltzmann network whose topology corresponds to that of the target Bayes Belief network.

(b) Train your Boltzmann network with the following patterns according to the probabilities listed:

Probability	x	y	z
0.2	100	0001	10
0.3	010	1000	10
0.4	100	0100	01
0.1	001	0010	01

That is, present to your network the given patterns for training according to the probabilities listed.

(c) Suppose **z** encodes two categories in a standard 1-of-2 representation. Use your trained network to classify the following two sets of inputs: **x** = 001, **y** = 0010 and **x** = 100, **y** = 000*, where * denotes a missing feature.

Section 7.5

8. Use genetic algorithms to evolve a two-category tree classifier using the representation in Fig. 7.15. (You may wish to look ahead to Chapter 8 for background on decision tree classifiers.) Let your population size be $L = 15$ chromosomes, and use fitness proportional reproduction to choose five chromosomes for reproduction, where the fitness score is the classification accuracy of the following eight points (marked as lists):

Pattern	x	ω_1	ω_2
1	$(1, 5, -1, 3)$	×	
2	$(-1, 5, 2, 2)$	×	
3	$(2, 3, -1, 0)$	×	
4	$(-3, 4, -2, -1)$	×	
5	$(-1, -3, 1, 2)$		×
6	$(-2, 4, -3, 0)$		×
7	$(-3, 5, 1, 1)$		×
8	$(1, -2, 0, 0)$		×

Use your final classifier to categorize the following points: $(-1, 4, 1, 1)$, $(-2, 4, -1, 1)$, and $(3, 3, 0, 1)$.

Section 7.6

9. Consider a two-category problem with four features in the bounded region $-1 \leq x_i \leq +1$ for $i = 1, 2, 3, 4$.

(a) Generate training points in each of two categories defined by

$$\omega_1 : x_1 + 0.5x_2 - 0.3x_3 - 0.1x_4 < 0.5$$

$$\omega_2 : - x_1 + 0.2x_2 + x_3 - 0.6x_4 < 0.2$$

by randomly selecting a point in the four-dimensional space. If it satisfies just one of the inequalities, label its category accordingly. If it satisfies both inequalities, randomly choose a label with probability 0.5. If it satisfies neither of the inequalities, discard the point. Continue in this way until you have 50 points for each category.

(b) Use genetic programming to evolve a standard *Lisp* expression whose value is positive for ω_1 and negative for ω_2. Let your population have 100 snippets of code, and use fitness proportional reproduction to select 10 snippets for reproduction at each generation. Allow the use of the insertion operator. Plot the population average classification error as a function of generation.

(**c**) Repeat parts (a) and (b) but for categories defined by

$$\omega_1 : 0.5x_1 - 0.3x_2^2 + x_1x_2 - 0.2x_3 - 0.4x_4 < 0.3$$
$$\omega_2 : \quad 0.2x_1 + x_1x_2 - 0.3x_3 + 0.6x_1x_4 < 0.7.$$

BIBLIOGRAPHY

[1] Emile Aarts and Jan Korst. *Simulated Annealing and Boltzmann Machines: A Stochastic Approach to Combinatorial Optimization and Neural Computing*. Wiley, New York, 1989.

[2] David H. Ackley, Geoffrey E. Hinton, and Terrence J. Sejnowski. A learning algorithm for Boltzmann machines. *Cognitive Science*, 9(1):147–169, 1985.

[3] Rudolf Ahlswede and Ingo Wegener. *Search Problems*. Wiley, New York, 1987.

[4] Frantzisko Xabier Albizuri, Alicia d'Anjou, Manuel Graña, Francisco Javier Torrealdea, and Mari Carmen Hernandez. The high-order Boltzmann machine: Learned distribution and topology. *IEEE Transactions on Neural Networks*, TNN-6(3):767–770, 1995.

[5] David Andre, Forrest H. Bennett III, and John R. Koza. Discovery by genetic programming of a cellular automata rule that is better than any known rule for the majority classification problem. In John R. Koza, David E. Goldberg, David B. Fogel, and Rick L. Riolo, editors, *Genetic Programming 1996: Proceedings of the First Annual Conference*, pages 3–11. MIT Press, Cambridge, MA, 1996.

[6] Thomas Bäck, Frank Hoffmeister, and Hans-Paul Schwefel. A survey of evolution strategies. In Rik K. Belew and Lashon B. Booker, editors, *Proceedings of the Fourth International Conference on Genetic Algorithms*, pages 2–9. Morgan Kaufmann, San Mateo, CA, 1991.

[7] J. Mark Baldwin. A new factor in evolution. *American Naturalist*, 30:441–451, 536–553, 1896.

[8] John S. Bridle and Roger K. Moore. Boltzmann machines for speech pattern processing. *Proceedings of the Institute of Acoustics*, 6(4):315–322, 1984.

[9] Lawrence Davis, editor. *Genetic Algorithms and Simulated Annealing*. Research Notes in Artificial Intelligence. Morgan Kaufmann, Los Altos, CA, 1987.

[10] Lawrence Davis, editor. *Handbook of Genetic Algorithms*. Van Nostrand Reinhold, New York, 1991.

[11] Michael de la Maza and Bruce Tidor. An analysis of selection procedures with particular attention paid to proportional and Boltzmann selection. In Stephanie Forrest, editor, *Proceedings of the 5th International Conference on Genetic Algorithms*, pages 124–131. Morgan Kaufmann, San Mateo, CA, 1993.

[12] David B. Fogel, editor. *Evolutionary Computation: The Fossil Record*. IEEE Press, Los Alamitos, CA, 1998.

[13] Lawrence J. Fogel, Alvin J. Owens, and Michael J. Walsh. *Artificial Intelligence through Simulated Evolution*. Wiley, New York, 1966.

[14] Lawrence J. Fogel, Alvin J. Owens, and Michael J. Walsh. *Intelligence through Simulated Evolution*. Wiley, New York, updated and expanded edition, 1999.

[15] Yoav Freund and David Haussler. Unsupervised learning of distributions on binary vectors using two layer networks. In John E. Moody, Stephen J. Hanson, and Richard P. Lippmann, editors, *Advances in Neural Information Processing Systems 4*, pages 912–919. Morgan Kaufmann, San Mateo, CA, 1992.

[16] Conrad C. Galland. The limitations of deterministic Boltzmann machine learning. *Network*, 4(3):355–379, 1993.

[17] Stewart Geman and Donald Geman. Stochastic relaxation, Gibbs distributions, and the Bayesian restoration of images. *IEEE Transactions on Pattern Analysis and Machine Intelligence*, PAMI-6(6):721–741, 1984.

[18] David E. Goldberg. *Genetic Algorithms in Search, Optimization and Machine Learning*. Addison-Wesley, Reading, MA, 1989.

[19] Richard W. Hamming. *Coding and Information Theory*. Prentice-Hall, Englewood Cliffs, NJ, second edition, 1986.

[20] Peter E. Hart, Nils Nilsson, and Bertram Raphael. A formal basis for the heuristic determination of minimum cost paths. *IEEE Transactions of Systems Science and Cybernetics*, SSC-4(2):100–107, 1968.

[21] John Hertz, Anders Krogh, and Richard G. Palmer. *Introduction to the Theory of Neural Computation*. Addison-Wesley, Redwood City, CA, 1991.

[22] Geoffrey E. Hinton. Deterministic Boltzmann learning performs steepest descent in weight space. *Neural Computation*, 1(1):143–150, 1989.

[23] Geoffrey E. Hinton and Stephen J. Nowlan. How learning can guide evolution. *Complex Systems*, 1(1):495–502, 1987.

[24] Yuzo Hirai. Hardware implementation of neural networks in Japan. *Neurocomputing*, 5(1):3–16, 1993.

[25] John H. Holland. *Adaptation in Natural and Artificial Systems: An Introductory Analysis with Applications to*

Biology, Control and Artificial Intelligence. MIT Press, Cambridge, MA, second edition, 1992.

[26] John J. Hopfield. Neural networks and physical systems with emergent collective computational abilities. *Proceedings of the National Academy of Sciences of the USA*, 79(8):2554–2558, 1982.

[27] Hugo Van Hove and Alain Verschoren. Genetic algorithms and trees I: Recognition trees (the fixed width case). *Computers and Artificial Intelligence*, 13(5):453–476, 1994.

[28] Michael I. Jordan, editor. *Learning in Graphical Models.* MIT Press, Cambridge, MA, 1999.

[29] Ron Keesing and David G. Stork. Evolution and learning in neural networks: The number and distribution of learning trials affect the rate of evolution. In Richard P. Lippmann, John E. Moody, and David S. Touretzky, editors, *Advances in Neural Information Processing Systems 3*, pages 804–810. Morgan Kaufmann, San Mateo, CA, 1991.

[30] James D. Kelly, Jr. and Lawrence Davis. A hybrid genetic algorithm for classification. In Raymond Reiter and John Myopoulos, editors, *Proceedings of the 12th International Joint Conference on Artificial Intelligence*, pages 645–650. Morgan Kaufmann, San Mateo, CA, 1991.

[31] Scott Kirkpatrick, C. Daniel Gelatt, Jr., and Mario P. Vecchi. Optimization by simulated annealing. *Science*, 220(4598):671–680, 1983.

[32] Charles Kittel and Herbert Kroemer. *Thermal Physics.* Freeman, San Francisco, CA, second edition, 1980.

[33] Donald E. Knuth. *The Art of Computer Programming*, volume 3. Addison-Wesley, Reading, MA, first edition, 1973.

[34] Richard E. Korf. Optimal path finding algorithms. In Laveen N. Kanal and Vipin Kumar, editors, *Search in Artificial Intelligence*, chapter 7, pages 223–267. Springer-Verlag, Berlin, Germany, 1988.

[35] John R. Koza. *Genetic Programming: On the Programming of Computers by Means of Natural Selection.* MIT Press, Cambridge, MA, 1992.

[36] John R. Koza. *Genetic Programming II: Automatic Discovery of Reusable Programs.* MIT Press, Cambridge, MA, 1994.

[37] Vipin Kumar and Laveen N. Kanal. The CDP: A unifying formulation for heuristic search, dynamic programming, and branch & bound procedures. In Laveen N. Kanal and Vipin Kumar, editors, *Search in Artificial Intelligence*, pages 1–27. Springer-Verlag, New York, 1988.

[38] David J. C. MacKay. Equivalence of Boltzmann chains and hidden Markov models. *Neural Computation*, 8(1):178–181, 1996.

[39] Robert E. Marmelstein and Gary B. Lamont. Pattern classification using a hybrid genetic program decision tree approach. In John R. Koza, Wolfgang Banzhaf, Kumar Chellapilla, Kalyanmoy Deb, Marco Dorigo, David B. Fogel, Max H. Garzon, David E. Goldberg, Hitoshi Iba, and Rick Riolo, editors, *Genetic Programming 1998: Proceedings of the Third Annual Conference*, pages 223–231. Morgan Kaufmann, San Mateo, CA, 1998.

[40] Robert J. McEliece, Edward C. Posner, Eugene R. Rodemich, and Santosh S. Venkatesh. The capacity of the Hopfield associative memory. *IEEE Transactions on Information Theory*, IT-33(4):461–482, 1985.

[41] Nicholas Metropolis, Arianna W. Rosenbluth, Marshall N. Rosenbluth, Augusta H. Teller, and Edward Teller. Equation of state calculations by fast computing machines. *Journal of Chemical Physics*, 21(6):1087–1092, 1953.

[42] Melanie Mitchell. *An Introduction to Genetic Algorithms.* MIT Press, Cambridge, MA, 1996.

[43] John Moussouris. Gibbs and Markov random systems with constraints. *Journal of Statistical Physics*, 10(1):11–33, 1974.

[44] Michael Murray, James B. Burr, David G. Stork, Ming-Tak Leung, Kan Boonyanit, Gregory J. Wolff, and Allen M. Peterson. Deterministic Boltzmann machine VLSI can be scaled using multi-chip modules. In Jose Fortes, Edward Lee, and Teresa Meng, editors, *Proceedings of the International Conference on Application-Specific Array Processors ASAP-92*, volume 9, pages 206–217. IEEE Press, Los Alamitos, CA, 1992.

[45] Michael Murray, Ming-Tak Leung, Kan Boonyanit, Kong Kritayakirana, James B. Burr, Greg Wolff, David G. Stork, Takahiro Watanabe, Ed Schwartz, and Allen M. Peterson. Digital Boltzmann VLSI for constraint satisfaction and learning. In Jack D. Cowan, Gerald Tesauro, and Joshua Alspector, editors, *Advances in Neural Information Processing Systems*, volume 6, pages 896–903. MIT Press, Cambridge, MA, 1994.

[46] Dana S. Nau, Vipin Kumar, and Laveen N. Kanal. General branch and bound, and its relation to A* and AO*. *Artificial Intelligence*, 23(1):29–58, 1984.

[47] Nils J. Nilsson. *Artificial Intelligence: A New Synthesis.* Morgan Kaufmann, San Mateo, CA, 1998.

[48] Mattias Ohlsson, Carsten Peterson, and Bo Söderberg. Neural networks for optimization problems with inequality constraints: The knapsack problem. *Neural Computation*, 5(2):331–339, 1993.

[49] Sankar K. Pal and Paul P. Wang, editors. *Genetic Algorithms for Pattern Recognition.* CRC Press, Boca Raton, FL, 1996.

[50] Morton With Pederson and David G. Stork. Pruning Boltzmann networks and Hidden Markov Models. In *Thirteenth Asilomar Conference on Signals, Systems and Computers*, volume 1, pages 258–261. IEEE Press, New York, 1997.

[51] Carsten Peterson and James R. Anderson. A mean-field theory learning algorithm for neural networks. *Complex Systems*, 1(5):995–1019, 1987.

[52] Richard W. Prager, Tim D. Harrison, and Frank Fallside. Boltzmann machines for speech recognition. *Computer Speech and Language*, 1(1):3–27, 1986.

[53] Ingo Rechenberg. Bionik, evolution und optimierung (in German). *Naturwissenschaftliche Rundschau*, 26(11):465–472, 1973.

[54] Ingo Rechenberg. Evolutionsstrategie – optimierung nach prinzipien der biologischen evolution (in German). In Jörg Albertz, editor, *Evolution und Evolutionsstrategien in Biologie, Technik und Gesellschaft*, pages 25–72. Freie Akademie, 1989.

[55] Stuart Russell and Peter Norvig. *Artificial Intelligence: A Modern Approach*. Prentice-Hall Series in Artificial Intelligence. Prentice-Hall, Englewood Cliffs, NJ, 1995.

[56] Lawrence K. Saul and Michael I. Jordan. Boltzmann chains and Hidden Markov Models. In Gerald Tesauro, David S. Touretzky, and Todd K. Leen, editors, *Advances in Neural Information Processing Systems*, volume 7, pages 435–442. MIT Press, Cambridge, MA, 1995.

[57] Oliver G. Selfridge. Pandemonium: A paradigm for learning. In *Mechanisation of Thought Processes: Proceedings of a Symposium held at the National Physical Laboratory*, pages 513–526. HMSO, London, 1958.

[58] Paul Smolensky. Information processing in dynamical systems: Foundations of Harmony theory. In David E. Rumelhart and James L. McClelland, and the PDP Research Group editors, *Parallel Distributed Processing*, volume 1, chapter 6, pages 194–281. MIT Press, Cambridge, MA, 1986.

[59] Karl Steinbuch. Die lernmatrix (in German). *Kybernetik (Biological Cybernetics)*, 1(1):36–45, 1961.

[60] Karl Steinbuch. *Automat und Mensch (in German)*. Springer, New York, 1971.

[61] David G. Stork, Bernie Jackson, and Scott Walker. Non-optimality via pre-adaptation in simple neural systems. In Christopher G. Langton, Charles Taylor, J. Doyne Farmer, and Steen Rasmussen, editors, *Artificial Life II*, pages 409–429. Addison Wesley, Reading, MA, 1992.

[62] David G. Stork, Bernie Jackson, and Scott Walker. Nonoptimality in a neurobiological systems. In Daniel S. Levine and Wesley R. Elsberry, editors, *Optimality in Biological and Artificial Networks?*, pages 57–75. Lawrence Erlbaum Associates, Mahwah, NJ, 1997.

[63] Harold Szu. Fast simulated annealing. In John S. Denker, editor, *Neural Networks for Computing*, pages 420–425. American Institute of Physics, New York, 1986.

[64] Vladimír Černý. Thermodynamical approach to the traveling salesman problem: An efficient simulation algorithm. *Journal of Optimization Theory and Applications*, 45:41–51, 1985.

[65] Santosh S. Venkatesh and Demetri Psaltis. Linear and logarithmic capacities in associative neural networks. *IEEE Transactions on Information Theory*, IT-35(3):558–568, 1989.

[66] Gérard Weisbuch and Françoise Fogelman-Soulié. Scaling laws for the attractors of Hopfield networks. *Journal of Physics Letters*, 46(14):623–630, 1985.

[67] Patrick Henry Winston. *Artificial Intelligence*. Addison-Wesley, Reading, MA, third edition, 1992.

NONMETRIC METHODS

8.1 INTRODUCTION

We have considered pattern recognition based on feature vectors of real-valued and discrete-valued numbers, and in all cases there has been a natural measure of distance between vectors. For instance in the nearest-neighbor classifier the notion figures conspicuously—indeed it is the core of the technique—while for neural networks the notion of similarity appears when two input vectors sufficiently "close" lead to similar outputs. Most practical pattern recognition methods address problems of this sort, where feature vectors are real-valued and there exists some notion of metric.

NOMINAL DATA

But suppose a classification problem involves *nominal* data—for instance descriptions that are discrete and without any natural notion of similarity or even ordering. Consider the use of information about teeth in the classification of fish and sea mammals. Some teeth are small and fine (as in baleen whales) for straining tiny prey from the sea. Others (as in sharks) come in multiple rows. Some sea creatures, such as walruses, have tusks. Yet others, such as squid, lack teeth altogether. There is no clear notion of similarity (or metric) for this information about teeth. For instance, the teeth of a baleen whale and the tusks of a walrus are no more "similar" than are the teeth of a shark to the teeth in a flounder.

PROPERTY
D-TUPLE

Thus in this chapter our attention turns away from describing patterns by vectors of real numbers and toward using *lists* of attributes. A common approach is to specify the values of a fixed number of properties by a *property d-tuple* For example, consider describing a piece of fruit by four properties: color, texture, taste and size. Then a particular piece of fruit might be described by the 4-tuple {*red*, *shiny*, *sweet*, *small*}, which is a shorthand for color = red, texture = shiny, taste = sweet and size = small. Another common approach is to de-

STRING

scribe the pattern by a variable length *string* of nominal attributes, such as a sequence of base pairs in a segment of DNA—for example, "AGCTTCAGATTCCA."* Such lists or strings might themselves be the output of other component classifiers of the type we have seen elsewhere. For instance, we might train a neural network to recognize different component brush strokes used in Chinese and Japanese characters (roughly a dozen basic forms); a classifier would then accept as inputs a list of these nominal attributes and make the final, full character classification.

---

*We often put strings between quotation marks, particularly if this will help to avoid ambiguities.

How can we best use such nominal data for classification? Most importantly, how can we efficiently *learn* categories using such nonmetric data? If there is structure in strings, how can it be represented? In considering such problems, we move beyond the notion of continuous probability distributions and metrics toward discrete problems that are addressed by rule-based or syntactic pattern recognition methods.

8.2 DECISION TREES

It is natural and intuitive to classify a pattern through a sequence of questions, in which the next question asked depends on the answer to the current question. This "20-questions" approach is particularly useful for nonmetric data, because all of the questions can be asked in a "yes/no" or "true/false" or "value(property) ∈ set_of_values" style that does not require any notion of metric.

ROOT NODE
LINK
BRANCH
LEAF

Such a sequence of questions is displayed in a directed *decision tree* or simply *tree*, where by convention the first or *root node* is displayed at the top, connected by successive (directional) *links* or *branches* to other nodes. These are similarly connected until we reach terminal or *leaf* nodes, which have no further links (Fig. 8.1). Sections 8.3 and 8.4 describe some generic methods for *creating* such trees, but let us first understand how they are used for classification. The classification of a particular pattern begins at the root node, which asks for the value of a particular property of the pattern. The different links from the root node correspond to the different possible values. Based on the answer, we follow the appropriate link to a subsequent or

DESCENDENT

descendent node. In the trees we shall discuss, the links must be mutually distinct and exhaustive; that is, one and only one link will be followed. The next step is to make the decision at the appropriate subsequent node, which can be considered the

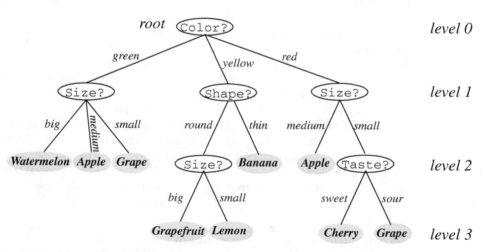

FIGURE 8.1. Classification in a basic decision tree proceeds from top to bottom. The questions asked at each node concern a particular property of the pattern, and the downward links correspond to the possible values. Successive nodes are visited until a terminal or leaf node is reached, where the category label is read. Note that the same question, Size?, appears in different places in the tree and that different questions can have different numbers of branches. Moreover, different leaf nodes, shown in pink, can be labeled by the same category (e.g., **Apple**).

root of a *subtree*. We continue this way until we reach a leaf node, which has no further question. Each leaf node bears a category label, and the test pattern is assigned the category of the leaf node reached.

The simple decision tree in Fig. 8.1 illustrates one benefit of trees over many other classifiers such as neural networks: interpretability. It is a straightforward matter to render the information in such a tree as logical expressions. Such interpretability has two manifestations. First, we can easily interpret the decision for any *particular* test pattern as the conjunction of decisions along the path to its corresponding leaf node. Thus if the properties are {taste, color, shape, size}, the pattern $\mathbf{x}$ = {sweet, yellow, thin, medium} is classified as **Banana** because it is (color = yellow) AND (shape = thin).* Second, we can occasionally get clear interpretations of the *categories* themselves, by creating logical descriptions using conjunctions and disjunctions (Problem 8). For instance, the tree shows **Apple** = (green AND medium) OR (red AND medium).

Rules derived from trees—especially large trees—are often quite complicated and must be reduced to aid interpretation. For our example, one simple rule describes **Apple** = (medium AND NOT yellow). Another benefit of trees is that they lead to rapid classification, employing a sequence of typically simple queries. Finally, we note that trees provide a natural way to incorporate prior knowledge from human experts. In practice, though, such expert knowledge if of greatest use when the classification problem is fairly simple and the training set is small.

8.3 CART

Now we turn to the matter of using training data to create or "grow" a decision tree. We assume that we have a set $\mathcal{D}$ of labeled training data and we have decided on a set of properties that can be used to discriminate patterns, but do not know how to organize the tests into a tree. Clearly, any decision tree will progressively split the set of training examples into smaller and smaller subsets. It would be ideal if all the samples in each subset had the same category label. In that case, we would say that each subset was *pure*, and could terminate that portion of the tree. Usually, however, there is a mixture of labels in each subset, and thus for each branch we will have to decide either to stop splitting and accept an imperfect decision, or instead select another property and grow the tree further.

This suggests an obvious recursive tree-growing process: Given the data represented at a node, either declare that node to be a leaf (and state what category to assign to it), or find another property to use to split the data into subsets. However, this is only one example of a more generic tree-growing methodology known as CART (classification and regression trees). CART provides a general framework that can be instantiated in various ways to produce different decision trees. In the CART approach, six general kinds of questions arise:

1. Should the properties be restricted to binary-valued or allowed to be multi-valued? That is, how many decision outcomes or *splits* will there be at a node?

---

*We retain our convention of representing patterns in boldface even though they need not be true vectors; that is, they might contain nominal data that cannot be added or multiplied the way vector components can. For this reason we use the terms "attribute" to represent both nominal data and real-valued data, and we reserve "feature" for real-valued data.

2. Which property should be tested at a node?

3. When should a node be declared a leaf?

4. If the tree becomes "too large," how can it be made smaller and simpler, that is, pruned?

5. If a leaf node is impure, how should the category label be assigned?

6. How should missing data be handled?

We consider each of these questions in turn.

8.3.1 Number of Splits

Each decision outcome at a node is called a *split*, because it corresponds to splitting a subset of the training data. The root node splits the full training set; each successive decision splits a proper subset of the data. The number of splits at a node is closely related to question 2, specifying *which* particular split will be made at a node. In general, the number of splits is set by the designer and could vary throughout the tree, as we saw in Fig. 8.1. The number of links descending from a node is sometimes called the node's *branching factor* or *branching ratio*, denoted B. However, every decision (and hence every tree) can be represented using just *binary* decisions (Problem 2). Thus the root node querying fruit color ($B = 3$) in our example could be replaced by two nodes: The first would ask `fruit = green?`, and at the end of its "no" branch, another node would ask `fruit = yellow?`. Because of the universal expressive power of binary trees and the comparative simplicity in training, we shall concentrate on such trees (Fig. 8.2).

BRANCHING
FACTOR

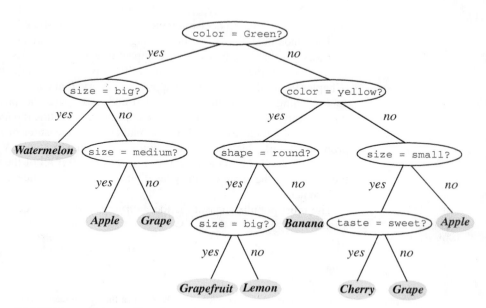

FIGURE 8.2. A tree with arbitrary branching factor at different nodes can always be represented by a functionally equivalent binary tree—that is, one having branching factor $B = 2$ throughout, as shown here. By convention the "yes" branch is on the left, the "no" branch on the right. This binary tree contains the same information and implements the same classification as that in Fig. 8.1.

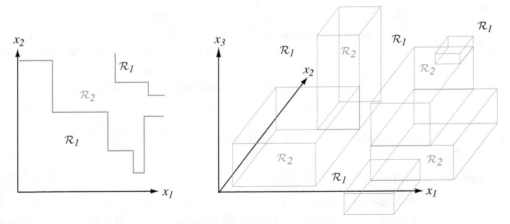

FIGURE 8.3. Monothetic decision trees create decision boundaries with portions perpendicular to the feature axes. The decision regions are marked $\mathcal{R}_1$ and $\mathcal{R}_2$ in these two-dimensional and three-dimensional two-category examples. With a sufficiently large tree, any decision boundary can be approximated arbitrarily well in this way.

8.3.2 Query Selection and Node Impurity

Much of the work in designing trees focuses on deciding which property test or query should be performed at each node.* With nonnumeric data, there is no geometrical interpretation of how the query at a node splits the data. However, for numerical data, there is a simple way to visualize the decision boundaries that are produced by decision trees. For example, suppose that the query at each node has the form "Is $x_i \leq x_{is}$?" This leads to hyperplane decision boundaries that are perpendicular to the coordinate axes, and to decision regions of the form illustrated in Fig. 8.3.

The fundamental principle underlying tree creation is that of simplicity: We prefer decisions that lead to a simple, compact tree with few nodes. This is a version of Occam's razor, that the simplest model that explains data is the one to be preferred (Chapter 9). To this end, we seek a property query T at each node N that makes the data reaching the immediate descendent nodes as "pure" as possible. In formalizing this notion, it turns out to be more convenient to define the *im*purity, rather than the purity of a node. Several different mathematical measures of impurity have been proposed, all of which have basically the same behavior. Let $i(N)$ denote the impurity of a node N. In all cases, we want $i(N)$ to be 0 if all of the patterns that reach the node bear the same category label, and to be large if the categories are equally represented.

The most popular measure is the *entropy impurity* (or occasionally *information impurity*):

PURITY

ENTROPY
IMPURITY

$$i(N) = -\sum_j P(\omega_j) \log_2 P(\omega_j), \tag{1}$$

---

*The problem is further complicated by the fact that there is no reason why the query at a node has to involve only one property. One might well consider logical combinations of properties, such as using (size = medium) AND (NOT (color = yellow))? as a query. Trees in which each query is based on a single property are called *monothetic*; if the query at any of the nodes involves two or more properties, the tree is called *polythetic*. For simplicity, we generally restrict our treatment to monothetic trees. In all cases, the key requirement is that the decision at a node be well-defined and unambiguous so that the response leads down one and only one branch.

where $P(\omega_j)$ is the fraction of patterns at node N that are in category ω_j.* By the well-known properties of entropy, if all the patterns are of the same category, the impurity is 0; otherwise it is positive, with the greatest value occuring when the different classes are equally likely.

Another definition of impurity is particularly useful in the two-category case. Given the desire to have zero impurity when the node represents only patterns of a single category, the simplest polynomial form is

$$i(N) = P(\omega_1)P(\omega_2). \tag{2}$$

VARIANCE
IMPURITY

GINI IMPURITY

This can be interpreted as a *variance impurity* because under reasonable assumptions it is related to the variance of a distribution associated with the two categories (Problem 10). A generalization of the variance impurity, applicable to two or more categories, is the *Gini impurity*:

$$i(N) = \sum_{i \neq j} P(\omega_i)P(\omega_j) = \frac{1}{2}\left[1 - \sum_j P^2(\omega_j)\right]. \tag{3}$$

This is just the expected error rate at node N if the category label is selected randomly from the class distribution present at N. This criterion is more strongly peaked at equal probabilities than is the entropy impurity (Fig. 8.4).

MISCLASSIFI-
CATION
IMPURITY

The *misclassification impurity* can be written as

$$i(N) = 1 - \max_j P(\omega_j), \tag{4}$$

and it measures the minimum probability that a training pattern would be misclassified at N. Of the impurity measures typically considered, this measure is the most strongly peaked at equal probabilities. It has a discontinuous derivative, though, and this can present problems when searching for an optimal decision over a continuous

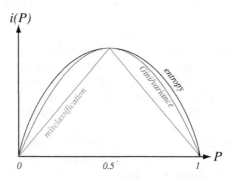

FIGURE 8.4. For the two-category case, the impurity functions peak at equal class frequencies and the variance and the Gini impurity functions are identical. The entropy, variance, Gini, and misclassification impurities (given by Eqs. 1–4, respectively) have been adjusted in scale and offset to facilitate comparison here; such scale and offset do not directly affect learning or classification.

---

*Here we are a bit sloppy with notation, because we normally reserve P for probability and $\hat{P}$ for frequency ratios. We could be even more precise by writing $\hat{P}(\mathbf{x} \in \omega_j | N)$—that is, the fraction of training patterns $\mathbf{x}$ at node N that are in category ω_j, given that they have survived all the previous decisions that led to the node N—but for the sake of simplicity we still avoid such notational overhead.

parameter space. Figure 8.4 shows these impurity functions for a two-category case, as a function of the probability of one of the categories.

We now come to the key question: Given a partial tree down to node N, what value s should we choose for the property test T? An obvious heuristic is to choose the query that decreases the impurity as much as possible. The drop in impurity is defined by

$$\Delta i(N) = i(N) - P_L i(N_L) - (1 - P_L)i(N_R), \qquad (5)$$

where N_L and N_R are the left and right descendent nodes, $i(N_L)$ and $i(N_R)$ are their impurities, and P_L is the fraction of patterns at node N that will go to N_L when property query T is used. Then the "best" query value s is the choice for T that maximizes $\Delta i(T)$. If the entropy impurity is used, then the impurity reduction corresponds to an information gain provided by the query. Because each query in a binary tree is a single "yes/no" one, the reduction in entropy impurity due to a split at a node cannot be greater than one bit (Problem 5).

The way to find an optimal decision for a node depends upon the general form of decision. Because the decision criteria are based on the extrema of the impurity functions, we are free to change such a function by an additive constant or overall scale factor and this will not affect which split is found. Designers typically choose functions that are easy to compute, such as those based on a *single* feature or attribute, giving a monothetic tree. If the form of the decisions is based on the nominal attributes, we may have to perform extensive or exhaustive search over all possible subsets of the training set to find the rule maximizing Δi. If the attributes are real-valued, one could use gradient descent algorithms to find a splitting hyperplane (Section 8.3.8), giving a polythetic tree. An important reason for favoring binary trees is that the decision at any node can generally be cast as a one-dimensional optimization problem. If the branching factor B were instead greater than 2, a two- or higher-dimensional optimization would be required; this is generally much more difficult.

Sometimes there will be several decisions s that lead to the same reduction in impurity, and the question arises how to choose among them. For example, if the features are real-valued and a split lying anywhere in a range $x_l < x_s < x_u$ for the x variable leads to the same (maximum) impurity reduction, it is traditional to choose either the midpoint or the weighted average—$x_s = (x_l + x_u)/2$ or $x_s = (1 - P)x_l + x_u P$, respectively—where P is the probability a pattern goes to the "left" under the decision. Computational simplicity may be the determining factor as there are rarely deep theoretical reasons to favor one over another.

Note too that the optimization of Eq. 5 is *local*—done at a single node. As with the vast majority of such *greedy methods*, there is no assurance that successive locally optimal decisions lead to the *global* optimum. In particular, there is no guarantee that after training we have the smallest tree. Nevertheless, for every reasonable impurity measure and learning method, we can always continue to split further to get the lowest possible impurity at the leafs. There is no assurance that the impurity at a leaf node will be the zero, however: if two patterns have the same attribute description yet come from different categories, the impurity will be greater than zero.

Occasionally during tree creation the misclassification impurity (Eq. 4) will not decrease whereas the Gini impurity would; thus although classification is our final goal, we may prefer the Gini impurity because it "anticipates" later splits that will be useful. Consider a case where at node N there are 90 patterns in ω_1 and 10 in ω_2. Thus the misclassification impurity at N is 0.1. Suppose there are no splits that guarantee a

ω_2 majority in either of the two descendent nodes. Then the misclassification remains at 0.1 for all splits. Now consider a split that sends 70 ω_1 patterns to the right along with 0 ω_2 patterns, and sends 20 ω_1 and 10 ω_2 to the left. This is an attractive split, but the misclassification impurity is still 0.1. On the other hand, the Gini impurity for this split is less than the Gini for the parent node. In short, the Gini impurity shows that this is a good split while the misclassification rate does not.

In multiclass binary tree creation, the *twoing criterion* may be useful. The overall goal is to find the split that best splits *groups* of the c categories—that is, a candidate "supercategory" $\mathcal{C}_1$ consisting of all patterns in some subset of the categories, and candidate "supercategory" $\mathcal{C}_2$ as all remaining patterns. Let the class of categories be $\mathcal{C} = \{\omega_1, \omega_2, \ldots, \omega_c\}$. At each node, the decision splits the categories into $\mathcal{C}_1 = \{\omega_{i_1}, \omega_{i_2}, \ldots, \omega_{i_k}\}$ and $\mathcal{C}_2 = \mathcal{C} - \mathcal{C}_1$. For every candidate split s, we compute a change in impurity $\Delta i(s, \mathcal{C}_1)$ as though it corresponded to a standard two-class problem. That is, we find the split $s^*(\mathcal{C}_1)$ that maximizes the change in impurity. Finally, we find the supercategory $\mathcal{C}_1^*$ that maximizes $\Delta i(s^*(\mathcal{C}_1), \mathcal{C}_1)$. The benefit of this impurity is that it is *strategic*—it may learn the largest scale structure of the overall problem (Problem 4).

It may be surprising, but the particular choice of an impurity function rarely seems to affect the final classifier and its accuracy. An entropy impurity is frequently used because of its computational simplicity and basis in information theory, though the Gini impurity has received significant attention as well. In practice, the stopping criterion and the pruning method—when to stop splitting nodes, and how to merge leaf nodes—are more important than the impurity function itself in determining final classifier accuracy, as we shall see.

Multiway Splits

Although we shall concentrate on binary trees, we briefly mention the matter of allowing the branching ratio at each node to be set during training, a technique we will return to in a discussion of the ID3 algorithm (Section 8.4.1). In such a case, it is tempting to use a multibranch generalization of Eq. 5 of the form

$$\Delta i(s) = i(N) - \sum_{k=1}^{B} P_k i(N_k), \qquad (6)$$

where P_k is the fraction of training patterns sent down the link to node N_k, and $\sum_{k=1}^{B} P_k = 1$. However, the drawback with Eq. 6 is that decisions with large B are inherently favored over those with small B whether or not the large-B splits in fact represent meaningful structure in the data. For instance, even in random data, a high-B split will reduce the impurity more than will a low-B split. To avoid this drawback, the candidate change in impurity of Eq. 6 must be scaled, according to

$$\Delta i_B(s) = \frac{\Delta i(s)}{-\sum_{k=1}^{B} P_k \log_2 P_k}. \qquad (7)$$

a method based on the *gain ratio impurity* (Problem 17). Just as before, the optimal split is the one maximizing $\Delta i_B(s)$.

8.3.3 When to Stop Splitting

Consider now the problem of deciding when to stop splitting during the training of a binary tree. If we continue to grow the tree fully until each leaf node corresponds to the lowest impurity, then the data have typically been overfit (Chapter 9). In the extreme but rare case, each leaf corresponds to a single training point and the full tree is merely a convenient implementation of a lookup table; it thus cannot be expected to generalize well in (noisy) problems having high Bayes error. Conversely, if splitting is stopped too early, then the error on the training data is not sufficiently low and hence performance may suffer.

How shall we decide when to stop splitting? One traditional approach is to use techniques we shall explore in Chapter 9, in particular validation and cross-validation. In validation, the tree is trained using a subset of the data (e.g., 90%), with the remaining (10%) kept as a validation set. We continue splitting nodes in successive layers until the error on the validation data is minimized. (Cross-validation relies on *several* independently chosen subsets.)

Another method is to set a (small) threshold value in the reduction in impurity; splitting is stopped if the best candidate split at a node reduces the impurity by less than that preset amount—that is, if $\max_s \Delta i(s) \leq \beta$. This method has two main benefits. First, unlike cross-validation, the tree is trained directly using *all* the training data. Second, leaf nodes can lie in different levels of the tree, which is desirable whenever the complexity of the data varies throughout the range of input. (Such an

BALANCED TREE

unbalanced tree requires a different number of decisions for different test patterns.) A fundamental drawback of the method, however, is that it is often difficult to know how to set the threshold because there is rarely a simple relationship between β and the ultimate performance (Computer exercise 2). A very simple method is to stop when a node represents fewer than some threshold number of points, say 10, or some fixed percentage of the total training set, say 5%. This has a benefit analogous to that in k-nearest-neighbor classifiers (Chapter 4); that is, the size of the partitions is small in regions where data is dense, but large where the data is sparse.

Yet another method is to trade complexity for test accuracy by splitting until a minimum in a new, global criterion function,

$$\alpha \cdot size + \sum_{leaf\ nodes} i(N), \tag{8}$$

is reached. Here *size* could represent the number of nodes or links and α is some positive constant. (This is analogous to regularization methods in neural networks that penalize connection weights or nodes.) If an impurity based on entropy is used

MINIMUM
DESCRIPTION
LENGTH

for $i(N)$, then Eq. 8 finds support from *minimum description length* (MDL), which we shall consider again in Chapter 9. The sum of the impurities at the leaf nodes is a measure of the uncertainty (in bits) in the training data given the model represented by the tree; the size of the tree is a measure of the complexity of the classifier itself (which also could be measured in bits). A difficulty, however, is setting α, because it is not always easy to find a simple relationship between α and the final classifier performance.

An alternative approach is to use a stopping criterion based on the statistical significance of the reduction of impurity. During tree construction, we estimate the distribution of all the Δi for the current collection of nodes; we assume this is the full distribution of Δi. For any candidate node split, we then determine whether it is statistically different from zero, for instance by a chi-squared test (cf. Appendix Section A.6.1). If a candidate split does not reduce the impurity *significantly*, splitting is stopped (Problem 15).

A variation in this technique of *hypothesis testing* can be applied even without strong assumptions on the distribution of Δi. We seek to determine whether a candidate split is "meaningful"—that is, whether it differs significantly from a random split. Suppose n patterns survive at node N (with n_1 in ω_1 and n_2 in ω_2); we wish to decide whether a candidate split s differs significantly from a random one. Suppose a particular candidate split s sends Pn patterns to the left branch and sends $(1-P)n$ patterns to the right branch. A random split having this probability (i.e., the null hypothesis) would place Pn_1 of the ω_1 patterns and Pn_2 of the ω_2 patterns to the left and would place the remaining ones to the right. We quantify the deviation of the results due to candidate split s from the (weighted) random split by means of the *chi-squared statistic*, which in this two-category case is

$$\chi^2 = \sum_{i=1}^{2} \frac{(n_{iL} - n_{ie})^2}{n_{ie}}, \tag{9}$$

where n_{iL} is the number of patterns in category ω_i sent to the left under decision s, and $n_{ie} = Pn_i$ is the number expected by the random rule. The chi-squared statistic vanishes if the candidate split s gives the same distribution as the random one, and it is larger the more s differs from the random one. When χ^2 is greater than a critical value, as given in a table (cf. Appendix Section A.6.1), we can reject the null hypothesis because s differs "significantly" at some probability or *confidence level*, such as 0.01 or 0.05. The critical values of the confidence depend upon the number of degrees of freedom, which in the case just described is 1, because for a given probability P the *single* value n_{1L} specifies all other values (n_{1R}, n_{2L}, and n_{2R}). If the "most significant" split at a node does not yield a χ^2 exceeding the chosen confidence level threshold, splitting should be stopped.

8.3.4 Pruning

Occasionally, stopped splitting suffers from the lack of sufficient look ahead, a phenomenon called the *horizon effect*. The determination of the optimal split at a node N is not influenced by decisions at N's descendent nodes—that is, those at subsequent levels. In stopped splitting, node N might be declared a leaf, cutting off the possibility of beneficial splits in subsequent nodes; as such, a stopping condition may be met "too early" for overall optimal recognition accuracy. Informally speaking, the stopped splitting biases the learning algorithm toward trees in which the greatest impurity reduction is near the root node.

The principal alternative approach to stopped splitting is *pruning*. In pruning, a tree is grown fully—that is, until leaf nodes have minimum impurity, beyond any putative "horizon." Then, all pairs of neighboring leaf nodes (i.e., ones linked to a common antecedent node, one level above) are considered for elimination. Any pair whose elimination yields a satisfactory (small) increase in impurity is eliminated, and the common antecedent node is declared a leaf. (This antecedent, in turn, could then itself be pruned.) Clearly, such *merging* or *joining* of the two leaf nodes is the inverse of splitting. It is not unusual that after such pruning, the leaf nodes lie in a wide range of levels and the tree is unbalanced.

Although it is most common to prune starting at the leaf nodes, this is not necessary: Cost-complexity pruning can replace a complex subtree with a leaf directly.

Furthermore, the C4.5 algorithm (Section 8.4.2) can eliminate an arbitrary test node, thereby replacing a subtree by one of its branches.

The benefits of pruning are that it avoids the horizon effect; furthermore, because there is no training data held out for cross-validation, it directly uses *all* information in the training set. Naturally, this comes at a greater computational expense than stopped splitting, and for problems with large training sets, the expense can be prohibitive. For small problems, though, these computational costs are low and pruning is generally to be preferred over stopped splitting. Incidentally, what we have been calling stopped training and pruning are sometimes called prepruning and postpruning, respectively.

A conceptually different pruning method is based on *rules*. Each leaf has an associated rule—the conjunction of the individual decisions from the root node, through the tree, to the particular leaf. Thus the full tree can be described by a large list of rules, one for each leaf. Occasionally, some of these rules can be simplified if a series of decisions is redundant. Eliminating the *irrelevant* precondition rules simplifies the description, but has no influence on the classifier function, including its generalization ability. The predominant reason to prune, however, is to improve generalization. In this case we therefore eliminate rules so as to improve accuracy on a validation set (Computer exercise 5). This technique may even allow the elimination of a rule corresponding to a node near the root.

One of the benefits of rule pruning is that it allows us to distinguish between the contexts in which any particular node N is used. For instance, for some test pattern $\mathbf{x}_1$ the decision rule at node N is necessary; for another test pattern $\mathbf{x}_2$, that rule is irrelevant and thus N could be pruned. In traditional node pruning, we must either keep N or prune it away. In rule pruning, however, we can eliminate it where it is not necessary (i.e., for patterns such as $\mathbf{x}_1$) and retain it for others (such as $\mathbf{x}_2$).

A final benefit is that the reduced rule set may give improved interpretability. Although rule pruning was not part of the original CART approach, such pruning can be easily applied to CART trees. We shall consider an example of rule pruning in Section 8.4.2.

8.3.5 Assignment of Leaf Node Labels

Assigning category labels to the leaf nodes is the simplest step in tree construction. If successive nodes are split as far as possible, and each leaf node corresponds to patterns in a single category (zero impurity), then of course this category label is assigned to the leaf. In the more typical case, where either stopped splitting or pruning is used and the leaf nodes have positive impurity, each leaf should be labeled by the category that has most points represented. An extremely small impurity is not necessarily desirable because it may be an indication that the tree is overfitting the training data.

Example 1 illustrates some of these steps.

EXAMPLE 1   A Simple Tree

Consider the following $n = 16$ points in two dimensions for training a binary CART tree ($B = 2$) using the entropy impurity (Eq. 1).

ω_1 (black)		ω_2 (red)	
x_1	x_2	x_1	x_2
.15	.83	.10	.29
.09	.55	.08	.15
.29	.35	.23	.16
.38	.70	.70	.19
.52	.44	.62	.47
.57	.73	.91	.27
.73	.75	.65	.90
.47	.06	.75	.36* (.32†)

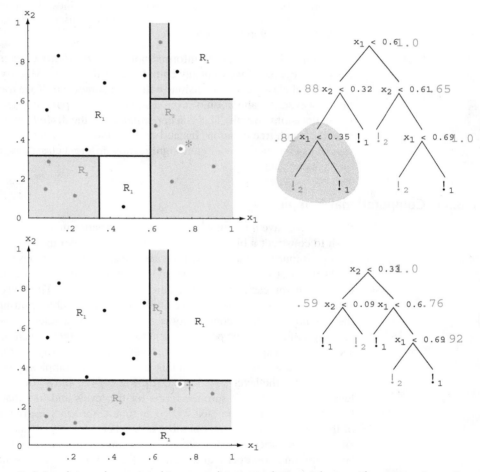

Training data and associated (unpruned) tree are shown at the top. The entropy impurity at nonterminal nodes is shown in red and the impurity at each leaf is 0. If the single training point marked * were instead slightly lower (marked †), the resulting tree and decision regions would differ significantly, as shown at the bottom.

The impurity of the root node is

$$i(N_{root}) = -\sum_{i=1}^{2} P(\omega_i)\log_2 P(\omega_i) = -[0.5\log_2 0.5 + 0.5\log_2 0.5] = 1.0.$$

For simplicity we consider candidate splits parallel to the feature axes—that is, of the form "Is $x_i < x_{is}$?". By exhaustive search of the $n-1$ positions for the x_1 feature and $n-1$ positions for the x_2 feature we find by Eq. 5 that the greatest reduction in the impurity occurs near $x_{1s} = 0.6$, and hence this becomes the decision criterion at the root node. We continue for each subtree until each final node represents a single category (and thus has the lowest impurity, 0), as shown in the figure. If pruning were invoked, the pair of leaf nodes at the left would be the first to be deleted (gray shading) because there the impurity is increased the least. In this example, stopped splitting with the proper threshold would also give the same final network. In general, however, with large trees and many pruning steps, pruning and stopped splitting need not lead to the same final tree.

This particular training set shows how trees can be sensitive to details of the training points. If the ω_2 point marked * in the top figure is moved slightly (marked †), the tree and decision regions differ significantly, as shown at the bottom. Such instability is due in large part to the discrete and greedy nature of decisions early in the tree learning.

STABILITY Example 1 illustrates the informal notion of *instability* or sensitivity to training points. Of course, if we train any common classifier with a slightly different training set the final classification decisions will differ somewhat. If we train a CART classifier, however, the alteration of even a single training point can lead to radically different decisions overall. This is a consequence of the discrete and inherently greedy nature of such tree creation. Instability often indicates that incremental and off-line versions of the method will yield significantly different classifiers, even when trained on the same data.

8.3.6 Computational Complexity

Suppose we have n training patterns in d dimensions in a two-category problem and wish to construct a binary tree based on splits parallel to the feature axes using an entropy impurity. What are the time and the space complexities?

At the root node (level 0) we must first sort the training data, an $O(n \log n)$ calculation for each of the d features or dimensions. The entropy calculation is $O(n) + (n-1)O(d)$ because we examine $n-1$ possible splitting points. Thus for the root node the time complexity is $O(dn \log n)$. Consider an average case, where roughly half the training points are sent to each of the two branches. The above analysis implies that splitting each node in level 1 has complexity $O(d\, n/2 \log(n/2))$; because there are two such nodes at that level, the total complexity is $O(dn \log (n/2))$. Similarly, for the level 2 we have $O(dn \log (n/4))$, and so on. The total number of levels is $O(\log n)$. We sum the terms for the levels and find that the total average time complexity is $O(dn (\log n)^2)$. The time complexity for recall is just the depth of the tree—that is, the total number of levels is $O(\log n)$. The space complexity is simply the number of nodes, which, given some simplifying assumptions (such as a single training point per leaf node), is $1 + 2 + 4 + \cdots + n/2 \approx n$, that is, $O(n)$, as explored in Problem 9.

We stress that these assumptions (e.g., equal splits at each node) rarely hold exactly; moreover, heuristics can be used to speed the search for splits during training. Nevertheless, the result that for fixed dimension d the training is $O(dn^2 \log n)$ and classification $O(\log n)$ is a good rule of thumb; it illustrates how training is far more computationally expensive than is classification, and that on average this discrepancy grows as the problem gets larger.

There are several techniques for reducing the complexity during the training of trees based on real-valued data. One of the simplest heuristics is to begin the search for splits x_{is} at the "middle" of the range of the training set, moving alternately to progressively higher and lower values. Optimal splits always occur for decision thresholds between adjacent points from *different* categories, and thus one should test only such ranges. These and related techniques generally provide only moderate reductions in computation. When the patterns consist of nominal data, candidate splits could be over every subset of attributes, or just a single entry, and the computational burden is best lowered using insight into features (Problem 3).

8.3.7 Feature Choice

As with most pattern recognition techniques, CART and other tree-based methods work best if the "proper" features are used (Fig. 8.5). For real-valued vector data, most standard preprocessing techniques can be used before creating a tree. Preprocessing by principal components (Chapter 10, Section 10.13) can be effective, be-

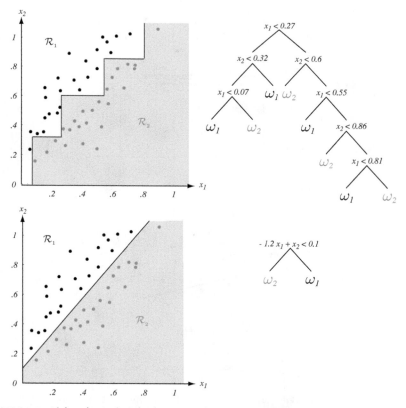

FIGURE 8.5. If the class of node decisions does not match the form of the training data, a very complicated decision tree will result, as shown at the top. Here decisions are parallel to the axes while in fact the data is better split by boundaries along another direction. If, however, "proper" decision forms are used (here, linear combinations of the features), the tree can be quite simple, as shown at the bottom.

cause it finds the "important" axes, and this generally leads to simple decisions at the nodes. If, however, the principal axes in one region differ significantly from those in another region, then no single choice of axes overall will suffice. In that case we may need to employ the techniques of Section 8.3.8—for instance, allowing splits to be at arbitrary orientation, often giving smaller and more compact trees.

8.3.8 Multivariate Decision Trees

If the "natural" splits of real-valued data do not fall parallel to the feature axes or the full training data set differs significantly from simple or accommodating distributions, then the above methods may be rather inefficient and lead to poor generalization (Fig. 8.6); even pruning may be insufficient to give a good classifier. The simplest solution is to allow splits that are not parallel to the feature axes, such as a general linear classifier trained via gradient descent on a classification or sum-squared-error criterion (Chapter 5). While such training may be slow for the nodes near the root if the training set is large, training will be faster at nodes closer to

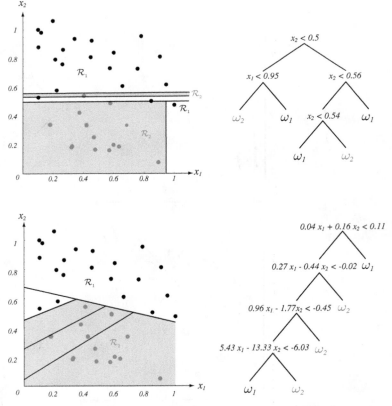

FIGURE 8.6. One form of multivariate tree employs general linear decisions at each node, giving splits along arbitrary directions in the feature space. In virtually all interesting cases the training data are not linearly separable, and thus the LMS algorithm is more useful than methods that require the data to be linearly separable, even though the LMS need not yield a minimum in classification error (Chapter 5). The tree at the bottom can be simplified by methods outlined in Section 8.4.2.

the leafs because fewer training samples are involved. Recall can remain quite fast because the linear functions at each node can be computed rapidly.

8.3.9 Priors and Costs

Up to now we have tacitly assumed that a category ω_i is represented with the same frequency in both the training and the test data. If this is not the case, we need a method for controlling tree creation so as to have lower error on the actual final classification task when the frequencies are different. The most direct method is to "weight" samples to correct for the prior frequencies (Problem 16). Furthermore, we may seek to minimize a general cost, rather than a strict misclassification or 0–1 cost. As in Chapter 2, we represent such information in a cost matrix λ_{ij}—the cost of classifying a pattern as ω_i when it is actually ω_j. Cost information is easily incorporated into a Gini impurity, giving the following weighted Gini impurity,

WEIGHTED GINI
IMPURITY

$$i(N) = \sum_{ij} \lambda_{ij} P(\omega_i) P(\omega_j), \tag{10}$$

which should be used during training. Costs can be incorporated into other impurity measures as well (Problem 11).

8.3.10 Missing Attributes

Classification problems might have missing attributes during training, during classification, or both. Consider first training a tree classifier despite the fact that some training patterns are missing attributes. A naive approach would be to delete from consideration any such *deficient patterns*; however, this is quite wasteful and should be employed only if there are many complete patterns. A better technique is to proceed as otherwise described above (Section 8.3.2), but instead calculate impurities at a node N using only the attribute information present. Suppose there are n training points at N and that each has three attributes, except one pattern that is missing attribute x_3. To find the best split at N, we calculate possible splits using all n points using attribute x_1, then all n points for attribute x_2, then the $n-1$ nondeficient points for attribute x_3. Each such split has an associated reduction in impurity, calculated as before, though here with different numbers of patterns. As always, the desired split is the one which gives the greatest decrease in impurity. The generalization of this procedure to more features, to multiple patterns with missing attributes, and even to patterns with several missing attributes is straightforward, as is its use in classifying nondeficient patterns (Problem 14).

DEFICIENT
PATTERN

Now consider how to create and use trees that can *classify* a deficient pattern. The trees described above cannot directly handle test patterns lacking attributes (but see Section 8.4.2), and thus if we suspect that such deficient test patterns will occur, we must modify the training procedure discussed in Section 8.3.2. The basic approach during classification is to use the traditional ("primary") decision at a node whenever possible (i.e., when the queries involve a feature that is present in the deficient test pattern) but to use alternative queries whenever the test pattern is missing that feature.

During training then, in addition to the primary split, each nonterminal node N is given an ordered set of *surrogate splits*, consisting of an attribute label and a rule. The first such surrogate split maximizes the "predictive association" with the primary split. A simple measure of the predictive association of two splits s_1 and s_2 is merely the numerical count of patterns that are sent to the "left" by both s_1 and s_2 plus the

SURROGATE
SPLIT
PREDICTIVE
ASSOCIATION

count of the patterns sent to the "right" by both the splits. The second surrogate split is defined similarly, being the one that uses another feature and best approximates the primary split in this way. Of course, during classification of a deficient test pattern, we use the first surrogate split that does not involve the test pattern's missing attributes. This missing value strategy corresponds to a linear model replacing the pattern's missing value by the value of the nonmissing attribute most strongly correlated with it. This strategy uses to maximum advantage the (local) associations among the attributes to decide the split when attribute values are missing. A method

VIRTUAL VALUE   closely related to surrogate splits is that of *virtual values*, in which the missing attribute is assigned its most likely value.

EXAMPLE 2   Surrogate Splits and Missing Attributes

Consider the creation of a monothetic tree using an entropy impurity and the following ten training points. Because the tree will be used to classify test patterns with missing features, we will give each node surrogate splits.

$$\omega_1: \quad \begin{matrix} \mathbf{x}_1 \\ \begin{pmatrix} 0 \\ 7 \\ 8 \end{pmatrix} \end{matrix}, \quad \begin{matrix} \mathbf{x}_2 \\ \begin{pmatrix} 1 \\ 8 \\ 9 \end{pmatrix} \end{matrix}, \quad \begin{matrix} \mathbf{x}_3 \\ \begin{pmatrix} 2 \\ 9 \\ 0 \end{pmatrix} \end{matrix}, \quad \begin{matrix} \mathbf{x}_4 \\ \begin{pmatrix} 4 \\ 1 \\ 1 \end{pmatrix} \end{matrix}, \quad \begin{matrix} \mathbf{x}_5 \\ \begin{pmatrix} 5 \\ 2 \\ 2 \end{pmatrix} \end{matrix}$$

$$\omega_2: \quad \begin{matrix} \mathbf{y}_1 \\ \begin{pmatrix} 3 \\ 3 \\ 3 \end{pmatrix} \end{matrix}, \quad \begin{matrix} \mathbf{y}_2 \\ \begin{pmatrix} 6 \\ 0 \\ 4 \end{pmatrix} \end{matrix}, \quad \begin{matrix} \mathbf{y}_3 \\ \begin{pmatrix} 7 \\ 4 \\ 5 \end{pmatrix} \end{matrix}, \quad \begin{matrix} \mathbf{y}_4 \\ \begin{pmatrix} 8 \\ 5 \\ 6 \end{pmatrix} \end{matrix}, \quad \begin{matrix} \mathbf{y}_5 \\ \begin{pmatrix} 9 \\ 6 \\ 7 \end{pmatrix} \end{matrix}.$$

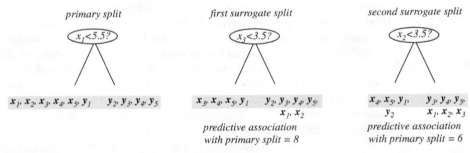

primary split   *first surrogate split*   *second surrogate split*

$x_1 < 5.5?$ $x_3 < 3.5?$ $x_2 < 3.5?$

$x_1, x_2, x_3, x_4, x_5, y_1$ y_2, y_3, y_4, y_5 x_3, x_4, x_5, y_1 $y_2, y_3, y_4, y_5,$ x_4, x_5, y_1 $y_3, y_4, y_5,$

                                                        x_1, x_2 y_2 x_1, x_2, x_3

                              predictive association   *predictive association*
                              with primary split = 8   *with primary split = 6*

Of all possible splits based on a single feature, the primary split, "Is $x_1 < 5.5$?", minimizes the entropy impurity of the full training set. The first surrogate split at the root node must use a feature other than x_1; its threshold is set in order to best approximate the action of the primary split. In this case "Is $x_3 < 3.5$?" is the first surrogate split. Likewise, here the second surrogate split must use the x_2 feature; its threshold is chosen to best approximate the action of the primary split. In this case "Is $x_2 < 3.5$?" is the second surrogate split. The pink shaded band marks those patterns sent to the matching direction as the primary split. The number of patterns in the shading is thus the predictive association with the primary split.

During classification, any test pattern containing feature x_1 would be queried using the primary split, "$x_1 \leq 5.5$?" Consider though the deficient test pattern $(*, 2, 4)^t$, where $*$ indicates that the x_1 feature is missing. Since the primary split cannot be used, we turn instead to the first surrogate split, "Is $x_3 < 3.5$?", which sends this point to the right. Likewise, the test pattern $(*, 2, *)^t$ would be queried by the second surrogate split, "Is $x_2 < 3.5$?", and sent to the left.

Through exhaustive search along all three features, we find the primary split at the root node should be "Is $x_1 < 5.5$?", which sends $\{\mathbf{x}_1, \mathbf{x}_2, \mathbf{x}_3, \mathbf{x}_4, \mathbf{x}_5, \mathbf{y}_1\}$ to the left and $\{\mathbf{y}_2, \mathbf{y}_3, \mathbf{y}_4, \mathbf{y}_5\}$ to the right, as shown in the figure.

We now seek the first surrogate split at the root node; such a split must be based on either the x_2 or the x_3 feature.

Through exhaustive search we find that the split "Is $x_3 < 3.5$?" has the highest predictive association with the primary split—a value of 8, because 8 patterns are sent to matching directions by each rule, as shown in the figure. The second surrogate split must be along the only remaining feature, x_2. We find that for this feature the rule "Is $x_2 < 3.5$?" has the highest predictive association with the primary split, a value of 6. (This, incidentally, is not the optimal x_2 split for impurity reduction—we use it because it best approximates the preferred, primary split.) While the above describes the training of the root node, training of other nodes is conceptually the same, though computationally less complex because fewer points need be considered.

Sometimes the fact that an attribute is missing can be informative. For instance, in medical diagnosis, the fact that an attribute (such as blood sugar level) is missing might imply that the physician had some reason not to measure it. As such, a missing attribute could be represented as a new feature and could be used in classification.

8.4 OTHER TREE METHODS

Virtually all tree-based classification techniques can incorporate the fundamental techniques described above. In fact, our discussion expanded beyond the core ideas in the earliest presentations of CART. While most tree-growing algorithms use an entropy impurity, there are many choices for stopping rules, for pruning methods, and for the treatment of missing attributes. Here we discuss just two other popular tree algorithms.

8.4.1 ID3

ID3 received its name because it was the third in a series of "interactive dichotomizer" procedures. It is intended for use with nominal (unordered) inputs only. If the problem involves real-valued variables, they are first binned into intervals, each interval being treated as an unordered nominal attribute. Every split has a branching factor B_j, where B_j is the number of discrete attribute bins of the variable j chosen for splitting. In practice these are seldom binary and thus a gain ratio impurity should be used (Section 8.3.2). Such trees have their number of levels equal to the number of input variables. The algorithm continues until all nodes are pure or there are no more variables on which to split. While there is thus no pruning in standard presentations of the ID3 algorithm, it is straightforward to incorporate pruning along the ideas presented above (Computer exercise 3).

8.4.2 C4.5

The C4.5 algorithm, the successor and refinement of ID3, is the most popular in a series of "classification" tree methods. In it, real-valued variables are treated the same as in CART. Multiway ($B > 2$) splits are used with nominal data, as in ID3 with a gain ratio impurity based on Eq. 7. The algorithm uses heuristics for pruning based on the statistical significance of splits.

A clear difference between C4.5 and CART involves classifying patterns with missing features. During training there are no special accommodations for subsequent classification of deficient patterns in C4.5; in particular, there are no surrogate splits precomputed. Instead, if node N with branching factor B queries the missing feature in a deficient test pattern, C4.5 follows *all* B possible answers to the descendent nodes and ultimately B leaf nodes. The final classification is based on the labels of the B leaf nodes, weighted by the decision probabilities at N. (These probabilities are simply those of decisions at N on the training data.) Each of N's immediate descendent nodes can be considered the root of a subtree implementing part of the full classification model. This missing-attribute scheme corresponds to weighting these submodels by the probability *any* training pattern at N would go to the corresponding outcome of the decision. This method does not exploit statistical correlations between different features of the training points, whereas the method of surrogate splits in CART does. Because C4.5 does not compute surrogate splits and hence does not need to store them, this algorithm may be preferred over CART if space complexity (storage) is a major concern.

The C4.5 algorithm has the provision for pruning based on the rules derived from the learned tree. Each leaf node has an associated rule—the conjunction of the decisions leading from the root node, through the tree, to that leaf. A technique called C4.5*Rules* deletes redundant antecedents in such rules. To understand this, consider the leftmost leaf in the tree at the bottom of Fig. 8.6, which corresponds to the rule

C4.5RULES

$$IF\big[\qquad (0.04x_1 + 0.16x_2 < 0.11)$$
$$AND\ (0.27x_1 - 0.44x_2 < -0.02)$$
$$AND\ (0.96x_1 - 1.77x_2 < -0.45)$$
$$AND(5.43x_1 - 13.33x_2 < -6.03)\big]$$
$$THEN \qquad \mathbf{x} \in \omega_1.$$

This rule can be simplified to give

$$IF\big[\qquad (0.04x_1 + 0.16x_2 < 0.11)$$
$$AND(5.43x_1 - 13.33x_2 < -6.03)\big]$$
$$THEN\ \mathbf{x} \in \omega_1,$$

as should be evident in that figure. Note especially that information corresponding to nodes near the root can be pruned by C4.5Rules. This is more general than impurity based pruning methods, which instead merge leaf nodes.

8.4.3 Which Tree Classifier Is Best?

In Chapter 9 we shall consider some general issues associated with comparing different classifiers. Here, rather than directly comparing typical implementations of CART, ID3, C4.5, and other numerous tree methods, it is more instructive to consider variations within the different component steps. After all, with care one can generate a tree using any reasonable feature processing, impurity measure, stopping criterion, or pruning method. Many of the basic principles applicable throughout pattern classification guide us here. Of course, if the designer has insight into feature preprocessing, this should be exploited. The binning of real-valued features used in early versions of ID3 does not take full advantage of order information, and thus ID3 should be applied to such data only if computational costs are otherwise too high. It has been found that an entropy impurity works acceptably in most cases and is a natural default. In general, pruning is to be preferred over stopped training and cross-validation because it takes advantage of more of the information in the training set; however, pruning large training sets can be computationally expensive. The pruning of rules is less useful for problems that have high noise and are statistical in nature, but such pruning can often simplify classifiers for problems where the data were generated by rules themselves. Likewise, decision trees are poor at inferring simple concepts—for instance, whether more than half of the binary (discrete) attributes have value +1. As with most classification methods, one gains expertise and insight through experimentation on a wide range of problems. No single tree algorithm dominates or is dominated by others.

In a wide range of applications, trees have yielded classifiers with accuracy comparable to other methods we have discussed, such as neural networks and nearest-neighbor classifiers, especially when specific prior information about the appropriate form of classifier is lacking. Tree-based classifiers are particularly useful with nonmetric data, and as such they are an important tool in pattern recognition research.

*8.5 RECOGNITION WITH STRINGS

Suppose the patterns are represented as ordered sequences or *strings* of discrete items, as in a sequence of letters in an English word or in DNA bases in a gene sequence, such as "AGCTTCGAATC." (The letters A, G, C, and T stand for the nucleic acids adenine, guanine, cytosine, and thymine.) Pattern classification based on such strings of discrete symbols differs in a number of ways from the more commonly used techniques we have addressed up to here. Because the string elements—called

CHARACTER

characters, letters or symbols—are nominal, there is no obvious notion of distance between strings. There is a further difficulty arising from the fact that strings need not be of the same length. While such strings are surely not vectors, we nevertheless broaden our familiar boldface notation to now apply to strings as well (e.g., $\mathbf{x} = $ "AGCTTC"), though we will often refer to them as patterns, strings, templates, or

WORD

general *words*. (Of course, there is no requirement that these be meaningful words in a natural language such as English or French.) A particularly long string is denoted

TEXT

text. Any contiguous string that is part of $\mathbf{x}$ is called a substring, segment, or more

FACTOR

frequently a *factor* of $\mathbf{x}$. For example, "GCT" is a factor of "AGCTTC."

There is a large number of problems in computations on strings. The ones that are of greatest importance in pattern recognition are as follows:

String Matching: Given $\mathbf{x}$ and *text*, determine whether $\mathbf{x}$ is a factor of *text*, and, if so, where it appears.

Edit Distance: Given two strings $\mathbf{x}$ and $\mathbf{y}$, compute the minimum number of basic operations—character insertions, deletions and exchanges—needed to transform $\mathbf{x}$ into $\mathbf{y}$.

String Matching with Errors: Given $\mathbf{x}$ and *text*, find the locations in *text* where the "cost" or "distance" of $\mathbf{x}$ to any factor of *text* is minimal.

String Matching with the "Don't–Care" Symbol: This is the same as basic string

DON'T CARE
SYMBOL

matching, but with a special symbol—∅, the *don't-care* symbol—which can match any other symbol.

We should begin by understanding the several ways in which these string operations are used in pattern classification. Basic string matching can be viewed as an extreme case of template matching, as in finding a particular English word within a large electronic corpus such as a novel or digital repository. Alternatively, suppose we have a large text such as Herman Melville's **Moby Dick**, and we want to classify it as either most relevant to the topic of fish or to the topic of hunting. Test strings or *keywords* for the fish topic might include "salmon," "whale," "fishing," "ocean," while those for hunting might include "gun," "bullet," "shoot," and so on. String matching would determine the number of occurrences of such keywords in the text. A simple count of the keyword occurrences could then be used to classify the text according to topic. (Other, more sophisticated methods for this latter stage would generally be preferable.)

The problem of string matching with the don't-care symbol is closely related to standard string matching, even though the best algorithms for the two types of problems differ, as we shall see. Suppose, for instance, that in DNA sequence analysis we have a segment of DNA, such as $\mathbf{x} =$ "AGCCG∅∅∅∅∅GACTG," where the first and last sections (called motifs) are important for coding a protein while the middle section, which consists of five characters, is nevertheless known to be inert and to have no function. If we are given an extremely long DNA sequence (the *text*), string matching with the don't care symbol using the pattern $\mathbf{x}$ containing $\varnothing$ symbols would determine if *text* is in the class of sequences that could yield the particular protein.

The string operation that finds greatest use in pattern classification is based on edit distance, and it is best understood in terms of the nearest-neighbor algorithm (Chapter 4). Recall that in that algorithm each training pattern or prototype is stored along with its category label; an unknown test pattern is then classified by its nearest prototype. Suppose now that the prototypes are strings and we seek to classify a novel test string by its "nearest" stored string. For instance an acoustic speech recognizer might label every 10-ms interval with the most likely phoneme present in an utterance, giving a string of discrete phoneme labels such as "tttoooonn." Edit distance would then be used to find the "nearest" stored training pattern, so that its category label can be read.

The difficulty in this approach is that there is no obvious notion of metric or distance between strings. In order to proceed, then, we must introduce some measure of distance between the strings. The resulting *edit distance* is the minimum number of fundamental operations needed to transform the test string into a prototype string, as we shall see.

The string-matching-with-errors problem contains aspects of both the basic string matching and the edit distance problems. The goal is to find all locations in *text* where $\mathbf{x}$ is "close" to the substring or factor of *text*. This measure of closeness is chosen to be an edit distance. Thus the string-matching-with-errors problem appears in the same types of problems as basic string matching, the only difference being that there is a certain "tolerance" for a match. String matching with errors finds use, for example, in searching digital texts for possibly misspelled versions of a given target word.

Naturally, deciding *which* strings to consider is highly problem-dependent. Nevertheless, given target strings and the relevance of tolerances, and so on, the string matching problems just outlined are conceptually very simple; the challenge arises when the problems are large, such as searching for a segment within the roughly 3×10^9 base pairs in the human genome, the 3×10^7 characters in an electronic version of **War and Peace**, or the more than 10^{13} characters in a very large digital repository. For such cases, the effort is in finding tricks and heuristics that make the problem computationally tractable.

We now consider these four string operations in greater detail.

8.5.1 String Matching

The most fundamental and useful operation in string matching is testing whether a candidate string $\mathbf{x}$ is a factor of *text*. Naturally we assume that the number of characters in *text*, denoted length[*text*] or $|text|$, is greater than that in $\mathbf{x}$, and for most computationally interesting cases $|text| \gg |\mathbf{x}|$. Each discrete character is taken from an *alphabet* $\mathcal{A}$—for example, binary or decimal numerals, the English letters, or four DNA bases, that is, $\mathcal{A} = \{0, 1\}$ or $\{0,1,2,\ldots,9\}$ or $\{a,b,c,\ldots,z\}$ or $\{A,G,C,T\}$, respectively. A *shift*, s, is an offset needed to align the first character of $\mathbf{x}$ with character number $s+1$ in *text*. The basic string matching problem is to find whether there exists a *valid shift*, that is, one where there is a perfect match between each character in $\mathbf{x}$ and the corresponding one in *text*. The general string-matching problem is to list *all* valid shifts (Fig. 8.7).

ALPHABET

SHIFT

VALID SHIFT

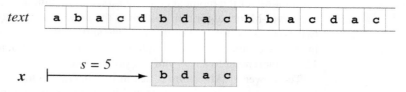

FIGURE 8.7. The general string-matching problem is to find all shifts s for which the pattern $\mathbf{x}$ appears in *text*. Any such shift is called *valid*. In this case $\mathbf{x}$ = "bdac" is indeed a factor of *text*, and $s = 5$ is the only valid shift.

The most straightforward approach in string matching is to test each possible shift s in turn, as given in the naive string-matching algorithm.

■ **Algorithm 1.   (Naive String Matching)**

1 **begin initialize** $\mathcal{A}$, $\mathbf{x}$, *text*, $n \leftarrow$ length[*text*], $m \leftarrow$ length[$\mathbf{x}$]
2         $s \leftarrow 0$
3   **while** $s \leq n - m$
4   **if** $\mathbf{x}[1 \ldots m] = text[s+1 \ldots s+m]$
5   **then** print "pattern occurs at shift" s
6                     $s \leftarrow s + 1$
7   **return**
8 **end**

Algorithm 1 is hardly optimal; it takes time $\Theta((n - m + 1)m)$ in the worst case; if $\mathbf{x}$ and *text* are random, however, the algorithm is efficient (Problem 18). The weakness in the naive string-matching algorithm is that information from one candidate shift s is not exploited when seeking a subsequent candidate shift. A more sophisticated method, the Boyer-Moore algorithm, uses such information in a clever way.

■ **Algorithm 2.** **(Boyer-Moore String Matching)**

1 **begin initialize** $\mathcal{A}$, $\mathbf{x}$, *text*, $n \leftarrow$ length[*text*], $m \leftarrow$ length[$\mathbf{x}$]
2          $\mathcal{F}(\mathbf{x}) \leftarrow$ last-occurrence function
3          $\mathcal{G}(\mathbf{x}) \leftarrow$ good-suffix function
4          $s \leftarrow 0$
5   **while** $s \leq n - m$
6   **do** $j \leftarrow m$
7   **while** $j > 0$ and $\mathbf{x}[j] = text[s + j]$
8   **do** $j \leftarrow j - 1$
9   **if** $j = 0$
10   **then** print "pattern occurs at shift" s
11                          $s \leftarrow s + \mathcal{G}(0)$
12   **else** $s \leftarrow s + \max[\mathcal{G}(j), j - \mathcal{F}(text[s + j])]$
13   **return**
14 **end**

Postponing for the moment considerations of the functions $\mathcal{F}$ and $\mathcal{G}$, we can see that the Boyer-Moore algorithm resembles the naive string-matching algorithm, but with two exceptions. First, at each candidate shift s, the character comparisons are done in reverse order, that is, from right to left (line 8). Second, according to lines 11 and 12, the increment to a new shift apparently need not be 1.

The power of Algorithm 2 lies in two heuristics that allow it to skip the examination of a large number shifts and hence character comparisons: The *good-suffix heuristic* and the *bad-character heuristic* operate independently and in parallel. After a mismatch is detected, each heuristic proposes an amount by which s can be safely increased without missing a valid shift; the larger of these proposed shifts is selected and s is increased accordingly.

BAD-CHARACTER
HEURISTIC
The *bad-character heuristic* utilizes the rightmost character in *text* that does not match the aligned character in $\mathbf{x}$. Because character comparisons proceed right-to-left, this "bad character" is found as efficiently as possible. Because the current shift s is invalid, no more character comparisons are needed and a shift increment can be made. The bad-character heuristic proposes incrementing the shift by an amount to align the rightmost occurrence of the bad character in $\mathbf{x}$ with the bad character identified in *text*. This guarantees that no valid shifts have been skipped (Fig. 8.8).

GOOD-SUFFIX
HEURISTIC
SUFFIX
PREFIX

GOOD SUFFIX
Now consider the *good-suffix heuristic*, which operates in parallel with the bad-character heuristic, and also proposes a safe shift increment. A general *suffix* of $\mathbf{x}$ is a factor or substring of $\mathbf{x}$ that contains the final character in $\mathbf{x}$. (Likewise, a *prefix* contains the initial character in $\mathbf{x}$.) At shift s the rightmost contiguous characters in *text* that match those in $\mathbf{x}$ are called the *good suffix*, or "matching suffix." As before, because character comparisons are made right-to-left, the good suffix is found with the minimum number of comparisons. Once a character mismatch has been found, the good-suffix heuristic proposes to increment the shift so as to align the next occurrence of the good suffix in $\mathbf{x}$ with that identified in *text*. This ensures that no valid shift has been skipped. Given the two shift increments proposed by the bad-character and the good-suffix heuristics, line 12 of the Boyer-Moore algorithm chooses the larger.

LAST-
OCCURRENCE
FUNCTION
These heuristics rely on the functions $\mathcal{F}$ and $\mathcal{G}$. The *last-occurrence function*, $\mathcal{F}(\mathbf{x})$, is merely a table containing every letter in the alphabet and the position of its

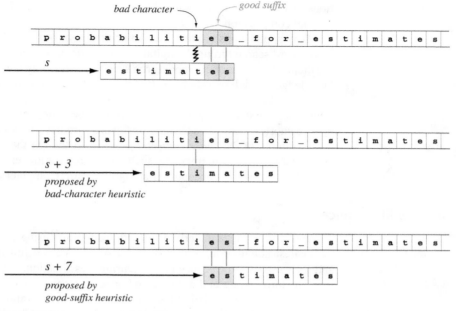

FIGURE 8.8. String matching by the Boyer-Moore algorithm takes advantage of information obtained at one shift *s* to propose the next shift; the algorithm is generally much less computationally expensive than naive string matching, which always increments shifts by a single character. The top figure shows the alignment of *text* and pattern **x** for an invalid shift *s*. Character comparisons proceed right to left, and the first two such comparisons are a match—the good suffix is "es." The first (rightmost) mismatched character in *text*, here "i," is called the *bad character*. The bad-character heuristic proposes incrementing the shift to align the rightmost "i" in **x** with the bad character "i" in *text*—a shift increment of 3, as shown in the middle figure. The bottom figure shows the effect of the good-suffix heuristic, which proposes incrementing the shift the least amount that will align the good suffix, "es" in **x**, with that in *text*—here an increment of 7. Lines 11 and 12 of the Boyer-Moore algorithm select the larger of the two proposed shift increments, i.e., 7 in this case. Although not shown in this figure, after the mismatch is detected at shift $s+7$, both the bad-character and the good-suffix heuristics propose an increment of yet another 7 characters, thereby finding a valid shift.

rightmost occurrence in **x**. For the pattern in Fig. 8.8, the table would contain: a, 6; e, 8; i, 4; m, 5; s, 9; and t, 8. All 20 other letters in the English alphabet are assigned a value 0, signifying that they do not appear in **x**. The construction of this table is simple (Problem 22) and need be done just once; it does not significantly affect the computational cost of the Boyer-Moore algorithm.

GOOD-SUFFIX
FUNCTION

The *good-suffix function*, $\mathcal{G}(\mathbf{x})$, creates a table that for each suffix gives the location of its second right-most occurrence in **x**. In the example in Fig. 8.8, the suffix s (the last character in "estimates") also occurs at position 2 in **x**. Furthermore, the suffix "es" occurs at position 1 in **x**. The suffix "tes" does not appear elsewhere in **x** and hence it, and all remaining suffixes, are assigned the value 0. In sum, then, the table of $\mathcal{G}(\mathbf{x})$ would have just two nonzero entries: s, 2 and es, 1.

In practice, these heuristics make the Boyer-Moore one of the most attractive string-matching algorithms on serial computers. Other powerful methods quickly become conceptually more involved and are generally based on precomputing functions of **x** that enable efficient shift increments, or dividing the problem for efficient parallel computation.

Many applications require a *text* to be searched for several strings, as in the case of keyword search through a digital text. Occasionally, some of these search strings are themselves factors of other search strings. Presumably we would not want to acknowledge a match of a short string if it were also part of a match for a longer string. Thus if our keywords included "beat," "eat," and "be," we would want our search to return only the string match of "beat" from *text* = "when_chris_beats_the_drum," not the shorter strings "eat" and "be," which are nevertheless "there" in *text*. This

is an example of the *subset-superset problem*. Although there may be much bookkeeping associated with imposing such a strict bias for longer sequences over shorter ones, the approach is conceptually straightforward (Computer exercise 8).

8.5.2 Edit Distance

The fundamental idea underlying pattern recognition using edit distance is based on the nearest-neighbor algorithm (Chapter 4). We store a full training set of strings and their associated category labels. During classification, a test string is compared to each stored string and a "distance" or score is computed; the test string is then assigned the category label of the "nearest" string in the training set.

Unlike the case using real-valued vectors discussed in Chapter 4, there is no single obvious measure of the similarity or difference between two strings. For instance, it is not clear whether "abbccc" is closer to "aabbcc" or to "abbcccb." To proceed, then, we introduce a measure of the difference between two strings. Such an *edit distance* between **x** and **y** describes how many fundamental operations are required to transform **x** into **y**. These fundamental operations are as follows:

Substitutions: A character in **x** is replaced by the corresponding character in **y**.

Insertions: A character in **y** is inserted into **x**, thereby increasing the length of **x** by one character.

Deletions: A character in **x** is deleted, thereby decreasing the length of **x** by one character.

INTERCHANGE Occasionally we also consider a fourth operation, *interchange*, or "twiddle," or *transposition*, which interchanges two neighboring characters in **x**. Thus, one could transform **x** = "asp" into **y** = "sap" with a single interchange. Because such an interchange can always be expressed as two substitutions, for simplicity we shall not consider the interchange operation.

Let **C** be an m-by-n matrix of integers associated with a cost or "distance" and let $\delta(\cdot, \cdot)$ denote a generalization of the Kronecker delta function, having value 1 if the two arguments (characters) match and 0 otherwise. The basic edit-distance algorithm is then as follows:

■ **Algorithm 3. (Edit Distance)**

```
1 begin initialize A, x, y, m ← length[x], n ← length[y]
2        C[0, 0] ← 0
3        i ← 0
4        do i ← i + 1
5            C[i, 0] ← i
6        until i = m
7        j ← 0
```

8   **do** $j \leftarrow j + 1$
9         $C[0, j] \leftarrow j$
10   **until** $j = n$
11     $i \leftarrow 0; \ j \leftarrow 0$
12   **do** $i \leftarrow i + 1$
13   **do** $j \leftarrow j + 1$
14

$$C[i, j] = \min\Big[\underbrace{C[i-1, j]+1}_{insertion}, \ \underbrace{C[i, j-1]+1}_{deletion}, \ \underbrace{C[i-1, j-1]+1-\delta(\mathbf{x}[i], \mathbf{y}[j])}_{no\ change/exchange} \Big]$$

15   **until** $j = n$
16   **until** $i = m$
17   **return** $C[m, n]$
18 **end**

Lines 4–10 initialize the left column and top row of $\mathbf{C}$ with the integer number of "steps" away from $i = 0, j = 0$. The core of this algorithm, line 14, finds the minimum cost in each entry of $\mathbf{C}$, column by column (Fig. 8.9). Algorithm 3 is thus greedy in that each column of the distance or cost matrix is filled using merely the costs in the previous column. Linear programming techniques can also be used to find a global minimum, though this nearly always requires greater computational effort (Problem 28).

If insertions and deletions are equally costly, then the symmetry property of a metric holds. However, we can broaden the applicability of the algorithm by allowing in line 14 different costs for the fundamental operations; for example, insertions might cost twice as much as substitutions. In such a broader case, properties of symmetry and the triangle inequality no longer hold, and edit distance is not a true metric (Problem 27).

As shown in Fig. 8.9, $\mathbf{x} =$ "excused" can be transformed to $\mathbf{y} =$ "exhausted" through one substitution and two insertions. The table shows the steps of this trans-

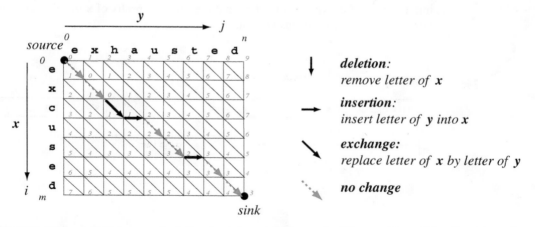

FIGURE 8.9. The edit distance calculation for strings $\mathbf{x}$ and $\mathbf{y}$ can be illustrated in a table. Algorithm 3 begins at *source*, $i = 0, j = 0$, and fills in the cost matrix $\mathbf{C}$, column by column (shown in red), until the full edit distance is placed at the *sink*, $C[i = m, j = n]$. The edit distance between "excused" and "exhausted" is thus 3.

formation, along with the computed entries of the cost matrix **C**. For the case shown, where each fundamental operation has a cost of 1, the edit distance is given by the value of the cost matrix at the sink, that is, **C**[7,9] = 3.

x	excused	source string	$C[0, 0] = 0$
	exhused	substitute h for c	$C[3, 3] = 1$
	exhaused	insert a	$C[3, 4] = 2$
	exhausted	insert t	$C[5, 7] = 3$
y	exhausted	target string	$C[7, 9] = 3$

8.5.3 Computational Complexity

Algorithm 3 is $O(mn)$ in time, of course; it is $O(m)$ in space (memory) because only the entries in the previous column need be stored when computing $C[i, j]$ for $i = 0$ to m. Because of the importance of string matching and edit distance throughout computer science, a number of algorithms have been proposed. We need not delve into the details here (but see the Bibliography) except to say that there are sophisticated string-matching algorithms with time complexity $O(m + n)$.

8.5.4 String Matching with Errors

There are several versions of the string-matching-with-errors problem; the one that concerns us is this: Given a pattern **x** and *text*, find the shift for which the edit distance between **x** and a factor of *text* is minimum. The algorithm for the string-matching-with-errors problem is very similar to that for edit distance. Let **E** be a matrix of costs, analogous to **C** in Algorithm 3. We seek a shift for which the edit distance to a factor of *text* is minimum, or formally min[**C**(**x**, **y**)] where **y** is any factor of *text*. To this end, the algorithm must compute its new cost **E** whose entries are $E[i, j] = \min[C(x[1 \ldots i], y[1 \ldots j])]$.

The principal difference between the algorithms for the two problems (i.e., with or without errors) is that we initialize E[0,j] to 0 in the string matching with errors problem, instead of to j in lines 4–10 of the basic string matching algorithm. This initialization of **E** expresses the fact that the "empty" prefix of **x** matches an empty factor of *text* and contributes no cost.

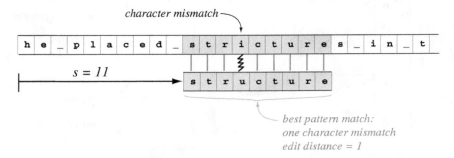

FIGURE 8.10. The string-matching-with-errors problem is to find the shift *s* for which the edit distance between **x** and an aligned factor of *text* is minimum. In this illustration, the minimum edit distance is 1, corresponding to the character exchange u → i, and the shift *s* = 11 is the location.

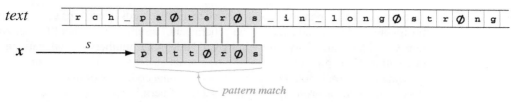

text: `r c h _ p a Ø t e r Ø s _ i n _ l o n g Ø s t r Ø n g`

x ——— *s* ——→ `p a t t Ø r Ø s`

pattern match

FIGURE 8.11. The problem of string matching with don't care symbol is the same as that in basic string matching except that the ∅ symbol—in either *text* or **x**—can match any character. The figure shows the only valid shift.

Two minor heuristics for reducing computational effort are relevant to the string-matching-with-errors problem. The first is that except in highly unusual cases, the lengths of the candidate factors of *text* that need be considered are roughly equal to length[**x**]. Second, for each candidate shift, the edit-distance calculation can be terminated if it already exceeds the current minimum. In practice, this latter heuristic can reduce the computational burden significantly. Otherwise, the algorithm for string matching with errors is virtually the same as that for edit distance.

8.5.5 String Matching with the "Don't-Care" Symbol

String matching with the "don't-care" symbol, ∅, is formally the same as basic string matching, but the ∅ in either **x** or *text* is said to match any character (Fig. 8.11). An obvious approach to string matching with the don't-care symbol is to modify the naive string-matching algorithm to include a condition for matching the don't care symbol. Such an approach, however, retains the computational inefficiencies of naive string matching (Problem 29). Furthermore, extending the Boyer-Moore algorithm to include ∅ is somewhat difficult and inefficient. The most effective methods are based on fundamental methods in computer arithmetic and, while fascinating, would take us away from our central concerns of pattern recognition (cf. Bibliography). The use of this technique in pattern recognition is the same as string matching, with a particular type of "tolerance."

While learning is a general and fundamental technique throughout pattern recognition, it has found limited use in recognition with basic string matching. This is because the designer typically knows precisely which strings are being sought—they do not need to be learned. Learning can, of course, be based on the *outputs* of a string-matching algorithm, as part of a larger pattern recognition system.

8.6 GRAMMATICAL METHODS

Up to here, we have not been concerned with any detailed models that might underlie the generation of the sequence of characters in a string. We now turn to the case where rules of a particular sort were used to generate the strings and thus where their structure is fundamental. Often this structure is hierarchical, where at the highest or most abstract level a sequence is very simple, but at subsequent levels there is greater and greater complexity. For instance, at its most abstract level, the string "The history book clearly describes several wars" is merely a sentence. At a somewhat more detailed level it can be described as a noun phrase followed by a verb phrase. The noun phrase can be expanded at yet a subsequent level, as can the verb phrase. The expansion ends when we reach the words "The," "history," and so forth—items that are considered the "characters," atomic and without further structure. Consider also strings representing valid telephone numbers—local, national and international. Such numbers conform to a strict structure: Either a country code is present or it is not; if not, then the domestic national code may or may not be present; if a country code is present, then there is a set of permissible city codes and for each city there is a set of permissible area codes and individual local numbers, and so on.

As we shall see, such structure is easily specified in a *grammar*, and when such structure is present, the use of a grammar for recognition can improve accuracy. For instance, grammatical methods can be used to provide constraints for a full system that uses a statistical recognizer as a component. Consider an optical character recognition system that recognizes and interprets mathematical equations based on a scanned pixel image. The mathematical symbols often have specific "slots" that can be filled with certain other symbols; this can be specified by a grammar. Thus an integral sign has two slots, for upper and lower limits, and these can be filled by only a limited set of symbols. (Indeed, a grammar is used in many mathematical typesetting programs in order to prevent authors from creating meaningless "equations.") A full system that recognizes the integral sign could use a grammar to limit the number of candidate categories for a particular slot, and this increases the accuracy of the full system. Similarly, consider the problem of recognizing phone numbers within acoustic speech in an automatic dialing application. A statistical or Hidden-Markov-Model acoustic recognizer might perform word spotting and pick out number words such as "eight" and "hundred." A subsequent stage based on a formal grammar would then exploit the fact that telephone numbers are highly constrained, as mentioned.

We shall study the case when crisp rules specify how the representation at one level leads to a more expanded and complicated representation at the next level. We sometimes call a string generated by a set of rules a *sentence*; the rules are specified by a grammar, denoted G. (Naturally, there is no requirement that these be related in any way to sentences in natural language such as English.) In pattern recognition, we are given a sentence and a grammar and seek to determine whether the sentence was generated by G.

SENTENCE

8.6.1 Grammars

The notion of a grammar is very general and powerful. Formally, a grammar G consists of four components:

Symbols: Every sentence consists of a string of characters (which are also called primitive symbols, terminal symbols or letters), taken from an alphabet $\mathcal{A}$. For bookkeeping, it is also convenient to include the *null* or *empty string* denoted ϵ, which has length zero; if ϵ is appended to any string $\mathbf{x}$, the result is again $\mathbf{x}$.

NULL STRING

Variables: These are also called nonterminal symbols, intermediate symbols, or occasionally internal symbols, and are taken from a set $\mathcal{I}$.

Root Symbol: The *root symbol* or starting symbol is a special internal symbol, the source from which all sequences are derived. The root symbol is taken from a set $\mathcal{S}$.

ROOT SYMBOL

Productions: The set of *production rules*, rewrite rules, or simply rules, denoted $\mathcal{P}$, specifies how to transform a set of variables and symbols into other variables and symbols. These rules determine the core structures that can be produced by the grammar. For instance if A is an internal symbol and c a terminal symbol, the

PRODUCTION RULE

rewrite rule cA → cc means that any time the segment cA appears in a string, it can be replaced by cc.

Thus we denote a general grammar by its alphabet, its variables, its particular root symbol, and the rewrite rules: $G = (\mathcal{A}, \mathcal{I}, \mathcal{S}, \mathcal{P})$. The *language* generated by grammar, denoted $\mathcal{L}(G)$, is the set of all strings (possibly infinite in number) that can be generated by G.

Consider two examples; the first is quite simple and abstract. Let $\mathcal{A} = \{a, b, c\}$, $\mathcal{S} = S, \mathcal{I} = \{A, B, C\}$, and

$$\mathcal{P} = \left\{ \begin{array}{ll} \mathbf{p}_1: & S \rightarrow aSBA \text{ OR } aBA \quad \mathbf{p}_2: \ AB \rightarrow BA \\ \mathbf{p}_3: & bB \rightarrow bb \qquad\qquad\qquad \mathbf{p}_4: \ bA \rightarrow bc \\ \mathbf{p}_5: & cA \rightarrow cc \qquad\qquad\qquad \mathbf{p}_6: \ aB \rightarrow ab \end{array} \right\}.$$

(In order to make the list of rewrite rules more compact, we shall condense rules having the same left-hand side by means of the *OR* on the right-hand side. Thus rule $\mathbf{p}_1$ is a condensation of the two rules $S \rightarrow aSBA$ and $S \rightarrow aBA$.) If we start with S and apply the rewrite rules in the following orders, we have the following two cases:

root	S
$\mathbf{p}_1$	a$B$$A$
$\mathbf{p}_6$	ab$A$
$\mathbf{p}_4$	abc

root	S
$\mathbf{p}_1$	aSBA
$\mathbf{p}_1$	aa$BABA$
$\mathbf{p}_6$	aabABA
$\mathbf{p}_2$	aabBAA
$\mathbf{p}_3$	aabbAA
$\mathbf{p}_4$	aabbc$A$
$\mathbf{p}_5$	aabbcc

After the rewrite rules have been applied in these sequences, no more symbols match the left-hand side of any rewrite rule, and the process is complete. Such a transformation from the root symbol to a final string is called a *production*. These two productions show that abc and aabbcc are in $\mathcal{L}(G)$, the language generated by G. In fact, it can be shown (Problem 38) that this grammar generates the language $\mathcal{L}(G) = \{a^n b^n c^n | n \geq 1\}$.

A much more complicated grammar underlies the English language, of course. The alphabet consists of all English words, $\mathcal{A} = \{$the, history, book, sold, over, 1000, copies, ...$\}$, and the intermediate symbols are the parts of speech: $\mathcal{I}$ = {⟨*noun*⟩, ⟨*verb*⟩, ⟨*noun phrase*⟩, ⟨*verb phrase*⟩, ⟨*adjective*⟩, ⟨*adverb*⟩, ⟨*adverbial phrase*⟩}. The root symbol here is $\mathcal{S} = $ ⟨*sentence*⟩. A restricted set of the production rules in English includes:

$$\mathcal{P} = \left\{ \begin{array}{l} \langle sentence \rangle \rightarrow \langle noun\ phrase \rangle \langle verb\ phrase \rangle \\ \langle noun\ phrase \rangle \rightarrow \langle adjective \rangle \langle noun\ phrase \rangle \\ \langle verb\ phrase \rangle \rightarrow \langle verb\ phrase \rangle \langle adverbial\ phrase \rangle \\ \langle noun \rangle \rightarrow \text{book } OR \text{ theorem } OR \ ... \\ \langle verb \rangle \rightarrow \text{describes } OR \text{ buys } OR \text{ holds } OR \ ... \\ \langle adverb \rangle \rightarrow \text{over } OR \ ... \end{array} \right\}.$$

This subset of the rules of English grammar does not prevent the generation of meaningless sentences, of course. For instance, the nonsense sentence "Squishy green dreams hop heuristically" can be derived in this subset of English grammar.

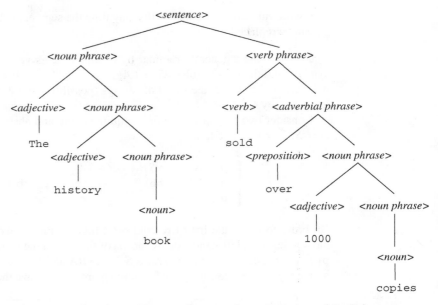

FIGURE 8.12. This derivation tree illustrates how a portion of English grammar can transform the root symbol, here ⟨*sentence*⟩, into a particular sentence or string of elements, here English words, which are read from left to right.

DERIVATION
TREE

Figure 8.12 shows the steps of a production in a *derivation tree*, where the root symbol is displayed at the top and the terminal symbols are shown at the bottom.

8.6.2 Types of String Grammars

There are four main types of grammar, arising from different types of structure in the productions. As we have seen, a rewrite rule is of the form $\alpha \rightarrow \beta$, where α and β are strings made up of intermediate and terminal symbols.

Type 0: Free or Unrestricted. Free grammars have no restrictions on the rewrite rules and thus they provide no constraints or structure on the strings they can produce. While in principle they can express an arbitrary set of rules, this generality comes at the tremendous expense of possibly unbounded learning time. Knowing that a string is derived from a type 0 grammar provides no information, and, as such, type 0 grammars have found but little use in pattern recognition.

Type 1: Context-Sensitive. A grammar is called context-sensitive if every rewrite rule is of the form

$$\alpha I \beta \rightarrow \alpha \gamma \beta$$

where α and β are any strings made up of intermediate and terminal symbols, I is an intermediate symbol, and γ is a string made up of intermediate or terminal symbols (other than ϵ). We say that "I can be rewritten as γ in the context of α on the left and β on the right."

Type 2: Context-Free. A grammar is called context free if every production is of the form

$$I \rightarrow \gamma$$

where I is an intermediate symbol and γ is a string made up of intermediate or terminal symbols (other than ϵ). Clearly, unlike a type 1 grammar, here there is no need for a "context" for the rewriting of I by γ.

Type 3: Finite State or Regular. A grammar is called regular if every rewrite rule is of the form

$$\alpha \rightarrow z\beta \qquad OR \qquad \alpha \rightarrow z$$

where α and β are made up of intermediate symbols and z is a terminal symbol (other than ϵ). Such grammars are also called finite state because they can be generated by a finite-state machine, which we shall see in Fig. 8.16

A language generated by a grammar of type i is called a type i language. It can be shown that the class of grammars of type i includes all grammars of type $i + 1$; thus there is a strict hierarchy in grammars.

CHOMSKY
NORMAL FORM

Any context-free grammar can be converted into one in *Chomsky normal form* (CNF). Such a grammar has all rules of the form

$$A \rightarrow BC \qquad \text{and} \qquad A \rightarrow z$$

where A, B, and C are intermediate symbols (i.e., they are in $\mathcal{I}$) and z is a terminal symbol. For every context-free grammar G, there is another G' in Chomsky normal form such that $\mathcal{L}(G) = \mathcal{L}(G')$, as explored in Problem 36.

EXAMPLE 3 A Grammar for Pronouncing Numbers

In order to understand these issues better, consider a grammar that yields pronunciation of any number between 1 and 999,999. The alphabet has 29 basic terminal symbols—that is, the spoken words $\mathcal{A} = \{one, two, \ldots, ten, eleven, \ldots, twenty, thirty, \ldots, ninety, hundred, thousand\}$.

There are six intermediate symbols, corresponding to general six-digit, three-digit, and two-digit numbers, the numbers between ten and nineteen, and so forth, as will be clear below:

$\mathcal{I} = \{digits6, digits3, digits2, digit1, teens, tys\}$.

The root node corresponds to a general number up to six digits in length:

$\mathcal{S} = digits6$.

The set of rewrite rules is based on a knowledge of English:

$$\mathcal{P} = \left\{ \begin{array}{l} digits6 \rightarrow digits3 \ thousand \ digits3 \\ digits6 \rightarrow digits3 \ thousand \ OR \ digits3 \\ digits3 \rightarrow digit1 \ hundred \ digits2 \\ digits3 \rightarrow digit1 \ hundred \ OR \ digits2 \\ digits2 \rightarrow teens \ OR \ tys \ OR \ tys \ digit1 \ OR \ digit1 \\ digit1 \rightarrow one \ OR \ two \ OR \ \ldots \ nine \\ teens \rightarrow ten \ OR \ eleven \ OR \ \ldots \ nineteen \\ tys \rightarrow twenty \ OR \ thirty \ OR \ \ldots \ OR \ ninety \end{array} \right\}$$

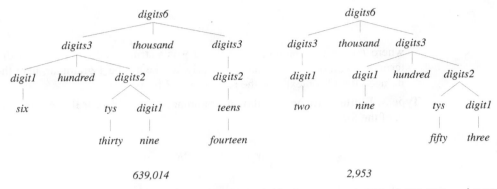

These two derivation trees show how the grammar G yields the pronunciation of 639,014 and 2,953. The final strings of terminal symbols are read from left to right.

This grammar takes *digits*6 and applies the productions until the elements in the final alphabet are produced, as shown in the figure. Because it contains rewrite rules such as *digits*6 → *digits*3 *thousand*, this grammar cannot be type 3. It is easy to confirm that it is type 2.

8.6.3 Recognition Using Grammars

Recognition using grammars is formally very similar to the general approaches used throughout pattern recognition. Suppose we suspect that a test sentence was generated by one of c different grammars, $G_1, G_2, \ldots, G_c$, which can be considered as different models or classes. A test sentence $\mathbf{x}$ is classified according to which grammar could have produced it, or, equivalently, the language $\mathcal{L}(G_i)$ of which $\mathbf{x}$ is a member.

Up to now we have worked forward—forming a derivation from a root node to a final sentence. For recognition, though, we must employ the inverse process: That is, given a particular $\mathbf{x}$, find a derivation in G that leads to $\mathbf{x}$. This process, called

PARSING

parsing, is virtually always much more difficult than forming a derivation. We now discuss one general approach to parsing, and we briefly mention two others.

Bottom-Up Parsing

Bottom-up parsing starts with the test sentence $\mathbf{x}$, and seeks to simplify it, so as to represent it as the root symbol. The basic approach is to use candidate productions from $\mathcal{P}$ "backwards," that is, find rewrite rules whose right hand side matches part of the current string, and replace that part with a segment that could have produced it. This is the general method in the Cocke-Younger-Kasami algorithm, which fills

PARSE TABLE

a *parse table* from the "bottom up." The grammar must be expressed in Chomsky normal form and thus the productions $\mathcal{P}$ must all be of the form $A \rightarrow BC$, a broad but not all inclusive category of grammars. Entries in the table are candidate strings in a portion of a valid derivation. If the table contains the source symbol S, then indeed we can work forward from S and derive the test sentence, and hence $\mathbf{x} \in \mathcal{L}(G)$. In the following, x_i (for $i = 1, \ldots n$) represents the individual terminal characters in the string to be parsed.

■ **Algorithm 4.   (Bottom-Up Parsing)**

```
 1  begin initialize G = (A, I, S, P), x = x₁x₂...xₙ
 2              i ← 0
 3              do  i ← i + 1
 4                  Vᵢ₁ ← {A | A → xᵢ}
 5              until i = n
 6              j ← 1
 7              do  j ← j + 1
 8                  i ← 0
 9                  do  i ← i + 1
10                      Vᵢⱼ ← ∅
11                      k ← 0
12                      do  k ← k + 1
13                          Vᵢⱼ ← Vᵢⱼ ∪ {A | A → BC ∈ P, B ∈ Vᵢₖ and C ∈ Vᵢ₊ₖ,ⱼ₋ₖ}
14                      until k = j − 1
15                  until i = n − j + 1
16              until j = n
17              if  S ∈ V₁ₙ then print "parse of" x "successful in G"
18          return
19  end
```

Consider the operation of Algorithm 4 in the following simple abstract example. Let the grammar G have two terminal and three intermediate symbols: $A = \{a, b\}$, and $I = \{A, B, C\}$. The root symbol is S, and there are just four production rules:

$$\mathcal{P} = \left\{ \begin{array}{ll} \mathbf{p}_1: & S \rightarrow AB \; OR \; BC \\ \mathbf{p}_2: & A \rightarrow BA \; OR \; \mathtt{a} \\ \mathbf{p}_3: & B \rightarrow CC \; OR \; \mathtt{b} \\ \mathbf{p}_4: & C \rightarrow AB \; OR \; \mathtt{a} \end{array} \right\}.$$

Figure 8.13 shows the parse table generated by Algorithm 4 for the input string $\mathbf{x} = $ "baaba." Along the bottom are the characters x_i of this string. Lines 2 through 5 of the algorithm fill in the first ($j = 1$) row with any internal symbols that derive the corresponding character in $\mathbf{x}$. The $i = 1$ and $i = 4$ entries of that bottom row are filled with B, since rewrite rule $\mathbf{p}_3$: $B \rightarrow \mathtt{b}$. Likewise, the remaining entries are filled with both A and C, as a result of rewrite rules $\mathbf{p}_2$ and $\mathbf{p}_4$.

The core computation in the algorithm is performed in line 13, which fills entries throughout the table with symbols that could produce segments in lower rows, and hence might be part of a valid derivation (if indeed one is found). For instance, the $i = 1, j = 2$ entries contain any symbols that could produce segments in the row beneath it. Thus this entry contains S because by rule $\mathbf{p}_1$ we have $S \rightarrow BC$, and it also contains A because by rule $\mathbf{p}_2$ we have $A \rightarrow BA$. According to the innermost loop over k (lines 12–14), we seek the left-hand side for rules that span a range. For instance, the $i = 3, j = 3$ entry contains B because for $k = 2$ and rule $\mathbf{p}_3$ we have $B \rightarrow CC$ (as shown in Fig. 8.14).

Figure 8.14 shows the cells that are searched when filling a particular cell in the parse table. The sequence sweeps vertically up to the cell in question, while diagonally down from the cell in question; this guarantees that the all paths from the top cell in a valid derivation can be found. If the top cell contains the root symbol S (and

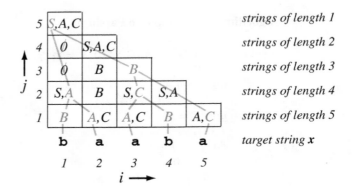

FIGURE 8.13. The bottom-up parsing algorithm fills the parse table with symbols that might be part of a valid derivation. The pink lines are not provided by the algorithm, but when read downward from the root symbol they confirm that a valid derivation exists.

possibly other symbols), then indeed the string is successfully parsed. That is, there exists a valid production leading from S to the target string $\mathbf{x}$.

To understand how this table is filled, consider first the $j = 1$ row. The $j = 4$, $i = 1$ cell contains B, because according to rewrite rule $\mathbf{p}_3$, B is the only intermediate symbol that could yield b in the query sentence, directly below. The same logic holds for the $i = 1$, $j = 1$ cell. The remaining three cells for $j = 1$ contain A and C, since these are the only intermediate variables that can derive a. Incidentally, the derivation in Fig. 8.15 confirms that the parse is valid.

The computational complexity of bottom-up parsing performed by Algorithm 4 is high. The innermost loop of line 13 is executed n or fewer times, while lines 7 and 9 are $O(n^2)$, which is also the space complexity. The time complexity is $O(n^3)$.

Top-Down and Other Methods of Parsing

As its name suggests, top-down parsing starts with the root node and successively applies productions from $\mathcal{P}$, with the goal of finding a derivation of the test sentence $\mathbf{x}$. Because it is rare indeed that the sentence is derived in the first production attempted, it is necessary to specify some criteria to guide the choice of which rewrite rule to apply. Such criteria could include beginning the parse at the first (left) character in the sentence (i.e., finding a small set of rewrite rules that yield the first character), then iteratively expanding the production to derive subsequent characters, or instead starting at the last (right) character in the sentence.

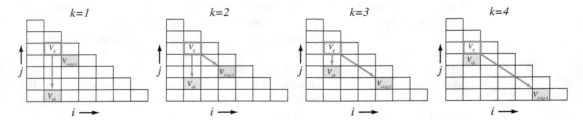

FIGURE 8.14. The innermost loop of Algorithm 4 seeks to fill a cell V_{ij} (outlined in red) by the left-hand side of any rewrite rule whose right-hand side corresponds to symbols in the two shaded cells. As k is incremented, the cells queried move vertically upward to the cell in question and move diagonally down from that cell. The shaded cells show the possible right-hand sides in a derivation, as illustrated by the pink lines in Fig. 8.13.

FIGURE 8.15. This valid derivation of "babaa" in G can be read from the pink lines in the parse table of Fig. 8.13 generated by the bottom-up parse algorithm.

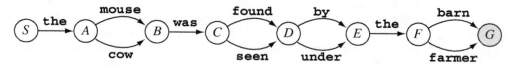

FIGURE 8.16. One type of finite-state machine consists of nodes that can emit terminal symbols ("the," "mouse," etc.) and transition to another node. Such operation can be described by a grammar. For instance, the rewrite rules for this finite-state machine include $S \rightarrow$ the$A$, $A \rightarrow$ mouse$B$ *OR* cow$B$, and so on. Clearly these rules imply this finite-state machine implements a type 3 grammar. The final internal node (shaded) would lead to the null symbol ϵ. Such finite-state machines are sometimes favored because of their clear interpretation and learning methods based on addition of nodes and links. In Section 8.7, though, we shall see general methods for grammatical learning that apply to a broader range of grammatical models.

FINITE-STATE
MACHINE

The bottom-up and top-down parsers just described are quite general and there are a number of parsing algorithms which differ in space and time complexities. Many parsing methods depend upon the model underlying the grammar. One popular such model is *finite-state machines*. Such a machine consists of nodes and transition links; each node can emit a symbol, as shown in Fig. 8.16.

8.7 GRAMMATICAL INFERENCE

In many applications, the grammar is designed by hand. Nevertheless, learning plays an extremely important role in pattern recognition research, and it is natural that we attempt to learn a grammar from example sentences it generates. When seeking to follow that general approach we are immediately struck by differences between the areas addressed by grammatical methods and those that can be described as statistical. First, for most languages there are many grammars (often an infinite number) that can produce it. If two grammars G_1 and G_2 generate the same language (and no other sentences), then this ambiguity is of no consequence; recognition will be the same. However, because learning is always based on a *finite* set of samples, the learning problem is underspecified; there is usually an infinite number of grammars consistent with the training data, and thus we cannot recover the source grammar uniquely.

There are two main techniques used to make the problem of inferring a grammar from instances tractable. The first is to use both *positive* and *negative* instances. That is, we use a set $\mathcal{D}^+$ of sentences known to be derivable in the grammar; we also use a set $\mathcal{D}^-$ of sentences that are known to be *not* derivable in the grammar. In a multicategory case, it is common to take the positive instances in G_i and use them for negative examples in G_j for $j \neq i$. Even with both positive and negative instances, a finite training set rarely specifies the grammar uniquely. Thus our second technique is to impose conditions and constraints. A trivial illustration is that we demand that the alphabet of the candidate grammar contain only those symbols that appear in the training sentences. Moreover, we demand that every production rule in the grammar be used. We seek the "simplest" grammar that explains the training instances where "simple" generally refers to the total number of rewrite rules, or the sum of their lengths, or other natural criteria. These are versions of Occam's razor, that the simplest explanation of the data is to be preferred (Chapter 9).

In broad overview, learning proceeds as follows. An initial grammar G^0 is guessed. Often it is useful to specify the type of grammar (1, 2, or 3), and thus place constraints on the forms of the candidate rewrite rules. In the absence of other prior information, it is traditional to make G^0 as simple as possible and gradually expand the set of productions as needed. Positive training sentences $\mathbf{x}_i^+$ are selected from $\mathcal{D}^+$ one by one. If $\mathbf{x}_i^+$ cannot be parsed by the grammar, then new rewrite rules are proposed for $\mathcal{P}$. A new rule is accepted if and only if it is used for a successful parse of $\mathbf{x}_i^+$ *and* does not allow any negative samples to be parsed.

In greater detail, then, an algorithm for inferring the grammar is as follows:

■ **Algorithm 5. (Grammatical Inference (Overview))**

```
 1  begin initialize 𝒟⁺, 𝒟⁻, G⁰
 2         n⁺ ← |𝒟⁺|  (number of instances in 𝒟⁺)
 3         𝒮 ← S
 4         𝒜 ← set of characters in 𝒟⁺
 5         i ← 0
 6         do i ← i + 1
 7             read xᵢ⁺ from 𝒟⁺
 8             if xᵢ⁺ cannot be parsed by G
 9                 then do  propose additional productions to 𝒫 and variables to ℐ
10                           accept updates if G parses xᵢ⁺ but no string in 𝒟⁻
11             until i = n⁺
12             eliminate redundant productions
13      return G ← {𝒜, ℐ, 𝒮, 𝒫}
14  end
```

Informally, Algorithm 5 continually adds new rewrite rules as required by the successive sentences selected from $\mathcal{D}^+$ so long as the candidate rewrite rule does not allow a sentence in $\mathcal{D}^-$ to be parsed. Line 9 does not state how to choose the specific candidate rewrite rule, but in practice the rule may be chosen from a predefined set (with simpler rules selected first), or based on specific knowledge of the underlying models generating the sentences.

EXAMPLE 4 Grammatical Inference

Consider inferring a grammar G from the following positive and negative examples: $\mathcal{D}^+ = \{\text{a, aaa, aaab, aab}\}$, and $\mathcal{D}^- = \{\text{ab, abc, abb, aabb}\}$. Clearly the alphabet of G is $\mathcal{A} = \{\text{a, b}\}$. We posit a single internal symbol for G^0, along with the simplest rewrite rule $\mathcal{P} = \{S \to A\}$.

i	$\mathbf{x}_i^+$	$\mathcal{P}$	$\mathcal{P}$ produces $\mathcal{D}^-$?
1	a	$S \to A$ $A \to$ a	No
2	aaa	$S \to A$ $A \to$ a $A \to$ aA	No
3	aaab	$S \to A$ $A \to$ a $A \to$ aA $A \to$ ab	Yes: ab $\in \mathcal{D}^-$
3	aaab	$S \to A$ $A \to$ a $A \to$ aA $A \to$ aab	No
4	aab	$S \to A$ $A \to$ a $A \to$ aA $A \to$ aab	No

The table shows the progress of the algorithm. The first positive instance, a, demands a rewrite rule $A \to$ **a**. This rule does not allow any sentences in $\mathcal{D}^-$ to be derived, and thus is accepted for $\mathcal{P}$. When $i = 3$, the proposed rule $A \to$ **ab** indeed allows $\mathbf{x}_3^+$ to be derived, but the rule is rejected because it also derives a sentence in $\mathcal{D}^-$. Instead, the next proposed rule, $A \to$ **aab** is accepted. The final grammar inferred has the four rewrite rules shown at the bottom of the table.

The method of grammatical inference just described is quite general. It is made more specialized by placing restrictions on the types of candidate rewrite rules, corresponding to the designer's assumptions about the type of grammar (1, 2, or 3). For a type 3 grammar, we can consider learning in terms of the finite-state machine. In that case, learning consists of adding nodes and links (cf. Bibliography).

*8.8 RULE-BASED METHODS

In problems where classes can be characterized by general *relationships* among entities, rather than by instances per se, it becomes attractive to build classifiers based on rules. Rule-based methods are integral to expert systems in artificial intelligence, but because they have found only modest use in pattern recognition, we shall give merely IF-THEN RULE a short overview. We shall focus on a broad class of *if-then rules* for representing and learning such relationships.

A very simple if-then rule is

IF Swims(x) *AND* HasScales(x) *THEN* Fish(x),

which means, of course, that if an object x has the property that it swims, and the property that it has scales, then it is a fish. Rules have the great benefits that they are easily interpreted and can be used in database applications where information is encoded in relations. A drawback is that there is no natural notion of probability and it is somewhat difficult, therefore, to use rules when there is high noise and a large Bayes error.

A *predicate*, such as Man(·), HasTeeth(·) and AreMarried(·, ·), is a test that returns a value of logical True or False.* Such predicates can apply to problems where the data are numerical, nonnumerical, linguistic, strings, or any of a broad class of types. The choice of predicates and their evaluation depend strongly on the problem, of course, and in practice these are generally more difficult tasks than learning the rules. For instance, Fig. 8.17 below illustrates the use of rules in categorizing a structure as an arch. Such a rule might involve predicates such as Touch(·, ·) or Supports(·, ·, ·) which address whether two blocks touch, or whether two blocks support a third. It is a very difficult problem in computer vision to evaluate such predicates based on a pixel image taken of the scene.

PROPOSITIONAL LOGIC
FIRST-ORDER LOGIC

There are two main types of if-then rules: *propositional* (variable-free) and *first-order*. A propositional rule describes a particular instance, as in

IF Male(Bill) *AND* IsMarried(Bill) *THEN* IsHusband(Bill),

CONSTANT

where Bill is a particular atomic item. Because its properties are fixed, Bill is an example of a (logical) *constant*. The deficiency of propositional logic is that it provides no general way to represent general relations among a large number of instances. For example, even if we knew Male(Edward) and IsMarried(Edward) are both True, the above rule would not allow us to infer that Edward is is a husband, because that rule applies only to the particular constant Bill.

This deficiency is overcome in first-order logic, which permits rules with *variables*, such as

VARIABLE

IF Eats(x,y) *AND* HasFlesh(x) *THEN* Carnivore(y),

where here x and y are the variables. This rule states that for any items x and y, if y eats x and x has flesh, then y is a carnivore. Clearly this is a very powerful summary of an enormous wealth of examples—first-order rules are far more expressive than classical propositional logic. The power of first-order logic is illustrated in the following rules:

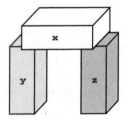 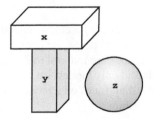 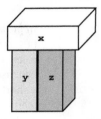

FIGURE 8.17. The rule in Eq. 11 identifies the figure on the left as an example of Arch, but not the other two figures. In practice, it is very difficult to develop subsystems that evaluate the propositions themselves, for instance Touch(x,y) and Supports(x,y,z).

---

*We shall ignore cases where a predicate is Undefined.

$$IF \text{ Male(x) } AND \text{ IsMarried(x,y) } THEN \text{ IsHusband(x)},$$

$$IF \text{ Parent(x,y) } AND \text{ Parent(y,z) } THEN \text{ GrandParent(x,z)}$$

and

$$IF \text{ Spouse(x,y) } THEN \text{ Spouse(y,x)}.$$

A rule from first-order logic can also apply to constants, for instance:

$$IF \text{ Eats(Mouse,Cat) } AND \text{ HasFlesh(Mouse) } THEN \text{ Carnivore(Cat)},$$

where Cat and Mouse are two particular constants.

FUNCTION

If-then rules can also incorporate *functions*, which return numerical values, as illustrated in the following:

$$IF \text{ Male(x) } AND \text{ (Age(x) < 16) } THEN \text{ Boy(x)},$$

TERM

where Age(x) is a function that returns a numerical age in years while the expression or *term* (Age(x) < 16) returns either logical True or False. In sum, the above rule states that a male younger than 16 years is a boy. If we were to use decision trees or statistical techniques, we would not be able to learn this rule perfectly, even given a tremendously large number of examples.

It is clear given a set of first-order rules how to use them in pattern classification: we merely present the unknown item and evaluate the propositions and rules. Thus consider the long rule

$$IF \text{ IsBlock(x) } AND \text{ IsBlock(y) } AND \text{ IsBlock(z)}$$

$$AND \text{ Touch(x,y) } AND \text{ Touch(x,z) } AND \text{ NotTouch(y,z)} \quad (11)$$

$$AND \text{ Supports(x,y,z) } THEN \text{ Arch(x,y,z)},$$

where Supports(x,y,z) means that x is supported by both y and z. We stress that designing algorithms to implement IsBlock(·), Supports(·, ·, ·), and so on, can be extremely difficult; there is little we can say about them here other than that nearly always building these component algorithms represents the greatest effort in designing the overall classifier. Nevertheless, given reliable such algorithms, the rule could be used to classify simple structures as an arch or nonarch (Fig. 8.17).

8.8.1 Learning Rules

Now we turn briefly to the learning of such if-then rules. We have already seen several ways to learn rules. For instance, we can train a decision tree via CART, ID3, C4.5, or other algorithm and then simplify the tree to extract rules (Section 8.4). For cases where the underlying data arise from a grammar, we can infer the particular rules via the methods in Section 8.7. A key distinction in the approach we now discuss is that it can learn sets of first-order rules containing variables. As in grammatical inference, our approach to learning rules from a set of positive and negative examples, $\mathcal{D}^+$ and $\mathcal{D}^-$, is to learn a single rule, delete the examples that it explains, and iterate. Such *sequential covering* learning algorithms lead to a disjunctive set of

SEQUENTIAL COVERING

rules that "cover" the training data. After such training, it is traditional to simplify the resulting logical rule by means of standard logical methods.

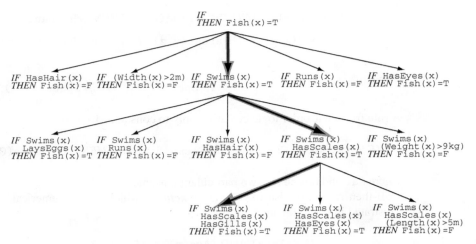

FIGURE 8.18. In sequential covering, candidate rules are searched through successive refinements. First, the "best" rule having a single conditional predicate is found—that is, the one explaining most training data. Next, other candidate predicates are added, the best compound rule is selected, and so forth.

The designer must specify the predicates and functions, based on a prior knowledge of the problem domain. The algorithm begins by considering the most general rules using these predicates and functions, and it finds the "best" simple rule. Here, "best" means that the rule describes the largest number of training examples. Then, the algorithm searches among all refinements of the best rule, choosing the refinement that too is "best." This process is iterated until no more refinements can be added, or when the number of items described is maximum. In this way a single, though possibly complex, if-then rule has been learned (Fig. 8.18). The sequential covering algorithm iterates this process and returns a set of rules.

A general approach is to search first through all rules having a *single* attribute. Next, consider the rule having a single conjunction of two predicates, then these conjunctions, and so on. Note that this greedy algorithm need not be optimal; that is, it need not yield the most compact rule.

SUMMARY

Nonmetric data consist of lists of nominal attributes; such lists might be unordered or ordered (strings). Tree-based methods such as CART, ID3, and C4.5, rely on answers to a series of questions (typically binary) for classification. The designer selects the form of question and the tree is grown, starting at the root node, by finding splits of data that make the representation more "pure." There are several acceptable impurity measures, such as misclassification, variance, and Gini; the entropy impurity, however, has found greatest use. To avoid overfitting and to improve generalization, one must either employ stopped splitting (declaring a node with nonzero impurity to be a leaf) or instead prune a tree trained to minimum impurity leafs. Tree classifiers are very flexible and can be used in a wide range of applications, including those with data that are metric, nonmetric, or in combination.

When comparing patterns that consist of strings of nonnumeric symbols, we use edit distance—a measure of the number of fundamental operations (insertions, deletions, exchanges) that must be performed to transform one string into another. While the general edit distance is not a true metric, edit distance can nevertheless be used for nearest-neighbor classification. String matching is finding whether a test string appears in a longer *text*. The requirement of a perfect match in basic string matching can be relaxed, as in string matching with errors or with the don't-care symbol. These basic string and pattern recognition ideas are simple and straightforward; addressing them in large problems requires algorithms that are computationally efficient.

Grammatical methods assume that the strings are generated from certain classes of rules, which can be described by an underlying grammar. A grammar G consists of an alphabet, intermediate symbols, a starting or root symbol, and most importantly a set of rewrite rules or productions. The four different types of grammars— free, context-sensitive, context-free, and regular—make different assumptions about the nature of the transformations. Parsing is the task of accepting a string $\mathbf{x}$, determining whether it is a member of the language generated by G, and, if so, finding a derivation. Grammatical methods find greatest use in highly structured environments, particularly where structure lies at many levels. Grammatical inference generally uses positive and negative example strings (i.e., ones in the language generated by G and ones not in that language), to infer a set of productions.

Rule-based systems use either propositional logic (variable-free) or first-order logic to describe a category. In broad overview, rules can be learned by sequentially "covering" elements in the training set by successively more complex compound rules.

BIBLIOGRAPHICAL AND HISTORICAL REMARKS

Most work on decision trees addresses problems in continuous features, though a key property of the method is that they apply to nominal data too. Some of the foundations of tree-based classifiers stem from the Concept Learning System described in reference [42], but the important book on CART [10] provided a strong statistics foundation and revived interest in the approach. Quinlan has been a leading exponent of tree classifiers, introducing ID3 [66], C4.5 [69], and the application of minimum description length for pruning [56, 71]. A good overview is reference [61], and a comparison of multivariate decision tree methods is given in reference [11]. Splitting and pruning criteria based on probabilities are explored in reference [53], and the use of an interesting information metric for this end is described in reference [52]. The Gini index was first used in analysis of variance in categorical data [47]. Incremental or on-line learning in decision trees is explored in reference [85]. The missing variable problem in trees is addressed in references [10] and [67], which describe methods more general than those presented here. An unusual parallel "neural" search through trees was presented in reference [78].

The use of edit distance began in the 1970s [64]; a key paper by Wagner and Fischer proposed the fundamental Algorithm 3 and showed that it was optimal [88]. The explosion of digital information, especially natural language text, has motivated work on string matching and related operations. An excellent survey is reference [5], and two thorough books are references [23] and [82]. The computational complexity of string algorithms is presented in reference [21, Chapter 34]. The fast string matching method of Algorithm 2 was introduced in reference [9]; its complexity and speedups and improvements were discussed in references [4, 18, 24, 35, 40] and [83]. String edit distance that permits block-level transpositions is discussed in reference [48]. Some sophisticated string operations—two-dimensional string matching, longest common subsequence, and graph matching—have found only occasionally use in pattern recognition. Statistical methods applied to strings are discussed in reference [26]. Finite-state automata have been applied to several problems in string matching [23, Chapter 7], as well as in time-series prediction and switching—for instance, converting from an alphanumeric representation to a binary representation [43]. String matching has been applied to the recognition of DNA sequences and text, and it is essential in most pattern recognition and template matching involving large databases of text [14]. There is a growing literature on special-purpose hardware for string operations, of which the Splash-2 system [12] is a leading example.

The foundations of a formal study of grammar, including the classification of grammars, began with the landmark book by Chomsky [16]. An early exposition of grammatical inference [39, Chapter 6] was the source for much of the discussion here. Recognition based on parsing (Latin *pars orationis* or "part of speech") has been fundamental in automatic language recognition. Some of the earliest work on three-dimensional object recognition relied on complex grammars that described the relationships of corners and edges, in block structures such arches and towers. It was found that such systems were very brittle; they failed whenever there were errors in feature extraction, due to occlusion, and even minor misspecifications of the model. For the most part, then, grammatical methods have been abandoned for object recognition and scene analysis [60, 25]. Grammatical methods have been applied to the recognition of some simple, highly structured diagrams, such as electrical circuits, simple maps, and even Chinese/Japanese kanji characters. For useful surveys of the basic ideas in syntactic pattern recognition see references [13, 14, 32, 33, 34] and [62]. Suveys of parsing appear in references [3] and [28], and for grammatical inference in reference [59]. The complexity of parsing type 3 grammars is linear in the length of the string, type 2 is low-order polynomial, type 1 is exponential; pointers to the relevant literature appear in [76]. There has been a great deal of work on parsing natural language and speech, and a good textbook on artificial intelligence addressing this topic and much more is reference [75]. There is much work on inferring grammars from instances, such as the Crespi-Reghizzi algorithm (context free) [22]. If queries can be presented interactively, the learning of a grammar can be accelerated, as was shown in reference [81].

The methods described in this chapter have been expanded to allow for stochastic grammars, where there are probabilities associated with rules [20]. A grammar can be considered a specification of a prior probability for a class—for instance, a uniform prior over all (legal) strings in the language $\mathcal{L}$. Error-correcting parsers have been used when random variations arise in an underlying stochastic grammar [50, 84]. One can also apply probability measures to languages [8].

Rule-based methods have formed the foundation of expert systems, and they have been applied extensively through many branches of artificial intelligence such as planning, navigation, and prediction; their use in pattern recognition has been modest, however. Early influential systems include DENDRAL, for inferring chemical structure from mass spectra [29], PROSPECTOR, for finding mineral deposits [38], and MYCIN, for medical diagnosis [79]. Early use of rule induction for pattern recognition include that of Michalski [57, 58]. Figure 8.17 was inspired by Winston's influential work on learning simple geometrical structures and relationships [91]. Learning rules can be called inductive logic programming; Clark and Niblett have made a number of contributions to the field, particularly their CN2 induction algorithm [17]. Quinlan, who has contributed much to the theory and application of tree-based classifiers, describes in reference [68] his FOIL algorithm, which uses a minimum description length criterion to stop the learning of first-order rules. Texts on inductive logic include references [46] and [63]; text on general machine learning include references [44] and [61].

PROBLEMS

Section 8.2

1. When a test pattern is classified by a decision tree, that pattern is subjected to a sequence of queries, corresponding to the nodes along a path from root to leaf. Prove that for any decision tree, there is a functionally equivalent tree in which every such path consists of *distinct* queries. That is, given an arbitrary tree, prove that it is always possible to construct an equivalent tree in which no test pattern is ever subjected to the same query twice.

Section 8.3

2. Consider classification trees that are nonbinary.
   - **(a)** Prove that for any arbitrary tree, with possibly unequal branching ratios throughout, there exists a binary tree that implements the same classification function.
   - **(b)** Consider a tree with just two levels—a root node connected to B leaf nodes ($B \geq 2$). What are the upper and the lower limits on the number of *levels* in a functionally equivalent binary tree, as a function of B?
   - **(c)** As in part (b), what are the upper and lower limits on the number of *nodes* in a functionally equivalent binary tree?

3. Compare the computational complexities of a monothetic and a polythetic tree classifier trained on the same data as follows. Suppose there are $n/2$ training patterns in each of two categories. Every pattern has d attributes, each of which can take on k discrete values. Assume that the best split evenly divides the set of patterns.
   - **(a)** How many levels will there be in the monothetic tree? The polythetic tree?
   - **(b)** In terms of the variables given, what is the complexity of finding the optimal split at the root of a monothetic tree? A polythetic tree?
   - **(c)** Compare the total complexities for training the two trees fully.

4. The task here is to find the computational complexity of training a tree classifier using the twoing impurity where candidate splits are based on a single feature. Suppose there are c classes, $\omega_1, \omega_2, \ldots, \omega_c$, each with n/c patterns that are d-dimensional. Proceed as follows:
   - **(a)** How many possible nontrivial divisions into two supercategories are there at the root node?
   - **(b)** For any one of these candidate supercategory divisions, what is the computational complexity of finding the split that minimizes the entropy impurity?
   - **(c)** Use your results from parts (a) and (b) to find the computational complexity of finding the split at the root node.
   - **(d)** Suppose for simplicity that each split divides the patterns into equal subsets and furthermore that each leaf node corresponds to a single pattern. In terms of the variables given, what will be the expected number of levels of the tree?

(e) Naturally, the number of classes represented at any particular node will depend upon the level in the tree; at the root all c categories must be considered, while at the level just above the leaves, only two categories must be considered. (The pairs of particular classes represented will depend upon the particular node.) State some natural simplifying assumptions, and determine the number of candidate classes at any node as a function of level. (You may need to use the floor or ceiling notation, $\lfloor x \rfloor$ or $\lceil x \rceil$, in your answer, as described in the Appendix Section A.1.)

(f) Use your results from part (e) and the number of patterns to find the computational complexity at an arbitrary level L.

(g) Use all your results to find the computational complexity of training the full tree classifier.

(h) Suppose there are $n = 2^{10}$ patterns, each of which is $d = 6$ dimensional, evenly divided among $c = 16$ categories. Suppose that on a uniprocessor a fundamental computation requires roughly 10^{-10} sec. Roughly how long will it take to train your classifier using the twoing criterion? How long will it take to classify a single test pattern?

5. Consider training a binary tree using the entropy impurity, and refer to Eqs. 1 and 5 for the case of an arbitrary number of categories.

(a) Prove that the decrease in entropy impurity provided by a single yes/no query can never be greater than one bit.

(b) For the two trees in Example 1, verify that each split reduces the impurity and that this reduction is never greater than 1 bit. Explain nevertheless why the impurity at a node *can* be lower than at its descendent, as occurs in that Example.

(c) Generalize your result from part (a) to the case with arbitrary branching ratio $B \geq 2$.

6. Let $P(\omega_1), \ldots, P(\omega_c)$ denote the probabilities of c classes at node N of a binary classification tree, and $\sum_{j=1}^{c} P(\omega_j) = 1$. Suppose the impurity $i(P(\omega_1), \ldots, P(\omega_c))$ at N is a strictly concave function of the probabilities. That is, for any probabilities

$$i_a = i(P^a(\omega_1), \ldots, P^a(\omega_c))$$
$$i_b = i(P^b(\omega_1), \ldots, P^b(\omega_c)) \tag{12}$$

and

$$i^*(\alpha) = i(\alpha P^a(\omega_1) + (1 - \alpha_1)P^b(\omega_1), \ldots, \alpha P^a(\omega_c) + (1 - \alpha_c)P^b(\omega_c)),$$

then we have $i^* \geq \alpha i_a + (1 - \alpha)i_b$.

(a) Prove that for any split, we have $\Delta i(s, t) \geq 0$, with equality if and only if $P(\omega_j|T_L) = P(\omega_j|T_R) = P(\omega_j|T)$, for $j = 1, \ldots, c$. In other words, for a concave impurity function, splitting never increases the impurity.

(b) Prove that entropy impurity (Eq. 1) is a concave function.

(c) Prove that Gini impurity (Eq. 3) is a concave function.

7. Show that the surrogate split method described in the text corresponds to the assumption that the missing feature (attribute) is the one most informative. State mathematically what "most informative" means in this context.

8. Consider a two-category problem and the following training patterns, each having four binary attributes:

ω_1	ω_2
0110	1011
1010	0000
0011	0100
1111	1110

   (a) Use the entropy impurity (Eq. 1) to create by hand an unpruned classifier for these data.

   (b) Apply simple logical reduction methods to your tree in order to express each category by the simplest logical expression—that is, with the fewest *ANDs* and *ORs*.

9. Show that the time complexity of recall in an unpruned, fully trained tree classifier with uniform branching ratio is $O(\log n)$ where n is the number of training patterns. For uniform branching factor, B, state the exact functional form of the number of queries applied to a test pattern as a function of B.

10. Consider impurity functions for a two-category classification problem as a function of $P(\omega_1)$ (and implicitly $P(\omega_2) = 1 - P(\omega_1)$). Show that the simplest reasonable polynomial form for the impurity is related to the sample variance as follows:

   (a) Consider restricting impurity functions to the family of polynomials in $P(\omega_1)$. Explain why i must be at least quadratic in $P(\omega_1)$.

   (b) Write the simplest quadratic form for $P(\omega_1)$ given the boundary conditions $i(P(\omega_1) = 0) = i(P(\omega_1) = 1) = 0$; show that your impurity function can be written $i \propto P(\omega_1)P(\omega_2)$.

   (c) Suppose all patterns in category ω_1 are assigned the value 1.0, while all those in ω_2 are assigned the value 0.0, thereby giving a bimodal distribution. Show that your impurity measure is proportional to the sample variance of this full distribution. Interpret your answer in words.

11. Show how general costs, represented in a cost matrix λ_{ij}, can be incorporated into the misclassification impurity (Eq. 4) during the training of a multicategory tree. Specifically, write pseudocode for a learning algorithm for such training where λ_{ij} is initialized at the beginning, and pseudocode for classification where the test pattern **x** is initialized at the beginning.

12. In this problem you are asked to create tree classifiers for a one-dimensional two-category problem in the limit of large number of points, where $P(\omega_1) = P(\omega_2) = 1/2$, $p(x|\omega_1) \sim N(0, 1)$ and $p(x|\omega_2) \sim N(1, 2)$, and all nodes have decisions of the form "Is $x \leq x_s$?" for some threshold x_s. Each binary tree will be small—just a root node plus two other (nonterminal) nodes and four leaf nodes. For each of the four impurity measures below, state the splitting criteria (i.e., the value x_s at each of the three nonterminal nodes) and the final test error. Whenever possible, express your answers functionally, possibly using the error function $erf(\cdot)$, as well as numerically.

(a) Entropy impurity (Eq. 1).

(b) Gini impurity (Eq. 3).

(c) Misclassification impurity (Eq. 4).

(d) Another splitting rule is based on the so-called *Kolmogorov-Smirnov test*. Let the cumulative distributions for a single variable x for each category be $F_i(x)$ for $i = 1, 2$. The splitting criterion is the maximum difference in the cumulative distributions, that is,

$$\max_{x_s} |F_1(x_s) - F_2(x_s)|.$$

(e) Using the methods of Chapter 2, calculate the Bayes decision rule and the Bayes error.

13. Repeat Problem 12 but for two one-dimensional Cauchy distributions,

$$p(x|\omega_i) = \frac{1}{\pi b_i} \cdot \frac{1}{1 + \left(\frac{x - a_i}{b_i}\right)^2}, \qquad i = 1, 2,$$

with $P(\omega_1) = P(\omega_2) = 1/2$, $a_1 = 0$, $b_1 = 1$, $a_2 = 1$ and $b_2 = 2$. (Here error functions are not needed.)

14. Generalize the missing attribute problem to the case of *several* missing features, and to several deficient patterns. Specifically, write pseudocode for creating a binary decision tree where each d-dimensional pattern can have multiple missing features.

15. During the growing of a decision tree, a node represents the following six-dimensional binary patterns:

ω_1	ω_2
110101	011100
101001	010100
100001	011010
101101	010000
010101	001000
111001	010100
100101	111000
011000	110101

Candidate decisions are based on single feature values.

(a) Which feature should be used for splitting?

(b) Recall the use of statistical significance for stopped splitting. What is the null hypothesis in this example?

(c) Calculate chi-squared for your decision in part (a). Does it differ significantly from the null hypothesis at the 0.01 confidence level? Should splitting be stopped?

(d) Repeat part (c) for the 0.05 level.

16. Consider the following patterns, each having four binary-valued attributes:

ω_1	ω_2
1100	1100
0000	1111
1010	1110
0011	0111

Note especially that the first patterns in the two categories are the same.

(a) Create by hand a binary classification tree for this data. Train your tree so that the leaf nodes have the lowest impurity possible.

(b) Suppose it is known that during testing, the prior probabilities of the two categories will not be equal, but instead $P(\omega_1) = 2P(\omega_2)$. Modify your training method and use the above data to form a tree for this case.

Section 8.4

17. Consider training a binary decision tree to classify two-component patterns from two categories. The first component is binary, 0 or 1, while the second component has six possible values, A through F:

ω_1: 1A	0E	0B	1B	1F	0D
ω_2: 0A	0C	1C	0F	0B	1D

Compare splitting the root node based on the first feature with splitting it on the second feature in the following way.

(a) Use an entropy impurity with a two-way split (i.e., $B = 2$) on the first feature and a six-way split on the second feature.

(b) Repeat (a) but using a gain ratio impurity.

(c) In light of your above answers, discuss the value of gain ratio impurity in cases where splits have different branching ratios.

Section 8.5

18. Consider strings $\mathbf{x}$ and *text*, of length m and n, respectively, from an alphabet $\mathcal{A}$ consisting of d characters. Assume that the naive string-matching algorithm (Algorithm 1) exits the implied loop in line 4 as soon as a mismatch occurs. Prove that the number of character-to-character comparisons made on average for random strings is

$$(n - m + 1)\frac{1 - d^{-m}}{1 - d^{-1}} \le 2(n - m + 1).$$

19. Consider string matching using the Boyer-Moore algorithm (Algorithm 2) based on the trinary alphabet $\mathcal{A} = \{a, b, c\}$. Apply the good-suffix function $\mathcal{G}$ and the last-occurrence function $\mathcal{F}$ to each of the following strings:

(a) "acaccacbac"

(b) "abababcbcbaaabcbaa"

(c) "cccaaababaccc"

(d) "abbabbabbcbbabbcbba"

20. Consider the string-matching problem illustrated in the top of Fig. 8.8. Assume *text* began at the first character of "probabilities."

    (a) How many basic character comparisons are required by the naive string-matching algorithm (Algorithm 1) to find a valid shift?

    (b) How many basic character comparisons are required by the Boyer-Moore string matching algorithm (Algorithm 2) to find a valid shift?

21. For each of the *text*s below, determine the number of fundamental character comparisons needed to find all valid shifts for the test string $\mathbf{x}$ = "abcca" using the naive string-matching algorithm (Algorithm 1) and the Boyer-Moore algorithm (Algorithm 2).

    (a) "abcccdabacabbca"

    (b) "dadadadadadadad"

    (c) "abcbcabcabcabc"

    (d) "accabcababacca"

    (e) "bbccacbccabbcca"

22. Write pseudocode for an efficient construction of the last-occurrence function $\mathcal{F}$ used in the Boyer-Moore algorithm (Algorithm 2). Let d be the number of elements in the alphabet $\mathcal{A}$, and m the length of string $\mathbf{x}$.

    (a) What is the time complexity of your algorithm in the worst case?

    (b) What is the space complexity of your algorithm in the worst case?

    (c) How many fundamental operations are required to compute $\mathcal{F}$ for the 26-letter English alphabet for $\mathbf{x}$ = "bonbon" ? For $\mathbf{x}$ = "marmalade" ? For $\mathbf{x}$ = "abcdabdabcaabcda" ?

23. Consider the training data from the trinary alphabet $\mathcal{A}$ = {a, b, c} in the table

ω_1	ω_2	ω_3
aabbc	bccba	caaaa
ababcc	bbbca	cbcaab
babbcc	cbbaaaa	baaca

Use the simple edit distance to classify each of the below strings. If there are ambiguities in the classification, state which two (or all three) categories are candidates.

    (a) "abacc"

    (b) "abca"

    (c) "ccbba"

    (d) "bbaaac"

24. Repeat Problem 23 using its training data but the following test data:

    (a) "ccab"

    (b) "abdca"

    (c) "abc"

    (d) "bacaca"

25. Repeat Problem 23 but assume that the cost of different string transformations are not equal. In particular, assume that an interchange is twice as costly as an insertion or a deletion.

26. Consider edit distance with positive but otherwise arbitrary costs associated with each of the fundamental operations of insertion, deletion and substitution.

(a) Which of the criteria for a metric are always obeyed and which not necessarily obeyed?

(b) For any criteria that are not always obeyed, construct a counterexample.

27. Consider the general edit distance with positive costs and prove whether it has the four properties of a true metric: nonnegativity, reflexivity, symmetry, and the triangle inequality.

28. Algorithm 3 employs a greedy heuristic for computing the edit distance between two strings $\mathbf{x}$ and $\mathbf{y}$; it need not give a global optimum. In the following, let $|\mathbf{x}| = n_1$ and $|\mathbf{y}| = n_2$.

(a) State the computational complexity of an exhaustive examination of all transformations of $\mathbf{x}$ into $\mathbf{y}$. (Assume that no transformation need be considered if it leads to a string shorter than $\min[n_1, n_2]$ or longer than $\max[n_1, n_2]$.)

(b) Recall from Chapter 5 the basic approach of linear programming. Write pseudocode that would apply linear programming to the calculation of edit distances.

29. Consider strings $\mathbf{x}$ and *text*, of length m and n, respectively, from an alphabet $\mathcal{A}$ consisting of d characters.

(a) Modify the pseudocode of the naive string-matching algorithm to include the don't-care symbol.

(b) Employ the assumptions of Problem 18 but also that $\mathbf{x}$ has exactly k don't care symbols while *text* has none. Find the number of character-to-character comparisons made on average for otherwise random strings.

(c) Show that in the limit of $k = 0$ your answer is closely related to that of Problem 18.

(d) What is your answer in part (b) in the limit $k = m$?

Section 8.6

30. Mathematical expressions in the computer language *Lisp* are of the form (operation *operand*₁ *operand*₂) where expressions can be nested, for example (quotient (plus 4 9) 6).

(a) Write a simple grammar for the four basic arithmetic operations plus, difference, times and quotient, applied to positive single-digit integers. Be sure to include parentheses in your alphabet.

(b) Determine by hand whether each of the following candidate *Lisp* expressions can be derived in your grammar and, if so, show a corresponding derivation tree.

- (times (plus (difference 5 9)(times 3 8))(quotient 2 6))
- (7 difference 2)
- (quotient (7 plus 2) (plus 6 3))
- ((plus) (6 2))
- (difference (plus 5 9) (difference 6 8)) .

31. Consider the language $\mathcal{L}(G) = \{a^n b | n \geq 1\}$.

(a) Construct by hand a grammar that generates this language.

(b) Use G to form derivation trees for the strings "ab" and "aaaaab."

32. Consider the grammar G: $\mathcal{A} = \{a, b, c\}$, $\mathcal{S} = S$, $\mathcal{I} = \{A, B\}$ and $\mathcal{P} = \{S \rightarrow cAb, A \rightarrow aBa, B \rightarrow aBa, B \rightarrow cb\}$.

    (a) What type of grammar is G?

    (b) Prove that this grammar generates the language $\mathcal{L}(G) = \{ca^n cba^n b | n \geq 1\}$.

    (c) Draw the derivation tree the following two strings: "caacbaab" and "cacbab."

33. A *palindrome* is a sequence of characters that reads the same forward and backward, such as "i," "tat," "boob," and "sitonapotatopanotis."

    (a) Write a grammar that generates all palindromes using 26 English letters (no spaces). Use your grammar to show a derivation tree for "noon" and "bib."

    (b) What type is your grammar (0, 1, 2, or 3)?

    (c) Write a grammar that generates all words that consist of a single letter followed by a palindrome. Use your grammar to show a derivation tree for "pi," "too," and "stat."

34. Consider the grammar G in Example 3.

    (a) How many possible derivations are there in G for numbers 1 through 999?

    (b) How many possible derivations are there for numbers 1 through 999,999?

    (c) Does the grammar allow any of the numbers (up to six digits) to be pronounced in more than one way?

35. Recall that ϵ is the empty string, defined to have zero length and to have no manifestation in a final string. Consider the following grammar G: $\mathcal{A} = \{a\}$, $\mathcal{S} = S$, $\mathcal{I} = \{A, B, C, D, E\}$ and eight rewrite rules:

$$\mathcal{P} = \left\{ \begin{array}{ll} S \rightarrow ACaB & Ca \rightarrow aaC \\ CB \rightarrow DB & CB \rightarrow E \\ aD \rightarrow Da & aD \rightarrow AC \\ aE \rightarrow Ea & AE \rightarrow \epsilon \end{array} \right\}.$$

    (a) Note how A and B mark the beginning and end of the sentence, respectively, and that C is a marker that doubles the number of as (while moving from left to right through the word). Prove that the language generated by this grammar is $\mathcal{L}(G) = \{a^{2^n} | n > 0\}$.

    (b) Show a derivation tree for "aaaa" and for "aaaaaaaa" (cf. Computer exercise 10).

36. Explore the notion of Chomsky normal form in the following way.

    (a) Show that the grammar G with $\mathcal{A} = \{a, b\}$, $\mathcal{S} = S$, $\mathcal{I} = \{A, B\}$ and rewrite rules:

$$\mathcal{P} = \left\{ \begin{array}{l} S \rightarrow bA \ OR \ aB \\ A \rightarrow bAA \ OR \ aS \ OR \ a \\ B \rightarrow aBB \ OR \ bS \ OR \ b \end{array} \right\},$$

    is not in Chomsky normal form.

    (b) Show that grammar G' with $\mathcal{A} = \{a, b\}$, $\mathcal{S} = S$, $\mathcal{I} = \{A, B, C_a, C_b, D_1, D_2\}$, and

$$\mathcal{P} = \begin{cases} S \rightarrow C_b A \text{ OR } C_a B & D_1 \rightarrow AA \\ A \rightarrow C_a S \text{ OR } C_b D_1 \text{ OR a} & D_2 \rightarrow BB \\ B \rightarrow C_b S \text{ OR } C_a D_2 \text{ OR b} & C_a \rightarrow \text{a} \\ & C_b \rightarrow \text{b} \end{cases}.$$

is in Chomsky normal form.

(c) Show that G and G' are equivalent by converting the rewrite rules of G into those of G' in the following way. Note that the rules $A \rightarrow$ a and $B \rightarrow$ b of G are already acceptable. Now convert other rules of G appropriate for Chomsky normal form. First replace $S \rightarrow bA$ in G by $S \rightarrow C_b A$ and $C_b \rightarrow$ b. Likewise, replace $A \rightarrow$ aS by $A \rightarrow C_a S$ and $C_a \rightarrow$ a. Continue in this way, keeping in mind the final form of the rewrite rules of G'.

(d) Attempt a derivation of "aabab" in G and in G'.

37. Prove that each of the following languages is not context-free.
    (a) $\mathcal{L}(G) = \{a^i b^j c^k | i < j < k\}$.
    (b) $\mathcal{L}(G) = \{a^i | i \text{ a prime}\}$.

38. Consider a grammar with $\mathcal{A} = \{\text{a, b, c}\}, \mathcal{S} = S, \mathcal{I} = \{A, B\}$, and

$$\mathcal{P} = \begin{cases} S \rightarrow \text{a}SBA \text{ OR a}BA & AB \rightarrow BA \\ bB \rightarrow \text{bb} & bA \rightarrow \text{bc} \\ cA \rightarrow \text{cc} & aB \rightarrow \text{ab} \end{cases}.$$

Prove that this grammar generates the language $\mathcal{L}(G) = \{a^n b^n c^n | n \geq 1\}$.

39. Try to parse by hand the following utterances using the grammar of Example 3. For each successful parse, show the corresponding derivation tree.
    (a) *three hundred forty two thousand six hundred nineteen*
    (b) *thirteen*
    (c) *nine hundred thousand*
    (d) *two thousand six hundred thousand five*
    (e) *one hundred sixty eleven*

Section 8.7

40. Let $\mathcal{D}_1 = \{\text{ab, abb, abbb}\}$ and $\mathcal{D}_2 = \{\text{ba, aba, babb}\}$ be positive training examples from two grammars, G_1 and G_2, respectively.
    (a) Suppose both grammars are of type 3. Generate some candidate rewrite rules.
    (b) Infer grammar G_1 using $\mathcal{D}_1$ as positive examples and $\mathcal{D}_2$ as negative examples.
    (c) Infer grammar G_2 using $\mathcal{D}_2$ as positive examples and $\mathcal{D}_1$ as negative examples.
    (d) Use your trained grammars to classify the following sentences; label any sentence that cannot be parsed in either grammar, or can be parsed in both grammars, as "ambiguous" : "bba," "abab," "bbb," and "abbbb."

Section 8.8

41. For each of the below, write a rule giving an equivalent relation using any of the following predicates: Male($\cdot$), Female($\cdot$), Parent($\cdot,\cdot$), Married($\cdot,\cdot$),

(a) Sister(·,·), where Sister(x,y) = True means that x is the sister of y.
(b) Father(·,·), where Father(x,y) = True means that x is the father of y.
(c) Grandmother(·,·), where Grandmother(x,y) = True means that x is the grandmother of y.
(d) Husband(·,·), where Husband(x,y) = True means that x is the husband of y.
(e) IsWife(·), where IsWife(x) = True means that simply that x is somebody's wife.
(f) Siblings(·,·)
(g) FirstCousins(·,·)

COMPUTER EXERCISES

Several exercises will make use of the following data sampled from three categories. Each of the five features takes on a discrete feature, indicated by the range listed at the along the top. Note particularly that there are different number of samples in each category, and that the number of possible values for the features is not the same. For instance, the first feature can take on four values ($A - D$, inclusive), while the last feature can take on just two ($M - N$).

Sample	Category	$A - D$	$E - G$	$H - J$	$K - L$	$M - N$
1	ω_1	A	E	H	K	M
2	ω_1	B	E	I	L	M
3	ω_1	A	G	I	L	N
4	ω_1	B	G	H	K	M
5	ω_1	A	G	I	L	M
6	ω_2	B	F	I	L	M
7	ω_2	B	F	J	L	N
8	ω_2	B	E	I	L	N
9	ω_2	C	G	J	K	N
10	ω_2	C	G	J	L	M
11	ω_2	D	G	J	K	M
12	ω_2	B	E	I	L	M
13	ω_3	D	E	H	K	N
14	ω_3	A	E	H	K	N
15	ω_3	D	E	H	L	N
16	ω_3	D	F	J	L	N
17	ω_3	A	F	H	K	N
18	ω_3	D	E	J	L	M
19	ω_3	C	F	J	L	M
20	ω_3	D	F	H	L	M

Section 8.3

1. Write a general program for growing a binary tree and use it to train a tree fully using the data from the three categories in the table, using an entropy impurity.

(a) Use the (unpruned) tree to classify the following patterns: $\{A, E, I, L, N\}$, $\{D, E, J, K, N\}$, $\{B, F, J, K, M\}$, and $\{C, D, J, L, N\}$.

(b) Prune one pair of leaf nodes so as to increase the entropy impurity as little as possible.

(c) Modify your program to allow for nonbinary splits, where the branching ratio B as is determined at each node during training. Train a new tree fully using a gain ratio impurity and then classify the points in (a).

2. Recall that one criterion for stopping the growing of a decision tree is to halt splitting when the best split reduces the impurity by less than some threshold value, that is, when $\max_s \Delta i(s) \leq \beta$, where s indicates the split and β is the threshold. Explore the relationship between classifier generalization and β through the following simulations.

(a) Generate 200 training points, 100 each for two two-dimensional Gaussian distributions:

$$p(\mathbf{x}|\omega_1) \sim N\left(\begin{pmatrix} -0.25 \\ 0 \end{pmatrix}, \mathbf{I}\right) \text{ and } p(\mathbf{x}|\omega_2) \sim N\left(\begin{pmatrix} +0.25 \\ 0 \end{pmatrix}, \mathbf{I}\right).$$

Also use your program to generate an independent test set of 100 points, 50 each of the categories.

(b) Write a program to grow a tree classifier, where a node is not split if $\max_s \Delta i(s) \leq \beta$.

(c) Plot the generalization error of your tree classifier versus β for $\beta = 0.01$, $0.02, \ldots$, as estimated on the test data generated in part (a).

(d) In light of your plot, discuss the relationship between β and generalization error.

Section 8.4

3. Write a program for training an ID3 decision tree in which the branching ratio B at each node is equal to the number of discrete "binned" possible values for each attribute. Use a gain ratio impurity.

(a) Use your program to train a tree fully with the ω_1 and ω_2 patterns in the table above.

(b) Use your tree to classify $\{B, G, I, K, N\}$ and $\{C, E, J, L, M\}$.

(c) Write a logical expression that describes the decisions in part (b). Simplify these expressions.

(d) Convert the information in your tree into a single logical expression which describes the ω_1 category. Repeat for the ω_2 category.

4. Consider the issue of tree-based classifiers and deficient patterns.

(a) Write a program to generate a binary decision tree for categories ω_1 and ω_2 using sample points $1 - 10$ from the table above and an entropy impurity. For each decision node store the primary split and four surrogate splits.

(b) Use your tree to classify the following patterns, where as usual * denotes a missing feature.
- $\{A, F, H, K, M\}$
- $\{*, G, H, K, M\}$
- $\{C, F, I, L, N\}$
- $\{B, *, *, K, N\}$

(c) Now write a program to train a tree using deficient points. Train with sample points 1–10 from the table, used in part (a), as well as with the following four points:

- ω_1: $\{*, F, I, K, N\}$
- ω_1: $\{B, G, H, K, *\}$
- ω_2: $\{C, G, *, L, N\}$
- ω_2: $\{*, F, I, K, N\}$

(d) Use your tree from part (c) to classify the test points in part (b).

5. Train a tree classifier to distinguish all three categories ω_i, $i = 1, 2, 3$, using all 20 sample points in the table above. Use an entropy criterion without pruning or stopping.

(a) Express your tree as a set of rules.

(b) Through exhaustive search, find the rule or rules, which, when deleted, lead to the smallest increase in classification error as estimated by the training data.

Section 8.5

6. Write a program to implement the naive string-matching algorithm (Algorithm 1). Insert a conditional branch so as to exit the innermost loop whenever a mismatch occurs (i.e., the shift is found to be invalid). Add a line to count the total number of character-to-character comparisons in the complete string search.

(a) Write a small program to generate a *text* of n characters, taken from an alphabet having d characters. Let $d = 5$ and use your program to generate a *text* of length $n = 1000$ and a test string $\mathbf{x}$ of length $m = 10$.

(b) Compare the number of character-to-character comparisons performed by your program with the theoretical result quoted in Problem 18 for all pairs of the following parameters: $m = \{10, 15, 20\}$ and $n = \{100, 1000, 10000\}$.

7. Write a program to implement the Boyer-Moore algorithm (Algorithm 2) in the following steps. Throughout let the alphabet have d characters.

(a) Write a routine for constructing the good-suffix function $\mathcal{G}$. Let $d = 3$ and apply your routine to the strings $\mathbf{x}_1 = $ "abcbab" and $\mathbf{x}_2 = $ "babab."

(b) Write a routine for constructing the last-occurrence function $\mathcal{F}$. Apply it to the strings $\mathbf{x}_1$ and $\mathbf{x}_2$ of part (a).

(c) Write an implementation of the Boyer-Moore algorithm incorporating your routines from parts (a) and (b). Generate a *text* of $n = 10000$ characters chosen from the alphabet $\mathcal{A} = \{a, b, c\}$. Use your program to search for $\mathbf{x}_1$ in *text*, and again $\mathbf{x}_2$ in *text*.

(d) Make some statistical assumptions to estimate the number of occurrences of $\mathbf{x}_1$ and $\mathbf{x}_2$ in *text*, and compare that number to your answers in part (c).

8. Write an algorithm for addressing the subset-superset problem in string matching. That is, search a *text* with several strings, some of which are factors of others.

(a) Let $\mathbf{x}_1 = $ "beats," $\mathbf{x}_2 = $ "beat," $\mathbf{x}_3 = $ "be," $\mathbf{x}_4 = $ "at," $\mathbf{x}_5 = $ "eat," $\mathbf{x}_6 = $ "sat." Search for these possible factors in

$$text = \underbrace{\text{``beats\_beats\_beats\_....\_beats''}}_{100\times},$$

but do not return any strings that are factors of other test strings found in *text*.

(b) Repeat with *text* consisting of 100 appended copies of "repeatable_," and the test items "repeatable," "pea," "table," "tab," "able," "peat," and "a."

Section 8.6

9. Write a parser for the grammar described in the text:

$$\mathcal{A} = \{a, b\}, \mathcal{I} = \{A, B\}, \mathcal{S} = S, \text{ and } \mathcal{P} = \left\{ \begin{array}{ll} \mathbf{p}_1: & S \to AB \text{ OR } BC \\ \mathbf{p}_2: & A \to BA \text{ OR } a \\ \mathbf{p}_3: & B \to CC \text{ OR } b \\ \mathbf{p}_4: & C \to AB \text{ OR } a \end{array} \right\}.$$

Use your program to attempt to parse each of the following strings. In all cases, show the parse tree; for each successful parse, show moreover the corresponding derivation tree.

- "aaaabbab"
- "ba"
- "baabab"
- "babab"
- "aaa"
- "baaa"

10. Consider the following grammar G described in the text: $\mathcal{A} = \{a\}, \mathcal{S} = S,$ $\mathcal{I} = \{A, B, C, D, E\}$ and eight rewrite rules:

$$\mathcal{P} = \left\{ \begin{array}{llll} S \to AC aB & & Ca \to aaC \\ CB \to DB & & CB \to E \\ aD \to Da & & aD \to AC \\ aE \to Ea & & AE \to \epsilon \end{array} \right\}.$$

Note how A and B mark the beginning and end of the sentence, respectively, and that C is a marker that doubles the number of as (while moving from left to right through the word).

(a) Prove that the language generated by this grammar is $\mathcal{L}(G) = \{a^{2^n} | n > 0\}$.

(b) Show a derivation tree for "aaaa" and for "aaaaaaaa."

Section 8.7

11. Write a program to infer a grammar G from the following positive and negative examples:

(a) $\mathcal{D}^+ = \{abc, aabbcc, aaabbbccc\}$
(b) $\mathcal{D}^- = \{abbc, abcc, aabcc\}$

Take the following as candidate rewrite rules:

$$\begin{array}{lll} S \to aSBA & AB \to BA & cB \to aC \\ S \to bSBA & BA \to AB & bA \to bc \\ S \to aBA & bB \to bb & bC \to bc \\ S \to aSB & bC \to ba & aB \to ab \\ S \to aSA & cA \to cc & aB \to ca \end{array}$$

Proceed as follows:

(a) Implement the general bottom-up parser of Algorithm 4.

(b) Implement the general grammatical inferencing method of Algorithm 5.

(c) Use your programs in conjunction to infer G from the data.

BIBLIOGRAPHY

[1] Alfred V. Aho, John E. Hopcroft, and Jeffrey D. Ullman. *The Design and Analysis of Computer Algorithms.* Addison-Wesley, Reading, MA, 1974.

[2] Alfred V. Aho, John E. Hopcroft, and Jeffrey D. Ullman. *Data Structures and Algorithms.* Addison-Wesley, Reading, MA, 1987.

[3] Alfred V. Aho and Jeffrey D. Ullman. *The Theory of Parsing, Translation, and Compiling, volume 1: Parsing.* Prentice-Hall, Englewood Cliffs, NJ, 1972.

[4] Alberto Apostolico and Raffaele Giancarlo. The Boyer-Moore-Galil string searching strategies revisited. *SIAM Journal of Computing,* 51(1):98–105, 1986.

[5] Ricardo Baeza-Yates. Algorithms for string searching: A survey. *SIGIR Forum,* 23(3,4):34–58, 1989.

[6] Alan A. Bertossi, Fabrizio Luccio, Elena Lodi, and Linda Pagli. String matching with weighted errors. *Theoretical Computer Science,* 73(3):319–328, 1990.

[7] Anselm Blumer, Janet A. Blumer, Andrzej Ehrenfeucht, David Haussler, Mu-Tian Chen, and Joel I. Seiferas. The smallest automaton recognizing the subwords of a text. *Theoretical Computer Science,* 40(1):31–55, 1985.

[8] Taylor L. Booth and Richard A. Thompson. Applying probability measures to abstract languages. *IEEE Transactions on Computers,* C-22(5):442–450, 1973.

[9] Robert S. Boyer and J. Strother Moore. A fast string-searching algorithm. *Communications of the ACM,* 20(10):762–772, 1977.

[10] Leo Breiman, Jerome H. Friedman, Richard A. Olshen, and Charles J. Stone. *Classification and Regression Trees.* Chapman & Hall, New York, 1993.

[11] Carla E. Brodley and Paul E. Utgoff. Multivariate decision trees. *Machine Learning,* 19(1):45–77, 1995.

[12] Duncan A. Buell, Jeffrey M. Arnold, and Walter J. Kleinfelder, editors. *Splash 2: FPGAs in a Custom Computing Machine.* IEEE Computer Society, Los Alamitos, CA, 1996.

[13] Horst Bunke, editor. *Advances in Structural and Syntactic Pattern Recognition.* World Scientific, River Edge, NJ, 1992.

[14] Horst Bunke and Alberto Sanfeliu, editors. *Syntactic and Structural Pattern Recognition: Theory and Applications.* World Scientific, River Edge, NJ, 1990.

[15] Wray L. Buntine. Decision tree induction systems: A Bayesian analysis. In Laveen N. Kanal, Tod S. Levitt, and John F. Lemmer, editors, *Uncertainty in Artificial Intelligence 3,* pages 109–127. North-Holland, Amsterdam, 1989.

[16] Noam Chomsky. *Syntactic Structures.* Mouton, The Hague, Netherlands, 1957.

[17] Peter Clark and Tim Niblett. The CN2 induction algorithm. *Machine Learning,* 3(4):261–284, 1989.

[18] Richard Cole. Tight bounds on the complexity of the Boyer-Moore string matching algorithm. In *Proceedings of the Second Annual ACM-SIAM Symposium on Discrete Algorithms,* pages 224–233, San Francisco, CA, 1991.

[19] Livio Colussi, Ziv Galil, and Raffaele Giancarlo. On the exact complexity of string matching. In *Proceedings of the 31st Annual Symposium on Foundations of Computer Science,* pages 135–144, 1990.

[20] Craig M. Cook and Azriel Rosenfeld. Some experiments in grammatical inference. In Jean-Claude Simon, editor, *Computer Oriented Learning Processes,* pages 157–171. Springer-Verlag, NATO ASI, New York, 1974.

[21] Thomas H. Cormen, Charles E. Leiserson, and Ronald L. Rivest. *Introduction to Algorithms.* MIT Press, Cambridge, MA, 1990.

[22] Stefano Crespi-Reghizzi. An effective model for grammatical inference. In C. V. Freiman, editor, *Proceedings of the International Federation for Information Processing (IFIP) Congress, Ljubljana Yugoslavia,* pages 524–529. North-Holland, Amsterdam, 1972.

[23] Maxime Crochemore and Wojciech Rytter. *Text Algorithms.* Oxford University Press, Oxford, UK, 1994.

[24] Min-Wen Du and Shih-Chio Chang. Approach to designing very fast approximate string matching algorithms. *IEEE Transactions on Knowledge and Data Engineering,* KDE-6(4):620–633, 1994.

[25] Richard O. Duda and Peter E. Hart. Experiments in scene analysis. *Proceedings of the First National Symposium on Industrial Robots,* pages 119–130, 1970.

[26] Richard M. Durbin, Sean R. Eddy, Anders Krogh, and Graeme J. Mitchison. *Biological Sequence Analysis: Probabilistic Models of Proteins and Nucleic Acids.* Cambridge University Press, Cambridge, UK, 1998.

[27] John Durkin. Induction via ID3. *AI Expert,* 7(4):48–53, 1992.

[28] Jay Earley. An efficient context-free parsing algorithm. *Communications of the ACM,* 13:94–102, 1970.

[29] Bruce G. Buchanan, Edward A. Feigenbaum and Joshua Lederberg. On generality and problem solving: A case study using the DENDRAL program. In Bernard Meltzer and Donald Michie, editors, *Machine Intelligence 6,* pages 165–190. Edinburgh University Press, Edinburgh, Scotland, 1971.

[30] Usama M. Fayyad and Keki B. Irani. On the handling of continuous-valued attributes in decision tree generation. *Machine Learning*, 8:87–102, 1992.

[31] Edward A. Feigenbaum and Bruce G. Buchanan. DEN-DRAL and Meta-DENDRAL: Roots of knowledge systems and expert system applications. *Artificial Intelligence*, 59(1–2):233–240, 1993.

[32] King-Sun Fu. *Syntactic Pattern Recognition and Applications*. Prentice-Hall, Englewood Cliffs, NJ, 1982.

[33] King-Sun Fu and Taylor L. Booth. Grammatical inference: Introduction and Survey–Part I. *IEEE Transactions on Pattern Analysis and Machine Intelligence*, PAMI-8(3):343–359, 1986.

[34] King-Sun Fu and Taylor L. Booth. Grammatical inference: Introduction and Survey–Part II. *IEEE Transactions on Pattern Analysis and Machine Intelligence*, PAMI-8(3):360–375, 1986.

[35] Zvi Galil and Joel I. Seiferas. Time-space-optimal string matching. *Journal of Computer and System Sciences*, 26(3):280–294, 1983.

[36] E. Mark Gold. Language identification in the limit. *Information and Control*, 10:447–474, 1967.

[37] Dan Gusfield. *Algorithms on Strings, Trees and Sequences*. Cambridge University Press, New York, 1997.

[38] Peter E. Hart, Richard O. Duda, and Marco T. Einaudi. PROSPECTOR–a computer-based consultation system for mineral exploration. *Mathematical Geology*, 10(5):589–610, 1978.

[39] John E. Hopcroft and Jeffrey D. Ullman. *Introduction to Automata Theory, Languages, and Computation*. Addison-Wesley, Reading, MA, 1979.

[40] R. Nigel Horspool. Practical fast searching in strings. *Software–Practice and Experience*, 10(6):501–506, 1980.

[41] Xiaoqiu Huang. A lower bound for the edit-distance problem under an arbitrary cost function. *Information Processing Letters*, 27(6):319–321, 1988.

[42] Earl B. Hunt, Janet Marin, and Philip J. Stone. *Experiments in Induction*. Academic Press, New York, 1966.

[43] Zvi Kohavi. *Switching and Finite Automata Theory*. McGraw-Hill, New York, second edition, 1978.

[44] Pat Langley. *Elements of Machine Learning*. Morgan Kaufmann, San Francisco, CA, 1996.

[45] Karim Lari and Stephen J. Young. Estimation of stochastic context-free grammars using the inside-out algorithm. *Computer Speech and Language*, 4(1):35–56, 1990.

[46] Nada Lavrač and Saso Džeroski. *Inductive Logic Programming: Techniques and Applications*. Ellis Horwood (Simon & Schuster), New York, 1994.

[47] Richard J. Light and Barry H. Margolin. An analysis of variance for categorical data. *Journal of the American Statistical Association*, 66(335):534–544, 1971.

[48] Dan Lopresti and Andrew Tomkins. Block edit models for approximate string matching. *Theoretical Computer Science*, 181(1):159–179, 1997.

[49] M. Lothaire, editor. *Combinatorics on Words*. Cambridge University Press, New York, second edition, 1997. M. Lothaire is a joint pseudonym for the following: Robert Cor, Dominque Perrin, Jean Berstel, Christian Choffrut, Dominque Foata, Jean Eric Pin, Guiseppe Pirillo, Christophe Reutenauer, Marcel P. Schutzenberger, Jadcques Sakaroovitch, and Imre Simon.

[50] Shin-Yee Lu and King-Sun Fu. Stochastic error-correcting syntax analysis and recognition of noisy patterns. *IEEE Transactions on Computers*, C-26(12):1268–1276, 1977.

[51] Maurice Maes. Polygonal shape recognition using string matching techniques. *Pattern Recognition*, 24(5):433–440, 1991.

[52] Ramon López De Mántaras. A distance-based attribute selection measure for decision tree induction. *Machine Learning*, 6:81–92, 1991.

[53] J. Kent Martin. An exact probability metric for decision tree splitting and stopping. *Machine Learning*, 28(2–3):257–291, 1997.

[54] Andres Marzal and Enrique Vidal. Computation of normalized edit distance and applications. *IEEE Transactions on Pattern Analysis and Machine Intelligence*, PAMI-15(9):926–932, 1993.

[55] Gerhard Mehlsam, Hermann Kaindl, and Wilhelm Barth. Feature construction during tree learning. In Klaus P. Jantke and Steffen Lange, editors, *Algorithmic Learning for Knowledge-Based Systems*, volume 961 of *Lecture Notes in Artificial Intelligence*, pages 391–403. Springer-Verlag, Berlin, 1995.

[56] Manish Mehta, Jorma Rissanen, and Rakesh Agrawal. MDL-based decision tree pruning. In *Proceedings of the First International Conference on Knowledge Discovery and Data Mining (KDD'95)*, pages 216–221, 1995.

[57] Ryszard S. Michalski. AQVAL/1: Computer implementation of a variable valued logic system VL1 and examples of its application to pattern recognition. In *Proceedings of the First International Conference on Pattern Recognition*, pages 3–17, 1973.

[58] Ryszard S. Michalski. Pattern recognition as rule-guided inductive inference. *IEEE Transactions on Pattern Analysis and Machine Intelligence*, PAMI-2(4):349–361, 1980.

[59] Laurent Miclet. Grammatical inference. In Horst Bunke and Alberto Sanfeliu, editors, *Syntactic and Structural Pattern Recognition: Theory and Applications*, chapter 9, pages 237–290. World Scientific, River Edge, NJ, 1990.

[60] William F. Miller and Alan C. Shaw. Linguistic methods in picture processing–A survey. *Proceedings of the Fall Joint Computer Conference FJCC*, pages 279–290, 1968.

[61] Tom M. Mitchell. *Machine Learning*. McGraw-Hill, New York, 1997.

[62] Roger Mohr, Theo Pavlidis, and Alberto Sanfeliu, editors. *Structural Pattern Recognition*. World Scientific, River Edge, NJ, 1990.

[63] Stephen Muggleton, editor. *Inductive Logic Programming*. Academic Press, London, 1992.

[64] Saul B. Needleman and Christian D. Wunsch. A general

method applicable to the search for similarities in the amino-acid sequence of two proteins. *Journal of Molecular Biology*, 48(3):443–453, 1970.

[65] Partha Niyogi. *The Informational Complexity of Learning: Perspectives on Neural Networks and Generative Grammar*. Kluwer, Boston, MA, 1998.

[66] J. Ross Quinlan. Learning efficient classification procedures and their application to chess end games. In Ryszard S. Michalski, Jaime G. Carbonell, and Tom M. Mitchell, editors, *Machine Learning: An Artificial Intelligence Approach*, pages 463–482. Morgan Kaufmann, San Francisco, CA, 1983.

[67] J. Ross Quinlan. Unknown attribute values in induction. In Alberto Maria Segre, editor, *Proceedings of the Sixth International Workshop on Machine Learning*, pages 164–168. Morgan Kaufmann, San Mateo, CA, 1989.

[68] J. Ross Quinlan. Learning logical definitions from relations. *Machine Learning*, 5(3):239–266, 1990.

[69] J. Ross Quinlan. *C4.5: Programs for Machine Learning*. Morgan Kaufmann, San Francisco, CA, 1993.

[70] J. Ross Quinlan. Improved use of continuous attributes in C4.5. *Journal of Artificial Intelligence*, 4:77–90, 1996.

[71] J. Ross Quinlan and Ronald L. Rivest. Inferring decision tress using the minimum description length principle. *Information and Computation*, 80(3):227–248, 1989.

[72] Stephen V. Rice, Horst Bunke, and Thomas A. Nartker. Classes of cost functions for string edit distance. *Algorithmica*, 18(2):271–180, 1997.

[73] Eric Sven Ristad and Peter N. Yianilos. Learning string-edit distance. *IEEE Transactions on Pattern Analysis and Machine Intelligence*, PAMI-20(5):522–532, 1998.

[74] Ron L. Rivest. Learning decision lists. *Machine Learning*, 2(3):229–246, 1987.

[75] Stuart Russell and Peter Norvig. *Artificial Intelligence: A Modern Approach*. Prentice-Hall Series in Artificial Intelligence. Prentice-Hall, Englewood Cliffs, NJ, 1995.

[76] Wojciech Rytter. On the complexity of parallel parsing of general context-free languages. *Theoretical Computer Science*, 47(3):315–321, 1986.

[77] David Sankoff and Joseph B. Kruskal. *Time Warps, String Edits, and Macromolecules: The Theory and Practice of Sequence Comparison*. Addison-Wesley, New York, 1983.

[78] Janet Saylor and David G. Stork. Parallel analog neural networks for tree searching. In John S. Denker, editor, *Neural Networks for Computing*, pages 392–397. American Institute of Physics, New York, 1986.

[79] Edward H. Shortliffe, editor. *Computer-Based Medical Consultations: MYCIN*. Elsevier/North-Holland, New York, 1976.

[80] Temple F. Smith and Michael S. Waterman. Identification of common molecular sequences. *Journal of Molecular Biology*, 147:195–197, 1981.

[81] Ray J. Solomonoff. A new method for discovering the grammars of phrase structure languages. In *Proceedings of the International Conference on Information Processing*, pages 285–290, 1959.

[82] Graham A. Stephen. *String Searching Algorithms*, volume 3. World Scientific, River Edge, NJ, Lecture Notes on Computing, 1994.

[83] Daniel Sunday. A very fast substring search algorithm. *Communications of the ACM*, 33(8):132–142, 1990.

[84] Eiichi Tanaka and King-Sun Fu. Error-correcting parsers for formal languages. *IEEE Transactions on Computers*, C-27(7):605–616, 1978.

[85] Paul E. Utgoff. Incremental induction of decision trees. *Machine Learning*, 4(2):161–186, 1989.

[86] Paul E. Utgoff. Perceptron trees: A case study in hybrid concept representations. *Connection Science*, 1(4):377–391, 1989.

[87] Uzi Vishkin. Optimal parallel pattern matching in strings. *Information and Control*, 67:91–113, 1985.

[88] Robert A. Wagner and Michael J. Fischer. The string-to-string correction problem. *Journal of the ACM*, 21(1):168–178, 1974.

[89] Sholom M. Weiss and Nitin Indurkhya. Decision tree pruning: Biased or optimal? In *Proceedings of the 12th National Conference on Artificial Intelligence*, volume 1, pages 626–632. AAAI Press, Menlo Park, CA, 1994.

[90] Allan P. White and Wei Zhong Liu. Bias in information-based measures in decision tree induction. *Machine Learning*, 15(3):321–329, 1994.

[91] Patrick Henry Winston. Learning structural descriptions from examples. In Patrick Henry Winston, editor, *The Psychology of Computer Vision*, pages 157–209. McGraw-Hill, New York, 1975.

ALGORITHM-INDEPENDENT MACHINE LEARNING

9.1 INTRODUCTION

In the previous chapters we have seen many learning algorithms and techniques for pattern recognition. When confronting such a range of algorithms, everyone has wondered at one time or another which one is "best." Of course, some algorithms may be preferred because of their lower computational complexity; others may be preferred because they take into account some prior knowledge of the form of the data (e.g., discrete, continuous, unordered list, string, ...). Nevertheless, there are classification problems for which such issues are of little or no concern, or we wish to compare algorithms that are equivalent in regard to them. In these cases we are left with the question, Are there any reasons to favor one algorithm over another? For instance, given two classifiers that perform equally well on the training set, it is frequently asserted that the *simpler* classifier can be expected to perform better on a test set. But is this version of *Occam's razor* really so evident? Likewise, we frequently prefer or impose *smoothness* on a classifier's decision functions. Do simpler or "smoother" classifiers generalize better, and, if so, why? In this chapter we address these and related questions concerning the foundations and philosophical underpinnings of statistical pattern classification. Now that the reader has intuition and experience with individual algorithms, these issues in the theory of learning may be better understood.

OCCAM'S RAZOR

In some fields there are strict conservation laws and constraint laws—such as the conservation of energy, charge, and momentum in physics, as well as the second law of thermodynamics, which states that the entropy of an isolated system can never decrease. These hold regardless of the number and configuration of the forces at play. Given the usefulness of such laws, we naturally ask, Are there analogous results in pattern recognition, ones that do not depend upon the particular choice of classifier or learning method? Are there any fundamental results that hold regardless of the cleverness of the designer, the number and distribution of the patterns, and the nature of the classification task?

Of course it is very valuable to know that there exists a constraint on classifier accuracy, the Bayes error rate, and it is sometimes useful to compare performance to this theoretical limit. Alas, in practice we rarely, if ever, know the Bayes error rate. Even if we did know this error rate, it would not help us much in designing a classifier except to tell us that further training and data collection is futile. Thus the Bayes error is generally of theoretical interest. What other fundamental principles and properties might be of greater use in designing classifiers?

Before we address such problems, we should clarify the meaning of the title of this chapter. "Algorithm-independent" here refers, first, to those mathematical foundations that do not depend upon the particular classifier or learning algorithm used. Our upcoming discussion of bias and variance is just as valid for methods based on neural networks or the nearest-neighbor or model-dependent maximum-likelihood. Second, we mean techniques that can be used in conjunction with different learning algorithms, or provide guidance in their use. For example, cross-validation and resampling techniques can be used with any of a large number of training methods. Of course by the very general notion of an algorithm, these too are algorithms, technically speaking, but we discuss them because of their breadth of applicability and their independence from details of the learning techniques used.

In this chapter we shall see, first, that no pattern classification method is inherently superior to any other, or even to random guessing; it is the type of problem, prior distribution, and other information that determine which form of classifier should provide the best performance. We shall then explore several ways to quantify and adjust the "match" between a learning algorithm and the problem it addresses. In any particular problem there are differences between classifiers, of course, and thus we show that with certain assumptions we can estimate their accuracies (even, for instance, before the candidate classifier is fully trained) and compare different classifiers. Finally, we shall see methods for integrating component or "expert" classifiers, which themselves might implement quite different algorithms.

We shall present the results that are most important for pattern recognition practitioners, generally skipping over mathematical proofs that can be found in the original research referenced in the Bibliographical and Historical Remarks section.

9.2 LACK OF INHERENT SUPERIORITY OF ANY CLASSIFIER

We now turn to the central question posed above: If we are interested solely in the generalization performance, are there any reasons to prefer one classifier or learning algorithm over another? If we make no prior assumptions about the nature of the classification task, can we expect any classification method to be superior or inferior overall? Can we even find an algorithm that is overall superior to (or inferior to) random guessing?

9.2.1 No Free Lunch Theorem

As summarized in the *No Free Lunch Theorem*, the answer to these and several related questions is "no." If the goal is to obtain good generalization performance, there are no context-independent or usage-independent reasons to favor one learning or classification method over another. If one algorithm seems to outperform another in a particular situation, it is a consequence of its fit to the particular pattern recognition problem, not the general superiority of the algorithm. When confronting a new pattern recognition problem, appreciation of this theorem reminds us to focus on the aspects that matter most—prior information, data distribution, amount of training data, and cost or reward functions. The theorem also justifies a healthy skepticism regarding studies that purport to demonstrate the overall superiority of a particular learning or recognition algorithm.

First we should consider more closely the method by which we judge the generalization performance of a classifer. Up to here, we have estimated such performance by means of a test data set, sampled independently, as is the training set. In some cases, this approach has some unexpected drawbacks when applied to comparing classifiers. In a discrete problem, for example, when the training set and test set are very large, they necessarily overlap, and we are testing on training patterns. Further, virtually any powerful algorithm such as the nearest-neighbor algorithm, unpruned decision trees, or neural networks with sufficient number of hidden nodes can learn the training set perfectly. Second, for low-noise or low-Bayes error cases, if we use an algorithm powerful enough to learn the training set, then the upper limit of the i.i.d. error decreases as the training set size increases.

OFF-TRAINING SET ERROR

Thus, in order to compare learning algorithms, we will use the *off-training set error*—the error on points *not* in the training set, as will become clear. If the training set is very large, then the maximum size of the off-training set data is necessarily small.

For simplicity consider a two-category problem, where the training set $\mathcal{D}$ consists of patterns $\mathbf{x}^i$ and associated category labels $y_i = \pm 1$ for $i = 1, \ldots, n$ generated by the unknown target function to be learned, $F(\mathbf{x})$, where $y_i = F(\mathbf{x}^i)$. In most cases of interest there is a random component in $F(\mathbf{x})$ and thus the same input could lead to different categories, giving nonzero Bayes error. At first we shall assume that the feature set is discrete; this simplifies notation and allows the use of summation and probabilities rather than integration and probability densities. The general conclusions hold in the continuous case as well, but the required technical details would cloud our discussion.

Let $\mathcal{H}$ denote the (discrete) set of hypotheses, or possible sets of parameters to be learned. A particular hypothesis $h(\mathbf{x}) \in \mathcal{H}$ could be described by quantized weights in a neural network, or parameters $\boldsymbol{\theta}$ in a functional model, or sets of decisions in a tree, and so on. Furthermore, $P(h)$ is the prior probability that the algorithm will produce hypothesis h after training; note that this is *not* the probability that h is correct. Next, $P(h|\mathcal{D})$ denotes the probability that the algorithm will yield hypothesis h when trained on the data $\mathcal{D}$. In deterministic learning algorithms such as the nearest-neighbor and decision trees, $P(h|\mathcal{D})$ will be everywhere zero except for a single hypothesis h. For stochastic methods (such as neural networks trained from random initial weights), or stochastic Boltzmann learning, $P(h|\mathcal{D})$ can be a broad distribution. Let E be the error for a zero-one or other loss function.

How shall we judge the generalization quality of a learning algorithm? Because we are not given the target function, the natural measure is the expected value of the error given $\mathcal{D}$, summed over all possible targets. This scalar value can be expressed as a weighted "inner product" between the distributions $P(h|\mathcal{D})$ and $P(F|\mathcal{D})$, as follows:

$$\mathcal{E}[E|\mathcal{D}] = \sum_{h,F} \sum_{\mathbf{x} \notin \mathcal{D}} P(\mathbf{x})[1 - \delta(F(\mathbf{x}), h(\mathbf{x}))] P(h|\mathcal{D}) P(F|\mathcal{D}), \tag{1}$$

where for the moment we assume there is no noise. The familiar Kronecker delta function, $\delta(\cdot, \cdot)$, has value 1 if its two arguments match, and value 0 otherwise. Equation 1 states that the expected error rate, given a fixed training set $\mathcal{D}$, is related to the sum over all possible inputs weighted by their probabilities, $P(\mathbf{x})$, as well as the "alignment" or "match" of the learning algorithm, $P(h|\mathcal{D})$, to the actual posterior $P(F|\mathcal{D})$. The important insight provided by this equation is that without prior knowledge concerning $P(F|\mathcal{D})$, we can prove little about any particular learning algorithm $P(h|\mathcal{D})$, including its generalization performance.

The expected off-training-set classification error when the true function is $F(\mathbf{x})$ and the probability for the kth candidate learning algorithm is $P_k(h(\mathbf{x})|\mathcal{D})$ is given by

$$\mathcal{E}_k(E|F, n) = \sum_{\mathbf{x} \notin \mathcal{D}} P(\mathbf{x})[1 - \delta(F(\mathbf{x}), h(\mathbf{x}))]P_k(h(\mathbf{x})|\mathcal{D}). \qquad (2)$$

Although we shall not provide a formal proof, we are now in a position to give a precise statement of the No Free Lunch Theorem.*

■ **Theorem 9.1. (No Free Lunch)** For any two learning algorithms $P_1(h|\mathcal{D})$ and $P_2(h|\mathcal{D})$, the following are true, independent of the sampling distribution $P(\mathbf{x})$ and the number n of training points:

1. Uniformly averaged over all target functions F, $\mathcal{E}_1(E|F, n) - \mathcal{E}_2(E|F, n) = 0$
2. For any fixed training set $\mathcal{D}$, uniformly averaged over F, $\mathcal{E}_1(E|F, \mathcal{D}) - \mathcal{E}_2(E|F, \mathcal{D}) = 0$
3. Uniformly averaged over all priors $P(F)$, $\mathcal{E}_1(E|n) - \mathcal{E}_2(E|n) = 0$
4. For any fixed training set $\mathcal{D}$, uniformly averaged over $P(F)$, $\mathcal{E}_1(E|\mathcal{D}) - \mathcal{E}_2(E|\mathcal{D}) = 0$

Part 1 says that uniformly averaged over all target functions the expected off-training set error for all learning algorithms is the same, that is,

$$\sum_F \sum_\mathcal{D} P(\mathcal{D}|F)[\mathcal{E}_1(E|F, n) - \mathcal{E}_2(E|F, n)] = 0, \qquad (3)$$

for any two learning algorithms. In short, no matter how clever we are at choosing a "good" learning algorithm $P_1(h|\mathcal{D})$ and a "bad" algorithm $P_2(h|\mathcal{D})$ (perhaps even random guessing, or a constant output), if all target functions are equally likely, then the "good" algorithm will not outperform the "bad" one. Stated more generally, there are no i and j such that for all $F(\mathbf{x})$, $\mathcal{E}_i(E|F, n) > \mathcal{E}_j(E|F, n)$. Furthermore, no matter what algorithm we use, there is at least one target function for which random guessing is a better algorithm.

With the assumption that the training set can be learned by all algorithms we consider, Part 2 states that even if we know $\mathcal{D}$, then averaged over all target functions no learning algorithm yields an off-training set error that is superior to any other, that is,

$$\sum_F [\mathcal{E}_1(E|F, \mathcal{D}) - \mathcal{E}_2(E|F, \mathcal{D})] = 0. \qquad (4)$$

Parts 3 and 4 concern nonuniform target function distributions and have related interpretations (Problems 2–5). Example 1 provides an elementary illustration.

---

*The clever name for the theorem was suggested by David Haussler.

EXAMPLE 1   No Free Lunch for Binary Data

Consider input vectors consisting of three binary features, and a particular target function $F(\mathbf{x})$, as given in the table. Suppose (deterministic) learning algorithm 1 assumes every pattern is in category ω_1 unless trained otherwise, and algorithm 2 assumes every pattern is in ω_2 unless trained otherwise. Thus when trained with $n = 3$ points in $\mathcal{D}$, each algorithm returns a single hypothesis, h_1 and h_2, respectively. In this case the expected errors on the off-training set data are $\mathcal{E}_1(E|F, \mathcal{D}) = 0.4$ and $\mathcal{E}_2(E|F, \mathcal{D}) = 0.6$.

	x	F	h_1	h_2
	000	1	1	1
$\mathcal{D}$	001	-1	-1	-1
	010	1	1	1
	011	-1	1	-1
	100	1	1	-1
	101	-1	1	-1
	110	1	1	-1
	111	1	1	-1

For this target function $F(\mathbf{x})$, clearly algorithm 1 is superior to algorithm 2. But note that the designer does not *know* $F(\mathbf{x})$—indeed, we assume we have no prior information about $F(\mathbf{x})$. The fact that all targets are equally likely means that $\mathcal{D}$ provides no information about $F(\mathbf{x})$. If we wish to compare the algorithms overall, we therefore must average over all such possible target functions consistent with the training data. Part 2 of Theorem 9.1 states that averaged over all possible target functions, there is no difference in off-training set errors between the two algorithms. For each of the 2^5 distinct target functions consistent with the $n = 3$ patterns in $\mathcal{D}$, there is exactly one other target function whose output is inverted for each of the patterns outside the training set, and this ensures that the performances of algorithms 1 and 2 will also be inverted, so that the contributions to the formula in Part 2 cancel. Thus indeed Part 2 of the Theorem as well as Eq. 4 are obeyed.

Figure 9.1 illustrates a result derivable from Part 1 of Theorem 9.1. Each of the six squares represents the set of all possible classification problems; note that this is *not* the standard feature space. If a learning system performs well—higher than average generalization accuracy—over some set of problems, then it must perform worse than average elsewhere, as shown in Fig. 9.1 a. No system can perform well throughout the full set of functions (Fig. 9.1 d); to do so would violate the No Free Lunch Theorem.

In sum, all statements of the form "learning/recognition algorithm 1 is better than algorithm 2" are ultimately statements about the relevant target functions. Hence there is a "conservation theorem" in generalization: For every possible learning algorithm for binary classification the sum of performance over all possible target functions is exactly zero. Thus we cannot achieve positive performance on some problems without getting an equal and opposite amount of negative performance on other problems. While we may hope that we never have to apply any particular algorithm to certain problems, all we can do is trade performance on problems we do not expect to encounter with those that we do expect to encounter. This, along with the other results from the No Free Lunch Theorem, stresses that it is the *assumptions*

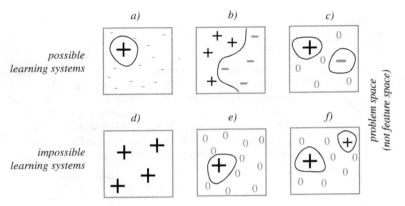

FIGURE 9.1. The No Free Lunch Theorem shows the generalization performance on the off-training set data that *can* be achieved (top row) and also shows the performance that *cannot* be achieved (bottom row). Each square represents all possible classification problems consistent with the training data—this is *not* the familiar feature space. A + indicates that the classification algorithm has generalization higher than average, a − indicates lower than average, and a 0 indicates average performance. The size of a symbol indicates the amount by which the performance differs from the average. For instance, part a shows that it is possible for an algorithm to have high accuracy on a small set of problems so long as it has mildly poor performance on all other problems. Likewise, part b shows that it is possible to have excellent performance throughout a large range of problem, but this will be balanced by very poor performance on a large range of other problems. It is impossible, however, to have good performance throughout the full range of problems, shown in part d. It is also impossible to have higher-than-average performance on some problems while having average performance everywhere else, shown in part e.

about the learning domains that are relevant. Another practical import of the theorem is that even popular and theoretically grounded algorithms will perform poorly on some problems, ones in which the learning algorithm and the posterior happen not to be "matched," as governed by Eq. 1. Practitioners must be aware of this possibility, which arises in real-world applications. Expertise limited to a small range of methods, even powerful ones such as neural networks, will not suffice for all classification problems. Experience with a broad range of techniques is the best insurance for solving arbitrary new classification problems.

★9.2.2 Ugly Duckling Theorem

While the No Free Lunch Theorem shows that in the absence of assumptions we should not prefer any learning or classification algorithm over another, an analogous theorem addresses features and patterns. Roughly speaking, the Ugly Duckling Theorem states that in the absence of assumptions there is no privileged or "best" feature representation, and that even the notion of similarity between patterns depends implicitly on assumptions that may or may not be correct.

Because we are using discrete representations, we can use logical expressions or "predicates" to describe a pattern, much as in Chapter 8. If we denote a binary feature attribute by f_i, then a particular pattern might be described by the predicate "f_1 AND f_2," another pattern might be described as "NOT f_2," and so on. Likewise

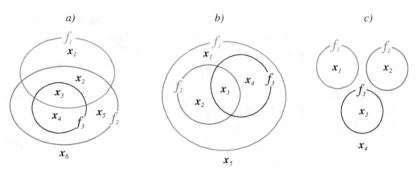

FIGURE 9.2. Patterns $\mathbf{x}_i$, represented as d-tuples of binary features f_i, can be placed in Venn diagram (here $d = 3$); the diagram itself depends upon the classification problem and its constraints. For instance, suppose f_1 is the binary feature attribute `has_legs`, f_2 is `has_right_arm` and f_3 the attribute `has_right_hand`. Thus in part a pattern $\mathbf{x}_1$ denotes a person who has legs but neither arm nor hand; $\mathbf{x}_2$ a person who has legs and an arm, but no hand; and so on. Notice that the Venn diagram expresses the biological constraints associated with real people: it is impossible for someone to have a right hand but no right arm. Part c expresses different constraints, such as the biological constraint of mutually exclusive eye colors. Thus attributes f_1, f_2 and f_3 might denote `brown`, `green`, and `blue`, respectively, and a pattern $\mathbf{x}_i$ describes a real person, whom we can assume cannot have eyes that differ in color.

we could have a predicate involving the patterns themselves, such as $\mathbf{x}_1$ *OR* $\mathbf{x}_2$. Figure 9.2 shows how patterns can be represented in a Venn diagram.

Below we shall need to count predicates, and for clarity it helps to consider a particular Venn diagram, such as that in Fig. 9.3. This is the most general Venn diagram based on two features, because for every configuration of f_1 and f_2 there is indeed a pattern. Here predicates can be as simple as "$\mathbf{x}_1$," or more complicated, such as "$\mathbf{x}_1$ *OR* $\mathbf{x}_2$ *OR* $\mathbf{x}_4$," and so on.

RANK

The *rank r* of a predicate is the number of the simplest or indivisible elements it contains. The tables below show the predicates of rank 1, 2, and 3 associated with the Venn diagram of Fig. 9.3.* Not shown is the fact that there is but one predicate of rank $r = 4$, the disjunction of the $\mathbf{x}_1, \ldots, \mathbf{x}_4$, which has the logical value True. If we

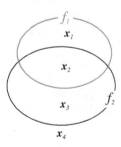

FIGURE 9.3. The Venn for a problem with no constraints on two features. Thus all four binary attribute vectors can occur.

---

*Technically speaking, we should use set operations rather than logical operations when discussing the Venn diagram, writing $\mathbf{x}_1 \cup \mathbf{x}_2$ instead of $\mathbf{x}_1$ OR $\mathbf{x}_2$. Nevertheless, we use logical operations here for consistency with the rest of the text.

let n be the total number of regions in the Venn diagram (i.e., the number of distinct possible patterns), then there are $\binom{n}{r}$ predicates of rank r, as shown at the bottom of the table.

rank $r = 1$

$\mathbf{x}_1$	f_1 AND NOT f_2
$\mathbf{x}_2$	f_1 AND f_2
$\mathbf{x}_3$	f_2 AND NOT f_1
$\mathbf{x}_4$	NOT $(f_1$ OR $f_2)$

$\binom{4}{1} = 4$

rank $r = 2$

$\mathbf{x}_1$ OR $\mathbf{x}_2$	f_1
$\mathbf{x}_1$ OR $\mathbf{x}_3$	f_1 XOR f_2
$\mathbf{x}_1$ OR $\mathbf{x}_4$	NOT f_2
$\mathbf{x}_2$ OR $\mathbf{x}_3$	f_2
$\mathbf{x}_2$ OR $\mathbf{x}_4$	NOT$(f_1$ AND $f_2)$
$\mathbf{x}_3$ OR $\mathbf{x}_4$	NOT f_1

$\binom{4}{2} = 6$

rank $r = 3$

$\mathbf{x}_1$ OR $\mathbf{x}_2$ OR $\mathbf{x}_3$	f_1 OR f_2
$\mathbf{x}_1$ OR $\mathbf{x}_2$ OR $\mathbf{x}_4$	f_1 OR NOT f_2
$\mathbf{x}_1$ OR $\mathbf{x}_3$ OR $\mathbf{x}_4$	NOT$(f_1$ AND $f_2)$
$\mathbf{x}_2$ OR $\mathbf{x}_3$ OR $\mathbf{x}_4$	f_2 OR NOTf_1

$\binom{4}{3} = 4$

The total number of predicates in the absence of constraints is

$$\sum_{r=0}^{n} \binom{n}{r} = (1 + 1)^n = 2^n,\tag{5}$$

and thus for the $d = 4$ case of Fig. 9.3, there are $2^4 = 16$ possible predicates (Problem 9). Note that Eq. 5 applies only to the case where there are no constraints; for Venn diagrams that do incorporate constraints, such as those in Fig. 9.2, the formula does not hold (Problem 10).

Now we turn to our central question: In the absence of prior information, is there a principled reason to judge any two distinct patterns as more or less similar than two other distinct patterns? A natural and familiar measure of similarity is the number of features or attributes shared by two patterns, but even such an obvious measure presents conceptual difficulties.

To appreciate such difficulties, consider first a simple example. Suppose attributes f_1 and f_2 represent blind_in_right_eye and blind_in_left_eye, respectively. If we base similarity on shared features, person $\mathbf{x}_1 = \{1, 0\}$ (blind only in the right eye) is maximally different from person $\mathbf{x}_2 = \{0, 1\}$ (blind only in the left eye). In particular, in this scheme $\mathbf{x}_1$ is more similar to a totally blind person and to a normally sighted person than he is to $\mathbf{x}_2$. But this result may prove unsatisfactory; we can easily envision many circumstances where we would consider a person blind in just the right eye to be "similar" to one blind in just the left eye. Such people might be permitted to drive automobiles, for instance. Furthermore, a person blind in just one eye would differ significantly from totally blind person who would not be able to drive.

A second, related point is that there are always multiple ways to represent vectors (or tuples) of attributes. For instance, in the above example, we might use alternative features f_1' and f_2' to represent blind_in_right_eye and same_in_both_eyes, respectively, and then the four types of people would be represented as shown in the tables.

	f_1	f_2		f_1'	f_2'
$\mathbf{x}_1$	0	0		0	1
$\mathbf{x}_2$	0	1		0	0
$\mathbf{x}_3$	1	0		1	0
$\mathbf{x}_4$	1	1		1	1

Of course there are other representations, each more or less appropriate to the particular problem at hand. In the absence of prior information, though, there is no principled reason to prefer one of these representations over another.

We must then still confront the problem of finding a principled measure of the similarity between two patterns, given some representation. The only plausible candidate measure in this circumstance would be the number of *predicates* (rather than the number of features) the patterns share. Consider two distinct patterns (in some representation) $\mathbf{x}_i$ and $\mathbf{x}_j$, where $i \neq j$. Regardless of the constraints in the problem (i.e., the Venn diagram), there are, of course, no predicates of rank $r = 1$ that are shared by the two patterns. There is but one predicate of rank $r = 2$; it is $\mathbf{x}_i \ OR \ \mathbf{x}_j$. A predicate of rank $r = 3$ must contain three patterns, two of which are $\mathbf{x}_i$ and $\mathbf{x}_j$. Because there are d patterns total, there are then $\binom{d-2}{1} = d - 2$ predicates of rank 3 that are shared by $\mathbf{x}_i$ and $\mathbf{x}_j$. Likewise, for an arbitrary rank r, there are $\binom{d-2}{r-2}$ predicates shared by the two patterns, where $2 \leq r \leq d$. The total number of predicates shared by the two patterns is thus the sum

$$\sum_{r-2}^{d} \binom{d-2}{r-2} = (1+1)^{d-2} = 2^{d-2}. \tag{6}$$

Note the key result: Equation 6 is *independent* of the choice of $\mathbf{x}_i$ and $\mathbf{x}_j$ (so long as they are distinct). Thus we conclude that the number of predicates shared by two distinct patterns is *constant*, and *independent* of the patterns themselves (Problem 11). We conclude that if we judge similarity based on the number of predicates that patterns share, then any two distinct patterns are "equally similar." This is stated formally as the following theorem:

▪ **Theorem 9.2. (Ugly Duckling)** Given that we use a finite set of predicates that enables us to distinguish any two patterns under consideration, the number of predicates shared by any two such patterns is constant and independent of the choice of those patterns. Furthermore, if pattern similarity is based on the total number of predicates shared by two patterns, then any two patterns are "equally similar." *

In summary, then, the Ugly Duckling Theorem states something quite basic yet often overlooked: There is no problem-independent or privileged or "best" set of features or feature attributes. Moreover, while the above was derived using d-tuples of binary values, it also applies to a continuous feature space too, if such a space is discretized (at any resolution). The theorem forces us to acknowledge that even the apparently simple notion of similarity between patterns is fundamentally based on implicit assumptions about the problem domain (Problem 12).

9.2.3 Minimum Description Length (MDL)

It is sometimes claimed that the minimum description length principle provides justification for preferring one type of classifier over another—specifically, "simpler" classifiers over "complex" ones. Briefly stated, the approach purports to find some irreducible, smallest representation of all members of a category (much like a "signal"); all variation among the individual patterns is then "noise." The argument is

---

*The theorem gets its fanciful name from the following counterintuitive statement: Assuming that similarity is based on the number of shared predicates, an ugly duckling A is as similar to beautiful swan B as beautiful swan C is to B, given that these items differ at all from one another.

that by simplifying recognizers appropriately, the signal can be retained while the noise is ignored. Because the principle is so often invoked, it is important to understand what properly derives from it, what does not, and how it relates to the No Free Lunch Theorem. To do so, however, we must first understand the notion of algorithmic complexity.

Algorithmic Complexity

Algorithmic complexity—also known as Kolmogorov complexity, Kolmogorov-Chaitin complexity, descriptional complexity, shortest program length, or algorithmic entropy—seeks the inherent complexity of a binary string. (We shall assume that both classifiers and patterns are described by such strings.) Algorithmic complexity can be explained by analogy to communication, the earliest application of information theory (Section A.7 of the Appendix). If the sender and receiver agree upon a specification method, such as an encoding or compression technique, then message x can then be transmitted as y and decoded given some fixed method L, denoted $L(y) = x$. The cost of transmission of x is the length of the transmitted message y, that is, $|y|$. The least such cost is hence the minimum length of such a message, denoted $\min_{y:L(y)=x} |y|$.

ABSTRACT
COMPUTER

Algorithmic complexity is defined by analogy to communication, where instead of a fixed decoding method L, we consider programs running on an *abstract computer*—that is, one whose functions (memory, processing, etc.) are described operationally and without regard to hardware implementation. Consider an abstract computer that takes as a program a binary string y and outputs a string x and halts. In such a case we say that y is an abstract encoding or description of x.

A *universal* description should be independent of the specification (up to some additive constant)—it should not depend upon whether our specification is programmed in *C++*, *Lisp*, *Java*, and so on. Such a description would then allow us to reliably compare the complexities of different binary strings x_1 and x_2. Such a method would provide a measure of the inherent information content, the amount of data that must be transmitted in the absence of any other prior knowledge. The Kolmogorov complexity of a binary string x, denoted $K(x)$, is defined as the size of the *shortest* program string y (where y's length is measured in bits), that, without additional data, computes the string x and halts. Formally, we write

$$K(x) = \min_{y:U(y)=x} |y|, \tag{7}$$

TURING
MACHINE

where U represents an abstract universal *Turing machine* or Turing computer. For our purposes it suffices to state that a Turing machine is "universal" in that it can implement any algorithm and compute any computable function. Kolmogorov complexity is a measure of the incompressibility of x and is analogous to minimal sufficient statistics, the optimally compressed representation of certain properties of a distribution (Chapter 3).

Consider the following examples. Suppose x consists solely of n 1s. This string is actually quite "simple." If we use some fixed number of bits k to specify a general program containing a loop for printing a string of 1s, we need merely $\log_2 n$ more bits to specify the iteration number n, the condition for halting. Thus the Kolmogorov complexity of a string of n 1s is $K(x) = O(\log_2 n)$. Next consider the transcendental number π, whose infinite sequence of seemingly random binary digits, $11.00100100001111110110101010001\ldots_2$, actually contains only a few bits of information: the size of the shortest program that can produce any arbitrarily large number of consecutive digits of π. Informally we say the algorithmic complexity of π is a constant; formally we write $K(\pi) = O(1)$, which means that $K(\pi)$ does not grow with increasing number of desired bits. Another example is a "truly" random binary string, which cannot be expressed as a shorter string; its algorithmic complexity is within a constant factor of its length. For such a string we write $K(x) = O(|x|)$, which means that $K(x)$ grows as fast as the length of x (Problem 13).

9.2.4 Minimum Description Length Principle

We now turn to a simple, "naive" version of the minimum description length principle and its application to pattern recognition. Given that all members of a category share some properties, yet differ in others, the recognizer should seek to learn the common or essential characteristics while ignoring the accidental or random ones. Kolmogorov complexity seeks to provide an objective measure of simplicity, and thus the description of the "essential" characteristics.

Suppose we seek to design a classifier using a training set $\mathcal{D}$. The *minimum description length (MDL) principle* states that we should minimize the sum of the model's algorithmic complexity and the description of the training data with respect to that model, that is,

$$K(h, \mathcal{D}) = K(h) + K(\mathcal{D} \text{ using } h). \tag{8}$$

Thus we seek the model h^* that obeys $h^* = \arg\min_h K(h, \mathcal{D})$, as explored in Problem 14. (Variations on the naive minimum description length principle use a *weighted* sum of the terms in Eq. 8.) In practice, determining the algorithmic complexity of a classifier depends upon a chosen class of abstract computers, and this means that the complexity can be specified only up to an additive constant.

A particularly clear application of the minimum description length principle is in the design of decision-tree classifiers (Chapter 8). In this case, a model h specifies the tree and the decisions at the nodes; thus the algorithmic complexity of the model is proportional to the number of nodes. The complexity of the data given the model could be expressed in terms of the entropy (in bits) of the data $\mathcal{D}$, the weighted sum of the entropies of the data at the leaf nodes. Thus if the tree is pruned based on an entropy criterion, there is an implicit global cost criterion that is equivalent to minimizing a measure of the general form in Eq. 8 (Computer exercise 1).

It can be shown theoretically that classifiers designed with a minimum description length principle are guaranteed to converge to the ideal or true model *in the limit of more and more data*. This is surely a very desirable property. However, such derivations cannot prove that the principle leads to superior performance in the *finite* data case; to do so would violate the No Free Lunch Theorem. Moreover, in practice it is often difficult to compute the minimum description length, because we may not be clever enough to find the "best" representation (Problem 17). Assume that there is some correspondence between a particular classifier and an abstract computer; in such a case it may be quite simple to determine the length of the string y necessary to create the classifier. But because finding the algorithmic complexity demands we find the *shortest* such string, we must perform a very difficult search through possible programs that could generate the classifier.

The minimum description length principle can be viewed from a Bayesian perspective. Using our current terminology, Bayes formula states

$$P(h|\mathcal{D}) = \frac{P(h)P(\mathcal{D}|h)}{P(\mathcal{D})} \tag{9}$$

for discrete hypotheses and data. The optimal hypothesis h^* is the one yielding the highest posterior probability, that is,

$$h^* = \arg\max_h[P(h)P(\mathcal{D}|h)]$$

$$= \arg\max_h[\log_2 P(h) + \log_2 P(\mathcal{D}|h)], \tag{10}$$

much as we saw in Chapter 3. We note that a string x can be communicated or represented at a cost bounded below by $-\log_2 P(x)$, as stated in Shannon's optimal coding theorem. Shannon's theorem thus provides a link between the minimum description length (Eq. 8) and the Bayesian approaches (Eq. 10). The minimum description length principle states that simple models (small $K(h)$) are to be preferred, and thus amounts to a bias toward "simplicity." It is often easier in practice to specify such a prior in terms of a description length than it is using functions of distributions (Problem 16). We shall revisit the issue of the trade-off between simplifying the model and fitting the data in the bias-variance dilemma in Section 9.3.

It is found empirically that classifiers designed using the minimum description length principle work well in many problems. As mentioned, the principle is effectively a method for biasing priors over models toward "simple" models. The reasons for the many empirical success of the principle are not trivial, as we shall see in Section 9.2.5. One of the greatest benefits of the principle is that it provides a computationally clear approach to balancing model complexity and the fit of the data. In somewhat more heuristic methods, such as pruning neural networks, it is difficult to compare the algorithmic complexity of the network (e.g., number of units or weights) with the entropy of the data with respect to that model.

9.2.5 Overfitting Avoidance and Occam's Razor

Throughout our discussions of pattern classifiers, we have mentioned the need to avoid overfitting by means of regularization, pruning, inclusion of penalty terms, minimizing a description length, and so on. The No Free Lunch results throw such techniques into question. If there are no problem-independent reasons to prefer one algorithm over another, why is overfitting avoidance nearly universally advocated? For a given training error, why do we generally advocate simple classifiers with fewer features and parameters?

In fact, techniques for avoiding overfitting or minimizing description length are not inherently beneficial; instead, such techniques amount to a preference, or "bias," over the forms or parameters of classifiers. They are only beneficial if they happen to address problems for which they work. It is the match of the learning algorithm to the *problem*—not the imposition of overfitting avoidance—that determines the empirical success. There are problems for which overfitting avoidance actually leads to worse performance. The effects of overfitting avoidance depend upon the choice of representation too; if the feature space is mapped to a new, formally equivalent one, overfitting avoidance has different effects.

In light of the negative results from the No Free Lunch theorems, we might probe more deeply into the frequent empirical "successes" of the minimum description length principle and the more general philosophical principle of Occam's razor. In its original form, Occam's razor stated merely that "entities" (or explanations) should not be multiplied beyond necessity, but it has come to be interpreted in pattern recognition as counseling that one should not use classifiers that are more complicated than are necessary, where "necessary" is determined by the quality of fit to the training data. Given the respective requisite assumptions, the No Free Lunch theorem proves that there is no benefit in "simple" classifiers (or "complex" ones, for that matter)— simple classifiers claim neither unique nor universal validity.

The frequent empirical "successes" of Occam's razor imply that the classes of problems addressed so far have certain properties. What might be the reason we explore problems that tend to favor simpler classifiers? A reasonable hypothesis is that through evolution, we have had strong selection pressure on our pattern recognition apparatuses to be computationally simple—require fewer neurons, less time, and so forth—and in general such classifiers tend to be "simple." We are more likely to ignore problems for which Occam's razor does not hold. Analogously, researchers naturally develop simple algorithms before more complex ones, as for instance in the progression from the Perceptron, to multilayer neural networks, to networks with pruning, to networks with topology learning, to hybrid neural net/rule-based methods, and so on—each more complex than its predecessor. Each method is found to work on some problems, but not ones that are "too complex." For instance the basic Perceptron is inadequate for optical character recognition; a simple three-layer neural network is inadequate for speaker-independent speech recognition. Hence our design methodology itself imposes a bias toward "simple" classifiers; we generally stop searching for a design when the classifier is "good enough." This principle of SATISFICING *satisficing*—creating an adequate though possibly nonoptimal solution—underlies much of practical pattern recognition as well as human cognition.

Another "justification" for Occam's razor derives from a property we might strongly desire or expect in a learning algorithm. If we assume that adding more training data does not, on average, degrade the generalization accuracy of a classifier, then a version of Occam's razor can in fact be derived. Note, however, that such a desired property amounts to a nonuniform prior over target functions; while this property is surely desirable, it is a premise and cannot be "proven." Finally, the No Free Lunch theorem implies that we cannot use training data to create a scheme by which we can with some assurance distinguish new problems for which the classifier will generalize well from new problems for which the classifier will generalize poorly (Problem 8).

9.3 BIAS AND VARIANCE

Given that there is no general best classifier unless the probability over the class of problems is restricted, practitioners must be prepared to explore a number of methods or models when solving any given classification problem. Below we will define two ways to measure the "match" or "alignment" of the learning algorithm to the classification problem: the bias and the variance. The bias measures the accuracy or quality of the match: high bias implies a *poor* match. The variance measures the precision or specificity of the match: a high variance implies a *weak* match. Designers can adjust the bias and variance of classifiers, but the important bias-variance relation shows that the two terms are not independent; in fact, for a given mean-square error, they obey a form of "conservation law." Naturally, though, with prior information or even mere luck, classifiers can be created that have a different mean-square error.

9.3.1 Bias and Variance for Regression

Bias and variance are most easily understood in the context of regression or curve fitting. Suppose there is a true (but unknown) function $F(\mathbf{x})$ with continuous valued output with noise, and we seek to estimate $F(\cdot)$ based on n samples in a set $\mathcal{D}$ generated by $F(\mathbf{x})$. The regression function estimated is denoted $g(\mathbf{x}; \mathcal{D})$ and we are interested in the dependence of this approximation on the training set $\mathcal{D}$. Due to random variations in data selection, for some data sets of finite size this approximation will be excellent, while for other data sets of the same size the approximation will be poor. The natural measure of the effectiveness of the estimator can be expressed as its mean-square deviation from the desired optimal. Thus we average over all training sets $\mathcal{D}$ of fixed size n and find (Problem 18)

$$
\mathcal{E}_{\mathcal{D}}\left[(g(\mathbf{x}; \mathcal{D}) - F(\mathbf{x}))^2\right]
$$
$$
= \underbrace{(\mathcal{E}_{\mathcal{D}}[g(\mathbf{x}; \mathcal{D}) - F(\mathbf{x})])^2}_{bias^2} + \underbrace{\mathcal{E}_{\mathcal{D}}\left[(g(\mathbf{x}; \mathcal{D}) - \mathcal{E}_{\mathcal{D}}[g(\mathbf{x}; \mathcal{D})])^2\right]}_{variance}. \quad (11)
$$

BIAS

VARIANCE

The first term on the right-hand side is the *bias* (squared)—the difference between the expected value and the true (but generally unknown) value—while the second term is the *variance*. Thus a low bias means that on average we will accurately estimate F from $\mathcal{D}$. Furthermore, a low variance means that the estimate of F does not change much as the training set varies. Even if an estimator is unbiased (i.e., the $bias = 0$ and its expected value is equal to the true value), there can nevertheless be a large mean-square error arising from a large variance term.

BIAS-VARIANCE DILEMMA

Equation 11 shows that the mean-square error can be expressed as the sum of a bias and a variance term. The *bias-variance dilemma* or *bias-variance trade-off* is a general phenomenon: Procedures with increased flexibility to adapt to the training data (e.g., have more free parameters) tend to have lower bias but higher variance. Different classes of regression functions $g(\mathbf{x}; \mathcal{D})$—linear, quadratic, sum of Gaussians, and so on—will have different overall errors; nevertheless, Eq. 11 will be obeyed.

Suppose, for example, that the true, target function $F(x)$ is a cubic polynomial of one variable, with noise, as illustrated in Fig. 9.4. We seek to estimate this function based on a sampled training set $\mathcal{D}$. Column a shows a very poor "estimate" $g(x)$—a fixed linear function, *independent* of the training data. For different training sets sampled from $F(x)$ with noise, $g(x)$ is unchanged. The histogram of this mean-square error of Eq. 11, shown at the bottom, reveals a spike at a fairly high error; because this estimate is so poor, it has a high bias. Furthermore, the variance of the constant model $g(x)$ is zero. The model in column b is also fixed, but happens to be a better estimate of $F(x)$. It too has zero variance, but a lower bias than the poor model in column a. Presumably the designer imposed some prior knowledge about $F(x)$ in order to get this improved estimate.

The model in column c is a cubic with trainable coefficients; it would learn $F(x)$ exactly if $\mathcal{D}$ contained infinitely many training points. Notice that the fit found for every training set is quite good. Thus the bias is low, as shown in the distribution at the bottom. The model in column d is linear in x, but its slope and intercept are determined from the training data. As such, the model in column d has a lower bias than the models in columns a and b.

In sum, for a given target function $F(x)$, if a model has many parameters (generally low bias), it will fit the data well but yield high variance. Conversely, if the

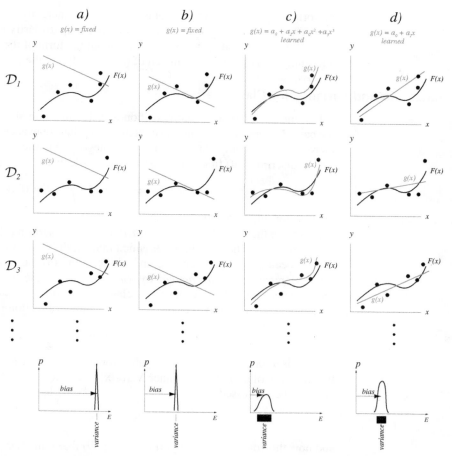

FIGURE 9.4. The bias-variance dilemma can be illustrated in the domain of regression. Each column represents a different model, and each row represents a different set of $n = 6$ training points, $\mathcal{D}_i$, randomly sampled from the true function $F(x)$ with noise. Probability functions of the mean-square error of $E = \mathcal{E}_\mathcal{D}[(g(x) - F(x))^2]$ of Eq. 11 are shown at the bottom. Column a shows a very poor model: a linear $g(x)$ whose parameters are held fixed, *independent* of the training data. This model has high bias and zero variance. Column b shows a somewhat better model, though it too is held fixed, independent of the training data. It has a lower bias than in column a and has the same zero variance. Column c shows a cubic model, where the parameters are trained to best fit the training samples in a mean-square-error sense. This model has low bias and a moderate variance. Column d shows a linear model that is adjusted to fit each training set; this model has intermediate bias and variance. If these models were instead trained with a very large number $n \to \infty$ of points, the bias in column c would approach a small value (which depends upon the noise), while the bias in column d would not; the variance of all models would approach zero.

model has few parameters (generally high bias), it may not fit the data particularly well, but this fit will not change much for different data sets (low variance). The best way to get low bias and low variance is the have prior information about the target function. We can virtually never get zero bias and zero variance; to do so would mean that there is only one learning problem to be solved, in which case the answer is already known. Furthermore, a large amount of training data will yield improved performance so long as the model is sufficiently general to represent the target function. These considerations of bias and variance help to clarify the reasons we seek to have as much accurate prior information about the form of the solution, and as large a training set as feasible; the match of the algorithm to the problem is crucial.

9.3.2 Bias and Variance for Classification

While the bias-variance decomposition and dilemma are simplest to understand in the case of regression, we are most interested in their relevance to classification; here there are a few complications. In a two-category classification problem we let the target (discriminant) function have value 0 or $+1$, that is,

$$F(\mathbf{x}) = \Pr[y = 1|\mathbf{x}] = 1 - \Pr[y = 0|\mathbf{x}]. \tag{12}$$

On first consideration, the mean-square error we saw for *regression* (Eq. 11) does not appear to be the proper one for *classification*. After all, even if the mean-square error fit is poor, we can have accurate classification, possibly even the lowest (Bayes) error. This is because the decision rule under a zero-one loss selects the higher posterior $P(\omega_i|\mathbf{x})$, regardless of the *amount* by which it is higher. Nevertheless, by considering the expected value of y, we can recast classification into the framework of regression we saw before. To do so, we consider a discriminant function

$$y = F(\mathbf{x}) + \epsilon, \tag{13}$$

where ϵ is a zero-mean, random variable, for simplicity here assumed to be a centered binomial distribution with variance $\text{Var}[\epsilon|\mathbf{x}] = F(\mathbf{x})(1 - F(\mathbf{x}))$. The target function can thus be expressed as

$$F(\mathbf{x}) = \mathcal{E}[y|\mathbf{x}], \tag{14}$$

and now the goal is to find an estimate $g(\mathbf{x}; \mathcal{D})$ that minimizes a mean-square error, such as in Eq. 11:

$$\mathcal{E}_{\mathcal{D}}[(g(\mathbf{x}; \mathcal{D}) - y)^2]. \tag{15}$$

In this way the regression methods of Section 9.3.1 can yield an estimate $g(\mathbf{x}; \mathcal{D})$ used for classification.

For simplicity we assume equal priors, $P(\omega_1) = P(\omega_2) = 0.5$, and thus the Bayes discriminant y_B has threshold $1/2$ and the Bayes decision boundary is the set of points for which $F(\mathbf{x}) = 1/2$. For a given training set $\mathcal{D}$, if the classification error rate $\Pr[g(\mathbf{x}; \mathcal{D}) = y]$ averaged over predictions at $\mathbf{x}$ agrees with the Bayes discriminant,

$$\Pr[g(\mathbf{x}; \mathcal{D}) = y] = \Pr[y_B(\mathbf{x}) \neq y] = \min[F(\mathbf{x}), 1 - F(\mathbf{x})], \tag{16}$$

then indeed we have the lowest error. If not, then the prediction yields an increased error

$$\Pr[g(\mathbf{x}; \mathcal{D})] = \max[F(\mathbf{x}), 1 - F(\mathbf{x})] \tag{17}$$

$$= |2F(\mathbf{x}) - 1| + \Pr[y_B(\mathbf{x}) = y].$$

We average over all data sets of size n and find

$$\Pr[g(\mathbf{x}; \mathcal{D}) \neq y] = |2F(\mathbf{x}) - 1|\Pr[g(\mathbf{x}; \mathcal{D}) \neq y_B] + \Pr[y_B \neq y]. \tag{18}$$

Equation 18 shows that classification error rate is linearly proportional to $\Pr[g(\mathbf{x}; \mathcal{D})$ $\neq y_B]$, which can be considered a *boundary error* in that it represents the incorrect estimation of the optimal (Bayes) boundary (Problem 19).

Because of random variations in training sets, the boundary error will depend upon $p(g(\mathbf{x}; \mathcal{D}))$, the probability density of obtaining a particular estimate of the discriminant given $\mathcal{D}$. This error is merely the area of the tail of $p(g(\mathbf{x}; \mathcal{D}))$ on the opposite side of the Bayes discriminant value $1/2$, much as we saw in Chapter 2:

$$\Pr[g(\mathbf{x}; \mathcal{D}) \neq y_B] = \begin{cases} \int_{1/2}^{\infty} p(g(\mathbf{x}; \mathcal{D})) \, dg & \text{if } F(\mathbf{x}) < 1/2 \\[2mm] \int_{-\infty}^{1/2} p(g(\mathbf{x}; \mathcal{D})) \, dg & \text{if } F(\mathbf{x}) \geq 1/2. \end{cases} \tag{19}$$

If we make the convenient assumption that $p(g(\mathbf{x}; \mathcal{D}))$ is a Gaussian, we find (Problem 20)

$$\Pr[g(\mathbf{x}; \mathcal{D}) \neq y_B] = \Phi\left[\text{Sgn}[F(\mathbf{x}) - 1/2]\frac{\mathcal{E}_{\mathcal{D}}[g(\mathbf{x}; \mathcal{D})] - 1/2}{\sqrt{\text{Var}[g(\mathbf{x}; \mathcal{D})]}}\right] \tag{20}$$

$$= \Phi\Big[\underbrace{\text{Sgn}[F(\mathbf{x}) - 1/2][\mathcal{E}_{\mathcal{D}}[g(\mathbf{x}; \mathcal{D})] - 1/2]}_{\textit{boundary bias}} \underbrace{\text{Var}[g(\mathbf{x}; \mathcal{D})]^{-1/2}}_{\textit{variance}}\Big],$$

where

$$\Phi[t] = \frac{1}{\sqrt{2\pi}} \int_{t}^{\infty} e^{-1/2u^2} \, du = \frac{1}{2}[1 - \text{erf}(t/\sqrt{2})] \tag{21}$$

and erf[·] is the familiar error function (Appendix Section A.5).

We have expressed this boundary error in terms of a *boundary bias*, in analogy with the simple bias-variance relation in regression (Eq. 11). Equation 20 shows that the effect of the variance term on the boundary error is highly nonlinear and depends on the value of the boundary bias. Furthermore, when the variance is small, this effect is particularly sensitive to the sign of the bias. In regression the estimation error is *additive* in *bias²* and *variance*, whereas for classification there is a nonlinear and *multiplicative* interaction. In classification the sign of the *boundary bias* affects the role of variance in the error. For this reason low *variance* is generally important for accurate classification, while low *boundary bias* need not be. Or said another way, in classification, *variance* generally dominates *bias*. In practical terms, this implies that we need not be particularly concerned if our estimation is biased, so long as the variance is kept low. Numerous specific methods of classifier adjustment—pruning neural networks or decision trees, varying the number of free parameters, and so on—affect the bias and variance of a classifier; in Section 9.5 we shall discuss some methods applicable to a broad range of classification methods. Much as we saw in the bias-variance dilemma for regression, classification procedures with increased flexibility to adapt to the training data (e.g., have more free parameters) tend to have lower bias but higher variance.

As an illustration of *boundary bias* and *variance* in classifiers, consider a simple two-class problem in which samples are drawn from two-dimensional Gaussian distributions, each parameterized by vectors $p(\mathbf{x}|\omega_i) \sim N(\boldsymbol{\mu}_i, \boldsymbol{\Sigma}_i)$, for $i = 1, 2$. Here the true distributions have diagonal covariances, as shown at the top of Fig. 9.5. We

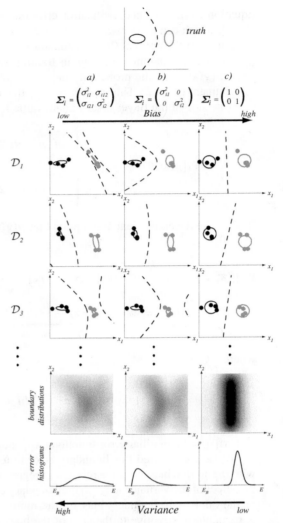

FIGURE 9.5. The (boundary) bias-variance trade-off in classification is illustrated with a two-dimensional Gaussian problem. The figure at the top shows the true distributions and the Bayes decision boundary. The nine figures in the middle show different learned decision boundaries. Each row corresponds to a different training set of $n = 8$ points selected randomly from the true distributions and labeled according to the true decision boundary. Column a shows case of a Gaussian model with fully general covariance matrices trained by maximum-likelihood. The learned boundaries differ significantly from one data set to the next; this learning algorithm has high variance. Column b shows the decision boundaries resulting from fitting a Gaussian model with diagonal covariances; in this case the decision boundaries vary less from one row to another. This learning algorithm has a lower variance than the one at the left. Finally, column c shows decision boundaries learning by fitting a Gaussian model with unit covariances (i.e., a linear model); notice that the decision boundaries are nearly identical from one data set to the next. This algorithm has low variance.

have just a few samples from each category and estimate the parameters in three different classes of models by maximum-likelihood. Column a shows the most general Gaussian classifiers; each component distribution can have arbitrary covariance matrix. Column b shows classifiers where each component Gaussian is constrained to have a diagonal covariance. Column c shows the most restrictive model: The covariances are equal to the identity matrix, yielding circular Gaussian distributions. Thus the left column corresponds to very low bias, and the right column corresponds to high bias.

Each row in Fig. 9.5 represents a different training set, randomly selected from the true distribution (shown at the top), and the resulting classifiers. Notice that most feature points in the high-bias cases retain their classification, regardless of the particular training set (i.e., such models have low variance), whereas the classification of a much larger range of points varies in the low-bias case (i.e., there is high variance). While in general a lower bias comes at the expense of higher variance, the relationship is nonlinear and multiplicative.

At the bottom of the figure, three density plots show how the location of the decision boundary varies across many different training sets. The leftmost density plot shows a very broad distribution (high variance). The rightmost plot shows a narrow, peaked distribution (low variance). To visualize the bias, imagine taking the spatial average of the decision boundaries obtained by running the learning algorithm on all possible data sets. The average of such boundaries for the leftmost algorithm will be equal to the true decision boundary—this algorithm has no bias. The rightmost average will be a vertical line, and hence there will be higher error—this algorithm has the highest bias of the three. Distributions of the generalization error are shown along the bottom.

For a given bias, the variance will decrease as n is increased. Naturally, if we had trained using a very large training set ($n \to \infty$), all error distributions become narrower and move to lower values of E. If a model is rich enough to express the optimal decision boundary, its error distribution for the large n case will approach a delta function at $E = E_B$, the Bayes error. As mentioned, to achieve the desired low generalization error it is more important to have low variance than to have low bias. The only way to get the ideal of zero bias and zero variance is to know the true model ahead of time (or be astoundingly lucky and guess it), in which case no learning was needed anyway. Bias and variance can be lowered with large training size n and accurate prior knowledge of the form of $F(\mathbf{x})$. Furthermore, as n grows, more parameters must be added to the model, g, so the data can be fit (reducing bias). For best classification based on a finite training set, it is desirable to match the form of the model to that of the (unknown) true distributions; this usually requires prior knowledge.

9.4 RESAMPLING FOR ESTIMATING STATISTICS

When we apply some learning algorithm to a new pattern recognition problem with unknown distribution, how can we determine the bias and variance? Figures 9.4 and 9.5 suggest a method using multiple samples, an inspiration for formal "resampling" methods, which we now discuss. Later we shall turn to our ultimate goal: using resampling and related techniques to improve classification (Section 9.5).

9.4.1 Jackknife

We begin with an example of how resampling can be used to yield a more informative estimate of a general statistic. Suppose we have a set $\mathcal{D}$ of n data points x_i ($i = 1, \ldots, n$), sampled from a one-dimensional distribution. The familiar estimate of the mean is, of course,

$$\hat{\mu} = \frac{1}{n} \sum_{i=1}^{n} x_i. \tag{22}$$

Likewise the estimate of the accuracy of the mean is the sample variance, given by

$$\hat{\sigma}^2 = \frac{(n-1)}{n} \sum_{i=1}^{n} (x_i - \hat{\mu})^2. \tag{23}$$

MEDIAN

Suppose we were instead interested in the *median*, the point for which half of the distribution is higher, half lower. Although we could determine the median explicitly, there does not seem to be a straightforward way to generalize Eq. 23 to give a measure of the *error* or spread of our estimate of the median. The same difficulty applies to estimating the *mode* (the most frequently represented point in a data set), the 25th percentile, or any of a large number of statistics other than the mean. The jackknife* and bootstrap (Section 9.4.2) are two of the most popular and theoretically grounded resampling techniques for extending the above approach (based on Eqs. 22 and 23) to *arbitrary* statistics, of which the mean is just one instance.

MODE

In resampling theory, we frequently use statistics in which a data point is eliminated from the data; we denote this by means of a special subscript. For instance, the *leave-one-out* mean is

LEAVE-ONE-OUT
MEAN

$$\mu_{(i)} = \frac{1}{n-1} \sum_{j \neq i}^{n} x_j = \frac{n\bar{x} - x_i}{n-1}, \tag{24}$$

that is, the sample average of the data set if the ith point is deleted. Next we define the jackknife estimate of the mean to be

$$\mu_{(\cdot)} = \frac{1}{n} \sum_{i=1}^{n} \mu_{(i)}, \tag{25}$$

that is, the mean of the leave-one-out means. It is simple to prove that the traditional estimate of the mean and the jackknife estimate of the mean are the same, that is, $\hat{\mu} = \mu_{(\cdot)}$ (Problem 23). Likewise, the jackknife estimate of the variance of the estimate obeys

$$\text{Var}[\hat{\mu}] = \frac{n-1}{n} \sum_{i=1}^{n} (\mu_{(i)} - \mu_{(\cdot)})^2 \tag{26}$$

and, applied to the mean, is equivalent to the traditional variance of Eq. 23 (Problem 26).

---

*The jackknife method, which also goes by the name of "leave one out," was due to Maurice Quenouille. The playful name was chosen by John W. Tukey to capture the impression that the method was handy and useful in lots of ways.

The benefit of expressing the variance in the form of Eq. 26 is that it can be generalized to any other estimator $\hat{\theta}$, such as the median or 25th percentile or mode. To do so we need to compute the statistic with one data point "left out." Thus we let

$$\hat{\theta}_{(i)} = \hat{\theta}(x_1, x_2, \cdots, x_{i-1}, x_{i+1}, \cdots, x_n) \tag{27}$$

take the place of $\mu_{(i)}$ and let $\hat{\theta}_{(\cdot)}$ take the place of $\mu_{(\cdot)}$ in Eqs. 25 and 26 above.

Jackknife Bias Estimate

BIAS

The notion of bias is more general than that described in Section 9.3; in fact it can be applied to the estimation of any statistic. The *bias* of an estimator θ is the difference between its true value and its expected value, that is,

$$bias = \theta - \mathcal{E}[\hat{\theta}]. \tag{28}$$

The jackknife method can be used estimate such a bias. The procedure is first to sequentially delete points x_i one at a time from $\mathcal{D}$ and compute the estimate $\hat{\theta}_{(\cdot)}$, where $\hat{\theta}_{(\cdot)} = \frac{1}{n} \sum_{i=1}^{n} \hat{\theta}_{(i)}$. Then the jackknife estimate of the bias is (Problem 21)

$$bias_{jack} = (n - 1)(\hat{\theta}_{(\cdot)} - \hat{\theta}). \tag{29}$$

We rearrange terms and thus see that the jackknife estimate of $\hat{\theta}$ is

$$\tilde{\theta} = \hat{\theta} - bias_{jack} = n\hat{\theta} - (n - 1)\hat{\theta}_{(\cdot)}. \tag{30}$$

The benefit of using Eq. 30 is that it is unbiased for estimating the true bias (Problem 25).

Jackknife Variance Estimate

Now we seek the jackknife estimate of the variance of an arbitrary statistic θ. First, recall that the traditional variance is defined as

$$\text{Var}[\hat{\theta}] = \mathcal{E}\big[[\hat{\theta}(x_1, x_2, \ldots, x_n) - \mathcal{E}[\hat{\theta}]]^2\big]. \tag{31}$$

The jackknife estimate of the variance, defined by analogy to Eq. 26, is

$$\text{Var}_{jack}[\hat{\theta}] = \frac{n - 1}{n} \sum_{i=1}^{n} [\hat{\theta}_{(i)} - \hat{\theta}_{(\cdot)}]^2. \tag{32}$$

EXAMPLE 2   Jackknife Estimate of Bias and Variance of the Mode

---

Consider an elementary example where we are interested in the mode of the following $n = 6$ points: $\mathcal{D} = \{0, 10, 10, 10, 20, 20\}$. It is clear from the histogram that the most frequently represented point is $\hat{\theta} = 10$. The jackknife estimate of the mode is

$$\hat{\theta}_{(\cdot)} = \frac{1}{n} \sum_{i=1}^{n} \hat{\theta}_{(i)} = \frac{1}{6}[10 + 15 + 15 + 15 + 10 + 10] = 12.5,$$

where for $i = 2, 3, 4$ we used the fact that the mode of a distribution having two equal peaks is the point midway between those peaks. The fact that $\hat{\theta}_{(\cdot)} > \hat{\theta}$ reveals immediately that the jackknife estimate takes into account more of the full (skewed) distribution than does the standard mode calculation.

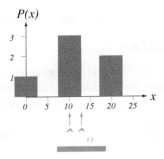

A histogram of $n = 6$ points whose mode is $\hat{\theta} = 10$ and whose jackknife estimate of the mode is $\hat{\theta}_{(\cdot)} = 12.5$. The square root of the jackknife estimate of the variance is a natural measure of the spread of probable values of the mode. This spread is indicated by the horizontal red bar.

The jackknife estimate of the bias of the estimate of the mode is given by Eq. 29:

$$bias_{jack} = (n - 1)(\hat{\theta}_{(\cdot)} - \hat{\theta}) = 5(12.5 - 10) = 12.5.$$

Likewise, the jackknife estimate of the variance is given by Eq. 32:

$$Var_{jack}[\hat{\theta}] = \frac{n-1}{n} \sum_{i=1}^{n} (\hat{\theta}_{(i)} - \hat{\theta}_{(\cdot)})^2$$

$$= \frac{5}{6}[(10 - 12.5)^2 + 3(15 - 12.5)^2 + 2(10 - 12.5)^2] = 31.25.$$

The square root of this variance, $\sqrt{31.25} \simeq 5.6$, serves as an effective standard deviation. A red bar of twice this width, shown below the histogram, reveals that the traditional mode lies within this tolerance to the jackknife estimate of the mode.

The jackknife resampling technique often gives us a more satisfactory estimate of a statistic such as the mode than do traditional methods, though it is more computationally complex (Problem 27).

9.4.2 Bootstrap

A "bootstrap" data set is one created by randomly selecting n points from the training set $\mathcal{D}$, with replacement. (Because $\mathcal{D}$ itself contains n points, there is nearly always duplication of individual points in a bootstrap data set.) In bootstrap estimation,* this selection process is independently repeated B times to yield B bootstrap data

---

*"Bootstrap" comes from subsequent versions of Rudolf Erich Raspe's wonderful stories "The adventures of Baron Munchhausen," in which the hero could pull himself up onto his horse by lifting his own bootstraps. A different but more common usage of the term applies to starting a computer, which must first run a program before it can run other programs.

sets, which are treated as independent sets. The bootstrap estimate of a statistic θ, denoted $\hat{\theta}^{*(\cdot)}$, is merely the mean of the B estimates on the individual bootstrap data sets:

$$\hat{\theta}^{*(\cdot)} = \frac{1}{B} \sum_{b=1}^{B} \hat{\theta}^{*(b)}, \tag{33}$$

where $\hat{\theta}^{*(b)}$ is the estimate on bootstrap sample b.

Bootstrap Bias Estimate

The bootstrap estimate of the bias is (Problem 28)

$$bias_{boot} = \frac{1}{B} \sum_{b=1}^{B} \hat{\theta}^{*(b)} - \hat{\theta} = \hat{\theta}^{*(\cdot)} - \hat{\theta}. \tag{34}$$

TRIMMED MEAN

Computer exercise 3 shows how the bootstrap can be applied to statistics that resist computational analysis, such as the "trimmed mean," in which the mean is calculated for a distribution in which some percentage (e.g., 5%) of the high and the low points in a distribution have been eliminated.

Bootstrap Variance Estimate

The bootstrap estimate of the variance is

$$\text{Var}_{boot}[\theta] = \frac{1}{B} \sum_{b=1}^{B} \left[\hat{\theta}^{*(b)} - \hat{\theta}^{*(\cdot)} \right]^2. \tag{35}$$

If the statistic θ is the mean, then in the limit of $B \to \infty$, the bootstrap estimate of the variance is the traditional variance of the mean (Problem 22). Generally speaking, the larger the number B of bootstrap samples, the more satisfactory is the estimate of a statistic and its variance. One of the benefits of bootstrap estimation is that B can be adjusted to the computational resources; if powerful computers are available for a long time, then B can be chosen large. In contrast, a jackknife estimate requires exactly n repetitions: Fewer repetitions gives a poorer estimate that depends upon the random points chosen; more repetitions merely duplicates information already provided by some of the first n leave-one-out calculations.

9.5 RESAMPLING FOR CLASSIFIER DESIGN

The previous section addressed the use of resampling in estimating statistics, including the accuracy of an existing classifier, but only indirectly referred to the design of classifiers themselves. We now turn to a number of general resampling methods that have proven effective when used in conjunction with any in a wide range of techniques for training classifiers. These are related to methods for estimating and comparing classifier models that we will discuss in Section 9.6.

9.5.1 Bagging

ARCING

The generic term *arcing*—adaptive reweighting and combining—refers to reusing or selecting data in order to improve classification. In Section 9.5.2 we shall consider the most popular arcing procedure, AdaBoost, but first we discuss briefly one of the simplest. Bagging—a name derived from "bootstrap aggregation"—uses multiple versions of a training set, each created by drawing $n' < n$ samples from $\mathcal{D}$ with replacement. Each of these bootstrap data sets is used to train a different *component*

COMPONENT
CLASSIFIER

classifier and the final classification decision is based on the vote of each component classifier.* Traditionally the component classifiers are of the same general form—for example, all hidden Markov models, or all neural networks, or all decision trees— merely the final parameter values differ among them due to their different sets of training patterns.

INSTABILITY

A classifier/learning algorithm combination is informally called *unstable* if "small" changes in the training data lead to significantly different classifiers and relatively "large" changes in accuracy. As we saw in Chapter 8, decision tree classifiers trained by a greedy algorithm can be unstable—a slight change in the position of a single training point can lead to a radically different tree. In general, bagging improves recognition for unstable classifiers because it effectively averages over such discontinuities. There are no convincing theoretical derivations or simulation studies showing that bagging will help all unstable classifiers, however.

Bagging is our first encounter with multiclassifier systems, where a final overall classifier is based on the outputs of a number of component classifiers. The global decision rule in bagging—a simple vote among the component classifiers—is the most elementary method of pooling or integrating the outputs of the component classifiers. We shall consider multiclassifier systems again in Section 9.7, with particular attention to forming a single decision rule from the outputs of the component classifiers.

9.5.2 Boosting

The goal of boosting is to improve the accuracy of any given learning algorithm. In boosting we first create a classifier with accuracy on the training set greater than average, and then add new component classifiers to form an ensemble whose joint decision rule has arbitrarily high accuracy on the training set. In such a case we say that the classification performance has been "boosted." In overview, the technique trains successive component classifiers with a subset of the training data that is "most informative" given the current set of component classifiers. Classification of a test point $\mathbf{x}$ is based on the outputs of the component classifiers, as we shall see.

For definiteness, consider creating three component classifiers for a two-category problem through boosting. First we randomly select a set of $n_1 < n$ patterns from the full training set $\mathcal{D}$ (without replacement); call this set $\mathcal{D}_1$. Then we train the first

WEAK LEARNER

classifier, C_1, with $\mathcal{D}_1$. Classifier C_1 need only be a *weak learner*—that is, have accuracy only slightly better than chance. (Of course, this is the minimum requirement; a weak learner could have high accuracy on the training set. In that case the benefit of boosting will be small.) Now we seek a second training set, $\mathcal{D}_2$, that is the "most informative" given component classifier C_1. Specifically, half of the patterns in $\mathcal{D}_2$ should be correctly classified by C_1, half incorrectly classified by C_1 (Problem 30). Such an informative set $\mathcal{D}_2$ is created as follows: We flip a fair coin. If the coin is heads, we select remaining samples from $\mathcal{D}$ and present them, one by one to C_1 until C_1 misclassifies a pattern. We add this misclassified pattern to $\mathcal{D}_2$. Next we flip the

---

*In Section 9.7 we shall come across other names for component classifiers. For the present purposes we simply note that these are not classifiers of *component features*, but are instead members in an ensemble of classifiers whose outputs are pooled so as to implement a single classification rule.

coin again. If heads, we continue through $\mathcal{D}$ to find another pattern misclassified by C_1 and add it to $\mathcal{D}_2$ as just described; if tails, we find a pattern that C_1 classifies correctly. We continue until no more patterns can be added in this manner. Thus half of the patterns in $\mathcal{D}_2$ are correctly classified by C_1, half are not. As such, $\mathcal{D}_2$ provides information complementary to that represented in C_1. Now we train a second component classifier C_2 with $\mathcal{D}_2$.

Next we seek a third data set, $\mathcal{D}_3$, which is not well classified by voting by C_1 and C_2. We randomly select a training pattern from those remaining in $\mathcal{D}$ and then classify that pattern with C_1 and with C_2. If C_1 and C_2 disagree, we add this pattern to the third training set $\mathcal{D}_3$; otherwise we ignore the pattern. We continue adding informative patterns to $\mathcal{D}_3$ in this way; thus $\mathcal{D}_3$ contains those not well represented by the combined decisions of C_1 and C_2. Finally, we train the last component classifier, C_3, with the patterns in $\mathcal{D}_3$.

Now consider the use of the ensemble of three trained component classifiers for classifying a test pattern $\mathbf{x}$. Classification is based on the votes of the component classifiers. Specifically, if C_1 and C_2 agree on the category label of $\mathbf{x}$, we use that label; if they disagree, then we use the label given by C_3 (Fig. 9.6).

We skipped over a practical detail in the boosting algorithm: how to choose the number of patterns n_1 to train the first component classifier. We would like the final system to be trained with *all* patterns in $\mathcal{D}$ of course; moreover, because the final decision is a simple vote among the component classifiers, we would like to have a

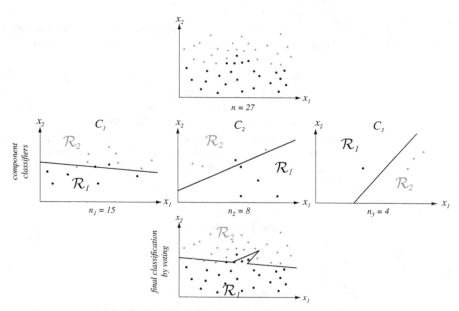

FIGURE 9.6. A two-dimensional two-category classification task is shown at the top. The middle row shows three component (linear) classifiers C_k trained by LMS algorithm (Chapter 5), where their training patterns were chosen through the basic boosting procedure. The final classification is given by the voting of the three component classifiers and yields a nonlinear decision boundary, as shown at the bottom. Given that the component classifiers are weak learners (i.e., each can learn a training set at least slightly better than chance), the ensemble classifier will have a lower training error on the full training set $\mathcal{D}$ than does any single component classifier. Of course, the ensemble classifier has lower error than a single linear classifier trained on the entire data set.

roughly equal number of patterns in each (i.e., $n_1 \simeq n_2 \simeq n_3 \simeq n/3$). A reasonable first guess is to set $n_1 \simeq n/3$ and create the three component classifiers. If the classification problem is very simple, however, component classifier C_1 will explain most of the data and thus n_2 (and n_3) will be much less than n_1, and not all of the patterns in the training set $\mathcal{D}$ will be used. Conversely, if the problem is extremely difficult, then C_1 will explain only a small amount of the data, and nearly all the patterns will be informative with respect to C_1; thus n_2 will be unacceptably large. Thus in practice we may need to run the overall boosting procedure a few times, adjusting n_1 in order to use the full training set and, if possible, get roughly equal partitions of the training set. A number of simple heuristics can be used to improve the partitioning of the training set as well (Computer exercise 6).

The above boosting procedure can be applied recursively to the component classifiers themselves, giving a 9-component or even 27-component full classifier. In this way, a very low training error can be achieved, even a vanishing training error if the problem is separable.

AdaBoost

There are a number of variations on basic boosting. The most popular, AdaBoost—from "adaptive boosting"—allows the designer to continue adding weak learners until some desired low training error has been achieved. In AdaBoost each training pattern receives a weight that determines its probability of being selected for a training set for an individual component classifier. If a training pattern is accurately classified, then its chance of being used again in a subsequent component classifier is reduced; conversely, if the pattern is not accurately classified, then its chance of being used again is raised. In this way, AdaBoost "focuses in" on the informative or "difficult" patterns. Specifically, we initialize the weights across the training set to be uniform. On each iteration k, we draw a training set at random according to these weights, and then we train component classifier C_k on the patterns selected. Next we increase weights of training patterns misclassified by C_k and decrease weights of the patterns correctly classified by C_k. Patterns chosen according to this new distribution are used to train the next classifier, C_{k+1}, and the process is iterated.

We again let the patterns and their labels in $\mathcal{D}$ be denoted $\mathbf{x}^i$ and y_i, respectively and let $W_k(i)$ be the kth (discrete) distribution over all these training samples. Thus the AdaBoost procedure is as follows:

■ **Algorithm 1. (AdaBoost)**

1 <u>**begin initialize**</u> $\mathcal{D} = \{\mathbf{x}^1, y_1, \ldots, \mathbf{x}^n, y_n\}, k_{max}, W_1(i) = 1/n, i = 1, \ldots, n$
2          $k \leftarrow 0$
3   <u>**do**</u> $k \leftarrow k + 1$
4   train weak learner C_k using $\mathcal{D}$ sampled according to $W_k(i)$
5                  $E_k \leftarrow$ training error of C_k measured on $\mathcal{D}$ using $W_k(i)$
6                  $\alpha_k \leftarrow \frac{1}{2}\ln[(1 - E_k)/E_k]$
7                  $W_{k+1}(i) \leftarrow \dfrac{W_k(i)}{Z_k} \times \begin{cases} e^{-\alpha_k} & \text{if } h_k(\mathbf{x}^i) = y_i \text{ (correctly classified)} \\ e^{\alpha_k} & \text{if } h_k(\mathbf{x}^i) \neq y_i \text{ (incorrectly classified)} \end{cases}$
8   <u>**until**</u> $k = k_{max}$
9   <u>**return**</u> C_k and α_k for $k = 1$ to k_{max} (ensemble of classifiers with weights)
10 <u>**end**</u>

Note that in line 5 the error for classifier C_k is determined with respect to the distribution $W_k(i)$ over $\mathcal{D}$ on which it was trained. In line 7, Z_k is simply a normalizing constant computed to ensure that $W_k(i)$ represents a true distribution, and $h_k(\mathbf{x}^i)$ is the category label ($+1$ or -1) given to pattern $\mathbf{x}^i$ by component classifier C_k. Naturally, the loop termination of line 8 could instead use the criterion of sufficiently low training error of the ensemble classifier.

The final classification decision of a test point $\mathbf{x}$ is based on a discriminant function that is merely the weighted sums of the outputs given by the component classifiers:

$$g(\mathbf{x}) = \left[\sum_{k=1}^{k_{max}} \alpha_k h_k(\mathbf{x}) \right]. \tag{36}$$

The classification decision for this two-category case is then simply $\text{Sgn}[g(\mathbf{x})]$.

Except in pathological cases, so long as each component classifier is a weak learner, the total training error of the ensemble can be made arbitrarily low by setting the number of component classifiers, k_{max}, sufficiently high. To see this, notice that the training error for weak learner C_k can be written as $E_k = 1/2 - G_k$ for some positive value G_k. Thus the ensemble training error is (Problem 32)

$$E = \prod_{k=1}^{k_{max}} \left[2\sqrt{E_k(1 - E_k)} \right] = \prod_{k=1}^{k_{max}} \sqrt{1 - 4G_k^2}$$

$$\leq \exp\left(-2 \sum_{k=1}^{k_{max}} G_k^2 \right), \tag{37}$$

as illustrated in Fig. 9.7. It is sometimes beneficial to increase k_{max} beyond the value needed for zero ensemble training error because this may improve generalization.

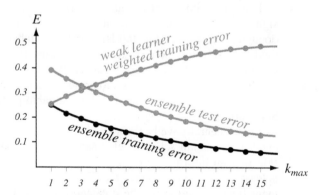

FIGURE 9.7. AdaBoost applied to a weak learning system can reduce the training error E exponentially as the number of component classifiers, k_{max}, is increased. Because AdaBoost "focuses on" *difficult* training patterns, the training error of each successive component classifier (measured on its own weighted training set) is generally larger than that of any previous component classifier (shown in gray). Nevertheless, so long as the component classifiers perform better than chance (e.g., have error less than 0.5 on a two-category problem), the weighted ensemble decision of Eq. 36 ensures that the training error will decrease, as given by Eq. 37. It is often found that the test error decreases in boosted systems as well, as shown in red.

While a large k_{max} could in principle lead to overfitting, simulation experiments have shown that overfitting rarely occurs, even when k_{max} is extremely large.

At first glance, it appears that boosting violates the No Free Lunch Theorem in that an ensemble classifier seems always to perform better than any single component classifier on the full training set. After all, according to Eq. 37 the training error drops exponentially fast with the number of component classifiers. The theorem is not violated, however: Boosting only improves classification if the component classifiers perform *better than chance*, but this cannot be guaranteed *a priori*. If the component classifiers cannot learn the task better than chance, then we do not have a strong match between the problem and model and should choose an alternative learning algorithm. Moreover, the exponential reduction in error on the training set does not ensure reduction of the *off-training set error* or generalization, as we saw in Section 9.2.1. Nevertheless, AdaBoost has proven effective in many real-world applications.

9.5.3 Learning with Queries

In the previous sections we assumed that there was a set of labeled training patterns $\mathcal{D}$ and employed resampling methods to reuse patterns to improve classification. In some applications, however, the patterns are *unlabeled*. In Chapter 10 we shall return to the problem of learning when no labels are available, but here we assume there exists some (possibly costly) way of labeling any pattern. Our current challenge is thus to determine which unlabeled patterns would be most informative (i.e., improve the classifier the most) if they were labeled and used as training patterns. These are the patterns we will present as a *query* to an *oracle*—a teacher who can label any pattern without error. This approach is called variously learning with queries, active learning, or interactive learning and is a special case of a resampling technique. A further refinement on this approach is *cost-based learning* in which there is a cost for obtaining new data. Here the task is to minimize an overall cost, which depends both on the classifier accuracy and the cost of data collection.

QUERY

ORACLE

COST-BASED
LEARNING

Learning with queries might be appropriate, for example, when we want to design a classifier for handwritten numerals using unlabeled pixel images scanned from documents from a corpus too large for us to label every pattern. We could start by randomly selecting some patterns, presenting them to an oracle, and then training the classifier with the returned labels. We then use learning with queries to select unlabeled patterns from our set to present to a human (the oracle) for labeling. Informally, we would expect that the most valuable patterns would be near the decision boundaries.

More generally we begin with a preliminary, weak classifier that has been developed with a small set of labeled samples. There are two related methods for then selecting an informative pattern—that is, a pattern for which the current classifier is least certain. In *confidence-based query selection* the classifier computes discriminant functions $g_i(\mathbf{x})$ for the c categories, $i = 1, \ldots, c$. An informative pattern $\mathbf{x}$ is one for which the two largest discriminant functions have nearly the same value; such patterns lie near the current decision boundaries. Several search heuristics can be used to find such points efficiently (Problem 31).

CONFIDENCE
BASED QUERY
SELECTION

The second method, *voting-based* or *committee-based query selection*, is similar to the previous method but is applicable to multiclassifier systems—that is, ones comprising several component classifiers (Section 9.7). Each unlabeled pattern is presented to each of the k component classifiers; the pattern that yields the greatest disagreement among the k resulting category labels is considered the most infor-

VOTING-
BASED QUERY
SELECTION

mative pattern, and is thus presented as a query to the oracle. Voting-based query selection can be used even if the component classifiers do not provide analog discriminant functions—for instance, decision trees, rule-based classifiers, or simple k-nearest neighbor classifiers. In both confidence-based and voting-based methods, the pattern labeled by the oracle is then used for training the classifier in the traditional way. (We shall return in Section 9.7 to training an ensemble of classifiers.)

Clearly such learning with queries does not directly exploit information about the prior distribution of the patterns. In particular, in most problems the distributions of query patterns will be large near the final decision boundaries (where patterns are informative) rather than at the region of highest prior probability (where they are typically less informative), as illustrated in Fig. 9.8. One benefit of learning with queries is that we need not guess the form of the underlying distribution, but can instead use nonparametric techniques, such as nearest-neighbor classification, that allow the decision boundary to be found directly.

If the set of unlabeled samples for queries is not large, we can nevertheless exploit learning with queries if there is a way to *generate* query patterns. Suppose we have a only small set of labeled handwritten characters. Suppose too we have image processing algorithms for altering these images to generate new, surrogate patterns for queries to an oracle. For instance, the pixel images might be rotated, scaled, sheared, be subject to random pixel noise, or have their lines thinned. Furthermore, we might be able to generate new patterns "in between" two labeled patterns by interpolating

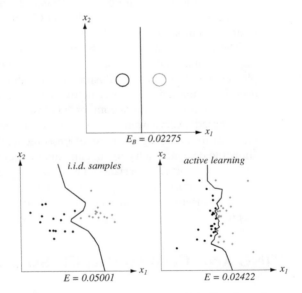

FIGURE 9.8. Active learning can be used to create classifiers that are more accurate than ones using i.i.d. sampling. The figure at the top shows a two-dimensional problem with two equal circular Gaussian priors; the Bayes decision boundary is a straight line and the Bayes error E_B equals 0.02275. The bottom figure on the left shows a nearest-neighbor classifier trained with $n = 30$ labeled points sampled i.i.d. from the true distributions. Note that most of these points are far from the decision boundary. The figure at the right illustrates active learning. The first four points were sampled near the extremes of the feature space. Subsequent query points were chosen midway between two points already used by the classifier, one randomly selected from each of the two categories. In this way, successive queries to the oracle "focused in" on the true decision boundary. The final generalization error of this classifier, $E = 0.02422$, is lower than the one trained using i.i.d. samples, $E = 0.05001$.

or somehow mixing them in a domain-specific way. With such generated query patterns the classifier can explore regions of the feature space about which it is least confident (Fig. 9.8).

9.5.4 Arcing, Learning with Queries, Bias and Variance

In Chapter 3 and many other places, we have stressed the need for training a classifier on samples drawn from the distribution on which it will be tested. Resampling in general—and learning with queries in particular—seems to violate this recommendation. Why can a classifier trained on a strongly weighted distribution of data be expected to do well—or better!—than one trained on the i.i.d. sample? Why doesn't resampling lead to *worse* performance, to the extent that the resampled distribution differs from the i.i.d. one?

Indeed, if we were to take a model of the true distribution and train it with a highly skewed distribution obtained by learning with queries, the final classifier accuracy might be unacceptably low. Consider, however, two interrelated points about resampling methods and altered distributions. The first is that resampling methods are generally used with techniques that do *not* attempt to model or fit the full category distributions. Thus even if we suspect we know that the prior distributions for two categories have a particular mathematical form, we might nevertheless use a nonparametric method such as nearest neighbor, radial basis function, or RCE classifiers when using learning with queries. Thus in learning with queries we are not fitting parameters in a model, as described in Chapter 3, but instead are seeking decision boundaries more directly.

The second point is that as the number of component classifiers is increased, techniques such as general boosting and AdaBoost effectively broaden that class of implementable functions, as illustrated in Fig. 9.6. While the final classifier might indeed be characterized as parametric, it is in an expanded space of parameters, one larger than that of the first component classifier.

In broad overview, resampling, boosting and related procedures are heuristic methods for adjusting the class of implementable decision functions. As such they allow the designer to try to "match" the final classifier to the problem by indirectly adjusting the bias and variance. The power of these methods is that they can be used with an arbitrary classification technique such as the two-layer Perceptron, which would otherwise prove extremely difficult to adjust to the complexity of an arbitrary problem.

9.6 ESTIMATING AND COMPARING CLASSIFIERS

There are at least two reasons for wanting to know the generalization rate of a classifier on a given problem. One is to see if the classifier performs well enough to be useful; another is to compare its performance with that of a competing design. Estimating the final generalization performance invariably requires making assumptions about the classifier or the problem or both, and can fail if the assumptions are not valid. We should stress, then, that all the following methods are heuristic. Indeed, if there were a foolproof method for choosing which of two classifiers would generalize better on an arbitrary new problem, we could incorporate such a method into the learning and violate the No Free Lunch Theorem. Occasionally our assumptions are explicit (as in parametric models), but more often than not they are implicit and difficult to identify or relate to the final estimation (as in empirical methods).

9.6.1 Parametric Models

One approach to estimating the generalization rate is to compute it from the assumed parametric model. For example, in the two-class multivariate normal case, we might estimate the probability of error using the Bhattacharyya or Chernoff bounds (Chapter 2), substituting estimates of the means and the covariance matrix for the unknown parameters. However, there are three problems with this approach. First, such an error estimate is often overly optimistic; characteristics that make the training samples peculiar or unrepresentative will not be revealed. Second, we should always suspect the validity of an assumed parametric model; a performance evaluation based on the same model cannot be believed unless the evaluation is unfavorable. Finally, in more general situations where the distributions are not simple, it is very difficult to compute the error rate exactly, even if the probabilistic structure is known completely.

9.6.2 Cross-Validation

VALIDATION SET

In simple validation we randomly split the set of labeled training samples $\mathcal{D}$ into two parts: one is used as the traditional training set for adjusting model parameters in the classifier. The other set—the *validation set*—is used to estimate the generalization error. Since our ultimate goal is low generalization error, we train the classifier until we reach a minimum of this validation error, as sketched in Fig. 9.9. It is essential that the validation (or the test) set not include points used for training the parameters in the classifier—a methodological error known as "testing on the training set." *

m-FOLD CROSS-VALIDATION

A simple generalization of the above method is *m-fold cross-validation*. Here the training set is randomly divided into m disjoint sets of equal size n/m, where n is

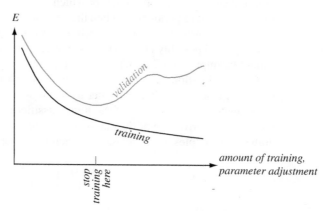

FIGURE 9.9. In validation, the data set $\mathcal{D}$ is split into two parts. The first (e.g., 90% of the patterns) is used as a standard training set for setting free parameters in the classifier model; the other (e.g., 10%) is the validation set and is meant to represent the full generalization task. For most problems, the training error decreases monotonically during training, as shown in black. Typically, the error on the validation set decreases, but then increases, an indication that the classifier may be overfitting the training data. In validation, training or parameter adjustment is stopped at the first minimum of the validation error. In the more general method of cross-validation, the performance is based on multiple independently formed validation sets.

---

*A related but less obvious problem arises when a classifier undergoes a long series of refinements guided by the results of repeated testing on the same test data. This form of "training on the test data" often escapes attention until new test samples are obtained.

again the total number of patterns in $\mathcal{D}$. The classifier is trained m times, each time with a different set held out as a validation set. The estimated performance is the mean of these m errors. In the limit where $m = n$, the method is in effect the leave-one-out approach to be discussed in Section 9.6.3.

Such techniques can be applied to virtually every classification method, where the specific form of learning or parameter adjustment depends upon the general training method. For example, in neural networks of a fixed topology (Chapter 6), the amount of training is the number of epochs or presentations of the training set. Alternatively, the number of hidden units can be set via cross-validation. Likewise, the width of the Gaussian window in Parzen windows (Chapter 4), and an optimal value of k in the k-nearest neighbor classifier can be set by validation or cross-validation.

Validation is heuristic and need not (indeed cannot) give improved classifiers in every case. Nevertheless, validation is extremely simple, and for many real-world problems it is found to improve generalization accuracy. There are several heuristics for choosing the portion γ of $\mathcal{D}$ to be used as a validation set ($0 < \gamma < 1$). Nearly always, a smaller portion of the data should be used as validation set ($\gamma < 0.5$) because the validation set is used merely to set a *single* global property of the classifier (i.e., when to stop adjusting parameters) rather than the large number of classifier parameters learned using the training set. If a classifier has a large number of free parameters or degrees of freedom, then a larger portion of $\mathcal{D}$ should be used as a training set, that is, γ should be reduced. A traditional default is to split the data with $\gamma = 0.1$, which has proven effective in many applications. Finally, when the number of degrees of freedom in the classifier is small compared to the number of training points, the predicted generalization error is relatively insensitive to the choice of γ.

We reiterate that cross-validation is a heuristic and need not work on every problem. Indeed, there are problems for which *anti-cross-validation* is effective—halting the adjustment of parameters when the validation error is the first local *maximum*. As such, in any particular problem, designers must be prepared to explore different values of γ, and possibly abandon the use of cross-validation altogether if performance cannot be improved (Computer exercise 7).

Cross-validation is, at base, an empirical approach that tests the classifier experimentally. Once we train a classifier using cross-validation, the validation error gives an estimate of the accuracy of the final classifier on the unknown test set. If the true but unknown error rate of the classifier is p and if k of the n' independent, randomly drawn test samples are misclassified, then k has the binomial distribution

$$P(k) = \binom{n'}{k} p^k (1 - p)^{n'-k}. \tag{38}$$

Thus, the fraction of test samples misclassified is exactly the maximum-likelihood estimate for p (Problem 40):

$$\hat{p} = \frac{k}{n'}. \tag{39}$$

The properties of this estimate for the parameter p of a binomial distribution are well known. In particular, Fig. 9.10 shows 95% confidence intervals as a function of $\hat{p}$ and n'. For a given value of $\hat{p}$, the probability is 0.95 that the true value of p lies in the interval between the lower and upper curves marked by the number n' of test samples (Problem 37). These curves show that unless n' is fairly large, the maximum-likelihood estimate must be interpreted with caution. For example, if no errors are

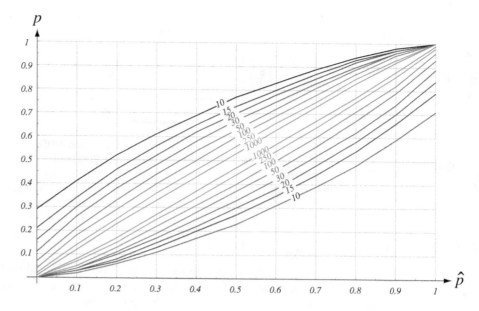

FIGURE 9.10. The 95% confidence intervals for a given estimated error probability $\hat{p}$ can be derived from a binomial distribution of Eq. 38. For each value of $\hat{p}$, the true probability has a 95% chance of lying between the curves marked by the number of test samples n'. The larger the number of test samples, the more precise the estimate of the true probability and hence the smaller the 95% confidence interval.

made on 50 test samples, with probability 0.95 the true error rate is between zero and 8%. The classifier would have to make no errors on more than 250 test samples to be reasonably sure that the true error rate is below 2%.

9.6.3 Jackknife and Bootstrap Estimation of Classification Accuracy

A method for comparing classifiers closely related to cross-validation is to use the jackknife or bootstrap estimation procedures (Sections 9.4.1 and 9.4.2). The application of the jackknife approach to classification is straightforward. We estimate the accuracy of a given algorithm by training the classifier n separate times, each time using the training set $\mathcal{D}$ from which a different single training point has been deleted. This is merely the $m = n$ limit of m-fold cross-validation. Each resulting classifier is tested on the single deleted point, and the jackknife estimate of the accuracy is then simply the mean of these leave-one-out accuracies. Here the computational complexity may be very high, especially for large n (Problem 29).

The jackknife, in particular, generally gives good estimates because each of the the n classifiers is quite similar to the classifier being tested (differing solely due to a single training point). Likewise, the jackknife estimate of the variance of this estimate is given by a simple generalization of Eq. 32. A particular benefit of the jackknife approach is that it can provide measures of confidence or statistical significance in the comparison between two classifier designs. Suppose that trained classifier C_1 has an accuracy of 80% while C_2 has accuracy of 85%, as estimated by the jackknife procedure. Is C_2 really better than C_1? To answer this, we calculate the jackknife estimate of the variance of the classification accuracies and use traditional hypothesis testing to see if C_1's apparent superiority is statistically significant (Fig. 9.11).

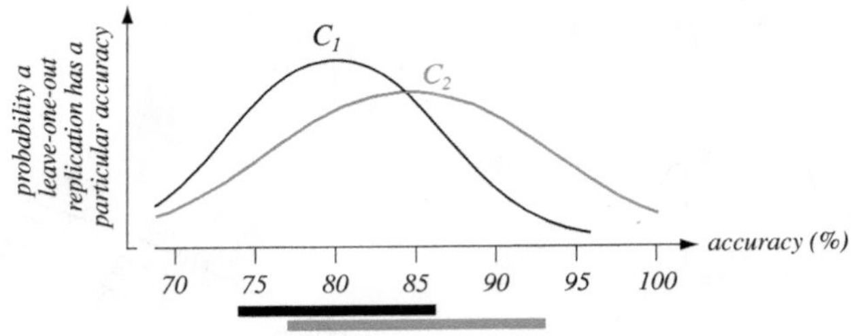

FIGURE 9.11. Jackknife estimation can be used to compare the accuracies of classifiers. The jackknife estimate of classifiers C_1 and C_2 are 80% and 85%, and full widths (twice the square root of the jackknife estimate of the variances) are 12% and 15%, as shown by the bars at the bottom. In this case, traditional hypothesis testing could show that the difference is not statistically significant at some confidence level.

There are several ways to generalize the bootstrap method to the problem of estimating the accuracy of a classifier. One of the simplest approaches is to train B classifiers, each with a different bootstrap data set, and test on other bootstrap data sets. The bootstrap estimate of the classifier accuracy is simply the mean of these bootstrap accuracies. In practice, the high computational complexity of bootstrap estimation of classifier accuracy is sometimes worth possible improvements in that estimate (Section 9.5.1).

9.6.4 Maximum-Likelihood Model Comparison

Recall first the maximum-likelihood parameter estimation methods discussed in Chapter 3. Given a model with unknown parameter vector $\boldsymbol{\theta}$, we find the value $\hat{\boldsymbol{\theta}}$ that maximizes the probability of the training data, that is, $p(\mathcal{D}|\hat{\boldsymbol{\theta}})$. Maximum-likelihood *model comparison* or maximum-likelihood *model selection*—sometimes called ML-II—is a direct generalization of those techniques. The goal here is to choose the *model* that best explains the training data, in a way that will become clear below.

ML-II

We again let $h_i \in \mathcal{H}$ represent a candidate hypothesis or model (assumed discrete for simplicity) and let $\mathcal{D}$ represent the training data. The posterior probability of any given model is given by Bayes rule:

$$P(h_i|\mathcal{D}) = \frac{P(\mathcal{D}|h_i)P(h_i)}{p(\mathcal{D})} \propto P(\mathcal{D}|h_i)P(h_i), \tag{40}$$

EVIDENCE

where we will rarely need the normalizing factor $p(\mathcal{D})$. The data-dependent term, $P(\mathcal{D}|h_i)$, is the *evidence* for h_i; the second factor, $P(h_i)$, is our subjective prior over the space of hypotheses—it rates our confidence in different models even before the data arrive. In practice, the data-dependent term dominates in Eq. 40, and hence the priors $P(h_i)$ are often neglected in the computation. In maximum-likelihood model comparison, we find the maximum-likelihood parameters for each of the candidate models, calculate the resulting likelihoods, and select the model with the largest such likelihood in Eq. 40 (Fig. 9.12).

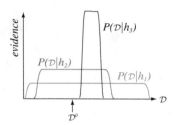

FIGURE 9.12. The evidence (i.e., probability of generating different data sets given a model) is shown for three models of different expressive power or complexity. Model h_1 is the most expressive, because with different values of its parameters the model can fit a wide range of data sets. Model h_3 is the most restrictive of the three. If the actual data observed is $\mathcal{D}^0$, then maximum-likelihood model selection states that we should choose h_2, which has the highest evidence. Model h_2 "matches" this particular data set better than do the other two models, and it should be selected.

9.6.5 Bayesian Model Comparison

Bayesian model comparison uses the full information over priors when computing posterior probabilities in Eq. 40. In particular, the evidence for a particular hypothesis is an integral,

$$P(\mathcal{D}|h_i) = \int p(\mathcal{D}|\boldsymbol{\theta}, h_i)p(\boldsymbol{\theta}|\mathcal{D}, h_i)\, d\boldsymbol{\theta}, \tag{41}$$

where, as before, $\boldsymbol{\theta}$ describes the parameters in the candidate model. It is common for the posterior $P(\boldsymbol{\theta}|\mathcal{D}, h_i)$ to be peaked at $\hat{\boldsymbol{\theta}}$, and thus the evidence integral can often be approximated as

$$P(\mathcal{D}|h_i) \simeq \underbrace{P(\mathcal{D}|\hat{\boldsymbol{\theta}}, h_i)}_{\substack{\text{best-fit} \\ \text{likelihood}}}\ \underbrace{p(\hat{\boldsymbol{\theta}}|h_i)\Delta\boldsymbol{\theta}}_{\text{Occam factor}}. \tag{42}$$

Before the data arrive, model h_i has some broad range of model parameters, denoted by $\Delta^0\boldsymbol{\theta}$ and shown in Fig. 9.13. After the data arrive, a smaller range is commensurate or compatible with $\mathcal{D}$, denoted $\Delta\boldsymbol{\theta}$. The *Occam factor* in Eq. 42,

OCCAM FACTOR

$$\text{Occam factor} = p(\hat{\boldsymbol{\theta}}|h_i)\Delta\boldsymbol{\theta} = \frac{\Delta\boldsymbol{\theta}}{\Delta^0\boldsymbol{\theta}} \tag{43}$$

$$= \frac{\text{param. vol. commensurate with } \mathcal{D}}{\text{param. vol. commensurate with any data}},$$

is the ratio of two volumes in parameter space: (1) the volume that can account for data $\mathcal{D}$ and (2) the prior volume, accessible to the model without regard to $\mathcal{D}$. The Occam factor has magnitude less than 1.0; it is simply the factor by which the hypothesis space collapses by the presence of data. The more the training data, the smaller the range of parameters that are commensurate with it, and thus the greater this collapse in the parameter space and the larger the Occam factor (Fig. 9.13).

Naturally, once the posteriors for different models have been calculated by Eqs. 40 and 42, we select the single one having the highest such posterior. (Ironically, the Bayesian model selection procedure is itself not truly Bayesian, because a Bayesian procedure would average over *all* possible models when making a decision.)

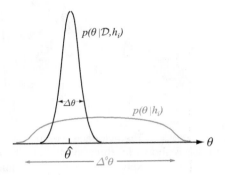

FIGURE 9.13. In the absence of training data, a particular model h_i has available a large range of possible values of its parameters, denoted $\Delta^0\theta$. In the presence of a particular training set $\mathcal{D}$, a smaller range is available. The Occam factor, $\Delta\theta/\Delta^0\theta$, measures the fractional decrease in the volume of the model's parameter space due to the presence of training data $\mathcal{D}$. In practice, the Occam factor can be calculated fairly easily if the evidence is approximated as a p-dimensional Gaussian, centered on the maximum-likelihood value $\hat{\boldsymbol{\theta}}$.

The evidence for h_i—that is, $P(\mathcal{D}|h_i)$—was ignored in a maximum-likelihood setting of parameters $\hat{\boldsymbol{\theta}}$; nevertheless, it is the central term in our comparison of models. As mentioned, in practice the evidence term in Eq. 40 dominates the prior term, and it is traditional to ignore such priors, which are often highly subjective or problematic anyway (Problem 39, Computer exercise 9). This procedure represents an inherent bias toward simple models (small $\Delta\boldsymbol{\theta}$); models that are overly complex (large $\Delta\boldsymbol{\theta}$) are automatically self-penalizing and "overly complex" is a data-dependent concept.

In the general case, the full integral of Eq. 41 is too difficult to calculate analytically or even numerically. Nevertheless, if $\boldsymbol{\theta}$ is p-dimensional and the posterior can be assumed to be a Gaussian, then the Occam factor can be calculated directly (Problem 38), yielding

$$P(\mathcal{D}|h_i) \simeq \underbrace{P(\mathcal{D}|\hat{\boldsymbol{\theta}}, h_i)}_{\substack{\text{best-fit}\\\text{likelihood}}} \underbrace{p(\hat{\boldsymbol{\theta}}|h_i)(2\pi)^{k/2}|\mathbf{H}|^{-1/2}}_{\text{Occam factor}}, \tag{44}$$

where

$$\mathbf{H} = \frac{\partial^2 \ln p(\boldsymbol{\theta}|\mathcal{D}, h_i)}{\partial \boldsymbol{\theta}^2} \tag{45}$$

is a Hessian matrix—a matrix of second-order derivatives—and measures how "peaked" the posterior is around the value $\hat{\boldsymbol{\theta}}$. Note that this Gaussian approximation does not rely on the fact that the underlying model of the distribution of the *data* in feature space is or is not Gaussian. Rather, it is based on the assumption that the evidence distribution arises from a large number of independent uncorrelated processes and is governed by the Law of Large Numbers. The integration inherent in Bayesian methods is simplified using this Gaussian approximation to the evidence. Because calculating the needed Hessian via differentiation is nearly always simpler than a high-dimensional numerical integration, the Bayesian method of model selection is not at a severe computational disadvantage relative to its maximum-likelihood counterpart.

There may be a problem due to degeneracies in a model: Several parameters could be relabeled and leave the classification rule (and hence the likelihood) unchanged. The resulting degeneracy leads, in essence, to an "overcounting" that alters the effective volume in parameter space. Degeneracies are especially common in neural network models where the parameterization comprises many equivalent weights (Chapter 6). For such cases, we must multiply the right-hand side of Eq. 42 by the degeneracy of $\hat{\theta}$ in order to scale the Occam factor, and thereby obtain the proper estimate of the evidence (Problem 43).

Bayesian Model Selection and the No Free Lunch Theorem

There seems to be a fundamental contradiction between two of the deepest ideas in the foundation of statistical pattern recognition. On the one hand, the No Free Lunch Theorem states that in the absence of prior information about the problem, there is no reason to prefer one classification algorithm over another. On the other hand, Bayesian model selection is theoretically well-founded and seems to show how to reliably choose the better of two algorithms.

Consider two "composite" algorithms—algorithm A and algorithm B—each of which employs two others (algorithm 1 and algorithm 2). For any problem, algorithm A uses Bayesian model selection and applies the "better" of algorithm 1 and algorithm 2. Algorithm B uses *anti*-Bayesian model selection and applies the "worse" of algorithm 1 and algorithm 2. It appears that algorithm A will reliably outperform algorithm B throughout the full class of problems—in contradiction with Part 1 of the No Free Lunch Theorem.

What is the resolution of this apparent contradiction? In Bayesian model selection we ignore the prior over the space of models, $\mathcal{H}$, effectively assuming that it is uniform. This assumption therefore does not take into account how those models correspond to underlying target functions—that is, mappings from input to category labels. Accordingly, Bayesian model selection usually corresponds to a nonuniform prior over target functions. Moreover, depending on the arbitrary choice of model, the precise nonuniform prior will vary. In fact, this arbitrariness is very well known in statistics, and good practitioners rarely apply the *principle of indifference*, assuming a uniform prior over models, as Bayesian model selection requires. Indeed, there are many "paradoxes" described in the statistics literature that arise from not being careful to have the prior over models be tailored to the choice of models (Problem 39). The No Free Lunch Theorem allows that for some particular nonuniform prior there may be a learning algorithm that gives better than chance—or even optimal—results. Apparently Bayesian model selection corresponds to nonuniform priors that seem to match many important real-world problems.

PRINCIPLE OF
INDIFFERENCE

9.6.6 The Problem-Average Error Rate

The examples we have given thus far suggest that the problem with having only a small number of samples is that the resulting classifier will not perform well on new data—it will not generalize well. Thus, we expect the error rate to be a function of the number n of training samples, typically decreasing to some minimum value as n approaches infinity. To investigate this analytically, we must carry out the following familiar steps:

1. Estimate the unknown parameters from samples.

2. Use these estimates to determine the classifier.

3. Calculate the error rate for the resulting classifier.

In general this analysis is very complicated. The answer depends on everything—on the particular training patterns, on the way they are used to determine the classifier, and on the unknown, underlying probability structure. However, by using histogram approximations to the unknown probability densities and averaging appropriately, it is possible to draw some illuminating conclusions.

Consider a case in which two categories have equal prior probabilities. Suppose that we partition the feature space into some number m of disjoint cells $C_1, \ldots, C_m$. If the conditional densities $p(\mathbf{x}|\omega_1)$ and $p(\mathbf{x}|\omega_2)$ do not vary appreciably within any cell, then instead of needing to know the actual value of $\mathbf{x}$, we need only know into which cell $\mathbf{x}$ falls. This reduces the problem to the discrete case. Let $p_i = P(\mathbf{x} \in C_i|\omega_1)$ and $q_i = P(\mathbf{x} \in C_i|\omega_2)$. Then, because we have assumed that $P(\omega_1) = P(\omega_2) = 1/2$, the vectors $\mathbf{p} = (p_1, \ldots, p_m)^t$ and $\mathbf{q} = (q_1, \ldots, q_m)^t$ determine the probability structure of the problem. If $\mathbf{x}$ falls in C_i, the Bayes decision rule is to decide ω_i if $p_i > q_i$. The resulting Bayes error rate is given by

$$P(E|\mathbf{p}, \mathbf{q}) = \frac{1}{2} \sum_{i=1}^{m} \min[p_i, q_i]. \tag{46}$$

When the parameters $\mathbf{p}$ and $\mathbf{q}$ are unknown and must be estimated from a set of training patterns, the resulting error rate will be larger than the Bayes rate. The exact error probability will depend on the set of training patterns and the way in which they are used to obtain the classifier. Suppose that half of the samples are labeled ω_1 and half are labeled ω_2, with n_{ij} being the number that fall in C_i and are labeled ω_j. Suppose further that we design the classifier by using the maximum-likelihood estimates $\hat{p}_i = 2n_{i1}/n$ and $\hat{q}_i = 2n_{i2}/n$ as if they were the true values. Then a new feature vector falling in C_i will be assigned to ω_1 if $n_{i1} > n_{i2}$. With all of these assumptions, it follows that the probability of error for the resulting classifier is given by

$$P(E|\mathbf{p}, \mathbf{q}, \mathcal{D}) = \frac{1}{2} \sum_{n_{i1} > n_{i2}} q_i + \frac{1}{2} \sum_{n_{i1} \le n_{i2}} p_i. \tag{47}$$

To evaluate this probability of error, we need to know the true conditional probabilities $\mathbf{p}$ and $\mathbf{q}$ and the set of training patterns, or at least the numbers n_{ij}. Different sets of n randomly chosen patterns will yield different values for $P(E|\mathbf{p}, \mathbf{q}, \mathcal{D})$. We can use the fact that the numbers n_{ij} have a multinomial distribution to average over all of the possible sets of n random samples and obtain an average probability of error $P(E|\mathbf{p}, \mathbf{q}, n)$. Roughly speaking, this is the typical error rate one should expect for n samples. However, evaluation of this average error rate still requires knowing the underlying problem—that is, the values for $\mathbf{p}$ and $\mathbf{q}$. If $\mathbf{p}$ and $\mathbf{q}$ are quite different, the average error rate will be near zero, while if $\mathbf{p}$ and $\mathbf{q}$ are quite similar, it will be near 0.5.

A sweeping way to eliminate this dependence of the answer upon the problem is to average the answer over all possible problems. That is, we assume some prior distribution for the unknown parameters $\mathbf{p}$ and $\mathbf{q}$, and we average $P(E|\mathbf{p}, \mathbf{q}, n)$ with respect to $\mathbf{p}$ and $\mathbf{q}$. The resulting *problem-average probability of error* $\bar{P}(E|m, n)$

will depend only on the number m of cells, the number n of samples, and the prior distributions.

Of course, choosing the prior distributions is a delicate matter. By favoring easy problems we can make $\bar{P}$ approach zero, and by favoring hard problems we can make $\bar{P}$ approach 0.5. We would like to choose a prior distribution corresponding to the class of problems we typically encounter, but there is no obvious way to do that. A bold approach is merely to assume that problems are "uniformly distributed"—that is, that the vectors **p** and **q** are distributed uniformly over the simplexes

$$p_i \geq 0, \qquad \sum_{i=1}^{m} p_i = 1, \tag{48}$$

$$q_i \geq 0, \qquad \sum_{i=1}^{m} q_i = 1.$$

Note that this uniform distribution over the space of **p** and **q** does not correspond to some purported uniform distribution over possible distributions or target functions, the issue pointed out in Section 9.6.5.

Figure 9.14 summarizes simulation experiments and shows curves of $\bar{P}$ as a function number of cells for fixed numbers of training patterns. With an infinite number of training patterns the maximum-likelihood estimates are perfect, and $\bar{P}$ is the average of the Bayes error rate over all problems. The corresponding curve for $\bar{P}(E|m, \infty)$ decreases rapidly from 0.5 at $m = 1$ to the asymptotic value of 0.25 as m approaches infinity. The fact that $\bar{P} = 0.5$ if $m = 1$ is not surprising, because if there is only one cell the decision must be based solely on the prior probabilities. The fact that $\bar{P}$ approaches 0.25 as m approaches infinity is aesthetically pleasing, because this value is halfway between the extremes of 0.0 and 0.5. The fact that the problem-average error rate is so high merely shows that many hopelessly difficult classification prob-

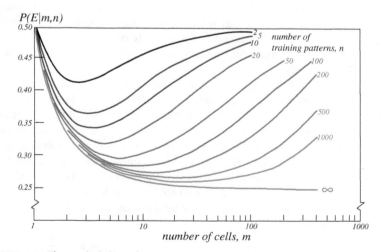

FIGURE 9.14. The probability of error E on a two-category problem for a given number of samples, n, can be estimated by splitting the feature space into m cells of equal size and classifying a test point according to the label of the most frequently represented category in the cell. The graphs show the average error of a large number of random problems having the given n and m indicated.

lems are included in this average. Clearly, it would be rash indeed to conclude that the "average" pattern recognition problem will have this error rate.

However, the most interesting feature of these curves is that for every curve involving a finite number of samples there is an optimal number of cells. This is directly related to the fact that with a finite number of samples the performance will worsen if too many features are used. In this case it is clear why there exists an optimal number of cells for any given n and m. At first, increasing the number of cells makes it easier to distinguish between the distributions represented by the vectors $\mathbf{p}$ and $\mathbf{q}$, thereby allowing improved performance. However, if the number of cells becomes too large, there will not be enough training patterns to fill them. Eventually, the number of patterns in most cells will be zero, and we must return to using just the ineffective *a priori* probabilities for classification. Thus, for any finite n, $\bar{P}(E|m, n)$ must approach 0.5 as m approaches infinity.

The value of m for which $\bar{P}(E|m, n)$ is minimum is quite small. For $n = 500$ samples, it is somewhere around $m = 200$ cells. Suppose that we were to form the cells by dividing each feature axis into l intervals. Then with d features we would have $m = l^d$ cells. If $l = 2$, which is extremely crude quantization, this implies that using more than four or five binary features will lead to worse rather than better performance. This is a very pessimistic result, but then so is the statement that the average error rate is 0.25. These numerical values are a consequence of the prior distribution chosen for the problems, and they are of no significance when one is facing a particular problem. The main thing to be learned from this analysis is that the performance of a classifier certainly does depend on the number of training patterns, and that if this number is fixed, increasing the number of features beyond a certain point raises the variance unacceptably and will be counterproductive.

9.6.7 Predicting Final Performance from Learning Curves

Training on very large data sets can be computationally intensive, requiring days, weeks, or even months on powerful machines. If we are exploring and comparing several different classification techniques, the total training time needed may be unacceptably long. What we seek, then, is a method to compare classifiers without the need of training all of them fully on the complete data set. If we can determine the most promising model quickly and efficiently, all we then must do is train *this* model fully.

One method is to use a classifier's performance on a relatively *small* training set to predict its performance on the ultimate large training set. Such performance is revealed in a type of learning curve in which the test error is plotted versus the size of the training set. Figure 9.15 shows the error rate on an independent test set after the classifier has been fully trained on $n' \leq n$ points in the training set. (Note that in this form of learning curve the training error decreases monotonically and does not show "overtraining" evident in curves such as Fig. 9.9.)

For many real-world problems, such learning curves decay monotonically and can be adequately described by a power-law function of the form

$$E_{test} = a + \frac{b}{n'^{\alpha}} \tag{49}$$

where a, b and $\alpha \geq 1$ depend upon the task and the classifier. In the limit of very large n', the training error equals the test error, because both the training and test sets represent the full problem space. Thus we also model the training error as a

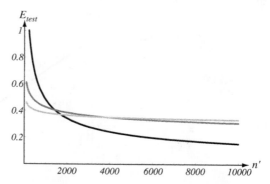

FIGURE 9.15. The test error for three classifiers, each fully trained on the given number n' of training patterns, decreases in a typical monotonic power-law function. Notice that the rank order of the classifiers trained on $n' = 500$ points differs from that for $n' = 10000$ points and the asymptotic case.

power-law function, having the same asymptotic error,

$$E_{train} = a - \frac{c}{n'^\beta}. \tag{50}$$

If the classifier is sufficiently powerful, this asymptotic error, a, is equal to the Bayes error. Furthermore, such a powerful classifier can learn perfectly the small training sets and thus the training error (measured on the n' points) will vanish at small n', as shown in Fig. 9.16.

Now we seek to estimate the asymptotic error, a, from the training and test errors on small and intermediate size training sets. From Eqs. 49 and 50 we find

$$E_{test} + E_{train} = 2a + \frac{b}{n'^\alpha} - \frac{c}{n'^\beta}, \tag{51}$$

$$E_{test} - E_{train} = \frac{b}{n^\alpha} + \frac{c}{n^\beta}.$$

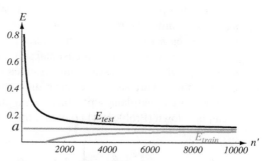

FIGURE 9.16. Test and training error of a classifier fully trained on data subsets of different size n' selected randomly from the full set $\mathcal{D}$. At low n', the classifier can learn the category labels of the points perfectly, and thus the training error vanishes there. In the limit $n' \to \infty$, both training and test errors approach the same asymptotic value, a. If the classifier is sufficiently powerful and the training data is sampled i.i.d., then a is the Bayes error rate, E_B.

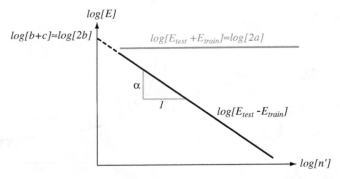

FIGURE 9.17. If the test and training errors versus training set size obey the power-law functions of Eqs. 49 and 50, then the log of the sum and log of the difference of these errors are straight lines on a log-log plot. The estimate of the asymptotic error rate a is then simply related to the height of the $\log[E_{test} + E_{train}]$ line, as shown.

If we make the assumption of $\alpha = \beta$ and $b = c$, then Eq. 51 reduces to

$$E_{test} + E_{train} = 2a, \tag{52}$$

$$E_{test} - E_{train} = \frac{2b}{n'^\alpha}.$$

Given this assumption, it is a simple matter to measure the training and test errors for small and intermediate values of n', plot them on a log-log scale, and estimate a, as shown in Fig. 9.17. Even if the approximations $\alpha = \beta$ and $b = c$ do not hold in practice, the difference $E_{test} - E_{train}$ nevertheless still forms a straight line on a log-log plot and the sum, $s = b+c$, can be found from the height of the $\log[E_{test}+E_{train}]$ curve. The weighted sum $cE_{test} + bE_{train}$ will be a straight line for some empirically set values of b and c, constrained to obey $b + c = s$, enabling a to be estimated (Problem 42). Once a has been estimated for each in the set of candidate classifiers, the one with the lowest a is chosen and must be trained on the full training set $\mathcal{D}$.

9.6.8 The Capacity of a Separating Plane

Consider the partitioning of a d-dimensional feature space by a hyperplane $\mathbf{w}^t\mathbf{x} + w_0 = 0$, as might be trained by the Perceptron algorithm (Chapter 5). Suppose that we are given n sample points in general position—that is, with no subset of $d + 1$ points falling in a $(d - 1)$-dimensional subspace. Assume each point is labeled either ω_1 or ω_2. Of the 2^n possible dichotomies of n points in d dimensions, a certain fraction $f(n, d)$ are said to be linear dichotomies. These are the labelings for which there exists a hyperplane separating the points labeled ω_1 from the points labeled ω_2. It can be shown (Problem 41) that this fraction is given by

$$f(n, d) = \begin{cases} 1 & \text{for } n \leq d + 1 \\ \frac{2}{2^n} \sum_{i=0}^{d} \binom{n-1}{i} & \text{for } n > d + 1, \end{cases} \tag{53}$$

as plotted in Fig. 9.18 for several values of d.

To understand the issue more fully, consider the one-dimensional case with four points; according to Eq. 53, we have $f(n = 4, d = 1) = 0.5$. The table shows schematically all 16 of the equally likely labels for four patterns along a line. (For

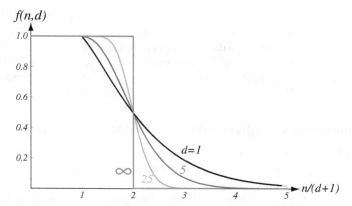

FIGURE 9.18. The fraction of dichotomies of n points in d dimensions that are linear, as given by Eq. 53.

instance, 0010 indicates that the labels are assigned $\omega_1\omega_1\omega_2\omega_1$.) The $\times$ marks those arrangements that are linearly separable—that is, in which a single point decision boundary can separate all ω_1 patterns from all ω_2 patterns. Indeed as given by Eq. 53, 8 of the 16—half—are linearly separable.

Labels	Linearly Separable?	Labels	Linearly Separable?
0000	$\times$	1000	$\times$
0001	$\times$	1001	
0010		1010	
0011	$\times$	1011	
0100		1100	$\times$
0101		1101	
0110		1110	$\times$
0111	$\times$	1111	$\times$

Note from Fig. 9.18 that all dichotomies of $d + 1$ or fewer points are linear. This means that a hyperplane is not overconstrained by the requirement of correctly classifying $d + 1$ or fewer points. In fact, if d is large, it is not until n is a sizable fraction of $2(d + 1)$ that the problem begins to become difficult. At $n = 2(d + 1)$, which is sometimes called the *capacity* of a hyperplane, half of the possible dichotomies are still linear. Thus, a linear discriminant is not effectively overdetermined until the number of samples is several times as large as the dimensionality of the feature space or subset of the problems. This is often expressed as follows: "Generalization begins only after learning ends." Alternatively, we cannot expect a linear classifier to "match" a problem, on average, if the dimension of the feature space is greater than $n/2 - 1$.

CAPACITY

9.7 COMBINING CLASSIFIERS

We have already mentioned classifiers whose decision is based on the outputs of component classifiers (Sections 9.5.1 and 9.5.2). Such full classifiers are variously

MIXTURE OF EXPERT

called *mixture-of-expert* models, ensemble classifiers, modular classifiers, or occasionally pooled classifiers. Such classifiers are particularly useful if each of its component classifiers is highly trained (i.e., an "expert") in a different region of the feature space. We first consider the case where each component classifier provides probability estimates. Later, in Section 9.7.2 we consider the case where component classifiers provide rank order information or one-of-c outputs.

9.7.1 Component Classifiers with Discriminant Functions

MIXTURE MODEL

We assume that each pattern is produced by a *mixture model*, in which first some fundamental process or function indexed by r (where $1 \leq r \leq k$) is randomly chosen according to distribution $P(r|\mathbf{x}, \boldsymbol{\theta}_0^0)$ where $\boldsymbol{\theta}_0^0$ is a parameter vector. Next, the selected process r emits an output $\mathbf{y}$ (e.g., a category label) according to $P(\mathbf{y}|\mathbf{x}, \boldsymbol{\theta}_r^0)$, where the parameter vector $\boldsymbol{\theta}_r^0$ describes the state of the process. (The superscript 0 indicates the properties of the generating model. Below, terms without this superscript refer to the parameters in a classifier.) The overall probability of producing output $\mathbf{y}$ is then the sum over all the processes according to

$$P(\mathbf{y}|\mathbf{x}, \boldsymbol{\Theta}^0) = \sum_{r=1}^{k} P(r|\mathbf{x}, \boldsymbol{\theta}_0^0) P(\mathbf{y}|\mathbf{x}, \boldsymbol{\theta}^0), \qquad (54)$$

MIXTURE DISTRIBUTION

where $\boldsymbol{\Theta}^0 = [\boldsymbol{\theta}_0^0, \boldsymbol{\theta}_1^0, \ldots, \boldsymbol{\theta}_k^0]^t$ represents the vector of all relevant parameters. Equation 54 describes a *mixture distribution*, which could be discrete or continuous.

Figure 9.19 shows the basic architecture of an ensemble classifier whose task is to classify a test pattern $\mathbf{x}$ into one of c categories; this architecture matches the assumed mixture model. A test pattern $\mathbf{x}$ is presented to each of the k component classifiers,

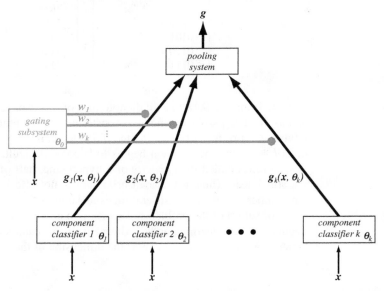

FIGURE 9.19. The mixture-of-experts architecture consists of k component classifiers or "experts," each of which has trainable parameters $\boldsymbol{\theta}_i$, $i = 1, \ldots, k$. For each input pattern $\mathbf{x}$, each component classifier i gives estimates of the category membership $g_{ir} = P(\omega_r|\mathbf{x}, \boldsymbol{\theta}_i)$. The outputs are weighted by the gating subsystem governed by parameter vector $\boldsymbol{\theta}_0$ and are pooled for ultimate classification.

each of which emits c scalar discriminant values, one for each category. The c discriminant values from component classifier r are grouped and marked $\mathbf{g}(\mathbf{x}, \boldsymbol{\theta}_r)$ in the figure, with

$$\sum_{j=1}^{c} g_{rj} = 1 \qquad \text{for all } r. \tag{55}$$

GATING
SUBSYSTEM

All discriminant values from component classifier r are multiplied by a scalar weight w_r, governed by the *gating subsystem*, which has a parameter vector $\boldsymbol{\theta}_0$. Below we shall use the conditional mean of the mixture density, which can be calculated from Eq. 54:

$$\boldsymbol{\mu} = \mathcal{E}[\mathbf{y}|\mathbf{x}, \boldsymbol{\Theta}] = \sum_{r=1}^{k} w_r \boldsymbol{\mu}_r, \tag{56}$$

where $\boldsymbol{\mu}_r$ is the conditional mean associated with $P(\mathbf{y}|\mathbf{x}, \boldsymbol{\theta}_r^0)$.

The mixture-of-experts architecture is trained so that each component classifier models a corresponding process in the mixture model, and the gating subsystem models the mixing parameters $P(r|\mathbf{x}, \boldsymbol{\theta}_0^0)$ in Eq. 54. The goal is to find parameters that maximize the log-likelihood for n training patterns $\mathbf{x}^1, \ldots, \mathbf{x}^n$ in set $\mathcal{D}$:

$$l(\mathcal{D}, \boldsymbol{\Theta}) = \sum_{i=1}^{n} \ln \left(\sum_{r=1}^{k} P(r|\mathbf{x}^i, \boldsymbol{\theta}_0) P(\mathbf{y}^i|\mathbf{x}^i, \boldsymbol{\theta}_r) \right). \tag{57}$$

A straightforward approach is to use gradient descent on the parameters, where the derivatives are (Problem 44)

$$\frac{\partial l(\mathcal{D}, \boldsymbol{\Theta})}{\partial \boldsymbol{\theta}_r} = \sum_{i=1}^{n} P(r|\mathbf{y}^i, \mathbf{x}^i) \frac{\partial}{\partial \boldsymbol{\theta}_r} \ln[P(\mathbf{y}^i|\mathbf{x}^i, \boldsymbol{\theta}_r)] \quad \text{for } r = 1, \ldots, k \tag{58}$$

and

$$\frac{\partial l(\mathcal{D}, \boldsymbol{\Theta})}{\partial g_r} = \sum_{i=1}^{n} (P(r|\mathbf{y}^i, \mathbf{x}^i) - w_r^i). \tag{59}$$

Here $P(r|\mathbf{y}^i, \mathbf{x}^i)$ is the posterior probability of process r conditional on the input and output being $\mathbf{x}^i$ and $\mathbf{y}^i$, respectively. Moreover, w_r^i is the prior probability $P(r|\mathbf{x}^i)$ that process r is chosen given the input is $\mathbf{x}^i$. Gradient descent according to Eq. 59 moves the prior probabilities to the posterior probabilities. The Expectation-Maximization (EM) algorithm can be used to train this architecture as well (Chapter 3).

The final decision rule is simply to choose the category corresponding to the maximum discriminant value after the pooling system. An alternative, *winner-take-all* method is to use the decision of the single component classifier that is "most confident"—that is, has the largest single discriminant value g_{rj}. While the winner-take-all method is provably suboptimal, it nevertheless is simple and can work well if the component classifiers are experts in separate regions of the input space.

WINNER-
TAKE-ALL

We have skipped over a problem: How many component classifiers should be used? Of course, if we have prior information about the number of component processes that generated the mixture density, this should guide our choice of k. In the absence of such information, we may have to explore different values of k, thereby tailoring the bias and variance of the full ensemble classifier. Typically, if the true number of components in the mixture density is k^*, a mixture-of-experts more than k^* component classifiers will generalize better than one with fewer than k^* component classifiers because the extra component classifiers learn to duplicate one another and hence become redundant.

9.7.2 Component Classifiers without Discriminant Functions

Occasionally we seek to form an ensemble classifier from highly trained component classifiers, some of which might not themselves compute discriminant functions. For instance, we might have four component classifiers—a k-nearest-neighbor classifier, a decision tree, a neural network, and a rule-based system—all addressing the same problem. While a neural network would provide analog values for each of the c categories, the rule-based system would give only a single category label (i.e., a one-of-c representation) and the k-nearest-neighbor classifier would give only rank order of the categories.

In order to integrate the information from the component classifiers we must convert the their outputs into discriminant values obeying the constraint of Eq. 55 so we can use the framework of Fig. 9.19. The simplest heuristics to this end are the following:

Analog: If the outputs of a component classifier are analog values $\tilde{g}_i$, we can use the *softmax* transformation,

SOFTMAX

$$g_i = \frac{e^{\tilde{g}_i}}{\sum_{j=1}^{c} e^{\tilde{g}_j}}. \tag{60}$$

to convert them to values g_i.

Rank order: If the output is a rank order list, we assume the discriminant function is linearly proportional to the rank order of the item on the list. Of course, the resulting g_i should then be properly normalized, and thus sum to 1.0.

One-of-c: If the output is a one-of-c representation, in which a single category is identified, we let $g_j = 1.0$ for the j corresponding to the chosen category, and 0.0 otherwise.

The table gives a simple illustration of these heuristics.

Analog value		Rank order		One-of-c	
$\tilde{g}_i$	g_i	$\tilde{g}_i$	g_i	$\tilde{g}_i$	g_i
0.4	0.158	3rd	$4/21 = 0.194$	0	0.0
0.6	0.193	6th	$1/21 = 0.048$	1	1.0
0.9	0.260	5th	$2/21 = 0.095$	0	0.0
0.3	0.143	1st	$6/21 = 0.286$	0	0.0
0.2	0.129	2nd	$5/21 = 0.238$	0	0.0
0.1	0.111	4th	$3/21 = 0.143$	0	0.0

Once the outputs of the component classifiers have been converted to effective discriminant functions in this way, the component classifiers are themselves held fixed, but the gating network is trained as described in Eq. 59. This method is particularly useful when several highly trained component classifiers are pooled to form a single decision.

SUMMARY

The No Free Lunch Theorem states that in the absence of prior information about the problem there are no reasons to prefer one learning algorithm or classifier model over another. Given that a finite set of feature values are used to distinguish the patterns under consideration, the Ugly Duckling Theorem states that the number of predicates shared by any two different patterns is constant and does not depend upon the choice of the two objects. Together, these theorems highlight the need for insight into proper features and matching the algorithm to the data distribution: There is no problem-independent "best" learning or pattern recognition system nor feature representation. In short, formal theory and algorithms taken alone are not enough; pattern classification is an empirical subject.

Two ways to describe the match between classifier and problem are the bias and variance. The bias measures the accuracy or quality of the match (high bias implies a *poor* match), and the variance measures the precision or specificity of the match (a high variance implies a *weak* match). The bias-variance dilemma states that learning procedures with increased flexibility to adapt to the training data (e.g., have more free parameters) tend to have lower bias but higher variance. In classification there is a nonlinear relationship between bias and variance, and low variance tends to be more important for classification than low bias. If classifier models can be expressed as binary strings, the minimum-description-length principle states that the best model is the one with the minimum sum of such a model description and the training data with respect to that model. This general principle can be extended to cover model-specific heuristics such as weight decay and pruning in neural networks, regularization in specific models, and so on.

The basic insight underlying resampling techniques—such as the bootstrap, jackknife, boosting, and bagging—is that multiple data sets selected from a given data set enable the value and ranges of arbitrary statistics to be computed. In classification, boosting techniques such as AdaBoost adjust the match of full classifier to the problem (and thus the bias and variance) even for an arbitrary basic classification method. In learning with queries, the classifier system presents query patterns to an oracle for labeling. Such learning is most efficient if *informative* patterns—ones for which the classifier is least certain—are presented as queries.

There are a number of methods for estimating the final accuracy of classifiers and thus comparing them. Each is based on assumptions—for example, that the parametric model is known or that the form of its learning curve is known. Cross-validation, jackknife, and bootstrap methods are closely related techniques that use subsets of the training data to estimate classifier accuracy. Maximum-likelihood (ML-II) and Bayesian methods—extensions of methods for setting *parameters*—can be used to compare and choose among *models*. A key term in Bayesian model selection is the Occam factor, which describes how the allowable volume in parameter space shrinks due to constraints imposed by the training data. The method penalizes "overly complex" models, where such complexity is a data-dependent property.

There are a number of methods for combining the outputs of separate component or "expert" classifiers, such as linear weighting, winner-takes-all, and so on. Overall classification is generally better when the decision rules of the component classifiers differ and provide complementary information.

BIBLIOGRAPHICAL AND HISTORICAL REMARKS

The No Free Lunch Theorem appears in reference [110] as well as in Wolpert's collection of contributions on the foundations of the theory of generalization [109]. Schaffer's "conservation law in generalization" is a reformulation of one of the parts of the theorem, and was the inspiration for Fig. 9.1 [83]. The Ugly Duckling Theorem was proven in reference [105], which also explores some of its philosophical implications [79].

The foundational work on Kolmogorov complexity appears in references [57, 58, 93] and [94], but a short elementary overview [14] and Chaitin's [15] and particularly Li and Vitányi's [66] books are far more accessible. Barron and Cover were the first to use a minimum description length (MDL) principle to estimate densities [7]. There are several versions of MDL [80, 81], such as the Akaike Information Criterion (AIC) [1, 2] and the Bayes Information Criterion (BIC) [86] (which differ from MDL by relative weighting of model penalty). Likewise, the Network Information Criterion (NIC) can be used to compare neural networks of the same architecture [73]. More generally, neural network pruning and general regularization methods can be cast as "minimum description" principles, but with different measures for model and fit of the data [65].

Convincing theoretical and philosophical justifications of Occam's razor have been elusive. Karl Popper has argued that Occam's razor is without operational value, because there is no clear criterion or measure of simplicity [76], a point echoed by other philosophers [92]. It is worth pointing out alternatives to Occam's razor, which Isaac Newton cast in **Principia** as "Natura enim simplex est, et rerum causis superfluis non luxuriat," or "for nature indeed is simple, and does not luxuriate in superfluous causes" [74]. The first alternative stems from Epicurus (342?–270? B.C.), who, in a letter to Pythocles, stated what we now call the principle of multiple explanations or principle of indifference: If several theories are consistent with the data, retain all such theories [29]. The second is a restatement of Bayes approach: The probability of a model or hypothesis being true is proportional to the designer's prior belief in the hypothesis multiplied by the conditional probability of the data given the hypothesis in question. Occam's razor, here favoring "simplicity" in classifiers, can be motivated by considering the cost (difficulty) of designing the classifier and the principle of bounded rationality—that we often settle for an adequate but not necessarily the optimal solution [89]. An empirical study showing that simple classifiers often work well can be found in reference [45].

The basic bias-variance decomposition and bias-variance dilemma [37] in regression appear in many statistics books [16, 41]. Geman et al. give a very clear presentation in the context of neural networks, but their discussion of *classification* is only indirectly related to their mathematical derivations for *regression* [35]. Our presentation for classification (zero-one loss) is based on Friedman's important paper [32]; the bias-variance decomposition has been explored in other nonquadratic cost functions as well [42].

Quenouille introduced the jackknife in 1956 [78]. The theoretical foundations of resampling techniques are presented in Efron's clear book [28], and practical guides to their use include references [25] and [36]. Papers on bootstrap techniques for error estimation include reference [48]. Breiman has been particularly active in introducing and exploring resampling methods for estimation and classifier design, such as bagging [11] and general arcing [13]. AdaBoost [31] builds upon Schapire's analysis of the strength of weak learnability [84] and Freund's early work in the theory of learning [30]. Boosting in multicategory problems is a bit more subtle than in two-category problems we discussed [85]. Angluin's early work on queries for concept learning [3] was generalized to active learning by Cohn and many others [18, 20] and is fundamental to some efforts in collecting large databases, as discussed in references [95, 96] and [100].

Cross-validation was introduced by Cover [23] and has been used extensively in conjunction with classification methods such as neural networks. Estimates of error under different conditions include references [34, 104] and [111], and an excellent paper that derives the size of test set needed for accurate estimation of classification accuracy is reference [39]. Bowyer and Phillip's book covers empirical evaluation techniques in computer vision [10], many of which apply to more general classification domains.

The roots of maximum-likelihood model selection stem from Bayes himself, but one of the earlier technical presentations is reference [38]. Interest in Bayesian model selection was revived in a series of papers by MacKay, whose primary interest was in applying the method to neural networks and interpolation [67, 68, 69, 70]. These model selection methods have subtle relationships to minimum-description-length (MDL) [80] and so-called maximum entropy approaches—topics that would take us a bit beyond our central concerns. Cortes and her colleagues pioneered the analysis of learning curves for estimating the final quality of a classifier in references [21] and [22]. No rate of convergence results can be made in the arbitrary case for finding the Bayes error, however [6]. Hughes [46] first carried out the required computations displayed in Fig. 9.14.

Books on techniques for combining general classifiers include references [55] and [56] and those for combining neural nets in particular include references [9] and [88]. Perrone and Cooper described the benefits that arise when expert classifiers disagree [75]. Dasarathy's book [24] has a nice mixture of theory (focusing more on sensor fusion than multiclassifier systems per se) and a collection of important original papers, including references [43, 61] and [97]. The simple heuristics for converting 1-of-c and rank order outputs to numerical values enabling integration were discussed in reference [63]. The hierarchical mixture-of-experts architecture and learning algorithm were first described in references [51] and [52]. A specific hierarchical multiclassifier technique is stacked generalization [12, 90, 91] and [108], where for instance Gaussian kernel estimates at one level are pooled by yet other Gaussian kernels at a higher level.

We have skipped over a great deal of work from the formal field of computational learning theory. Such work is generally preoccupied with convergence properties, asymptotics, and computational complexity, and usually relies on simplified or general models. Anthony and Biggs' short, clear and elegant book is an excellent introduction to the field [5]; broader texts include references [49, 53] and [72]. Perhaps the work from the field most useful for pattern recognition practitioners comes from weak learnability and boosting, mentioned above. The probably approximately correct (PAC) framework, introduced by Valiant [99], has been very influential in computation learning theory, but has had only minor influence on the development of practical pattern recognition systems. A somewhat broader formulation, probably almost Bayes (PAB), is described in reference [4]. The work by Vapnik and Chervonenkis on structural risk minimization [103], and later Vapnik-Chervonenkis (VC) theory [101, 102], derives (among other things) expected error bounds; it too has proven influential to the theory community. Alas, the bounds derived are somewhat loose in practice, as presented in references [19] and [107].

PROBLEMS

Section 9.2

1. One of the "conservations laws" for generalization states that the positive generalization performance of an algorithm in some learning situations must be offset by negative performance elsewhere. Consider a very simple learning algorithm that seems to contradict this law. For each test pattern, the prediction of the *majority learning algorithm* is merely the category most prevalent in the training data.

   (a) Show that averaged over all two-category problems of a given number of features the off-training set error is 0.5.

   (b) Repeat (a) but for the *minority learning algorithm*, which always predicts the category label of the category *least* prevalent in the training data.

   (c) Use your answers from (a) and (b) to illustrate Part 2 of the No Free Lunch Theorem (Theorem 9.1).

2. Prove Part 1 of Theorem 9.1, that is, uniformly averaged over all target functions F, $\mathcal{E}_1(E|F, n) - \mathcal{E}_2(E|F, n) = 0$. Summarize and interpret this result in words.

3. Prove Part 2 of Theorem 9.1, that is, for any fixed training set $\mathcal{D}$, uniformly averaged over F, $\mathcal{E}_1(E|F, \mathcal{D}) - \mathcal{E}_2(E|F, \mathcal{D}) = 0$. Summarize and interpret this result in words.

4. Prove Part 3 of Theorem 9.1, that is, uniformly averaged over all priors $P(F)$, $\mathcal{E}_1(E|n) - \mathcal{E}_2(E|n) = 0$. Summarize and interpret this result in words.

5. Prove Part 4 of Theorem 9.1, i.e., for any fixed training set $\mathcal{D}$, uniformly averaged over $P(F)$, $\mathcal{E}_1(E|\mathcal{D}) - \mathcal{E}_2(E|\mathcal{D}) = 0$. Summarize and interpret this result in words.

6. Suppose you call an algorithm better if it performs slightly better than average over *most* problems, but very poorly on a small number of problems. Explain why the NFL Theorem does not preclude the existence of algorithms "better" in this sense.

7. Show by simple counterexamples that the averaging in the different Parts of the No Free Lunch Theorem (Theorem 9.1) must be "uniformly." For instance imagine that the sampling distribution is a Dirac delta distribution centered on a single target function, and algorithm 1 guesses the target function exactly while algorithm 2 disagrees with algorithm 1 on every prediction.

   (a) Part 1

   (b) Part 2

   (c) Part 3

   (d) Part 4

8. State how the No Free Lunch theorems imply that you cannot use training data to distinguish between new problems for which you generalize well from those for which you generalize poorly. Argue by *reductio ad absurdum*—that is, that if you could distinguish such problems, then the No Free Lunch Theorem would be violated.

9. Prove the relation $\sum_{r=0}^{n} \binom{n}{r} = (1+1)^n = 2^n$ of Eq. 5 two ways:

   (a) State the polynomial expansion of $(x + y)^n$ as a summation of coefficients and powers of x and y. Then, make a simple substitution for x and y.

   (b) Prove the relation by induction. Let $K(n) = \sum_{r=0}^{n} \binom{n}{r}$. First confirm that the relation is valid for $n = 1$—that is, that $K(1) = 2^1$. Now prove that $K(n+1) = 2K(n)$ for arbitrary n.

10. Consider the number of different Venn diagrams for k binary features $f_1, \ldots, f_k$. (Figure 9.2 shows several of these configurations for the $k = 3$ case.)

    (a) How many functionally different Venn diagrams exist for the $k = 2$ case? Sketch all of them. For each case, state how many different regions exists—that is, how many functionally different patterns can be represented.

    (b) Repeat part (a) for the $k = 3$ case.

    (c) How many functionally different Venn diagrams exist for the arbitrary k case?

11. While the text outlined a proof of the Ugly Duckling Theorem (Theorem 9.2), this problem asks you to fill in some of the details and explain some of its implications.

    (a) The discussion in the text assumed that the classification problem had no constraints, and thus could be described by the most general Venn diagram, in which all predicates of a given rank r were present. How do the derivations change, if at all, if we know that there are constraints provided by the problem, as in Fig. 9.2 (b) and (c)?

    (b) Someone sees two cars, A and B, made by the same manufacturer in the same model year, both are four-door and have the same engine type, but they differ solely in that one is red while the other is green. Car C is made by a different manufacturer, has a different engine, is two-door, and is blue. Explain in as much detail as possible why, even in this seemingly clear case, that in fact there are no prior reasons to view cars A and B as any "more similar" than cars B and C.

12. Suppose we describe patterns by means of predicates of a particular rank r^*. Show that the Ugly Duckling Theorem (Theorem 9.2) applies to any single level r^* and thus applies to all predicates up to an arbitrary maximum level.

13. Make some simple assumptions and state, using $O(\cdot)$ notation, the Kolmogorov complexity of the following binary strings:

    (a) $\underbrace{010110111011110\ldots}_{n}$

    (b) $\underbrace{000\ldots00100\ldots000}_{n}$

    (c) $e = 10.10110111111000010\ldots_2$

    (d) $2e = 101.01101111110000101\ldots_2$

(e) The binary digits of π, but where every 100th digit is changed to the numeral 1.

(f) The binary digits of π, but where every nth digit is changed to the numeral 1.

14. Recall the notation from our discussion of the No Free Lunch Theorem and of Kolmogorov complexity. Suppose we use a learning algorithm with uniform $P(h|\mathcal{D})$. In that case $K(h, \mathcal{D}) = K(\mathcal{D})$ in Eq. 8. Explain and interpret this result.

15. Consider two binary strings x_1 and x_2. Explain why the Kolmogorov complexity of the pair obeys $K(x_1, x_2) \le K(x_1) + K(x_2) + c$ for some positive constant c.

16. Consider designing a tree-based classifier of the general type explored in Chapter 8 employing the minimum-description length principle and alternatively the imposition of priors over the form of classifier.

   (a) Suppose we design a tree classifier using a minimum-description length principle where the total entropy (in bits) consists of two terms: the entropy of the data with respect to the tree and the number of nodes in the tree. This is formally equivalent to a training the tree by maximum-likelihood techniques with a prior that favors smaller trees. Determine the functional form of this corresponding prior $P(K)$, where K is the total number of nodes in the tree. State any assumptions you must make.

   (b) Suppose a tree classifier is trained by maximum-likelihood techniques where the prior on the number of nodes decreases with the number of nodes according to and exponential, that is, $P(K) \propto e^{-K}$. Determine the functional form of an equivalent minimum-description length criterion that will lead to the same final classifier.

BERRY PARADOX   17. The *Berry paradox* is related to the famous liar's paradox ("this statement is false"), as well as to a number of paradoxes in set theory explored by Bertrand Russell and Kurt Gödel. The Berry paradox shows indirectly how the notion of Kolmogorov complexity can be difficult and subtle. Consider describing positive integers by means of sentences—for instance, "the number of fingers on a human hand" or "the number of primes less than a million." Explain why the definition "the least number that cannot be defined in less than twenty words" is paradoxical, and explain informally how this relates to difficulties in computing Kolmogorov complexity.

Section 9.3

18. Expand the left-hand side of Eq. 11 to get the right-hand side, which expresses the mean-square error as a sum of a *bias*^{2} and *variance*. Can *bias* ever be negative? Can *variance* ever be negative?

19. Fill in the steps leading to Eq. 18, that is,

$$\Pr[g(\mathbf{x}; \mathcal{D}) \ne y] = |2F(\mathbf{x}) - 1|\Pr[g(\mathbf{x}; \mathcal{D}) \ne y_B] + \Pr[y_B \ne y]$$

where the target function is $F(\mathbf{x})$, $g(\mathbf{x}; \mathcal{D})$ is the computed discriminant value, and y_B is the Bayes discriminant value.

20. Assume that the probability of obtaining a particular discriminant value for pattern $\mathbf{x}$ for a training algorithm trained with data $\mathcal{D}$, denoted $p(g(\mathbf{x}; \mathcal{D}))$, is a Gaussian. Use this Gaussian assumption and Eq. 19 to derive Eq. 20.

Section 9.4

21. Derive the jackknife estimate of the *bias* in Eq. 29.

22. Prove that in the limit of $B \to \infty$, the bootstrap estimate of the variance of the mean is the same as the standard estimate of the variance of the mean.

23. Prove that Eq. 24 for the average of the leave one out means, $\mu_{(\cdot)}$, is equivalent to Eq. 22 for the sample mean, $\hat{\mu}$.

CONSISTENT

24. We say that an estimator is *consistent* if it converges to the true value in the limit of infinite data. Prove that the standard mean of Eq. 22 is *not* consistent for the distribution $p(x) \sim \tan^{-1}(x - a)$ for any finite real constant a.

25. Prove that the jackknife estimate of an arbitrary statistic θ given in Eq. 30 is unbiased for estimating the true bias.

26. Verify that Eq. 26 for the jackknife estimate of the variance of the mean is formally equivalent to the variance implied by the traditional estimate given in Eq. 23.

27. Consider n points in one dimension. Use $O(\cdot)$ notation to express the computational complexity associated with each of the following estimations.

   (a) The jackknife estimate of the mean

   (b) The jackknife estimate of the median

   (c) The jackknife estimate of the standard deviation

   (d) The bootstrap estimate of the mean

   (e) The bootstrap estimate of the median

   (f) The bootstrap estimate of the standard deviation

28. Derive Eq. 34 for the bootstrap estimate of the bias.

Section 9.5

29. What is the computational complexity of full jackknife estimate of accuracy and variance for an unpruned nearest-neighbor classifier (Chapter 4)?

30. In standard boosting applied to a two-category problem, we must create a data set that is "most informative" with respect to the current state of the classifier. Why does this imply that *half* of its patterns should be classified correctly, rather than *none* of them? In a c-category problem, what portion of the patterns should be misclassified in a "most informative" set?

31. In active learning, learning can be speeded by creating patterns that are "informative"—that is, those for which the two largest discriminants are approximately equal. Consider the two-category case where for any point $\mathbf{x}$ in feature space, discriminant values g_1 and g_2 are returned by the classifier. Write pseudocode that takes two points—$\mathbf{x}_1$ classified as ω_1 and $\mathbf{x}_2$ classified as ω_2—and rapidly finds a new point $\mathbf{x}_3$ that is "near" the current decision boundary and hence is "informative." Assume only that the discriminant functions are monotonic along the line linking $\mathbf{x}_1$ and $\mathbf{x}_2$.

32. Consider AdaBoost with an arbitrary number of component classifiers.

   (a) State clearly any assumptions you make, and derive Eq. 37 for the ensemble training error of the full boosted system.

(b) Recall that the training error for a weak learner applied to a two-category problem can be written $E_k = 1/2 - G_k$ for some positive value G_k. The training error for the first component classifier is $E_1 = 0.25$. Suppose that $G_k = 0.05$ for all $k = 1$ to k_{max}. Plot the upper bound on the ensemble test error given by Eq. 37, such as shown in Fig. 9.7.

(c) Suppose that G_k decreases as a function of k. Specifically, repeat part (b) with the assumption $G_k = 0.05/k$ for $k = 1$ to k_{max}.

Section 9.6

33. The No Free Lunch Theorem implies that if all problems are equally likely, then cross-validation must fail as often as it succeeds. Show this as follows: Consider algorithm 1 to be standard cross-validation and consider algorithm 2 to be *anti-cross-validation*, which advocates choosing the model that does *worst* on a validation set. Argue that if cross-validation were better than anti-cross-validation overall, the No Free Lunch Theorem would be violated.

34. Suppose we believe that the data for a pattern classification task from one category comes either from a uniform distribution $p(x) \sim U(x_l, x_u)$ or from a normal distribution, $p(x) \sim N(\mu, \sigma^2)$, but we have no reason to prefer one over the other. Our sample data is $\mathcal{D} = \{0.2, 0.5, 0.4, 0.3, 0.9, 0.7, 0.6\}$.

(a) Find the maximum-likelihood values of x_l, and x_u for the uniform model.

(b) Find the maximum-likelihood values of μ and σ for the Gaussian model.

(c) Use maximum-likelihood model selection to decide which model should be preferred.

35. Suppose we believe that the data for a pattern classification task from one category comes either from a uniform distribution bounded below by 0—that is, $p(x) \sim U(0, x_u)$—or from a normal distribution, $p(x) \sim N(\mu, \sigma^2)$, but we have no reason to prefer one over the other. Our sample data is $\mathcal{D} = \{0.2, 0.5, 0.4, 0.3, 0.9, 0.7, 0.6\}$.

(a) Find the maximum-likelihood values of x_u for the uniform model.

(b) Find the maximum-likelihood values of μ and σ for the Gaussian model.

(c) Use maximum-likelihood model selection to decide which model should be preferred.

(d) State qualitatively the difference between your solution here and that to Problem 34, without necessarily having to solve that problem. In particular, what are the implications from the fact that the two candidate models have different numbers of parameters?

36. Consider three candidate one-dimensional distributions, each parameterized by an unknown value for its "center":

- Gaussian: $p(x) \sim N(\mu, 1)$
- Triangle: $p(x) \sim T(\mu, 1) = \begin{cases} 1 - |x - \mu| & \text{for } |x - \mu| < 1 \\ 0 & \text{otherwise} \end{cases}$

- Uniform: $p(x) \sim U(\mu - 1, \mu + 1)$

We are given the data $\mathcal{D} = \{-0.9, -0.1, 0., 0.1, 0.9\}$, and thus, clearly the maximum-likelihood solution $\hat{\mu} = 0$ applies to each model.

   (a) Use maximum-likelihood model selection to determine the best model for this data. State clearly any assumptions you make.

   (b) Suppose we are sure for each model that the center must lie in the range $-1 \le \mu \le 1$. Calculate the Occam factor for each model and the data given.

   (c) Use Bayesian model selection to determine the "best" model given $\mathcal{D}$.

37. Use Eq. 38 and generate curves of the form shown in Fig. 9.10. Prove analytically that the curves are symmetric with respect to the interchange $\hat{p} \to (1 - \hat{p})$ and $p \to (1 - p)$. Explain the reasons for this symmetry.

38. Let model h_i be described by a k-dimensional parameter vector $\boldsymbol{\theta}$. State your assumptions and show that the Occam factor can be written as

$$p(\hat{\boldsymbol{\theta}}|h_i)(2\pi)^{k/2}|\mathbf{H}|^{-1/2},$$

as given in Eq. 44, where the Hessian $\mathbf{H}$ is a matrix of second-order derivatives defined in Eq. 45.

BERTRAND'S PARADOX   **39.** *Bertrand's paradox* shows how the notion of "uniformly distributed" models can be problematic, and it leads us to question the principle of indifference (cf. Computer exercise 9). Consider the following problem: Given a circle, find the probability that a "randomly selected" chord has length greater than the side of an inscribed equilateral triangle.

   Here are three possible solutions to this problem and their justifications, illustrated in the figure:

   1. By definition, a chord strikes the circle at two points. We can arbitrarily rotate the figure so as to place one of those points at the top. The other point is equally likely to strike any other point on the circle. As shown in the left-hand side of the figure, one-third of such points (red) will yield that a chord with length greater than that of the side of an inscribed equilateral triangle. Thus the probability a chord is longer than the length of the side of an inscribed equilateral triangle is $P = 1/3$.

   2. A chord is uniquely determined by the location of its midpoint. Any such midpoint that lies in the circular disk whose radius is half that of the full circle will yield a chord with length greater than that of an inscribed equilateral triangle. Because the area of this red disk is one-fourth that of the full circular disk, the probability is $P = 1/4$.

   3. We can arbitrarily rotate the circle such that the midpoint of a chord lies on a vertical line. If the midpoint lies closer to the center than half the radius of the circle, the chord will be longer than the side of the inscribed equilateral triangle. Thus the probability is $P = 1/2$.

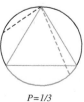
$P=1/3$

$P=1/4$

$P=1/2$

Explain why there is little or no reason to prefer one of the solution methods over another, and thus the solution to the problem itself is ill-defined. Use your answer to reconcile Bayesian model selection with the No Free Lunch Theorem (Theorem 9.1).

40. If k of n' independent, randomly chosen test patterns are misclassified, then as given in Eq. 38, k has a binomial distribution

$$P(k) = \binom{n'}{k} p^k (1 - p)^{n'-k}$$

Prove that the maximum-likelihood estimate for p is then $\hat{p} = k/n'$, as given in Eq. 39.

41. Derive the relation for $f(n, d)$, the fraction of dichotomies of n randomly chosen points in d dimensions that are linearly separable, given by Eq. 53. Explain why $f(n, d) = 1$ for $n \leq d + 1$.

42. Write pseudocode for an algorithm to determine the large n' limit of the test error, given the assumption of a power-law decrease in error described by Eq. 52 and illustrated in Fig. 9.17.

43. Suppose a standard three-layer neural network having d input units, n_H hidden units, a single bias unit and $c = 2$ output units is trained on a two-category problem (Chapter 6). What is the degeneracy of the final assignment of weights? That is, how many ways can the weights be relabeled with the decision rule being unchanged? Explain how this degeneracy would need to be incorporated into a Bayesian model selection.

Section 9.7

44. Let $\mathbf{x}^i$ and $\mathbf{y}^i$ denote input and output vectors, respectively, and r index component processes ($1 \leq r \leq k$) in a mixture model. Use Bayes theorem,

$$P(r|\mathbf{y}^i, \mathbf{x}^i) = \frac{P(r|\mathbf{x}^i) P(\mathbf{y}^i|\mathbf{x}^i, r)}{\sum_{q=1}^{k} P(q|\mathbf{x}^i) P(\mathbf{y}^i|\mathbf{x}^i, q)},$$

to derive the derivatives Eqs. 58 and 59 used for gradient descent learning in the mixture of experts model.

45. Suppose a mixture-of-experts classifier has k Gaussians component classifiers of arbitrary mean and covariance in d dimensions, $N(\boldsymbol{\mu}_r, \boldsymbol{\Sigma}_r)$. Derive the specific learning rules for the parameters of each component classifier and for the gating subsystem, special cases of Eqs. 58 and 59.

COMPUTER EXERCISES

Several exercises will make use of the following three-dimensional data sampled from four categories, denoted ω_i.

Sample	ω_1			ω_2			ω_3			ω_4		
	x_1	x_2	x_3	x_1	x_2	x_3	x_1	x_2	x_3	x_1	x_2	x_3
1	2.5	3.4	7.9	4.2	4.9	11.3	2.9	15.5	4.6	16.9	12.4	0.2
2	4.3	4.4	7.1	11.7	5.3	10.5	3.6	13.9	9.8	12.1	16.8	2.1
3	7.1	0.8	6.3	8.4	11.1	6.6	10.3	6.1	12.3	13.7	12.1	5.5
4	1.4	−0.2	2.5	8.2	10.4	4.9	8.2	5.5	7.1	11.9	13.4	3.4
5	3.9	4.3	3.4	5.3	7.7	8.8	13.3	4.7	11.7	14.5	15.5	2.8
6	3.2	6.8	5.1	7.9	4.5	9.5	6.6	8.1	16.7	15.6	14.9	4.4
7	7.3	6.5	7.1	10.7	6.9	10.9	12.2	5.1	5.9	16.2	12.3	3.2
8	−0.7	3.1	8.1	9.6	9.7	7.3	15.6	3.3	10.7	12.2	16.3	3.2
9	2.8	5.9	2.2	8.2	11.2	6.3	4.6	10.1	13.8	14.5	12.9	−0.9
10	6.1	7.6	4.3	5.3	10.1	4.9	9.1	4.4	8.9	15.8	15.6	4.5

Section 9.2

1. Consider the use of the minimum description length principle for the design of a binary decision tree classifier (Chapter 4). Each question at a node is of the form "Is $x_i > \theta$?" (or alternatively "Is $x_i < \theta$?"). Specify each such question with five bits: Two bits specify the feature queried (x_1, x_2, or x_3), a single bit specifies whether the comparison is $>$ or $<$, and four bits specify each θ as an integer $0 \leq \theta \leq 16$. Assume that the Kolmogorov complexity of the classifier is, up to an additive constant, the sum of the bits of all questions. Assume too that the Kolmogorov complexity of the data, given the tree classifier, is merely the entropy of the data at the leaves, also measured in bits.

   (a) Train your tree with the data from the four-category problem in the table above. Starting at the root, grow your tree a single node at a time, continuing until each node is as pure as possible. Plot as a function of the total number of nodes the Kolmogorov complexity of (1) the classifier, (2) the data with respect to the classifier, and (3) their sum (Eq. 8). Show the tree (including the questions at its nodes) having the minimum description length.

   (b) The minimum description length principle gives a principled method for comparing classifiers in which the resolution of parameters (e.g., weights, thresholds, etc.) can be altered. Repeat part (a) but using only three bits to specify each threshold θ in the nodes.

   (c) Assume that the additive constants for the Kolmogorov complexities of your above classifiers are equal. Which of all the classifiers has the minimum description length?

Section 9.3

2. Illustrate the bias-variance decomposition and the bias-variance dilemma for regression through simulations. Let the target function be $F(x) = x^2$ with Gaussian noise of variance 0.1. First, randomly generate 100 data sets, each of size $n = 10$, by selecting a value of x uniformly in the range $-1 \leq x \leq 1$ and then applying $F(x)$ with noise. Train any free parameters a_i (by minimum square error criterion) in each of the regression functions in parts (a) – (d), one data set at a time. Then make a histogram of the sum-square error of Eq. 11 (cf. Fig. 9.4). For each model use your results to estimate the bias and the variance.

   (a) $g(x) = 0.5$
   (b) $g(x) = 1.0$
   (c) $g(x) = a_0 + a_1 x$

(d) $g(x) = a_0 + a_1 x + a_2 x^2 + a_3 x^3$

(e) Repeat parts (a)–(d) for 100 data sets of size $n = 100$.

(f) Summarize all your above results, with special consideration of (i) the bias-variance decomposition and dilemma, and (ii) the effect of the size of the data set.

Section 9.4

TRIMMED MEAN

3. The *trimmed mean* of a distribution is merely the sample mean of the distribution from which some portion α (e.g., 0.1) of the highest and of the lowest points have been deleted. The trimmed mean is, of course, less sensitive to the presence of outliers than is the traditional sample mean.

(a) Show how in the limit $\alpha \to 0.5$, the trimmed mean of a distribution is the median.

(b) Let the data $\mathcal{D}$ be the x_3 values of the 10 patterns in category ω_2 in the table above. Write a program to determine (i) the jackknife estimate of the median of $\mathcal{D}$ and (ii) the jackknife estimate of the variance of this estimate.

(c) Repeat part (b) but for the $\alpha = 0.1$ trimmed mean and its variance.

(d) Repeat part (b) but for the $\alpha = 0.2$ trimmed mean and its variance.

(e) Repeat parts (b)–(d) but where $\mathcal{D}$ has an additional ("outlier") point at $x_3 = 20$.

(f) Interpret your results, with special attention to the sensitivity of the trimmed mean to outliers.

Section 9.5

4. Write a program to implement the AdaBoost procedure (Algorithm 1) with component classifiers whose linear discriminants are trained by the basic LMS algorithm (Chapter 5).

(a) Apply your system to the problem of discriminating the 10 points in ω_1 from the 10 points in ω_2 in the table above. Plot your training error as a function of the number of component classifiers. Be sure the graph extends to a k_{max} sufficiently high that the training error vanishes.

(b) Define a "supercategory" consisting of all the patterns in ω_1 and ω_2 in the table, and define another supercategory for the ω_3 and ω_4 patterns. Repeat part (a) for discriminating these supercategories.

(c) Compare and interpret your graphs in (a) and (b), paying particular attention to the relative difficulties of the classification problems.

5. Explore the value of active learning in a two-dimensional two-category problem in which the priors are Gaussians, $p(\mathbf{x}|\omega_i) \sim N(\boldsymbol{\mu}_i, \boldsymbol{\Sigma}_i)$ with $\boldsymbol{\mu}_1 = \binom{+5}{+5}$, $\boldsymbol{\mu}_2 = \binom{-5}{-5}$, $\boldsymbol{\Sigma}_1 = \boldsymbol{\Sigma}_2 = \binom{20\ \ 0}{0\ \ 20}$, and $P(\omega_1) = P(\omega_2) = 0.5$. Throughout this problem, restrict data to be in the domain $-10 \le x_i \le +10$, for $i = 1, 2$.

(a) State by inspection the Bayes classifier. This will be the decision used by the oracle in part (c).

(b) Generate a training set of 100 points, 50 labeled ω_1 sampled according to $p(\mathbf{x}|\omega_1)$ and likewise 50 patterns according to $p(\mathbf{x}|\omega_2)$. Train a nearest-neighbor classifier (Chapter 4) using your data, and plot the decision boundary in two dimensions.

(c) Now assume there is an oracle, which can label any query pattern according to your answer in part (a), which we exploit through a particular form of active learning. To begin the learning, choose 10 points according to a uniform distribution in the domain $-10 \le x_i \le +10$, for $i = 1, 2$. Apply labels to these points according to the oracle to get $\mathcal{D}_1$ and $\mathcal{D}_2$, for each category. Now generate new query points as follows. Randomly choose a point from $\mathcal{D}_1$ and a point from $\mathcal{D}_2$; create a query point midway between these two points. Label the point according to the oracle and add it to the appropriate $\mathcal{D}_j$. Continue until the total number of labeled points is 100. Now create a nearest-neighbor classifier using all points, and plot the decision boundary in two dimensions.

(d) Compare qualitatively your classifiers from parts (a), (b) and (c), and discuss your results.

6. In simple boosting using three component classifiers we would like to use all n training points and to have a roughly equal number of patterns used in each component classifier (i.e., $n_1 \simeq n_2 \simeq n_3 \simeq n/3$).

(a) Generate a training set $\mathcal{D}$ consisting of $n = 300$ two-dimensional points, 150 in each of two categories. Draw your samples from uniform distributions in squares defined by their opposite corners, in particular

- $p(\mathbf{x}|\omega_1) \sim U\left(\binom{0}{0}, \binom{3}{3}\right)$
- $p(\mathbf{x}|\omega_2) \sim U\left(\binom{1}{1}, \binom{4}{4}\right)$

(b) Assume each component classifier is a simple monothetic binary decision tree consisting of a root node, two offspring nodes and four leaf nodes and trained on an entropy impurity (Chapter 8). Train a three-component classifier system using boosting. Start with $n_1 = n/3 = 100$ patterns.

(c) Perform a simple heuristic search for an appropriate for a good value of n_1 based on the values of n_2 and n_3 from your simulation in part (b). In particular, if you find n_1 is too low, re-initialize n_1 to be halfway between the maximum possible value $n_1 = n = 300$ and the current value. Alternatively, if you find n_1 is too high, re-initialize n_1 to be halfway between the minimum value $n_1 = 0$ and the current value.

(d) How many full applications of the boosting algorithm are needed to achieve an "acceptable" starting value for n_1? Be sure to state what you mean by "acceptable."

Section 9.6

7. Explore a case where validation need not yield an improved classifier. Throughout, the classifier will be k-nearest-neighbor (Chapter 4), where k will be set by validation. Consider a two-category problem in two dimensions, with uniform prior distributions throughout the range $0 \le x_i \le 1$ for $i = 1, 2$.

(a) First form a test set $\mathcal{D}_{test}$ of 20 points—10 points in ω_1 and 10 in ω_2— randomly chosen according to a uniform distribution.

(b) Next generate 100 points—50 patterns in each category. Set $\gamma = 0.1$ and split this set into a training set $\mathcal{D}_{train}$ (90 points) and validation set $\mathcal{D}_{val}$ (10 points).

(c) Now form a k-nearest-neighbor classifier in which k is increased until the first *minimum* of the validation error is found. (Restrict k to odd values, to avoid ties.) Now determine the error of your classifier using the test set.

(d) Repeat part (c), but instead find the k that is the first *maximum* of the validation error.

(e) Repeat parts (c) and (d) five times, noting the test error in all 10 cases.

(f) Discuss your results—in particular, how they depend or do not depend upon the fact that the data were all uniformly distributed.

8. Consider three candidate one-dimensional distributions, each parameterized by an unknown value for its "center":

  • Gaussian: $p(x) \sim N(\mu, \sigma^2)$

  • Triangle: $p(x) \sim T(\mu, 1) = \begin{cases} 1 - |x - \mu| & \text{for } |x - \mu| < 1 \\ 0 & \text{otherwise} \end{cases}$

  • Uniform: $p(x) \sim U(\mu - 2, \mu + 2)$

Suppose we are sure that for each model the center must lie in the range $-1 < \mu < 1$, and for the Gaussian that $0 \le \sigma^2 \le 1$. Suppose too that we are given the data $\mathcal{D} = \{-0.9, -0.1, 0.0, 0.1, 0.9\}$. Clearly, the maximum-likelihood solution $\hat{\mu} = 0$ applies to each model.

(a) Estimate the Occam factor in each case.

(b) Use Bayesian model selection to choose the best of these models.

9. Problem 39 describes Bertrand's paradox, which involves the probability that a circle's chord "randomly chosen" will have length greater than that of an inscribed equilateral triangle.

(a) Write a program to generate chords according to the logic of solution (1) in Problem 39. Generate 1000 such chords and estimate empirically the probability that a chord has length greater than that of an inscribed equilateral triangle.

(b) Repeat part (a), assuming the logic underlying solution (2).

(c) Repeat part (a), assuming the logic underlying solution (3).

(d) Explain why there is little or no reason to prefer one of the solution methods over another, and thus the solution to the stated problem is ill-defined.

(e) Relate your answers above to the No Free Lunch Theorem (Theorem 9.1) and Bayesian model selection.

Section 9.7

10. Create a multiclassifier system for the data in the table above. As in Computer exercise 4, define two supercategories, where the 20 points in ω_1 and ω_2 form one category, ω_A, and the remaining 20 points form ω_B.

(a) Let the first component classifier be based on Gaussian priors, where the mean $\boldsymbol{\mu}_i$ is arbitrary and the covariance is estimated by maximum-likelihood (Chapter 3). What is training error measured using ω_A and ω_B?

(b) Let the second component classifier also be based on Gaussian priors, but where the covariance is arbitrary. Now what is the training error measured using ω_A and ω_B?

(c) Train your two-component classifier by gradient descent (Eqs. 58 and 59). What is the training error of the full system?

BIBLIOGRAPHY

[1] Hirotugu Akaike. On entropy maximization principle. In Paruchuri R. Krishnaiah, editor, *Applications of Statistics*, pages 27–42. North-Holland, Amsterdam, 1977.

[2] Hirotugu Akaike. A Bayesian analysis of the minimum AIC procedure. *Annals of the Institute of Statistical Mathematics*, 30A(1):9–14, 1978.

[3] Dana Angluin. Queries and concept learning. *Machine Learning*, 2(4):319–342, 1988.

[4] Svetlana Anoulova, Paul Fischer, Stefan Pölt, and Hans-Ulrich Simon. Probably almost Bayes decisions. *Information and Computation*, 129(1):63–71, 1996.

[5] Martin Anthony and Norman Biggs, editors. *Computational Learning Theory: An Introduction*. Cambridge University Press, Cambridge, UK, 1992.

[6] András Antos, Luc Devroye, and László Györfi. Lower bounds for Bayes error estimation. *IEEE Transactions on Pattern Analysis and Applications*, PAMI-21(7):643–645, 1999.

[7] Andrew R. Barron and Thomas M. Cover. Minimum complexity density estimation. *IEEE Transactions on Information Theory*, IT-37(4):1034–1054, 1991.

[8] Charles H. Bennett, Péter Gács, Ming Li, Paul M. B. Vitányi, and Wojciech H. Zurek. Information distance. *IEEE Transactions on Information Theory*, IT-44(4):1407–1423, 1998.

[9] Christopher M. Bishop. *Neural Networks for Pattern Recognition*. Oxford University Press, Oxford, UK, 1995.

[10] Kevin W. Bowyer and P. Jonathon Phillips, editors. *Empirical Evaluation Techniques in Computer Vision*. IEEE Computer Society, Los Alamitos, CA, 1998.

[11] Leo Breiman. Bagging predictors. *Machine Learning*, 26(2):123–140, 1996.

[12] Leo Breiman. Stacked regressions. *Machine Learning*, 24(1):49–64, 1996.

[13] Leo Breiman. Arcing classifiers. *The Annals of Statistics*, 26(3):801–824, 1998.

[14] Gregory J. Chaitin. Information-theoretic computational complexity. *IEEE Transactions on Information Theory*, IT-20(1):10–15, 1974.

[15] Gregory J. Chaitin. *Algorithmic Information Theory*. Cambridge University Press, Cambridge, UK, 1987.

[16] Vladimir Cherkassky and Filip Mulier. *Learning from Data: Concepts, Theory, and Methods*. Wiley, New York, 1998.

[17] Bertrand S. Clarke and Andrew R. Barron. Information theoretic asymptotics of Bayes methods. *IEEE Transactions on Information Theory*, IT-36(3):453–471, 1990.

[18] David Cohn, Les Atlas, and Richard Ladner. Improving generalization with active learning. *Machine Learning*, 15(2):201–221, 1994.

[19] David Cohn and Gerald Tesauro. How tight are the Vapnik-Chervonenkis bounds? *Neural Computation*, 4(2):249–269, 1992.

[20] David A. Cohn, Zoubin Ghahramani, and Michael I. Jordan. Active learning with statistical models. In Gerald Tesauro, David S. Touretzky, and Todd K. Leen, editors, *Advances in Neural Information Processing Systems*, volume 7, pages 705–712. MIT Press, Cambridge, MA, 1995.

[21] Corinna Cortes, Larry D. Jackel, and Wan-Ping Chiang. Limits on learning machine accuracy imposed by data quality. In Gerald Tesauro, David S. Touretzky, and Todd K. Leen, editors, *Advances in Neural Information Processing Systems*, volume 7, pages 239–246. MIT Press, Cambridge, MA, 1995.

[22] Corinna Cortes, Larry D. Jackel, Sara A. Solla, Vladimir Vapnik, and John S. Denker. Learning curves: Asymptotic values and rate of convergence. In Jack D. Cowan, Gerald Tesauro, and Joshua Alspector, editors, *Advances in Neural Information Processing Systems*, volume 6, pages 327–334. Morgan Kaufmann, San Francisco, CA, 1994.

[23] Thomas M. Cover. Learning in pattern recognition. In Satoshi Watanabe, editor, *Methodologies of Pattern Recognition*, pages 111–132. Academic Press, New York, 1969.

[24] Belur V. Dasarathy, editor. *Decision Fusion*. IEEE Computer Society, Washington, DC, 1994.

[25] Anthony Christopher Davison and David V. Hinkley, editors. *Bootstrap Methods and Their Application*. Cambridge University Press, Cambridge, UK, 1997.

[26] Tom G. Dietterich. Overfitting and undercomputing in machine learning. *Computing Surveys*, 27(3):326–327, 1995.

[27] Robert P. W. Duin. A note on comparing classifiers. *Pattern Recognition Letters*, 17(5):529–536, 1996.

[28] Bradley Efron. *The Jackknife, the Bootstrap and Other Resampling Plans*. Society for Industrial and Applied Mathematics (SIAM), Philadelphia, PA, 1982.

[29] Epicurus and Eugene Michael O'Connor (Editor). *The Essential Epicurus: Letters, Principal Doctrines, Vatican Sayings, and Fragments*. Prometheus Books, New York, 1993.

[30] Yoav Freund. Boosting a weak learning algorithm by majority. *Information and Computation*, 121(2):256–285, 1995.

[31] Yoav Freund and Robert E. Schapire. A decision-theoretic generalization of on-line learning and an application to boosting. *Journal of Computer and System Sciences*, 55(1):119–139, 1995.

[32] Jerome H. Friedman. On bias, variance, 0/1-loss, and the curse-of-dimensionality. *Data Mining and Knowledge Discovery*, 1(1):55–77, 1997.

[33] Kenji Fukumizu. Active learning in multilayer perceptrons. In David S. Touretzky, Michael C. Mozer, and Michael E. Hasselmo, editors, *Advances in Neural Information Processing Systems*, volume 8, pages 295–301. MIT Press, Cambridge, MA, 1996.

[34] Keinosuke Fukunaga and Raymond R. Hayes. Effects of sample size in classifier design. *IEEE Transactions on Pattern Analysis and Machine Intelligence*, PAMI-11:873–885, 1989.

[35] Stewart Geman, Elie Bienenstock, and René Doursat. Neural networks and the bias/variance dilemma. *Neural Networks*, 4(1):1–58, 1992.

[36] Phillip I. Good. *Resampling Methods: A Practical Guide to Data Analysis*. Birkhauser, Boston, MA, 1999.

[37] Ulf Grenander. On empirical spectral analysis of stochastic processes. *Arkiv Matematiki*, 1(35):503–531, 1951.

[38] Stephen F. Gull. Bayesian inductive inference and maximum entropy. In Gary L. Ericson and C. Ray Smith, editors, *Maximum Entropy and Bayesian Methods in Science and Engineering 1: Foundations*, volume 1, pages 53–74. Kluwer, Boston, MA, 1988.

[39] Isabelle Guyon, John Makhoul, Richard Schwartz, and Vladimir Vapnik. What size test set gives good error rate estimates? *IEEE Transactions on Pattern Analysis and Machine Intelligence*, PAMI-20(1):52–64, 1998.

[40] Lars Kai Hansen and Peter Salamon. Neural network ensembles. *IEEE Transactions on Pattern Analysis and Machine Intelligence*, PAMI-12(10):993–1001, 1990.

[41] Trevor J. Hastie and Robert J. Tibshirani. *General Additive Models*. Chapman & Hall, London, 1990.

[42] Thomas Heskes. Bias/variance decompositions for likelihood-based estimators. *Neural Computation*, 10(6):1425–1433, 1998.

[43] Imad Y. Hoballah and Pramod K. Varshney. An information theoretic approach to the distributed detection problem. *IEEE Transactions on Information Theory*, IT-35(5):988–994, 1989.

[44] Wassily Hoeffding. Probability inequalities for sums of bounded random variables. *Journal of the American Statistical Association*, 58(301):13–30, 1963.

[45] Robert C. Holte. Very simple classification rules perform well on most commonly used data sets. *Machine Learning*, 11(1):63–91, 1993.

[46] Gordon F. Hughes. On the mean accuracy of statistical pattern recognizers. *IEEE Transactions on Information Theory*, IT-14(1):55–63, 1968.

[47] Robert A. Jacobs. Bias/variance analyses of mixtures-of-experts architectures. *Neural Computation*, 9(2):369–383, 1997.

[48] Anil K. Jain, Richard C. Dubes, and Chaur-Chin Chen. Bootstrap techniques for error estimation. *IEEE Transactions on Pattern Analysis and Machine Intelligence*, PAMI-9(5):628–633, 1998.

[49] Sanjay Jain, Daniel Osherson, James S. Royer, and Arun Sharma. *Systems that Learn: An Introduction to Learning Theory*. MIT Press, Cambridge, MA, second edition, 1999.

[50] Chuanyi Ji and Sheng Ma. Combinations of weak classifiers. *IEEE Transactions on Neural Networks*, 8(1):32–42, 1997.

[51] Michael I. Jordan and Robert A. Jacobs. Hierarchies of adaptive experts. In John E. Moody, Stephen J. Hanson, and Richard P. Lippmann, editors, *Advances in Neural Information Processing Systems 4*, pages 985–992. Morgan Kaufmann, San Mateo, CA, 1992.

[52] Michael I. Jordan and Robert A. Jacobs. Hierarchical mixtures of experts and the EM algorithm. *Neural Computation*, 6(2):181–214, 1994.

[53] Michael J. Kearns. *The Computational Complexity of Machine Learning*. MIT Press, Cambridge, MA, 1990.

[54] Gary D. Kendall and Trevor J. Hall. Optimal network construction by minimum description length. *Neural Computation*, 5(2):210–212, 1993.

[55] Josef Kittler. Combining classifiers: A theoretical framework. *Pattern Analysis and Applications*, 1(1):18–27, 1998.

[56] Josef Kittler, Mohamad Hatef, Robert P. W. Duin, and Jiri Matas. On combining classifiers. *IEEE Transactions on Pattern Analysis and Applications*, PAMI-20(3):226–239, 1998.

[57] Andreæi N. Kolmogorov. Three approaches to the quantitative definition of information. *Problems of Information and Transmission*, 1(1):3–11, 1965.

[58] Andreæi N. Kolmogorov and Vladimir A. Uspensky. On the definition of an algorithm. *American Mathematical Society Translations, Series 2*, 29:217–245, 1963.

[59] Eun Bae Kong and Thomas G. Dietterich. Error-correcting output coding corrects bias and variance. In *Proceedings of the Twelfth International Conference on Machine Learning*, pages 313–321. Morgan Kaufmann, San Francisco, CA, 1995.

[60] Anders Krogh and Jesper Vedelsby. Neural network ensembles, cross validation, and active learning. In Gerald Tesauro, David S. Touretzky, and Todd K. Leen, editors, *Advances in Neural Information Processing Systems*, volume 7, pages 231–238. MIT Press, Cambridge, MA, 1995.

[61] Roman Krzysztofowicz and Dou Long. Fusion of detection probabilities and comparison of multisensor systems. *IEEE Transactions on Systems, Man and Cybernetics*, SMC-20(3):665–677, 1990.

[62] Sanjeev R. Kulkarni, Gábor Lugosi, and Santosh S. Venkatesh. Learning pattern classification – A survey. *IEEE Transactions on Information Theory*, IT-44(6):2178–2206, 1998.

[63] Dar-Shyang Lee and Sargur N. Srihari. A theory of classifier combination: The neural network approach. In *Proceedings of the Third International Conference on Document Analysis and Recognition (ICDAR)*, volume 1, pages 14–16. IEEE Press, Los Alamitos, CA, 1995.

[64] Sik K. Leung-Yan-Cheong and Thomas M. Cover. Some equivalences between Shannon entropy and Kolmogorov complexity. *IEEE Transactions on Information Theory*, IT-24(3):331–338, 1978.

[65] Ming Li and Paul M. B. Vitányi. Inductive reasoning and Kolmogorov complexity. *Journal of Computer and System Sciences*, 44(2):343–384, 1992.

[66] Ming Li and Paul M. B. Vitányi. *Introduction to Kol-*

mogorov Complexity and Its Applications. Springer, New York, second edition, 1997.

[67] David J. C. MacKay. Bayesian interpolation. *Neural Computation*, 4(3):415–447, 1992.

[68] David J. C. MacKay. Bayesian model comparison and backprop nets. In John E. Moody, Stephen J. Hanson, and Richard P. Lippmann, editors, *Advances in Neural Information Processing Systems*, volume 4, pages 839–846. Morgan Kaufmann, San Mateo, CA, 1992.

[69] David J. C. MacKay. The evidence framework applied to classification networks. *Neural Computation*, 4(5):698–714, 1992.

[70] David J. C. MacKay. A practical Bayesian framework for backpropagation networks. *Neural Computation*, 4(3):448–472, 1992.

[71] Armand Maurer. *The Philosophy of William of Ockham in the Light of Its Principles.* Pontifical Institute of Mediaval Studies, Toronto, Ontario, 1999.

[72] Tom M. Mitchell. *Machine Learning.* McGraw-Hill, New York, 1997.

[73] Noboru Murata, Shuji Hoshizawa, and Shun-ichi Amari. Learning curves, model selection and complexity in neural networks. In Stephen José Hanson, Jack D. Cowan, and C. Lee Giles, editors, *Advances in Neural Information Processing Systems*, volume 5, pages 607–614. MIT Press, Cambridge, MA, 1993s.

[74] Isaac Newton and Alexander Koyre (Editor). *Isaac Newton's Philosophiae Naturalis Principia Mathematica.* Harvard University Press, Cambridge, MA, third edition, 1972.

[75] Michael Perrone and Leon N Cooper. When networks disagree: Ensemble methods for hybrid neural networks. In Richard J. Mammone, editor, *Artificial Neural Networks for Speech and Vision*, pages 126–142. Chapman & Hall, London, 1993.

[76] Karl Raimund Popper. *Conjectures and Refutations: The Growth of Scientific Knowledge.* Routledge Press, New York, fifth edition, 1992.

[77] Maurice H. Quenouille. Approximate tests of correlation in time series. *Journal of the Royal Statistical Society B*, 11:18–84, 1949.

[78] Maurice H. Quenouille. Notes on bias in estimation. *Biometrika*, 43:353–360, 1956.

[79] Willard V. Quine. *Ontological Relativity and other Essays.* Columbia University Press, New York, 1969.

[80] Jorma Rissanen. Modelling by shortest data description. *Automatica*, 14(5):465–471, 1978.

[81] Jorma Rissanen. *Stochastic Complexity in Statistical Inquiry.* World Scientific, Singapore, 1989.

[82] Steven Salzberg. On comparing classifiers: Pitfalls to avoid and a recommended approach. *Data Mining and Knowledge Discovery*, 1(3):317–327, 1997.

[83] Cullen Schaffer. A conservation law for generalization performance. In William W. Cohen and Haym Hirsh, editors, *Proceeding of the Eleventh International Conference on Machine Learning*, pages 259–265. Morgan Kaufmann, San Francisco, CA, 1994.

[84] Robert E. Schapire. The strength of weak learnability. *Machine Learning*, 5(2):197–227, 1990.

[85] Robert E. Schapire. Using output codes to boost multiclass learning problems. In *Machine Learning: Proceedings of the Fourteenth International Conference*, pages 313–321. Morgan Kaufmann, San Mateo, CA, 1997.

[86] Gideon Schwarz. Estimating the dimension of a model. *Annals of Statistics*, 6(2):461–464, 1978.

[87] Holm Schwarze and John Hertz. Discontinuous generalization in large committee machines. In Jack D. Cowan, Gerald Tesauro, and Joshua Alspector, editors, *Advances in Neural Information Processing Systems*, volume 6, pages 399–406. Morgan Kaufmann, San Mateo, CA, 1994.

[88] Amanda J. C. Sharkey, editor. *Combining Artificial Neural Nets: Ensemble and Modular Multi-Net Systems.* Springer-Verlag, London, 1999.

[89] Herbert Simon. Theories of bounded rationality. In C. Bartlett McGuire and Roy Radner, editors, *Decision and Organization: A Volume in Honor of Jacob Marschak*, chapter 8, pages 161–176. North-Holland, Amsterdam, 1972.

[90] Padhraic Smyth and David Wolpert. Stacked density estimation. In Michael I. Jordan, Michael J. Kearns, and Sara A. Solla, editors, *Advances in Neural Information Processing Systems*, volume 10, pages 668–674. MIT Press, Cambridge, MA, 1998.

[91] Padhraic Smyth and David Wolpert. An evaluation of linearly combining density estimators via stacking. *Machine Learning*, 36(1/2):59–83, 1999.

[92] Elliott Sober. *Simplicity.* Oxford University Press, New York, 1975.

[93] Ray J. Solomonoff. A formal theory of inductive inference. Part I. *Information and Control*, 7(1):1–22, 1964.

[94] Ray J. Solomonoff. A formal theory of inductive inference. Part II. *Information and Control*, 7(2):224–254, 1964.

[95] David G. Stork. Document and character research in the Open Mind Initiative. In *Proceedings of the International Conference on Document Analysis and Recognition (ICDAR99)*, pages 1–12, Bangalore, India, September 1999.

[96] David G. Stork. The Open Mind Initiative. *IEEE Expert Systems and Their Application*, 14(3):19–20, 1999.

[97] Stelios C. A. Thomopoulos, Ramanarayanan Viswanathan, and Dimitrios C. Bougoulias. Optimal decision fusion in multiple sensor systems. *IEEE Transactions on Aerospace and Electronic Systems*, AES-23(5):644–653, 1987.

[98] Godfried T. Toussaint. Bibliography on estimation of misclassification. *IEEE Transactions on Information Theory*, IT-20(4):472–279, 1974.

[99] Les Valiant. Theory of the learnable. *Communications of the ACM*, 27(11):1134–1142, 1984.

[100] Jean-Marc Valin and David G. Stork. Open Mind speech recognition. In *Proceedings of the Automatic*

Speech Recognition and Understanding Workshop (ASRU99), Keystone, CO, 1999.

[101] Vladimir Vapnik. *Estimation of Dependences Based on Empirical Data*. Springer-Verlag, New York, 1995.

[102] Vladimir Vapnik. *The Nature of Statistical Learning Theory*. Springer-Verlag, New York, 1995.

[103] Vladimir Vapnik and Aleksei Chervonenkis. *Theory of Pattern Recognition*. Nauka, Moscow, Russian edition, 1974.

[104] Šarūnas Raudys and Anil K. Jain. Small sample size effects in statistical pattern recognition: Recommendations for practitioners. *IEEE Transactions on Pattern Analysis and Applications*, PAMI-13(3):252–264, 1991.

[105] Satoshi Watanabe. *Pattern Recognition: Human and Mechanical*. Wiley, New York, 1985.

[106] Ole Winther and Sara A. Solla. Optimal Bayesian online learning. In Kwok-Yee Michael Wong, Irwin King, and Dit-Yan Yeung, editors, *Theoretical Aspects of Neural Computation: A Multidisciplinary Perspective*, pages 61–70. Springer, New York, 1998.

[107] Greg Wolff, Art Owen, and David G. Stork. Empirical error-confidence curves for neural network and Gaussian classifiers. *International Journal of Neural Systems*, 7(3):363–371, 1996.

[108] David H. Wolpert. On the connection between in-sample testing and generalization error. *Complex Systems*, 6(1):47–94, 1992.

[109] David H. Wolpert, editor. *The Mathematics of Generalization*. Addison-Wesley, Reading, MA, 1995.

[110] David H. Wolpert. The relationship between PAC, the statistical physics framework, the Bayesian framework, and the VC framework. In David H. Wolpert, editor, *The Mathematics of Generalization*, pages 117–214. Addison-Wesley, Reading, MA, 1995.

[111] Frank J. Wyman, Dean M. Young, and Danny W. Turner. A comparison of asymptotic error rate for the sample linear discriminant function. *Pattern Recognition*, 23(7):775–785, 1990.

UNSUPERVISED LEARNING AND CLUSTERING

10.1 INTRODUCTION

Until now we have assumed that the training samples used to design a classifier were labeled by their category membership. Procedures that use labeled samples are said to be supervised. Now we shall investigate a number of *unsupervised* procedures, which use unlabeled samples. That is, we shall see what can be done when all one has is a collection of samples without being told their categories.

One might wonder why anyone is interested in such an unpromising problem, and whether or not it is possible even in principle to learn anything of value from unlabeled samples. There are at least five basic reasons for interest in unsupervised procedures. First, collecting and labeling a large set of sample patterns can be surprisingly costly. For instance, recorded speech is virtually free, but accurately *labeling* the speech—marking what word or phoneme is being uttered at each instant—can be very expensive and time consuming. If a classifier can be crudely designed on a small set of labeled samples, and then "tuned up" by allowing it to run without supervision on a large, unlabeled set, much time and trouble can be saved. Second, one might wish to proceed in the reverse direction: Train with large amounts of (less expensive) unlabeled data, and only then use supervision to label the groupings found. This may be appropriate for large "data mining" applications, where the contents of a large database are not known beforehand. Third, in many applications the characteristics of the patterns can change slowly with time—for example, in automated food classification as the seasons change. If these changes can be tracked by a classifier running in an unsupervised mode, improved performance can be achieved. Fourth, we can use unsupervised methods to find *features* that will then be useful for categorization. There are unsupervised methods that provide a form of data-dependent "smart preprocessing" or "smart feature extraction." Lastly, in the early stages of an investigation it may be valuable to perform exploratory data analysis and thereby gain some insight into the nature or structure of the data. The discovery of distinct subclasses—clusters or groups of patterns whose members are more similar to each other than they are to other patterns—or of major departures from expected characteristics may suggest we significantly alter our approach to designing the classifier.

The answer to the question of whether or not it is possible in principle to learn anything from unlabeled data depends upon the assumptions one is willing to accept—theorems cannot be proved without premises. We shall begin with the very restrictive assumption that the functional forms for the underlying probability densities are known and that the only thing that must be learned is the value of an unknown parameter vector. Interestingly enough, the formal solution to this problem will turn out to be almost identical to the solution for the problem of supervised learning given in Chapter 3. Unfortunately, in the unsupervised case the solution suffers from the usual problems associated with parametric assumptions without providing any of the benefits of computational simplicity. This will lead us to various attempts to reformulate the problem as one of partitioning the data into subgroups or clusters. While some of the resulting clustering procedures have no known significant theoretical properties, they are still among the more useful tools for pattern recognition problems.

10.2 MIXTURE DENSITIES AND IDENTIFIABILITY

We begin by assuming that we know the complete probability structure for the problem with the sole exception of the values of some parameters. To be more specific, we make the following assumptions:

1. The samples come from a known number c of classes.
2. The prior probabilities $P(\omega_j)$ for each class are known, $j = 1, \ldots, c$.
3. The forms for the class-conditional probability densities $p(\mathbf{x}|\omega_j, \boldsymbol{\theta}_j)$ are known, $j = 1, \ldots, c$.
4. The values for the c parameter vectors $\boldsymbol{\theta}_1, \ldots, \boldsymbol{\theta}_c$ are unknown.
5. The category labels are unknown.

We assume the samples were obtained by selecting a state of nature ω_j with probability $P(\omega_j)$ and then selecting an $\mathbf{x}$ according to the probability law $p(\mathbf{x}|\omega_j, \boldsymbol{\theta}_j)$. Thus, the probability density function for the samples is given by

$$p(\mathbf{x}|\boldsymbol{\theta}) = \sum_{j=1}^{c} p(\mathbf{x}|\omega_j, \boldsymbol{\theta}_j) P(\omega_j), \tag{1}$$

COMPONENT
DENSITIES
MIXING
PARAMETERS

where $\boldsymbol{\theta} = (\boldsymbol{\theta}_1, \ldots, \boldsymbol{\theta}_c)^t$. For obvious reasons, a density function of this form is called a *mixture density*. The conditional densities $p(\mathbf{x}|\omega_j, \boldsymbol{\theta}_j)$ are called the *component densities*, and the prior probabilities $P(\omega_j)$ are called the *mixing parameters*. The mixing parameters can also be included among the unknown parameters, but for the moment we shall assume that only $\boldsymbol{\theta}$ is unknown.

Our basic goal will be to use samples drawn from this mixture density to estimate the unknown parameter vector $\boldsymbol{\theta}$. Once we know $\boldsymbol{\theta}$, we can decompose the mixture into its components and use a maximum *a posteriori* classifier on the derived densities, if indeed classification is our final goal. Before seeking explicit solutions to this problem, however, let us ask whether or not it is possible in principle to recover $\boldsymbol{\theta}$ from the mixture. Suppose that we had an unlimited number of samples and that we used one of the nonparametric methods of Chapter 4 to determine the value of $p(\mathbf{x}|\boldsymbol{\theta})$ for every $\mathbf{x}$. If there is only one value of $\boldsymbol{\theta}$ that will produce the observed values for $p(\mathbf{x}|\boldsymbol{\theta})$, then a solution is at least possible in principle. However, if several different values of $\boldsymbol{\theta}$ can produce the same values for $p(\mathbf{x}|\boldsymbol{\theta})$, then there is no hope of obtaining a unique solution.

These considerations lead us to the following definition: A density $p(\mathbf{x}|\boldsymbol{\theta})$ is said to be *identifiable* if $\boldsymbol{\theta} \neq \boldsymbol{\theta}'$ implies that there exists an $\mathbf{x}$ such that $p(\mathbf{x}|\boldsymbol{\theta}) \neq p(\mathbf{x}|\boldsymbol{\theta}')$. Or put another way, a density $p(\mathbf{x}|\boldsymbol{\theta})$ is *not* identifiable if we cannot recover a unique $\boldsymbol{\theta}$, even from an infinite amount of data. In the discouraging situation where we cannot infer *any* of the individual parameters (i.e., components of $\boldsymbol{\theta}$), the density is *completely unidentifiable*. Note that the identifiability of $\boldsymbol{\theta}$ is a property of the *model*, irrespective of any procedure we might use to determine its value. As one might expect, the study of unsupervised learning is greatly simplified if we restrict ourselves to identifiable mixtures. Fortunately, most mixtures of commonly encountered density functions are identifiable, as are most complex or high-dimensional density functions encountered in real-world problems.

COMPLETE
UNIDENTIFI-
ABILITY

Mixtures of discrete distributions are not always so obliging. As a simple example consider the case where x is binary and $P(x|\boldsymbol{\theta})$ is the mixture

$$P(x|\boldsymbol{\theta}) = \frac{1}{2}\theta_1^x(1 - \theta_1)^{1-x} + \frac{1}{2}\theta_2^x(1 - \theta_2)^{1-x}$$

$$= \begin{cases} \frac{1}{2}(\theta_1 + \theta_2) & \text{if } x = 1 \\ 1 - \frac{1}{2}(\theta_1 + \theta_2) & \text{if } x = 0. \end{cases}$$

Suppose, for example, that we know for our data that $P(x = 1|\boldsymbol{\theta}) = 0.6$ and hence that $P(x = 0|\boldsymbol{\theta}) = 0.4$. Then we know the function $P(x|\boldsymbol{\theta})$, but we cannot determine $\boldsymbol{\theta}$, and hence cannot extract the component distributions. The most we can say is that $\theta_1 + \theta_2 = 1.2$. Thus, here we have a case in which the mixture distribution is completely unidentifiable, and hence a case for which unsupervised learning is impossible in principle. Related situations may permit us to determine one or *some* parameters, but not all (Problem 3).

This kind of problem commonly occurs with discrete distributions. If there are too many components in the mixture, there may be more unknowns than independent equations, and identifiability can be a serious problem. For the continuous case, the problems are less severe, although certain minor difficulties can arise due to the possibility of special cases. While it can be shown that mixtures of normal densities are usually identifiable, the parameters in the simple mixture density

$$p(x|\boldsymbol{\theta}) = \frac{P(\omega_1)}{\sqrt{2\pi}} \exp\left[-\frac{1}{2}(x - \theta_1)^2\right] + \frac{P(\omega_2)}{\sqrt{2\pi}} \exp\left[-\frac{1}{2}(x - \theta_2)^2\right] \qquad (2)$$

cannot be uniquely identified if $P(\omega_1) = P(\omega_2)$, because then θ_1 and θ_2 can be interchanged without affecting $p(x|\boldsymbol{\theta})$. To avoid such irritations, we shall acknowledge that identifiability can be a problem, but shall henceforth assume that the mixture densities we are working with are identifiable.

10.3 MAXIMUM-LIKELIHOOD ESTIMATES

Suppose now that we are given a set $\mathcal{D} = \{\mathbf{x}_1, \ldots, \mathbf{x}_n\}$ of n unlabeled samples drawn independently from the mixture density

$$p(\mathbf{x}|\boldsymbol{\theta}) = \sum_{j=1}^{c} p(\mathbf{x}|\omega_j, \boldsymbol{\theta}_j)P(\omega_j), \qquad (1)$$

where the full parameter vector $\boldsymbol{\theta}$ is fixed but unknown. The likelihood of the observed samples is, by definition, the joint density

$$p(\mathcal{D}|\boldsymbol{\theta}) \equiv \prod_{k=1}^{n} p(\mathbf{x}_k|\boldsymbol{\theta}). \tag{3}$$

The maximum-likelihood estimate $\hat{\boldsymbol{\theta}}$ is that value of $\boldsymbol{\theta}$ that maximizes $p(\mathcal{D}|\boldsymbol{\theta})$.

If we assume that $p(\mathcal{D}|\boldsymbol{\theta})$ is a differentiable function of $\boldsymbol{\theta}$, then we can derive some interesting necessary conditions for $\hat{\boldsymbol{\theta}}$. Let l be the logarithm of the likelihood, and let $\boldsymbol{\nabla}_{\boldsymbol{\theta}_i} l$ be the gradient of l with respect to $\boldsymbol{\theta}_i$. Then ·

$$l = \sum_{k=1}^{n} \ln p(\mathbf{x}_k|\boldsymbol{\theta}) \tag{4}$$

and

$$\boldsymbol{\nabla}_{\boldsymbol{\theta}_i} l = \sum_{k=1}^{n} \frac{1}{p(\mathbf{x}_k|\boldsymbol{\theta})} \boldsymbol{\nabla}_{\boldsymbol{\theta}_i} \left[\sum_{j=1}^{c} p(\mathbf{x}_k|\omega_j, \boldsymbol{\theta}_j) P(\omega_j) \right]. \tag{5}$$

If we assume that the elements of $\boldsymbol{\theta}_i$ and $\boldsymbol{\theta}_j$ are functionally independent if $i \neq j$, and if we introduce the posterior probability

$$P(\omega_i|\mathbf{x}_k, \boldsymbol{\theta}) = \frac{p(\mathbf{x}_k|\omega_i, \boldsymbol{\theta}_i) P(\omega_i)}{p(\mathbf{x}_k|\boldsymbol{\theta})}, \tag{6}$$

we see that the gradient of the log-likelihood can be written in the interesting form

$$\boldsymbol{\nabla}_{\boldsymbol{\theta}_i} l = \sum_{k=1}^{n} P(\omega_i|\mathbf{x}_k, \boldsymbol{\theta}) \boldsymbol{\nabla}_{\boldsymbol{\theta}_i} \ln p(\mathbf{x}_k|\omega_i, \boldsymbol{\theta}_i). \tag{7}$$

Because the gradient must vanish at the value of $\boldsymbol{\theta}_i$ that maximizes l, the maximum-likelihood estimate $\hat{\boldsymbol{\theta}}_i$ must satisfy the conditions

$$\sum_{k=1}^{n} P(\omega_i|\mathbf{x}_k, \hat{\boldsymbol{\theta}}) \boldsymbol{\nabla}_{\boldsymbol{\theta}_i} \ln p(\mathbf{x}_k|\omega_i, \hat{\boldsymbol{\theta}}_i) = 0 \qquad i = 1, \dots, c. \tag{8}$$

Among the solutions to these equations for $\hat{\boldsymbol{\theta}}_i$ we will find the maximum-likelihood solution.

It is not hard to generalize these results to include the prior probabilities $P(\omega_i)$ among the unknown quantities. In this case the search for the maximum value of $p(\mathcal{D}|\boldsymbol{\theta})$ extends over $\boldsymbol{\theta}$ and $P(\omega_i)$, subject to the constraints

$$P(\omega_i) \geq 0 \qquad i = 1, \dots, c \tag{9}$$

and

$$\sum_{i=1}^{c} P(\omega_i) = 1. \tag{10}$$

Let $\hat{P}(\omega_i)$ be the maximum-likelihood estimate for $P(\omega_i)$, and let $\hat{\boldsymbol{\theta}}_i$ be the maximum-likelihood estimate for $\boldsymbol{\theta}_i$. It can be shown (Problem 6) that if the likelihood function is differentiable and if $\hat{P}(\omega_i) \neq 0$ for any i, then $\hat{P}(\omega_i)$ and $\hat{\boldsymbol{\theta}}_i$ must satisfy

$$\hat{P}(\omega_i) = \frac{1}{n} \sum_{k=1}^{n} \hat{P}(\omega_i | \mathbf{x}_k, \hat{\boldsymbol{\theta}}) \tag{11}$$

and

$$\sum_{k=1}^{n} \hat{P}(\omega_i | \mathbf{x}_k, \hat{\boldsymbol{\theta}}) \nabla_{\boldsymbol{\theta}_i} \ln p(\mathbf{x}_k | \omega_i, \hat{\boldsymbol{\theta}}_i) = 0, \tag{12}$$

where

$$\hat{P}(\omega_i | \mathbf{x}_k, \hat{\boldsymbol{\theta}}) = \frac{p(\mathbf{x}_k | \omega_i, \hat{\boldsymbol{\theta}}_i) \hat{P}(\omega_i)}{\sum_{j=1}^{c} p(\mathbf{x}_k | \omega_j, \hat{\boldsymbol{\theta}}_j) \hat{P}(\omega_j)}. \tag{13}$$

These equations have the following interpretation. Equation 11 states that the maximum-likelihood estimate of the probability of a category is the average over the entire data set of the estimate derived from each sample—each sample is weighted equally. Equation 13 is ultimately related to Bayes Theorem, but notice that in estimating the probability for class ω_i, the numerator on the right-hand side depends on $\hat{\boldsymbol{\theta}}_i$ and not the full $\hat{\boldsymbol{\theta}}$ directly. While Eq. 12 is a bit subtle, we can understand it clearly in the trivial $n = 1$ case. Because $\hat{P} \neq 0$, this case states merely that the probability density is maximized as a function of $\boldsymbol{\theta}_i$—surely what is needed for the maximum-likelihood solution.

10.4 APPLICATION TO NORMAL MIXTURES

It is enlightening to see how these general results apply to the case where the component densities are multivariate normal, $p(\mathbf{x}|\omega_i, \boldsymbol{\theta}_i) \sim N(\boldsymbol{\mu}_i, \boldsymbol{\Sigma}_i)$. The following table illustrates a few of the different cases that can arise depending upon which parameters are known ($\times$) and which are unknown (?):

Case	$\boldsymbol{\mu}_i$	$\boldsymbol{\Sigma}_i$	$P(\omega_i)$	c
1	?	$\times$	$\times$	$\times$
2	?	?	?	$\times$
3	?	?	?	?

Case 1 is the simplest, and it will be considered in detail because of its pedagogical value. Case 2 is more realistic, though somewhat more involved. Case 3 represents the problem we face on encountering a completely unknown set of data; unfortunately, it cannot be solved by maximum-likelihood methods. We shall postpone discussion of what can be done when the number of classes is unknown until Section 10.10.

10.4.1 Case 1: Unknown Mean Vectors

If the only unknown quantities are the mean vectors $\boldsymbol{\mu}_i$, then of course $\boldsymbol{\theta}_i$ consists of the components of $\boldsymbol{\mu}_i$. Equation 8 can then be used to obtain necessary conditions on the maximum-likelihood estimate for $\boldsymbol{\mu}_i$. Because the likelihood is

$$\ln p(\mathbf{x}|\omega_i, \boldsymbol{\mu}_i) = -\ln \left[(2\pi)^{d/2} |\boldsymbol{\Sigma}_i|^{1/2} \right] - \frac{1}{2} (\mathbf{x} - \boldsymbol{\mu}_i)^t \boldsymbol{\Sigma}_i^{-1} (\mathbf{x} - \boldsymbol{\mu}_i), \qquad (14)$$

its derivative is

$$\nabla_{\boldsymbol{\mu}_i} \ln p(\mathbf{x}|\omega_i, \boldsymbol{\mu}_i) = \boldsymbol{\Sigma}_i^{-1}(\mathbf{x} - \boldsymbol{\mu}_i). \qquad (15)$$

Thus according to Eq. 8, the maximum-likelihood estimate $\hat{\boldsymbol{\mu}}_i$ must satisfy

$$\sum_{k=1}^{n} P(\omega_i|\mathbf{x}_k, \hat{\boldsymbol{\mu}}) \boldsymbol{\Sigma}_i^{-1}(\mathbf{x}_k - \hat{\boldsymbol{\mu}}_i) = 0, \qquad \text{where } \hat{\boldsymbol{\mu}} = (\hat{\boldsymbol{\mu}}_1, \dots, \hat{\boldsymbol{\mu}}_c)^t. \qquad (16)$$

After multiplying by $\boldsymbol{\Sigma}_i$ and rearranging terms, we obtain the solution

$$\hat{\boldsymbol{\mu}}_i = \frac{\sum_{k=1}^{n} P(\omega_i|\mathbf{x}_k, \hat{\boldsymbol{\mu}}) \mathbf{x}_k}{\sum_{k=1}^{n} P(\omega_i|\mathbf{x}_k, \hat{\boldsymbol{\mu}})}. \qquad (17)$$

This equation is intuitively very satisfying. It shows that the maximum-likelihood estimate for $\boldsymbol{\mu}_i$ is merely a weighted average of the samples; the weight for the kth sample is an estimate of how likely it is that $\mathbf{x}_k$ belongs to the ith class. If $P(\omega_i|\mathbf{x}_k, \hat{\boldsymbol{\mu}})$ happened to be 1.0 for some of the samples and 0.0 for the rest, then $\hat{\boldsymbol{\mu}}_i$ would be the mean of those samples estimated to belong to the ith class. More generally, suppose that $\hat{\boldsymbol{\mu}}_i$ is sufficiently close to the true value of $\boldsymbol{\mu}_i$ that $P(\omega_i|\mathbf{x}_k, \hat{\boldsymbol{\mu}})$ is essentially the true posterior probability for ω_i. If we think of $P(\omega_i|\mathbf{x}_k, \hat{\boldsymbol{\mu}})$ as the fraction of those samples having value $\mathbf{x}_k$ that come from the ith class, then we see that Eq. 17 essentially gives $\hat{\boldsymbol{\mu}}_i$ as the average of the samples coming from the ith class.

Unfortunately, Eq. 17 does not give $\hat{\boldsymbol{\mu}}_i$ explicitly, and if we substitute

$$P(\omega_i|\mathbf{x}_k, \hat{\boldsymbol{\mu}}) = \frac{p(\mathbf{x}_k|\omega_i, \hat{\boldsymbol{\mu}}_i) P(\omega_i)}{\sum_{j=1}^{c} p(\mathbf{x}_k|\omega_j, \hat{\boldsymbol{\mu}}_j) P(\omega_j)}$$

with $p(\mathbf{x}|\omega_i, \hat{\boldsymbol{\mu}}_i) \sim N(\hat{\boldsymbol{\mu}}_i, \boldsymbol{\Sigma}_i)$, we obtain a tangled snarl of coupled simultaneous nonlinear equations. These equations usually do not have a unique solution, and we must test the solutions we get to find the one that actually maximizes the likelihood.

If we have some way of obtaining fairly good initial estimates $\hat{\boldsymbol{\mu}}_i(0)$ for the unknown means, Eq. 17 suggests the following iterative scheme for improving the estimates:

$$\hat{\boldsymbol{\mu}}_i(j+1) = \frac{\sum_{k=1}^{n} P(\omega_i|\mathbf{x}_k, \hat{\boldsymbol{\mu}}(j)) \mathbf{x}_k}{\sum_{k=1}^{n} P(\omega_i|\mathbf{x}_k, \hat{\boldsymbol{\mu}}(j))}. \qquad (18)$$

This is basically a gradient ascent or hill-climbing procedure for maximizing the log-likelihood function. If the overlap between component densities is small, then the coupling between classes will be small and convergence will be fast. However, when convergence does occur, all that we can be sure of is that the gradient is zero. Like all hill-climbing procedures, this one carries no guarantee of yielding the global maximum. Note too that if the model is misspecified (e.g., we assume the "wrong" number of clusters), then the estimate of the log-likelihood given by Eq. 18 can actually decrease.

To illustrate the kind of behavior that can occur in gradient searches in unsupervised learning, consider the simple two-component one-dimensional normal mixture:

$$p(x|\mu_1, \mu_2) = \underbrace{\frac{1}{3\sqrt{2\pi}} \exp\left[-\frac{1}{2}(x - \mu_1)^2\right]}_{\omega_1} + \underbrace{\frac{2}{3\sqrt{2\pi}} \exp\left[-\frac{1}{2}(x - \mu_2)^2\right]}_{\omega_2}, \quad (19)$$

where ω_i denotes a Gaussian component. The 25 samples shown in the table below were drawn sequentially from this mixture with $\mu_1 = -2$ and $\mu_2 = 2$. Let us use these samples to compute the log-likelihood function

$$l(\mu_1, \mu_2) = \sum_{k=1}^{n} \ln p(x_k|\mu_1, \mu_2) \qquad (20)$$

for various values of μ_1 and μ_2. The bottom of Fig. 10.1 shows how l varies with μ_1 and μ_2. The maximum value of l occurs at $\hat{\mu}_1 = -2.130$ and $\hat{\mu}_2 = 1.668$, which is in the rough vicinity of the true values $\mu_1 = -2$ and $\mu_2 = 2$. However, l reaches another peak of comparable height at $\hat{\mu}_1 = 2.085$ and $\hat{\mu}_2 = -1.257$.

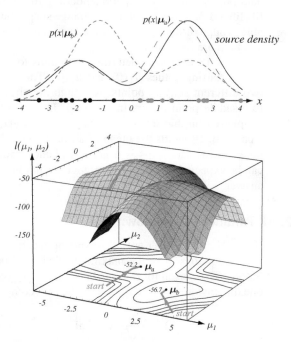

FIGURE 10.1. (Above) The source mixture density used to generate sample data, and two maximum-likelihood estimates based on the data in the table. (Bottom) Log-likelihood of a mixture model consisting of two univariate Gaussians as a function of their means, for the data in the table. Trajectories for the iterative maximum-likelihood estimation of the means of a two-Gaussian mixture model based on the data are shown as red lines. Two local optima (with log-likelihoods −52.2 and −56.7) correspond to the two density estimates shown above.

Roughly speaking, this solution corresponds to interchanging μ_1 and μ_2. Note that had the prior probabilities been equal, interchanging μ_1 and μ_2 would have produced no change in the log-likelihood function. Thus, as we mentioned before, when the mixture density is not identifiable, the maximum-likelihood solution is not unique.

k	x_k	ω_1	ω_2	k	x_k	ω_1	ω_2	k	x_k	ω_1	ω_2
1	0.608		×	9	0.262		×	17	−3.458	×	
2	−1.590	×		10	1.072		×	18	0.257		×
3	0.235		×	11	−1.773	×		19	2.569		×
4	3.949		×	12	0.537		×	20	1.415		×
5	−2.249	×		13	3.240		×	21	1.410		×
6	2.704		×	14	2.400		×	22	−2.653	×	
7	−2.473	×		15	−2.499	×		23	1.396		×
8	0.672		×	16	2.608		×	24	3.286		×
								25	−0.712	×	

Additional insight into the nature of these multiple solutions can be obtained by examining the resulting estimates for the mixture density. The figure at the top shows the true (source) mixture density and the estimates obtained by using the two maximum-likelihood estimates as if they were the true parameter values. The 25 sample values are shown as a scatter of points along the abscissa—ω_1 points in black, ω_2 points in red. Note that the peaks of both the true mixture density and the maximum-likelihood solutions are located so as to encompass two major groups of data points. The estimate corresponding to the smaller local maximum of the log-likelihood function has a mirror-image shape, but its peaks also encompass reasonable groups of data points. To the eye, neither of these solutions is clearly superior, and both are interesting.

If Eq. 18 is used to determine solutions to Eq. 17 iteratively, the results depend on the starting values $\hat{\mu}_1(0)$ and $\hat{\mu}_2(0)$. The bottom figure shows trajectories from two different starting points. Although not shown, if $\hat{\mu}_1(0) = \hat{\mu}_2(0)$, convergence to a saddle point occurs in one step. This is not a coincidence; it happens for the simple reason that for this starting point $P(\omega_i|x_k, \hat{\mu}_1(0), \hat{\mu}_2(0)) = P(\omega_i)$. In such a case Eq. 18 yields the mean of all of the samples for $\hat{\mu}_1$ and $\hat{\mu}_2$ for all successive iterations. Clearly, this is a general phenomenon, and such saddle-point solutions can be expected if the starting point does not bias the search away from a symmetric answer.

10.4.2 Case 2: All Parameters Unknown

If $\boldsymbol{\mu}_i$, $\boldsymbol{\Sigma}_i$, and $P(\omega_i)$ are all unknown and if no constraints are placed on the covariance matrix, then the maximum-likelihood principle yields useless singular solutions. The reason for this can be appreciated from the following simple example in one dimension. Let $p(x|\mu, \sigma^2)$ be the two-component normal mixture:

$$p(x|\mu, \sigma^2) = \frac{1}{2\sqrt{2\pi}\sigma} \exp\left[-\frac{1}{2}\left(\frac{x-\mu}{\sigma}\right)^2 \right] + \frac{1}{2\sqrt{2\pi}} \exp\left[-\frac{1}{2}x^2 \right].$$

The likelihood function for n samples drawn from this probability density is merely the product of the n densities $p(x_k|\mu, \sigma^2)$. Suppose that we make the choice $\mu = x_1$, so that

$$p(x_1|\mu, \sigma^2) = \frac{1}{2\sqrt{2\pi}\sigma} + \frac{1}{2\sqrt{2\pi}} \exp\left[-\frac{1}{2}x_1^2 \right].$$

Clearly, for the rest of the samples

$$p(x_k | \mu, \sigma^2) \geq \frac{1}{2\sqrt{2\pi}} \exp\left[-\frac{1}{2} x_k^2\right],$$

so that

$$p(x_1, \ldots, x_n | \mu, \sigma^2) \geq \left\{\frac{1}{\sigma} + \exp\left[-\frac{1}{2} x_1^2\right]\right\} \frac{1}{(2\sqrt{2\pi})^n} \exp\left[-\frac{1}{2}\sum_{k=2}^{n} x_k^2\right].$$

Thus, the first factor at the right shows that by letting σ approach zero we can make the likelihood arbitrarily large, and the maximum-likelihood solution is singular.

Ordinarily, singular solutions are of no interest, and we are forced to conclude that the maximum-likelihood principle fails for this class of normal mixtures. However, it is an empirical fact that meaningful solutions can still be obtained if we restrict our attention to the largest of the finite local maxima of the likelihood function. Assuming that the likelihood function is well-behaved at such maxima, we can use Eqs. 11–13 to obtain estimates for μ_i, Σ_i, and $P(\omega_i)$. When we include the elements of Σ_i in the elements of the parameter vector θ_i, we must remember that only half of the off-diagonal elements are independent. In addition, it turns out to be much more convenient to let the independent elements of Σ_i^{-1} rather than Σ_i be the unknown parameters. With these observations, the actual differentiation of

$$\ln p(\mathbf{x}_k | \omega_i, \theta_i) = \ln \frac{|\Sigma_i^{-1}|^{1/2}}{(2\pi)^{d/2}} - \frac{1}{2}(\mathbf{x}_k - \mu_i)^t \Sigma_i^{-1}(\mathbf{x}_k - \mu_i) \tag{21}$$

with respect to the elements of μ_i and Σ_i^{-1} is relatively routine. Let $x_p(k)$ be the pth element of $\mathbf{x}_k$, $\mu_p(i)$ be the pth element of μ_i, $\sigma_{pq}(i)$ be the pqth element of Σ_i, and $\sigma^{pq}(i)$ be the pqth element of Σ_i^{-1}. Then differentiation gives

$$\nabla_{\mu_i} \ln p(\mathbf{x}_k | \omega_i, \theta_i) = \Sigma_i^{-1}(\mathbf{x}_k - \mu_i) \tag{22}$$

and

$$\frac{\partial \ln p(\mathbf{x}_k | \omega_i, \theta_i)}{\partial \sigma^{pq}(i)} = \left(1 - \frac{\delta_{pq}}{2}\right)\left[\sigma_{pq}(i) - (x_p(k) - \mu_p(i))(x_q(k) - \mu_q(i))\right], \tag{23}$$

where δ_{pq} is the Kronecker delta. We substitute these results into Eq. 12 and perform a small amount of algebraic manipulation (Problem 17) and thereby obtain the following equations for the local-maximum-likelihood estimates $\hat{\mu}_i$, $\hat{\Sigma}_i$, and $\hat{P}(\omega_i)$:

$$\hat{P}(\omega_i) = \frac{1}{n}\sum_{k=1}^{n} \hat{P}(\omega_i | \mathbf{x}_k, \hat{\theta}) \tag{24}$$

$$\hat{\mu}_i = \frac{\sum_{k=1}^{n} \hat{P}(\omega_i | \mathbf{x}_k, \hat{\theta})\mathbf{x}_k}{\sum_{k=1}^{n} \hat{P}(\omega_i | \mathbf{x}_k, \hat{\theta})} \tag{25}$$

$$\hat{\Sigma}_i = \frac{\sum_{k=1}^{n} \hat{P}(\omega_i | \mathbf{x}_k, \hat{\theta})(\mathbf{x}_k - \hat{\mu}_i)(\mathbf{x}_k - \hat{\mu}_i)^t}{\sum_{k=1}^{n} \hat{P}(\omega_i | \mathbf{x}_k, \hat{\theta})} \tag{26}$$

where

$$\hat{P}(\omega_i|\mathbf{x}_k, \hat{\boldsymbol{\theta}}) = \frac{p(\mathbf{x}_k|\omega_i, \hat{\boldsymbol{\theta}}_i)\hat{P}(\omega_i)}{\sum_{j=1}^{c} p(\mathbf{x}_k|\omega_j, \hat{\boldsymbol{\theta}}_j)\hat{P}(\omega_j)}$$

$$= \frac{|\hat{\boldsymbol{\Sigma}}_i|^{-1/2} \exp\left[-\frac{1}{2}(\mathbf{x}_k - \hat{\boldsymbol{\mu}}_i)^t \hat{\boldsymbol{\Sigma}}_i^{-1}(\mathbf{x}_k - \hat{\boldsymbol{\mu}}_i)\right]\hat{P}(\omega_i)}{\sum_{j=1}^{c} |\hat{\boldsymbol{\Sigma}}_j|^{-1/2} \exp\left[-\frac{1}{2}(\mathbf{x}_k - \hat{\boldsymbol{\mu}}_j)^t \hat{\boldsymbol{\Sigma}}_j^{-1}(\mathbf{x}_k - \hat{\boldsymbol{\mu}}_j)\right]\hat{P}(\omega_j)}. \tag{27}$$

While the notation may make these equations appear to be rather formidable, their interpretation is actually quite simple. In the extreme case where $\hat{P}(\omega_i|\mathbf{x}_k, \hat{\boldsymbol{\theta}})$ is 1.0 when $\mathbf{x}_k$ is from Class ω_i and 0.0 otherwise, $\hat{P}(\omega_i)$ is the fraction of samples from ω_i, $\hat{\boldsymbol{\mu}}_i$ is the mean of those samples, and $\hat{\boldsymbol{\Sigma}}_i$ is the corresponding sample covariance matrix. More generally, $\hat{P}(\omega_i|\mathbf{x}_k, \hat{\boldsymbol{\theta}})$ is between 0.0 and 1.0, and all of the samples play some role in the estimates. However, the estimates are basically still frequency ratios, sample means, and sample covariance matrices.

The problems involved in solving these implicit equations are similar to the problems discussed in Section 10.4.1, with the additional complication of having to avoid singular solutions. Of the various techniques that can be used to obtain a solution, the most obvious approach is to use initial estimates to evaluate Eq. 27 for $\hat{P}(\omega_i|\mathbf{x}_k, \hat{\boldsymbol{\theta}})$ and then to use Eqs. 24–26 to update these estimates. If the initial estimates are very good, having perhaps been obtained from a fairly large set of labeled samples, convergence can be quite rapid. However, the results do depend upon the starting point, and the problem of multiple solutions is always present. Furthermore, the repeated computation and inversion of the sample covariance matrices can be quite time consuming.

Considerable simplification can be obtained if it is possible to assume that the covariance matrices are diagonal. This has the added virtue of reducing the number of unknown parameters, which is very important when the number of samples is not large. If this assumption is too strong, it still may be possible to obtain some simplification by assuming that the c covariance matrices are equal, which also may eliminate the problem of singular solutions (Problem 17).

10.4.3 *k*-Means Clustering

Of the various techniques that can be used to simplify the computation and accelerate convergence, we shall briefly consider one elementary but very popular approximate method. We are tempted to call it the *c*-means procedure, since its goal is to find the c mean vectors $\boldsymbol{\mu}_1, \boldsymbol{\mu}_2, \ldots, \boldsymbol{\mu}_c$. However, it is much more widely known as *k*-means clustering, where k (which is the same as our c) is the number of cluster centers. Thus we shall follow common practice and call it by that name.

From Eq. 27, it is clear that the probability $\hat{P}(\omega_i|\mathbf{x}_k, \hat{\boldsymbol{\theta}})$ is large when the squared Mahalanobis distance $(\mathbf{x}_k - \hat{\boldsymbol{\mu}}_i)^t \hat{\boldsymbol{\Sigma}}_i^{-1}(\mathbf{x}_k - \hat{\boldsymbol{\mu}}_i)$ is small. Suppose that we merely compute the squared Euclidean distance $\|\mathbf{x}_k - \hat{\boldsymbol{\mu}}_i\|^2$, find the mean $\hat{\boldsymbol{\mu}}_m$ nearest to $\mathbf{x}_k$, and approximate $\hat{P}(\omega_i|\mathbf{x}_k, \hat{\boldsymbol{\theta}})$ as

$$\hat{P}(\omega_i|\mathbf{x}_k, \hat{\boldsymbol{\theta}}) \simeq \begin{cases} 1 & \text{if } i = m \\ 0 & \text{otherwise.} \end{cases} \tag{28}$$

Then the iterative application of Eq. 25 leads to the following procedure for finding $\hat{\boldsymbol{\mu}}_1, \ldots, \hat{\boldsymbol{\mu}}_c$. In the absence of other information, we may need to guess the "proper" number of clusters, c. Likewise, we may assign c based on the final application. For example we might set $c = 26$ in a handwritten English letter application, whether or not the data fall into 26 natural clusters. We shall return in Section 10.10 to this problem of cluster validity.

Here and throughout, we denote the known number of patterns as n, and the desired number of clusters c. It is traditional to let c samples randomly chosen from the data set serve as initial cluster centers. The algorithm is then:

■ **Algorithm 1.** (*k-Means Clustering*)

1 **begin initialize** n, c, $\boldsymbol{\mu}_1, \boldsymbol{\mu}_2, \ldots, \boldsymbol{\mu}_c$
2   **do** classify n samples according to nearest $\boldsymbol{\mu}_i$
3   recompute $\boldsymbol{\mu}_i$
4   **until** no change in $\boldsymbol{\mu}_i$
5 **return** $\boldsymbol{\mu}_1, \boldsymbol{\mu}_2, \ldots, \boldsymbol{\mu}_c$
6 **end**

The computational complexity of the algorithm is $O(ndcT)$ where d is the number of features and T is the number of iterations (Problem 16). In practice, the number of iterations is generally much less than the number of samples.

This is typical of a class of procedures that are known as *clustering* procedures or algorithms. Later on we shall place it in the class of iterative optimization procedures, because the means tend to move so as to minimize a squared-error criterion function. For the moment we view it merely as an approximate way to obtain maximum-likelihood estimates for the means. The values obtained can be accepted as the answer, or can be used as starting points for the more exact computations.

It is interesting to see how this procedure behaves on the example data we saw in Fig. 10.1. Figure 10.2 shows the sequence of values for $\hat{\mu}_1$ and $\hat{\mu}_2$ obtained for several different starting points. Since interchanging $\hat{\mu}_1$ and $\hat{\mu}_2$ merely interchanges

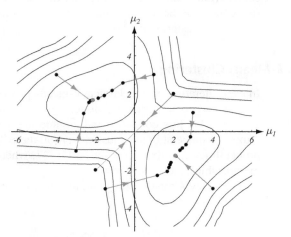

FIGURE 10.2. The k-means clustering procedure is a form of stochastic hill climbing in the log-likelihood function. The contours represent equal log-likelihood values for the one-dimensional data in Fig. 10.1. The dots indicate parameter values after different iterations of the k-means algorithm. Six of the starting points shown lead to local maxima, whereas two (i.e., $\mu_1(0) = \mu_2(0)$) lead to a saddle point near $\boldsymbol{\mu} = \mathbf{0}$.

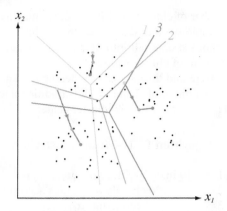

FIGURE 10.3. Trajectories for the means of the k-means clustering procedure applied to two-dimensional data. The final Voronoi tesselation (for classification) is also shown—the means correspond to the "centers" of the Voronoi cells. In this case, convergence is obtained in three iterations.

the labels assigned to the data, the trajectories are symmetric about the line $\hat{\mu}_1 = \hat{\mu}_2$. The trajectories lead either to the point $\hat{\mu}_1 = -2.176$, $\hat{\mu}_2 = 1.684$ or to its symmetric image. This is close to the solution found by the maximum-likelihood method (namely, $\hat{\mu}_1 = -2.130$ and $\hat{\mu}_2 = 1.688$), and the trajectories show a general resemblance to those shown in Fig. 10.1. In general, when the overlap between the component densities is small the maximum-likelihood approach and the k-means procedure can be expected to give similar results.

Figure 10.3 shows a two-dimensional example, with the assumption of $c = 3$ clusters. The three initial cluster centers, chosen randomly from the training points, and their associated Voronoi tesselation, are shown in pink. According to the algorithm, the points in each of the three Voronoi cells are used to calculate new cluster centers (pink), and so on. Here, after the third iteration the algorithm has converged (red). Because the k-means algorithm is very simple and works well in practice, it is a staple of clustering methods.

*10.4.4 Fuzzy k-Means Clustering

In every iteration of the classical k-means procedure, each data point is assumed to be in exactly one cluster, as implied by Eq. 28 and by lines 2 and 3 of Algorithm 1. We can relax this condition and assume that each sample $\mathbf{x}_j$ has some graded or "fuzzy" membership in a cluster. At root, these "memberships" are equivalent to the probabilities $\hat{P}(\omega_i|\mathbf{x}_j, \hat{\boldsymbol{\theta}})$ given in Eq. 27, where $\hat{\boldsymbol{\theta}}$ is the parameter vector for the membership functions.

The fuzzy k-means clustering algorithm seeks a minimum of a heuristic global cost function

$$J_{fuz} = \sum_{i=1}^{c} \sum_{j=1}^{n} [\hat{P}(\omega_i|\mathbf{x}_j, \hat{\boldsymbol{\theta}})]^b ||\mathbf{x}_j - \boldsymbol{\mu}_i||^2, \qquad (29)$$

where b is a free parameter chosen to adjust the "blending" of different clusters. If b is set to 0, J_{fuz} is merely a sum-of-squared errors criterion with each pattern assigned to a single cluster, which we shall see again in Eq. 54. For $b > 1$, the criterion allows each pattern to belong to multiple clusters.

The probabilities of cluster membership for each point are normalized as

$$\sum_{i=1}^{c} \hat{P}(\omega_i|\mathbf{x}_j) = 1, \qquad j = 1, \ldots, n, \tag{30}$$

where, for simplicity, we have not explicitly shown the dependence on $\hat{\boldsymbol{\theta}}$. We let $\hat{P}_j$ denote the prior probability of $\hat{P}(\omega_j)$, then at the solution (i.e., the minimum of J_{fuz}), we have

$$\partial J_{fuz}/\partial \boldsymbol{\mu}_i = 0 \qquad \text{and} \qquad \partial J_{fuz}/\partial \hat{P}_j = 0. \tag{31}$$

These lead (Problem 15) to the solutions

$$\boldsymbol{\mu}_j = \frac{\sum_{j=1}^{n} [\hat{P}(\omega_i|\mathbf{x}_j)]^b \mathbf{x}_j}{\sum_{j=1}^{n} [\hat{P}(\omega_i|\mathbf{x}_j)]^b} \tag{32}$$

and

$$\hat{P}(\omega_i|\mathbf{x}_j) = \frac{(1/d_{ij})^{1/(b-1)}}{\sum_{r=1}^{c} (1/d_{rj})^{1/(b-1)}} \qquad \text{and} \quad d_{ij} = ||\mathbf{x}_j - \boldsymbol{\mu}_i||^2. \tag{33}$$

In general, the J_{fuz} criterion is minimized when the cluster centers $\boldsymbol{\mu}_j$ are near those points that have high estimated probability of being in cluster j. Because Eqs. 32 and 33 rarely have analytic solutions, the cluster means and point probabilities are estimated iteratively according to the following algorithm:

■ **Algorithm 2. (Fuzzy *k*-Means Clustering)**

1 **begin initialize** $n, c, b, \boldsymbol{\mu}_1, \ldots, \boldsymbol{\mu}_c, \hat{P}(\omega_i \mid \mathbf{x}_j), i = 1 \ldots, c; \; j = 1, \ldots, n$
2   normalize $\hat{P}(\omega_i \mid \mathbf{x}_j)$ by Eq. 30
3   **do** recompute $\boldsymbol{\mu}_i$ by Eq. 32
4   recompute $\hat{P}(\omega_i \mid \mathbf{x}_j)$ by Eq. 33
5   **until** small change in $\boldsymbol{\mu}_i$ and $\hat{P}(\omega_i \mid \mathbf{x}_j)$
6   **return** $\boldsymbol{\mu}_1, \boldsymbol{\mu}_2, \ldots, \boldsymbol{\mu}_c$
7 **end**

Figure 10.4 illustrates the algorithm. At early iterations the means lie near the center of the full data set because each point has a nonnegligible "membership" (i.e., probability) in each cluster. At later iterations the means separate and each membership tends toward the value 1.0 or 0.0. Clearly, the classical *k*-means algorithm is just a special case where the memberships for all points obey

$$P(\omega_i|\mathbf{x}_j) = \begin{cases} 1 & \text{if } ||\mathbf{x}_j - \boldsymbol{\mu}_i|| < ||\mathbf{x}_j - \boldsymbol{\mu}_{i'}|| \text{ for all } i' \neq i \\ 0 & \text{otherwise,} \end{cases} \tag{34}$$

as given by Eq. 17.

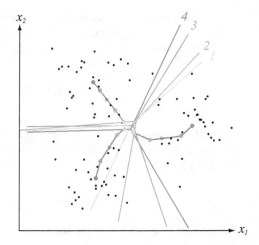

FIGURE 10.4. At each iteration of the fuzzy k-means clustering algorithm, the probability of category memberships for each point are adjusted according to Eqs. 32 and 33 (here $b = 2$). While most points have nonnegligible memberships in two or three clusters, we nevertheless draw the boundary of a Voronoi tesselation to illustrate the progress of the algorithm. After four iterations, the algorithm has converged to the red cluster centers and associated Voronoi tesselation.

The incorporation of probabilities (as graded memberships) sometimes improves convergence of k-means over its classical counterpart. One drawback of the method is that according to Eq. 30 the probability of "membership" of a point $\mathbf{x}_j$ in a cluster i depends implicitly on the number of clusters, and when this number is specified incorrectly, serious problems may arise (Computer exercise 4).

10.5 UNSUPERVISED BAYESIAN LEARNING

10.5.1 The Bayes Classifier

As we saw in Chapter 3, maximum-likelihood methods do not assume the parameter vector $\boldsymbol{\theta}$ to be random—it is just unknown. In such methods, prior knowledge about the likely values for $\boldsymbol{\theta}$ is not directly relevant, although in practice such knowledge may be used in choosing good starting points for hill-climbing procedures. In this section, however, we shall take a Bayesian approach to unsupervised learning. That is, we shall assume that $\boldsymbol{\theta}$ is a random variable with a known prior distribution $p(\boldsymbol{\theta})$, and we shall use the training samples to compute the posterior density $p(\boldsymbol{\theta}|\mathcal{D})$. Interestingly enough, the analysis will closely parallel the analysis of supervised Bayesian learning in Chapter 3, showing that the two problems are formally very similar.

We begin with an explicit statement of our basic assumptions. We assume the following:

1. The number c of classes is known.
2. The prior probabilities $P(\omega_j)$ for each class are known, $j = 1, \ldots, c$.
3. The forms for the class-conditional probability densities $p(\mathbf{x}|\omega_j, \boldsymbol{\theta}_j)$ are known, $j = 1, \ldots, c$, but the full parameter vector $\boldsymbol{\theta} = (\boldsymbol{\theta}_1, \ldots, \boldsymbol{\theta}_c)^t$ is not known.

4. Part of our knowledge about $\boldsymbol{\theta}$ is contained in a known prior density $p(\boldsymbol{\theta})$.

5. The rest of our knowledge about $\boldsymbol{\theta}$ is contained in a set $\mathcal{D}$ of n samples $\mathbf{x}_1, \ldots, \mathbf{x}_n$ drawn independently from the familiar mixture density

$$p(\mathbf{x}|\boldsymbol{\theta}) = \sum_{j=1}^{c} p(\mathbf{x}|\omega_j, \boldsymbol{\theta}_j) P(\omega_j). \tag{35}$$

At this point we could go directly to the calculation of $p(\boldsymbol{\theta}|\mathcal{D})$. However, let us first see how this density is used to determine the Bayes classifier. Suppose that a state of nature is selected with probability $P(\omega_i)$ and a feature vector $\mathbf{x}$ is selected according to the probability law $p(\mathbf{x}|\omega_i, \boldsymbol{\theta}_i)$. To derive the Bayes classifier we must use all of the information at our disposal to compute the posterior probability $P(\omega_i|\mathbf{x})$. We exhibit the role of the samples explicitly by writing this as $P(\omega_i|\mathbf{x}, \mathcal{D})$. By Bayes formula, we have

$$P(\omega_i|\mathbf{x}, \mathcal{D}) = \frac{p(\mathbf{x}|\omega_i, \mathcal{D}) P(\omega_i|\mathcal{D})}{\sum_{j=1}^{c} p(\mathbf{x}|\omega_j, \mathcal{D}) P(\omega_j|\mathcal{D})}. \tag{36}$$

Because the selection of the state of nature ω_i was done independently of the previously drawn samples, $P(\omega_i|\mathcal{D}) = P(\omega_i)$, and we obtain

$$P(\omega_i|\mathbf{x}, \mathcal{D}) = \frac{p(\mathbf{x}|\omega_i, \mathcal{D}) P(\omega_i)}{\sum_{j=1}^{c} p(\mathbf{x}|\omega_j, \mathcal{D}) P(\omega_j)}. \tag{37}$$

Central to the Bayesian approach is the introduction of the unknown parameter vector $\boldsymbol{\theta}$ via

$$p(\mathbf{x}|\omega_i, \mathcal{D}) = \int p(\mathbf{x}, \boldsymbol{\theta}|\omega_i, \mathcal{D}) \, d\boldsymbol{\theta}$$

$$= \int p(\mathbf{x}|\boldsymbol{\theta}, \omega_i, \mathcal{D}) p(\boldsymbol{\theta}|\omega_i, \mathcal{D}) \, d\boldsymbol{\theta}. \tag{38}$$

Because the selection of $\mathbf{x}$ is independent of the samples, we have $p(\mathbf{x}|\boldsymbol{\theta}, \omega_i, \mathcal{D}) = p(\mathbf{x}|\omega_i, \boldsymbol{\theta}_i)$. Similarly, because knowledge of the state of nature when $\mathbf{x}$ is selected tells us nothing about the distribution of $\boldsymbol{\theta}$, we have $p(\boldsymbol{\theta}|\omega_i, \mathcal{D}) = p(\boldsymbol{\theta}|\mathcal{D})$, and thus

$$p(\mathbf{x}|\omega_i, \mathcal{D}) = \int p(\mathbf{x}|\omega_i, \boldsymbol{\theta}_i) p(\boldsymbol{\theta}|\mathcal{D}) \, d\boldsymbol{\theta}. \tag{39}$$

That is, our best estimate of $p(\mathbf{x}|\omega_i)$ is obtained by averaging $p(\mathbf{x}|\omega_i, \boldsymbol{\theta}_i)$ over $\boldsymbol{\theta}_i$. Whether or not this is a good estimate depends on the nature of $p(\boldsymbol{\theta}|\mathcal{D})$, and thus our attention turns at last to that density.

10.5.2 Learning the Parameter Vector

We can use Bayes formula to write

$$p(\boldsymbol{\theta}|\mathcal{D}) = \frac{p(\mathcal{D}|\boldsymbol{\theta}) p(\boldsymbol{\theta})}{\int p(\mathcal{D}|\boldsymbol{\theta}) p(\boldsymbol{\theta}) \, d\boldsymbol{\theta}}, \tag{40}$$

where the independence of the samples yields the likelihood

$$p(\mathcal{D}|\boldsymbol{\theta}) = \prod_{k=1}^{n} p(\mathbf{x}_k|\boldsymbol{\theta}). \tag{41}$$

Alternatively, letting $\mathcal{D}^n$ denote the set of n samples, we can write Eq. 40 in the recursive form

$$p(\boldsymbol{\theta}|\mathcal{D}^n) = \frac{p(\mathbf{x}_n|\boldsymbol{\theta})p(\boldsymbol{\theta}|\mathcal{D}^{n-1})}{\int p(\mathbf{x}_n|\boldsymbol{\theta})p(\boldsymbol{\theta}|\mathcal{D}^{n-1})\,d\boldsymbol{\theta}}. \tag{42}$$

These are the basic equations for unsupervised Bayesian learning. Equation 40 emphasizes the relation between the Bayesian and the maximum-likelihood solutions. If $p(\boldsymbol{\theta})$ is essentially uniform over the region where $p(\mathcal{D}|\boldsymbol{\theta})$ peaks, then $p(\boldsymbol{\theta}|\mathcal{D})$ peaks at the same place. If the only significant peak occurs at $\boldsymbol{\theta} = \hat{\boldsymbol{\theta}}$ and if the peak is very sharp, then Eqs. 37 and 39 yield

$$p(\mathbf{x}|\omega_i, \mathcal{D}) \simeq p(\mathbf{x}|\omega_i, \hat{\boldsymbol{\theta}}) \tag{43}$$

and

$$P(\omega_i|\mathbf{x}, \mathcal{D}) \simeq \frac{p(\mathbf{x}|\omega_i, \hat{\boldsymbol{\theta}}_i)P(\omega_i)}{\sum_{j=1}^{c} p(\mathbf{x}|\omega_j, \hat{\boldsymbol{\theta}}_j)P(\omega_j)}. \tag{44}$$

That is, these conditions justify the use of the maximum-likelihood estimate $\hat{\boldsymbol{\theta}}$ as if it were the true value of $\boldsymbol{\theta}$ in designing the Bayes classifier.

As we saw in Chapter 3, in the limit of large amounts of data, maximum-likelihood and the Bayes methods will agree (or nearly agree). While in many *small* sample size problems they will agree, there exist small problems where the approximations are poor (Fig. 10.5). As we saw in the analogous case in supervised learning, whether one chooses to use the maximum-likelihood or the Bayes method depends not only on how confident one is of the prior distributions, but also on computational considerations; maximum-likelihood techniques are often easier to implement than Bayesian ones.

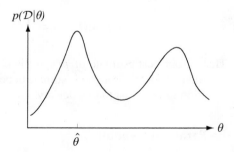

FIGURE 10.5. In a highly skewed or multiple peak posterior distribution such as illustrated here, the maximum-likelihood solution $\hat{\theta}$ will yield a density very different from a Bayesian solution, which requires the integration over the full range of parameter space θ.

Of course, if $p(\boldsymbol{\theta})$ has been obtained by supervised learning using a large set of labeled samples, it will be far from uniform, and it will have a dominant influence on $p(\boldsymbol{\theta}|\mathcal{D}^n)$ when n is small. Equation 42 shows how the observation of an additional unlabeled sample modifies our opinion about the true value of $\boldsymbol{\theta}$, and it highlights the ideas of updating and learning. If the mixture density $p(\mathbf{x}|\boldsymbol{\theta})$ is identifiable, then each additional sample tends to sharpen $p(\boldsymbol{\theta}|\mathcal{D}^n)$, and under fairly general conditions $p(\boldsymbol{\theta}|\mathcal{D}^n)$ can be shown to converge (in probability) to a Dirac delta function centered at the true value of $\boldsymbol{\theta}$ (Problem 9). Thus, even though we do not know the categories of the samples, identifiability assures us that we can learn the unknown parameter vector $\boldsymbol{\theta}$, and thereby learn the component densities $p(\mathbf{x}|\omega_i, \boldsymbol{\theta})$.

This, then, is the formal Bayesian solution to the problem of unsupervised learning. In retrospect, the fact that unsupervised learning of the parameters of a mixture density is so similar to supervised learning of the parameters of a component density is not at all surprising. Indeed, if the component density is itself a mixture, there would appear to be no essential difference between the two problems.

There are, however, some significant differences between supervised and unsupervised learning. One of the major differences concerns the issue of identifiability. With supervised learning, the lack of identifiability merely means that instead of obtaining a unique parameter vector we obtain an equivalence class of parameter vectors. Because all of these yield the same component density, lack of identifiability presents no theoretical difficulty. A lack of identifiability is much more serious in unsupervised learning. When $\boldsymbol{\theta}$ cannot be determined uniquely, the mixture cannot be decomposed into its true components. Thus, while $p(\mathbf{x}|\mathcal{D}^n)$ may still converge to $p(\mathbf{x})$, $p(\mathbf{x}|\omega_i, \mathcal{D}^n)$ given by Eq. 39 will not in general converge to $p(\mathbf{x}|\omega_i)$, and a theoretical barrier to learning exists. It is here that a few *labeled* training samples would be valuable for "decomposing" the mixture into its components.

Another serious problem for unsupervised learning is computational complexity. With supervised learning, the possibility of finding sufficient statistics allows solutions that are analytically pleasing and computationally feasible. With unsupervised learning, there is no way to avoid the fact that the samples are obtained from a mixture density,

$$p(\mathbf{x}|\boldsymbol{\theta}) = \sum_{j=1}^{c} p(\mathbf{x}|\omega_j, \boldsymbol{\theta}_j) P(\omega_j), \tag{1}$$

and this gives us little hope of ever finding simple exact solutions for $p(\boldsymbol{\theta}|\mathcal{D})$. Such solutions are tied to the existence of a simple sufficient statistic (Chapter 3), and the factorization theorem requires the ability to factor $p(\mathcal{D}|\boldsymbol{\theta})$ as

$$p(\mathcal{D}|\boldsymbol{\theta}) = g(\mathbf{s}, \boldsymbol{\theta})h(\mathcal{D}). \tag{45}$$

But from Eqs. 41 and 1, we see that the likelihood can be written as

$$p(\mathcal{D}|\boldsymbol{\theta}) = \prod_{k=1}^{n} \left[\sum_{j=1}^{c} p(\mathbf{x}_k|\omega_j, \boldsymbol{\theta}_j) P(\omega_j) \right]. \tag{46}$$

Thus, $p(\mathcal{D}|\boldsymbol{\theta})$ is the sum of c^n products of component densities. Each term in this sum can be interpreted as the joint probability of obtaining the samples $\mathbf{x}_1, \ldots, \mathbf{x}_n$ bearing a particular labeling, with the sum extending over all of the ways that the samples could be labeled. Clearly, this results in a thorough mixture of $\boldsymbol{\theta}$ and the $\mathbf{x}$'s, and no simple factoring should be expected. An exception to this statement arises if the component densities do not overlap; thus as $\boldsymbol{\theta}$ varies, only one term in the mixture density is nonzero. In that case, $p(\mathcal{D}|\boldsymbol{\theta})$ is the product of the n nonzero terms and may possess a simple sufficient statistic. However, because that case allows the class of any sample to be determined, it actually reduces the problem to one of supervised learning and thus is not a significant exception.

Another way to compare supervised and unsupervised learning is to substitute the mixture density for $p(\mathbf{x}_n|\boldsymbol{\theta})$ in Eq. 42 and obtain

$$p(\boldsymbol{\theta}|\mathcal{D}^n) = \frac{\sum_{j=1}^{c} p(\mathbf{x}_n|\omega_j, \boldsymbol{\theta}_j) P(\omega_j)}{\sum_{j=1}^{c} \int p(\mathbf{x}_n|\omega_j, \boldsymbol{\theta}_j) P(\omega_j) p(\boldsymbol{\theta}|\mathcal{D}^{n-1}) \, d\boldsymbol{\theta}} p(\boldsymbol{\theta}|\mathcal{D}^{n-1}). \quad (47)$$

If we consider the special case where $P(\omega_1) = 1$ and all the other prior probabilities are zero, corresponding to the supervised case in which all samples come from Class ω_1, then Eq. 47 simplifies to

$$p(\boldsymbol{\theta}|\mathcal{D}^n) = \frac{p(\mathbf{x}_n|\omega_1, \boldsymbol{\theta}_1)}{\int p(\mathbf{x}_n|\omega_1, \boldsymbol{\theta}_1) p(\boldsymbol{\theta}|\mathcal{D}^{n-1}) \, d\boldsymbol{\theta}} p(\boldsymbol{\theta}|\mathcal{D}^{n-1}). \quad (48)$$

Let us compare Eqs. 47 and 48 to see how observing an additional sample changes our estimate of $\boldsymbol{\theta}$. In each case we can ignore the normalizing denominator, which is independent of $\boldsymbol{\theta}$. Thus, the only significant difference is that in the supervised case we multiply the "prior" density for $\boldsymbol{\theta}$ by the component density $p(\mathbf{x}_n|\omega_1, \boldsymbol{\theta}_1)$, while in the unsupervised case we multiply it by the mixture density $\sum_{j=1}^{c} p(\mathbf{x}_n|\omega_j, \boldsymbol{\theta}_j) P(\omega_j)$. Assuming that the sample really did come from Class ω_1, we see that the effect of not knowing this category membership in the unsupervised case is to diminish the influence of $\mathbf{x}_n$ on changing $\boldsymbol{\theta}$. Because $\mathbf{x}_n$ could have come from any of the c classes, we cannot use it with full effectiveness in changing the component(s) of $\boldsymbol{\theta}$ associated with any one category. Rather, we must distributed its effect over the various categories in accordance with the probability that it arose from each category.

EXAMPLE 1   Unsupervised Learning of Gaussian Data

As an example, consider the one-dimensional, two-component mixture with $p(x|\omega_1) \sim N(\mu, 1), p(x|\omega_2, \theta) \sim N(\theta, 1)$, where μ, $P(\omega_1)$ and $P(\omega_2)$ are known. Here we have

$$p(x|\theta) = \underbrace{\frac{P(\omega_1)}{\sqrt{2\pi}} \exp\left[-\frac{1}{2}(x - \mu)^2\right]}_{\omega_1} + \underbrace{\frac{P(\omega_2)}{\sqrt{2\pi}} \exp\left[-\frac{1}{2}(x - \theta)^2\right]}_{\omega_2},$$

and we seek the mean of the second component.

Viewed as a function of x, this mixture density is a superposition of two normal densities—one peaking at $x = \mu$ and the other peaking at $x = \theta$. Viewed as a function of θ, $p(x|\theta)$ has a single peak at $\theta = x$. Suppose that the prior density $p(\theta)$ is uniform from a to b. Then after one observation $(x = x_1)$ we have

$$p(\theta|x_1) = \alpha p(x_1|\theta) p(\theta) = \left\{ \begin{array}{ll} \alpha'\{P(\omega_1) \exp[-\frac{1}{2}(x_1 - \mu)^2] + \\ \quad P(\omega_2) \exp[-\frac{1}{2}(x_1 - \theta)^2]\} & a \le \theta \le b \\ 0 & \text{otherwise} \end{array} \right\},$$

where α and α' are normalizing constants that are independent of θ. If the sample x_1 is in the range $a \leq x_1 \leq b$, then $p(\theta|x_1)$ peaks at $\theta = x_1$, of course. Otherwise it peaks either at $\theta = a$ if $x_1 < a$ or at $\theta = b$ if $x_1 > b$. Note that the additive constant $\exp[-(1/2)(x_1 - \mu)^2]$ is large if x_1 is near μ, and thus the peak of $p(\theta|x_1)$ is less pronounced if x_1 is near μ. This corresponds to the fact that if x_1 is near μ, it is more likely to have come from the $p(x|\omega_1)$ component, and hence its influence on our estimate for θ is diminished.

With the addition of a second sample x_2, $p(\theta|x_1)$ changes to

$$p(\theta|x_1, x_2) = \beta p(x_2|\theta)p(\theta|x_1)$$

$$= \begin{cases} \beta' \Big\{ P(\omega_1)P(\omega_1) \exp\left[-\tfrac{1}{2}(x_1 - \mu)^2 - \tfrac{1}{2}(x_2 - \mu)^2\right] \\ \quad + P(\omega_1)P(\omega_2) \exp\left[-\tfrac{1}{2}(x_1 - \mu)^2 - \tfrac{1}{2}(x_2 - \theta)^2\right] \\ \quad + P(\omega_2)P(\omega_1) \exp\left[-\tfrac{1}{2}(x_1 - \theta)^2 - \tfrac{1}{2}(x_2 - \mu)^2\right] \\ \quad + P(\omega_2)P(\omega_2) \exp\left[-\tfrac{1}{2}(x_1 - \theta)^2 - \tfrac{1}{2}(x_2 - \theta)^2\right] \Big\} \\ \qquad\qquad\qquad\qquad\qquad\qquad\qquad\qquad a \leq \theta \leq b \\ 0 \qquad\qquad\qquad\qquad\qquad\qquad\qquad\qquad \text{otherwise.} \end{cases}$$

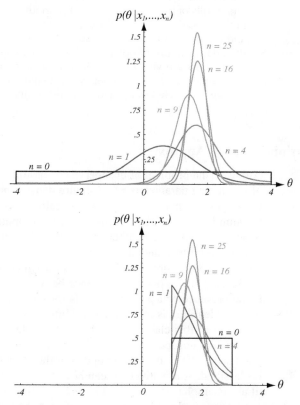

In unsupervised Bayesian learning of the parameter θ, the density becomes more peaked as the number of samples increases. The top figures uses a wide uniform prior $p(\theta) = 1/8$, for $-4 \leq \theta \leq 4$ while the bottom figure uses a narrower one, $p(\theta) = 1/2$, for $1 \leq \theta \leq 3$. Despite these different prior distributions, after all 25 samples have been used, the posterior densities are virtually identical in the two cases—the information in the samples overwhelms the prior information.

Unfortunately, the primary thing we learn from this expression is that $p(\theta|\mathcal{D}^n)$ is already complicated when $n = 2$. The four terms in the sum correspond to the four ways in which the samples could have been drawn from the two component populations. With n samples there will be 2^n terms, and no simple sufficient statistics can be found to facilitate understanding or to simplify computations.

It is possible to use the relation

$$p(\theta|\mathcal{D}^n) = \frac{p(x_n|\theta)p(\theta|\mathcal{D}^{n-1})}{\int p(x_n|\theta)p(\theta|\mathcal{D}^{n-1})\,d\theta}$$

and numerical integration to obtain an approximate numerical solution for $p(\theta|\mathcal{D}^n)$. This was done for the data in Fig. 10.1 using the values $\mu = 2$, $P(\omega_1) = 1/3$, and $P(\omega_2) = 2/3$. A prior density $p(\theta)$ uniform from -4 to $+4$ encompasses the data in the table. When this was used to start the recursive computation of $p(\theta|\mathcal{D}^n)$, the results are shown in the figure were obtained. As n goes to infinity, we can confidently expect $p(\theta|\mathcal{D}^n)$ to approach an impulse centered at $\theta = 2$. This graph gives some idea of the rate of convergence.

One of the main differences between the Bayesian and the maximum-likelihood approaches to unsupervised learning is the presence of the prior density $p(\theta)$. The figure shows how $p(\theta|\mathcal{D}^n)$ changes when $p(\theta)$ is assumed to be uniform from 1 to 3, corresponding to more certain initial knowledge about θ. The results of this change are most pronounced when n is small. It is here (just as in the classification case in Chapter 3) that the differences between the Bayesian and the maximum-likelihood solutions are most significant. As n increases, the importance of prior knowledge diminishes, and in the particular case the curves for $n = 25$ are virtually identical. In general, one would expect the difference to be small when the number of unlabeled samples is several times the effective number of labeled samples used to determine $p(\theta)$.

10.5.3 Decision-Directed Approximation

Although the problem of unsupervised learning can be stated as merely the problem of estimating parameters of a mixture density, neither the maximum-likelihood nor the Bayesian approach yields analytically simple results. Exact solutions for even the simplest nontrivial examples lead to computational requirements that grow exponentially with the number of samples. The problem of unsupervised learning is too important to abandon just because exact solutions are hard to find, however, and numerous procedures for obtaining approximate solutions have been suggested.

Because the important difference between supervised and unsupervised learning is the presence or absence of labels for the samples, an obvious approach to unsupervised learning is to use the prior information to design a classifier and to use the decisions of this classifier to label the samples. This is called the *decision-directed* approach to unsupervised learning, and it is subject to many variations. It can be applied sequentially on-line by updating the classifier each time an unlabeled sample is classified. Alternatively, it can be applied in parallel (batch mode) by waiting until all n samples are classified before updating the classifier. If desired, this process can be repeated until no changes occur in the way the samples are labeled. Various heuristics can be introduced to make the extent of any corrections depend upon the confidence of the classification decision.

There are some obvious dangers associated with the decision-directed approach. If the initial classifier is not reasonably good or if an unfortunate sequence of samples is encountered, the errors in classifying the unlabeled samples can drive the classifier the wrong way, resulting in a solution corresponding roughly to one of the lesser peaks of the likelihood function. Even if the initial classifier is optimal, in general the resulting labeling will not be the same as the true class membership; the act of classification will exclude samples from the tails of the desired distribution and will include samples from the tails of the other distributions. Thus, if there is significant overlap between the component densities, one can expect biased estimates and less-than-optimal results.

Despite these drawbacks, the simplicity of decision-directed procedures makes the Bayesian approach computationally feasible, and a flawed solution is often better than none. If conditions are favorable, performance that is nearly optimal can be achieved at far less computational expense. In practice it is found that most of these procedures work well if the parametric assumptions are valid, if there is little overlap between the component densities, and if the initial classifier design is at least roughly correct (Computer exercise 7).

10.6 DATA DESCRIPTION AND CLUSTERING

Let us reconsider our original problem of learning the structure of multidimensional patterns from a set of unlabeled samples. Viewed geometrically, these samples may form clouds of points in a d-dimensional space. Suppose that we knew, somehow, that these points came from a single normal distribution. Then the most we could learn form the data would be contained in the sufficient statistics—the sample mean and the sample covariance matrix. In essence, these statistics constitute a compact description of the data. The sample mean locates the center of gravity of the cloud; it can be thought of as the single point $\mathbf{m}$ that best represents all of the data in the sense of minimizing the sum of squared distances from $\mathbf{m}$ to the samples. The sample covariance matrix describes the amount the data scatters along various directions around $\mathbf{m}$. If the data points are actually normally distributed, then the cloud has a simple hyperellipsoidal shape, and the sample mean tends to fall in the region where the samples are most densely concentrated.

Of course, if the samples are not normally distributed these statistics can give a very misleading description of the data. Figure 10.6 shows four different data sets that all have the same mean and covariance matrix. Obviously, second-order statistics are incapable of revealing all of the structure in an arbitrary set of data.

If we assume that the samples come from a mixture of c normal distributions, we can approximate a greater variety of situations. In essence, this corresponds to assuming that the samples fall in hyperellipsoidally shaped clouds of various sizes and orientations. If the number of component densities is sufficiently high, we can approximate virtually any density function as a mixture model in this way and use the parameters of the mixture to describe the data. Alas, we have seen that the problem of estimating the parameters of a mixture density is not trivial. Furthermore, in situations where we have relatively little prior knowledge about the nature of the data, the assumption of particular parametric forms may lead to poor or meaningless results. Instead of finding structure in the data, we would be imposing structure on it.

One alternative is to use one of the nonparametric methods described in Chapter 4 to estimate the unknown mixture density. If accurate, the resulting estimate is certainly a complete description of what we can learn from the data. Regions of high local density, which might correspond to significant subclasses in the population, can be found from the peaks or modes of the estimated density.

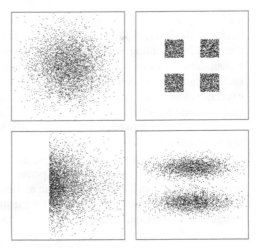

FIGURE 10.6. These four data sets have identical statistics up to second-order—that is, the same mean μ and covariance Σ. In such cases it is important to include in the model more parameters to represent the structure more completely.

CLUSTERING
PROCEDURE
If the goal is to find subclasses, a more direct alternative is to use a *clustering procedure*. Roughly speaking, clustering procedures yield a data description in terms of clusters or groups of data points that possess strong internal similarities. Formal clustering procedures use a criterion function, such as the sum of the squared distances from the cluster centers, and seek the grouping that extremizes the criterion function. Because even this can lead to unmanageable computational problems, other procedures have been proposed that are intuitively appealing but that lead to solutions having few, if any, established properties. Their use is usually justified on the grounds that they are easy to apply and often yield interesting results that may guide the application of more rigorous procedures.

10.6.1 Similarity Measures

Once we describe the clustering problem as one of finding natural groupings in a set of data, we are obliged to define what we mean by a natural grouping. In what sense are we to say that the samples in one cluster are more like one another than like samples in other clusters? This question actually involves two separate issues:

- How should one measure the similarity between samples?
- How should one evaluate a partitioning of a set of samples into clusters?

In this section we address the first of these issues.

The most obvious measure of the similarity (or dissimilarity) between two samples is the distance between them. One way to begin a clustering investigation is to define a suitable metric (Section 4.6) and compute the matrix of distances between all pairs of samples. If distance is a good measure of dissimilarity, then one would expect the distance between samples in the *same* cluster to be significantly less than the distance between samples in *different* clusters.

Suppose for the moment that we say that two samples belong to the same cluster if the Euclidean distance between them is less than some threshold distance d_0. It is immediately obvious that the choice of d_0 is very important. If d_0 is very large, all of

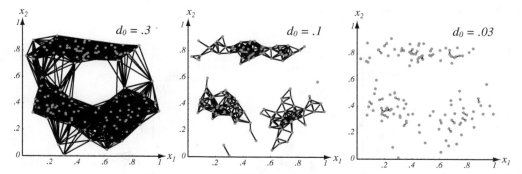

FIGURE 10.7. The distance threshold affects the number and size of clusters in similarity based clustering methods. For three different values of distance d_0, lines are drawn between points closer than d_0—the smaller the value of d_0, the smaller and more numerous the clusters.

the samples will be assigned to one cluster. If d_0 is very small, each sample will form an isolated, singleton cluster. To obtain "natural" clusters, d_0 will have to be greater than the typical within-cluster distances and less than the typical between-cluster distances (Fig. 10.7).

Less obvious perhaps is the fact that the results of clustering depend on the choice of Euclidean distance as a measure of dissimilarity. That particular choice is generally justified if the feature space is isotropic and the data are spread roughly evenly all directions. Clusters defined by Euclidean distance will be invariant to translations or rotations in feature space—rigid-body motions of the data points. However, they will not be invariant to linear transformations in general, or to other transformations that distort the distance relationships. Thus, as Fig. 10.8 illustrates, a simple scaling of the coordinate axes can result in a different grouping of the data into clusters. Of course, this is of no concern for problems in which arbitrary rescaling is an unnatural or meaningless transformation. However, if clusters are to mean anything, they should be invariant to transformations natural to the problem.

One way to achieve invariance is to normalize the data prior to clustering. For example, to obtain invariance to displacement and scale changes, one might translate and scale the axes so that all of the features have zero mean and unit variance—standardize the data. To obtain invariance to rotation, one might rotate the axes so that they coincide with the eigenvectors of the sample covariance matrix. This transformation to *principal components* (Section 10.13.1) can be preceded or followed by normalization for scale.

However, we should not conclude that this kind of normalization is necessarily desirable. Consider, for example, the matter of translating and scaling the axes so that each feature has zero mean and unit variance. The rationale usually given for this normalization is that it prevents certain features from dominating distance calculations merely because they have large numerical values, much as we saw in networks trained with backpropagation (Chapter 6). Subtracting the mean and dividing by the standard deviation is an appropriate normalization if this spread of values is due to normal random variation; however, it can be quite inappropriate if the spread is due to the presence of subclasses (Fig. 10.9). Thus, this routine normalization may be less than helpful in the cases of greatest interest.

Instead of scaling axes, we can change the metric in interesting ways. For instance, one broad class of metrics is of the form

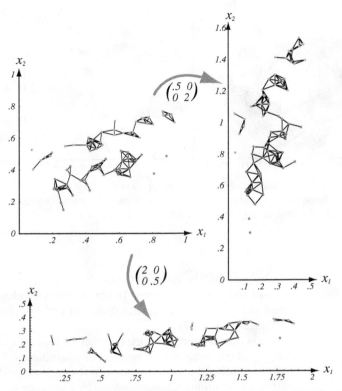

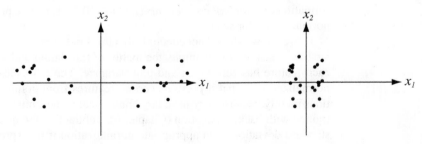

FIGURE 10.8. Scaling axes affects the clusters in a minimum distance cluster method. The original data and minimum-distance clusters are shown in the upper left; points in one cluster are shown in red, while the others are shown in gray. When the vertical axis is expanded by a factor of 2.0 and the horizontal axis shrunk by a factor of 0.5, the clustering is altered (as shown at the right). Alternatively, if the vertical axis is shrunk by a factor of 0.5 and the horizontal axis is expanded by a factor of 2.0, smaller more numerous clusters result (shown at the bottom). In both these scaled cases, the assignment of points to clusters differ from that in the original space.

FIGURE 10.9. If the data fall into well-separated clusters (left), normalization by scaling for unit variance for the full data may reduce the separation, and hence be undesirable (right). Such a normalization may in fact be appropriate if the full data set arises from a single fundamental process (with noise), but inappropriate if there are several different processes, as shown here.

$$d(\mathbf{x}, \mathbf{x}') = \left(\sum_{k=1}^{d} |x_k - x_k'|^q \right)^{1/q}, \tag{49}$$

MINKOWSKI
METRIC

CITY BLOCK
METRIC

where $q \geq 1$ is a selectable parameter—the general *Minkowski metric* we considered in Chapter 4. Setting $q = 2$ gives the familiar Euclidean metric while setting $q = 1$ gives the Manhattan or *city block* metric—the sum of the absolute distances along each of the d coordinate axes. Note that only $q = 2$ is invariant to an arbitrary rotation or translation in feature space. Another alternative is to use some kind of metric based on the data itself, such as the Mahalanobis distance.

SIMILARITY
FUNCTION

More generally, one can abandon the use of distance altogether and introduce a nonmetric *similarity function* $s(\mathbf{x}, \mathbf{x}')$ to compare two vectors $\mathbf{x}$ and $\mathbf{x}'$. Conventionally, this is a symmetric function whose value is large when $\mathbf{x}$ and $\mathbf{x}'$ are somehow "similar." For example, when the angle between two vectors is a meaningful measure of their similarity, then the normalized inner product

$$s(\mathbf{x}, \mathbf{x}') = \frac{\mathbf{x}^t \mathbf{x}'}{\|\mathbf{x}\| \, \|\mathbf{x}'\|} \tag{50}$$

may be an appropriate similarity function. This measure, which is the cosine of the angle between $\mathbf{x}$ and $\mathbf{x}'$, is invariant to rotation and dilation, though it is not invariant to translation and general linear transformations.

When the features are binary-valued (0 or 1), the similarity function of Eq. 50 has a simple nongeometrical interpretation in terms of shared features or shared attributes. Let us say that a sample $\mathbf{x}$ *possesses* the ith attribute if $x_i = 1$. Then $\mathbf{x}^t \mathbf{x}'$ is merely the number of attributes possessed by both $\mathbf{x}$ and $\mathbf{x}'$, and $\|\mathbf{x}\| \, \|\mathbf{x}'\| = (\mathbf{x}^t \mathbf{x} \mathbf{x}'^t \mathbf{x}')^{1/2}$ is the geometric mean of the number of attributes possessed by $\mathbf{x}$ and the number possessed by $\mathbf{x}'$. Thus, $s(\mathbf{x}, \mathbf{x}')$ is a measure of the relative possession of common attributes. Some simple variations are

$$s(\mathbf{x}, \mathbf{x}') = \frac{\mathbf{x}^t \mathbf{x}'}{d}, \tag{51}$$

the fraction of attributes shared, and

$$s(\mathbf{x}, \mathbf{x}') = \frac{\mathbf{x}^t \mathbf{x}'}{\mathbf{x}^t \mathbf{x} + \mathbf{x}'^t \mathbf{x}' - \mathbf{x}^t \mathbf{x}'}, \tag{52}$$

TANIMOTO
DISTANCE

the ratio of the number of shared attributes to the number possessed by $\mathbf{x}$ or $\mathbf{x}'$. This latter measure (sometimes known as the Tanimoto coefficient or *Tanimoto distance*) is frequently encountered in the fields of information retrieval and biological taxonomy. Related measures of similarity arise in other applications, with the variety of measures testifying to the diversity of problem domains.

Fundamental issues in measurement theory are involved in the use of any distance or similarity function. The calculation of the similarity between two vectors always involves combining the values of their components. Yet in many pattern recognition applications the components of the feature vector measure seemingly noncomparable quantities, such as meters and kilograms. Recall our example of classifying fish: How can we compare the lightness of the skin to the length or weight of the fish? Should the comparison depend on whether the length is measured in meters or inches? How do we treat vectors whose components have a mixture of nominal, ordinal, interval and ratio scales? Ultimately, there are rarely clear methodological answers to these questions. When a designer selects a particular similarity function or normalizes the data in a particular way, information is introduced that gives the procedure meaning. We have given examples of some alternatives that have proved to be useful. Beyond that we can do little more than alert the unwary to these pitfalls of clustering.

Amidst all this discussion of clustering, we must not lose sight of the fact that often the clusters found will later be labeled (e.g., by resorting to a teacher or small number of labeled samples) and that the clusters can then be used for classification. In that case, the same similarity (or metric) should be used for classification as was used for forming the clusters (Computer exercise 8).

10.7 CRITERION FUNCTIONS FOR CLUSTERING

We have just considered the first major issue in clustering: how to measure "similarity." Now we turn to the second major issue: the criterion function to be optimized. Suppose that we have a set $\mathcal{D} = \{\mathbf{x}_1, \ldots, \mathbf{x}_n\}$ of n samples that we want to partition into exactly c disjoint subsets $\mathcal{D}_1, \ldots, \mathcal{D}_c$. Each subset is to represent a cluster, with samples in the same cluster being somehow more similar to each other than they are to samples in other clusters. One way to make this into a well-defined problem is to define a criterion function that measures the clustering quality of any partition of the data. Then the problem is one of finding the partition that extremizes the criterion function. In this section we examine the characteristics of several basically similar criterion functions, postponing until later the question of how to find an optimal partition.

10.7.1 The Sum-of-Squared-Error Criterion

The simplest and most widely used criterion function for clustering is the sum-of-squared-error criterion. Let n_i be the number of samples in $\mathcal{D}_i$ and let $\mathbf{m}_i$ be the mean of those samples,

$$\mathbf{m}_i = \frac{1}{n_i} \sum_{\mathbf{x} \in \mathcal{D}_i} \mathbf{x}. \tag{53}$$

Then the sum-of-squared errors is defined by

$$J_e = \sum_{i=1}^{c} \sum_{\mathbf{x} \in \mathcal{D}_i} \|\mathbf{x} - \mathbf{m}_i\|^2. \tag{54}$$

This criterion function has a simple interpretation: For a given cluster $\mathcal{D}_i$, the mean vector $\mathbf{m}_i$ is the best representative of the samples in $\mathcal{D}_i$ in the sense that it minimizes the sum of the squared lengths of the "error" vectors $\mathbf{x} - \mathbf{m}_i$ in $\mathcal{D}_i$. Thus, J_e measures the total squared error incurred in representing the n samples $\mathbf{x}_1, \ldots, \mathbf{x}_n$ by the c cluster centers $\mathbf{m}_1, \ldots, \mathbf{m}_c$. The value of J_e depends on how the samples are grouped into clusters and the number of clusters; the optimal partitioning is defined as one that minimizes J_e. Clusterings of this type are often called *minimum variance* partitions.

MINIMUM
VARIANCE

What kind of clustering problems are well-suited to a sum-of-squared-error criterion? Basically, J_e is an appropriate criterion when the clusters form compact clouds that are rather well-separated from one another. A less obvious problem arises when there are great differences in the number of samples in different clusters. In that case

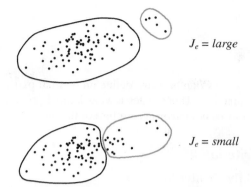

J_e = large

J_e = small

FIGURE 10.10. When two natural groupings have very different numbers of points, the clusters minimizing a sum-squared-error criterion J_e of Eq. 54 may not reveal the true underlying structure. Here the criterion is smaller for the two clusters at the bottom than for the more natural clustering at the top.

it can happen that a partition that splits a large cluster is favored over one that maintains the integrity of the natural clusters, as illustrated in Fig. 10.10. This situation frequently arises because of the presence of "outliers" or "wild shots" and brings up the problem of interpreting and evaluating the results of clustering. Because little can be said about that problem, we shall merely observe that if additional considerations render the results of minimizing J_e unsatisfactory, then these considerations should be used, if possible, in formulating a better criterion function.

10.7.2 Related Minimum Variance Criteria

By some simple algebraic manipulation (Problem 20) we can eliminate the mean vectors from the expression for J_e and obtain the equivalent expression

$$J_e = \frac{1}{2} \sum_{i=1}^{c} n_i \bar{s}_i, \tag{55}$$

where

$$\bar{s}_i = \frac{1}{n_i^2} \sum_{\mathbf{x} \in \mathcal{D}_i} \sum_{\mathbf{x}' \in \mathcal{D}_i} \|\mathbf{x} - \mathbf{x}'\|^2. \tag{56}$$

Equation 56 leads us to interpret $\bar{s}_i$ as the average squared distance between points in the ith cluster, and it emphasizes the fact that the sum-of-squared-error criterion uses Euclidean distance as the measure of similarity. It also suggests an obvious way of obtaining other criterion functions. For example, one can replace $\bar{s}_i$ by the average, the median, or perhaps the maximum distance between points in a cluster. More generally, one can introduce an appropriate similarity function $s(\mathbf{x}, \mathbf{x}')$ and replace $\bar{s}_i$ by functions such as

$$\bar{s}_i = \frac{1}{n_i^2} \sum_{\mathbf{x} \in \mathcal{D}_i} \sum_{\mathbf{x}' \in \mathcal{D}_i} s(\mathbf{x}, \mathbf{x}') \tag{57}$$

or

$$\bar{s}_i = \min_{\mathbf{x}, \mathbf{x}' \in \mathcal{D}_i} s(\mathbf{x}, \mathbf{x}'). \tag{58}$$

As in Chapter 4, we define an optimal partition as one that extremizes the criterion function. This creates a well-defined problem, and the hope is that its solution discloses the intrinsic structure of the data.

10.7.3 Scatter Criteria

The Scatter Matrices

Another interesting class of criterion functions can be derived from the scatter matrices used in multiple discriminant analysis. The definitions in Table 10.1 directly parallel those given in Chapter 3.

Table 10.1. Mean Vectors and Scatter Matrices Used in Clustering Criteria

	Depend on cluster center?			
	Yes	No		
Mean vector for the ith cluster		×	$\mathbf{m}_i = \dfrac{1}{n_i} \sum_{\mathbf{x} \in \mathcal{D}_i} \mathbf{x}$	(59)
Total mean vector		×	$\mathbf{m} = \dfrac{1}{n} \sum_{\mathcal{D}} \mathbf{x} = \dfrac{1}{n} \sum_{i=1}^{c} n_i \mathbf{m}_i$	(60)
Scatter matrix for the ith cluster	×		$\mathbf{S}_i = \sum_{\mathbf{x} \in \mathcal{D}_i} (\mathbf{x} - \mathbf{m}_i)(\mathbf{x} - \mathbf{m}_i)^t$	(61)
Within-cluster scatter matrix	×		$\mathbf{S}_W = \sum_{i=1}^{c} \mathbf{S}_i$	(62)
Between-cluster scatter matrix	×		$\mathbf{S}_B = \sum_{i=1}^{c} n_i (\mathbf{m}_i - \mathbf{m})(\mathbf{m}_i - \mathbf{m})^t$	(63)
Total scatter matrix		×	$\mathbf{S}_T = \sum_{\mathbf{x} \in \mathcal{D}} (\mathbf{x} - \mathbf{m})(\mathbf{x} - \mathbf{m})^t$	(64)

As before, it follows from the definitions in Table 10.1 that the total scatter matrix is the sum of the within-cluster scatter matrix and the between-cluster scatter matrix:

$$\mathbf{S}_T = \mathbf{S}_W + \mathbf{S}_B. \tag{65}$$

Note that the total scatter matrix does not depend on how the set of samples is partitioned into clusters; it depends only on the total set of samples. The within-cluster and between-cluster scatter matrices taken separately do depend on the partitioning, of course. Roughly speaking, there is an exchange between these two matrices: The between-cluster scatter goes up as the within-cluster scatter goes down. This is fortunate, because by trying to minimize the within-cluster scatter we will also tend to maximize the between-cluster scatter.

To be more precise in talking about the amount of within-cluster or between-cluster scatter, we need a scalar measure of the "size" of a scatter matrix. The two measures that we shall consider are the *trace* and the *determinant*. In the univariate case, these two measures are equivalent, and we can define an optimal partition as one that minimizes $\mathbf{S}_W$ or maximizes $\mathbf{S}_B$. In the multivariate case, things are somewhat more complicated, and a number of related but distinct optimality criteria have been suggested.

The Trace Criterion

Perhaps the simplest scalar measure of a scatter matrix is its trace—the sum of its diagonal elements. Roughly speaking, the trace measures the square of the scattering radius, because it is proportional to the sum of the variances in the coordinate directions. Thus, an obvious criterion function to minimize is the trace of $\mathbf{S}_W$. In fact, this criterion is nothing more or less than the sum-of-squared-error criterion, because the definitions of scatter matrices (Eqs. 61 and 62) yield

$$\text{tr}[\mathbf{S}_W] = \sum_{i=1}^{c} \text{tr}[\mathbf{S}_i] = \sum_{i=1}^{c} \sum_{\mathbf{x} \in \mathcal{D}_i} \|\mathbf{x} - \mathbf{m}_i\|^2 = J_e. \tag{66}$$

Because $\text{tr}[\mathbf{S}_T] = \text{tr}[\mathbf{S}_W] + \text{tr}[\mathbf{S}_B]$ and $\text{tr}[\mathbf{S}_T]$ is independent of how the samples are partitioned, we see that no new results are obtained by trying to maximize $\text{tr}[\mathbf{S}_B]$. However, it is comforting to know that in seeking to minimize the within-cluster criterion $J_e = \text{tr}[\mathbf{S}_W]$ we are also maximizing the between-cluster criterion

$$\text{tr}[S_B] = \sum_{i=1}^{c} n_i \|\mathbf{m}_i - \mathbf{m}\|^2. \tag{67}$$

The Determinant Criterion

We have used the determinant of the scatter matrix to obtain a scalar measure of scatter. Roughly speaking, the determinant measures the square of the scattering volume, since it is proportional to the product of the variances in the directions of the principal axes. Because $\mathbf{S}_B$ will be singular if the number of clusters is less than or equal to the dimensionality, $|\mathbf{S}_B|$ is obviously a poor choice for a criterion function. Furthermore, $\mathbf{S}_B$ may become singular, and will certainly be so if $n - c$ is less than the dimensionality d (Problem 29). However, if we assume that $\mathbf{S}_W$ is nonsingular, we are led to consider the determinant criterion function

$$J_d = |\mathbf{S}_W| = \left| \sum_{i=1}^{c} \mathbf{S}_i \right|. \tag{68}$$

The partition that minimizes J_d is often similar to the one that minimizes J_e, but the two need not be the same, as shown in Example 2. We observed before that the minimum-squared-error partition might change if the axes are scaled, though this does not happen with J_d (Problem 27). Thus J_d is to be favored under conditions where there may be unknown or irrelevant linear transformations of the data.

Invariant Criteria

It is not particularly hard to show that the eigenvalues $\lambda_1, \ldots, \lambda_d$ of $\mathbf{S}_W^{-1}\mathbf{S}_B$ are invariant under nonsingular linear transformations of the data (Problem 28). Indeed, these eigenvalues are the basic linear invariants of the scatter matrices. Their numerical values measure the ratio of between-cluster to within-cluster scatter in the direction of the eigenvectors, and partitions that yield large values are usually desirable. Of course, as we have seen before, the fact that the rank of $\mathbf{S}_B$ can not exceed $c - 1$ means that no more than $c - 1$ of these eigenvalues can be nonzero. Nevertheless, good partitions are ones for which the nonzero eigenvalues are large.

One can invent a great variety of invariant clustering criteria by composing appropriate functions of these eigenvalues. Some of these follow naturally from standard matrix operations. For example, because the trace of a matrix is the sum of its eigenvalues, we might elect to maximize the criterion function

$$\text{tr}[\mathbf{S}_W^{-1}\mathbf{S}_B] = \sum_{i=1}^{d} \lambda_i. \tag{69}$$

By using the relation $\mathbf{S}_T = \mathbf{S}_W + \mathbf{S}_B$, we can derive the following invariant relatives of $\text{tr}[\mathbf{S}_W]$ and $|\mathbf{S}_W|$, as asked in Problem 26:

$$J_f = \text{tr}[\mathbf{S}_T^{-1}\mathbf{S}_W] = \sum_{i=1}^{d} \frac{1}{1 + \lambda_i} \tag{70}$$

and

$$\frac{|\mathbf{S}_W|}{|\mathbf{S}_T|} = \prod_{i=1}^{d} \frac{1}{1 + \lambda_i}. \tag{71}$$

Because all of these criterion functions are invariant to linear transformations, the same is true of the partitions that extremize them. In the special case of two clusters, only one eigenvalue is nonzero, and all of these criteria yield the same clustering. However, when the samples are partitioned into more than two clusters, the optimal partitions, though often similar, need not be the same, as shown in Example 2.

EXAMPLE 2   Clustering Criteria

We can gain some intuition by considering these clustering criteria applied to the below data set. All of the clusterings seem reasonable, and there is no strong argument to favor one over the others. For the case $c = 2$, the clusters minimizing the J_e indeed tend to favor clusters of roughly equal numbers of points; in contrast, J_d favors one large and one fairly small cluster. Because the full data set happens to be spread horizontally more than vertically, the eigenvalue in the horizontal direction is greater than that in the vertical direction. As such, the clusters are "stretched" horizontally somewhat. In general, the differences between

Sample	x_1	x_2	Sample	x_1	x_2
1	−1.82	0.24	11	0.41	0.91
2	−0.38	−0.39	12	1.70	0.48
3	−0.13	0.16	13	0.92	−0.49
4	−1.17	0.44	14	2.41	0.32
5	−0.92	0.16	15	1.48	−0.23
6	−1.69	−0.01	16	−0.34	1.88
7	0.33	−0.17	17	0.83	0.23
8	−0.71	−0.21	18	0.62	0.81
9	1.27	−0.39	19	−1.42	−0.51
10	−0.16	−0.23	20	0.67	−0.55

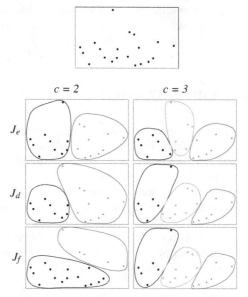

The raw data shown at the top does not exhibit any obvious clusters. The clusters found by minimizing a criterion depends upon the criterion function as well as the assumed number of clusters. The sum-of-squared-error criterion J_e (Eq. 54), the determinant criterion J_d (Eq. 68) and the more subtle trace criterion J_f (Eq. 70) were applied to the 20 points in the table with the assumption of $c = 2$ and $c = 3$ clusters. (Each point in the table is shown, with bounding boxes defined by $-1.8 < x_1 < 2.5$ and $-0.6 < x_2 < 1.9$.)

the cluster criteria become less pronounced for large numbers of clusters. For the $c = 3$ case, for instance, the clusters depend only mildly upon the cluster criterion—indeed, two of the clusterings are identical.

With regard to the criterion function involving $\mathbf{S}_T$, note that $\mathbf{S}_T$ does not depend on how the samples are partitioned into clusters. Thus, the clusterings that minimize $|\mathbf{S}_W|/|\mathbf{S}_T|$ are exactly the same as the ones that minimize $|\mathbf{S}_W|$. If we rotate and scale the axes so that $\mathbf{S}_T$ becomes the identity matrix, we see that minimizing $\text{tr}[\mathbf{S}_T^{-1}\mathbf{S}_W]$ is equivalent to minimizing the sum-of-squared-error criterion $\text{tr}[\mathbf{S}_W]$ after performing this normalization. Clearly, this criterion suffers from the very defects that we warned about in Section 10.6.1, and it is probably the least desirable of these criteria.

One final warning about invariant criteria is in order. If different apparent clusters can be obtained by scaling the axes or by applying any other linear transformation, then all of these groupings will be exposed by invariant procedures. Thus, invariant criterion functions are more likely to possess multiple local extrema and are correspondingly more difficult to optimize.

The variety of the criterion functions we have discussed and the somewhat subtle differences between them should not be allowed to obscure their essential similarity. In every case the underlying model is that the samples form c fairly well separated clouds of points. The within-cluster scatter matrix $\mathbf{S}_W$ is used to measure the compactness of these clouds, and the basic goal is to find the most compact grouping. While this approach has proved useful for many problems, it is not universally applicable. For example, it will not extract a very dense cluster embedded in the center of a diffuse cluster, or separate intertwined line-like clusters. When the minimum of the cluster criterion is not sufficiently low, and the cluster structure is not inferred by the algorithm, we must devise other criterion functions that are better matched to the structure present or being sought.

*10.8 ITERATIVE OPTIMIZATION

Once a criterion function has been selected, clustering becomes a well-defined problem in discrete optimization: Find those partitions of the set of samples that extremize the criterion function. Because the sample set is finite, there are only a finite number of possible partitions. Thus, in theory the clustering problem can always be solved by exhaustive enumeration. However, the computational complexity renders such an approach unthinkable for all but the simplest problems; there are approximately $c^n/c!$ ways of partitioning a set of n elements into c subsets, and this exponential growth with n is overwhelming (Problem 18). For example, an exhaustive search for the best set of 5 clusters in 100 samples would require considering more than 10^{67} partitionings. Simply put, in most applications an exhaustive search is completely infeasible.

The approach most frequently used in seeking optimal partitions is iterative optimization. The basic idea is to find some reasonable initial partition and to "move" samples from one group to another if such a move will improve the value of the criterion function. Like hill-climbing procedures in general, these approaches guarantee local but not global optimization. Different starting points can lead to different solutions, and one never knows whether or not the best solution has been found. Despite these limitations, the fact that the computational requirements are bearable makes this approach attractive.

Let us consider the use of iterative improvement to minimize the sum-of-squared-error criterion J_e, written as

$$J_e = \sum_{i=1}^{c} J_i,$$ (72)

where an effective error per cluster is defined to be

$$J_i = \sum_{\mathbf{x} \in \mathcal{D}_i} \|\mathbf{x} - \mathbf{m}_i\|^2$$ (73)

and the mean of each cluster is, as before,

$$\mathbf{m}_i = \frac{1}{n_i} \sum_{\mathbf{x} \in \mathcal{D}_i} \mathbf{x}. \tag{53}$$

Suppose that a sample $\hat{\mathbf{x}}$ currently in cluster $\mathcal{D}_i$ is tentatively moved to $\mathcal{D}_j$. Then $\mathbf{m}_j$ changes to

$$\mathbf{m}_j^* = \mathbf{m}_j + \frac{\hat{\mathbf{x}} - \mathbf{m}_j}{n_j + 1} \tag{74}$$

and J_j increases to

$$\begin{aligned} J_j^* &= \sum_{\mathbf{x} \in \mathcal{D}_j} \|\mathbf{x} - \mathbf{m}_j^*\|^2 + \|\hat{\mathbf{x}} - \mathbf{m}_j^*\|^2 \\ &= \left(\sum_{\mathbf{x} \in \mathcal{D}_i} \left\| \mathbf{x} - \mathbf{m}_j - \frac{\hat{\mathbf{x}} - \mathbf{m}_j}{n_j + 1} \right\|^2 \right) + \left\| \frac{n_j}{n_j + 1} (\hat{\mathbf{x}} - \mathbf{m}_j) \right\|^2 \\ &= J_j + \frac{n_j}{n_j + 1} \|\hat{\mathbf{x}} - \mathbf{m}_j\|^2. \end{aligned} \tag{75}$$

Under the assumption that $n_i \neq 1$ (singleton clusters should not be destroyed), a similar calculation (Problem 31) shows that $\mathbf{m}_i$ changes to

$$\mathbf{m}_i^* = \mathbf{m}_i - \frac{\hat{\mathbf{x}} - \mathbf{m}_i}{n_i - 1} \tag{76}$$

and J_i decreases to

$$J_i^* = J_i - \frac{n_i}{n_i - 1} \|\hat{\mathbf{x}} - \mathbf{m}_i\|^2. \tag{77}$$

These equations greatly simplify the computation of the change in the criterion function. The transfer of $\hat{\mathbf{x}}$ from $\mathcal{D}_i$ to $\mathcal{D}_j$ is advantageous if the decrease in J_i is greater than the increase in J_j. This is the case if

$$\frac{n_i}{n_i - 1} \|\hat{\mathbf{x}} - \mathbf{m}_i\|^2 > \frac{n_j}{n_j + 1} \|\hat{\mathbf{x}} - \mathbf{m}_j\|^2, \tag{78}$$

which typically happens whenever $\hat{\mathbf{x}}$ is closer to $\mathbf{m}_j$ than $\mathbf{m}_i$. If reassignment is profitable, the greatest decrease in sum of squared error is obtained by selecting the cluster for which $n_j \|\hat{\mathbf{x}} - \mathbf{m}_j\|^2 / (n_j + 1)$ is minimum. This leads to the following clustering procedure:

■ **Algorithm 3. (Basic Iterative Minimum-Squared-Error Clustering)**

1 <u>**begin initialize**</u> $n, c, \mathbf{m}_1, \mathbf{m}_2, \ldots, \mathbf{m}_c$
2   <u>**do**</u> randomly select a sample $\hat{\mathbf{x}}$
3              $i \leftarrow \arg\min_{i'} \|\mathbf{m}_{i'} - \hat{\mathbf{x}}\|$ (classify $\hat{\mathbf{x}}$)

4   **if** $n_i \neq 1$ **then** compute

5
$$\rho_j = \begin{cases} \frac{n_j}{n_j+1}\|\hat{\mathbf{x}} - \mathbf{m}_j\|^2 & j \neq i \\ \frac{n_j}{n_j-1}\|\hat{\mathbf{x}} - \mathbf{m}_i\|^2 & j = i \end{cases}$$

6   **if** $\rho_k \leq \rho_j$ for all j **then** transfer $\hat{\mathbf{x}}$ to $\mathcal{D}_k$

7   recompute $J_e, \mathbf{m}_i, \mathbf{m}_k$

8   **until** no change in J_e in n attempts

9   **return** $\mathbf{m}_1, \mathbf{m}_2, \ldots, \mathbf{m}_c$

10   **end**

A moment's consideration will show that this procedure is essentially a sequential version of the k-means procedure (Algorithm 1) described in Section 10.4.3. Where the k-means procedure waits until all n samples have been reclassified before updating, the Basic Iterative Minimum-Squared-Error procedure updates after each sample is reclassified. It has been experimentally observed that this procedure is more susceptible to being trapped in local minima, and it has the further disadvantage of making the results depend on the order in which the candidates are selected. However, it is at least a stepwise optimal procedure, and it can be easily modified to apply to problems in which samples are acquired sequentially and clustering must be done on-line.

One question that plagues all hill-climbing procedures is the choice of the starting point. Unfortunately, there is no simple, universally good solution to this problem. One approach is to select c samples randomly for the initial cluster centers, using them to partition the data on a minimum-distance basis. Repetition with different random selections can give some indication of the sensitivity of the solution to the starting point. Yet another approach is to find the c-cluster starting point from the solutions to the $(c-1)$-cluster problem. The solution for the one-cluster problem is the total sample mean; the starting point for the c-cluster problem can be the final means for the $(c-1)$-cluster problem plus the sample that is farthest from the nearest cluster center. This approach leads us directly to the so-called hierarchical clustering procedures, which are simple methods that can provide very good starting points for iterative optimization.

10.9 HIERARCHICAL CLUSTERING

Up to now, our methods have formed disjoint clusters—in computer science terminology, we would say that the data description is "flat." However, there are many times when clusters have subclusters, these have subsubclusters, and so on. In biological taxonomy, for instance, kingdoms are split into phyla, which are split into subphyla, which are split into orders, suborders, families, subfamilies, genus and species, and so on, all the way to a particular individual organism. Thus we might have kingdom = animal, phylum = Chordata, subphylum = Vertebrata, class = Osteichthyes, subclass = Actinopterygii, order = Salmoniformes, family = Salmonidae, genus = Oncorhynchus, species = Oncorhynchus kisutch, and finally the individual being the particular Coho salmon caught in our net. Organisms that lie in the animal kingdom—such as a salmon and a moose—share important attributes that are not present in organisms in the plant kingdom, such as redwood trees. In fact, this kind of hierarchical clustering permeates classifactory activities in the sciences. Thus we now turn to clustering methods which will lead to representations that are "hierarchical," rather than flat.

10.9.1 Definitions

Let us consider a sequence of partitions of the n samples into c clusters. The first of these is a partition into n clusters, each cluster containing exactly one sample. The next is a partition into $n-1$ clusters, the next a partition into $n-2$, and so on until the nth, in which all the samples form one cluster. We shall say that we are at level k in the sequence when $c = n - k + 1$. Thus, level one corresponds to n clusters and level n corresponds to one cluster. Given any two samples $\mathbf{x}$ and $\mathbf{x}'$, at *some* level they will be grouped together in the same cluster. If the sequence has the property that whenever two samples are in the same cluster at level k they remain together at all higher levels, then the sequence is said to be a *hierarchical clustering*.

DENDROGRAM

The most natural representation of hierarchical clustering is a corresponding tree, called a *dendrogram*, which shows how the samples are grouped. Figure 10.11 shows a dendrogram for a simple problem involving eight samples. Level $k = 1$ shows the eight samples as singleton clusters. At level 2, samples $\mathbf{x}_6$ and $\mathbf{x}_7$ have been grouped to form a cluster, and they stay together at all subsequent levels. If it is possible to measure the similarity between clusters, then the dendrogram is usually drawn to scale to show the similarity between the clusters that are grouped. In Fig. 10.11, for example, the similarity between the two groups of samples that are merged at level 5 has a value of roughly 60.

We shall see shortly how such similarity values can be obtained, but first note that the similarity values can be used to help determine whether groupings are natural or forced. If the similarity values for the levels are roughly evenly distributed throughout the range of possible values, then there is no principled argument that any particular number of clusters is better or "more natural" than another. Conversely, suppose that there is a unusually large gap between the similarity values for the levels corresponding to $c = 3$ and to $c = 4$ clusters. In such a case, one can argue that $c = 3$ is the most natural number of clusters (Problem 37).

Another representation for hierarchical clustering is based on sets, in which each level of cluster may contain sets that are subclusters, as shown in Fig. 10.12.

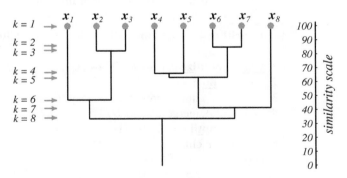

FIGURE 10.11. A dendrogram can represent the results of hierarchical clustering algorithms. The vertical axis shows a generalized measure of similarity among clusters. Here, at level 1 all eight points lie in singleton clusters; each point in a cluster is highly similar to itself, of course. Points $\mathbf{x}_6$ and $\mathbf{x}_7$ happen to be the most similar, and are merged at level 2, and so forth.

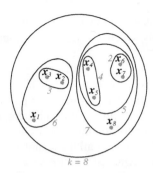

FIGURE 10.12. A set or Venn diagram representation of two-dimensional data (which was used in the dendrogram of Fig. 10.11) reveals the hierarchical structure but not the quantitative distances between clusters. The levels are numbered by k, in red.

Yet another, textual, representation uses brackets, such as $\{\{\mathbf{x}_1, \{\mathbf{x}_2, \mathbf{x}_3\}\}, \{\{\{\mathbf{x}_4, \mathbf{x}_5\}, \{\mathbf{x}_6, \mathbf{x}_7\}\}, \mathbf{x}_8\}\}$. While such representations may reveal the hierarchical structure of the data, they do not naturally represent the similarities *quantitatively*. For this reason, dendrograms are generally preferred.

Because of their conceptual simplicity, hierarchical clustering procedures are among the best known of unsupervised methods. The procedures themselves can be divided according to two distinct approaches: agglomerative and divisive. *Agglomerative* (bottom-up, clumping) procedures start with n singleton clusters and form the sequence by successively merging clusters. *Divisive* (top-down, splitting) procedures start with all of the samples in one cluster and form the sequence by successively splitting clusters. The computation needed to go from one level to another is usually simpler for the agglomerative procedures. However, when there are many samples and one is interested in only a small number of clusters, this computation will have to be repeated many times. For simplicity, we shall concentrate on agglomerative procedures, and we shall merely touch on some divisive methods in Section 10.12.

AGGLOMERATIVE

DIVISIVE

10.9.2 Agglomerative Hierarchical Clustering

The major steps in agglomerative clustering are contained in the following procedure, where c is the desired number of final clusters:

■ **Algorithm 4. (Agglomerative Hierarchical Clustering)**

1 **begin initialize** $c, \hat{c} \leftarrow n, \mathcal{D}_i \leftarrow \{\mathbf{x}_i\}, i = 1, \ldots, n$
2   **do** $\hat{c} \leftarrow \hat{c} - 1$
3   find nearest clusters, say, $\mathcal{D}_i$ and $\mathcal{D}_j$
4   merge $\mathcal{D}_i$ and $\mathcal{D}_j$
5   **until** $c = \hat{c}$
6   **return** c clusters
7 **end**

As described, this procedure terminates when the specified number of clusters has been obtained and returns the clusters, described as set of points (rather than as mean or representative vectors). If we continue until $c = 1$ we can produce a dendrogram like that in Fig. 10.11. At any level the "distance" between nearest clusters can provide the dissimilarity value for that level. Note that we have not said how to measure the distance between two clusters, and hence how to find the "nearest" clusters, required by line 3 of Algorithm 4. The considerations here are much like those involved in selecting a general clustering criterion function. For simplicity, we shall generally restrict our attention to the following distance measures:

$$d_{min}(\mathcal{D}_i, \mathcal{D}_j) = \min_{\substack{\mathbf{x} \in \mathcal{D}_i \\ \mathbf{x}' \in \mathcal{D}_j}} \|\mathbf{x} - \mathbf{x}'\| \tag{79}$$

$$d_{max}(\mathcal{D}_i, \mathcal{D}_j) = \max_{\substack{\mathbf{x} \in \mathcal{D}_i \\ \mathbf{x}' \in \mathcal{D}_j}} \|\mathbf{x} - \mathbf{x}'\| \tag{80}$$

$$d_{avg}(\mathcal{D}_i, \mathcal{D}_j) = \frac{1}{n_i n_j} \sum_{\mathbf{x} \in \mathcal{D}_i} \sum_{\mathbf{x}' \in \mathcal{D}_j} \|\mathbf{x} - \mathbf{x}'\| \tag{81}$$

$$d_{mean}(\mathcal{D}_i, \mathcal{D}_j) = \|\mathbf{m}_i - \mathbf{m}_j\|. \tag{82}$$

All of these measures have a minimum-variance flavor, and they usually yield the same results if the clusters are compact and well separated. However, if the clusters are close to one another, or if their shapes are not basically hyperspherical, quite different results can be obtained. Below we shall illustrate some of the differences.

But first let us consider the computational complexity of a particularly simple agglomerative clustering algorithm. Suppose we have n patterns in d-dimensional space, and we seek to form c clusters using $d_{min}(\mathcal{D}_i, \mathcal{D}_j)$ defined in Eq. 79. We will, once and for all, need to calculate $n(n-1)$ interpoint distances—each of which is an $O(d)$ calculation—and place the results in an interpoint distance table. The space complexity is, then, $O(n^2)$. Finding the minimum distance pair (for the first merging) requires that we step through the complete list, keeping the index of the smallest distance. Thus for the first agglomerative step, the complexity is $O(n(n-1) (d+1)) = O(n^2 d)$. For an arbitrary agglomeration step (i.e., from $\hat{c}$ to $\hat{c}-1$), we need merely step through the $n(n-1) - \hat{c}$ "unused" distances in the list and find the smallest for which $\mathbf{x}$ and $\mathbf{x}'$ lie in different clusters. This is, again, $O(n(n-1) - \hat{c})$. The full time complexity is thus $O(cn^2 d)$, and in typical conditions $n \gg c$.*

The Nearest-Neighbor Algorithm

When $d_{min}(\cdot, \cdot)$ is used to measure the distance between clusters (Eq. 79) the algorithm is sometimes called the nearest-neighbor cluster algorithm, or *minimum algorithm* Moreover, if it is terminated when the distance between nearest clusters exceeds an arbitrary threshold, it is called the *single-linkage algorithm*. Suppose that we think of the data points as being nodes of a graph, with edges forming a path between the nodes in the same subset $\mathcal{D}_i$. When $d_{min}(\cdot, \cdot)$ is used to measure the distance between subsets, the nearest-neighbor nodes determine the nearest subsets. The merging of $\mathcal{D}_i$ and $\mathcal{D}_j$ corresponds to adding an edge between the nearest pair of nodes in $\mathcal{D}_i$ and $\mathcal{D}_j$. Because edges linking clusters always go between distinct clusters, the resulting graph never has any closed loops or circuits; in the terminology of graph theory, this procedure generates a *tree*. If it is allowed to continue until all of the subsets are linked, the result is a *spanning tree*—a tree with a path from any

---

*There are methods for sorting or arranging the entries in the interpoint distance table so as to easily avoid inspection of points in the same cluster, but these typically do not improve the complexity results significantly.

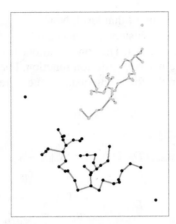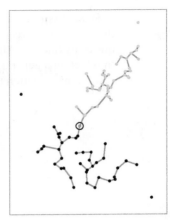

FIGURE 10.13. Two Gaussians were used to generate two-dimensional samples, shown in pink and black. The nearest-neighbor clustering algorithm gives two clusters that well approximate the generating Gaussians (left). If, however, another particular sample is generated (circled red point at the right) and the procedure is restarted, the clusters do not well approximate the Gaussians. This illustrates how the algorithm is sensitive to the details of the samples.

node to any other node. Moreover, it can be shown that the sum of the edge lengths of the resulting tree will not exceed the sum of the edge lengths for any other spanning tree for that set of samples (Problem 39). Thus, with the use of $d_{min}(\cdot, \cdot)$ as the distance measure, the agglomerative clustering procedure becomes an algorithm for generating a *minimal spanning tree*.

Figure 10.13 shows the results of applying this procedure to Gaussian data. In both cases the procedure was stopped, giving two large clusters (plus three singleton outliers); a minimal spanning tree can be obtained by adding the shortest possible edge between the two clusters. In the first case where the clusters are fairly well separated, the obvious clusters are found. In the second case, the presence of a point located so as to produce a bridge between the clusters results in a rather unexpected grouping into one large, elongated cluster and one small, compact cluster. This behavior is often called the "chaining effect" and is sometimes considered to be a defect of this distance measure. To the extent that the results are very sensitive to noise or to slight changes in position of the data points, this is certainly a valid criticism.

The Farthest-Neighbor Algorithm

When $d_{max}(\cdot, \cdot)$ of Eq. 80 is used to measure the distance between subsets, the algorithm is sometimes called the farthest-neighbor clustering algorithm, or *maximum algorithm*. If it is terminated when the distance between nearest clusters exceeds an arbitrary threshold, it is called the *complete-linkage algorithm*. The farthest-neighbor algorithm discourages the growth of elongated clusters. Application of the procedure can be thought of as producing a graph in which edges connect all of the nodes in a cluster. In the terminology of graph theory, every cluster constitutes a *complete* subgraph. The distance between two clusters is determined by the most distant nodes in the two clusters. When the nearest clusters are merged, the graph is changed by adding edges between every pair of nodes in the two clusters.

If we define the diameter of a partition as the largest diameter for clusters in the partition, then each iteration increases the diameter of the partition as little as possi-

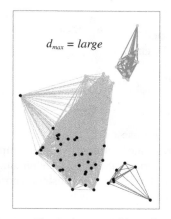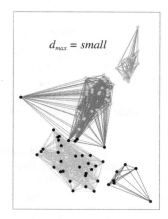

FIGURE 10.14. The farthest-neighbor clustering algorithm uses the separation between the most distant points as a criterion for cluster membership. If this distance is set very large, then all points lie in the same cluster. In the case shown at the left, a fairly large d_{max} leads to three clusters; a smaller d_{max} gives four clusters, as shown at the right.

ble. As Fig. 10.14 illustrates, this is advantageous when the true clusters are compact and roughly equal in size. Nevertheless, when this is not the case—as happens with the two elongated clusters—the resulting groupings can be meaningless. This is another example of imposing structure on data rather than finding structure in it.

Compromises

The minimum and maximum measures represent two extremes in measuring the distance between clusters. Like all procedures that involve minima or maxima, they tend to be overly sensitive to "outliers" or "wildshots." The use of averaging is an obvious way to ameliorate these problems, and $d_{avg}(\cdot, \cdot)$ and $d_{mean}(\cdot, \cdot)$ in Eqs. 81 and 82 are natural compromises between $d_{min}(\cdot, \cdot)$ and $d_{max}(\cdot, \cdot)$. Computationally, $d_{mean}(\cdot, \cdot)$ is the simplest of all of these measures, because the others require computing all $n_i n_j$ pairs of distances $\|\mathbf{x} - \mathbf{x}'\|$. However, a measure such as $d_{avg}(\cdot, \cdot)$ can be used when the distances $\|\mathbf{x} - \mathbf{x}'\|$ are replaced by similarity measures, where the similarity between mean vectors may be difficult or impossible to define.

10.9.3 Stepwise-Optimal Hierarchical Clustering

We observed earlier that if clusters are grown by merging the nearest pair of clusters, then the results have a minimum variance flavor. However, when the measure of distance between clusters is chosen arbitrarily, we can rarely assert that the resulting partition extremizes any particular criterion function. In effect, hierarchical clustering defines a cluster as whatever results from applying the clustering procedure. Nevertheless, with a simple modification it is possible to obtain a stepwise-optimal procedure for extremizing a criterion function. This is done merely by replacing line 3 of the Basic Iterative Agglomerative Clustering Procedure (Algorithm 4) by a more general form to get the following algorithm:

■ **Algorithm 5. (Stepwise Optimal Hierarchical Clustering)**

1 <u>**begin initialize**</u> $c, \hat{c} \leftarrow n, \mathcal{D}_i \leftarrow \{\mathbf{x}_i\}, i = 1, \ldots, n$
2   <u>**do**</u> $\hat{c} \leftarrow \hat{c} - 1$

3	find clusters whose merger changes the criterion the least, say, $\mathcal{D}_i$ and $\mathcal{D}_j$
4	merge $\mathcal{D}_i$ and $\mathcal{D}_j$
5	**until** $c = \hat{c}$
6	**return** c clusters
7	**end**

We saw earlier that the use of $d_{max}(\cdot, \cdot)$ causes the smallest possible stepwise increase in the diameter of the partition. Another simple example is provided by the sum-of-squared-error criterion function J_e. By an analysis very similar to that used in Section 10.8, we find that the pair of clusters whose merger increases J_e as little as possible is the pair for which the "distance"

$$d_e(\mathcal{D}_i, \mathcal{D}_j) = \sqrt{\frac{n_i n_j}{n_i + n_j}} \|\mathbf{m}_i - \mathbf{m}_j\| \tag{83}$$

is minimum (Problem 36). Thus, in selecting clusters to be merged, this criterion takes into account the number of samples in each cluster as well as the distance between clusters. In general, the use of $d_e(\cdot, \cdot)$ tends to favor growth by merging singletons or small clusters with large clusters over merging medium-sized clusters. While the final partition may not minimize J_e, it usually provides a very good starting point for further iterative optimization.

10.9.4 Hierarchical Clustering and Induced Metrics

DISSIMILARITY

Suppose that we are unable to supply a metric for our data, but that we can measure a *dissimilarity* value $\delta(\mathbf{x}, \mathbf{x}')$ for every pair of samples, where $\delta(\mathbf{x}, \mathbf{x}') \geq 0$, with equality holding if and only if $\mathbf{x} = \mathbf{x}'$. Then agglomerative clustering can still be used, with the understanding that the nearest pair of clusters is the least dissimilar pair. Interestingly enough, if we define the dissimilarity between two clusters by

$$\delta_{min}(\mathcal{D}_i, \mathcal{D}_j) = \min_{\substack{\mathbf{x} \in \mathcal{D}_i \\ \mathbf{x}' \in \mathcal{D}_j}} \delta(\mathbf{x}, \mathbf{x}') \tag{84}$$

or

$$\delta_{max}(\mathcal{D}_i, \mathcal{D}_j) = \max_{\substack{\mathbf{x} \in \mathcal{D}_i \\ \mathbf{x}' \in \mathcal{D}_j}} \delta(\mathbf{x}, \mathbf{x}'), \tag{85}$$

then the hierarchical clustering procedure will induce a distance function for the given set of n samples. Furthermore, the ranking of the distances between samples will be invariant to any monotonic transformation of the dissimilarity values (Problem 19).

We can now define a generalized *distance* $d(\mathbf{x}, \mathbf{x}')$ between $\mathbf{x}$ and $\mathbf{x}'$ as the value of the lowest level clustering for which $\mathbf{x}$ and $\mathbf{x}'$ are in the same cluster. To show that this is a legitimate distance function, or *metric*, we need to show four things: For all vectors $\mathbf{x}$, $\mathbf{x}'$, and $\mathbf{x}''$,

METRIC

nonnegativity: $d(\mathbf{x}, \mathbf{x}') \geq 0$

reflexivity: $d(\mathbf{x}, \mathbf{x}') = 0$ if and only if $\mathbf{x} = \mathbf{x}'$

symmetry: $d(\mathbf{x}, \mathbf{x}') = d(\mathbf{x}', \mathbf{x})$

triangle inequality: $d(\mathbf{x}, \mathbf{x}') + d(\mathbf{x}', \mathbf{x}'') \geq d(\mathbf{x}, \mathbf{x}'')$.

It is easy to see that these requirements are satisfied and hence that dissimilarity can induce a metric. For our formula for dissimilarity, we have moreover that

$$d(\mathbf{x}, \mathbf{x}'') \leq \max[d(\mathbf{x}, \mathbf{x}'), d(\mathbf{x}', \mathbf{x}'')] \quad \text{for any } \mathbf{x}' \tag{86}$$

ULTRAMETRIC

in which case we say that $d(\cdot, \cdot)$ is an *ultrametric* (Problem 33). Ultrametric criteria can be more immune to local minima problems since stricter ordering of distances among clusters is maintained.

*10.10 THE PROBLEM OF VALIDITY

With almost all of the procedures considered thus far we have assumed that the number of clusters is known. That is a reasonable assumption if we are upgrading a classifier that has been designed on a small labeled set, or if we are tracking slowly time-varying patterns in an environment with known number of clusters. However, it may be an unjustified assumption if we are exploring a data set whose properties are unknown. Thus, a recurring problem in cluster analysis is that of deciding just how many clusters are present.

When clustering is done by extremizing a criterion function, a common approach is to repeat the clustering procedure for $c = 1$, $c = 2$, $c = 3$, and so on, and to see how the criterion function changes with c. For example, it is clear that the sum-of-squared-error criterion J_e must decrease monotonically with c, because the squared error can be reduced each time c is increased merely by transferring a single sample to a new singleton cluster. If the n samples are really grouped into $\hat{c}$ compact, well-separated clusters, one would expect to see J_e decrease rapidly until $\hat{c} = c$, decreasing much more slowly thereafter until it reaches zero at $c = n$. Similar arguments have been advanced for hierarchical clustering procedures and can be apparent in a dendrogram, the usual assumption being that a large disparity in the levels at which clusters merge indicates the presence of natural groupings.

A more formal approach to this problem is to devise some measure of goodness of fit that expresses how well a given c-cluster description matches the data. The chi-squared and Kolmogorov-Smirnov statistics are the traditional measures of goodness of fit, but the curse of dimensionality usually demands the use of simpler measures, some criterion function, which we denote $J(c)$. Because we expect a description in terms of $c + 1$ clusters to give a better fit than a description in terms of c clusters, we would like to know what constitutes a statistically significant improvement in $J(c)$.

A formal way to proceed is first, to advance the *null hypothesis* that there are exactly c clusters present and, next, to compute the sampling distribution for $J(c+1)$ under this hypothesis. This distribution tells us what kind of apparent improvement to expect when a c-cluster description is actually correct. The decision procedure would be to accept the null hypothesis if the observed value of $J(c + 1)$ falls within limits corresponding to an acceptable probability of false rejection.

Unfortunately, it is usually very difficult to do anything more than crudely estimate the sampling distribution of $J(c + 1)$. The resulting solutions are not above suspicion, and the statistical problem of testing cluster validity is still essentially unsolved. However, under the assumption that a suspicious test is better than none, we include the following approximate analysis for the simple sum-of-squared-error criterion that closely parallels our discussion in Chapter 8.

Suppose that we have a set $\mathcal{D}$ of n samples and we want to decide whether or not there is any justification for assuming that they form more than one cluster. Let us advance the null hypothesis that all n samples come from a normal population with mean $\boldsymbol{\mu}$ and covariance matrix $\sigma^2\mathbf{I}$.* If this hypothesis were true, multiple clusters found would have to have been formed by chance, and any observed decrease in the sum-of-squared error obtained by clustering would have no significance.

The sum of squared error $J_e(1)$ is a random variable, because it depends on the particular set of samples:

$$J_e(1) = \sum_{\mathbf{x}\in\mathcal{D}} \|\mathbf{x} - \mathbf{m}\|^2, \tag{87}$$

where $\mathbf{m}$ is the sample mean of the full data set. Under the null hypothesis, the distribution for $J_e(1)$ is approximately normal with mean $nd\sigma^2$ and variance $2nd\sigma^4$ (Problem 40). Suppose now that we partition the set of samples into two subsets $\mathcal{D}_1$ and $\mathcal{D}_2$ so as to minimize $J_e(2)$, where

$$J_e(2) = \sum_{i=1}^{2} \sum_{\mathbf{x}\in\mathcal{D}_i} \|\mathbf{x} - \mathbf{m}_i\|^2, \tag{88}$$

$\mathbf{m}_i$ being the mean of the samples in $\mathcal{D}_i$. Under the null hypothesis, this partitioning is spurious, but it nevertheless results in a value for $J_e(2)$ that is smaller than $J_e(1)$. If we knew the sampling distribution for $J_e(2)$, we could determine how small $J_e(2)$ would have to be before we were forced to abandon a one-cluster null hypothesis. Lacking an analytical solution for the optimal partitioning, we cannot derive an exact solution for the sampling distribution. However, we can obtain a rough estimate by considering the suboptimal partition provided by a hyperplane through the sample mean. For large n, it can be shown that the sum of squared error for this partition is approximately normal with mean $n(d - 2/\pi)\sigma^2$ and variance $2n(d - 8/\pi^2)\sigma^4$.

This result agrees with our statement that $J_e(2)$ is smaller than $J_e(1)$, because the mean of $J_e(2)$ for the suboptimal partition—$n(d - 2/\pi)\sigma^2$—is less than the mean for $J_e(1)$—$nd\sigma^2$. To be considered significant, the reduction in the sum-of-squared error must certainly be greater than this. We can obtain an approximate critical value for $J_e(2)$ by assuming that the suboptimal partition is nearly optimal, by using the normal approximation for the sampling distribution, and by estimating σ^2 according to

$$\hat{\sigma}^2 = \frac{1}{nd} \sum_{\mathbf{x}\in\mathcal{D}} \|\mathbf{x} - \mathbf{m}\|^2 = \frac{1}{nd} J_e(1). \tag{89}$$

The final result can be stated as follows (Problem 41): Reject the null hypothesis at the p-percent significance level if

$$\frac{J_e(2)}{J_e(1)} < 1 - \frac{2}{\pi d} - \alpha\sqrt{\frac{2(1 - 8/\pi^2 d)}{nd}}, \tag{90}$$

---

*We could of course assume a different cluster form, but in the absence of further information the Gaussian can be justified on the grounds we have discussed before.

where α is determined by

$$p = 100 \int_{\alpha}^{\infty} \frac{1}{\sqrt{2\pi}} e^{-u^2/2} \, du = 50(1 - \text{erf}(\alpha/\sqrt{2})), \tag{91}$$

ERROR
FUNCTION

and $\text{erf}(\cdot)$ is the standard *error function*. This provides us with a test for deciding whether or not the splitting of a cluster is justified. Clearly the c-cluster problem can be treated by applying the same test to all clusters found.

*10.11 ON-LINE CLUSTERING

All of the clustering procedures that we have considered up to this point either explicitly or implicitly attempt to optimize a global criterion function with a known or assumed number of clusters. While this leads to crisp formulations of the problem of unsupervised learning, it does not always produce appropriate or expected results. It frequently leads to cluster structures that are sensitive to small changes in the criterion function, sensitive in ways that may be difficult to appreciate. In particular, when these kinds of procedures are used for on-line learning, it occasionally happens that the cluster structures are not stable, but instead continually wander and drift. Of course, to gain the advantages of being able to learn from new data, a system must be adaptive or exhibit "plasticity," possibly allowing the creation of new clusters, if the data warrant it. On the other hand, if the cluster structures are unstable and the most recently acquired piece of information can cause major reorganization, then it is difficult to ascribe much significance to any particular clustering description. This general problem has been called the *stability/plasticity dilemma*.

STABILITY/
PLASTICITY
DILEMMA

One source of this problem is that with clustering based on a global criterion, every sample can have an influence on the location of a cluster center, regardless of how remote it might be. This observation has led to so-called *competitive learning* procedures in which learning adjustments are confined to the cluster that is most similar to the pattern currently being presented. As a result, the characterizations of previously discovered clusters that are unrelated to the current pattern are not disrupted. Competitive learning developed from neural network research, and we shall use neural network terminology to describe these procedures. We begin with a simple procedure that can be thought of as a modification of a sequential k-means algorithm, for reasons that will become clear.

Competitive learning is related to decision-directed versions of k-means (Algorithm 1) and is based on neural network learning rules (Chapter 6). In both procedures, the number of desired clusters and their centers are initialized, and during clustering each pattern is provisionally classified into one of the clusters. The methods of updating the cluster centers differ, however. In the decision-directed method, each cluster center is calculated as the mean of the current provisional members. In competitive learning, the adjustment is confined to the single cluster center most similar to the pattern presented. As a result, in competitive learning, clusters that are "far away" from the current pattern tend not to be altered (but see Section 10.11.2)—sometimes considered a desirable property. The drawback is that the solution need not minimize a single easily defined global cost or criterion function.

We now turn to the specific competitive learning algorithm. For reasons that will become clear, each d-dimensional pattern is augmented (with $x_0 = 1$) and normalized to have length $\|\mathbf{x}\| = 1$; thus all patterns lie on the surface of a $(d + 1)$-

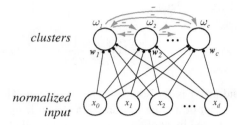

FIGURE 10.15. The two-layer network that implements the competitive learning algorithm consists of $d + 1$ input units and c output or cluster units. Each augmented input pattern is normalized to unit length (i.e., $\|\mathbf{x}\| = 1$), as is the set of weights at each cluster unit. When a pattern is presented, each of the cluster units computes its net activation $net_j = \mathbf{w}_j^t \mathbf{x}$; only the weights at the most active cluster unit are modified. (The suppression of activity in all but the most active cluster units can be implemented by competition among these units, as indicated by the red arrows.) The weights of the most active unit are then modified to be more similar to the pattern presented.

dimensional hypersphere. The competitive learning algorithm can be understood by its neural network implementation (Fig. 10.15), which resembles a Perceptron network (Chapter 5), with input units fully connected to c output or cluster units.

Each of the c cluster centers is initialized with a randomly chosen weight vector, also normalized $\|\mathbf{w}_j\| = 1$, $j = 1, \ldots, c$. It is traditional, but not required, to initialize cluster centers to be c points randomly selected from the data. When a new pattern is presented, each of the cluster units computes its net activation, $net_j = \mathbf{w}_j^t \mathbf{x}$. Only the unit with the largest net (i.e., whose weight vector is closest to the new pattern), is permitted to update its weights. While this selection of the most active unit is algorithmically trivial, it appears to be an additional operation, external to the network. However, if desired, it can be implemented in a winner-take-all network, where each cluster unit j inhibits others by an amount proportional to net_j, as shown by the red arrows in Fig. 10.15. It is this competition between cluster units, along with the resulting suppression of activity in all but the one with the largest net, that gives the algorithm its name.

As mentioned, learning is confined to the weights at the most active unit. The weight vector at this unit is updated to be more like the pattern:

$$\mathbf{w}(t + 1) = \mathbf{w}(t) + \eta \mathbf{x}, \qquad (92)$$

where η is a learning rate. The weights are then normalized to ensure $\sum_{i=0}^{d} w_i^2 = 1$. This normalization guarantees that the net activation $\mathbf{w}^t \mathbf{x}$ depends only on the angle between $\mathbf{w}$ and $\mathbf{x}$, and does not depend on the magnitude of $\mathbf{w}$; without such weight normalization, a single weight, say $\mathbf{w}_j$, could grow in magnitude and forever give the greatest value net_j, and through competition thereby prevent other clusters from learning. Figure 10.16 shows the trajectories of three cluster centers in response to a sequence of patterns chosen randomly from the set shown. If we let k denote a stopping criterion, the algorithm is as follows:

■ **Algorithm 6. (Competitive Learning)**

1 <u>**begin initialize**</u> $\eta, n, c, k, \mathbf{w}_1, \mathbf{w}_2, \ldots, \mathbf{w}_c$
2      $\mathbf{x}_i \leftarrow \{1, \mathbf{x}_i\}, \ i = 1, \ldots, n$ (augment all patterns)

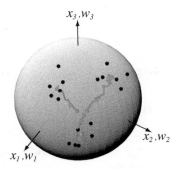

FIGURE 10.16. All of the two-dimensional patterns have been augmented and normalized and hence lie on a two-dimensional sphere in three dimensions. Likewise, the weights of the three cluster centers have been normalized. The red curves show the trajectory of the weight vectors, which start at the red points and end at the center of a cluster.

$$
\begin{array}{ll}
3 & \mathbf{x}_i \leftarrow \mathbf{x}_i / \|\mathbf{x}_i\|, \ i = 1, \dots, n \quad \text{(normalize all patterns)} \\
4 & \underline{\textbf{do}} \text{ randomly select a pattern } \mathbf{x} \\
5 & \quad j \leftarrow \arg\max_{j'} \mathbf{w}_{j'}^t \mathbf{x} \quad \text{(classify } \mathbf{x}) \\
6 & \quad \mathbf{w}_j \leftarrow \mathbf{w}_j + \eta \mathbf{x} \quad \text{(weight update)} \\
7 & \quad \mathbf{w}_j \leftarrow \mathbf{w}_j / \|\mathbf{w}_j\| \quad \text{(weight normalization)} \\
8 & \underline{\textbf{until}} \text{ no significant change in } \mathbf{w} \text{ in } k \text{ attempts} \\
9 & \underline{\textbf{return}} \ \mathbf{w}_1, \mathbf{w}_2, \dots, \mathbf{w}_c \\
10 & \underline{\textbf{end}}
\end{array}
$$

A drawback of Algorithm 6 is that there is no guarantee that it will terminate, even for a finite, nonpathological data set. For instance, the termination condition in line 8 may never be satisfied, and thus the weights may vary forever. A simple heuristic is to let the learning rate in line 6 decrease over time, for instance by letting $\eta(t) = \eta(0)\alpha^t$ for $0 < \alpha < 1$, where t is an iteration number. If the initial cluster centers are representative of the full data set, and the rate of decay is set so that the full data set is presented at least several times before the learning is reduced to very small values, then good results can be expected. However, if a novel pattern is added, it cannot be learned, because η is too small. Likewise, such a learning decay scheme is inappropriate if we seek to track gradual but ongoing changes in the data.

10.11.1 Unknown Number of Clusters

We have mentioned the problem of unknown number c of cluster centers. When c is unknown, we can proceed in one of two general ways. In the first, we solve the problem repeatedly for many different values of c, and compare some criterion for each clustering. If there is a large gap in the criterion values, it suggests a "natural" number of clusters. A second approach is to state a threshold for the creation of a new cluster. This latter approach is particularly useful in on-line cases. The drawback is that it depends more strongly on the order of data presentation.

Whereas clustering algorithms such as k-means and hierarchical clustering typically have all data present before clustering begins (i.e., are off-line), there are occasionally situations in which clustering must be performed on-line as the data streams

FIGURE 10.17. In leader-follower clustering, the number of clusters and their centers depend upon the random sequence of presentations of the points. The three simulations shown employed the same learning rate η, threshold θ, and number of presentations of each point (50), but differ in the random sequence of presentations. Notice that in the simulation on the left, three clusters are generated, whereas only two are generated in the other simulations.

in—for instance, when there is inadequate memory to store all the patterns themselves, or in a time-critical situation where the clusters need to be used even before the full data are present. Our graph-theoretic methods can be performed on-line—one merely links the new pattern to an existing cluster based on some similarity measure.

In order to make on-line versions of methods such as k-means, we will have to be a bit more careful. Under these conditions, the best approach generally is to represent clusters by their "centers" (e.g., means) and update a center based solely on its current value and the incoming pattern.

Suppose we currently have c cluster centers; they may have been placed initially at random positions, or at positions corresponding to the first c patterns presented, or in the current state after any number of patterns have been presented. The simplest approach is to alter only the cluster center most similar to a new pattern being presented, and the cluster center is changed to be somewhat more like the pattern (Fig. 10.17), a generic approach called *leader-follower clustering*.

LEADER-FOLLOWER CLUSTERING

If we let $\mathbf{w}_i$ represent the current center for cluster i, let η represent a learning rate, and let θ represent a threshold, then a Basic leader-follower clustering algorithm is as follows:

■ **Algorithm 7. (Basic Leader-Follower Clustering)**

1 <u>**begin initialize**</u> η, θ
2          $\mathbf{w}_1 \leftarrow \mathbf{x}$
3   <u>**do**</u> accept new $\mathbf{x}$
4              $j \leftarrow \arg\min_{j'} \|\mathbf{x} - \mathbf{w}_{j'}\|$ (find nearest cluster)
5   <u>**if**</u> $\|\mathbf{x} - \mathbf{w}_j\| < \theta$
6   <u>**then**</u> $\mathbf{w}_j \leftarrow \mathbf{w}_j + \eta\mathbf{x}$
7   <u>**else**</u> add new $\mathbf{w} \leftarrow \mathbf{x}$
8                  $\mathbf{w} \leftarrow \mathbf{w}/\|\mathbf{w}\|$ (normalize weight)
9   <u>**until**</u> no more patterns
10   <u>**return**</u> $\mathbf{w}_1, \mathbf{w}_2, \ldots$
11 <u>**end**</u>

Naturally, for a given problem the threshold θ determines implicitly the number of clusters that will be formed. A large threshold leads to a small number of large clusters, while a small threshold leads to a large number of small clusters. In the absence of information about the data, there are no firm guidelines to setting the threshold so that the "proper" number of clusters are found. It should be noted that Algorithm 7 has no provision for *reducing* the number of clusters, by, for instance, merging two clusters that are sufficiently similar.

Before we discuss further properties of such a leader-follower clustering algorithm, let us consider one popular neural approach based on it.

10.11.2 Adaptive Resonance

Leader-follower clustering is at the heart of a general approach to designing self-organizing neural networks that is known as adaptive resonance theory or ART. This theory was developed primarily to model how biological neural networks might be able to recognize unexpected patterns and to remember them for future use. If we think of one of these unexpected patterns as being a new cluster center, then one of the goals of ART is to make sure that even if slightly different new instances of the pattern called for some adjustments to this new cluster center, it would retain its basic characteristics in a stable fashion (Fig. 10.18). Another goal is to show how *expectations* can influence the response of a network. This leads to feedback connections and interesting dynamic behavior.

Adaptive resonance theory can be applied to a number of different network structures. For simplicity, we shall consider only the simple two-layer structure in Fig. 10.19. We shall provide only a general outline of its structure and behavior, suppressing many details that would be needed for a complete working implementation.

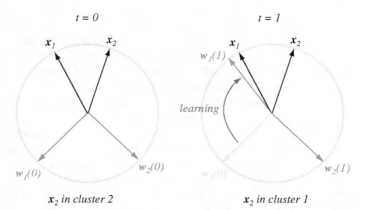

FIGURE 10.18. Instability and recoding can occur during competitive learning, as illustrated in this simple case of two patterns and two cluster centers. Two patterns, x_1 and x_2, are presented to a 2-2 network of Fig. 10.15 represented by two weight vectors. At $t = 0$, w_1 happens to be most aligned with x_1 and hence this pattern belongs to cluster 1; likewise, x_2 is most aligned with w_2 and hence it belongs to cluster 2, as shown at the left. Next, suppose pattern x_1 is presented several times; through the competitive learning weight update rule, w_1 moves to become closer to x_1. Now x_2 is most aligned with w_1, and thus it has changed from class 2 to class 1. Surprisingly, this recoding of x_2 occurs even though x_2 was not used for weight update. It is theoretically possible that such recoding will occur numerous times in response to particular sequences of pattern presentations.

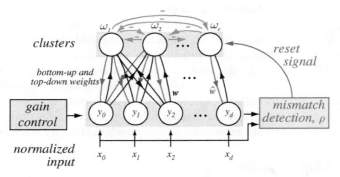

FIGURE 10.19. A generic adaptive resonance network has inputs and cluster units, much like a network for performing competitive learning. However, the input and the category layers are fully interconnected by *both* bottom-up and top-down connections with weights. The bottom-up weights, denoted **w**, learn the cluster centers while the top-down weights, **ŵ**, learn expected input patterns. If a match between input and a learned cluster is poor (where the quality of the match is specified by a user-specified vigilance parameter ρ), then the active cluster unit is suppressed by a reset signal, and a new cluster center can be recruited.

As Fig. 10.19 shows, the network has an input layer, fully interconnected by bottom-up weights to units in a cluster layer, much like a network implementation of competitive learning. In addition, an ART network has top-down weights, denoted **ŵ** in the figure; these will give "priming signals" back to units in the input layer. The units in the bottom layer of the network receive three kinds of inputs: (a) the usual input pattern **x**, (b) feedback signals from the top layer through the top-down weights, and (c) a time-varying bias signal from the so-called gain-control unit. The bottom-up weights $\mathbf{w}_i$ going into the ith output unit are normalized to unit length. As with the basic leader-follower algorithm, the number of clusters can increase over time, so the number c of units in the top layer is variable.

When the ART network is working properly, if the inputs to the bottom layer are close to a familiar pattern, the outputs of the bottom layer are supposed to be a "cleaned-up" version of the input **x**. To be more precise, if the input vector **x** is close to a cluster center $\mathbf{w}_i$, then the output **y** of the bottom layer would ideally be $\mathbf{w}_i$ itself. What is it that drives the bottom-layer responses toward a cluster center? Two things: (a) feedback from the top layer, and (b) a "gain-control" signal. In the absense of either feedback or gain control, the outputs of the bottom layer would merely be a copy of the inputs. The gain-control system is just a mechanism to keep a constant level of activity in the bottom layer, i.e., to hold $\|\mathbf{y}\|$ constant. The feedback signals come from the most active cluster unit, through top-down weights **ŵ**.

The weights going to each unit in the top layer represent the long-term memory of a cluster center. The top layer employs a "winner-take-all" competition, so only the unit i for which $\mathbf{x}^t\mathbf{w}_i$ is large will have a strong reponse. In clustering terms, the output unit that responds most strongly identifies the cluster center $\mathbf{w}_i$ that is nearest to the cleaned-up input **y** emerging from the bottom layer.

Clearly, for all vectors **y** of a given length, the input that will cause output unit i to have the strongest response is proportional to $\mathbf{w}_i$, just as in standard competitive learning. The top-down weights leading from that output unit provide an *expectation* for what the response that it most wants to see from the bottom layer. When an input **x** is first presented, the output **y** of the bottom layer is just the input vector **x**, which may or may not be particularly close to the weight vector $\mathbf{w}_i$ for the most strongly responding output unit. However, the network uses top-down signals to provide *additional input* to the bottom layer through the feedback connections from the top layer. After some time delay and adjustment by the gain control, this has the effect of pushing the responses of the bottom-layer units closer to $\mathbf{w}_i$, which in turn stimulates the top unit more strongly, which in turn stimulates the input, and so on. In ART terminology, this convergence of the feedback sequence is called a "resonance," although the behavior is unrelated to the resonant response of a driven oscillator.

But what might this network actually do? One possibility is that it might actually arrive at a stable state in which (a) the input $\mathbf{x}$ is close to $\mathbf{w}_i$, (b) the output of the bottom-layer units is very close to $\mathbf{w}_i$, and (c) only the ith top-layer unit is responding strongly. That represents the desired result when the input is indeed close to $\mathbf{w}_i$. As with the basic leader-follower algorithm, the value of $\mathbf{w}_i$ can be adjusted slightly to be closer to the new input.

However, another possibility is that the feedback is so powerful that even though the input vector $\mathbf{x}$ is very different from all of the cluster centers, one of the output units grabs control, and the output of the first layer is $\mathbf{w}_i$ anyway. This is not a desirable situation. In the leader-follower algorithm, it corresponds to the situation in which a new cluster center should be introduced. In ART terminology, the fact that there is a large difference between $\mathbf{x}$ and $\mathbf{w}_i$ is detected by the so-called *orienting subsystem*, which allows a new cluster unit to form or be recruited and initialized to $\mathbf{x}$ whenever the input and the activations at the input layer differ such that $\mathbf{x}^t\mathbf{y} < \rho$.

ORIENTING SUBSYSTEM

VIGILANCE

Here ρ is a user-determined parameter called the *vigilance*. The role of the vigilance is analogous to the role of the threshold θ in the leader-follower algorithm. If the vigilance is low, there can be a poor "match" between the input and the closest learned cluster and the network will accept it anyway. However, if the vigilance is high, the network will frequently generate new cluster centers. In ART networks this is implemented by a "reset wave," whose biologically inspired properties need not concern us. For the same input data set, a low vigilance leads to a small number of large coarse clusters, while a high vigilance leads to a large number of fine clusters.

This description of adaptive resonance is incomplete, and needs to be more precisely stated before the behavior of an actual network could be simulated. In many ways, one can view ART as being merely a neural-network implementation of leader-follower clustering. However, the network realization suggests interesting generalizations, such as using multi-layer architectures and allowing higher-level or "cross-modality" expectations to influence the activation of lower-level units, which are explored in the references in the Bibliography.

10.11.3 Learning with a Critic

Most of this text has addressed the problem of supervised learning, where a teacher provides the category labels of each training pattern. This chapter concerns unsupervised where no such label information is available. There is an intermediate case, in which the teacher knows only that the category provided by the classifier is correct or incorrect. Thus, the teacher serves as a *critic* of the system as it learns.

It is simple to incorporate learning with a critic in competitive learning and adaptive resonance networks. For instance, if the tentative cluster assignment for a pattern is judged to be correct by the critic, then standard learning is permitted. If, however, the cluster is judged to be incorrect by the critic, then no learning (weight update) is permitted.

*10.12 GRAPH-THEORETIC METHODS

Where the mathematics of normal mixtures and minimum-variance partitions leads us to picture clusters as isolated clumps, the language and concepts of graph theory lead us to consider much more intricate structures. Unfortunately, there is no uniform way of posing clustering problems as problems in graph theory. Thus, the effective use of these ideas is still largely an art, and the reader who wants to explore the possibilities should be prepared to be creative.

We begin our brief look into graph-theoretic methods by reconsidering the simple procedures that produce the graphs shown in Fig. 10.7. Here a threshold distance d_0 was selected, and two points are placed in the same cluster if the distance between them is less than d_0. This procedure can easily be generalized to apply to arbitrary similarity measures. Suppose that we pick a threshold value s_0 and say that $\mathbf{x}_i$ is

SIMILARITY
MATRIX

similar to $\mathbf{x}_j$ if $s(\mathbf{x}_i, \mathbf{x}_j) > s_0$. This defines an $n \times n$ *similarity matrix* $\mathbf{S} = [s_{ij}]$, with binary element

$$s_{ij} = \begin{cases} 1 & \text{if } s(\mathbf{x}_i, \mathbf{x}_j) > s_0 \\ 0 & \text{otherwise.} \end{cases} \tag{93}$$

SIMILARITY
GRAPH

Furthermore, this matrix induces a *similarity graph*, dual to $\mathbf{S}$, in which nodes correspond to points and an edge joins node i and node j if and only if $s_{ij} = 1$.

The clusterings produced by the single-linkage algorithm and by a modified version of the complete-linkage algorithm are readily described in terms of this graph. With the single-linkage algorithm, two samples $\mathbf{x}$ and $\mathbf{x}'$ are in the same cluster if and only if there exists a chain $\mathbf{x}, \mathbf{x}_1, \mathbf{x}_2, \ldots, \mathbf{x}_k, \mathbf{x}'$ such that $\mathbf{x}$ is similar to $\mathbf{x}_1$, $\mathbf{x}_1$ is similar to $\mathbf{x}_2$, and so on for the whole chain. Thus, this clustering corresponds to

CONNECTED
COMPONENT

the *connected components* of the similarity graph. With the complete-linkage algorithm, all samples in a given cluster must be similar to one another, and no sample can be in more than one cluster. If we drop this second requirement, then this clus-

MAXIMAL
COMPLETE
SUBGRAPH

tering corresponds to the *maximal complete subgraphs* of the similarity graph—the "largest" subgraphs with edges joining all pairs of nodes. (In general, the clusters of the complete-linkage algorithm will be found among the maximal complete subgraphs, but they cannot be determined without knowing the unquantized similarity values.)

We have noted above that the nearest-neighbor algorithm could be viewed as an algorithm for finding a minimal spanning tree. Conversely, given a minimal spanning tree we can find the clusterings produced by the nearest-neighbor algorithm. Removal of the longest edge produces the two-cluster grouping, removal of the next longest edge produces the three-cluster grouping, and so on. This amounts to a divisive hierarchical procedure, and it suggests other ways of dividing the graph into subgraphs. For example, in selecting an edge to remove, we can compare its length to the lengths of other edges incident upon its nodes. Let us say that an edge is *in-*

INCONSISTENT
EDGE

consistent if its length l is significantly larger than $\bar{l}$, the average length of all other edges incident on its nodes. Figure 10.20 shows a minimal spanning tree for a two-dimensional point set and the clusters obtained by systematically removing all edges for which $l > 2\bar{l}$ in this way. This criterion is sensitive to local conditions, giving results that are quite different from merely removing the two longest edges.

When the data points are strung out into long chains, a minimal spanning tree

DIAMETER PATH

forms a natural skeleton for the chain. If we define the *diameter path* as the longest path through the tree, then a chain will be characterized by the shallow depth of the branching off the diameter path. In contrast, for a large, uniform cloud of data

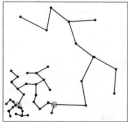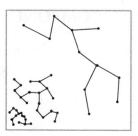

FIGURE 10.20. The removal of inconsistent edges—ones with length significantly larger than the average incident upon a node—may yield natural clusters. The original data are shown at the left, and its minimal spanning tree is shown in the middle. At virtually every node, incident edges are of nearly the same length. Each of the two nodes shown in red are exceptions: their incident edges are of very different lengths. When the two such inconsistent edges are removed, three clusters are produced, as shown at the right.

points, the tree will usually not have an obvious diameter path, but rather several distinct, near-diameter paths. For any of these, an appreciable number of nodes will be off the path. While slight changes in the locations of the data points can cause major rerouting of a minimal spanning tree, they typically have little effect on such statistics.

One of the useful statistics that can be obtained from a minimal spanning tree is the edge length distribution. Figure 10.21 shows a situation in which two dense clusters are embedded in a sparse set of points; the lengths of the edges of the minimal spanning tree exhibit two distinct clusters that would easily be detected by a minimum-variance procedure. By deleting all edges longer than some intermediate value, we can extract the dense cluster as the largest connected component of the remaining graph. While more complicated configurations cannot be disposed of this easily, the flexibility of the graph-theoretic approach suggests that it is applicable to a wide variety of clustering problems.

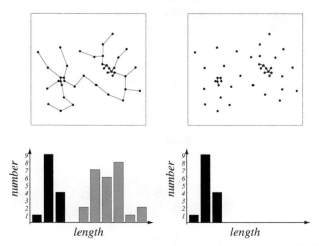

FIGURE 10.21. A minimal spanning tree is shown at the left; its bimodal edge length distribution is evident in the histogram below. If all links of intermediate or high length are removed (red), the two natural clusters are revealed (right).

10.13 COMPONENT ANALYSIS

Component analysis is an unsupervised approach to finding the "right" features from the data. We shall discuss several leading methods, each having a somewhat different goal. We have seen in Chapter 3 how principal component analysis (PCA) projects d-dimensional data onto a lower-dimensional subspace in a way that is optimal in a sum-squared error sense. Nonlinear component analysis (NLCA), typically implemented by neural techniques, is a nonlinear generalization of PCA. In independent component analysis (ICA) we seek those directions in feature space that show the independence of signals. This method is particularly helpful for segmenting signals from multiple sources.

10.13.1 Principal Component Analysis (PCA)

KARHUNEN-LOÉVE TRANSFORM

For completeness, we reiterate the basic approach in principal components or *Karhunen-Loéve transform* presented in Chapter 3. First, the d-dimensional mean vector $\boldsymbol{\mu}$ and $d \times d$ covariance matrix $\boldsymbol{\Sigma}$ are computed for the full data set. Next, the eigenvectors and eigenvalues are computed (cf. Section A.2.7 of the Appendix), and sorted according to decreasing eigenvalue. Call these eigenvectors $\mathbf{e}_1$ with eigenvalue λ_1, $\mathbf{e}_2$ with eigenvalue λ_2, and so on, and choose the k eigenvectors having the largest eigenvalues. Often there will be just a few large eigenvalues, and this implies that k is the inherent dimensionality of the subspace governing the "signal" while the remaining $d - k$ dimensions generally contain noise. Next we form a $d \times k$ matrix $\mathbf{A}$ whose columns consist of the k eigenvectors. The representation of data by principal components consists of projecting the data onto the k-dimensional subspace according to

$$\mathbf{x}' = \mathbf{F}_1(\mathbf{x}) = \mathbf{A}^t(\mathbf{x} - \boldsymbol{\mu}). \tag{94}$$

AUTO-ENCODER

A simple three-layer linear neural network, trained as an auto-encoder can form such a representation, as shown in Fig. 10.22. Each pattern of the data set is presented to *both* the input and output layers and the full network trained by gradient descent on a sum-squared-error criterion, for instance by backpropagation. It can be shown

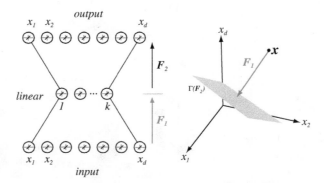

FIGURE 10.22. A three-layer neural network with linear hidden units, trained to be an auto-encoder, develops an internal representation that corresponds to the principal components of the full data set. The transformation $\mathbf{F}_1$ is a linear projection onto a k-dimensional subspace denoted $\Gamma(\mathbf{F}_2)$.

that this representation minimizes a squared error criterion (Problem 44). After the network is trained, the top layer is discarded, and the linear hidden layer provides the principal components.

10.13.2 Nonlinear Component Analysis (NLCA)

Principal component analysis yields a k-dimensional linear subspace of feature space that best represents the full data according to a minimum-square-error criterion. If the data represent complicated interactions of features, then the linear subspace may be a poor representation and nonlinear components may be needed.

A neural network approach to such nonlinear component analysis employs a network with five layers of units, as shown in Fig. 10.23. The middle layer consists of $k < d$ linear units, and it is here that the nonlinear components will be revealed. It is important that the two other internal layers have nonlinear units (Problem 46). The entire network is trained using the techniques of Chapter 6 as an auto-encoder or auto-associator. That is, each d-dimensional pattern is presented as both the input and as the target or desired output. When trained using a sum-squared error criterion, such a network readily learns the auto-encoder problem. The top two layers of the trained network are discarded, and the rest used for nonlinear component analysis. For each input pattern $\mathbf{x}$, the outputs of the k units of the three-layer network correspond to the nonlinear components.

We can understand the function of the full five-layer network in terms of two successive mappings, $\mathbf{F}_1$ followed by $\mathbf{F}_2$. As Fig. 10.23 illustrates, $\mathbf{F}_1$ is a projection from the d-dimensional input onto a k-dimensional nonlinear subspace and $\mathbf{F}_2$ is a mapping from that subspace back to the full d-dimensional space.

There are often multiple local minima in the error surface associated with the five-layer network, and we must take care to set an appropriate number k of units. Recall that in (linear) principal component analysis, the number of components k could be chosen based on the spectrum of eigenvectors. If the eigenvalues are ordered by magnitude, any significant drop between successive values indicates a "natural" number dimension to the subspace. Likewise, suppose five-layer networks are trained, with

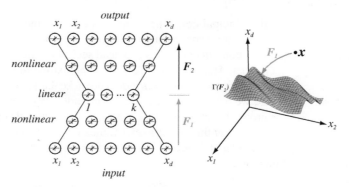

FIGURE 10.23. A five-layer neural network with two layers of nonlinear units (e.g., sigmoidal), trained to be an auto-encoder, develops an internal representation that corresponds to the nonlinear components of the full data set. The process can be viewed in feature space (at the right). The transformation $\mathbf{F}_1$ is a nonlinear projection onto a k-dimensional subspace, denoted $\Gamma(\mathbf{F}_2)$. Points in $\Gamma(\mathbf{F}_2)$ are mapped via $\mathbf{F}_2$ back to the the d-dimensional space of the original data. After training, the top two layers of the net are removed and the remaining three-layer network maps inputs $\mathbf{x}$ to the space $\Gamma(\mathbf{F}_2)$.

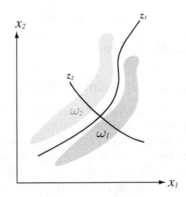

FIGURE 10.24. Features from two classes are as shown, along with nonlinear components of the full data set. Apparently, these classes are well-separated along the line marked z_2, but the large noise gives the largest nonlinear component to be along z_1. Preprocessing by keeping merely the largest nonlinear component would retain the "noise" and discard the "signal," giving poor recognition. The same defect can arise in linear principal components, where the coordinates are linear and orthogonal.

different numbers k of units in the middle layer. Assuming poor local minima have been avoided, the training error will surely decrease for successively larger values of k. If the improvement $k + 1$ over k is small, this may indicate that k is the "natural" dimension of the subspace at the network's middle layer.

We should not conclude that principal component analysis or nonlinear component analysis is always beneficial for classification. If the noise is large compared to the difference between categories, then component analysis will find the directions of the noise, rather than the signal, as illustrated in Fig. 10.24. In such cases, we seek to ignore the noise, and instead we extract the directions that are indicative of the categories—a technique we consider next.

*10.13.3 Independent Component Analysis (ICA)

While principal component analysis and nonlinear component analysis seek directions in feature space that best represent the data in a sum-squared error sense, independent component analysis instead seeks directions that are most *independent* from each other. This goal of ICA can be understood in the domain of *blind source sepa-*
BLIND SOURCE
SEPARATION
ration. Suppose there are d independent scalar source signals $x_i(t)$ for $i = 1, \ldots, d$, where we can consider t to be a time index $1 \leq t \leq T$. For notational convenience we group the d values at an instant into a vector $\mathbf{x}(t)$ and assume for simplicity that averaged over the full data set, the mean of $\mathbf{x}$ vanishes. Because of our assumptions of source independence and the absense of noise, we can write the multivariate density function as

$$p[\mathbf{x}(t)] = \prod_{i=1}^{d} p[x_i(t)]. \tag{95}$$

Suppose that a k-dimensional data (or sensor) vector is observed at each moment,

$$\mathbf{s}(t) = \mathbf{A}\mathbf{x}(t), \tag{96}$$

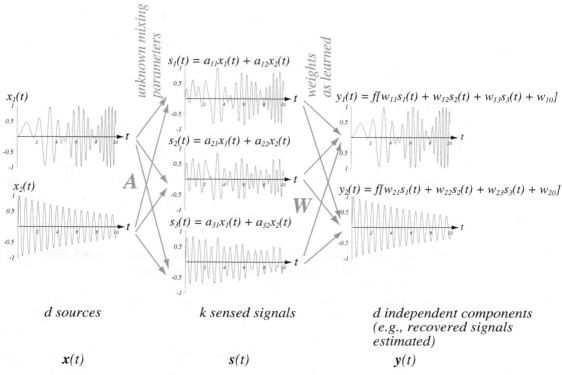

d sources

k sensed signals

d independent components
(e.g., recovered signals
estimated)

$x(t)$

$s(t)$

$y(t)$

FIGURE 10.25. Independent component analysis (ICA) is an unsupervised method that can be applied to the problem of blind source separation. In such problems, two or more source signals (assumed independent) $x_1(t)$, $x_2(t)$, $\cdots$, $x_d(t)$ are mixed linearly to yield sum signals $s_1(t)$, $s_2(t)$, $\cdots$, $s_k(t)$, where $k \geq d$. (This figure illustrates the case $d = 2$ and $k = 3$.) Given merely the sensed signals $\mathbf{x}(t)$ and an assumed number of components, d, the task of ICA is to find independent components in $\mathbf{s}$. In a blind source separation application, these are merely the source signals.

where $\mathbf{A}$ is a $k \times d$ matrix. If $\mathbf{x}$ represents acoustic sources, and $\mathbf{s}$ the signals in k microphones, then the matrix $\mathbf{A}$ depends upon the attenuations due to the source-sensor separations. The goal of ICA is to extract d components in $\mathbf{s}$ that are independent. If the components of $\mathbf{x}$ are independent as described by Eq. 95, then the components found by ICA will allow the discovery of the source signals themselves (Fig. 10.25). (We shall ignore any effects of delays or room echoes which greatly complicate the practical application of this particular example.)

For simplicity, we will treat the case where the number of sensor signals is the same as the number of desired independent components, i.e., we let $k = d$ in the below. (Problem 49 asks you to generalize the following results to the case where $k > d$.) The distribution in the outputs is related to the distribution

$$p_\mathbf{y}(\mathbf{y}) = \frac{p_\mathbf{s}(\mathbf{s})}{|\mathbf{J}|}, \tag{97}$$

where $\mathbf{J}$ is the Jacobean matrix

$$\mathbf{J} = \begin{pmatrix} \dfrac{\partial y_1}{\partial s_1} & \cdots & \dfrac{\partial y_d}{\partial s_1} \\ \vdots & \ddots & \vdots \\ \dfrac{\partial y_1}{\partial s_d} & \cdots & \dfrac{\partial y_d}{\partial s_d} \end{pmatrix} \tag{98}$$

and

$$|\mathbf{J}| = \left| |\mathbf{W}| \prod_{i=1}^{d} \frac{\partial y_i}{\partial s_i} \right|. \tag{99}$$

We model the final stage as a linear transform of the source signals, followed by a static nonlinearity, i.e.,

$$\mathbf{y} = f[\mathbf{Ws} + \mathbf{w}_0] \tag{100}$$

where $\mathbf{w}_0$ is a bias vector and $f[\cdot]$ is typically chosen to be a sigmoid. The central goal in ICA is to find the parameters $\mathbf{W}$ and $\mathbf{w}_0$ so as to make the outputs y_i as independent from one another as possible. The natural measure of such independence—and hence the criterion to be maximized—is the *joint entropy*

JOINT ENTROPY

$$H(\mathbf{y}) = -\mathcal{E}[\ln p_\mathbf{y}(\mathbf{y})] \tag{101}$$
$$= \mathcal{E}[\ln |\mathbf{J}|] - \underbrace{\mathcal{E}[\ln p_\mathbf{s}(\mathbf{s})]}_{\substack{independent \\ of\ weights}},$$

where the expectation is over the the full samples $t = 1, \ldots, T$.

Thus the learning rule for the weight matrix, based on gradient descent of the criterion in Eq 101, is

$$\Delta \mathbf{W} \propto \frac{\partial H(\mathbf{y})}{\partial \mathbf{W}} = \frac{\partial}{\partial \mathbf{W}} \ln |\mathbf{J}| = \frac{\partial}{\partial \mathbf{W}} \ln |\mathbf{W}| + \frac{\partial}{\partial \mathbf{W}} \ln \prod_{i=1}^{d} \left| \frac{\partial y_i}{\partial s_i} \right|, \tag{102}$$

where we used Eq. 99. In component form, the first term on the right-hand side of Eq. 102 is

$$\frac{\partial}{\partial W_{ij}} \ln |\mathbf{W}| = \frac{\text{cof}[W_{ij}]}{|\mathbf{W}|} \tag{103}$$

where the cofactor of W_{ij}, that is, $(-1)^{i+j}$ times the determinant of $(d-1)$-by-$(k-1)$-dimensional matrix obtained by deleting the ith row and jth column of $\mathbf{W}$ (cf. Section A.2.6 of the Appendix). Thus we have

$$\frac{\partial}{\partial \mathbf{W}} \ln |\mathbf{W}| = [\mathbf{W}^t]^{-1} \tag{104}$$

Assuming a sigmoidal nonlinearity, Eq. 103 then gives our weight update rule for the matrix $\mathbf{W}$

$$\Delta \mathbf{W} \propto [\mathbf{W}^t]^{-1} + (\mathbf{1} - 2\mathbf{y})\mathbf{s}_g^t, \tag{105}$$

where $\mathbf{1}$ is a d-component vector of 1s.

Problem 48 asks you to make the same assumptions as above, and follow similar arguments, and show that the learning rule for the bias weights is

$$\Delta \mathbf{w}_0 \propto \mathbf{1} - 2\mathbf{y}. \tag{106}$$

Equations 105 and 106 give the learning rule for ICA. It is often difficult to know ahead of time the proper number of components to seek. If ICA is being used in a pattern recognition application with known number of categories, then in the absence of other information d should be set to equal to the number of categories. If, however, this number is fairly large, then ICA may be overly sensitive to numerical simulation and give unreliable components.

Generally speaking, when used as preprocessing for *classification*, ICA has several characteristics that make it more desirable than linear or nonlinear PCA. As we saw in Fig. 10.24, such principal components need not be effective in separating classes. If the different sources arise from different models, then we can expect them to be independent, and ICA should be able to extract them.

10.14 LOW-DIMENSIONAL REPRESENTATIONS AND MULTIDIMENSIONAL SCALING (MDS)

Part of the problem of deciding whether or not a given clustering means anything stems from our inability to visualize the structure of multidimensional data. This problem is further aggravated when similarity or dissimilarity measures are used that lack the familiar properties of distance. One way to attack this problem is to try to represent the data points as points in some lower-dimensional space in such a way that the distances between points in that space correspond to the dissimilarities between points in the original space. If acceptably accurate representations can be found in two or perhaps three dimensions, this can be an extremely valuable way to gain insight into the structure of the data. The general process of finding a configuration of points whose interpoint distances correspond to similarities or dissimilarities is often called *multidimensional scaling*.

Let us begin with the simpler case where it is meaningful to talk about the distances between the n samples $\mathbf{x}_1, \ldots, \mathbf{x}_n$. Let $\mathbf{y}_i$ be the lower-dimensional *image* of $\mathbf{x}_i$, let δ_{ij} be the distance between $\mathbf{x}_i$ and $\mathbf{x}_j$, and let d_{ij} be the distance between $\mathbf{y}_i$ and $\mathbf{y}_j$ (Fig. 10.26). Then we are looking for a *configuration* of image points $\mathbf{y}_1, \ldots, \mathbf{y}_n$ for which the $n(n-1)/2$ distances d_{ij} between image points are as close as possible to the corresponding original distances δ_{ij}. Because it will usually not be possible to find a configuration for which $d_{ij} = \delta_{ij}$ for all i and j, we need some criterion for deciding whether or not one configuration is better than another. The following sum-of-squared-error functions are all reasonable candidates:

$$J_{ee} = \frac{\sum_{i<j}(d_{ij} - \delta_{ij})^2}{\sum_{i<j}\delta_{ij}^2} \tag{107}$$

$$J_{ff} = \sum_{i<j}\left(\frac{d_{ij} - \delta_{ij}}{\delta_{ij}}\right)^2 \tag{108}$$

$$J_{ef} = \frac{1}{\sum_{i<j}\delta_{ij}} \sum_{i<j}\frac{(d_{ij} - \delta_{ij})^2}{\delta_{ij}}. \tag{109}$$

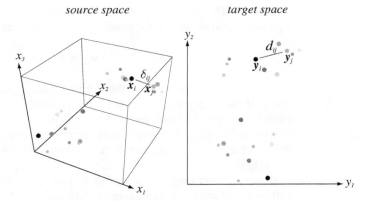

FIGURE 10.26. The figure shows an example of points in a three-dimensional space being mapped to a two-dimensional space. The size and color of each point $\mathbf{x}_i$ matches that of its image, $\mathbf{y}_i$. Here we use simple Euclidean distance, that is, $\delta_{ij} = \|\mathbf{x}_i - \mathbf{x}_j\|$ and $d_{ij} = \|\mathbf{y}_i - \mathbf{y}_j\|$. In typical applications, the source space usually has high dimensionality, but to allow easy visualization the target space is only two- or three-dimensional.

Because these criterion functions involve only the distances between points, they are invariant to rigid-body motions of the configurations. Moreover, they have all been normalized so that their minimum values are invariant to dilations of the sample points. While J_{ee} emphasizes large errors (regardless of whether the distances δ_{ij} are large or small), J_{ff} emphasizes the large fractional errors (regardless of whether the errors $|d_{ij} - \delta_{ij}|$ are large or small). A useful compromise is J_{ef}, which emphasizes large products of error and fractional error.

Once a criterion function has been selected, an optimal configuration $\mathbf{y}_1, \ldots, \mathbf{y}_n$ is defined as one that minimizes that criterion function. Such a configuration can be sought by a standard gradient-descent procedure, starting with some initial configuration and changing the $\mathbf{y}_i$'s in the direction of greatest rate of decrease in the criterion function. Because the distance in the lower-dimensional space is given by $d_{ij} = \|\mathbf{y}_i - \mathbf{y}_j\|$, the gradient of d_{ij} with respect to $\mathbf{y}_i$ is merely a unit vector in the direction of $\mathbf{y}_i - \mathbf{y}_j$. Thus, the gradients of the criterion functions are easy to compute:

$$\nabla_{\mathbf{y}_k} J_{ee} = \frac{2}{\sum_{i<j} \delta_{ij}^2} \sum_{j \neq k} (d_{kj} - \delta_{kj}) \frac{\mathbf{y}_k - \mathbf{y}_j}{d_{kj}}$$

$$\nabla_{\mathbf{y}_k} J_{ff} = 2 \sum_{j \neq k} \frac{d_{kj} - \delta_{kj}}{\delta_{kj}^2} \frac{\mathbf{y}_k - \mathbf{y}_j}{d_{kj}}$$

$$\nabla_{\mathbf{y}_k} J_{ef} = \frac{2}{\sum_{i<j} \delta_{ij}} \sum_{j \neq k} \frac{d_{kj} - \delta_{kj}}{\delta_{kj}} \frac{\mathbf{y}_k - \mathbf{y}_j}{d_{kj}}.$$

The starting configuration can be chosen randomly, or in any convenient way that spreads the image points about. If the image points lie in a $\hat{d}$-dimensional space, then a simple and effective starting configuration can be found by selecting those $\hat{d}$ coordinates of the samples that have the largest variance.

The following example illustrates the kind of results that can be obtained by these techniques. The data consist of 30 points spaced at unit intervals along a spiral in three dimensions:

$$x_1(k) = \cos(k/\sqrt{2})$$
$$x_2(k) = \sin(k/\sqrt{2})$$
$$x_3(k) = k/\sqrt{2}, \qquad k = 0, 1, \ldots, 29.$$

Figure 10.27 shows the three-dimensional data. When the J_{ef} criterion was used, 20 iterations of a gradient descent procedure produced the two-dimensional configuration shown at the right. Of course, translations, rotations, and reflections of this configuration would be equally good solutions.

In nonmetric multidimensional scaling problems, the quantities δ_{ij} are dissimilarities whose numerical values are not as important as their rank order. An ideal configuration would be one for which the rank order of the distances d_{ij} is the same as the rank order of the dissimilarities δ_{ij}. Let us order the $m = n(n-1)/2$ dissimilarities so that $\delta_{i_1 j_1} \leq \cdots \leq \delta_{i_m j_m}$, and let $\hat{d}_{ij}$ be *any* m numbers satisfying the *monotonicity constraint*

MONOTONICITY CONSTRAINT

$$\hat{d}_{i_1 j_1} \leq \hat{d}_{i_2 j_2} \leq \cdots \leq \hat{d}_{i_m j_m}. \tag{110}$$

In general, the distances d_{ij} will not satisfy this constraint, and the numbers $\hat{d}_{ij}$ will not be distances. However, the degree to which the d_{ij} satisfy this constraint is measured by

$$\hat{J}_{mon} = \min_{\hat{d}_{ij}} \sum_{i<j} (d_{ij} - \hat{d}_{ij})^2, \tag{111}$$

where it is always to be understood that the $\hat{d}_{ij}$ must satisfy the monotonicity constraint. Thus, $\hat{J}_{mon}$ measures the degree to which the configuration of points $\mathbf{y}_1, \ldots, \mathbf{y}_n$ represents the original dissimilarities. Unfortunately, $\hat{J}_{mon}$ cannot be used to define an optimal configuration because it can be made to vanish by collapsing

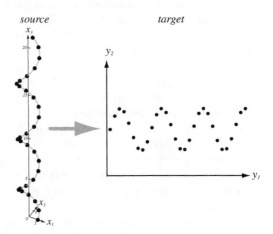

FIGURE 10.27. Thirty points of the form $\mathbf{x} = (\cos(k/\sqrt{2}), \sin(k/\sqrt{2}), k/\sqrt{2})^t$ for $k = 0, 1, \ldots, 29$ are shown at the left. Multidimensional scaling using the J_{ef} criterion (Eq. 109) and a two-dimensional target space leads to the image points shown at the right. This lower-dimensional representation shows clearly the fundamental sequential nature of the points in the original source space.

the configuration to a single point. However, this defect is easily removed by a normalization such as the following:

$$J_{mon} = \frac{\hat{J}_{mon}}{\sum_{i<j} d_{ij}^2}.$$

(112)

Thus, J_{mon} is invariant to translations, rotations, and dilations of the configuration, and an optimal configuration can be defined as one that minimizes this criterion function. It has been observed experimentally that when the number of points is larger than dimensionality of the image space, the monotonicity constraint is actually quite confining. This might be expected from the fact that the number of constraints grows as the square of the number of points, and it is the basis for the frequently encountered statement that this procedure allows the recovery of metric information from nonmetric data. The quality of the representation generally improves as the dimensionality of the image space is increased, and it may be necessary to go beyond three dimensions to obtain an acceptably small value of J_{mon}. However, this may be a small price to pay to allow the use of the many clustering procedures available for data points in metric spaces.

10.14.1 Self-Organizing Feature Maps

A method closely related to multidimensional scaling is that of self-organizing feature maps, sometimes called topologically ordered maps or *Kohonen self-organizing feature maps*. As before, the goal is to represent all points in the source space by points in a target space, such that distance and proximity relationships are preserved as much as possible. The self-organizing map algorithm we shall discuss does not require the storage of a large number of samples, and thus it has much lower space complexity than multidimensional scaling. (In practice, both methods have high time complexities.) Moreover, the method is particularly useful when there is a nonlinear mapping inherent in the problem itself, as we shall see.

KOHONEN MAPS

It is simplest to explain self-organizing maps by means of an example. Suppose we seek to learn a mapping from a circular disk region (the source space) to a linear target space, as shown in Fig. 10.28. The source space is sensed by a movable two-joint arm of fixed segment lengths; thus each point (x_1, x_2) in the disk area leads to a pair of angles (ϕ_1, ϕ_2), which we denote as a two-component vector $\boldsymbol{\phi}$. The algorithm uses a sequence of $\boldsymbol{\phi}$ values—but not the (x_1, x_2) values themselves, because they and their nonlinear transformation are not directly accessible. In our illustration the nonlinearity involves inverse trigonometric functions, but in most applications it is more complicated and not even known.

The task is this: Given a sequence of $\boldsymbol{\phi}$'s (corresponding to points sampled in the source space), create a mapping from $\boldsymbol{\phi}$ to $\mathbf{y}$ such that points neighboring in the source space are mapped to points that are neighboring in the target space. It is this goal of preserving neighborhoods that gives the resulting "topologically ordered maps" their name.

The mapping is learned by a simple two-layer neural network, here with two inputs (ϕ_1 and ϕ_2), fully connected to a large number of outputs, corresponding to points along the target line. When a pattern $\boldsymbol{\phi}$ is presented, each node in the target space computes its net activation, $net_k = \boldsymbol{\phi}^t \mathbf{w}_k$. One of the units is most activated; call it $\mathbf{y}^*$. The weights to this unit and those in its immediate neighborhood are updated according to:

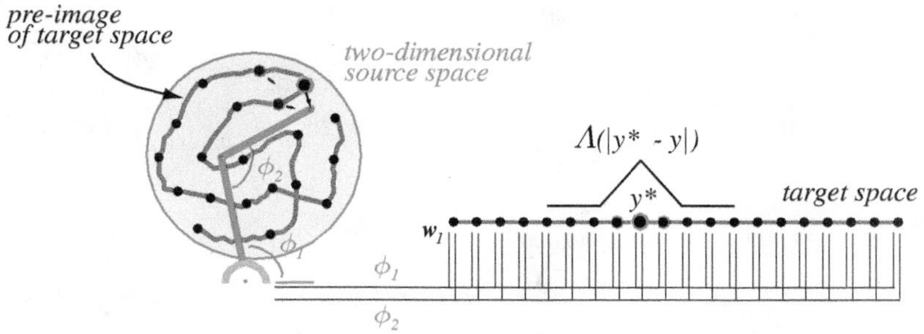

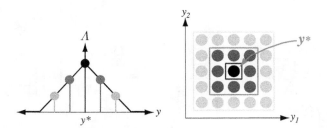

FIGURE 10.28. A self-organizing map from the (two-dimensional) disk source space to the (one-dimensional) line of the target space can be learned as follows. For each point y in the target line, there exists a corresponding point in the source space that, if sensed, would lead to y being most active. For clarity, then, we can link theses points in the source; it is as if the image line is placed in the source space. We call this the pre-image of the target space. At the state shown, the particular sensed point leads to y^* begin most active. The learning rule (Eq. 113) makes its source point move toward the sensed point, as shown by the small arrow. Because of the window function $\Lambda(|y^*-y|)$, the pre-image of points adjacent to y^* are also moved toward the sensed point, thought not as much. If such learning is repeated many times as the arm randomly senses the whole source space, a topologically correct map is learned.

$$w_{ki}(t+1) = w_{ki}(t) + \eta(t)\Lambda(|\mathbf{y} - \mathbf{y}^*|)(\boldsymbol{\phi}_i - w_{ki}(t)), \tag{113}$$

where $\eta(t)$ is a learning rate which depends upon the iteration number t. The function $\Lambda(|\mathbf{y} - \mathbf{y}^*|)$ is called the "window function" and has value 1.0 for $\mathbf{y} = \mathbf{y}^*$ and smaller for large values of $|\mathbf{y} - \mathbf{y}^*|$. The window function is vital to the success of the algorithm: It ensures that neighboring points in the target space have weights that are similar, and thus correspond to neighboring points in the source space, thereby ensuring topological neighborhoods (Fig. 10.29). Next, every weight vector is normalized such that $\|\mathbf{w}\| = 1$. (Naturally, only those weight vectors that have been altered during the learning trial need be renormalized.) The learning rate $\eta(t)$ decreases slowly as a function of iteration number (i.e., as patterns are presented) to ensure that learning will ultimately stop.

 Equation 113 has a particularly straightforward interpretation. For each pattern presentation, the "winning" unit in the target space is adjusted so that it is more like the particular pattern. Others in the neighborhood of $\mathbf{y}^*$ are also adjusted so that their

WINDOW FUNCTION

FIGURE 10.29. Typical window functions for self-organizing maps for target spaces in one dimension (left) and two dimensions (right). In each case, the weights at the maximally active unit, $\mathbf{y}^*$, in the target space get the largest weight update while units more distant get smaller update.

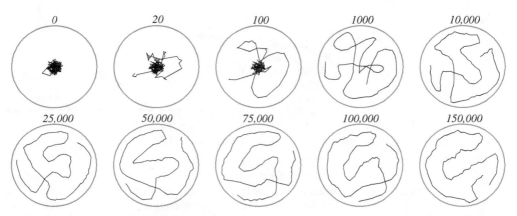

FIGURE 10.30. If a large number of pattern presentations are made using the setup of Fig. 10.28, a topologically ordered map develops. The number of pattern presentations is listed.

weights more nearly match that of the input pattern (though not quite as much as for y^*, as given by the window function). In this way, neighboring points in the input space lead to neighboring points being active.

After a large number of pattern presentations, learning according to Eq. 113 ensures that neighboring points in the source space lead to neighboring points in the target space. Informally speaking, it is as if the target space line has been placed on the source space; learning pulls and stretches the line to fill the source space, as illustrated in Fig. 10.30, which shows the development of the map. After 150,000 training presentations, a topological map has been learned.

The learning of such self-organizing maps is very general and can be applied to virtually any source space, target space, and continuous nonlinear mapping. Figure 10.31 shows the development of a self-organizing map from a square source space to a square (grid) target space.

There are generally inherent ambiguities in the maps learned by this algorithm. For instance, a mapping from a square to a square could have eight possible orientations, corresponding to the four rotation and two flip symmetries. Such ambiguity

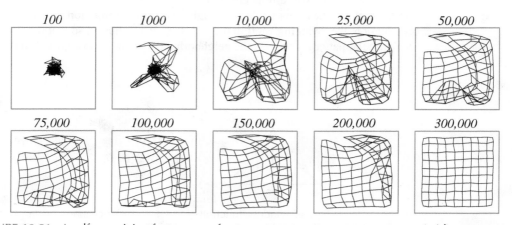

FIGURE 10.31. A self-organizing feature map from a square source space to a square (grid) target space. As in Fig. 10.28, each grid point of the target space is shown atop the point in the source space that maximally excites that target point.

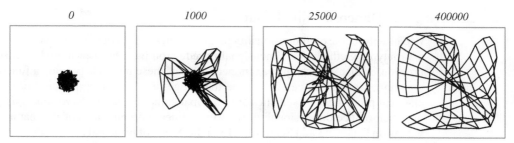

0   *1000*   *25000*   *400000*

FIGURE 10.32. Some initial (random) weights and the particular sequence of patterns (randomly chosen) lead to kinks in the map; even extensive further training does not eliminate the kink. In such cases, learning should be restarted with randomized weights and possibly a wider window function and slower decay in learning.

is generally irrelevant for subsequent clustering or classification in the target space. Nevertheless, the mapping ambiguities are related to a more significant drawback— the possibility of "kinks" in the map. A particular initial condition can lead to *part* of the map learning one of the orientations, while a different part learns another one, as shown in Fig. 10.32. When this occurs, it is generally best to reinitialize the weights randomly and restart the learning with perhaps a wider window function or slower decay in the learning rate.

One of the benefits of this learning algorithm is that it naturally takes account of the probability of sampling in the source space, that is, $p(\mathbf{x})$. Regions of high such probability attract more of the points in the target space, and this means that more of the image points will be mapped there, as shown in Fig. 10.33.

Such self-organizing feature maps can be used in a number of systems. For instance, in a signal-processing application we can use the outputs of a bank of filters to map a waveform to a two-dimensional target space. When such an approach is applied to spoken vowel sounds, similar utterances such as /ee/ and /eh/ will be close together, while others, such as /ee/ and /oo/ will be far apart—just as we had in multidimensional scaling. Subsequent supervised learning can label regions in this target space and thus lead to a full classifier, but one formed using only a small amount of supervised training.

0   *1000*   *400,000*   *800,000*

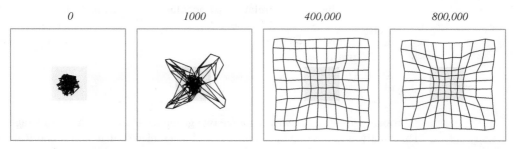

FIGURE 10.33. As in Fig. 10.31 except that the sampling of the input space was not uniform. In particular, the probability density for sampling a point in the central square region (pink) was 20 times greater than elsewhere. Notice that the final map devotes more nodes to this central region than in Fig. 10.31.

10.14.2 Clustering and Dimensionality Reduction

Because the curse of dimensionality plagues so many pattern recognition procedures, a variety of methods for dimensionality reduction have been proposed. Unlike the procedures that we have just examined, most of these methods provide a functional mapping, so that one can determine the image of an arbitrary feature vector. The classical procedures of statistics are principal components and *factor analysis*, both of which reduce dimensionality by forming linear combinations of the features. As we saw in Chapter 3 and Section 10.13.1, the object of principal components analysis is to find a lower-dimensional representation that accounts for the *variance* of the features. The object of factor analysis is to find a lower-dimensional representation that accounts for the *correlations* among the features. If we think of the problem as one of removing or combining (i.e., grouping) highly correlated features, then it becomes clear that the techniques of clustering are applicable to this problem. In terms of the *data matrix*, whose n rows are the d-dimensional samples, ordinary clustering can be thought of as a grouping of the rows, with a smaller number of *cluster centers* being used to represent the data, whereas dimensionality reduction can be thought of as a grouping of the columns, with *combined features* being used to represent the data.

FACTOR
ANALYSIS

DATA MATRIX

Let us consider a simple modification of hierarchical clustering to reduce dimensionality. In place of an n-by-n matrix of distances between samples, we consider a d-by-d *correlation matrix* $\mathbf{R} = [\rho_{ij}]$, where the correlation coefficient ρ_{ij} is related to the covariances (or sample covariances) by

CORRELATION
MATRIX

$$\rho_{ij} = \frac{\sigma_{ij}}{\sqrt{\sigma_{ii}\sigma_{jj}}}. \tag{114}$$

Because $0 \leq \rho_{ij}^2 \leq 1$, with $\rho_{ij}^2 = 0$ for uncorrelated features and $\rho_{ij}^2 = 1$ for completely correlated features, ρ_{ij}^2 plays the role of a similarity function for features. Two features for which ρ_{ij}^2 is large are clearly good candidates to be merged into one feature, thereby reducing the dimensionality by one. Repetition of this process leads to the following hierarchical procedure:

■ **Algorithm 8.   (Hierarchical Dimensionality Reduction)**

1 **begin initialize** d', $\mathcal{D}_i \leftarrow \{\mathbf{x}_i\}$, $i = 1, \ldots, d$
2 $\quad\quad \hat{d} \leftarrow d + 1$
3 $\quad\quad$ **do** $\hat{d} \leftarrow \hat{d} - 1$
4 $\quad\quad\quad\quad$ compute $\mathbf{R}$ by Eq. 114
5 $\quad\quad\quad\quad$ find most correlated distinct clusters, say $\mathcal{D}_i$ and $\mathcal{D}_j$
6 $\quad\quad\quad\quad$ $\mathcal{D}_i \leftarrow \mathcal{D}_i \cup \mathcal{D}_j$ merge
7 $\quad\quad\quad\quad$ delete $\mathcal{D}_j$
8 $\quad\quad\quad$ **until** $\hat{d} = d'$
9 $\quad$ **return** d' clusters
10 **end**

Probably the simplest way to merge two groups of features is just to average them. (This tacitly assumes that the features have been scaled so that their numerical ranges are comparable.) With this definition of a new feature, there is no problem in defining the correlation matrix for groups of features. It is not hard to think of variations on this general theme, but we shall not pursue this topic further.

For the purposes of pattern *classification*, the most serious criticism of all of the approaches to dimensionality reduction that we have mentioned is that they are overly concerned with faithful *representation* of the data. Greatest emphasis is usually placed on those features or groups of features that have the greatest variability. But for classification, we are interested in *discrimination*—not representation. While it is a truism that the ideal representation is the one that makes classification easy, it is not always so clear that clustering without explicitly incorporating classification criteria will find such a representation. Of course, even if an algorithm finds clear, isolated clusters, there is no guarantee that these will be useful for classification; after all, each isolated cluster might consist of points from the different categories highly interspersed. Roughly speaking, the most interesting features are the ones for which the difference in the class means is large relative to the standard deviations, not the ones for which merely the standard deviations are large. In short, for classification we are interested in something more like the method of multiple discriminant analysis described in Chapter 3.

There is a large body of theory on methods of dimensionality reduction for pattern classification. Some of these methods seek to form new features out of linear combinations of old ones. Others seek merely a smaller subset of the original features. A major problem confronting this theory is that the division of pattern recognition into feature extraction followed by classification is theoretically artificial. A completely optimal feature extractor can never be anything but an optimal classifier. It is only when constraints are placed on the classifier or limitations are placed on the size of the set of samples that one can formulate nontrivial problems. Various ways of circumventing this problem that may be useful under the proper circumstances can be found in the literature. When it is possible to exploit knowledge of the problem domain to obtain more informative features, that is usually the most profitable course of action.

SUMMARY

Unsupervised learning and clustering seek to extract information from unlabeled samples. If the underlying distribution comes from a mixture of component densities described by a set of unknown parameters θ, then θ can be estimated by Bayesian or maximum-likelihood methods. A more general approach is to define some measure of similarity between two clusters, as well as a global criterion such as a sum-squared-error or trace of a scatter matrix. Because there are only occasionally analytic methods for computing the clustering which optimizes the criterion, a number of greedy (*locally* stepwise optimal) iterative algorithms can be used, such as k-means and fuzzy k-means clustering.

If we seek to reveal structure in the data at many levels—that is, clusters with subclusters and subsubclusters—then hierarchical methods are needed. Agglomerative or bottom-up methods start with each sample as a singleton cluster and iteratively merge clusters that are "most similar" according to some chosen similarity or distance measure. Conversely, divisive or top-down methods start with a single cluster representing the full data set and iteratively split them into smaller clusters, each time seeking the subclusters that are most dissimilar. The resulting hierarchical structure is revealed in a dendrogram. A large disparity in the similarity measure for successive cluster levels in a dendrogram usually indicates the "natural" number of clusters. Alternatively, the problem of cluster validity—knowing the proper number of clusters—can also be addressed by hypothesis testing. In that case the null hypothesis is that there are some number c of clusters; we then determine if the reduction of the cluster criterion due to an additional cluster is statistically significant.

Competitive learning is a neural network clustering algorithm in which the cluster center most similar to an input pattern is modified to become more like that pattern. In order to guarantee that learning stops, the learning rate must decay. Competitive learning can be modified to allow for the creation of new cluster centers if no center is sufficiently similar to a particular input pattern, as in leader-follower clustering and adaptive resonance. While these methods have many advantages, such as computational ease and ability to track gradual variations in the data, they rarely optimize an easily specified global criterion such as sum-of-squared error. Unsupervised learning and clustering algorithms are often sensitive to a number of critical user-selected parameters.

Graph-theoretic methods in clustering treat the data as points, to be linked based on a number of heuristics and distance measures. The clusters produced by these methods can exhibit chaining or other intricate structures, but they rarely optimize an easily specified global cost function. Graph methods are, moreover, generally more sensitive to details of the data.

Component analysis seeks directions or axes in feature space that provide an improved, lower-dimensional representation for the full data space. In principal component analysis, which is a linear process, such directions are merely the eigenvectors of the covariance matrix having the largest eigenvalues; this projection optimizes a sum-squared-error criterion. Nonlinear component analysis, for instance as learned in an internal layer an auto-encoder neural network, yields curved surfaces embedded in the full d-dimensional feature space, onto which an arbitrary pattern $\mathbf{x}$ is projected. The goal in independent component analysis—which is developed through gradient descent in an entropy criterion—is to determine the directions in feature space that are statistically most independent. Such directions may reveal the true sources (if they are indeed independent) and can be used for segmentation and blind source separation.

Self-organizing feature maps and multidimensional scaling are two general methods for dimensionality reduction. Self-organizing feature maps can be highly nonlinear, and represents points close in the source space by points close in the lower-dimensional target space. In preserving neighborhoods in this way, such maps are also said to be "topologically correct." The source and target spaces can be of very general shapes, and the mapping will depend upon the the distribution of samples within the source space. Multidimensional scaling similarly learns a correspondence between points that preserves neighborhoods, and is often used for data visualization. Because the basic method requires all the inter-point distances for minimizing a global criterion function, its space complexity limits the usefulness of multidimensional scaling to problems of moderate size.

BIBLIOGRAPHICAL AND HISTORICAL REMARKS

Historically, the literature on unsupervised learning and clustering can be traced back to Karl Pearson, who in 1894 used sample moments to determine the parameters in a mixture of two univariate Gaussians. While most books on pattern classification address unsupervised learning, there are several modern books and review articles on unsupervised learning that go into great detail, such as references [1] and [23]. Much of the mathematical analysis of unsupervised methods comes from the signal compression community, where vector quantization (VQ) seeks to represent an arbitrary vector by one of c "prototype" vectors, which correspond to our cluster centers, as described in reference [18].

A clear book on mixture models is reference [34] and the issue of identifiability in unsupervised learning is covered in reference [46]. In reference [20], Hasselblad showed how the parameters of one-dimensional normals could be learned in an unsupervised environment. The k-means algorithm was introduced in a paper by Lloyd [32], which inspired many variations, such as the use of the Mahalanobis distance [33] or use of "fuzzy" measures [5, 6]. Efficient agglomerative methods for hierarchical clustering are summarized in reference [12]. Cladistics, the classificatory foundation of biology (from the Greek *klados*, branch), provides useful background for the use of classification in all scientific fields [27]. A good introduction to methods of hierarchical clustering is reference [26].

The key mathematical concepts underlying principal component analysis appear in references [9, 24, 13] and [30]. Independent component analysis was introduced by Jutten and Herault [25], and the maximum-likelihood approach was introduced by Gaeta and Lacoume [16]. Generalizations and a maximum-likelihood approach are given in references [39] and [40]. Bell and Sejnowski [4] showed a neural network implementation of independent component analysis, and in reference [36] explained the very close relationship between ICA and information maximization. A good compendium on ICA is reference [48]; several studies have shown the benefits of the technique for classification, such as reference [15]. Multidimensional scaling, occassionally called nonlinear projection in distinction to linear projection, is discussed in references [8] and [43] and its relationship to clustering is explored in reference [31].

Kohonen's long series of papers on self-organizing feature maps began in the early 1980s [28], and good compendia can be found in references [29] and [42]; convergence properties of algorithms for self-organizing feature maps are proved in reference [49]. A good comparison of self-organizing maps with other methods such as principal component analysis and discriminant analysis is reference [33]. There have been numerous applications of the method, from speech to finding patterns of poverty in the world.

The main emphasis of research on Adaptive Resonance has been to explore pattern recognition and clustering in biological systems, as discussed in reference [10]. A wonderfully clear exposition of the central algorithmic ideas is reference [35]; an attempt to translate the ideas and terminology of adaptive resonance to standard engineering terminology, complete with a glossary, is reference [45].

PROBLEMS

Section 10.2

1. Suppose that x can assume the values $0, 1, \ldots, m$ and that $P(x|\boldsymbol{\theta})$ is a mixture of c binomial distributions

$$P(x|\boldsymbol{\theta}) = \sum_{j=1}^{c} \binom{m}{x} \theta_j^m (1 - \theta_j)^{m-x} P(\omega_j),$$

where $\boldsymbol{\theta}$ is a vector of length c representing the parameters in the distributions.

   (a) Assuming that the prior probabilities $P(\omega_j)$ are known, explain why this mixture is not identifiable if $m < c$.

(b) Under these conditions, is the mixture *completely* unidentifiable?

(c) How do your answers above change if the prior probabilities are also unknown?

2. Consider a mixture distribution of two triangle distributions, where component density ω_i is centered on μ_i and has "halfwidth" w_i, according to

$$p(x|\omega_i) \sim T(\mu_i, w_i) = \begin{cases} (w_i - |x - \mu_i|)/w_i^2 & \text{for } |x - \mu_i| < w_i \\ 0 & \text{otherwise.} \end{cases}$$

(a) Assume $P(\omega_1) = P(\omega_2) = 0.5$ and derive the equations for the maximum-likelihood values $\hat{\mu}_i$ and $\hat{w}_i$, $i = 1, 2$.

(b) Under the conditions in part (a), is the distribution identifiable?

(c) Assume that both widths w_i are known, but the centers are not. Assume, too, that there exist values for the centers that give nonzero probability to each of the samples. Derive a formula for the maximum-likelihood value of the centers.

(d) Under the conditions in part (c), is the distribution identifiable?

3. Suppose there is a one-dimensional mixture density consisting of two Gaussian components, each centered on the origin:

$$p(x|\boldsymbol{\theta}) = P(\omega_1)\frac{1}{\sqrt{2\pi}\sigma_1}e^{-x^2/(2\sigma_1^2)} + (1 - P(\omega_1))\frac{1}{\sqrt{2\pi}\sigma_2}e^{-x^2/(2\sigma_2^2)},$$

and $\boldsymbol{\theta} = (P(\omega_1), \sigma_1, \sigma_2)^t$ describes the parameters.

(a) Show that under these conditions this density is completely unidentifiable.

(b) Suppose the value $P(\omega_1)$ is fixed and known. Is the model identifiable?

(c) Suppose σ_1 and σ_2 are known, but $P(\omega_1)$ is unknown. Is this resulting model identifiable? That is, can $P(\omega_1)$ be identified using data?

Section 10.3

4. Let $\mathbf{x}$ be a d-component binary vector $(0,1)$ and let $P(\mathbf{x}|\boldsymbol{\theta})$ be a mixture of c multivariate Bernoulli distributions,

$$P(\mathbf{x}|\boldsymbol{\theta}) = \sum_{i=1}^{c} P(\mathbf{x}|\omega_i, \boldsymbol{\theta}_i)P(\omega_i)$$

where

$$P(\mathbf{x}|\omega_i, \boldsymbol{\theta}_i) = \prod_{j=1}^{d} \theta_{ij}^{x_j}(1 - \theta_{ij})^{1-x_j}.$$

(a) Derive the formula for the partial derivative:

$$\frac{\partial \ln P(\mathbf{x}|\omega_i, \boldsymbol{\theta}_i)}{\partial \theta_{ij}} = \frac{x_i - \theta_{ij}}{\theta_{ij}(1 - \theta_{ij})}.$$

(b) Using the general equations for maximum-likelihood estimates, show that the maximum-likelihood estimate $\hat{\boldsymbol{\theta}}_i$ for $\boldsymbol{\theta}_i$ must satisfy

$$\hat{\boldsymbol{\theta}}_i = \frac{\sum_{k=1}^n \hat{P}(\omega_i|\mathbf{x}_k, \hat{\boldsymbol{\theta}})\mathbf{x}_k}{\sum_{k=1}^n \hat{P}(\omega_i|\mathbf{x}_k, \hat{\boldsymbol{\theta}})}.$$

(c) Interpret your answer to part (b) in words.

5. Let $p(\mathbf{x}|\boldsymbol{\theta})$ be a c-component normal mixture with $p(\mathbf{x}|\omega_i, \boldsymbol{\theta}_i) \sim N(\boldsymbol{\mu}_i, \sigma_i^2\mathbf{I})$. Using the results of Section 10.3, show that the maximum-likelihood estimate for σ_i^2 must satisfy

$$\hat{\sigma}_i^2 = \frac{1/d \sum_{k=1}^n \hat{P}(\omega_i|\mathbf{x}_k, \hat{\boldsymbol{\theta}}_i)\|\mathbf{x}_k - \hat{\boldsymbol{\mu}}_i\|^2}{\sum_{k=1}^m \hat{P}(\omega_i|\mathbf{x}_k, \hat{\boldsymbol{\theta}}_i)},$$

where $\hat{\boldsymbol{\mu}}_i$ and $\hat{P}(\omega_i|\mathbf{x}_k, \hat{\boldsymbol{\theta}}_i)$ are given by Eqs. 25 and 27, respectively.

6. Consider a c-component mixture model with unknown parameter vector $\boldsymbol{\theta}$ and unknown prior probabilities $P(\omega_i)$. Let $\hat{P}(\omega_i)$ be the maximum-likelihood estimate for $P(\omega_i)$, and let $\hat{\boldsymbol{\theta}}_i$ be the maximum-likelihood estimate for $\boldsymbol{\theta}_i$. Show that if the likelihood function is differentiable and if $\hat{P}(\omega_i) \neq 0$ for any i, then $\hat{P}(\omega_i)$ and $\hat{\boldsymbol{\theta}}_i$ must satisfy Eqs. 11 and 12, that is,

$$\hat{P}(\omega_i) = \frac{1}{n}\sum_{k=1}^n \hat{P}(\omega_i|\mathbf{x}_k, \hat{\boldsymbol{\theta}})$$

and

$$\sum_{k=1}^n \hat{P}(\omega_i|\mathbf{x}_k, \hat{\boldsymbol{\theta}})\boldsymbol{\nabla}_{\boldsymbol{\theta}_i} \ln p(\mathbf{x}_k|\omega_i, \hat{\boldsymbol{\theta}}_i) = 0,$$

where

$$\hat{P}(\omega_i|\mathbf{x}_k, \hat{\boldsymbol{\theta}}) = \frac{p(\mathbf{x}_k|\omega_i, \hat{\boldsymbol{\theta}}_i)\hat{P}(\omega_i)}{\sum_{j=1}^c p(\mathbf{x}_k|\omega_j, \hat{\boldsymbol{\theta}}_j)\hat{P}(\omega_j)}.$$

7. The derivation of the equations for maximum-likelihood estimation of parameters of a mixture density was made under the assumption that the parameters in each component density are functionally independent. Suppose instead that

$$p(\mathbf{x}|\alpha) = \sum_{j=1}^c p(\mathbf{x}|\omega_j, \alpha)P(\omega_j),$$

where α is a parameter that appears in *several* (and possibly all) of the component densities. Let l be the n-sample log-likelihood function. Show that the derivative of l with respect to that parameter is

$$\frac{\partial l}{\partial \alpha} = \sum_{k=1}^n \sum_{j=1}^c P(\omega_j|\mathbf{x}_k, \alpha)\frac{\partial \ln p(\mathbf{x}_k|\omega_j, \alpha)}{\partial \alpha},$$

where

$$P(\omega_j|\mathbf{x}_k, \alpha) = \frac{p(\mathbf{x}_k|\omega_j, \alpha)P(\omega_j)}{p(\mathbf{x}_k|\alpha)}.$$

8. Let θ_1 and θ_2 be unknown parameters for the component densities $p(x|\omega_1, \theta_1)$ and $p(x|\omega_2, \theta_2)$, respectively. Assume that θ_1 and θ_2 are initially statistically independent, so that $p(\theta_1, \theta_2) = p_1(\theta_1)p_2(\theta_2)$.

(a) Show that after one sample x_1 from the mixture density is observed, $p(\theta_1, \theta_2|x_1)$ can no longer be factored as $p_1(\theta_1|x_1)p_2(\theta_2|x_1)$ if

$$\frac{\partial p(x|\omega_i, \theta_i)}{\partial \theta_i} \neq 0, \qquad i = 1, 2.$$

(b) What does this imply in general about the statistical dependence of parameters in unsupervised learning?

9. Assume that a mixture density $p(\mathbf{x}|\boldsymbol{\theta})$ is identifiable. Prove that under very general conditions, $p(\boldsymbol{\theta}|\mathcal{D}^n)$ converges (in probability) to a Dirac delta function centered at the true value of $\boldsymbol{\theta}$ as the number of samples becomes very large.

10. Assume that the likelihood function of Eq. 3 is differentiable and derive the maximum-likelihood conditions of Eqs. 11–13.

Section 10.4

11. Let $p(\mathbf{x}|\omega_i, \boldsymbol{\theta}_i) \sim N(\boldsymbol{\mu}_i, \boldsymbol{\Sigma})$, where $\boldsymbol{\Sigma}$ is a common covariance matrix for the c component densities. Let σ_{pq} be the pqth element of $\boldsymbol{\Sigma}$, let σ^{pq} be the pqth element of $\boldsymbol{\Sigma}^{-1}$, let $x_p(k)$ be the pth element of $\mathbf{x}_k$, and let $\mu_p(i)$ be the pth element of $\boldsymbol{\mu}_i$.

(a) Show that

$$\frac{\partial \ln p(\mathbf{x}_k|\omega_i, \boldsymbol{\theta}_i)}{\partial \sigma^{pq}} = \left(1 - \frac{\delta_{pq}}{2}\right)\left[\sigma_{pq} - (x_p(k) - \mu_p(i))(x_q(k) - \mu_q(i))\right],$$

where

$$\delta_{pq} = \begin{cases} 1 & \text{if } p = q \\ 0 & \text{if } p \neq q \end{cases}$$

is the Kroneker delta symbol.

(b) Use this result and the results of Problem 7 to show that the maximum-likelihood estimate for $\boldsymbol{\Sigma}$ must satisfy

$$\hat{\boldsymbol{\Sigma}} = \frac{1}{n}\sum_{k=1}^{n}\mathbf{x}_k\mathbf{x}_k^t - \sum_{i=1}^{c}\hat{P}(\omega_i)\hat{\boldsymbol{\mu}}_i\hat{\boldsymbol{\mu}}_i^t,$$

where $\hat{P}(\omega_i)$ and $\hat{\boldsymbol{\mu}}_i$ are the maximum-likelihood estimates given by Eqs. 24 and 25.

12. Show that the maximum-likelihood estimate of a prior probability can be zero by considering the following special case. Let $p(x|\omega_1) \sim N(0, 1)$ and $p(x|\omega_2) \sim N(0, 1/2)$, so that $P(\omega_1)$ is the only unknown parameter in the mixture

$$p(x) = \frac{P(\omega_1)}{\sqrt{2\pi}}e^{-x^2/2} + \frac{(1 - P(\omega_1))}{\sqrt{\pi}}e^{-x^2}.$$

(a) Show that the maximum-likelihood estimate $\hat{P}(\omega_1)$ of $P(\omega_1)$ is zero if one sample x_1 is observed and if $x_1^2 < \ln 2$.

(b) What is the value of $\hat{P}(\omega_1)$ if $x_1^2 > \ln 2$?

(c) Summarize and interpret your answer in words.

13. Consider the univariate normal mixture

$$p(x|\mu_1, \ldots, \mu_c) = \sum_{j=1}^{c} \frac{P(\omega_j)}{\sqrt{2\pi}\sigma} \exp\left[-\frac{1}{2}\left(\frac{x - \mu_j}{\sigma}\right)^2\right]$$

in which all of the c components have the same known variance σ^2. Suppose that the means are so far apart compared to σ that for any observed x all but one of the terms in this sum are negligible. Use a heuristic argument to show that the value of

$$\max_{\mu_1, \ldots, \mu_c} \left\{\frac{1}{n} \ln p(x_1, \ldots, x_n|\mu_1, \ldots, \mu_c)\right\}$$

ought to be approximately

$$\sum_{j=1}^{c} P(\omega_j) \ln P(\omega_j) - \frac{1}{2} \ln[2\pi\sigma e]$$

when the number n of independently drawn samples is large. (Here e is the base of the natural logarithms.)

14. Let $\mathbf{x}_1, \ldots, \mathbf{x}_n$ be n d-dimensional samples and $\mathbf{\Sigma}$ be any nonsingular d-by-d matrix. Show that the vector $\mathbf{x}$ that minimizes

$$\sum_{k=1}^{m} (\mathbf{x}_k - \mathbf{x})^t \mathbf{\Sigma}^{-1} (\mathbf{x}_k - \mathbf{x})$$

is the sample mean, $\bar{\mathbf{x}} = 1/n \sum_{k=1}^{n} \mathbf{x}_k$.

15. Perform the differentiation in Eq. 31 to derive Eqs. 32 and 33.

16. Show that the computational complexity of Algorithm 1 is $O(ndcT)$, where n is the number of d-dimensional patterns, c is the assumed number of clusters, and T is the number of iterations.

17. Fill in the steps of the derivation of Eqs. 24–26. Be sure to state any assumptions you invoke.

Section 10.5

18. Consider the combinatorics of exhaustive inspection of clusters of n samples into c clusters.

(a) Show that there are exactly

$$\frac{1}{c!} \sum_{i=1}^{c} \binom{c}{i} (-1)^{c-i} i^n$$

such distinct clusterings.

(b) How many clusters are there for $n = 100$ and $c = 5$?

(c) Find an approximation for your answer to (a) for the case $n \gg c$. Use your answer to estimate the number of clusterings of 1000 points into 10 clusters.

Section 10.6

19. Prove that the ranking of distances between samples discussed in Section 10.6 is invariant to any monotonic transformation of the dissimilarity values. Do this as follows:

(a) Define the *value* v_k for the clustering at level k, and for level 1 let $v_1 = 0$. For all higher levels, v_k is the minimum dissimilarity between pairs of distinct clusters at level $k - 1$. Explain why with both δ_{min} and δ_{max} the value v_k either stays the same or increases as k increases.

(b) Assume that no two of the n samples are identical, so that $v_2 > 0$. Use this to prove monotonicity—that is, that $0 = v_1 \leq v_2 \leq v_3 \leq \cdots \leq v_n$.

Section 10.7

20. Derive Eq. 55 from Eq. 54 using the definition given in Eq. 56.

21. If a set of n samples $\mathcal{D}$ is partitioned into c disjoint subsets $\mathcal{D}_1, \ldots, \mathcal{D}_c$, the sample mean $\mathbf{m}_i$ for samples in $\mathcal{D}_i$ is undefined if $\mathcal{D}_i$ is empty. In such a case, the sum-of-squared errors involves only the nonempty subsets:

$$J_e = \sum_{\mathcal{D}_i \neq \emptyset} \sum_{\mathbf{x} \in \mathcal{D}_i} \|\mathbf{x} - \mathbf{m}_i\|^2.$$

Assuming that $n \geq c$, show there are no empty subsets in a partition that minimizes J_e. Explain your answer in words.

22. Consider a set of $n = 2k + 1$ samples, k of which coincide at $x = -2$, k at $x = 0$, and one at $x = a > 0$.

(a) Show that the two-cluster partitioning that minimizes J_e groups the k samples at $x = 0$ with the one at $x = a$ if $a^2 < 2(k + 1)$.

(b) What is the optimal grouping if $a^2 > 2(k + 1)$?

23. Let $\mathbf{x}_1 = \binom{4}{5}$, $\mathbf{x}_2 = \binom{1}{4}$, $\mathbf{x}_3 = \binom{0}{1}$, and $\mathbf{x}_4 = \binom{5}{0}$, and consider the following three partitions:

1. $\mathcal{D}_1 = \{\mathbf{x}_1, \mathbf{x}_2\}$, $\mathcal{D}_2 = \{\mathbf{x}_3, \mathbf{x}_4\}$
2. $\mathcal{D}_1 = \{\mathbf{x}_1, \mathbf{x}_4\}$, $\mathcal{D}_2 = \{\mathbf{x}_2, \mathbf{x}_3\}$
3. $\mathcal{D}_1 = \{\mathbf{x}_1, \mathbf{x}_2, \mathbf{x}_3\}$, $\mathcal{D}_2 = \{\mathbf{x}_4\}$

Show that by the sum-of-square error J_e criterion (Eq. 54), the third partition is favored, whereas by the invariant J_d (Eq. 68) criterion the first two partitions are favored.

24. Let $\mathbf{x}_1 = \binom{0}{0}$, $\mathbf{x}_2 = \binom{1}{1}$, $\mathbf{x}_3 = \binom{1}{0}$, and $\mathbf{x}_4 = \binom{2}{0.5}$, and consider the following three partitions:

1. $\mathcal{D}_1 = \{\mathbf{x}_1, \mathbf{x}_2\}$, $\mathcal{D}_2 = \{\mathbf{x}_3, \mathbf{x}_4\}$
2. $\mathcal{D}_1 = \{\mathbf{x}_1, \mathbf{x}_4\}$, $\mathcal{D}_2 = \{\mathbf{x}_2, \mathbf{x}_3\}$
3. $\mathcal{D}_1 = \{\mathbf{x}_1, \mathbf{x}_2, \mathbf{x}_3\}$, $\mathcal{D}_2 = \{\mathbf{x}_4\}$

(a) Find the clustering that minimizes the sum-of-squared error criterion, J_e (Eq. 54).

(b) Find the clustering that minimizes the determinant criterion, J_d (Eq. 68).

25. Consider the problem of invariance to transformation of the feature space.

(a) Show that the eigenvalues $\lambda_1, \ldots, \lambda_d$ of $\mathbf{S}_W^{-1}\mathbf{S}_B$ are invariant to nonsingular linear transformations of the data.

(b) Show that the eigenvalues $\nu_1, \ldots, \nu_d$ of $\mathbf{S}_T^{-1}\mathbf{S}_W$ are related to those of $\mathbf{S}_W^{-1}\mathbf{S}_B$ by $\nu_i = 1/(1 + \lambda_i)$.

(c) Use your above results to show that $J_d = |\mathbf{S}_W|/|\mathbf{S}_T|$ is invariant to nonsingular linear transformations of the data.

26. Recall the definitions of the within-cluster and the between-cluster scatter matrices (Eqs. 62 and 63). Define the total scatter matrix to be $\mathbf{S}_T = \mathbf{S}_W + \mathbf{S}_B$. Show that the following measures (Eqs. 70 and 71) are invariant to linear transformations of the data.

(a) $\mathrm{tr}[\mathbf{S}_T^{-1}\mathbf{S}_W] = \displaystyle\sum_{i=1}^{d} \frac{1}{1 + \lambda_i}$

(b) $|\mathbf{S}_W|/|\mathbf{S}_T| = \displaystyle\prod_{i=1}^{d} \frac{1}{1 + \lambda_i}$

(c) $|\mathbf{S}_W^{-1}\mathbf{S}_B| = \displaystyle\prod_{i=1}^{d} \lambda_i$

(d) What is the typical value of the criterion in (c)? Why, therefore, is that criterion not very useful?

27. Show that the clustering criterion J_d in Eq. 68 is invariant to linear transformations of the space as follows. Let $\mathbf{T}$ be a nonsingular matrix and consider the change of variables $\mathbf{x}' = \mathbf{Tx}$.

(a) Write the new mean vectors $\mathbf{m}'_i$ and scatter matrices $\mathbf{S}'_i$ in terms of the old values and $\mathbf{T}$.

(b) Calculate J'_d in terms of the (old) J_d and show that they differ solely by an overall scalar factor.

(c) Because this factor is the same for all partitions, argue that J_d and J'_d rank the partitions in the same way, and hence that the optimal clustering based on J_d is invariant to nonsingular linear transformations of the data.

28. The eigenvalues $\lambda_1, \ldots, \lambda_d$ of $\mathbf{S}_W^{-1}\mathbf{S}_B$ are the basic linear invariants of the scatter matrices. Show that the eigenvalues are indeed invariant under nonsingular linear transformations of the data.

29. Consider the problems that might arise when using the determinant criterion for clustering.

(a) Show that the rank of the within-cluster scatter matrix $\mathbf{S}_i$ cannot exceed $n_i - 1$, and thus the rank of $\mathbf{S}_W$ cannot exceed $\sum_{i=1}^{c}(n_i - 1) = n - c$.

(b) Use your answer to explain why the between-cluster scatter matrix $\mathbf{S}_B$ may become singular. (Of course, if the samples are confined to a lower-dimensional subspace, it is possible to have $\mathbf{S}_W$ be singular even though $n - c \geq d$.)

Section 10.8

30. One way to generalize the basic-minimum-squared-error procedure is to define the criterion function

$$J_T = \sum_{i=1}^{c} \sum_{\mathbf{x} \in \mathcal{D}_i} (\mathbf{x} - \mathbf{m}_i)^t \mathbf{S}_T^{-1} (\mathbf{x} - \mathbf{m}_i),$$

where $\mathbf{m}_i$ is the mean of the n_i samples in $\mathcal{D}_i$ and $\mathbf{S}_T$ is the total scatter matrix.

(a) Show that J_T is invariant to nonsingular linear transformations of the data.

(b) Show that the transfer of a sample $\hat{\mathbf{x}}$ from $\mathcal{D}_i$ to $\mathcal{D}_j$ causes J_T to change to

$$J_T^* = J_T + \left[\frac{n_j}{n_j + 1} (\hat{\mathbf{x}} - \mathbf{m}_j)^t \mathbf{S}_T^{-1} (\hat{\mathbf{x}} - \mathbf{m}_j) - \frac{n_i}{n_i - 1} (\hat{\mathbf{x}} - \mathbf{m}_i)^t \mathbf{S}_T^{-1} (\hat{\mathbf{x}} - \mathbf{m}_i) \right].$$

(c) Using this result, write pseudocode for an iterative procedure for minimizing J_T.

31. Consider how the transfer of a single point from one cluster to another affects the mean and sum-squared error, and thereby derive Eqs. 76 and 77.

Section 10.9

32. Let a similarity measure be defined as $s(\mathbf{x}, \mathbf{x}') = \mathbf{x}^t \mathbf{x}' / (\|\mathbf{x}\| \|\mathbf{x}'\|)$.

(a) Interpret this similarity measure if the d features have binary values, where $x_i = 1$ if $\mathbf{x}$ possesses the ith feature and $x_i = -1$ if it does not.

(b) Show that for this case the squared Euclidean distance satisfies

$$\|\mathbf{x} - \mathbf{x}'\|^2 = 2d(1 - s(\mathbf{x}, \mathbf{x}')).$$

33. Let patterns $\mathbf{x}$ and $\mathbf{x}'$ be arbitrary points in a d-dimensional space, and let q be a scalar parameter ($q > 1$). For each of the measures shown, state whether or not it represents a metric, and whether or not it represents an ultrametric.

(a) $s(\mathbf{x}, \mathbf{x}') = \|\mathbf{x} - \mathbf{x}'\|^2$ (squared Euclidean)

(b) $s(\mathbf{x}, \mathbf{x}') = \|\mathbf{x} - \mathbf{x}'\|$ (Euclidean)

(c) $s(\mathbf{x}, \mathbf{x}') = \left(\sum_{k=1}^{d} |x_k - x_k'|^q \right)^{1/q}$ (Minkowski)

(d) $s(\mathbf{x}, \mathbf{x}') = \mathbf{x}^t \mathbf{x}' / (\|\mathbf{x}\| \|\mathbf{x}'\|)$ (cosine)

(e) $s(\mathbf{x}, \mathbf{x}') = \mathbf{x}^t \mathbf{x}'$ (dot product)

34. Let cluster $\mathcal{D}_i$ contain n_i samples, and let d_{ij} be some measure of the distance between two clusters $\mathcal{D}_i$ and $\mathcal{D}_j$. In general, one might expect that if $\mathcal{D}_i$ and $\mathcal{D}_j$ are merged to form a new cluster $\mathcal{D}_k$, then the distance from $\mathcal{D}_k$ to some other cluster $\mathcal{D}_h$ is not simply related to d_{hi} and d_{hj}. However, consider the equation

$$d_{hk} = \alpha d_{hi} + \alpha_i d_{hj} + \beta d_{ij} + \gamma |d_{hi} - d_{hj}|.$$

Show that the following choices for the coefficients α_i, α_j, β, and γ lead to the distance functions indicated.

(a) d_{min} : $\alpha_i = \alpha_j = 0.5$, $\beta = 0$, $\gamma = -0.5$.

(b) d_{max} : $\alpha_i = \alpha_j = 0.5$, $\beta = 0$, $\gamma = +0.5$.

(c) $d_{avg}: \alpha_i = \dfrac{n_i}{n_i + n_j}, \alpha_j = \dfrac{n_j}{n_i + n_j}, \beta = \gamma = 0.$

(d) $d_{mean}^2: \alpha_i = \dfrac{n_i}{n_i + n_j}, \alpha_j = \dfrac{n_j}{n_i + n_j}, \beta = -\alpha_i \alpha_j, \gamma = 0.$

35. Consider a hierarchical clustering procedure in which clusters are merged so as to produce the smallest increase in the sum-of-squared error at each step. If the ith cluster contains n_i samples with sample mean $\mathbf{m}_i$, show that the smallest increase results from merging the pair of clusters for which

$$\frac{n_i n_j}{n_i + n_j} \|\mathbf{m}_i - \mathbf{m}_j\|^2$$

is minimum.

36. Assume we are clustering using the sum-of-squared error criterion J_e (Eq. 54). Show that a "distance" measure between clusters can be derived, Eq. 83, such that merging the "closest" such clusters increases J_e as little as possible.

37. Create by hand a dendrogram for the following eight points in one dimension: $\{-5.5, -4.1, -3.0, -2.6, 10.1, 11.9, 12.3, 13.6\}$. Define the similarity between two clusters to be $20 - d_{min}(\mathcal{D}_i, \mathcal{D}_j)$, where $d_{min}(\mathcal{D}_i, \mathcal{D}_j)$ is given in Eq. 79. Based on your dendrogram, argue that two is the natural number of clusters.

38. Create by hand a dendrogram for the following 10 points in one dimension: $\{-2.2, -2.0, -0.3, 0.1, 0.2, 0.4, 1.6, 1.7, 1.9, 2.0\}$. Define the similarity between to clusters to be $20 - d_{min}(\mathcal{D}_i, \mathcal{D}_j)$, where $d_{min}(\mathcal{D}_i, \mathcal{D}_j)$ is given in Eq. 79. Based on your dendrogram, argue that three is the natural number of clusters.

39. Assume that the nearest-neighbor cluster algorithm has been allowed to continue fully, thereby giving a tree with a path from any node to any other node. Show that the sum of the edge lengths of this resulting tree will not exceed the sum of the edge lengths for any other spanning tree for that set of samples.

Section 10.10

40. Assume that a large number n of d-dimensional samples has been chosen from a multidimensional Gaussian—that is, $p(\mathbf{x}) \sim N(\mathbf{m}, \boldsymbol{\Sigma})$, where $\boldsymbol{\Sigma}$ is an arbitrary positive-definite covariance matrix.

   (a) Prove that the distribution of the criterion function $J_e(1)$ given in Eq. 87 is normal with mean $nd\sigma^2$. Express σ in terms of $\boldsymbol{\Sigma}$.

   (b) Prove that the variance of this distribution is $2nd\sigma^4$.

   (c) Consider a suboptimal partition of the Gaussian by a hyperplane through the sample mean. Show that for large n, the sum of squared error for this partition is approximately normal with mean $n(d - 2/\pi)\sigma^2$ and variance $2n(d - 8/\pi^2)\sigma^4$, where σ is given in part (a).

41. Derive Eqs. 90 and 91 used in rejecting the null hypothesis associated with determining the number of clusters for a data set. Be sure to state any conditions you need to impose.

Section 10.11

42. Consider a simple greedy algorithm for creating a spanning tree based on the Euclidean distance between points.

(a) Write pseudocode for creating a minimal spanning tree linking n points in d dimension.

(b) Let k denote the average linkage per node. What is the average space complexity of your algorithm?

(c) What is the average time complexity?

Section 10.12

43. Consider the adaptive resonance clustering algorithm.

(a) Show that the standard ART algorithm of Fig. 10.19 cannot learn the XOR problem.

(b) Explain how the number of clusters generated by the adaptive resonance algorithm depends upon the order of presentation of the samples.

(c) Discuss the benefits and drawbacks of adaptive resonance in stationary and in nonstationary environments.

Section 10.13

44. Show that minimizing a mean-squared error criterion for d-dimensional data leads to the k-dimensional representation ($k < d$) of the Karhunen-Loéve transform (Eq. 94) as follows. For simplicity, assume that the data set has zero mean. (If the mean is not zero, we can always subtract off the mean from each vector to define new vectors.)

(a) The (scalar) projection of a vector $\mathbf{x}$ onto a unit vector $\mathbf{e}$, $a(\mathbf{e}) = \mathbf{x}^t \mathbf{e}$, is, of course, a random variable. Define the variance of a to be $\sigma^2 = \mathcal{E}_{\mathbf{x}}[a^2]$. Show that $\sigma^2 = \mathbf{e}^t \mathbf{\Sigma} \mathbf{e}$, where $\mathbf{\Sigma} = \mathcal{E}_{\mathbf{x}}[\mathbf{x}\mathbf{x}^t]$ is the correlation matrix.

(b) A vector $\mathbf{e}$ that yields an extremal or stationary value of this variance must obey $\sigma^2(\mathbf{e} + \delta\mathbf{e}) = \sigma^2(\mathbf{e})$, where $\delta\mathbf{e}$ is a small perturbation. Show that this condition implies $(\delta\mathbf{e})^t \mathbf{\Sigma} \mathbf{e} = 0$ at such a stationary point.

(c) Consider small variations $\delta\mathbf{e}$ that do not change the length of the vector, that is, ones in which $\delta\mathbf{e}$ is perpendicular to $\mathbf{e}$. Use this condition and your above results to show that $(\delta\mathbf{e})^t \mathbf{\Sigma} \mathbf{e} - \lambda(\delta\mathbf{e})^t \mathbf{e} = 0$, where λ is a scalar. Show that the necessary and sufficient solution is $\mathbf{\Sigma}\mathbf{e} = \lambda\mathbf{e}$—as based on Eq. 94.

(d) Define a sum-squared-error criterion for a set of points in d-dimensional space and their projections onto a k-dimensional linear subspace ($k < d$). Use your results above to show that in order to minimize your criterion, the subspace shoud be spanned by the k largest eigenvectors of the correlation matrix.

45. Consider a linear neural net network consisting of $d - k - d$ units ($k < d$). Show that such a neural net, when trained to autoassociate patterns, performs principal component analysis by considering the minimization it solves.

46. Consider the use of neural networks for nonlinear principal component analysis.

(a) Prove that if all units in the five-layer network of Fig. 10.23 are linear, and the network trained to serve as an autoencoder, then the representation learned at the middle layer corresponds to the linear principal component of the data.

(b) State briefly why this also implies that a three-layer network (input, hidden, output) cannot be used for nonlinear principal component analysis, even if the middle layer consists of nonlinear units.

47. Use the fact that the sum of samples from two Gaussians is again a Gaussian to show why independent component analysis cannot isolate sources perfectly if two or more components are Gaussian.

48. Follow arguments similar to those leading to Eq. 102 to show that the learning rule for bias weights in independent component analysis using sigmoidal non-linearity is the one given by Eq. 106.

49. Generalize the learning rules of Eqs. 102 and 106 to the case where the number of sensed signals k is strictly greater than the number of independent components d, as is illustrated in Fig. 10.25. Be sure to note that $\mathbf{W}$ is not square.

Section 10.14

50. Consider the use of multidimensional scaling for representing the points $\mathbf{x}_1 = (1, 0)^t$, $\mathbf{x}_2 = (0, 0)^t$, and $\mathbf{x}_3 = (0, 1)^t$ in one dimensions. To obtain a unique solution, assume that the image points satisfy $0 = y_1 < y_2 < y_3$.

  (a) Show that the criterion function J_{ee} is minimized by the configuration with $y_2 = (1 + \sqrt{2})/3$ and $y_3 = 2y_2$.

  (b) Show that the criterion function J_{ff} is minimized by the configuration with $y_2 = (2 + \sqrt{2})/4$ and $y_3 = 2y_2$.

COMPUTER EXERCISES

Several exercises make use of the data in the following table.

Sample	x_1	x_2	x_3	Sample	x_1	x_2	x_3
1	−7.82	−4.58	−3.97	11	6.18	2.81	5.82
2	−6.68	3.16	2.71	12	6.72	−0.93	−4.04
3	4.36	−2.19	2.09	13	−6.25	−0.26	0.56
4	6.72	0.88	2.80	14	−6.94	−1.22	1.13
5	−8.64	3.06	3.50	15	8.09	0.20	2.25
6	−6.87	0.57	−5.45	16	6.81	0.17	−4.15
7	4.47	−2.62	5.76	17	−5.19	4.24	4.04
8	6.73	−2.01	4.18	18	−6.38	−1.74	1.43
9	−7.71	2.34	−6.33	19	4.08	1.30	5.33
10	−6.91	−0.49	−5.68	20	6.27	0.93	−2.78

Section 10.4

1. Consider the univariate normal mixture

$$p(\mathbf{x}|\boldsymbol{\theta}) = \frac{P(\omega_1)}{\sqrt{2\pi}\sigma_1} \exp\left[-\frac{1}{2}\left(\frac{x - \mu_1}{\sigma_1}\right)^2\right] + \frac{1 - P(\omega_1)}{\sqrt{2\pi}\sigma_2} \exp\left[-\frac{1}{2}\left(\frac{x - \mu_2}{\sigma_2}\right)^2\right].$$

Write a general program for computing the maximum-likelihood values of the parameters, and apply it to the 20 x_1 points in the table above under the following assumptions of what is known and what is unknown:

  (a) Known: $P(\omega_1) = 0.5$, $\sigma_1 = \sigma_2 = 1$; Unknown: μ_1 and μ_2.

  (b) Known: $P(\omega_1) = 0.5$; Unknown: $\sigma_1 = \sigma_2 = \sigma$, μ_1 and μ_2.

(c) Known: $P(\omega_1) = 0.5$; Unknown: σ_1, σ_2, μ_1 and μ_2.

(d) Unknown: $P(\omega_1)$, σ_1, σ_2, μ_1 and μ_2.

2. Write a program to implement k-means clustering (Algorithm 1), and apply it to the three-dimensional data in the table for the following assumed numbers of clusters, and starting points.

(a) Let $c = 2$, $\mathbf{m}_1(0) = (1, 1, 1)^t$, and $\mathbf{m}_2(0) = (-1, 1, -1)^t$.

(b) Let $c = 2$, $\mathbf{m}_1(0) = (0, 0, 0)^t$, and $\mathbf{m}_2(0) = (1, 1, -1)^t$. Compare your final solution with that from part (a), and explain any differences, including the number of iterations for convergence.

(c) Let $c = 3$, $\mathbf{m}_1(0) = (0, 0, 0)^t$, $\mathbf{m}_2(0) = (1, 1, 1)^t$, and $\mathbf{m}_3(0) = (-1, 0, 2)^t$.

(d) Let $c = 3$, $\mathbf{m}_1(0) = (-0.1, 0, 0.1)^t$, $\mathbf{m}_2(0) = (0, -0.1, 0.1)^t$, and $\mathbf{m}_3(0) = (-0.1, -0.1, .1)^t$. Compare your final solution with that from part (c), and explain any differences, including the number of iterations for convergence.

3. Repeat Computer exercise 2, but use instead a fuzzy k-means algorithm with the "blending" set by $b = 2$ (Eqs. 32 and 33).

4. Explore the problems that can come with misspecifying the number of clusters in the fuzzy k-means algorithm (Algorithm 2) using the following one-dimensional data: $\mathcal{D} = \{-5.0, -4.5, -4.1, -3.9, 2.5, 2.8, 3.1, 3.9, 4.5\}$.

(a) Use your program in the four conditions defined by $c = 2$, $c = 3$, $b = 1$, and $b = 4$. In each case initialize the cluster centers to distinct values, but ones near $x = 0$.

(b) Compare your solutions to the $c = 3$, $b = 4$ case to the $c = 3$, $b = 1$ case, and discuss any sources of the differences.

5. Show how a few labeled samples in a k-means algorithm can improve clustering of unlabeled samples in the following somewhat extreme case.

(a) Generate 50 two-dimensional samples for each of four spherical Gaussians, $p(\mathbf{x}|\omega_i) \sim N(\boldsymbol{\mu}_i, \mathbf{I})$, where $\boldsymbol{\mu}_1 = \left(\begin{smallmatrix}-2\\-2\end{smallmatrix}\right)$, $\boldsymbol{\mu}_2 = \left(\begin{smallmatrix}-2\\2\end{smallmatrix}\right)$, $\boldsymbol{\mu}_3 = \left(\begin{smallmatrix}2\\2\end{smallmatrix}\right)$, and $\boldsymbol{\mu}_4 = \left(\begin{smallmatrix}2\\-2\end{smallmatrix}\right)$.

(b) Choose $c = 4$ initial positions for the cluster means randomly from the full 200 samples. What is the probability that your random selection yields exactly one cluster center for each component density? (Make the simplifying assumption that the component densities do not overlap significantly.)

(c) Using the four samples selected in part (b), run a k-means clusterer on the full 200 points. (If the four points in fact come from different components, reselect samples to ensure that at least two come from the same component density before using your clusterer.)

(d) Now assume you have some label information, in particular four samples known to come from distinct component densities. Using these as your initial cluster centers, run a k-means clusterer on the full 200 points.

(e) Discuss the value of a few labeled samples for clustering in light of the final clusters given in parts (c) and (d).

Section 10.5

6. Explore unsupervised Bayesian learning of the mean of a Gaussian distribution in the following way.

(a) Generate a data set $\mathcal{D}$ of 30 points, uniformly distributed in the interval $-10 \leq x \leq +10$.

(b) Assume (incorrectly) that the data in $\mathcal{D}$ were drawn from a normal distribution with known variance but unknown mean, namely, $p(x) \sim N(\mu, 2)$. Thus the unknown parameter $\boldsymbol{\theta}$ in Eq. 42 is simply the scalar μ. Assume a wide prior for the parameter: $p(\mu)$ is uniform in the range $-10 \leq \mu \leq +10$. Plot posterior probabilities for $k = 0, 1, 2, 3, 4, 5, 10, 15, 20, 25$, and 30 points from $\mathcal{D}$.

(c) Now assume instead a narrow prior—that is, $p(\mu)$ uniform in the range $-1 \leq \mu \leq +1$—and repeat part (b) using the same order of data presentation.

(d) Are your curves for part (b) and part (c) the same for small number of points? For large number of points? Explain.

7. Write a decision-directed clusterer related to k-means in the following way.

(a) First, generate a set $\mathcal{D}$ of $n = 1000$ three-dimensional points in the unit cube, $0 \leq x_i \leq 1$, $i = 1, 2, 3$.

(b) Randomly choose $c = 4$ of these points as the initial cluster centers $\mathbf{m}_j$, $j = 1, 2, 3, 4$.

(c) The core of the algorithm operates as follows: First, each sample $\mathbf{x}_i$ is classified by the nearest cluster center $\mathbf{m}_j$. Next, each mean $\mathbf{m}_j$ is calculated to be the mean of the samples in ω_j. If there is no change in the centers after n presentations, halt.

(d) Use your algorithm to plot four trajectories of the position of the cluster centers.

(e) What is the space and the time complexities of this algorithm? State any assumptions you invoke.

Section 10.6

8. Explore the role of metrics, similarity measures, and thresholds on cluster formation in the following way.

(a) First, generate a two-dimensional data set consisting of two parts: $\mathcal{D}_1$ contains 1000 points whose distance from the origin is chosen uniformly in the range $3 \leq r \leq 5$, and angular position uniform in the range $0 \leq \phi < 2\pi$; likewise, $\mathcal{D}_2$ consists of 50 points of distance $0 \leq r \leq 2$ and angle $0 \leq \phi < 2\pi$. The full data set used below is $\mathcal{D} = \mathcal{D}_1 \cup \mathcal{D}_2$.

(b) Write a simple clustering algorithm that links any two points $\mathbf{x}$ and $\mathbf{x}'$ if $d(\mathbf{x}, \mathbf{x}') < \theta$, where θ is a threshold selected by the user, and distance is calculated by means of a general Minkowski metric (Eq. 49),

$$d(\mathbf{x}, \mathbf{x}') = \left(\sum_{k=1}^{d} |x_k - x_k'|^q \right)^{1/q}.$$

Let $q = 2$ (Euclidean distance) and apply your algorithm to the data $\mathcal{D}$ for the following thresholds: $\theta = 0.01, 0.05, 0.1, 0.5, 1, 5$. In each case, plot all 150 points and differentiate the clusters by linking their points or other plotting convention.

(c) Repeat part (b) with $q = 1$ (city block distance).

    (d) Repeat part (b) with $q = 4$.

    (e) Discuss how the metric affects the "natural" number of clusters implied by your results.

Section 10.7

9. Explore different clustering criteria by exhaustive search in the following way. Let $\mathcal{D}$ be the first seven three-dimensional points in the table above.

    (a) If we assume that any cluster must have at least one point, how many cluster configurations are possible for the seven points?

    (b) Write a program to search through each of the cluster configurations, and for each compute the following criteria: J_e (Eq. 54), J_d (Eq. 68), $\sum_{i=1}^{d} \lambda_i$ (Eq. 69), $J_f = \mathrm{tr}[S_T^{-1}S_W]$ (Eq. 70) and $|S|/|S_T|$ (Eq. 71). Show the optimal clusters for each of your four criteria.

    (c) Perform a whitening transformation on your points and repeat part (b).

    (d) In light of your results, discuss which of the criteria are invariant to the whitening transformation.

Section 10.8

10. Show that the Basic Iterative Least-Squares clustering algorithm gives solutions and final criterion values that depend upon starting conditions in the following way. Implement Algorithm 3 for $c = 3$ clusters and apply it to the data in the table above. For each simulation, list the final clusters as sets of points (identified by their index in the table), along with the corresponding value of the criterion function.

    (a) $\mathbf{m}_1(0) = (1, 1, 1)^t$, $\mathbf{m}_2(0) = (-1, -1, -1)^t$, and $\mathbf{m}_3(0) = (0, 0, 0)^t$.

    (b) $\mathbf{m}_1(0) = (0.1, 0.1, 0.1)^t$, $\mathbf{m}_2(0) = (-0.1, -0.1, -0.1)^t$, and $\mathbf{m}_3(0) = (0, 0, 0)^t$.

    (c) $\mathbf{m}_1(0) = (2, 0, 2)^t$, $\mathbf{m}_2(0) = (-2, 0, -2)^t$, and $\mathbf{m}_3(0) = (1, 1, 1)^t$.

    (d) $\mathbf{m}_1(0) = (0.5, 1, 0.2)^t$, $\mathbf{m}_2(0) = (0.2, -1, 0.5)^t$, and $\mathbf{m}_3(0) = (0.2, 0.4, 0.6)^t$.

    (e) Explain why your final answers differ.

Section 10.9

11. Implement the basic hierarchical agglomerative clustering algorithm (Algorithm 4), as well as a method for drawing dendrograms based on its results. Apply your algorithm and draw dendrograms to the data in the table above using the distance measures indicated below. Define the similarity between two clusters to be linear in distance, with similarity = 100 for singleton clusters ($c = 20$) and similarity = 0 for the single cluster ($c = 1$).

    (a) d_{min} (Eq. 79)

    (b) d_{max} (Eq. 80)

    (c) d_{avg} (Eq. 81)

    (d) d_{mean} (Eq. 82)

12. Explore the use of cluster dendrograms for selecting the "most natural" number of clusters.

    (a) Write a program to perform hierarchical clustering and display a dendrogram, using measure of distance to be selected from Eqs. 79–82.

(b) Write a program to generate n/c points from each of c one-dimensional Gaussians, $p(x|\omega_i) \sim N(\mu_i, \sigma_i^2)$, $i = 1, \ldots, c$. Use your program to generate $n = 50$ points, 25 in each of two clusters, with $\mu_1 = 0$, $\mu_2 = 1$, and $\sigma_1^2 = \sigma_2^2 = 1$. Repeat with $\mu_2 = 4$.

(c) Use your program from (a) to generate dendrograms for each of the two data sets generated in (b).

(d) The difference in similarity values for successive levels is a random variable, which we can model as a normal distribution with some mean and variance. Suppose we define the "most natural" number of clusters according to the largest gap in similarity values, and that this largest gap is *significant* if it differs "significantly" from the distribution. State your criterion analytically, and show that one of the cases in (b) indeed has two clusters.

Section 10.11

13. Implement a basic Competitive Learning clustering algorithm (Algorithm 6) and apply it to the three-dimensional data in the table above as follows.

(a) First, preprocess the data by augmenting each vector with $x_0 = 1$ and normalizing to unit length. In this way, each point lies on the surface of a hypersphere.

(b) Set $c = 2$, and let the initial (normalized) weght vectors correspond to patterns 1 and 2. Let the learning rate be $\eta = 0.1$. Present the patterns in cyclic order, $1, 2, \ldots, 20, 1, 2, \ldots, 20, 1, 2, \ldots$.

(c) Modify your program so as to reduce the learning rate by multiplying by the constant factor $\alpha < 1$ after each pattern presentation, so the learning rate approaches zero exponentially. Repeat your simulation of part (b) with such decay, where $\alpha = 0.99$. Compare your final clusterings with those from using $\alpha = 0.5$.

(d) Repeat part (c) but with the patterns chosen in a random order—that is, with the probability of presenting any given pattern being 1/20 per trial. Discuss the role of random versus sequenced pattern presentation on the final clusterings.

Section 10.12

14. Consider the application of graph-theoretic clustering methods applied to the 20 points in the table above.

(a) Write a program to form the similarity matrix $\mathbf{S} = [s_{ij}]$, of Eq. 93 where two points are similar if their Euclidean distance is less than a threshold d_0.

(b) Apply your program to the $x_1 x_2$-components of the data in the table. Specifically, for $d_0 = 0.01, 0.05, 0.1, 0.5, 1.0$ show the number of distinct clusters, including singleton clusters.

(c) Repeat part (b) for the full three-dimensional data in the table above.

Section 10.13

15. Use principal component analysis to represent all the three-dimensional data in the table above in two dimensions. What are the eigenvectors and eigenvalues?

16. Explore the use of independent component analysis for blind source separation in the following example.

(a) Generate 100 points for $t = 1, \ldots, 100$ for $x_1(t) = \cos(t)$, and $x_2(t) = e^{-t} - 5e^{-t/5}$ and two sensor signals:

$$s_1(t) = 0.5x_1(t) + 0.2x_2(t)$$

$$s_2(t) = 0.1x_1(t) + 0.4x_2(t).$$

(b) Write a program for independent component analysis, based on Eqs. 105 and 48 to find $\mathbf{W}$ and $\mathbf{w}_0$, as defined in the text. Let your starting matrix be $\mathbf{W} = \begin{pmatrix} 0.1 & 0.3 \\ 1.0 & 0.2 \end{pmatrix}$ and bias vector be $\mathbf{w}_0 = \begin{pmatrix} 0.01 \\ -0.02 \end{pmatrix}$.

Section 10.14

17. Write a program to implement multidimensional scaling.

(a) Use your program to represent the three-dimensional data in the table above into a two-dimensional space minimizing the J_{ee} criterion of Eq. 107. Use a numeral, 1–20, to label each of your points.

(b) Repeat with J_{ff} of Eq. 108.

(c) Repeat with J_{ef} of Eq. 109.

(d) In light of your above three plots, discuss the relationships among your three different clustering criteria.

BIBLIOGRAPHY

[1] Phipps Arabie, Lawrence J. Hubert, and Geert De Soete, editors. *Clustering and Classification*. World Scientific, River Edge, NJ, 1998.

[2] Thomas A. Bailey and Richard C. Dubes. Cluster validity profiles. *Pattern Recognition*, 15:61–83, 1982.

[3] Geoffrey H. Ball and David J. Hall. Some implications of interactive graphic computer systems for data analysis and statistics. *Technometrics*, 12:17–31, 1970.

[4] Anthony J. Bell and Terrence J. Sejnowski. An information-maximization approach to blind separation and blind deconvolution. *Neural Computation*, 7(6):1129–1159, 1996.

[5] James C. Bezdek. *Fuzzy Mathematics in Pattern Classification*. Ph.D. thesis, Cornell University, Applied Mathematics Center, Ithaca, NY, 1973.

[6] James C. Bezdek. *Pattern Recognition with Fuzzy Objective Function Algorithms*. Plenum Press, New York, 1981.

[7] Christopher M. Bishop. Bayesian PCA. In Michael S. Kearns, Sara A. Solla, and David A. Cohn, editors, *Advances in Neural Information Processing Systems 3*, pages 382–387. MIT Press, Cambridge, MA, 1999.

[8] Ingwer Borg and Patrick J. F. Groenen. *Modern Multidimensional Scaling: Theory and Applications*. Springer-Verlag, New York, 1997.

[9] Hervé Bourlard and Yves Kamp. Auto-association by multilayer perceptrons and singular value decomposition. *Biological Cybernetics*, 59:291–294, 1988.

[10] Gail A. Carpenter and Stephen Grossberg, editors. *Pat-tern Recognition by Self-Organizing Neural Networks*. MIT Press, Cambridge, MA, 1991.

[11] Michael A. Cohen, Stephen Grossberg, and David G. Stork. Speech perception and production by a self-organizing neural network. In Gail A. Carpenter and Stephen Grossberg, editors, *Pattern Recognition by Self-Organizing Neural Networks*, pages 615–633. MIT Press, Cambridge, MA, 1991.

[12] William H. E. Day and Herbert Edelsbrunner. Efficient algorithms for agglomerative hierarchical clustering methods. *Journal of Classification*, 1(1):7–24, 1984.

[13] Konstantinos I. Diamantaras and Sun-Yuang Kung. *Principal Component Neural Networks: Theory and Applications*. Wiley-Interscience, New York, 1996.

[14] Thomas Eckes. An error variance approach to 2-mode hierarchical-clustering. *Journal of Classification*, 10(1):51–74, 1993.

[15] Ze'ev Roth and Yoram Baram. Multidmensional density shaping by sigmoids. *IEEE Transactions on Neural Networks*, TNN-7(5):1291–1298, 1996.

[16] Michel Gaeta and Jean-Louis Lacoume. Sources separation without *a priori* knowledge: The maximum likelihood solution. In Luis Torres, Enrique Masgrau, and Miguel A. Lagunas, editors, *European Association for Signal Processing, Eusipco 90*, pages 621–624, Barcelona, Spain, 1990. Elsevier.

[17] Selvanayagam Ganesalingam. Classification and mixture approaches to clustering via maximum likelihood. *Applied Statistics*, 38(3):455–466, 1989.

[18] Allen Gersho and Robert M. Gray. *Vector Quantization and Signal Processing*. Kluwer Academic Publishers, Boston, MA, 1992.

[19] John C. Gower and Gavin J. S. Ross. Minimum spanning trees and single-linkage cluster analysis. *Applied Statistics*, 18:54–64, 1969.

[20] Victor Hasselblad. Estimation of parameters for a mixture of normal distributions. *Technometrics*, 8:431–444, 1966.

[21] Geoffrey Hinton and Terrence J. Sejnowski, editors. *Unsupervised Learning: Foundations of Neural Computation*. MIT Press, Cambridge, MA, 1999.

[22] Lawrence J. Hubert. Min and max hierarchical clustering using asymmetric similarity measures. *Psychometrika*, 38(1):63–72, 1973.

[23] Anil K. Jain and Richard C. Dubes. *Algorithms for Clustering Data*. Prentice-Hall, Englewood Cliffs, NJ, 1988.

[24] Ian T. Jolliffe. *Principal Component Analysis*. Springer-Verlag, New York, 1986.

[25] Christian Jutten and Jeanny Herault. Blind separation of sources 1: An adaptive algorithm based on neuromimetic architecture. *Signal Processing*, 24(1):1–10, 1991.

[26] Leonard Kaufman and Peter J. Rousseeuw. *Finding Groups in Data: An Introduction to Cluster Analysis*. Wiley, New York, 1990.

[27] Ian J. Kitching, Peter L. Forey, Christopher J. Humphries, and David M. Williams. *Cladistics: A Practical Course in Systematics*. Clarendon Press, Oxford, UK, second edition, 1998.

[28] Teuvo Kohonen. Self-organizing formation of topologically correct feature maps. *Biological Cybernetics*, 43(1):59–69, 1982.

[29] Tuevo Kohonen. *Self-Organization and Associative Memory*. Springer-Verlag, Berlin, third edition, 1989.

[30] Mark A. Kramer. Nonlinear principal component analysis using autoassociative neural networks. *AIChE Journal*, 37(2):233–243, 1991.

[31] Joseph B. Kruskal. The relationship between multidimensional scaling and clustering. In John Van Ryzin, editor, *Classification and Clustering: Proceedings of an Advanced Seminar Conducted by the Mathematics Research Center, the University of Wisconsin–Madison, May 3-5, 1976*, pages 7–44. Academic Press, New York, 1977.

[32] Stuart P. Lloyd. Least squares quantization in PCM. *IEEE Transactions on Information Theory*, IT-2:129–137, 1982.

[33] Jianchang Mao and Anil K. Jain. A self-organizing network for hyperellipsoidal clustering (HEC). *IEEE Transactions on Neural Networks*, TNN-7(1):16–29, 1996.

[34] Geoffrey J. McLachlan and Kaye E. Basford. *Mixture Models*. Dekker, New York, 1988.

[35] Barbara Moore. ART1 and pattern clustering. In David Touretzky, Geoffrey Hinton, and Terrence Sejnowski, editors, *Proceedings of the 1988 Connectionist Models Summer School*, pages 174–185. Morgan Kaufmann, San Mateo, CA, 1988.

[36] Dragan Obradovic and Gustavo Deco. Information maximization and independent component analysis: Is there a difference? *Neural Computation*, 10(8):2085–2101, 1998.

[37] Erkki Oja. *Subspace Methods of Pattern Recognition*. John Wiley, New York, 1983.

[38] Erkki Oja and Samuel Kaski, editors. *Kohonen Maps*. Elsevier, Amsterdam, The Netherlands, 1999.

[39] Barak A. Pearlmutter and Lucas C. Parra. A context-sensitive generalization of ICA. In *International Conference on Neural Information Processing*, pages 151–157. Springer-Verlag, Hong Kong, 1996.

[40] Barak Perlmutter and Lucas C. Parra. Maximum-likelihood blind source separation: A context-sensitive generalization of ICA. In Michael C. Mozer, Michael I. Jordan, and Thomas Petsche, editors, *Advances in Neural Information Processing Systems*, volume 9, pages 613–619. MIT Press, Cambridge, MA, 1997.

[41] James O. Ramsay. Maximum-likelihood estimation in multidimensional scaling. *Psychometrika*, 42(2):241–266, 1977.

[42] Helge Ritter, Thomas Martinetz, and Klaus Schulten. *Neural Computation and Self-Organizing Maps: An Introduction*. Addison-Wesley, New York, English translation from the German edition, 1992.

[43] Susan S. Schiffman, Mark Lance Reynolds, and Forrest W. Young. *Introduction to Multidimensional Scaling: Theory, Methods and Applications*. Academic Press, New York, 1981.

[44] Stephen P. Smith and Anil K. Jain. Testing for uniformity in multidimensional data. *IEEE Transaction on Pattern Analysis and Machine Intelligence*, PAMI-6(1):73–81, 1984.

[45] David G. Stork. Self-organization, pattern recognition and adaptive resonance networks. *Journal of Neural Network Computing*, 1(1):26–42, 1989.

[46] Henry Teicher. Identifiability of mixtures. *Annals of Mathematical Statistics*, 32(1):244–248, 1961.

[47] John H. Wolfe. Pattern clustering by multivariate mixture analysis. *Multivariate Behavioral Research*, 5:329–350, 1970.

[48] Te Won Lee. *Independent Component Analysis: Theory and Applications*. Kluwer Academic Publishers, Dordrecht, 1998.

[49] Hujun Yin and Nigel M. Allinson. On the distribution and convergence of feature space in self-organizing maps. *Neural Computation*, 7(6):1178–1187, 1998.

MATHEMATICAL FOUNDATIONS

Our goal here is to present the basic results and definitions from linear algebra, probability theory, information theory, and computational complexity that serve as the mathematical foundations for pattern recognition. We will try to give intuitive insight whenever appropriate, but do not attempt to prove these results; systematic expositions can be found in the references.

A.1 NOTATION

Here are the terms and notation used throughout the book. In addition, there are numerous specialized variables and functions whose definitions and usage should be clear from the text.

Variables, Symbols, and Operations

$\simeq$	approximately equal to
$\equiv$	equivalent to (or defined to be)
$\propto$	proportional to
∞	infinity
$x \rightarrow a$	x approaches a
$t \leftarrow t + 1$	in an algorithm: assign to variable t the new value $t + 1$
$\lim_{x \to a} f(x)$	the value of $f(x)$ in the limit as x approaches a
$\arg\max_x f(x)$	the value of x that leads to the maximum value of $f(x)$
$\arg\min_x f(x)$	the value of x that leads to the minimum value of $f(x)$
$\lceil x \rceil$	ceiling of x—that is, the least integer not smaller than x (e.g., $\lceil 3.5 \rceil = 4$)

$\lfloor x \rfloor$	floor of x—that is, the greatest integer not larger than x (e.g., $\lfloor 3.5 \rfloor = 3$)
$m \bmod n$	m modulo n—that is, the remainder when m is divided by n (e.g., $7 \bmod 5 = 2$)
$\text{Rand}[l, u)$	in a computer program, a routine that returns a real number x, randomly chosen in the range $l \leq x < u$
$\ln(x)$	logarithm base e, or natural logarithm of x
$\log(x)$	logarithm base 10 of x
$\log_2(x)$	logarithm base 2 of x
$\exp[x]$ or e^x	exponential of x—that is, e raised to the power of x
$\partial f(x)/\partial x$	partial derivative of f with respect to x
$\int_a^b f(x)dx$	the integral of $f(x)$ between a and b. If no limits are written, the full space is assumed
$F(x; \theta)$	function of x, with implied dependence upon θ
■	Q.E.D., quod erat demonstrandum ("which was to be proved")—used to signal the end of a proof

Mathematical Operations

$\bar{x}$	mean or average value of x
$\mathcal{E}[f(x)]$	the expected value of function $f(x)$ where x is a random variable
$\mathcal{E}_y[f(x, y)]$	the expected value of function over several variables, $f(x, y)$, taken over a subset y of them
$\text{Var}[f(\cdot)]$	the variance—that is, $\mathcal{E}[(f(x) - \mathcal{E}[f(x)])^2]$
$\text{Var}_f[\cdot]$	the variance—that is, $\mathcal{E}_f[(x - \mathcal{E}_f[x])^2]$
$\sum_{i=1}^n a_i$	the sum from $i = 1$ to n—that is, $a_1 + a_2 + \cdots + a_n$
$\prod_{i=1}^n a_i$	the product from $i = 1$ to n—that is, $a_1 \times a_2 \times \cdots \times a_n$
$f(t) \star g(t)$	convolution of $f(t)$ and $g(t)$, $\int_{-\infty}^{\infty} f(\tau)g(t - \tau)\,d\tau$

Vectors and Matrices

$\mathbf{R}^d$	d-dimensional Euclidean space
$\mathbf{x}, \mathbf{A}, \ldots$	boldface is used for (column) vectors and matrices
$\mathbf{f}(x)$	vector-valued function (note the boldface) of a scalar argument
$\mathbf{f}(\mathbf{x})$	vector-valued function (note the boldface) of a vector argument
$\mathbf{I}$	identity matrix, a square matrix having 1's on the diagonal and 0 everywhere else
$\mathbf{1}_i$	vector of length i consisting solely of 1's
$diag(a_1, a_2, \ldots, a_d)$	matrix whose diagonal elements are $a_1, a_2, \ldots, a_d$, and off-diagonal elements are 0

$\mathbf{x}^t$	transpose of vector $\mathbf{x}$
$\|\mathbf{x}\|$	Euclidean norm of vector $\mathbf{x}$
$\boldsymbol{\Sigma}$	covariance matrix
$\text{tr}[\mathbf{A}]$	the trace of $\mathbf{A}$—that is, the sum of its diagonal elements
$\mathbf{A}^{-1}$	the inverse of matrix $\mathbf{A}$
$\mathbf{A}^{\dagger}$	pseudoinverse of matrix $\mathbf{A}$
$|\mathbf{A}|$ or $\text{Det}[\mathbf{A}]$	determinant of $\mathbf{A}$
λ	eigenvalue
$\mathbf{e}$	eigenvector
$\mathbf{u}_i$	unit vector in the ith direction in Euclidean space

Sets

$\mathcal{A}, \mathcal{B}, \mathcal{C}, \mathcal{D}, \ldots$	"Calligraphic" font generally denotes sets or lists—for example, a data set $\mathcal{D} = \{\mathbf{x}_1, \ldots, \mathbf{x}_n\}$
$\mathbf{x} \in \mathcal{D}$	$\mathbf{x}$ is an element of set $\mathcal{D}$
$\mathbf{x} \notin \mathcal{D}$	$\mathbf{x}$ is not an element of set $\mathcal{D}$
$\mathcal{A} \cup \mathcal{B}$	union of two sets—that is, the set containing all elements in either $\mathcal{A}$ or $\mathcal{B}$
$\mathcal{A} \cap \mathcal{B}$	intersection of two sets—that is, the set containing all elements that are in both $\mathcal{A}$ and $\mathcal{B}$
$|\mathcal{D}|$	the cardinality of set $\mathcal{D}$—that is, the number of (possibly nondistinct) discrete elements in it

Probability, Distributions, and Complexity

ω	state of nature
$P(\cdot)$	probability mass
$p(\cdot)$	probability density
$P(a, b)$	the joint probability—that is, the probability of having both a and b
$p(a, b)$	the joint probability density—that is, the probability density of having both a and b
$\text{Pr}[\cdot]$	the probability of a condition being met—for example, $\text{Pr}[x < x_0]$ means the probability that x is less than x_0
$p(\mathbf{x}|\boldsymbol{\theta})$	the conditional probability density of $\mathbf{x}$ given $\boldsymbol{\theta}$
$\mathbf{w}$	weight vector
$\lambda(\cdot, \cdot)$	loss function
$\nabla = \begin{pmatrix} \frac{\partial}{\partial x_1} \\ \frac{\partial}{\partial x_2} \\ \vdots \\ \frac{\partial}{\partial x_d} \end{pmatrix}$	gradient operator in $\mathbf{R}^d$, sometimes written $grad[\cdot]$

$$\nabla_{\boldsymbol{\theta}} = \begin{pmatrix} \frac{\partial}{\partial \theta_1} \\ \frac{\partial}{\partial \theta_2} \\ \vdots \\ \frac{\partial}{\partial \theta_d} \end{pmatrix}$$ gradient operator in $\boldsymbol{\theta}$ coordinates, sometimes written $grad_{\boldsymbol{\theta}}[\cdot]$

$\hat{\boldsymbol{\theta}}$ maximum-likelihood estimate of $\boldsymbol{\theta}$

$\sim$ "has the distribution"—for example, $p(x) \sim N(\mu, \sigma^2)$ means that the density of x is normal, with mean μ and variance σ^2

$N(\mu, \sigma^2)$ normal or Gaussian distribution with mean μ and variance σ^2

$N(\boldsymbol{\mu}, \boldsymbol{\Sigma})$ multidimensional normal or Gaussian distribution with mean vector $\boldsymbol{\mu}$ and covariance matrix $\boldsymbol{\Sigma}$

$U(x_l, x_u)$ a one-dimensional uniform distribution between x_l and x_u

$U(\mathbf{x}_l, \mathbf{x}_u)$ a d-dimensional uniform density—that is, uniform density within the smallest axes-aligned bounding box that contains both $\mathbf{x}_l$ and $\mathbf{x}_u$, and 0 elsewhere

$T(\mu, \delta)$ triangle distribution, having center μ and full half-width δ

$\delta(x)$ Dirac delta function, which has value 0 for $x \neq 0$, and integrates to unity

δ_{ij} Kronecker delta symbol, which has value 1 if its two indexes match, and 0 otherwise

$\Gamma(\cdot)$ Gamma function

$n!$ n factorial—that is, $n \times (n-1) \times (n-2) \times \cdots \times 1$

$\binom{n}{k} = \frac{n!}{k!(n-k)!}$ binomial coefficient, read "n choose k," for n and k integers

$O(h(x))$ big oh order of $h(x)$

$\Theta(h(x))$ big theta order of $h(x)$

$\Omega(h(x))$ big omega order of $h(x)$

$\sup_x f(x)$ the supremum value of $f(x)$—the least upper bound or global maximum of $f(x)$ over all values of x

A.2 LINEAR ALGEBRA

A.2.1 Notation and Preliminaries

A d-dimensional column vector $\mathbf{x}$ and its transpose $\mathbf{x}^t$ can be written as

$$\mathbf{x} = \begin{pmatrix} x_1 \\ x_2 \\ \vdots \\ x_d \end{pmatrix} \quad \text{and} \quad \mathbf{x}^t = (x_1 \ x_2 \ \dots \ x_d), \tag{1}$$

where all components can take on real values. We denote an $n \times d$ (rectangular) matrix $\mathbf{M}$ and its $d \times n$ transpose $\mathbf{M}^t$ as

$$\mathbf{M} = \begin{pmatrix} m_{11} & m_{12} & m_{13} & \cdots & m_{1d} \\ m_{21} & m_{22} & m_{23} & \cdots & m_{2d} \\ \vdots & \vdots & \vdots & \ddots & \vdots \\ m_{n1} & m_{n2} & m_{n3} & \cdots & m_{nd} \end{pmatrix} \tag{2}$$

and

$$\mathbf{M}^t = \begin{pmatrix} m_{11} & m_{21} & \cdots & m_{n1} \\ m_{12} & m_{22} & \cdots & m_{n2} \\ m_{13} & m_{23} & \cdots & m_{n3} \\ \vdots & \vdots & \ddots & \vdots \\ m_{1d} & m_{2d} & \cdots & m_{nd} \end{pmatrix}. \tag{3}$$

In other words, the jith entry of $\mathbf{M}^t$ is the ijth entry of $\mathbf{M}$.

A square ($d \times d$) matrix is called symmetric if its entries obey $m_{ij} = m_{ji}$; it is called skew-symmetric (or anti-symmetric) if $m_{ij} = -m_{ji}$. A general matrix is called nonnegative if $m_{ij} \geq 0$ for all i and j. A particularly important matrix is the IDENTITY MATRIX *identity matrix*, $\mathbf{I}$—a $d \times d$ (square) matrix whose diagonal entries are 1's, and all other entries 0. The *Kronecker delta* function or Kronecker symbol, defined as KRONECKER DELTA

$$\delta_{ij} = \begin{cases} 1 & \text{if } i = j \\ 0 & \text{otherwise,} \end{cases} \tag{4}$$

can serve to define the entries of an identity matrix. A general diagonal matrix (i.e., one having 0 for all off diagonal entries) is denoted $diag(m_{11}, m_{22}, \ldots, m_{dd})$, the entries being the successive elements $m_{11}, m_{22}, \ldots, m_{dd}$. Addition of vectors and of matrices is component by component.

We can multiply a vector by a matrix, $\mathbf{Mx} = \mathbf{y}$, that is,

$$\begin{pmatrix} m_{11} & m_{12} & \cdots & m_{1d} \\ m_{21} & m_{22} & \cdots & m_{2d} \\ \vdots & \vdots & \ddots & \vdots \\ m_{n1} & m_{n2} & \cdots & m_{nd} \end{pmatrix} \begin{pmatrix} x_1 \\ x_2 \\ \vdots \\ \vdots \\ x_d \end{pmatrix} = \begin{pmatrix} y_1 \\ y_2 \\ \vdots \\ y_n \end{pmatrix}, \tag{5}$$

where

$$y_i = \sum_{j=1}^{d} m_{ij} x_j. \tag{6}$$

Note that the number of columns of $\mathbf{M}$ must equal the number of rows of $\mathbf{x}$. Also, if $\mathbf{M}$ is not square, the dimensionality of $\mathbf{y}$ differs from that of $\mathbf{x}$.

A.2.2 Inner Product

The *inner product* of two vectors having the same dimensionality will be denoted here as $\mathbf{x}^t \mathbf{y}$ and yields a scalar:

$$\mathbf{x}^t\mathbf{y} = \sum_{i=1}^{d} x_i y_i = \mathbf{y}^t\mathbf{x}. \tag{7}$$

EUCLIDEAN
NORM

It is sometimes also called the *scalar product* or *dot product* and denoted $\mathbf{x} \bullet \mathbf{y}$, or more rarely (x, y). The *Euclidean norm* or length of the vector is

$$\|\mathbf{x}\| = \sqrt{\mathbf{x}^t\mathbf{x}}. \tag{8}$$

We call a vector "normalized" if $\|\mathbf{x}\| = 1$. The angle θ between two d-dimensional vectors obeys

$$\cos\theta = \frac{\mathbf{x}^t\mathbf{y}}{\|\mathbf{x}\| \, \|\mathbf{y}\|}, \tag{9}$$

and thus the inner product is a measure of the colinearity of two vectors—a natural indication of their similarity. In particular, if $\mathbf{x}^t\mathbf{y} = 0$, then the vectors are orthogonal; if $\|\mathbf{x}^t\mathbf{y}\| = \|\mathbf{x}\| \, \|\mathbf{y}\|$, then the vectors are colinear. From Eq. 9, we have immediately the Cauchy-Schwarz inequality, which states

$$|\mathbf{x}^t\mathbf{y}| \le \|\mathbf{x}\| \, \|\mathbf{y}\|. \tag{10}$$

LINEAR
INDEPENDENCE

We say a set of vectors $\{\mathbf{x}_1, \mathbf{x}_2, \ldots, \mathbf{x}_n\}$ is *linearly independent* if no vector in the set can be written as a linear combination of any of the others. Informally, a set of d linearly independent vectors spans an d-dimensional vector space; that is, any vector in that space can be written as a linear combination of such spanning vectors.

A.2.3 Outer Product

MATRIX
PRODUCT

The outer product (sometimes called *matrix product* or more rarely *dyadic product*) of two vectors yields a matrix

$$\mathbf{M} = \mathbf{xy}^t = \begin{pmatrix} x_1 \\ x_2 \\ \vdots \\ x_d \end{pmatrix} (y_1 \; y_2 \; \cdots \; y_n) = \begin{pmatrix} x_1 y_1 & x_1 y_2 & \cdots & x_1 y_n \\ x_2 y_1 & x_2 y_2 & \cdots & x_2 y_n \\ \vdots & \vdots & \ddots & \vdots \\ x_d y_1 & x_d y_2 & \cdots & x_d y_n \end{pmatrix}, \tag{11}$$

and thus the components of $\mathbf{M}$ are $m_{ij} = x_i y_j$. Of course, if the dimensions of $\mathbf{x}$ and $\mathbf{y}$ are not the same, then $\mathbf{M}$ is not square.

A.2.4 Derivatives of Matrices

Suppose $f(\mathbf{x})$ is a scalar-valued function of d variables x_i, $i = 1, 2, \ldots, d$, which we represent as the vector $\mathbf{x}$. Then the derivative or gradient of $f(\cdot)$ with respect to this vector is computed component by component, that is,

$$\nabla f(\mathbf{x}) = \operatorname{grad} f(\mathbf{x}) = \frac{\partial f(\mathbf{x})}{\partial \mathbf{x}} = \begin{pmatrix} \frac{\partial f(\mathbf{x})}{\partial x_1} \\ \frac{\partial f(\mathbf{x})}{\partial x_2} \\ \vdots \\ \frac{\partial f(\mathbf{x})}{\partial x_d} \end{pmatrix}. \tag{12}$$

If we have an n-dimensional vector-valued function $\mathbf{f}$ (note the use of boldface), of a d-dimensional vector $\mathbf{x}$, we calculate the derivatives and represent them as the *Jacobian matrix*

JACOBIAN MATRIX

$$\mathbf{J}(\mathbf{x}) = \frac{\partial \mathbf{f}(\mathbf{x})}{\partial \mathbf{x}} = \begin{pmatrix} \frac{\partial f_1(\mathbf{x})}{\partial x_1} & \cdots & \frac{\partial f_1(\mathbf{x})}{\partial x_d} \\ \vdots & \ddots & \vdots \\ \frac{\partial f_n(\mathbf{x})}{\partial x_1} & \cdots & \frac{\partial f_n(\mathbf{x})}{\partial x_d} \end{pmatrix}. \tag{13}$$

If this matrix is square, its determinant (Section A.2.5) is called simply the *Jacobian* or occasionally the *Jacobian determinant*.

If the entries of $\mathbf{M}$ depend upon a scalar parameter θ, we can take the derivative of $\mathbf{M}$ component by component, to get another matrix, as

$$\frac{\partial \mathbf{M}}{\partial \theta} = \begin{pmatrix} \frac{\partial m_{11}}{\partial \theta} & \frac{\partial m_{12}}{\partial \theta} & \cdots & \frac{\partial m_{1d}}{\partial \theta} \\ \frac{\partial m_{21}}{\partial \theta} & \frac{\partial m_{22}}{\partial \theta} & \cdots & \frac{\partial m_{2d}}{\partial \theta} \\ \vdots & \vdots & \ddots & \vdots \\ \frac{\partial m_{n1}}{\partial \theta} & \frac{\partial m_{n2}}{\partial \theta} & \cdots & \frac{\partial m_{nd}}{\partial \theta} \end{pmatrix}. \tag{14}$$

In Section A.2.6 we shall discuss matrix inversion, but for convenience we give here the derivative of the inverse of a matrix, $\mathbf{M}^{-1}$:

$$\frac{\partial}{\partial \theta} \mathbf{M}^{-1} = -\mathbf{M}^{-1} \frac{\partial \mathbf{M}}{\partial \theta} \mathbf{M}^{-1}. \tag{15}$$

Consider a matrix $\mathbf{M}$ and a vector $\mathbf{y}$ that are independent of $\mathbf{x}$. The following vector derivative identities for such a matrix can be verified by writing out the components:

$$\frac{\partial}{\partial \mathbf{x}}[\mathbf{M}\mathbf{x}] = \mathbf{M} \tag{16}$$

$$\frac{\partial}{\partial \mathbf{x}}[\mathbf{y}^t\mathbf{x}] = \frac{\partial}{\partial \mathbf{x}}[\mathbf{x}^t\mathbf{y}] = \mathbf{y} \tag{17}$$

$$\frac{\partial}{\partial \mathbf{x}}[\mathbf{x}^t\mathbf{M}\mathbf{x}] = [\mathbf{M} + \mathbf{M}^t]\mathbf{x}. \tag{18}$$

In the case where $\mathbf{M}$ is symmetric (as for instance a covariance matrix, cf. Section A.4.10), then Eq. 18 simplifies to

$$\frac{\partial}{\partial \mathbf{x}}[\mathbf{x}^t\mathbf{M}\mathbf{x}] = 2\mathbf{M}\mathbf{x}. \tag{19}$$

We first recall the use of second derivatives of a scalar function of a scalar x in writing a Taylor series (or Taylor expansion) about a point:

$$f(x) = f(x_0) + \frac{df(x)}{dx}\bigg|_{x=x_0} (x - x_0) + \frac{1}{2!}\frac{d^2 f(x)}{dx^2}\bigg|_{x=x_0} (x - x_0)^2 + O((x - x_0)^3). \tag{20}$$

Analogously, if our scalar-valued f is instead a function of a vector $\mathbf{x}$, we can expand $f(\mathbf{x})$ in a Taylor series around a point $\mathbf{x}_0$:

$$f(\mathbf{x}) = f(\mathbf{x}_0) + \underbrace{\left[\frac{\partial f}{\partial \mathbf{x}}\right]^t_{\mathbf{x}=\mathbf{x}_0}}_{\mathbf{J}} (\mathbf{x} - \mathbf{x}_0) + \frac{1}{2!}(\mathbf{x} - \mathbf{x}_0)^t \underbrace{\left[\frac{\partial^2 f}{\partial \mathbf{x}^2}\right]^t_{\mathbf{x}=\mathbf{x}_0}}_{\mathbf{H}} (\mathbf{x} - \mathbf{x}_0) + O(||\mathbf{x} - \mathbf{x}_0||^3),$$

(21)

HESSIAN MATRIX   where $\mathbf{H}$ is the *Hessian matrix*, the matrix of second-order derivatives of $f(\cdot)$, here evaluated at $\mathbf{x}_0$. (We shall return in Section A.8 to consider the $O(\cdot)$ notation and the order of a function used in Eq. 21 and below.)

A.2.5 Determinant and Trace

The determinant of a $d \times d$ (square) matrix is a scalar, denoted $|\mathbf{M}|$, and reveals properties of the matrix. For instance, suppose we consider the columns of $\mathbf{M}$ as vectors; if these vectors are not linearly independent, then the determinant vanishes. In pattern recognition, we have particular interest in the covariance matrix $\boldsymbol{\Sigma}$, which contains the second moments of a set of data. In this case the absolute value of the determinant of a covariance matrix is a measure of the d-dimensional hypervolume of the data that yielded $\boldsymbol{\Sigma}$. (It can be shown that the determinant is equal to the product of the eigenvalues of a matrix, as mentioned in Section A.2.7.) If the data lie in a subspace of the full d-dimensional space, then the columns of $\boldsymbol{\Sigma}$ are not linearly independent, and the determinant vanishes. Furthermore, the determinant must be nonzero for the inverse of a matrix to exist (Section A.2.6).

The calculation of the determinant is simple in low dimensions, but a bit more involved in high dimensions. If $\mathbf{M}$ is itself a scalar (i.e., a 1×1 matrix M), then $|M| = M$. If $\mathbf{M}$ is 2×2, then $|\mathbf{M}| = m_{11}m_{22} - m_{21}m_{12}$. The determinant of a general EXPANSION BY   square matrix can be computed by a method called *expansion by minors*, and this MINORS   leads to a recursive definition. If $\mathbf{M}$ is our $d \times d$ matrix, we define $\mathbf{M}_{i|j}$ to be the $(d-1) \times (d-1)$ matrix obtained by deleting the ith row and the jth column of $\mathbf{M}$:

$$i \begin{pmatrix} m_{11} & m_{12} & \cdots & \otimes & \cdots & \cdots & m_{1d} \\ m_{21} & m_{22} & \cdots & \otimes & \cdots & \cdots & m_{2d} \\ \vdots & \vdots & \ddots & \otimes & \cdots & \cdots & \vdots \\ \vdots & \vdots & \cdots & \otimes & \cdots & \cdots & \vdots \\ \otimes & \otimes & \otimes & \otimes & \otimes & \otimes & \otimes \\ \vdots & \vdots & \cdots & \otimes & \cdots & \ddots & \vdots \\ m_{d1} & m_{d2} & \cdots & \otimes & \cdots & \cdots & m_{dd} \end{pmatrix} = \mathbf{M}_{i|j}.$$

(22)

Given the determinants $|\mathbf{M}_{i|j}|$, we can now compute the determinant of $\mathbf{M}$, with the expansion by minors on the first column giving

$$|\mathbf{M}| = m_{11}|\mathbf{M}_{1|1}| - m_{21}|\mathbf{M}_{2|1}| + m_{31}|\mathbf{M}_{3|1}| - \cdots \pm m_{d1}|\mathbf{M}_{d|1}|,$$

(23)

where the signs alternate. This process can be applied recursively to the successive (smaller) matrixes in Eq. 23.

Only for a 3×3 matrix, this determinant calculation can be represented by "sweeping" the matrix—that is, taking the sum of the products of matrix terms along a diagonal, where products from upper left to lower right are added with a positive sign, and those from the lower left to upper right with a minus sign. That is,

$$|\mathbf{M}| = \begin{vmatrix} m_{11} & m_{12} & m_{13} \\ m_{21} & m_{22} & m_{23} \\ m_{31} & m_{32} & m_{33} \end{vmatrix} \tag{24}$$

$$= m_{11}m_{22}m_{33} + m_{13}m_{21}m_{32} + m_{12}m_{23}m_{31}$$

$$- m_{13}m_{22}m_{31} - m_{11}m_{23}m_{32} - m_{12}m_{21}m_{33}.$$

Again, this "sweeping" rule does not work for matrices larger than 3×3. For any matrix we have $|\mathbf{M}| = |\mathbf{M}^t|$. Furthermore, for two square matrices of equal size $\mathbf{M}$ and $\mathbf{N}$, we have $|\mathbf{MN}| = |\mathbf{M}|\,|\mathbf{N}|$.

The *trace* of a $d \times d$ (square) matrix, denoted tr$[\mathbf{M}]$, is the sum of its diagonal elements:

$$\text{tr}[\mathbf{M}] = \sum_{i=1}^{d} m_{ii}. \tag{25}$$

Both the determinant and trace of a matrix are invariant with respect to rotations of the coordinate system.

A.2.6 Matrix Inversion

So long as its determinant does not vanish, the inverse of a $d \times d$ matrix $\mathbf{M}$, denoted $\mathbf{M}^{-1}$, is the $d \times d$ matrix such that

$$\mathbf{MM}^{-1} = \mathbf{I}. \tag{26}$$

COFACTOR We call the scalar $C_{ij} = (-1)^{i+j}|M_{i|j}|$ the i, j *cofactor* or, equivalently, the cofactor of the i, j entry of $\mathbf{M}$. As defined in Eq. 22, $\mathbf{M}_{i|j}$ is the $(d-1)$-by-$(d-1)$ matrix

ADJOINT formed by deleting the ith row and jth column of $\mathbf{M}$. The *adjoint* of $\mathbf{M}$, written Adj$[\mathbf{M}]$, is the matrix whose i, j entry is the j, i cofactor of $\mathbf{M}$. Given these definitions, we can write the inverse of a matrix as

$$\mathbf{M}^{-1} = \frac{\text{Adj}[\mathbf{M}]}{|\mathbf{M}|}. \tag{27}$$

If $\mathbf{M}$ is not square (or if $\mathbf{M}^{-1}$ in Eq. 27 does not exist because the columns of $\mathbf{M}$ are

PSEUDO-INVERSE not linearly independent), we typically use instead the *pseudoinverse* $\mathbf{M}^{\dagger}$. If $\mathbf{M}^t\mathbf{M}$ is nonsingular, the pseudoinverse is defined as

$$\mathbf{M}^{\dagger} = [\mathbf{M}^t\mathbf{M}]^{-1}\mathbf{M}^t. \tag{28}$$

The pseudoinverse ensures $\mathbf{M}^{\dagger}\mathbf{M} = \mathbf{I}$ and is very useful in solving least squares problems.

A.2.7 Eigenvectors and Eigenvalues

The inverse of the product of two square matrices obeys $[\mathbf{MN}]^{-1} = \mathbf{N}^{-1}\mathbf{M}^{-1}$, as can be verified by multiplying on the right or the left by $\mathbf{MN}$. Given a d-by-d matrix $\mathbf{M}$, a very important class of linear equations is of the form

$$\mathbf{Mx} = \lambda\mathbf{x} \tag{29}$$

for scalar λ, which can be rewritten

$$(\mathbf{M} - \lambda\mathbf{I})\mathbf{x} = \mathbf{0}, \tag{30}$$

where $\mathbf{I}$ the identity matrix and $\mathbf{0}$ is the zero vector. The solution vector $\mathbf{x} = \mathbf{e}_i$ and corresponding scalar $\lambda = \lambda_i$ are called the *eigenvector* and associated *eigenvalue*, respectively. If $\mathbf{M}$ is real and symmetric, there are d (possibly nondistinct) solution vectors $\{\mathbf{e}_1, \mathbf{e}_2, \ldots, \mathbf{e}_d\}$, each with an associated eigenvalue $\{\lambda_1, \lambda_2, \ldots, \lambda_d\}$. Under multiplication by $\mathbf{M}$ the eigenvectors are changed only in magnitude, not direction:

$$\mathbf{M}\mathbf{e}_j = \lambda_j\mathbf{e}_j. \tag{31}$$

If $\mathbf{M}$ is diagonal, then the eigenvectors are parallel to the coordinate axes.

One method of finding the eigenvectors and eigenvalues is to solve the *characteristic equation* (or *secular equation*,)

$$|\mathbf{M} - \lambda\mathbf{I}| = \lambda^d + a_1\lambda^{d-1} + \cdots + a_{d-1}\lambda + a_d = 0, \tag{32}$$

for each of its d (possibly nondistinct) roots λ_j. For each such root, we then solve a set of linear equations to find its associated eigenvector $\mathbf{e}_j$.

Finally, it can be shown that the trace of a matrix is just the sum of the eigenvalues and the determinant of a matrix is just the product of its eigenvalues:

$$\text{tr}[\mathbf{M}] = \sum_{i=1}^{d}\lambda_i \qquad \text{and} \qquad |\mathbf{M}| = \prod_{i=1}^{d}\lambda_i. \tag{33}$$

If a matrix is diagonal, then its eigenvalues are simply the nonzero entries on the diagonal, and the eigenvectors are the unit vectors parallel to the coordinate axes.

A.3 LAGRANGE OPTIMIZATION

Suppose we seek the position $\mathbf{x}_0$ of an extremum of a scalar-valued function $f(\mathbf{x})$, subject to some constraint. If a constraint can be expressed in the form $g(\mathbf{x}) = 0$, then we can find the extremum of $f(\mathbf{x})$ as follows. First we form the Lagrangian function

$$L(\mathbf{x}, \lambda) = f(\mathbf{x}) + \lambda\underbrace{g(\mathbf{x})}_{=0}, \tag{34}$$

where λ is a scalar called the Lagrange *undetermined multiplier*. We convert this constrained optimization problem into an unconstrained problem by taking the derivative,

$$\frac{\partial L(\mathbf{x}, \lambda)}{\partial \mathbf{x}} = \frac{\partial f(\mathbf{x})}{\partial \mathbf{x}} + \lambda\frac{\partial g(\mathbf{x})}{\partial \mathbf{x}} = 0, \tag{35}$$

and using standard methods from calculus to solve the resulting equations for λ and the extremizing value of $\mathbf{x}$. (Note that the term $\lambda\partial g/\partial\mathbf{x}$ does not vanish, in general.) The solution gives the $\mathbf{x}$ position of the extremum, and it is a simple matter of substitution to find the extreme value of $f(\cdot)$ under the constraints.

A.4 PROBABILITY THEORY

A.4.1 Discrete Random Variables

Let x be a discrete random variable that can assume any of the finite number m of different values in the set $\mathcal{X} = \{v_1, v_2, \ldots, v_m\}$. We denote by p_i the probability that x assumes the value v_i:

$$p_i = \Pr[x = v_i], \qquad i = 1, \ldots, m. \tag{36}$$

Then the probabilities p_i must satisfy the following two conditions:

$$p_i \geq 0 \quad \text{and} \quad \sum_{i=1}^{m} p_i = 1. \tag{37}$$

PROBABILITY
MASS FUNCTION

Sometimes it is more convenient to express the set of probabilities $\{p_1, p_2, \ldots, p_m\}$ in terms of the *probability mass function* $P(x)$, which must satisfy the following conditions:

$$P(x) \geq 0, \quad \text{and} \quad \sum_{x \in \mathcal{X}} P(x) = 1. \tag{38}$$

A.4.2 Expected Values

MEAN

The *expected value*, *mean*, or *average* of the random variable x is defined by

$$\mathcal{E}[x] = \mu = \sum_{x \in \mathcal{X}} x P(x) = \sum_{i=1}^{m} v_i p_i. \tag{39}$$

If one thinks of the probability mass function as defining a set of point masses, with p_i being the mass concentrated at $x = v_i$, then the expected value μ is just the center of mass. Alternatively, we can interpret μ as the arithmetic average of the values in a large random sample. More generally, if $f(x)$ is any function of x, the expected value of f is defined by

$$\mathcal{E}[f(x)] = \sum_{x \in \mathcal{X}} f(x) P(x). \tag{40}$$

Note that the process of forming an expected value is *linear*, in that if α_1 and α_2 are arbitrary constants, then we have

$$\mathcal{E}[\alpha_1 f_1(x) + \alpha_2 f_2(x)] = \alpha_1 \mathcal{E}[f_1(x)] + \alpha_2 \mathcal{E}[f_2(x)]. \tag{41}$$

EXPECTATION
OPERATOR

SECOND
MOMENT
VARIANCE

It is sometimes convenient to think of $\mathcal{E}$ as an operator—the (linear) *expectation operator*. Two important special-case expectations are the *second moment* and the *variance*:

$$\mathcal{E}[x^2] = \sum_{x \in \mathcal{X}} x^2 P(x) \tag{42}$$

$$\text{Var}[x] = \sigma^2 = \mathcal{E}[(x - \mu)^2] = \sum_{x \in \mathcal{X}} (x - \mu)^2 P(x), \tag{43}$$

STANDARD
DEVIATION

where σ is the *standard deviation* of x. The variance can be viewed as the moment of inertia of the probability mass function. The variance is never negative, and it is zero if and only if all of the probability mass is concentrated at one point.

The standard deviation is a simple but valuable measure of how far values of x are likely to depart from the mean. Its very name suggests that it is the standard or typical amount one should expect a randomly drawn value for x to deviate or differ from μ. *Chebyshev's inequality* (or the Bienaymé-Chebyshev inequality) provides a mathematical relation between the standard deviation and $|x - \mu|$:

CHEBYSHEV'S
INEQUALITY

$$\Pr[|x - \mu| > n\sigma] \leq \frac{1}{n^2}. \tag{44}$$

This inequality is not a tight bound (and it is useless for $n < 1$); a more practical rule of thumb, which strictly speaking is true only for the normal distribution, is that 68% of the values will lie within one, 95% within two, and 99.7% within three standard deviations of the mean (cf. Fig. A.1, ahead). Nevertheless, Chebyshev's inequality shows the strong link between the standard deviation and the spread of a distribution. In addition, it suggests that $|x - \mu|/\sigma$ is a meaningful normalized measure of the distance from x to the mean (cf. Section A.4.12).

By expanding the quadratic in Eq. 43, it is easy to prove the useful formula

$$\text{Var}[x] = \mathcal{E}[x^2] - (\mathcal{E}[x])^2. \tag{45}$$

Note that, unlike the mean, the variance is *not* linear. In particular, if $y = \alpha x$, where α is a constant, then $\text{Var}[y] = \alpha^2 \text{Var}[x]$. Moreover, the variance of the sum of two random variables is usually *not* the sum of their variances. However, as we shall see below, variances do add when the variables involved are statistically independent.

In the simple but important special case in which x is binary-valued (say, $v_1 = 0$ and $v_2 = 1$), we can obtain simple formulas for μ and σ. If we let $p = \Pr[x = 1]$, then it is easy to show that

$$\mu = p \quad \text{and} \quad \sigma = \sqrt{p(1 - p)}. \tag{46}$$

A.4.3 Pairs of Discrete Random Variables

Let x and y be random variables which can take on values in $\mathcal{X} = \{v_1, v_2, \ldots, v_m\}$, and $\mathcal{Y} = \{w_1, w_2, \ldots, w_n\}$, respectively. We can think of (x, y) as a vector or a point in the *product space* of x and y. For each possible pair of values (v_i, w_j) we have a *joint probability* $p_{ij} = \Pr[x = v_i, y = w_j]$. These mn joint probabilities p_{ij} are nonnegative and sum to 1. Alternatively, we can define a *joint probability mass function* $P(x, y)$ for which

PRODUCT SPACE

$$P(x, y) \geq 0 \quad \text{and} \quad \sum_{x \in \mathcal{X}} \sum_{y \in \mathcal{Y}} P(x, y) = 1. \tag{47}$$

The joint probability mass function is a complete characterization of the pair of random variables (x, y); that is, everything we can compute about x and y, individually or together, can be computed from $P(x, y)$. In particular, we can obtain the separate *marginal distributions* for x and y by summing over the unwanted variable:

MARGINAL
DISTRIBUTION

$$P_x(x) = \sum_{y \in \mathcal{Y}} P(x, y)$$

$$P_y(y) = \sum_{x \in \mathcal{X}} P(x, y). \tag{48}$$

We will occasionally use subscripts, as in Eq. 48, to emphasize the fact that $P_x(x)$ has a different functional form than $P_y(y)$. It is common to omit them and write simply $P(x)$ and $P(y)$ whenever the context makes it clear that these are in fact two different functions—rather than the same function merely evaluated with different values for the argument.

A.4.4 Statistical Independence

Variables x and y are said to be *statistically independent* if and only if

$$P(x, y) = P_x(x)P_y(y). \tag{49}$$

We can understand such independence as follows. Suppose that $p_i = \Pr[x = v_i]$ is the fraction of the time that $x = v_i$, and $q_j = \Pr[y = w_j]$ is the fraction of the time that $y = w_j$. Consider those situations where $x = v_i$. If it is still true that the fraction of those situations in which $y = w_j$ is the same value q_j, it follows that knowing the value of x did not give us any additional knowledge about the possible values of y; in that sense y is independent of x. Finally, if x and y are statistically independent, it is clear that the fraction of the time that the specific pair of values (v_i, w_j) occurs must be the product of the fractions $p_i q_j = P(v_i)P(w_j)$ as we shall explore in Section A.4.6.

A.4.5 Expected Values of Functions of Two Variables

In the natural extension of Section A.4.2, we define the expected value of a function $f(x, y)$ of two random variables x and y by

$$\mathcal{E}[f(x, y)] = \sum_{x \in \mathcal{X}} \sum_{y \in \mathcal{Y}} f(x, y)P(x, y), \tag{50}$$

and as before the expectation operator $\mathcal{E}$ is linear:

$$\mathcal{E}[\alpha_1 f_1(x, y) + \alpha_2 f_2(x, y)] = \alpha_1 \mathcal{E}[f_1(x, y)] + \alpha_2 \mathcal{E}[f_2(x, y)]. \tag{51}$$

The means (first moments) and variances (second moments) are

$$\mu_x = \mathcal{E}[x] = \sum_{x \in \mathcal{X}} \sum_{y \in \mathcal{Y}} x P(x, y)$$

$$\mu_y = \mathcal{E}[y] = \sum_{x \in \mathcal{X}} \sum_{y \in \mathcal{Y}} y P(x, y)$$

$$\sigma_x^2 = \text{Var}[x] = \mathcal{E}[(x - \mu_x)^2] = \sum_{x \in \mathcal{X}} \sum_{y \in \mathcal{Y}} (x - \mu_x)^2 P(x, y)$$

$$\sigma_y^2 = \text{Var}[y] = \mathcal{E}[(y - \mu_y)^2] = \sum_{x \in \mathcal{X}} \sum_{y \in \mathcal{Y}} (y - \mu_y)^2 P(x, y). \tag{52}$$

COVARIANCE

An important new "cross-moment" can now be defined, the *covariance* of x and y:

$$\sigma_{xy} = \mathcal{E}[(x - \mu_x)(y - \mu_y)] = \sum_{x \in \mathcal{X}} \sum_{y \in \mathcal{Y}} (x - \mu_x)(y - \mu_y)P(x, y). \tag{53}$$

Using vector notation, we can summarize Eqs. 52 and 53 as

$$\boldsymbol{\mu} = \mathcal{E}[\mathbf{x}] = \sum_{\mathbf{x} \in \{\mathcal{XY}\}} \mathbf{x} P(\mathbf{x}) \tag{54}$$

$$\boldsymbol{\Sigma} = \mathcal{E}[(\mathbf{x} - \boldsymbol{\mu})(\mathbf{x} - \boldsymbol{\mu})^t], \tag{55}$$

where $\{\mathcal{XY}\}$ respresents the space of all possible values for all components of $\mathbf{x}$ and $\boldsymbol{\Sigma}$ is the covariance matrix (cf., Section A.4.9).

The covariance is one measure of the degree of statistical dependence between x and y. If x and y are statistically independent, then $\sigma_{xy} = 0$. If $\sigma_{xy} = 0$, the variables x and y are said to be *uncorrelated*. It does *not* follow that uncorrelated variables must be statistically independent—covariance is just one measure of dependence. However, it is a fact that uncorrelated variables are statistically independent if they have a multivariate normal distribution, and in practice statisticians often treat uncorrelated variables as if they were statistically independent. If α is a constant and $y = \alpha x$, which is a case of strong statistical dependence, it is also easy to show that $\sigma_{xy} = \alpha \sigma_x^2$. Thus, the covariance is positive if x and y both increase or decrease together, and is negative if y decreases when x increases.

UNCORRELATED

CAUCHY-SCHWARZ INEQUALITY

There is an important *Cauchy-Schwarz inequality* for the variances σ_x and σ_y and the covariance σ_{xy}. It can be derived by observing that the variance of a random variable is never negative, and thus the variance of $\lambda x + y$ must be nonnegative no matter what the value of the scalar λ. This leads to the famous inequality

$$\sigma_{xy}^2 \leq \sigma_x^2 \sigma_y^2, \tag{56}$$

which is analogous to the vector inequality $(\mathbf{x}^t \mathbf{y})^2 \leq \|\mathbf{x}\|^2 \|\mathbf{y}\|^2$ given in Eq. 8.

CORRELATION COEFFICIENT

The *correlation coefficient*, defined as

$$\rho = \frac{\sigma_{xy}}{\sigma_x \sigma_y}, \tag{57}$$

is a normalized covariance, and must always be between -1 and $+1$. If $\rho = +1$, then x and y are maximally positively correlated, while if $\rho = -1$, they are maximally negatively correlated. If $\rho = 0$, the variables are uncorrelated. It is common practice to consider variables to be uncorrelated for practical purposes if the magnitude of their correlation coefficient is below some threshold, such as 0.05, although the threshold that makes sense does depend on the actual situation.

If x and y are statistically independent, then for any two functions f and g we obtain

$$\mathcal{E}[f(x)g(y)] = \mathcal{E}[f(x)]\mathcal{E}[g(y)], \tag{58}$$

a result which follows from the definition of statistical independence and expectation. Note that if $f(x) = x - \mu_x$ and $g(y) = y - \mu_y$, this theorem again shows that $\sigma_{xy} = \mathcal{E}[(x - \mu_x)(y - \mu_y)]$ is zero if x and y are statistically independent.

A.4.6 Conditional Probability

When two variables are statistically dependent, knowing the value of one of them lets us get a better estimate of the value of the other one. This is expressed by the following definition of the *conditional probability* of x given y:

$$\Pr[x = v_i | y = w_j] = \frac{\Pr[x = v_i, y = w_j]}{\Pr[y = w_j]}, \tag{59}$$

or, in terms of mass functions,

$$P(x|y) = \frac{P(x, y)}{P(y)}. \tag{60}$$

Note that if x and y are statistically independent, this gives $P(x|y) = P(x)$. That is, when x and y are independent, knowing the value of y gives you no information about x that you didn't already know from its marginal distribution $P(x)$.

Consider a simple illustration of a two-variable binary case where both x and y are either 0 or 1. Suppose that a large number n of pairs of xy-values are randomly produced. Let n_{ij} be the number of pairs in which we find $x = i$ and $y = j$, that is, we see the $(0, 0)$ pair n_{00} times, the $(0, 1)$ pair n_{01} times, and so on, where $n_{00} + n_{01} + n_{10} + n_{11} = n$. Suppose we pull out those pairs where $y = 1$—that is, the $(0, 1)$ pairs and the $(1, 1)$ pairs. Clearly, the fraction of those cases in which x is also 1 is

$$\frac{n_{11}}{n_{01} + n_{11}} = \frac{n_{11}/n}{(n_{01} + n_{11})/n}. \tag{61}$$

Intuitively, this is what we would like to get for $P(x|y)$ when $y = 1$ and n is large. And, indeed, this is what we do get, because n_{11}/n is approximately $P(x, y)$ and $(n_{01} + n_{11})/n$ is approximately $P(y)$ for large n.

A.4.7 The Law of Total Probability and Bayes Rule

The *Law of Total Probability* states that if an event A can occur in m different ways $A_1, A_2, \ldots, A_m$ and if these m subevents are *mutually exclusive*—that is, cannot occur at the same time—then the probability of A occurring is the sum of the probabilities of the subevents A_i. In particular, the random variable y can assume the value y in m different ways—with $x = v_1, x = v_2, \ldots$, and $x = v_m$. Because these possibilities are mutually exclusive, it follows from the Law of Total Probability that $P(y)$ is the sum of the joint probability $P(x, y)$ over all possible values for x. Formally we have

$$P(y) = \sum_{x \in \mathcal{X}} P(x, y). \tag{62}$$

But from the definition of the conditional probability $P(y|x)$ we have

$$P(x, y) = P(y|x)P(x), \tag{63}$$

and after rewriting Eq. 63 with x and y exchanged and some simple algebra, we obtain

$$P(x|y) = \frac{P(y|x)P(x)}{\sum_{x \in \mathcal{X}} P(y|x)P(x)}, \tag{64}$$

or in words we have

$$\text{posterior} = \frac{\text{likelihood} \times \text{prior}}{\text{evidence}},$$

where these terms are discussed more fully in Chapter 2.

Equation 64 is called *Bayes rule*. Note that the denominator, which is just $P(y)$, is obtained by summing the numerator over all x values. By writing the denominator in this form we emphasize the fact that everything on the right-hand side of the equation is conditioned on x. If we think of x as the important variable, then we can say that the shape of the distribution $P(x|y)$ depends only on the numerator $P(y|x)P(x)$; the

EVIDENCE

denominator is just a normalizing factor, sometimes called the *evidence*, needed to ensure that the $P(x|y)$ sum to one.

The standard interpretation of Bayes rule is that it "inverts" statistical connections, turning $P(y|x)$ into $P(x|y)$. Suppose that we think of x as a "cause" and y as an "effect" of that cause. That is, we assume that if the cause x is present, it is easy to determine the probability of the effect y being observed; the conditional probability

LIKELIHOOD

function $P(y|x)$—the *likelihood*—specifies this probability explicitly. If we observe the effect y, it might not be so easy to determine the cause x, because there might be several different causes, each of which could produce the same observed effect. However, Bayes rule makes it easy to determine $P(x|y)$, provided that we know both

PRIOR

$P(y|x)$ and the so-called *prior probability* $P(x)$, the probability of x before we make any observations about y. Said slightly differently, Bayes rule shows how the probability distribution for x changes from the *prior distribution* $P(x)$ before anything

POSTERIOR
DISTRIBUTION

is observed about y to the *posterior distribution* $P(x|y)$ once we have observed the value of y.

A.4.8 Vector Random Variables

To extend these results from two variables x and y to d variables $x_1, x_2, \ldots, x_d$, it is convenient to employ vector notation. As given by Eq. 47, the joint probability mass function $P(\mathbf{x})$ satisfies $P(\mathbf{x}) \geq 0$ and $\sum P(\mathbf{x}) = 1$, where the sum extends over all possible values for the vector $\mathbf{x}$. Note that $P(\mathbf{x})$ is a function of d variables and can be a very complicated, multidimensional function. However, if the random variables x_i are statistically independent, it reduces to the product

$$P(\mathbf{x}) = P_{x_1}(x_1) P_{x_2}(x_2) \cdots P_{x_d}(x_d)$$

$$= \prod_{i=1}^{d} P_{x_i}(x_i), \tag{65}$$

where we have used the subscripts just to emphasize the fact that the marginal distributions will generally have a different form. Here the separate marginal distributions $P_{x_i}(x_i)$ can be obtained by summing the joint distribution over the other variables. In addition to these univariate marginals, other marginal distributions can be obtained by this use of the Law of Total Probability. For example, suppose we have $P(x_1, x_2, x_3, x_4, x_5)$ and we want $P(x_1, x_4)$; we merely calculate

$$P(x_1, x_4) = \sum_{x_2} \sum_{x_3} \sum_{x_5} P(x_1, x_2, x_3, x_4, x_5). \tag{66}$$

One can define many different conditional distributions, such as $P(x_1, x_2|x_3)$ or $P(x_2|x_1, x_4, x_5)$. For example,

$$P(x_1, x_2|x_3) = \frac{P(x_1, x_2, x_3)}{P(x_3)}, \tag{67}$$

where all of the joint distributions can be obtained from $P(\mathbf{x})$ by summing out the unwanted variables. If instead of scalars we have vector variables, then these conditional distributions can also be written as

$$P(\mathbf{x}_1|\mathbf{x}_2) = \frac{P(\mathbf{x}_1, \mathbf{x}_2)}{P(\mathbf{x}_2)}, \tag{68}$$

and likewise, in vector form, Bayes rule becomes

$$P(\mathbf{x}_1|\mathbf{x}_2) = \frac{P(\mathbf{x}_2|\mathbf{x}_1)P(\mathbf{x}_1)}{\sum\limits_{\mathbf{x}_1} P(\mathbf{x}_2|\mathbf{x}_1)P(\mathbf{x}_1)}. \tag{69}$$

A.4.9 Expectations, Mean Vectors and Covariance Matrices

The expected value of a vector is defined to be the vector whose components are the expected values of the original components. Thus, if $\mathbf{f}(\mathbf{x})$ is an n-dimensional, vector-valued function of the d-dimensional random vector $\mathbf{x}$,

$$\mathbf{f}(\mathbf{x}) = \begin{bmatrix} f_1(\mathbf{x}) \\ f_2(\mathbf{x}) \\ \vdots \\ f_n(\mathbf{x}) \end{bmatrix}, \tag{70}$$

then the expected value of $\mathbf{f}$ is defined by

$$\mathcal{E}[\mathbf{f}] = \begin{bmatrix} \mathcal{E}[f_1(\mathbf{x})] \\ \mathcal{E}[f_2(\mathbf{x})] \\ \vdots \\ \mathcal{E}[f_n(\mathbf{x})] \end{bmatrix} = \sum_{\mathbf{x}} \mathbf{f}(\mathbf{x})P(\mathbf{x}). \tag{71}$$

MEAN VECTOR   In particular, the d-dimensional *mean vector* $\boldsymbol{\mu}$ is defined by

$$\boldsymbol{\mu} = \mathcal{E}[\mathbf{x}] = \begin{bmatrix} \mathcal{E}[x_1] \\ \mathcal{E}[x_2] \\ \vdots \\ \mathcal{E}[x_d] \end{bmatrix} = \begin{bmatrix} \mu_1 \\ \mu_2 \\ \vdots \\ \mu_d \end{bmatrix} = \sum_{\mathbf{x}} \mathbf{x}P(\mathbf{x}). \tag{72}$$

COVARIANCE MATRIX   Similarly, the *covariance matrix* $\boldsymbol{\Sigma}$ is defined as the (square) matrix whose ijth element σ_{ij} is the covariance of x_i and x_j:

$$\sigma_{ij} = \sigma_{ji} = \mathcal{E}[(x_i - \mu_i)(x_j - \mu_j)] \qquad i, j = 1 \ldots d, \tag{73}$$

as we saw in the two-variable case of Eq. 53. Therefore, in expanded form we have

$$
\boldsymbol{\Sigma} = \begin{bmatrix}
\mathcal{E}[(x_1 - \mu_1)(x_1 - \mu_1)] & \mathcal{E}[(x_1 - \mu_1)(x_2 - \mu_2)] & \cdots & \mathcal{E}[(x_1 - \mu_1)(x_d - \mu_d)] \\
\mathcal{E}[(x_2 - \mu_2)(x_1 - \mu_1)] & \mathcal{E}[(x_2 - \mu_2)(x_2 - \mu_2)] & \cdots & \mathcal{E}[(x_2 - \mu_2)(x_d - \mu_d)] \\
\vdots & \vdots & \ddots & \vdots \\
\mathcal{E}[(x_d - \mu_d)(x_1 - \mu_1)] & \mathcal{E}[(x_d - \mu_d)(x_2 - \mu_2)] & \cdots & \mathcal{E}[(x_d - \mu_d)(x_d - \mu_d)]
\end{bmatrix}
$$

$$
= \begin{bmatrix}
\sigma_{11} & \sigma_{12} & \cdots & \sigma_{1d} \\
\sigma_{21} & \sigma_{22} & \cdots & \sigma_{2d} \\
\vdots & \vdots & \ddots & \vdots \\
\sigma_{d1} & \sigma_{d2} & \cdots & \sigma_{dd}
\end{bmatrix}
= \begin{bmatrix}
\sigma_1^2 & \sigma_{12} & \cdots & \sigma_{1d} \\
\sigma_{21} & \sigma_2^2 & \cdots & \sigma_{2d} \\
\vdots & \vdots & \ddots & \vdots \\
\sigma_{d1} & \sigma_{d2} & \cdots & \sigma_d^2
\end{bmatrix}. \tag{74}
$$

We can use the vector product $(\mathbf{x} - \boldsymbol{\mu})(\mathbf{x} - \boldsymbol{\mu})^t$ to write the covariance matrix as

$$
\boldsymbol{\Sigma} = \mathcal{E}[(\mathbf{x} - \boldsymbol{\mu})(\mathbf{x} - \boldsymbol{\mu})^t]. \tag{75}
$$

Thus, $\boldsymbol{\Sigma}$ is symmetric, and its diagonal elements are just the variances of the individual elements of $\mathbf{x}$, which can never be negative; the off-diagonal elements are the covariances, which can be positive or negative. If the variables are statistically independent, the covariances are zero, and the covariance matrix is diagonal. The analog to the Cauchy-Schwarz inequality comes from recognizing that if $\mathbf{w}$ is any d-dimensional vector, then the variance of $\mathbf{w}^t \mathbf{x}$ can never be negative. This leads to the requirement that the quadratic form $\mathbf{w}^t \boldsymbol{\Sigma} \mathbf{w}$ never be negative. Matrices for which this is true are said to be *positive semidefinite*; thus, the covariance matrix $\boldsymbol{\Sigma}$ must be positive semidefinite. It can be shown that this is equivalent to the requirement that none of the eigenvalues of $\boldsymbol{\Sigma}$ can be negative.

A.4.10   Continuous Random Variables

When the random variable x can take values in the continuum, it no longer makes sense to talk about the probability that x has a particular value, such as 2.5136, because the probability of any particular exact value will almost always be zero. Rather, we talk about the probability that x falls in some interval (a, b); instead of having a probability mass function $P(x)$, we have a *probability density function* $p(x)$. The density has the property that

PROBABILITY
DENSITY

$$
\Pr[x \in (a, b)] = \int_a^b p(x)\, dx. \tag{76}
$$

The name *density* comes by analogy with material density. If we consider a small interval $(a, a + \Delta x)$ over which $p(x)$ is essentially constant, having value $p(a)$, we see that $p(a) = \Pr[x \in (a, a + \Delta x)]/\Delta x$. That is, the probability density at $x = a$ is the probability mass $\Pr[x \in (a, a + \Delta x)]$ per unit distance. It follows that the probability density function must satisfy

$$
p(x) \geq 0 \quad \text{and} \quad \int_{-\infty}^{\infty} p(x)\, dx = 1. \tag{77}
$$

In general, most of the definitions and formulas for discrete random variables carry over to continuous random variables with sums replaced by integrals. In particular, the expected value, mean, and variance for a continuous random variable are defined by

$$\mathcal{E}[f(x)] = \int_{-\infty}^{\infty} f(x)p(x)\,dx$$

$$\mu = \mathcal{E}[x] = \int_{-\infty}^{\infty} xp(x)\,dx \tag{78}$$

$$\text{Var}[x] = \sigma^2 = \mathcal{E}[(x-\mu)^2] = \int_{-\infty}^{\infty} (x-\mu)^2 p(x)\,dx,$$

and, as in Eq. 45, the variance obeys $\sigma^2 = \mathcal{E}[x^2] - (\mathcal{E}[x])^2$.

The multivariate situation is similarly handled with continuous random vectors $\mathbf{x}$. The probability density function $p(\mathbf{x})$ must satisfy

$$p(\mathbf{x}) \geq 0 \quad \text{and} \quad \int_{-\infty}^{\infty} p(\mathbf{x})\,d\mathbf{x} = 1, \tag{79}$$

where the integral is understood to be a d-fold, multiple integral and where $d\mathbf{x}$ is the element of d-dimensional volume $d\mathbf{x} = dx_1 dx_2 \cdots dx_d$. The corresponding moments for a general n-dimensional vector-valued function are

$$\mathcal{E}[\mathbf{f}(\mathbf{x})] = \int_{-\infty}^{\infty}\int_{-\infty}^{\infty} \cdots \int_{-\infty}^{\infty} \mathbf{f}(\mathbf{x})p(\mathbf{x})\,dx_1 dx_2 \cdots dx_d = \int_{-\infty}^{\infty} \mathbf{f}(\mathbf{x})p(\mathbf{x})\,d\mathbf{x} \tag{80}$$

and for the particular d-dimensional functions as above, we have

$$\boldsymbol{\mu} = \mathcal{E}[\mathbf{x}] = \int_{-\infty}^{\infty} \mathbf{x}p(\mathbf{x})\,d\mathbf{x} \tag{81}$$

$$\boldsymbol{\Sigma} = \mathcal{E}[(\mathbf{x}-\boldsymbol{\mu})(\mathbf{x}-\boldsymbol{\mu})^t] = \int_{-\infty}^{\infty} (\mathbf{x}-\boldsymbol{\mu})(\mathbf{x}-\boldsymbol{\mu})^t\,p(\mathbf{x})\,d\mathbf{x}.$$

If the components of $\mathbf{x}$ are statistically independent, then the joint probability density function factors as

$$p(\mathbf{x}) = \prod_{i=1}^{d} p_{x_i}(x_i) \tag{82}$$

and the covariance matrix is diagonal.

Conditional probability density functions are defined just as conditional mass functions. Thus, for example, the density for x given y is given by

$$p(x|y) = \frac{p(x, y)}{p(y)} \tag{83}$$

and Bayes rule for density functions is

$$p(x|y) = \frac{p(y|x)p(x)}{\displaystyle\int_{-\infty}^{\infty} p(y|x)p(x)\,dx}, \tag{84}$$

and likewise for the vector case.

Occasionally we will need to take the expectation with respect to a subset of the variables, and in that case we must show this as a subscript—for instance,

$$\mathcal{E}_{x_1}[f(x_1, x_2)] = \int_{-\infty}^{\infty} f(x_1, x_2)p(x_1)\,dx_1. \tag{85}$$

A.4.11 Distributions of Sums of Independent Random Variables

It frequently happens that we know the densities for two independent random variables x and y, and we need to know the density of their sum $z = x + y$. It is easy to obtain the mean and the variance of this sum:

$$\mu_z = \mathcal{E}[z] = \mathcal{E}[x + y] = \mathcal{E}[x] + \mathcal{E}[y] = \mu_x + \mu_y,$$
$$\sigma_z^2 = \mathcal{E}[(z - \mu_z)^2] = \mathcal{E}[(x + y - (\mu_x + \mu_y))^2] = \mathcal{E}[((x - \mu_x) + (y - \mu_y))^2]$$
$$= \mathcal{E}[(x - \mu_x)^2] + 2\underbrace{\mathcal{E}[(x - \mu_x)(y - \mu_y)]}_{=0} + \mathcal{E}[(y - \mu_y)^2] \tag{86}$$
$$= \sigma_x^2 + \sigma_y^2,$$

where we have used the fact that the cross-term factors into $\mathcal{E}[x - \mu_x]\mathcal{E}[y - \mu_y]$ when x and y are independent; in this case the product is manifestly zero, because each of the component expectations vanishes. Thus, the mean of the sum of two independent random variables is the sum of their means, and the variance of their sum is the sum of their variances. If the variables are random *yet not independent*—for instance $y = -x$, where x is a random variable—then the variance is not the sum of the component variances.

It is only slightly more difficult to work out the exact probability density function for $z = x + y$ from the separate density functions for x and y. The probability that z is between ζ and $\zeta + \Delta z$ can be found by integrating the joint density $p(x, y) = p_x(x)p_y(y)$ over the thin strip in the xy-plane between the lines $x + y = \zeta$ and $x + y = \zeta + \Delta z$. It follows that, for small Δz,

$$\Pr[\zeta < z < \zeta + \Delta z] = \left[\int_{-\infty}^{\infty} p(x)p(\zeta - x)\,dx\right]\Delta z, \tag{87}$$

CONVOLUTION and hence that the probability density function for the sum is the *convolution* of the probability density functions for the components:

$$p(z) = p_x(x) \star p_y(y) = \int\limits_{-\infty}^{\infty} p_x(x) p_y(z - x) \, dx. \tag{88}$$

As one would expect, these results generalize. It is not hard to show that:

- The mean of the sum of d independent random variables $x_1, x_2, \ldots, x_d$ is the sum of their means. (In fact the variables need not be independent for this to hold.)
- The variance of the sum is the sum of their variances.
- The probability density function for the sum is the convolution of the separate density functions:

$$p(z) = p(x_1) \star p(x_2) \star \cdots \star p(x_d). \tag{89}$$

A.4.12 Normal Distributions

CENTRAL LIMIT
THEOREM

GAUSSIAN

One of the most important results of probability theory is the *Central Limit Theorem*, which states that, under various conditions, the distribution for the sum of d independent random variables approaches a particular limiting form known as the *normal distribution*. As such, the *normal* or *Gaussian* probability density function is very important, both for theoretical and practical reasons. In one dimension, it is defined by

$$p(x) = \frac{1}{\sqrt{2\pi}\sigma} e^{-1/2((x-\mu)^2/\sigma^2)}. \tag{90}$$

The normal density is traditionally described as a "bell-shaped curve"; it is completely determined by the numerical values for two parameters, the mean μ and the variance σ^2. This is often emphasized by writing $p(x) \sim N(\mu, \sigma^2)$, which is read as "$x$ is distributed normally with mean μ and variance σ^2." The distribution is symmetrical about the mean, the peak occurring at $x = \mu$ and the width of the "bell" is proportional to the standard deviation σ. The parameters of a normal density in Eq. 90 satisfy the following equations:

$$\mathcal{E}[1] = \int\limits_{-\infty}^{\infty} p(x) \, dx = 1$$

$$\mathcal{E}[x] = \int\limits_{-\infty}^{\infty} x \, p(x) \, dx = \mu \tag{91}$$

$$\mathcal{E}[(x - \mu)^2] = \int\limits_{-\infty}^{\infty} (x - \mu)^2 p(x) \, dx = \sigma^2.$$

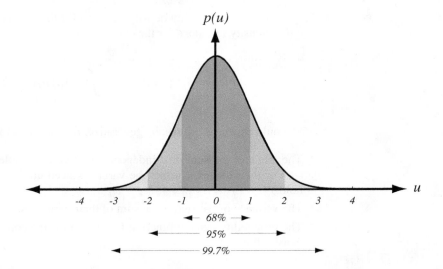

FIGURE A.1. A one-dimensional Gaussian distribution, $p(u) \sim N(0, 1)$, has 68% of its probability mass in the range $|u| \leq 1$, 95% in the range $|u| \leq 2$, and 99.7% in the range $|u| \leq 3$.

Normally distributed data points tend to cluster about the mean. Numerically, the probabilities obey

$$\Pr[|x - \mu| \leq \sigma] \simeq 0.68$$
$$\Pr[|x - \mu| \leq 2\sigma] \simeq 0.95 \tag{92}$$
$$\Pr[|x - \mu| \leq 3\sigma] \simeq 0.997,$$

as shown in Fig. A.1.

A natural measure of the distance from x to the mean μ is the distance $|x - \mu|$ measured in units of standard deviations:

$$r = \frac{|x - \mu|}{\sigma}, \tag{93}$$

MAHALANOBIS
DISTANCE

the *Mahalanobis distance* from x to μ. (In the one-dimensional case, this is sometimes called the z-score.) Thus for instance the probability is 0.95 that the Mahalanobis distance from x to μ will be less than 2. If a random variable x is modified by (a) subtracting its mean and (b) dividing by its standard deviation, it is said to be

STANDARDIZED

standardized. Clearly, a standardized normal random variable $u = (x - \mu)/\sigma$ has zero mean and unit standard deviation—that is,

$$p(u) = \frac{1}{\sqrt{2\pi}} e^{-u^2/2}, \tag{94}$$

which can be written as $p(u) \sim N(0, 1)$. Table A.1 shows the probability that a value, chosen at random according to $p(u) \sim N(0, 1)$, differs from the mean value by less than a criterion z.

Table A.1. The Probability a Sample Drawn from a Standardized Gaussian has Absolute Value Less Than a Criterion (i.e., Pr[$|u| \leq z$])

z	Pr[$|u| \leq z$]	z	Pr[$|u| \leq z$]	z	Pr[$|u| \leq z$]
0.0	0.0	1.0	0.683	2.0	0.954
0.1	0.080	1.1	0.729	2.1	0.964
0.2	0.158	1.2	0.770	2.326	0.980
0.3	0.236	1.3	0.806	2.5	0.989
0.4	0.311	1.4	0.838	2.576	0.990
0.5	0.383	1.5	0.866	3.0	0.9974
0.6	0.452	1.6	0.890	3.090	0.9980
0.7	0.516	1.7	0.911	3.291	0.999
0.8	0.576	1.8	0.928	3.5	0.9995
0.9	0.632	1.9	0.943	4.0	0.99994

A.5 GAUSSIAN DERIVATIVES AND INTEGRALS

Because of the prevalence of Gaussian functions throughout statistical pattern recognition, we often have occasion to integrate and differentiate them. The first three derivatives of a one-dimensional (standardized) Gaussian are

$$
\frac{\partial}{\partial x}\left[\frac{1}{\sqrt{2\pi}\sigma}e^{-x^2/(2\sigma^2)}\right] = \frac{-x}{\sqrt{2\pi}\sigma^3}e^{-x^2/(2\sigma^2)} = \frac{-x}{\sigma^2}p(x)
$$

$$
\frac{\partial^2}{\partial x^2}\left[\frac{1}{\sqrt{2\pi}\sigma}e^{-x^2/(2\sigma^2)}\right] = \frac{1}{\sqrt{2\pi}\sigma^5}\left(-\sigma^2 + x^2\right)e^{-x^2/(2\sigma^2)} = \frac{-\sigma^2 + x^2}{\sigma^4}p(x) \tag{95}
$$

$$
\frac{\partial^3}{\partial x^3}\left[\frac{1}{\sqrt{2\pi}\sigma}e^{-x^2/(2\sigma^2)}\right] = \frac{1}{\sqrt{2\pi}\sigma^7}\left(3x\sigma^2 - x^3\right)e^{-x^2/(2\sigma^2)} = \frac{-3x\sigma^2 - x^3}{\sigma^6}p(x),
$$

and are shown in Fig. A.2.

ERROR
FUNCTION

An important finite integral of the Gaussian is the so-called *error function*, defined as

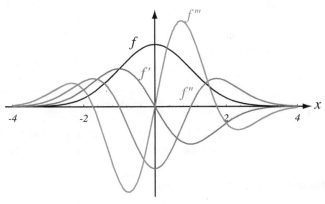

FIGURE A.2. A one-dimensional Gaussian distribution and its first three derivatives, shown for $f(x) \sim N(0, 1)$.

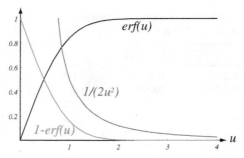

FIGURE A.3. The error function erf(u) corresponds to the area under a standardized Gaussian between $-\sqrt{2}u$ and $\sqrt{2}u$; that is, if x is a standardized Gaussian random variable, $\Pr[|x| \leq \sqrt{2}u] = \mathrm{erf}(u)$. Thus, the complementary probability, $1 - \mathrm{erf}(u)$, is the probability that a sample is chosen with $|x| > \sqrt{2}u$. Chebyshev's inequality states that for an arbitrary distribution having zero mean and unit standard deviation, $\Pr[|x| > \epsilon]$ is less than $1/\epsilon^2$, so that the lower curve is bounded by $1/(2u^2)$. As shown, this bound is quite loose for a Gaussian.

$$\mathrm{erf}(u) = \frac{2}{\sqrt{\pi}} \int_0^u e^{-x^2} dx. \tag{96}$$

As can be seen from Fig. A.1, $\mathrm{erf}(0) = 0$, and $\mathrm{erf}(1) = 0.84$. There is no closed analytic form for the error function, and thus we typically use tables, approximations, or numerical integration for its evaluation (Fig. A.3).

In calculating moments of Gaussians, we need the general integral of powers of x weighted by a Gaussian. Recall first the definition of a *gamma function*

GAMMA
FUNCTION

$$\Gamma(n + 1) = \int_0^\infty x^n e^{-x} dx, \tag{97}$$

where the gamma function obeys

$$\Gamma(n) = (n - 1)\Gamma(n - 1) \tag{98}$$

and $\Gamma(1/2) = \sqrt{\pi}$. For n an integer we have $\Gamma(n+1) = n \times (n-1) \times (n-2) \cdots \times 1 = n!$, read "$n$ factorial."

FACTORIAL

Changing variables in Eq. 97, we find the moments of a (normalized) Gaussian distribution as

$$2 \int_0^\infty x^n \frac{e^{-x^2/(2\sigma^2)}}{\sqrt{2\pi}\sigma} dx = \frac{2^{n/2}\sigma^n}{\sqrt{\pi}} \Gamma\left(\frac{n+1}{2}\right), \tag{99}$$

where again we have used a prefactor of 2 and lower integration limit of 0 in order give nontrivial (i.e., nonvanishing) results for odd n.

A.5.1 Multivariate Normal Densities

Normal random variables have many desirable theoretical properties. For example, it turns out that the convolution of two Gaussian functions is again a Gaussian function, and thus the distribution for the sum of two independent normal random variables is again normal. In fact, sums of dependent normal random variables also have normal distributions. Suppose that each of the d random variables x_i is normally distributed, each with its own mean and variance: $p_{x_i}(x_i) \sim N(\mu_i, \sigma_i^2)$. If these variables are independent, their joint density has the form

$$p(\mathbf{x}) = \prod_{i=1}^{d} p(x_i) = \prod_{i=1}^{d} \frac{1}{\sqrt{2\pi}\,\sigma_i} e^{-1/2((x_i - \mu_i)/\sigma_i)^2}$$

$$= \frac{1}{(2\pi)^{d/2} \prod\limits_{i=1}^{d} \sigma_i} \exp\left[-\frac{1}{2} \sum_{i=1}^{d} \left(\frac{x_i - \mu_i}{\sigma_i} \right)^2 \right]. \tag{100}$$

This can be written in a compact matrix form if we observe that for this case the covariance matrix is diagonal, that is,

$$\boldsymbol{\Sigma} = \begin{bmatrix} \sigma_1^2 & 0 & \cdots & 0 \\ 0 & \sigma_2^2 & \cdots & 0 \\ \vdots & \vdots & \ddots & \vdots \\ 0 & 0 & \cdots & \sigma_d^2 \end{bmatrix}, \tag{101}$$

and hence the inverse of the covariance matrix is easily written as

$$\boldsymbol{\Sigma}^{-1} = \begin{bmatrix} 1/\sigma_1^2 & 0 & \cdots & 0 \\ 0 & 1/\sigma_2^2 & \cdots & 0 \\ \vdots & \vdots & \ddots & \vdots \\ 0 & 0 & \cdots & 1/\sigma_d^2 \end{bmatrix}. \tag{102}$$

Thus, the exponent in Eq. 100 can be rewritten using

$$\sum_{i=1}^{d} \left(\frac{x_i - \mu_i}{\sigma_i} \right)^2 = (\mathbf{x} - \boldsymbol{\mu})^t \boldsymbol{\Sigma}^{-1} (\mathbf{x} - \boldsymbol{\mu}). \tag{103}$$

Finally, by noting that the determinant of $\boldsymbol{\Sigma}$ is just the product of the variances, we can write the joint density compactly in terms of the quadratic form

$$p(\mathbf{x}) = \frac{1}{(2\pi)^{d/2} |\boldsymbol{\Sigma}|^{1/2}} \exp\left[-\frac{1}{2} (\mathbf{x} - \boldsymbol{\mu})^t \boldsymbol{\Sigma}^{-1} (\mathbf{x} - \boldsymbol{\mu}) \right]. \tag{104}$$

MULTIVARIATE
NORMAL
DENSITY

This is the general form of a *multivariate normal density function*, where the covariance matrix $\boldsymbol{\Sigma}$ is no longer required to be diagonal. With a little linear algebra, it can be shown that if $\mathbf{x}$ obeys this probability law, then

$$\boldsymbol{\mu} = \mathcal{E}[\mathbf{x}] = \int_{-\infty}^{\infty} \mathbf{x} \, p(\mathbf{x}) \, d\mathbf{x}$$

$$\boldsymbol{\Sigma} = \mathcal{E}[(\mathbf{x} - \boldsymbol{\mu})(\mathbf{x} - \boldsymbol{\mu})^t] = \int_{-\infty}^{\infty} (\mathbf{x} - \boldsymbol{\mu})(\mathbf{x} - \boldsymbol{\mu})^t \, p(\mathbf{x}) \, d\mathbf{x}, \tag{105}$$

just as one would expect. Multivariate normal data tend to cluster about the mean vector, $\boldsymbol{\mu}$, falling in an ellipsoidally shaped cloud whose principal axes are the eigenvectors of the covariance matrix. The natural measure of the distance from $\mathbf{x}$ to the mean $\boldsymbol{\mu}$ is provided by the quantity

$$r^2 = (\mathbf{x} - \boldsymbol{\mu})^t \boldsymbol{\Sigma}^{-1} (\mathbf{x} - \boldsymbol{\mu}), \tag{106}$$

MAHALANOBIS
DISTANCE

which is the square of the *Mahalanobis distance* from $\mathbf{x}$ to $\boldsymbol{\mu}$. It is not as easy to standardize a vector random variable (reduce it to zero mean and unit covariance matrix) as it is in the univariate case. The expression analogous to $u = (x - \mu)/\sigma$ is $\mathbf{u} = \boldsymbol{\Sigma}^{-1/2}(\mathbf{x} - \boldsymbol{\mu})$, which involves the "square root" of the inverse of the covariance matrix. The process of obtaining $\boldsymbol{\Sigma}^{-1/2}$ requires finding the eigenvalues and eigenvectors of $\boldsymbol{\Sigma}$, and it is just a bit beyond the scope of this Appendix.

A.5.2 Bivariate Normal Densities

It is illuminating to look at the bivariate normal density—that is, the case of two normally distributed random variables x_1 and x_2. It is convenient to define $\sigma_1^2 = \sigma_{11}, \sigma_2^2 = \sigma_{22}$ and to introduce the correlation coefficient ρ defined by

$$\rho = \frac{\sigma_{12}}{\sigma_1 \sigma_2}. \tag{107}$$

With this notation, the covariance matrix becomes

$$\boldsymbol{\Sigma} = \begin{bmatrix} \sigma_{11} & \sigma_{12} \\ \sigma_{21} & \sigma_{22} \end{bmatrix} = \begin{bmatrix} \sigma_1^2 & \rho\sigma_1\sigma_2 \\ \rho\sigma_1\sigma_2 & \sigma_2^2 \end{bmatrix}, \tag{108}$$

and its determinant simplifies to

$$|\boldsymbol{\Sigma}| = \sigma_1^2 \sigma_2^2 (1 - \rho^2). \tag{109}$$

Thus, the inverse covariance matrix is given by

$$
\begin{aligned}
\boldsymbol{\Sigma}^{-1} &= \frac{1}{\sigma_1^2 \sigma_2^2 (1 - \rho^2)} \begin{bmatrix} \sigma_2^2 & -\rho\sigma_1\sigma_2 \\ -\rho\sigma_1\sigma_2 & \sigma_1^2 \end{bmatrix} \\
&= \frac{1}{1 - \rho^2} \begin{bmatrix} 1/\sigma_1^2 & -\rho/(\sigma_1\sigma_2) \\ -\rho/(\sigma_1\sigma_2) & 1/\sigma_2^2 \end{bmatrix}.
\end{aligned}
\tag{110}
$$

Next we explicitly expand the quadratic form in the normal density:

$$(\mathbf{x} - \boldsymbol{\mu})^t \boldsymbol{\Sigma}^{-1} (\mathbf{x} - \boldsymbol{\mu}) \tag{111}$$

$$= [(x_1 - \mu_1)\ (x_2 - \mu_2)] \frac{1}{1 - \rho^2} \begin{bmatrix} 1/\sigma_1^2 & -\rho/(\sigma_1\sigma_2) \\ -\rho/(\sigma_1\sigma_2) & 1/\sigma_2^2 \end{bmatrix} \begin{bmatrix} x_1 - \mu_1 \\ x_2 - \mu_2 \end{bmatrix}$$

$$= \frac{1}{1 - \rho^2} \left[\left(\frac{x_1 - \mu_1}{\sigma_1} \right)^2 - 2\rho \left(\frac{x_1 - \mu_1}{\sigma_1} \right) \left(\frac{x_2 - \mu_2}{\sigma_2} \right) + \left(\frac{x_2 - \mu_2}{\sigma_2} \right)^2 \right].$$

Thus, the general bivariate normal density has the form

$$p_{x_1 x_2}(x_1, x_2) = \frac{1}{2\pi \sigma_1 \sigma_2 \sqrt{1 - \rho^2}} \qquad (112)$$

$$\times \exp\left[-\frac{1}{2(1 - \rho^2)} \left[\left(\frac{x_1 - \mu_1}{\sigma_1}\right)^2 - 2\rho\left(\frac{x_1 - \mu_1}{\sigma_1}\right)\left(\frac{x_2 - \mu_2}{\sigma_2}\right) + \left(\frac{x_2 - \mu_2}{\sigma_2}\right)^2 \right] \right].$$

As we can see from Fig. A.4, $p(x_1, x_2)$ is a hill-shaped surface over the $x_1 x_2$ plane. The peak of the hill occurs at the point $(x_1, x_2) = (\mu_1, \mu_2)$—that is, at the mean vector $\boldsymbol{\mu}$. The shape of the hump depends on the two variances σ_1^2 and σ_2^2, and the correlation coefficient ρ. If we slice the surface with horizontal planes parallel to the $x_1 x_2$ plane, we obtain the so-called *level curves*, defined by the locus of points where the quadratic form

$$\left(\frac{x_1 - \mu_1}{\sigma_1}\right)^2 - 2\rho\left(\frac{x_1 - \mu_1}{\sigma_1}\right)\left(\frac{x_2 - \mu_2}{\sigma_2}\right) + \left(\frac{x_2 - \mu_2}{\sigma_2}\right)^2 \qquad (113)$$

is constant. It is not hard to show that $|\rho| \leq 1$ and that this implies that the level curves are ellipses. The x and y extent of these ellipses are determined by the variances σ_1^2 and σ_2^2, and their eccentricity is determined by ρ. More specifically, the PRINCIPAL AXES *principal axes* of the ellipse are in the direction of the eigenvectors $\mathbf{e}_i$ of $\boldsymbol{\Sigma}$, and the different widths in these directions are $\sqrt{\lambda_i}$. For instance, if $\rho = 0$, the principal axes of the ellipses are parallel to the coordinate axes, and the variables are statistically independent. In the special cases where $\rho = 1$ or $\rho = -1$, the ellipses collapse to straight lines. Indeed, the joint density becomes singular in this situation, because there is really only one independent variable. We shall avoid this degeneracy by assuming that $|\rho| < 1$.

One of the important properties of the multivariate normal density is that all conditional and marginal probabilities are also normal. To find such a density explicitly, which we denote $p_{x_2|x_1}(x_2|x_1)$, we substitute our formulas for $p_{x_1 x_2}(x_1, x_2)$ and

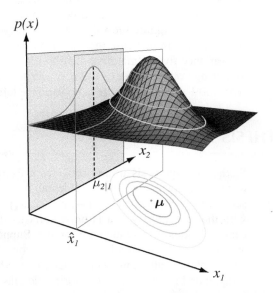

FIGURE A.4. A two-dimensional Gaussian having mean $\boldsymbol{\mu}$ and nondiagonal covariance $\boldsymbol{\Sigma}$. If the value on one variable is known, for instance $x_1 = \hat{x}_1$, the distribution over the other variable is Gaussian with mean $\mu_{2|1}$.

$p_{x_1}(x_1)$ in the defining equation

$$p_{x_2|x_1}(x_2|x_1) = \frac{p_{x_1 x_2}(x_1, x_2)}{p_{x_1}(x_1)}$$

$$= \left[\frac{1}{2\pi\sigma_1\sigma_2\sqrt{1-\rho^2}} e^{-\frac{1}{2(1-\rho^2)}\left[\left(\frac{x_1-\mu_1}{\sigma_1}\right)^2 - 2\rho\left(\frac{x_1-\mu_1}{\sigma_1}\frac{x_2-\mu_2}{\sigma_2}\right)\left(\frac{x_2-\mu_2}{\sigma_2}\right) + \left(\frac{x_2-\mu_2}{\sigma_2}\right)^2\right]} \right]$$

$$\times \left[\sqrt{2\pi}\sigma_1 e^{\frac{1}{2}\left(\frac{x_1-\mu_1}{\sigma_1}\right)^2} \right]$$

$$= \frac{1}{\sqrt{2\pi}\sigma_2\sqrt{1-\rho^2}} \exp\left[-\frac{1}{2(1-\rho^2)}\left[\frac{x_2-\mu_2}{\sigma_2} - \rho\frac{x_1-\mu_1}{\sigma_1} \right]^2 \right]$$

$$= \frac{1}{\sqrt{2\pi}\sigma_2\sqrt{1-\rho^2}} \exp\left[-\frac{1}{2}\left(\frac{x_2 - [\mu_2 + \rho\frac{\sigma_2}{\sigma_1}(x_1-\mu_1)]}{\sigma_2\sqrt{1-\rho^2}} \right)^2 \right]. \qquad (114)$$

CONDITIONAL MEAN

Thus, we have verified that the conditional density $p_{x_1|x_2}(x_1|x_2)$ is a normal distribution. Moreover, we have explicit formulas for the *conditional mean* $\mu_{2|1}$ and the conditional variance $\sigma_{2|1}^2$:

$$\mu_{2|1} = \mu_2 + \rho\frac{\sigma_2}{\sigma_1}(x_1 - \mu_1) \qquad \text{and} \qquad \sigma_{2|1}^2 = \sigma_2^2(1-\rho^2), \qquad (115)$$

as illustrated in Fig. A.4.

These formulas provide some insight into the question of how knowledge of the value of x_1 helps us to estimate x_2. Suppose that we know the value of x_1. Then a natural estimate for x_2 is the conditional mean, $\mu_{2|1}$. In general, $\mu_{2|1}$ is a linear function of x_1; if the correlation coefficient ρ is positive, the larger the value of x_1, the larger the value of $\mu_{2|1}$. If it happens that x_1 is the mean value μ_1, then the best we can do is to guess that x_2 is equal to μ_2. Also, if there is no correlation between x_1 and x_2, we ignore the value of x_1, whatever it is, and we always estimate x_2 by μ_2. Note that in that case the variance of x_2, given that we know x_1, is the same as the variance for the marginal distribution, that is, $\sigma_{2|1}^2 = \sigma_2^2$. If there is correlation, knowledge of the value of x_1, whatever the value is, reduces the variance. Indeed, with 100% correlation there is no variance left in x_2 when the value of x_1 is known.

A.6 HYPOTHESIS TESTING

Statistical hypothesis testing provides a formal way to decide if the results of an experiment are significant or accidental. It is standard statistical terminology to call a set of n measurements $\mathcal{X}_n = \{x_1, x_2, \ldots, x_n\}$ a *sample of size n*. However, in keeping with the terminology that is universally used in pattern recognition, we shall call each individual measurement a sample. Suppose that we have a set of samples that are drawn either from a known distribution D_0 or from some other distribution. In pattern classification, we seek to determine which distribution was the source of any sample; and if it is indeed D_0, we would classify the point accordingly. Hypothesis testing addresses a somewhat different but related problem. We assume initially that distribution D_0 is the source of the patterns; this is called the *null hypothesis* and is often denoted H_0. Based on the value of any observed sample, we ask whether we can *reject the null hypothesis*—that is, state with some degree of confidence (expressed as a probability) that the sample did *not* come from D_0.

For instance, D_0 might be a standardized Gaussian, $p(x) \sim N(0, 1)$, and hence our null hypothesis is that a sample comes from a Gaussian with mean $\mu = 0$. If the value of a particular sample is small (e.g., $x = 0.3$), it is likely that it came from the D_0; after all, 68% of the samples drawn from that distribution have absolute value less than $x = 1.0$ (cf. Fig. A.1). If a sample's value is large (e.g., $x = 5$), then we would be more confident that it did *not* come from D_0. At such a situation we merely conclude that (with some probability) the sample was drawn from a standardized Gaussian distribution with $\mu \neq 0$.

Viewed another way, for any confidence—expressed as a probability—there exists a criterion value such that if the sampled value differs from $\mu = 0$ by more than that criterion, we reject the null hypothesis. (It is traditional to use confidences of .01 or .05.) We then say that the difference of the sample from 0 is *statistically* STATISTICAL *significant*. For instance, if our null hypothesis is a standardized Gaussian, then if SIGNIFICANCE our sample differs from the value $x = 0$ by more than 2.576, we could reject the null hypothesis "at the .01 confidence level," as can be deduced from Table A.1. A more sophisticated analysis could be applied if *several* samples are all drawn from D_0 or if the null hypothesis involved a distribution other than a Gaussian. Of course, this usage of "significance" applies only to the statistical properties of the problem—it implies nothing about whether the results are "important." Hypothesis testing is of great generality, and it is useful when we seek to know whether something other than the assumed case (the null hypothesis) is likely to be the case.

A.6.1 Chi-Squared Test

Hypothesis testing can be applied to discrete problems too. Suppose we have n patterns—n_1 of which are known to be in ω_1, and n_2 in ω_2—and we are interested in determining whether a particular decision rule is useful or informative. In this case, the null hypothesis is that a random decision rule is present—one that selects a pattern and with some probability P places it in a category which we will call the "left" category, and otherwise in the "right" category. We say that a candidate rule is informative if it differs signficantly from such a random decision.

What we need is a clear mathematical definition of statistical significance under these conditions. The random rule (the null hypothesis) would place Pn_1 patterns from ω_1 and Pn_2 from ω_2 independently in the left category and the remainder in the right category. Our candidate decision rule would differ significantly from the random rule if the proportions differed significantly from those given by the random rule. Formally, we let n_{iL} denote the number of patterns from category ω_i placed in the left category by our candidate rule. The so-called *chi-squared* statistic for this case is

$$\chi^2 = \sum_{i=1}^{2} \frac{(n_{iL} - n_{ie})^2}{n_{ie}}, \tag{116}$$

where, according to the null hypothesis, the number of patterns in category ω_i that we expect to be placed in the left category is $n_{ie} = Pn_i$. Clearly χ^2 is nonnegative, and it is zero if and only if all the observed numbers n_{iL} match the expected numbers n_{ie}. The higher the value of χ^2, the less likely it is that the null hypothesis is true. Thus, for a sufficiently high χ^2, the difference between the expected and observed distributions is statistically significant, we can reject the null hypothesis, and we

Table A.2. Critical Values of Chi-Square (at Two Confidence levels) for Different Degrees of Freedom (df)

df	.05	.01	df	.05	.01	df	.05	.01
1	3.84	6.64	11	19.68	24.72	21	32.67	38.93
2	5.99	9.21	12	21.03	26.22	22	33.92	40.29
3	7.82	11.34	13	22.36	27.69	23	35.17	41.64
4	9.49	13.28	14	23.68	29.14	24	36.42	42.98
5	11.07	15.09	15	25.00	30.58	25	37.65	44.31
6	12.59	16.81	16	26.30	32.00	26	38.88	45.64
7	14.07	18.48	17	27.59	33.41	27	40.11	46.96
8	15.51	20.09	18	28.87	34.80	28	41.34	48.28
9	16.92	21.67	19	30.14	37.57	29	42.56	49.59
10	18.31	23.21	20	31.41	37.57	30	43.77	50.89

can consider our candidate decision rule is "informative." For any desired level of significance—such as .01 or .05—a table gives the critical values of χ^2 that allow us to reject the null hypothesis (Table A.2).

There is one detail that must be addressed: the number of degrees of freedom (df). In the situation described above, once the probability P is known, there is only one free variable needed to describe a candidate rule. For instance, once the number of patterns from ω_1 placed in the left category are known, all other values are determined uniquely. Hence in this case the number of degrees of freedom is 1. If there were more categories, or if the candidate decision rule had more possible outcomes, then the number of degrees of freedom df would be greater than 1. The higher the number of degrees of freedom, the higher must be the computed χ^2 to meet a desired level of significance.

We denote the critical values as, for instance, $\chi^2_{.01(1)} = 6.64$, where the subscript denotes the significance, here .01, and the integer in parentheses is the number of degrees of freedom. (In Table A.2, we conform to the usage in statistics, where this positive integer is denoted df, despite the possible confusion in calculus where it denotes an infinitesimal real number.) Thus if we have one degree of freedom and the observed χ^2 is greater than 6.64, then we can reject the null hypothesis and say that at the .01 confidence level our results did not come from a (weighted) random decision.

A.7 INFORMATION THEORY

A.7.1 Entropy and Information

Assume we have a discrete set of symbols $\{v_1, v_2, \ldots, v_m\}$ with associated probabilities P_i. The entropy of the discrete distribution—a measure of the randomness or unpredictability of a sequence of symbols drawn from it—is

$$H = -\sum_{i=1}^{m} P_i \log_2 P_i, \tag{117}$$

BIT

where entropy is measured in *bits* when we use the logarithm base 2. (For continuous distributions, we often use base-e or natural logarithm, denoted ln, in which case entropy is said to be measured in *nats*. In case any of the probabilities vanish, we use the fact that $\lim_{P \to 0} P \log P = 0$ to define $0 \log 0 = 0$.) One bit corresponds to the uncertainty that can be resolved by the answer to a single yes/no question. The expectation operator (cf. Eq. 40) can be used to write $H = \mathcal{E}[\log 1/P]$, where we think of P as being a random variable whose possible values are $P_1, P_2, \ldots, P_m$. The term $\log_2 1/P$ is sometimes called the *surprise*: If $P_i = 0$ except for one i, then there is no surprise when the corresponding symbol occurs.

SURPRISE

Note that the entropy does not depend on the symbols themselves, just on their probabilities. For a given number of symbols m, the uniform distribution, in which each symbol is equally likely, is the *maximum entropy distribution* (and $H = \log_2 m$ bits)—we have the maximum uncertainty about the identity of each symbol that will be chosen. Clearly if x is equally likely to take on integer values $0, 1, \ldots, 7$, we need 3 bits to describe the outcome and $H = -\sum_{i=0}^{7} \frac{1}{2^3} \log_2 \frac{1}{2^3} = \log_2 2^3 = 3$ bits. Conversely, if all the p_i are 0 except one, we have the *minimum entropy distribution* ($H = 0$ bits)—we are certain as to the symbol that will appear.

For a continuous distribution, the entropy is

$$H = -\int_{-\infty}^{\infty} p(x) \ln p(x) \, dx, \tag{118}$$

and again $H = \mathcal{E}[\ln 1/p]$. It is worth mentioning that among all continuous density functions having a given mean μ and variance σ^2, it is the Gaussian that has the maximum entropy ($H = 0.5 + \log_2 (\sqrt{2\pi}\sigma)$ bits). We can let σ approach zero to find that a probability density in the form of a *Dirac delta* function, that is,

DIRAC DELTA

$$\delta(x - a) = \begin{cases} 0 & \text{if } x \neq a \\ \infty & \text{if } x = a, \end{cases} \quad \text{with}$$

$$\int_{-\infty}^{\infty} \delta(x) \, dx = 1, \tag{119}$$

has the minimum entropy ($H = -\infty$ bits). For this Dirac density, we are sure that the value a will be selected each time.

Our use of entropy in continuous functions, such as in Eq. 118, belies some subtle issues which are worth pointing out. If x had units, such as meters, then the probability density $p(x)$ would have to have units of $1/x$. There would be something fundamentally wrong in taking the logarithm of $p(x)$—the argument of the logarithm function should be dimensionless. What we should really be dealing with is a dimensionless quantity, say $p(x)/p_0(x)$, where $p_0(x)$ is some reference density function (cf. Section A.7.2).

For discrete variable x and arbitrary function $f(\cdot)$, we have $H(f(x)) \leq H(x)$, that is, processing never increases entropy. In particular, if $f(x)$ is a constant, the entropy will vanish. Another key property of the entropy of a discrete distribution is that it is invariant to "shuffling" the event labels. The related question with continuous variables concerns what happens when one makes a change of variables. In general, if we make a change of variables, such as $y = x^3$ or even $y = 10x$, we will get a different value for the integral of $\int q(y) \log q(y) \, dy$, where q is the induced density for y. If entropy is supposed to measure the intrinsic disorganization, it doesn't make sense that y would have a different amount of intrinsic disorganization than x, because one is always derivable from the other; only if there were some randomness (e.g., shuffling) incorporated into the mapping could we say that one is more disorganized than the other.

Fortunately, in practice these concerns do not present important stumbling blocks since relative entropy and differences in entropy are more fundamental than H taken by itself. Nevertheless, questions of the foundations of entropy measures for continuous variables are addressed in books listed in Bibliographical Remarks.

A.7.2 Relative Entropy

KULLBACK-
LEIBLER
DISTANCE

Suppose we have two discrete distributions over the same variable x, $p(x)$, and $q(x)$. The relative entropy or *Kullback-Leibler distance* (which is closely related to cross entropy, information divergence and information for discrimination) is a measure of the "distance" between these distributions:

$$D_{KL}(p(x), q(x)) = \sum_x q(x) \ln \frac{q(x)}{p(x)}. \tag{120}$$

The continuous version is

$$D_{KL}(p(x), q(x)) = \int_{-\infty}^{\infty} q(x) \ln \frac{q(x)}{p(x)} \, dx. \tag{121}$$

Although $D_{KL}(p(\cdot), q(\cdot)) \geq 0$ and $D_{KL}(p(\cdot), q(\cdot)) = 0$ if and only if $p(\cdot) = q(\cdot)$, the relative entropy is not a true metric because D_{KL} is not necessarily symmetric in the interchange $p \leftrightarrow q$. Furthermore, $D_{KL}(\cdot, \cdot)$ need not satisfy the triangle inequality.

A.7.3 Mutual Information

Now suppose we have two distributions over possibly *different* variables—for example, $p(x)$ and $q(y)$. The mutual information is the reduction in uncertainty about one variable due to the knowledge of the other variable

$$I(p; q) = H(p) - H(p|q) = \sum_{x,y} r(x, y) \log_2 \frac{r(x, y)}{p(x)q(y)}, \tag{122}$$

where $r(x, y)$ is the joint distribution of finding value x and y. Mutual information is simply the relative entropy between the joint distribution $r(x, y)$ and the product distribution $p(x)q(y)$ and as such it measures how much the distributions of the variables differ from statistical independence. Mutual information does not obey all the properties of a metric. In particular, the metric requirement that if $p(x) = q(y)$ then $I(x; y) = 0$ need not hold, in general. As an example, suppose we have two binary random variables with $r(0, 0) = r(1, 1) = 1/2$, so $r(0, 1) = r(1, 0) = 0$. According to Eq. 122, the mutual information between $p(x)$ and $q(y)$ is $\log_2 2 = 1$.

The relationships among the entropy, relative entropy and mutual information are summarized in Fig. A.5. The figure shows, for instance, that the joint entropy $H(p, q)$ is never smaller than individual entropies $H(p)$ and $H(q)$, that $H(p) = H(p|q) + I(p; q)$, and so on.

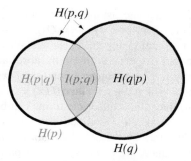

FIGURE A.5. For two distributions p and q, this figure shows the mathematical relationships among the entropy, mutual information $I(p; q)$, and conditional entropies $H(p|q)$ and $H(q|p)$. For instance $I(p; p) = H(p)$; if $I(p; q) = 0$, then $H(q|p) = H(q)$; $H(p, q) = H(p|q) + H(q)$; and so forth.

A.8 COMPUTATIONAL COMPLEXITY

In order to analyze and describe the difficulty of problems and the algorithms designed to solve such problems, we turn now to the technical notion of computational complexity. For instance, calculating the covariance matrix for a set of samples is somehow "harder" than calculating the mean. Furthermore, some algorithms for computing some function may be faster or take less memory than other algorithms. We seek to specify such differences, independent of the current computer hardware (which is always changing anyway).

To this end we use the concept of the order of a function and the asymptotic notations "big oh," "big omega," and "big theta." The three asymptotic bounds most often used are as follows (Fig. A.6):

Asymptotic upper bound: $O(g(x)) = \{f(x): \text{There exist positive constants } c \text{ and } x_0 \text{ such that } 0 \leq f(x) \leq cg(x) \text{ for all } x \geq x_0\}$.

Asymptotic lower bound: $\Omega(g(x)) = \{f(x): \text{There exist positive constants } c \text{ and } x_0 \text{ such that } 0 \leq cg(x) \leq f(x) \text{ for all } x \geq x_0\}$.

Asymptotically tight bound: $\Theta(g(x)) = \{f(x): \text{There exist positive constants } c_1, c_2, \text{ and } x_0 \text{ such that } 0 \leq c_1 g(x) \leq f(x) \leq c_2 g(x) \text{ for all } x \geq x_0\}$.

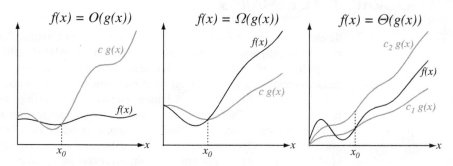

FIGURE A.6. Three types of asymptotic bounds are useful in describing the computational complexities arising in pattern recognition research.

Consider the asymptotic upper bound. We say that $f(x)$ is "of order big oh of $g(x)$"—written $f(x) = O(g(x))$—if there exist constants c_0 and x_0 such that $f(x) \leq cg(x)$ for all $x > x_0$. (We shall assume that all our functions are positive and dispense with taking absolute values.) This means simply that for sufficiently large x an upper bound on $f(x)$ grows no worse than $g(x)$. For instance, if $f(x) = a + bx + cx^2$, then $f(x) = O(x^2)$ because for sufficiently large x, the constant, linear, and quadratic terms can be "overcome" by proper choice of c and x_0. The generalization to functions of two or more variables is straightforward. It should be clear that by the definition above, the (big oh) order of a function is not unique. For instance, we can describe our particular $f(x)$ as being $O(x^2)$, $O(x^3)$, $O(x^4)$, $O(x^2 \ln x)$, and so forth. We use big omega notation, $\Omega(\cdot)$, for lower bounds, and little omega, $\omega(\cdot)$, for the tightest lower bound. Of these, the big oh notation has proven to be most useful because we generally want an *upper* bound on the resources when solving a problem.

The lower bound on the complexity of the *problem* is denoted $\Omega(g(x))$, and it is therefore the lower bound on any algorithm algorithm that solves that problem. Similarly, if the complexity of an algorithm is $O(g(x))$, it is an upper bound on the complexity of the problem it solves. The complexity of some problems—such as computing the mean of a discrete set—is known, and thus once we have found an algorithm having equal complexity, the only possible improvement could be on lowering the constants of proportionality.

Such a rough analysis does not tell us the constants c and x_0. For a finite size problem it is possible that a particular $O(x^3)$ algorithm is simpler than a particular $O(x^2)$ algorithm, and it is occasionally necessary for us to determine these constants to find which of several implemementations is the simplest. Nevertheless, for our purposes the big oh notation as just described is generally the best way to describe the computational complexity of an algorithm.

Suppose we have a set of n vectors, each of which is d-dimensional and we want to calculate the mean vector. Clearly, this requires $O(nd)$ multiplications. Sometimes we stress space and time complexities, which are particularly relevant when contemplating parallel hardware implementations. For instance, the d-dimensional sample mean could be calculated with d separate processors, each adding n sample values. Thus we can describe this implementation as $O(d)$ in *space* (i.e., the amount of memory or possibly the number of processors) and $O(n)$ in *time* (i.e., number of sequential steps). Of course for any particular algorithm there may be a number of time-space tradeoffs.

BIBLIOGRAPHICAL REMARKS

There are several good books on linear systems and matrix computations, such as references [15] and [8]. Lagrange optimization and related techniques are covered in a definitive book, reference [2]. While references [14] and [3] are of historic interest, readers seeking clear presentations of the central ideas in probability should consult references [6, 7, 11] and [22]. A handy reference to terms in probability and statistics is reference [21]. There are many books on hypothesis testing and statistical significance; an elementary book that can be recommended is reference [25]; advanced books include references [19] and [26]. Shannon's landmark paper [23] should be read by all students of pattern recognition. It and many other historically important papers on information theory can be found in reference [24]. An excellent textbook on information theory at the level needed for practical work in pattern recognition is reference [5], and readers seeking a more abstract and formal treatment should consult reference [9]. The study of time complexity of algorithms began with reference [13], and the study of space complexity began with references [12] and [20]. Knuth's classic volumes [16, 17, 18] contain a description of computational complexity, the big oh and other asymptotic notations. Somewhat more accessible treatments can be found in references [1] and [4].

BIBLIOGRAPHY

[1] Alfred V. Aho, John E. Hopcroft, and Jeffrey D. Ullman. *The Design and Analysis of Computer Algorithms.* Addison-Wesley, Reading, MA, 1974.

[2] Dimitri P. Bertsekas. *Constrained Optimization and Lagrange Multiplier Methods.* Athena Scientific, Belmont, MA, 1996.

[3] Patrick Billingsley. *Probability and Measure.* Wiley, New York, second edition, 1986.

[4] Thomas H. Cormen, Charles E. Leiserson, and Ronald L. Rivest. *Introduction to Algorithms.* MIT Press, Cambridge, MA, 1990.

[5] Thomas M. Cover and Joy A. Thomas. *Elements of Information Theory.* Wiley-Interscience, New York, 1991.

[6] Alvin W. Drake. *Fundamentals of Applied Probability Theory.* McGraw-Hill, New York, 1967.

[7] William Feller. *An Introduction to Probability Theory and Its Applications,* volume 1. Wiley, New York, 1968.

[8] Gene H. Golub and Charles F. Van Loan. *Matrix Computations.* Johns Hopkins University Press, Baltimore, MD, third edition, 1996.

[9] Robert M. Gray. *Entropy and Information Theory.* Springer-Verlag, New York, 1990.

[10] Daniel H. Greene and Donald E. Knuth. *Mathematics for the Analysis of Algorithms.* Springer-Verlag, New York, 1990.

[11] Richard W. Hamming. *The Art of Probability for Scientists and Engineers.* Addison-Wesley, New York, 1991.

[12] Juris Hartmanis, Philip M. Lewis II, and Richard E. Stearns. Hierarchies of memory limited computations. *Proceedings of the Sixth Annual IEEE Symposium on Switching Circuit Theory and Logical Design,* pages 179–190, 1965.

[13] Juris Hartmanis and Richard E. Stearns. On the computational complexity of algorithms. *Transactions of the American Mathematical Society,* 117:285–306, 1965.

[14] Harold Jeffreys. *Theory of Probability.* Oxford University Press, Oxford, UK, 1939, 1961 reprint edition.

[15] Thomas Kailath. *Linear Systems.* Prentice-Hall, Englewood Cliffs, NJ, 1980.

[16] Donald E. Knuth. *The Art of Computer Programming,* volume 1. Addison-Wesley, Reading, MA, first edition, 1973.

[17] Donald E. Knuth. *The Art of Computer Programming,* volume 3. Addison-Wesley, Reading, MA, first edition, 1973.

[18] Donald E. Knuth. *The Art of Computer Programming,* volume 2. Addison-Wesley, Reading, MA, first edition, 1981.

[19] Erich L. Lehmann. *Testing Statistical Hypotheses.* Springer, New York, 1997.

[20] Philip M. Lewis II, Richard E. Stearns, and Juris Hartmanis. Memory bounds for recognition of context-free and context-sensitive languages. *Proceedings of the Sixth Annual IEEE Symposium on Switching Circuit Theory and Logical Design,* pages 191–202, 1965.

[21] Francis H. C. Marriott. *A Dictionary of Statistical Terms.* Longman Scientific & Technical, Essex, UK, fifth edition, 1990.

[22] Yuri A. Rozanov. *Probability Theory: A Concise Course.* Dover, New York, 1969.

[23] Claude E. Shannon. A mathematical theory of communication. *Bell Systems Technical Journal,* 27:379–423, 623–656, 1948.

[24] David Slepian, editor. *Key Papers in the Development of Information Theory.* IEEE Press, New York, 1974.

[25] Richard C. Sprinthall. *Basic Statistical Analysis.* Allyn & Bacon, Needham Heights, MA, fifth edition, 1996.

[26] Rand R. Wilcox. *Introduction to Robust Estimation and Hypotheses Testing.* Academic Press, New York, 1997.

INDEX